GASEOUS
DIELECTRICS
VI

GASEOUS DIELECTRICS VI

Edited by
Loucas G. Christophorou
and
Isidor Sauers

Oak Ridge National Laboratory
Oak Ridge, Tennessee

SPRINGER SCIENCE+BUSINESS MEDIA, LLC

Library of Congress Cataloging-in-Publication Data

International Symposium on Gaseous Dielectrics (6th : 1990: Knoxville, Tenn.)
   Gaseous dielectrics VI / edited by Loucas G. Christophorou and Isidor Sauers.
      p.   cm.
   "Proceedings of the Sixth International Symposium on Gaseous Dielectrics held September 23-27, 1990, in Knoxville, Tennessee."--T.p. verso.
   Includes bibliographical references and index.
   ISBN 978-1-4613-6648-5   ISBN 978-1-4615-3706-9 (eBook)
   DOI 10.1007/978-1-4615-3706-9
   1. Dielectrics, Gaseous--Congresses.   I. Christophorou, L. G. II. Sauers, Isidor. III. Title: Gaseous dielectrics six.
QC5850.8.G38I58 1990
537'.24--dc20   91-11648
                                       CIP

Proceedings of the Sixth International Symposium on Gaseous Dielectrics, held September 23-27, 1990, in Knoxville, Tennessee

ISBN 978-1-4613-6648-5

© 1991 Springer Science+Business Media New York
Originally published by Plenum Press, New York in 1991
Softcover reprint of the hardcover 1st edition 1991

PREFACE

The Sixth International Symposium on Gaseous Dielectrics was held in Knoxville, Tennessee, U.S.A., on September 23-27, 1990. The symposium continued the transdisciplinary character and comprehensive approach of the preceding five symposia.

Gaseous Dielectrics VI is a detailed record of the symposium proceedings. It covers recent advances and developments in a wide range of basic, applied and industrial areas of gaseous dielectrics. It is hoped that *Gaseous Dielectrics VI* will aid future research and development in and encourage wider industrial use of gaseous dielectrics.

The Organizing Committee of the Sixth International Symposium on Gaseous Dielectrics consisted of L. G. Christophorou (U.S.A.), F. Y. Chu (Canada), A. H. Cookson (U.S.A.), D. L. Damsky (U.S.A.), O. Farish (U.K.), I. Gallimberti (Italy), A. Garscadden (U.S.A.), E. Marode (France), T. Nitta (Japan), W. Pfeiffer (Germany), I. Sauers (U.S.A.), R. J. Van Brunt (U.S.A.), and W. Zaengl (Switzerland). The local arrangements committee consisted of members of the Health and Safety Research Division and personnel of the Conference Office of the Oak Ridge National Laboratory, and staff of the University of Tennessee (UTK). The contributions of each member of these committees, the work of the Session Chairmen, the interest of the participants, and the advice of innumerable colleagues are gratefully acknowledged. I am especially indebted to Dr. Dennis L. McCorkle, Mrs. Joan E. Carrington, and Ms. Jo Ann Cripps for their assistance during the symposium and for their help with the manuscripts.

The symposium was hosted by the Oak Ridge National Laboratory and the University of Tennessee and was sponsored by the U.S. Department of Energy, the UTK/ORNL Science Alliance, and the Aero Propulsion and Power Laboratories of the Wright Research and Development Center; it was organized in cooperation with the Institute of Electrical and Electronics Engineers, Inc., the Power Engineering Society, and the Dielectrics and Electrical Insulation Society. The continued support of the Oak Ridge National Laboratory and the financial assistance of the sponsors are acknowledged with gratitude.

L. G. Christophorou, Symposium Chairman

Oak Ridge, Tennessee
December, 1990

CONTENTS

CHAPTER 1: BASIC PHYSICS OF GASEOUS DIELECTRICS

CHAPTER 2: BASIC MECHANISMS

CHAPTER 11: GAS INSULATED SUBSTATIONS

CHAPTER 12: RELIABILITY OF GIS/FAILURE MECHANISMS

CHAPTER 1: BASIC PHYSICS OF GASEOUS DIELECTRICS

COLLISIONAL ELECTRON DETACHMENT IN DIELECTRIC GASES

R. L. Champion

Department of Physics
College of William and Mary
Williamsburg, VA 23185

ABSTRACT

The collisional dynamics of anions common to the field of gaseous dielectrics is discussed. In particular, the results from recent measurements of absolute, two-body cross sections for detachment and other inelastic processes for collisions of SF_6^- with specific targets is emphasized. These results illustrate that although the "additional" electron attached to SF_6^- is only very weakly bound (or perhaps not bound at all unless some relaxation mechanism has occurred after the anion has formed), the probability of collisionally detaching that electron is extremely small even for collision energies which exceed the electron's binding energy by a factor as large as one hundred. The dominant collisional decomposition mechanism at low collision energies for sulphur hexafluoride anions is observed to be dissociation into, e.g., $SF_5 + F^-$. Measurements of this type can be used to provide insight into macroscopic phenomena such as discharge inception; it is clear however, that our current data base is inadequate to be used to make definitive predictions for such macroscopic phenomena.

INTRODUCTION

An important role for a molecule used as a gaseous dielectric is to attach and "hold-on" to any free electrons which might otherwise be instrumental in initiating a discharge process. Sulphur hexafluoride is an attractive dielectric gas specifically because of its large cross sections for direct and associative attachment of electrons; these properties have been well-documented[1]. Less well understood are the mechanisms whereby a molecular anion such as SF_6^- decomposes in subsequent collisions at high E/N conditions. There have been several studies[2] which provide indirect evidence suggesting that collisional detachment of SF_6^- is, in fact, a relatively unimportant process when compared to its collisional decomposition into SF_5 + F^-. The purpose of this report is to summarize the results of recent direct measurements in which absolute cross sections for collisional decomposition (and charge transfer) of SF_6^- have been determined.[3] The range of collision energies which are relevant to gaseous dielectrics is from the threshold for the inelastic process up to several eV above that threshold. An energy level diagram indicating the energetic thresholds for collisionally decomposing ground state SF_6^- into:

$$\begin{array}{lll}
SF_6^- + X \rightarrow & SF_6 + X + e & \text{(a)} \\
& SF_5 + F^- + X & \text{(b)} \\
& SF_5^- + F + X & \text{(c)}
\end{array} \qquad \text{(1)}$$

is given in Fig. 1.

The fundamental quantity to be measured is the cross section, $\sigma_i(E)$, where E is the relative collision energy (i.e., energy in the center-of-mass reference frame) and i stands for some inelastic channel. A target gas of principal interest to gaseous dielectrics is SF_6, but other targets can be used to probe the collisional dynamics of SF_6^- in a way which is precluded with the SF_6 target; we shall return to this point later.

The cross sections are related to the rate constants by

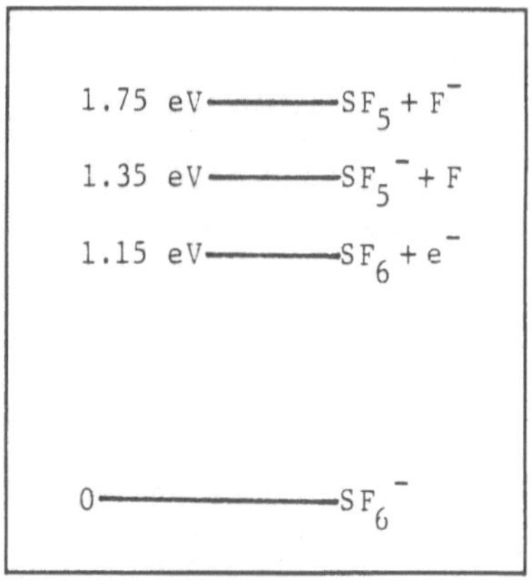

$$k_i = \int_0^\infty \sigma_i(v) f(v) dv \qquad (2)$$

1.75 eV ——— $SF_5 + F^-$

1.35 eV ——— $SF_5^- + F$

1.15 eV ——— $SF_6 + e^-$

0 ——— SF_6^-

Fig. 1.   Energy levels for SF_6^- and some products[4].

where f(v) is the normalized velocity distribution function. For the case where the anion speed can be described by a drift velocity, v_d, the reaction coefficients, κ_i, can be determined from the cross sections by

$$\frac{\kappa_i}{N} = \frac{1}{m v_d} \int_0^\infty \sigma_i(E_{lab}) \, f(E_{lab}) \, dE_{lab} \qquad (3)$$

where E_{lab} is the projectile energy in the lab frame, m is the projectile mass, and $f(E_{lab})$ is the kinetic energy distribution function descriptive of the system.[5] It is the reaction coefficients given by (3) that can be compared to model calculations based upon drift-tube data.

EXPERIMENTAL METHOD

The cross sections σ_i are defined in the usual manner; viz.

$$\sigma_i = \frac{dI_i}{dz} \frac{1}{NI} \qquad (4)$$

where dI_i is a particular decomposition current resulting from a primary beam of current I, dz is a path length and N is the number density of scattering centers. The way to measure these cross sections is to simply determine all of the experimental parameters on the right hand side of (4). This can be done in an ion beam, gas target configuration by separating and trapping the products from different channels with the use of electric and/or magnetic fields.[6] The separation of the charged products is possible because of either the large mass difference (e.g., e vs. F^-) or the substantial difference in the laboratory kinetic energies of the dissociation products. If the target species is not naturally available as a gas at room temperature (e.g., if the target is to be atomic hydrogen), then one must resort to a crossed-beam experiment to measure the cross sections.[7]

Such experiments are fairly straight-forward except when (i) low collision energies - such as those relevant to gaseous dielectrics - are involved or (ii) the projectile is a complex molecular anion which can be formed in any number of excited vibro-rotational states. It is useful to comment on each problem briefly.

In ion beam experiments, laboratory collision energies below about 50 eV are quite difficult to achieve owing to space charge limitations and to stray electric fields within the beam optics system. These effects result in a rapidly decreasing primary ion beam intensity as E_{lab} decreases below about 50 eV as well as to increased uncertainties in the trapping efficiencies for the different products and finally to a concomitant reduction in the signal-to-noise ratio in the cross section measurement. One way to attempt to circumvent this problem in the case of SF_6^- is to examine its collisional dynamics for the case where the target is not SF_6, but rather is a species with a small mass, say helium. In this case the relative collision energy is only 4/150 of the laboratory energy, allowing one to sample extremely low relative collision energies. Although helium is clearly not too similar to SF_6, we shall see that it is possible to gain insight into the collisional dynamics of SF_6^- for low relative collision energies by examining the reactants $SF_6^- +$ He.

The second problem specifically related to SF_6^- is the lack of information concerning the distribution of internal energy in the primary ion beam. Any SF_6^- ion formed by electron attachment must be relaxed collisionally or by a radiative transition if it is to survive. In high pressure environments, collisional relaxation insures that most SF_6^- ions stabilize quickly; this is not the case for pressures typical to ion sources however. Consequently the SF_6^- ion beam has a wide, almost continuous, distribution of internal energies, including those leading to autodetachment. It is with this ill-defined primary ion beam that the current ion-beam, gas target experiments have been performed. The next generation of experiments hopefully will be designed such that the internal energy of the ion beam is better defined than is presently the case.

RESULTS
Collisional detachment

Prior to examining the results for $SF_6^- + SF_6$, let us examine the results for several less complicated reactants. For collisional detachment of atomic anions, it is often the case that the energetic threshold for

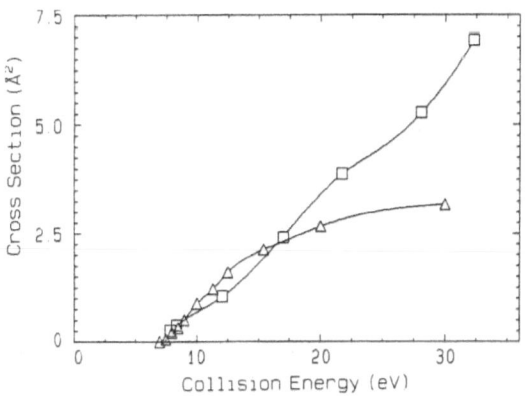

Fig. 2.   Detachment cross sections for F⁻ + Ne (triangles) and SF_6 (squares) as a function of relative collision energy.

3

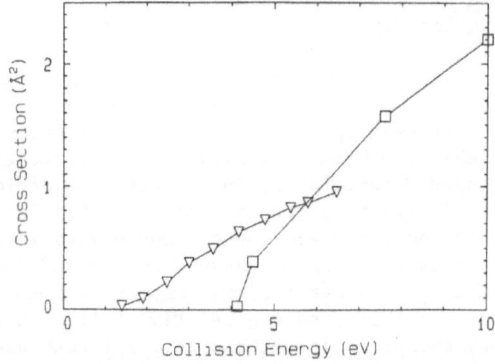

Fig. 3. Detachment cross sections for O_2^- + O_2 (squares) and He (triangles).

detachment exceeds the electron affinity of the atom by a factor ranging from two to one hundred.[8] The example of collisional detachment of F^- (with an electron affinity [EA] of 3.4 eV) by neon and SF_6 is shown in Fig. 2, as a function of the relative collision energy. The thing to note is that the energetic thresholds for detachment exceed the EA by a factor of two for both the simple (Ne) and complex (SF_6) target. This is a common feature in the detachment of atomic anions by atomic targets and is well understood in terms of curve crossings of intermolecular potentials.[8] The collisional properties of molecular anions have not been studied extensively, however. In Fig. 3, results for the detachment of O_2^- (possibly not all of which is in the ground vibrational state) by He[9] and O_2[10] are given. Again it is observed that the energetic threshold exceeds the EA of O_2 by a factor of three for He and a factor of almost ten for the O_2 target. The threshold for detaching O_2^- by O_2 may be high owing to the competing charge transfer process at low collision energies.

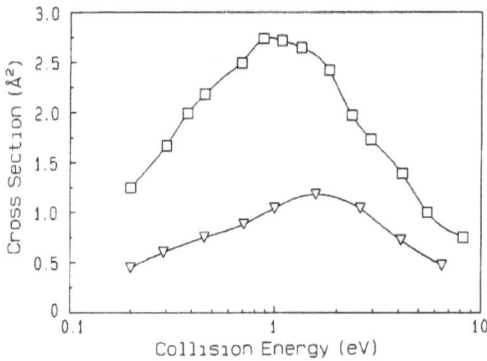

Fig. 4. Detachment cross sections for SF_6^- + He; the upper curve is for "hot" SF_6^- .

4

When we look at the results for the electron detachment of SF_6^- by He, as presented in Fig. 4, we immediately see that excited ions - i.e., $(SF_6^-)^*$ - are clearly in the primary ion beam. Instead of the energetic threshold for detachment lying <u>above</u> the adiabatic EA of SF_6 (~1.1 eV), electron detachment is observed to occur for relative collision energies <u>below</u> the EA, down to as low as 0.2 eV. As may be seen in Fig. 4, the effect of "cooling" the primary ions by altering the source conditions is unambiguous.

The detachment cross sections for SF_6^- ("hot" source conditions) + He, Ne and Ar are shown in Fig. 5. The rapid increase in the cross sections seen at $E \approx 30$ eV may be due to the detachment of "cool" SF_6^-: these cross sections (for E > 30 eV) do not depend upon the ion source conditions.

Now let us turn to the problem relevant to gaseous dielectrics; viz., the collisional detachment of SF_6^- by SF_6. This experimentally determined cross section is also given in Fig. 5. Unlike the case for rare gas targets, the detachment of SF_6^- by SF_6 is seen to be very small for E < 60 eV, even though the SF_6^- ion beam is produced under "hot" source conditions. In fact the measurement is consistent with the detachment cross section being zero to within the experimental uncertainty in the measurement, namely $\pm 0.1 \text{Å}^2$. Although the detachment cross section is tiny for E < 60 eV, we shall see that other inelastic scattering channels are highly probable in collisions of SF_6^- with SF_6.

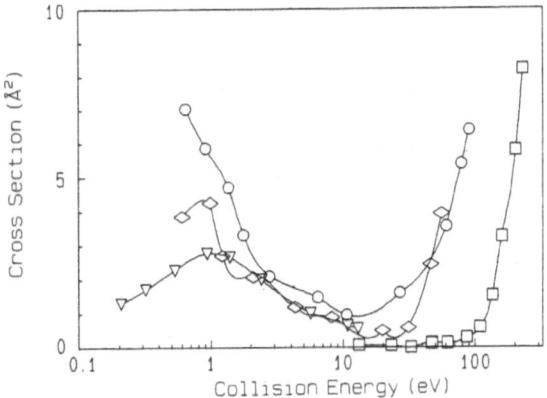

Fig. 5. Detachment cross sections for SF_6^- + He(triangles), Ne(diamonds), Ar(circles) and SF_6(squares).

Other inelastic channels

From the above discussion, it seems probable that collisional electron detachment of SF_6^- by SF_6 is not an important channel in modeling the dielectric properties of SF_6. Under any circumstances, it is clear that charge transfer and collision-induced dissociation are the dominant inelastic channels at low collision energies, as may be seen in Fig. 6, where the cross sections for (1b) and (1c) along with that for charge transfer are displayed. Also shown in Fig. 6 is the cross section for collision-induced dissociation by the target neon; this gives some idea of how this dissociation channel probably behaves for the SF_6 target for E < 10 eV.

It is likely that SF_5^- is formed in discharges by dissociative attachment to SF_6. Thus it is of interest to understand the collisional dynamics of the reactants SF_5^- + SF_6. For this system it is found that the cross sections for electron detachment, dissociation (producing F⁻) and charge transfer are remarkably similar in all respects to those illustrated in Fig. 6 for the SF_6^- projectile.

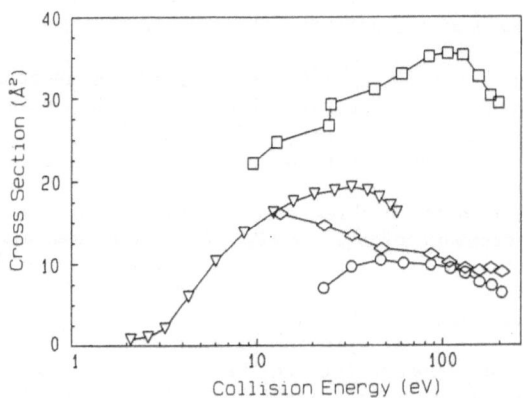

Fig. 6.   Cross sections for collisional decomposition of SF_6^-: F⁻/SF_6 (squares), F⁻/Ne (triangles), SF_5^-/SF_6 (diamonds), and for charge transfer in SF_6^- + SF_6 (circles).

SUMMARY

Absolute total cross sections for electron detachment and collision-induced dissociation for collisions of SF_6^- and SF_5^- with rare gases and SF_6 reveal that the cross sections for electron detachment remain surprisingly small for the SF_6 target until collision energies in excess of tens of electron volts are reached. This observation is clearly important for the development of an understanding of electrical breakdown when using SF_6 as a gaseous dielectric; this has been discussed in some detail recently.[5]

It is also clear that, in the present experiments, excited states of SF_6^- in the primary ion beam play an important - perhaps dominant - role in electron detachment by inert gases for collision energies in the vicinity of the electron affinity of SF_6. A more complete delineation of the role of excitation in the collisional decomposition of SF_6^- will unfortunately be a formidable experimental challenge.

For collision energies in excess of about 30 eV, the initial internal energy of SF_6^- is found <u>not</u> to influence the various decomposition cross sections. This is reasonable when the decomposition is viewed as a two-step process in which the reactant $SF_6^{-\prime}$ with initial internal energy U_1 is collisionally excited to a higher internal energy, U_2, which leaves the product $(SF_6^-)^*$ unstable with respect to the various decomposition channels. In reality, U_1 and U_2 may only be defined by distribution functions, but we may make some broad generalizations based upon their average values, $\langle U_1 \rangle$ and $\langle U_2 \rangle$. For large E, $\langle U_1 \rangle / \langle U_2 \rangle$ may be small and the magnitude of $\langle U_1 \rangle$ should not be important in the collisional dynamics. This is the observation in the present experiments.

A reasonable extension of the two-step model discussed above would be to treat the unimolecular decomposition of the collisionally-excited $SF_6^-(U_2)$ in a standard way in which the probability of decomposing into a particular product channel is taken to be proportional to the density of states for that channel.[11] Such a model will tend to favor the autodetachment channel when $1.15 \leq \langle U_2 \rangle \leq 1.35$ eV (see Fig. 1) but, because of phase space considerations will rapidly favor the autodissociation channels as $\langle U_2 \rangle$ is increased above 1.35 eV. The exact manner in which $\langle U_2 \rangle$ depends upon E is, of course, unknown.

If experiments such as these are to be of optimal use to the field of gaseous dielectrics, then it is obvious that improvements must be made in two areas. First, experiments must be designed and executed to specifically probe the low collision energies (i.e., a few eV) which are of paramount importance in determining the rate coefficients. Second, the internal energy distribution in the primary ion beam must be such it better reflects the conditions found in discharge applications.

ACKNOWLEDGEMENTS
This work was supported in part by the U. S. Dept of Energy, Office of Basic Energy Sciences.

REFERENCES

1. D.L. McCorkle, A.A. Christodoulides, L.G. Christophorou and I. Szamrej, J. Chem. Phys. 72, 4049 (1980); L.E. Kline, D.K. Davies, C.L. Chen and P.J. Chantry, J. Appl. Phys. 50, 6789 (1979); A. Chutjian and S.H. Alajajian, Phys. Rev A31, 2885 (1985); O.J. Orient and A. Chutjian, Phys Rev. A34, 1841 (1986).

2. N. Wiegart, IEEE Trans. Electr. Insul. EI-20, 587 (1985);   R.J. Van Brunt, J. Appl. Phys. 59, 2314 (1986).

3. For more details, see: Yicheng Wang, R.L. Champion, L.D. Doverspike, J.K. Olthoff and R.J. Van Brunt, J. Chem. Phys. 91, 2254 (1989).

4. Edward C.M. Chen, Lih-Ren Shuie, Ela Desai D'sa, C.F. Batten and W.E. Wentworth, J. Chem. Phys. 88, 4711 (1988).

5. See: J.K. Olthoff, R.J. Van Brunt, Yicheng Wang, R.L. Champion and L.D. Doverspike, J. Chem. Phys 91, 2261 (1989), and references cited therein.

6. N.R. White, D. Scott, M.S. Huq, L.D. Doverspike and R.L. Champion, J. Chem Phys. 80, 1108 (1984).

7. M.A. Huels, R.L. Champion, L.D. Doverspike and Yicheng Wang, Phys. Rev. A41, 4809 (1990).

8. See, e.g., L.D. Doverspike, B.T. Smith and R.L. Champion, Phys. Rev. A22, 393 (1980); D. Scott, M.S. Huq, R.L. Champion and L.D. Doverspike, Phys. Rev. A 32, 144 (1985).

9. Unpublished results, this laboratory.

10. A.E. Roche and C.C. Goodyear, J. Phys. B 2, 191 (1969).

11. C.E. Klots, Jour. Phys. Chem. 75, 1526 (1971) and Chem. Phys. Lett. 38, 61 (1976); S.E. Haywood, L.D. Doverspike, R.L. Champion, E. Herbst, B.K. Annis and S. Datz, J. Chem. Phys. 74, 2845 (1981).

DISCUSSION

A. E. D. HEYLEN: What practical application do UF_6 and SF_6 mixed with rare gases have with regards to gaseous insulation?

R. L. CHAMPION: Insofar as I know, UF_6^- is not relevant to gaseous insulation; I mentioned its interesting collisional dynamics simply for illustrative purposes. As far as the reactants SF_6^- and rare gas atoms are concerned, they are perhaps relevant to gaseous insulation for the following reason: It is very difficult to measure cross sections for SF_6^- and SF_6 at low collision energies (e.g., for $E_{lab} \leq 10$ eV). On the other hand, very low _relative_ collision energies may be sampled if experiments on SF_6^- and He are performed. It is hoped that these latter experiments give some insight into the dynamics for SF_6^- and SF_6 at relative energies below about 5 eV.

L. G. CHRISTOPHOROU: I would wish to make two comments: 1. It seems that these studies are important in understanding the basic physics of gaseous dielectrics, particularly with respect to collisional detachment as one of the ways of creating free electrons. You are saying that collisional detachment from SF_6^- is difficult unless SF_6^- is "hot". 2. Of more direct interest to gaseous dielectrics from the stand point of these studies is to consider SF_6/N_2 mixtures.

R. J. VAN BRUNT: One can understand why the direct collisional detachment of SF_6^- does not occur at lower kinetic energies ($E_{cm} < 30$ eV) down to the theoretical threshold defined by the electron affinity; namely there is a strong competition from other energetically allowed decay channels that result in dissociation. How does one explain the unusually high apparent collisional detachment threshold for the $C_s^- +$ Ar interaction?

R. L. CHAMPION: For these (simple) systems, detachment does not occur until the intermolecular potential for the molecular anion, in this case $C_s^- +$ Ar, rises above that for the corresponding neutral system ($C_s +$ Ar). This does not happen until $V(R) \sim$ 40–50 eV for the alkali anion–rare gas systems. Hence the energetic threshold for detachment may be 100 x the EA of the anion for these particular systems.

A TECHNIQUE FOR THE MEASUREMENT OF ELECTRON ATTACHMENT

TO SHORT–LIVED EXCITED SPECIES[1]

L. G. Christophorou, L. A. Pinnaduwage, and A. P. Bitouni

Atomic, Molecular, and High Voltage Physics Group
Health and Safety Research Division
Oak Ridge National Laboratory
Oak Ridge, Tennessee 37831 USA

and

Department of Physics
The University of Tennessee
Knoxville, Tennessee 37996 USA

ABSTRACT

A technique is described for the measurement of electron attachment to short–lived ($\lesssim 10^{-9}$ s) excited species. Preliminary results are presented for photoenhanced electron attachment to short–lived electronically–excited states of triethylamine molecules produced by laser two–photon excitation. The attachment cross sections for these excited states are estimated to be $> 10^{-11}$ cm^2 and are $\sim 10^7$ larger compared to those for the unexcited (ground–state) molecules.

INTRODUCTION

The significance of the basic physics of electron–ground state molecule collision processes in understanding the properties of gaseous dielectrics has been amply documented.[1] However, the role of electronically–excited species has not as yet been fully addressed, although it is considered to be important in its effects on the dielectric and "switching" properties of gases.[2-4] The principal reason for this is the lack of knowledge on electron attachment to, electron scattering from, and electron–impact ionization of electronically–excited species, which in turn is due to experimental difficulties encountered in producing sufficient numbers of specific excited species under controlled experimental conditions. Recently, we successfully produced indirectly via laser excitation, sufficient numbers of long–lived electronically–excited molecules under swarm conditions and observed 5 to 6 orders of magnitude increases in the attachment of slow electrons to them compared to the ground electronic states. In Fig. 1a is shown the laser/electrode arrangement used in these studies. The laser pulse which produced the excited species injected at the end of its path a pulse of photoelectrons in the gas from the cathode; these electrons attached to the laser–produced excited species en route to the anode. Figure 1b

---

[1]Research sponsored in part by the Office of Health and Environmental Research, U.S. Department of Energy, under Contract No. DE–AC05–84OR21400 with Martin Marietta Energy Systems, Inc., and in part by the National Science Foundation under Contract No. CHE–8813466 with the University of Tennessee, Knoxville.

shows the measured electron attachment coefficient for the first excited triplet and the ground singlet state of the thiophenol molecule.[5,6] Besides their intrinsic value these initial observations demonstrated the feasibility of optical switching of the dielectric properties of gases by laser–induced changes in electron attachment involving electronically–excited molecules. In this regard, it is desirable to achieve optically–enhanced electron attachment to short–lived ($\lesssim 10^{-8}$ s) excited electronic states since this would allow optical switching at higher repetition rates.

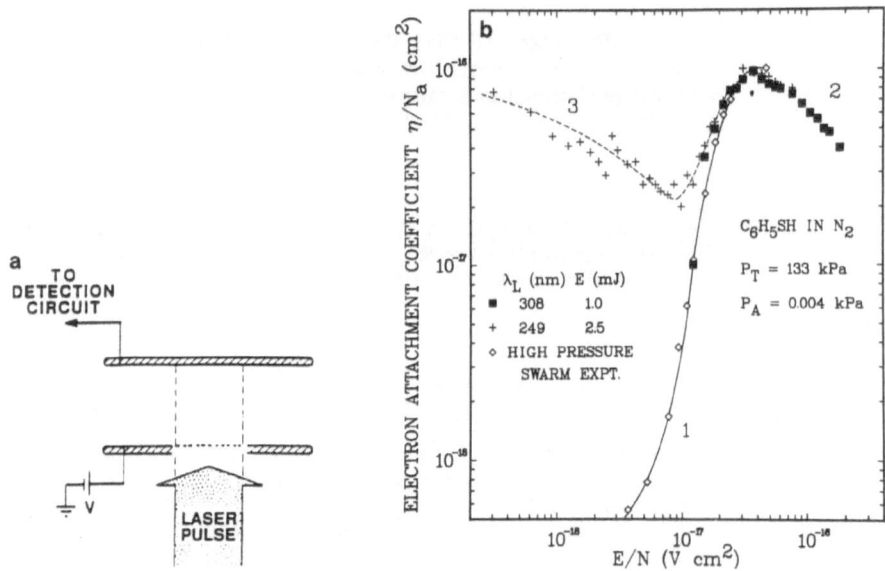

Fig. 1(a).   Schematic drawing of the two–electrode configuration employed for measurement of electron attachment to the first–excited triplet state of thiophenol. The detection region is the same as the interaction region and is located between the two electrodes; the same laser pulse irradiates the interaction region (producing excited attaching species) and also produces a pulse of attaching electrons at the cathode via photoelectric emission. At the laser fluences employed, laser photoionization is negligible.

(b)   The electron attachment coefficient, η/N_A vs. E/N for thiophenol in N_2 buffer gas. Curve 1 was obtained in a separate high pressure swarm experiment without laser irradiation, and depicts electron attachment to the ground state. Curves 2 and 3 were obtained in the present experiment with XeCl and KrF laser lines respectively. The photon energy of the XeCl line is not sufficient to excite electronically the molecule monophotonically and therefore only the ground state attachment is observed; however, electronic excitation and enhanced electron attachment occurs at the KrF line. Note that η/N_A^* is ~ 100 times larger than η/N_A since the excited molecule number density N_A^* is about one percent of N_A.

In this paper, we describe for the first time a technique for measuring photoenhanced electron attachment to short–lived ($\lesssim 10^{-8}$ s) excited electronic states of molecules and present preliminary results on low–energy electron attachment to high–lying excited electronic states of the triethylamine molecule.

TECHNIQUE

Description

In Fig. 2a is shown the laser/electrode configuration of the electron swarm apparatus[6] we employed. The gas Ar or N_2 was used as the buffer medium. A single excimer–laser pulse generated (via multiphoton ionization) the attaching electrons volumetrically over the laser–irradiated region (Fig. 2a) and, also, the electronically–excited molecules. Since the excited species and the electrons were produced concomitantly and in close proximity, electron attachment can occur within the short lifetimes of the electronically–excited molecules. The attaching electrons were produced by photoionization of the same gas that produced the electron–attaching excited species, although a suitable gas additive could be added to the binary mixture for this purpose. Since in this technique positive ions are also produced, the negative charges (electrons and anions) must be separated and detected unambiguously. This was accomplished (Fig. 2a) by separating the detection region (located between electrodes 2 and 3) from the interaction region (located between electrodes 1 and 2) via a three–electrode arrangement. Charge transmission between the two regions was through a fine grid. The same electric field was maintained in both regions and—depending on the direction of the field—either negative or positive charges produced in the interaction region were extracted into the detection region.

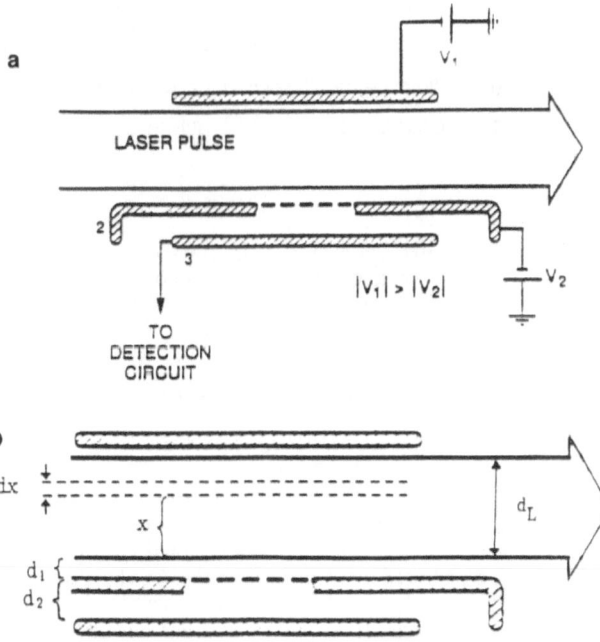

Fig. 2(a).   Schematic diagram of the experimental arrangement for the measurement of electron attachment to short–lived excited species.

(b)   Subdivision of electrons initially produced via laser photoionization in to "planar swarms" (see text).

Analysis

The data analysis depends on the electron attachment mechanisms involved and the relative values of four time parameters: the life time τ of the excited species; the time τ_{ss} for the photoionization electrons to reach steady state; the time τ_a for an electron to be attached; and the time τ_d for the electrons to drift through the laser irradiated region.

τ: can vary from ms to sub–ns depending on the excited state involved.

τ_{ss}: under our conditions [N_2, Ar pressures in the range 7 to 70 kPa] is in the range 10^{-6} to 10^{-8} s.

τ_a: depends on the electron attachment rate constant k_a and the number density N_a of the electron attaching species; when these are known, τ_a can be estimated from $\tau_a = (k_a N_a)^{-1}$. Since, moreover, electron attachment has to occur before the decay of the excited species,

$$\tau_a < \max\left\{\tau, \tau_L\right\} \tag{1}$$

where τ_L is the duration of the laser pulse.

τ_d: d_L/w, where d_L is the "height" of the laser pulse (Fig. 2a) and w is the electron drift velocity; under our experimental conditions τ_d varies from 10^{-5} to 10^{-6} s.

Two extreme cases can be distinguished:

(i) $\tau > \tau_d$. This is the case of "long–lived" ($\tau \gtrsim 10^{-5}$ s) species for which conventional electron swarm relationships apply in the present experiments. In conventional electron swarm studies $\tau_{ss} \ll \tau_d \simeq \tau_a < \tau$.

(ii) $\tau < \tau_L$. From Eq. (1), it follows that $\tau_a < \tau_L$ ($\approx 10^{-8}$ s in our experiments). This is the case of "short–lived" species ($\tau < 10^{-8}$ s). In this situation, $\tau_a \ll \tau_{ss}$ and the electrons are attached before a steady–state condition is reached.

Long–Lived Species ($\tau \gtrsim 10^{-5}$ s)

Molecules in metastable excited states or in their ground electronic state satisfy conditions (i) above. Under such conditions, the photoionization electrons quickly reach a steady–state and drift through the laser–irradiated region toward the anode; electron attachment occurs while the electrons drift. Conventional electron swarm relationships are applicable; for example, the k_a is related to w and the density–normalized electron attachment coefficient η/N_a by

$$k_a = (\eta/N_a)w \tag{2}$$

The present swarm technique differs from the conventional ones however in that instead of a planar electron swarm in the conventional, the electrons are produced over the entire volume irradiated by the laser pulse in the present technique. We can, however, subdivide this volume into an infinite number of "planar swarms" (Fig. 2b) and sum their contributions. If n_0 is the number density of electrons produced via laser photoionization, A the area of the grid in the middle electrode (Fig. 2b), and η the electron attachment coefficient of the attaching species located within the boundaries of the laser pulse, the number of unattached electrons crossing the lower laser boundary (Fig. 2b) is

$$N_E = \int_0^{d_L} n_0 \, A dx \, e^{-\eta x} = \frac{n_0 A}{\eta} [1 - e^{-\eta d_L}] \tag{3}$$

If no further electron attachment occurred during the traversal of d_1 and d_2, the measured voltage ratio, R_v, of the total signal, V_T, to the signal component, V_E, due to the unattached electrons would be

$$R_v = \frac{V_T}{V_E} = \frac{N_I + N_E}{N_E} = \frac{n_0 \, A d_L}{N_E} \tag{4}$$

From (3) and (4),

$$R_v = \frac{\eta d_L}{[1 - e^{-\eta d_L}]} \tag{5}$$

and η is, then, determined from the measured R_v by an iterative procedure.

When electron attachment occurs outside the laser–irradiated region, as in the case of electron attachment to ground state molecules, the measured V_E, $(V_E)_m$ would be smaller than one would expect if the attachment occurred only within the laser irradiated region. If η' is the attachment coefficient outside of the irradiated region, the electron component, $(V_E)_{c_1}$, corrected for electron attachment during the drift d_2 (inside the detection region where the applied electric field is the same as that in the interaction region) is

$$\frac{(V_E)_{c_1}}{(V_E)_m} = \frac{\eta' d_2}{1 - e^{-\eta d_2}} \tag{6}$$

When, in addition, correction is made for electron attachment during the drift d_1, the final corrected value $(V_E)_c$ is

$$(V_E)_c = (V_E)_{c_1} \, e^{\eta' d_1} = (V_E)_m \, e^{\eta' d_1} \left\{ \frac{\eta' d_2}{1 - e^{-\eta' d_2}} \right\} \tag{7}$$

In the case of measurement of electron attachment to ground–state molecules, which we have made to only verify the technique, $\eta' = \eta$ and published values of η were used in (7) to obtain $(V_E)_c$, which in turn was used in (5) to determine the electron attachment coefficient within the irradiated volume.

Short–Lived Species ($\tau < 10^{-8}$ s)

Unlike the long–lived species case just discussed, in this case electron attachment does not occur while the electrons drift but rather immediately after they are produced and before they begin to drift. In this case, then, the number density of negative ions formed within the laser irradiated volume is

$$N_I = \int_0^{\tau_L} N_e(t) \, N^*(t) \, k_a \, dt \tag{8}$$

where $N_e(t)$ and $N^*(t)$ are the number densities of electrons and excited species at time t and k_a is now the electron attachment rate constant of the excited species.

The number density of electrons initially produced by laser photoionization is

$$N_e \, (t = \tau_L) = N_T = N_I + N_E \qquad (9)$$

where N_I is the number density of negative ions and N_E is the number density of unattached electrons. The expressions for $N_e(t)$ and $N^*(t)$ will depend on the particular case under consideration. Disregarding any possible electron attachment to ground state molecules during the drift to and within the detection region, the measured voltage ratio, R_v, is given by,

$$R_v = \frac{V_I}{V_T} = \frac{N_I}{N_T} \qquad (10)$$

The laser fluence F, the partial pressure of the gas which yields the short—lived electron attaching species, and the electron attachment cross section $\sigma_a(\epsilon)$ are among the parameters which determine the value of R_v. Measurements of the R_v vs F, along with assumed or established electron attachment mechanisms can yield[7] the value of—or a quantity related to—$k_a(\epsilon)$ and $\sigma_a(\epsilon)$.

MEASUREMENTS

Electron Attachment to Ground—State SF$_6$

In order to test the present technique, we first measured electron attachment to ground state SF$_6$ molecules at room temperature using N$_2$ as the buffer gas. The attaching electrons were generated by XeF (250 nm; 3.5 eV) laser two—photon ionization of tetrakis—dimethylaminoethylene (TMAE) [adiabatic ionization potential, $I \lesssim 6$ eV] which was added to the SF$_6$/N$_2$ mixture. The laser fluence, was $\lesssim 0.3$ mJ cm^{-2} and the only effect of the laser irradiation was to photoionize TMAE. The cross sectional area of the laser pulse was 0.8 x 0.5 cm^2. Correction for electron attachment occurring outside of the laser—irradiated region was carried out (with Eq. (7)) using published values[8] of η' ($= \eta$). It can be seen from Fig. 3 that the present measurements of η/N_a (E/N) in the laser—irradiated region are in excellent agreement with the published data[8] obtained on SF$_6$/N$_2$ by conventional electron swarm techniques.

Electron Attachment to Short—Lived Electronically—Excited Triethylamine Molecules

We employed the present technique to measure electron attachment to electronically—excited states of triethylamine (TEA) which lie energetically above the first ionization—threshold energy I and are presumably Rydberg in character. Excited states of molecules lying above I are called superexcited states (SES); they normally decay rapidly (lifetimes $< 10^{-9}$ s) by preionization and/or predissociation. These studies were conducted at room temperature in N$_2$ or Ar buffer—gas mixtures at total pressures of ~ 6 to 60 kPa.

Figure 4 shows the measured total signal (proportional to the number density of electrons initially produced via photoionization; Eq. (8)), $(V_T)_m$, and the measured negative ion signal, $(V_I)_m$, as a function of the laser fluence, F. At low laser fluences,

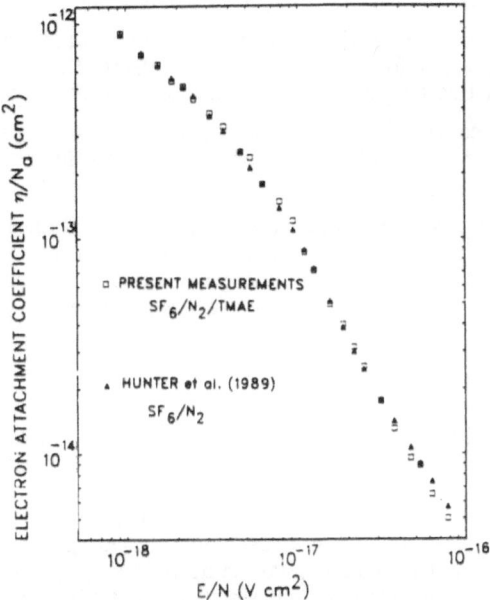

Fig. 3.   Comparison of electron attachment measurements for ground state SF_6 obtained using the present technique with published data[8] obtained using a conventional swarm technique.

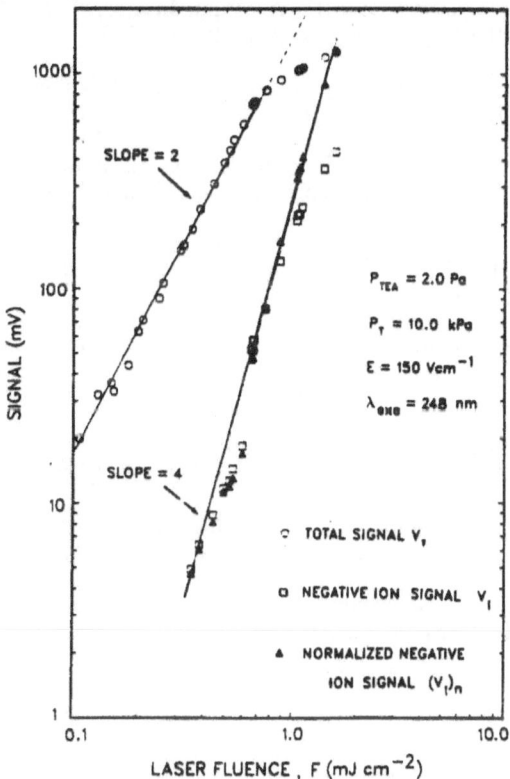

Fig. 4.   Laser fluence dependence of the measured total and negative ion signals and the normalized negative ion signal (see text) shown on a log–log plot for the experimental parameters indicated in the figure.

$(V_T)_m$ varies as F^2 indicating two–photon ionization. The deviation from this quadratic dependence at high F (> 0.75 mJ cm^{-2} in this case) is due to space–charge effects. Once corrected for the space–charge effects, the normalized negative ion signal, $(V_I)_n$ varied as F^4 (since four photons are needed to form a negative ion; two photons to produce the SES and two more photons to produce the attaching electron). The $(V_T)_n$ and $(V_I)_n$ were shown to have linear and quadratic dependences, respectively, on the triethylamine pressure. Using these findings and measurements along with the available photophysical studies on triethylamine we concluded[7] that the observed photoenhanced electron attachment is due to the SES of TEA and that their electron attachment cross sections are enormous ($> 10^{-11}$ cm^2). These cross sections are over 10^7 times larger than those for the ground state TEA molecules.

CONCLUSIONS

A technique has been described which allows measurement of electron attachment to short–lived electronically–excited species produced by pulsed–laser light. The technique provided the first results of this nature which indicate that slow electrons attach to short–lived electronically–excited molecules with enormous cross sections. More studies are needed—and are in progress—to establish the photoenhanced electron attachment mechanisms involved, determine accurately the rate constants and cross sections, and assess the potential of these findings for fast optical switching.

REFERENCES

1. L. G. Christophorou and L. A. Pinnaduwage, Basic Physics of Gaseous Dielectrics, IEEE Trans. Electr. Insul. 25:55 (1990).

2. J. M. Meek and J. D. Craggs (eds.), "Electrical Breakdown of Gases," Wiley, New York (1978).

3. L. G. Christophorou and S. R. Hunter, From Basic Research to Application, in: "Electron–Molecule Interactions and Their Applications," (L. G.Christophorou, ed.), Academic Press 2, New York (1990).

4. A. Guenther, M. Kristiansen, and T. Martin (eds.), "Opening Switches," Plenum Press, New York (1987).

5. L. G. Christophorou, S. R. Hunter, L. A. Pinnaduwage, J. G. Carter, A. A. Christodoulides, and S. M. Spyrou, Optically Enhanced Electron Attachment, Phys. Rev. Lett. 58:1316 (1987).

6. L. A. Pinnaduwage, L. G. Christophorou, and S. R. Hunter, Optically Enhanced Electron Attachment to Thiophenol, J. Chem. Phys. 90:6275 (1989).

7. L. A. Pinnaduwage and L. G. Christophorou, Enhanced Electron Attachment to Superexcited Molecular States, submitted to J. Chem. Phys. (1990).

8. S. R. Hunter, J. G. Carter, and L. G. Christophorou, Low–Energy Electron Attachment to SF$_6$ in N$_2$, Ar, and Xe Buffer Gases, J. Chem. Phys. 90:4879 (1989).

DISCUSSION

V. H. GEHMAN, JR.: How will excited species attach high—energy electrons, like those found in some pulsed—power—application switches?

L. G. CHRISTOPHOROU: Electron attachment is a resonant process, occurring over a narrow energy range. The cross section generally increases as the electron energy decreases. For pulsed power applications, some of the electrons, produced by ionization, near excited species, will have low energy and will attach very strongly. The high attachment cross section in the case of excited species may be explained as being due to an electron—dipole interaction. Because of these very high cross sections even if the concentration of excited species is a million times lower than that of the ground state species, electron attachment to the excited species will be important.

B. MARODE: Once attached to the excited species, is the negative ion produced left in an excited state?

L. G. CHRISTOPHOROU: Initially yes. This process is similar to an electron excited Feshbach resonance in which a fast electron excites the molecule leaving a low energy electron in the field of an excited species, and attachment occurs with a high cross section. In this case a laser beam is used to produce both the excited species and the electron (by photoionization). Following electron attachment, the transient anion will normally decay by dissociative attachment or autodetachment.

TOTAL CROSS SECTIONS FOR ELECTRON SCATTERING AND ATTACHMENT FOR SF₆ AND ITS ELECTRICAL-DISCHARGE BY-PRODUCTS

J. K. Olthoff and R. J. Van Brunt

National Institute of Standards and Technology
Gaithersburg, MD 20899

H.-X. Wan, J. H. Moore and J. A. Tossell

Department of Chemistry and Biochemistry
University of Maryland
College Park, MD 20742

INTRODUCTION

Sulfur hexafluoride (SF_6), either pure or mixed with other gases, is commonly used as an insulator in high voltage-equipment. Consequently, many studies have been performed to investigate the decomposition of SF_6 in various electrical discharges including corona,[1] sparks,[2] and arcs.[3] These studies have shown that large quantities of toxic and corrosive by-products such as SO_2, SOF_2, SO_2F_2, SOF_4, SF_4, and S_2F_{10} are produced when SF_6 is dissociated in the discharge. Additionally, recent studies of SF_6 as an etching gas for semiconductor processing have indicated that stable sulfur oxyfluoride by-products can account for more than 10% of the neutral molecules in the plasma.[4]

A full understanding of the physical processes occurring in SF_6 discharges and of the electron attaching processes in decomposed SF_6 requires a detailed knowledge of the interaction of free electrons with SF_6 and its by-products. In this paper we present absolute cross sections for electron scattering and for negative-ion formation through electron attachment to SF_6 and to several by-products produced by electrical discharges in SF_6 (SO_2, SOF_2, and SO_2F_2). These results are compared with previous data where available, and calculations of electron attachment energies are presented to aid in the interpretation of the cross section data.

EXPERIMENT

An electron transmission spectrometer employing a trochoidal monochromator[5] forms the basis of these experiments. This instrument consists of a thermionic electron source followed by the trochoidal monochromator, an accelerating lens, a gas cell, and a retarding lens which permits only unscattered electrons to be transmitted to an electron collector. The instrument is immersed in a uniform magnetic field of about 70 gauss. The electron energy resolution was about 100 meV and the temperature was maintained at 328 K. Total electron scattering cross sections are obtained by measuring the attenuation of the transmitted current due to the introduction of a sample into the gas cell. Cross sections for electron attachment (lifetimes > 10 μs) and dissociative attachment processes are determined from a measurement of the product negative ion flux to the walls of the gas cell.

The presence of the magnetic field introduces uncertainty in the length of the electron trajectories through the gas cell,[6] as well as uncertainty in the acceptance angle defined by the retarding lens which precedes the collector.[7] Additional uncertainty is associated with the measurement of the target gas pressure in the 0.2 to 1.0 mtorr (0.03 to 0.13 Pa) range at which the cross sections were determined. Overall, the cross sections reported are believed to be accurate to within 15% for electron energies above 1 eV. Below this energy, the uncertainty

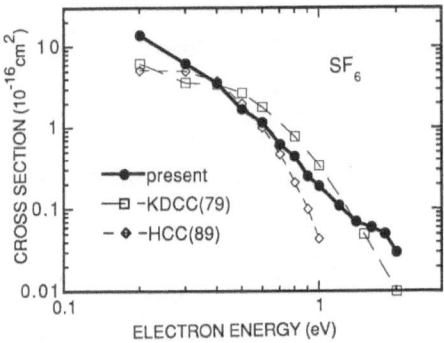

Fig. 1. Cross sections for electron attachment or dissociative attachment to SF_6 from 0.2 to 2 eV. Previous data from references 10 (HCC) and 13 (KDCC) are presented for comparison.

increases to as much as 50% at the lowest energies (≤ 0.2 eV). The limit of sensitivity in the dissociative attachment cross section measurements is about 2×10^{-18} cm^2. The precision of the measurements deteriorates as this limit is approached.

EXPERIMENTAL RESULTS AND DISCUSSION

SF$_6$

The total cross sections for electron scattering by SF_6 determined in the present experiment are not shown here but agree with previously reported values[8,9] to within the uncertainties discussed above.

Negative-ion formation from SF_6 by electron attachment and dissociative attachment has received considerable study. Christophorou and co-workers have performed several swarm studies[10] of electron attachment to SF_6, and Fenzloff *et al.*[11] have published a detailed study of the relative ion yields for dissociative attachment to SF_6. At very low energies (0–2 meV), Chutjian and co-workers[12] have measured absolute attachment cross sections using threshold photoelectron spectroscopy, while Kline and co-workers[13] have measured absolute cross sections for attachment and dissociative attachment from 0.01 eV to 15 eV in a beam experiment.

Absolute cross sections for electron attachment and dissociative attachment to SF_6 as measured by the present experiment are presented in Figure 1 for electron energies from 0.2 eV to 2.0 eV. At energies greater than 2 eV, the cross section was too small to measure in this experiment. Attachment and dissociative-attachment cross sections measured by Kline *et al.*[13] and calculated by Hunter *et al.*[10] from swarm data are shown for comparison. The cross sections in Figure 1 for Kline *et al.*[13] and for Hunter *et al.*[10] are the sum of their cross sections for SF_6^- and SF_5^- production. Note that our cross section values fall significantly below the values of Kline *et al.*[13] from about 0.4 eV to about 1.4 eV. This is in general agreement with analyses[10,14,15] of swarm data for which the experimentally determined electron-collision cross sections for SF_6 were adjusted downward in order to derive accurate transport, ionization, attachment, and dissociation coefficients of SF_6. At energies greater than 1.2 eV our results

20

appear to agree well with Kline *et al.*.[13] It must be noted, however, that at these energies the magnitude of the cross section approaches the detection limits of the experiment (2×10^{-18} cm^2). At lower energies (~ 0.2 eV) we appear to be in agreement with the attachment cross section (1.2×10^{-15} cm^2) published by Chutjian and co-workers.[12]

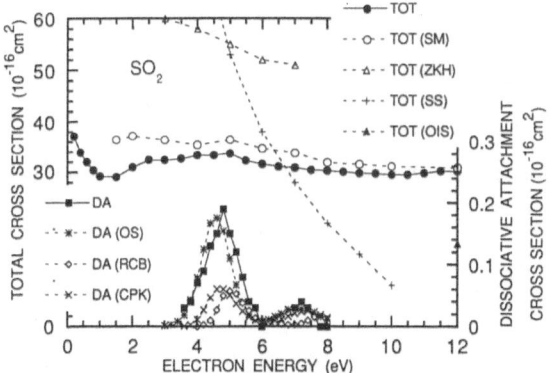

Fig. 2. Total electron-scattering cross sections and dissociative-attachment cross sections for SO$_2$. Previously published total cross sections from references 16 (ZKH), 17 (SS), 18 (SM), and 19 (OIS, total elastic-scattering cross section), and previously published dissociative-attachment data from references 21 (RCB), 22 (CPK), and 23 (OS) are shown for comparison.

SO$_2$

To date three conflicting experimental measurements of the total cross section for electron scattering by SO$_2$ have been published.[16-18] These three data sets are shown in Figure 2 along with the measurements from the present experiment. A single measurement[19] of the elastic scattering cross section at 12 eV is also shown. Our results are in closest agreement with the recent results of Szmytkowski and Maciag,[18] although discrepancies exceeding 20% are observed, especially at lower energies. Broad maxima observed near 2.5 eV and 5 eV in the cross sections measured here and in those of Szmytkowski and Maciag[18] correspond to the resonances observed by Sanche and Schulz[20] in derivative electron transmission spectra.

Previous measurements[21-24] of the cross sections for dissociative attachment to SO$_2$ differ in magnitude by as much as 70%. Figure 2 shows the measured dissociative-attachment cross sections from the present experiment and from Refs. 21-23. Qualitative agreement between these measurements is good with each experiment showing peaks near 4.7 eV and 7.2 eV. Mass spectrometric studies[24] have shown that the peak near 4.7 eV is composed primarily of O$^-$ and SO$^-$ while the peak near 7.2 eV is almost solely O$^-$. The peak near 4.7 eV corresponds to the broad maximum in the total cross section data near 5 eV. Although the dissociative-attachment data from our experiment are near the experimental detection limits, and therefore have fairly large error limits ($\pm 2 \times 10^{-18}$ cm^2), the present data are clearly in agreement with the values reported in Refs. 23 and 24, both of which show peak values near 18×10^{-18} cm^2.

SOF$_2$

The total electron-scattering cross sections for thionyl fluoride (SOF$_2$) from the present experiment are shown in Figure 3. A prominent resonance is observable at 0.6 eV with a weaker resonance appearing as a shoulder near 2.0 eV.

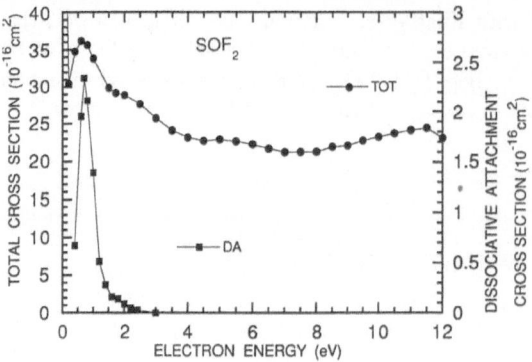

Fig. 3. Total electron-scattering cross sections and dissociative-attachment cross sections for SOF_2.

The dissociative-attachment cross section data from the present experiment are also shown in Figure 3. Note that the peak near 0.7 eV and the shoulder near 2 eV correspond to the resonances observed in the total cross section for electron scattering. Mass spectrometric studies of negative-ion formation[24] show an F^- peak near 0.6 eV and a shoulder near 2 eV, in agreement with the present data. Sauers *et al.*[24] also observed the formation of SOF_2^- at threshold electron energies but at peak intensities approximately 200 times smaller than for F^-. This small current would be undetectable in the present experiment.

SO_2F_2

Figure 4 shows the total cross sections for electron scattering by SO_2F_2. It is interesting to note that no prominent resonance peaks are observed. Additionally, the total electron-scattering cross section for SO_2F_2 is the lowest of any of the compounds investigated here.

The cross section for dissociative attachment to SO_2F_2 is also shown in Figure 4. The magnitude of the dissociative-attachment cross sections for SO_2F_2 is much smaller than for SOF_2, probably because there are no corresponding resonances in the total electron-scattering cross section. The peak in the dissociative-attachment cross section near 3.4 eV is in agreement with previous mass spectrometric studies by Wang and Franklin[25] and by Sauers and co-workers.[24] These studies indicate that this peak is produced by the formation of SO_2F^-, F_2^-, and F^-, and that the increase in the cross section at low energies is evidently due to the formation of the parent ion, $SO_2F_2^-$, by electron attachment. The cross section for dissociative attachment has been calculated from recent swarm studies[26] of SO_2F_2 at room temperature to be 1.06×10^{-16} cm^2 for 0.22 eV electrons. However, this value is more than an order of magnitude larger than the dissociative-attachment value measured by this experiment.

THEORY

In previous work we have found a high degree of correlation between the energies of shape resonances observed in electron transmission spectroscopy and those observed near inner-shell ionization edges in electron energy-loss or x-ray absorption spectroscopy.[6,27] The former involve temporary capture of low-energy electrons into low-lying, unfilled molecular orbits; the latter involve transitions of inner-shell electrons to analogous orbitals. For the inner-shell electron excitation process, the resonant state is stabilized by the positive core that is created. These energies differ by a stabilization energy, SE, given by the sum of the attachment energy, AE, which characterizes a resonance feature in the electron transmission spectrum (or total electron-

22

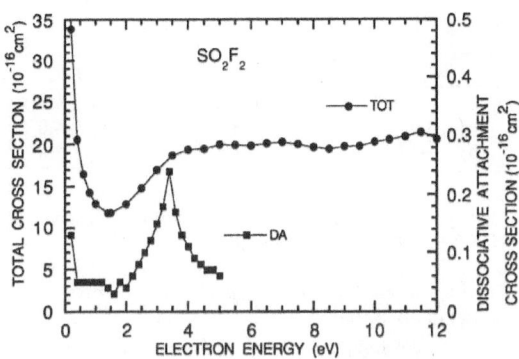

Fig. 4. Total electron-scattering cross sections and dissociative-attachment cross sections for SO_2F_2. The apparent increase in the dissociative attachment cross section at low energies is due to the formation of long-lived parent ions (see Ref. 24).

Table 1. Calculated term values (TV), attachment energies (AE), and stabilization energies (SE) for SO_2 and SF_4 in electron volts.

	$SO_2\ b_2$	$SF_4\ b_2$
TV	10.07	7.99
AE	-0.65	1.02
SE	9.42	9.01

scattering cross section) and the term value, TV, which is the difference between the inner-shell ionization edge and the inner-shell excitation energy to the state analogous to the resonant state observed in low-energy electron scattering: $SE = AE + TV$. The stabilization energy has been found to be relatively constant in a series of similar molecules, thus if SE can be estimated and TV is available for a particular unfilled orbital, it is possible to make reasonable assignments of features observed in low-energy electron scattering and dissociative-attachment cross section measurements.

To aid in the interpretation of our measurements we have, as in previous work, carried out an extensive series of *ab initio* Hartree-Fock calculations on both neutral and core-ionized sulfur fluorides and oxyfluorides within the approximation of the equivalent ion core virtual orbital model.[6,28] In particular, in order to establish the relation (for the series of molecules under investigation here) between term values from inner-shell excitation spectroscopy and attachment energies from electron transmission measurements, we have calculated TV and AE at the ΔSCF level[29] for the lowest virtual b_2 orbitals of SO_2 and SF_4. The calculated term values agree with experiment to within 1 eV or better. For SO_2, such a procedure is well defined for the calculation of the AE, since the $^2B_2$ negative ion state is stable (that is the AE is negative). The calculated attachment energy agrees within 0.5 eV with the measured[30] electron affinity of SO_2. For SF_4 the anion state is unbound at the neutral geometry and thus the calculated vertical attachment energy is unstable to the addition of diffuse functions and would indeed go to zero if a sufficient number of such functions were employed. However, using the same type basis as for the TV calculation, an attachment energy of 1.02 eV is calculated. As shown in Table 1, the values of SE implied by these calculations is 9.4 eV for SO_2 and 9.0 eV for SF_4. An average value for SE of 9.2 eV has been used in Table 2 to predict, from measured

Table 2. Projected values of AE's (in eV's) based on experimentally determined TV + 9.2 eV. Values derived from calculated TV's are in parenthesis.

	SF$_6$		SO$_2$		SOF$_2$		SO$_2$F$_2$		SF$_4$
		b_1	5.9	a'	(6.3)	a_1	6.4	b_1	3.2
$t1_u$	4.9	a_1	3.5	a'	(3.6)	a_1	4.0	a_1	3.2
$a1_g$	1.9	b_2	−1.0	a''	1.5	b_2	3.0	b_2	0.2

term values,[8,31-33] the energies and assignments of resonances observed in low energy electron scattering for other sulfur fluorides and oxyfluorides.

This approach suggests: (1) dissociative attachment to SF$_6$ proceeds through a threshold electron capture process, (2) dissociative attachment to SO$_2$ is associated with electron capture into the b_1 and higher unfilled molecular orbitals, (3) dissociative attachment to SOF$_2$ proceeds through the two lowest anion states, and (4) dissociative attachment to SO$_2$F$_2$ is primarily associated with electron capture in the lowest unoccupied molecular orbital.

ACKNOWLEDGMENT

This work was supported by NSF Grant No. CHE-87-21744 and the U. S. Department of Energy.

REFERENCES

1. R. J. Van Brunt, Production Rates for Oxyfluorides SOF$_2$, SO$_2$F$_2$, and SOF$_4$ in SF$_6$ Corona Discharges, *J. of Res. Nat. Bur. Stand.* **90**, 229 (1985).

2. I. Sauers, By-product Formation in Spark Breakdown of SF$_6$/O$_2$ Mixtures, *Plasma Chem. Plasma Process*, **8**, 247 (1988).

3. F. J. J. G. Janssen, Measurements at the Sub-ppm Level of Sulphur-fluorine Compounds Resulting from the Decomposition of SF$_6$ by Arc Discharge, *Kema Sci. Tech. Rep.* **2**, 9 (1984).

4. G. Turban and M. Rapeaux, Dry Etching of Polyimide in O$_2$/CF$_4$ and O$_2$/SF$_6$ Plasmas, *J. Electrochem. Soc.* **130**, 2231 (1983).

5. A. Stamatovic and G. J. Schulz, Characteristics of the Trochoidal Electron Monochromator, *Rev. Sci. Instrum.* **41**, 423 (1970); M. R. McMillan and J. H. Moore, Optimization of the Trochoidal Electron Monochromator, *ibid.* **51**, 944 (1980); G. J. Schulz, Electron Transmission Spectroscopy: Rare Gases, *Phys. Rev. A* **5**, 1672 (1972).

6. H.-X. Wan, J. H. Moore, and J. A. Tossell, Electron Scattering Cross Sections and Negative Ion States of Silane and Halide Derivatives of Silane, *J. Chem. Phys.* **91**, 7340 (1989).

7. A. R. Johnston and P. D. Burrow, Scattered-Electron Rejection in Electron Transmission Spectroscopy, *J. Electron Spectrosc. Relat. Phenom.* **25**, 119 (1982).

8. R. E. Kennerly, R. A. Bonham, and M. McMillan, The Total Absolute Electron Scattering Cross Sections for SF$_6$ for Incident Energies Between 0.5 and 100 eV Including Resonance Structure, *J. Chem. Phys.* **70**, 2039 (1979).

9. M. S. Dababneh, Y. -F. Hsieh, W. E. Kaupilla, C. K. Kwan, S. J. Smith, T. S. Stein, and M. N. Uddin, Total-cross-section Measurements for Positron and Electron Scattering by O$_2$, CH$_4$, and SF$_6$, *Phys. Rev.* **A38**, 1207 (1988).

10. S. R. Hunter, J. G. Carter, and L. G. Christophorou, Low Energy Electron Attachment to SF$_6$ in N$_2$, Ar, and Xe Buffer Gases, *J. Chem. Phys.* **90**, 4879 (1989).

11. M. Fenzloff, R. Gerhard, and E. Illenberger, Associative and Dissociative Electron Attachment by SF$_6$ and SF$_5$Cl, *J. Chem. Phys.* **88**, 149 (1988).

12. A. Chutjian and S. H. Alajajian, s-wave Threshold in Electron Attachment: Observations and Cross Sections in CCl$_4$ and SF$_6$ at Ultralow Electron Energies *Phys. Rev. A* **31**, 2885 (1985); O. J. Orient and A. Chutjian, Comparison of Calculated and Experimental Thermal Attachment Rate Constants for SF$_6$ in the Temperature Range 200-600 K, *ibid.* **34**, 1841 (1986).

13. L. E. Kline, D. K. Davis, C. L. Chen, and P. J. Chantry, Dielectric Properties for SF$_6$ and SF$_6$ Mixtures Predicted from Basic Data, *J. Appl. Phys.* **50**, 6789 (1979).

14. A. V. Phelps and R. J. Van Brunt, Electron-transport, Ionization, Attachment, and Dissociation Coefficients in SF$_6$ and Its Mixtures, *J. Appl. Phys.* **64**, 4269 (1988).

15. J. P. Novak and M. F. Fréchette, Transport Coefficients of SF$_6$ and SF$_6$/N$_2$ Mixtures from Revised Data, *J. Appl. Phys.* **55**, 107 (1984).

16. M. Zubek, S. Kadifachi and J. B. Hasted, in *Book of Abstracts of the European Conference on*

Atomic Physics, eds. J. Kowalski, G. zu Putlitz and H. G. Weber (Heidelberg, 1981) p. 763.

17. Cz. Szmytkowski, in *Book of Abstracts of the 2nd European Conference on Atomic and Molecular Physics*, eds. A. E. de Vries and M. J. van der Wiel (Amsterdam, 1985) p. 66.

18. C. Szmytkowski and K. Maciag, Absolute Total Electron-scattering Cross Section of SO_2, *Chem. Phys. Lett.* **124**, 463 (1986).

19. O. J. Orient, I. Iger, and S. K. Srivastava, Elastic Scattering of Electrons from SO_2, *J. Chem. Phys.* **77**, 3523 (1982).

20. L. Sanche and G. J. Schulz, Electron Transmission Spectroscopy: Resonances in Triatomic and Hydrocarbons, *J. Chem. Phys.* **58**, 79 (1973).

21. J. Rademacher, L. G. Christophorou, and R. P. Blaunstein, Electron Attachment to Sulphur Dioxide in High Pressure Gases, *J. Chem. Soc. Faraday Trans. II*, **71**, 1212 (1975).

22. I. M. Čadež, V. M. Pejčev, and M. V. Kurepa, Electron-sulphur Dioxide Total Ionization and Electron Attachment Cross Sections, *J. Phys. D: Appl. Phys.* **16**, 305 (1983).

23. O. J. Orient and S. K. Srivastava, Production of Negative Ions by Dissociative Electron Attachment to SO_2, *J. Chem. Phys.* **78**, 2949 (1983).

24. I. Sauers, L. G. Christophorou, and S. M. Spyrou, Negative Ion Formation in SF_6 Spark By-products, in *Gaseous Dielectrics IV* (L. G. Christophorou and M. O. Pace, Eds.), Pergamon Press, New York, pp. 261–272 (1984).

25. J. S. Wang and J. L. Franklin, Reactions and Energy Distribution in Dissociative Electron Capture Processes in Sulfuryl Halides, *Int. J. Mass. Spectrom. Ion Phys.* **36**, 233 (1980).

26. P. G. Datskos and L. G. Christophorou, Variation with Temperature of the Electron Attachment to SO_2F_2, *J. Chem. Phys.* **90**, 2626 (1989).

27. A. Benitez, J. H. Moore, and J. A. Tossell, The Correlation Between Electron Transmission and Inner Shell Electron Excitation Spectra, *J. Chem. Phys.* **88**, 6691 (1988).

28. W. H. E. Schwarz, Interpretation of the Core Electron Excitation Spectra of Hydride Molecules and the Properties of Hydride Radicals, *Chem. Phys.* **11**, 217 (1975); H. Friedrich, B. Pittel, P. Rabe, W. H. E. Schwarz, and B. Sonntag, Overlapping Core to Valence and Core to Rydberg Transitions and Resonances in the XUV Spectra of SiF_4, *J. Phys. B.* **13**, 25 (1980).

29. I. N. Levine, *Quantum Chemistry*, Allyn and Bacon, Boston (1983).

30. G. Piccardi, *Z. Phys.* **43**, 899 (1927); K. Krauss, W. Müller-Dunsing, and H. Nevert, *Z. Naturforsch* **169**, 1385 (1961); D. Feldman, Photoablosung von Elektronen bei Einigen Stabilen Negativen Ionen, *ibid.* **25a**, 621 (1970).

31. A. A. Krasnoperova, E. S. Gluskin, L. N. Mazalov, and V. A. Kochubel, The Fine Structure of the $L_{II,III}$ Absorption Edge of Sulfur in the SO_2 Molecule, *J. Struct. Chem.* **17**, 947 (1976).

32. J. L. Dehmer, Evidence of Effective Potential Barriers in the X-ray Absorption Spectra of Molecules, *Chem. Phys.* **56**, 4496 (1988).

33. A. Hitchcock and M. Tronc, Ionization Current Detection of Soft X-ray Photoabsorption: Sulfur and Chlorine K-Shell Spectra of SO_2F_2, SO_2FCl, and SO_2Cl_2, *Chem. Phys.* **121**, 265 (1988).

DISCUSSION

S. R. HUNTER: What is the uncertainty in the cross section mesurement in SF_6 at 0.2 eV? If this cross section were used in a Boltzmann analysis to obtain the thermal electron attachment rate in SF_6, the value would be considerably larger than the generally accepted value of 2.27×10^{-7} cm^3 s^{-1} (see ref. 11 of the paper).

J. K. OLTHOFF: The measured cross section for electron attachment at 0.2 eV in SF_6 must be considered to be an upper limit. The actual cross section may be as much as 50% lower due to uncertainties in the path length of the electron beam through the collision gas while immersed in the axial magnetic field.

L. G. CHRISTOPHOROU: Are your results, especially those for S_2F_{10}, a function of temperature?

J. K. OLTHOFF: At this point in time we have not had the opportunity to investigate the variation of the cross sections with temperature. However, results from other experiments indicate that the variations will be significant. In fact, if one wants to use these measured cross sections to model RF—etching plasmas in SF_6, then the temperature dependence must be measured and we plan to do these measurements in the future.

E. MARODE: Are the measured cross sections consistant with the negative ion spectra found in SF_6 discharges?

J. K. OLTHOFF: That is a very interesting question but we have not yet had the opportunity to investigate posssible correlations between the ion spectra observed directly from SF_6 discharges and the cross section data presented here.

ELECTRON LOCALIZATION EFFECTS AND RESONANT ATTACHMENT

TO O_2 IMPURITIES IN HIGHLY COMPRESSED NEON GAS

A.F.Borghesani and M.Santini

Dept. of Physics, University of Padua and
G.N.S.M./I.N.F.M., Padua, ITALY

INTRODUCTION

In a previous experiment [1] we measured the mobility of excess electrons injected in neon gas at moderately high density (up to $40 \times 10^{20}\,\mathrm{cm}^{-3}$) in the temperature range 25-300 K. The results of those measurements put into evidence the inadequacy of the theory of multiple scattering in its actual form as far as the description of the density dependence of the electron mobility is concerned. A heuristic model was therefore proposed in that paper in order to fit the mobility data without any adjustable parameters at all temperatures in the density range explored. The model relied on the assumption that the momentum transfer scattering cross section $\sigma_{\mathrm{MT}}(\epsilon)$ is shifted to $\sigma_{\mathrm{MT}}(\epsilon + \epsilon_0)$ by an energy shift ϵ_0 which depends on the gas density. The energy ϵ_0, calculated by exploiting an iterated Wigner-Seitz model, approaches for low densities the Fermi shift [2] $\epsilon_F = (2\pi\hbar^2/m)Na$, where N is the gas number density, m is the electron mass, $\hbar$ is the Planck constant and a is the scattering length of the Ne atom.

The idea of introducing this shift comes from the analysis of the measurements of electron resonant attachment to O_2 impurities in He gas [3,4]. This is a well known two-stage process [5], where the electron colliding on a O_2 molecule in its ground state may form an unstable O_2^- ion in the v=4 vibrationally excited state. The ion may be eventually stabilized into its ground state by collisions with the host gas atoms. The first step is a resonant one and takes place only if the electron energy lies in a restricted range close to $\simeq 90\,\mathrm{meV}$, where the capture cross section $\sigma_c(\epsilon)$ is sharply peaked. The electron energy can be varied usually by changing the temperature or the reduced electric field strength E/N. [5] However, it was found in He [3,4] that the attachment rate ν_A changes with density passing through a well defined peak at a density $N \simeq 30 \times 10^{20}\,\mathrm{cm}^{-3}$. This fact brings about the consequence that the electron energy ϵ increases with density, and that the capture cross section is shifted to $\sigma_c(\epsilon + \epsilon_0)$. Putting this idea of a density dependent energy shift into the model for the calculation of the electron mobility, we were able to fit our data at all temperatures in the full density range without introducing any adjustable parameters.

In this paper we present the extension of the mobility measurements to higher density (N up to $\simeq 180 \times 10^{20}\,\mathrm{cm}^{-3}$) at low temperature (45-49 K). We also present here measurements of the attachment frequency in a larger temperature range (up to 100 K). Main results of our research are: 1) The new measurements confirm the previous ones; 2) At a density $N^* \simeq 95 \times 10^{20}\,\mathrm{cm}^{-3}$ there is the onset of electron localization indicated by an even steeper decrease of the mobility with density and by a change of the curvature of the μ vs. N curve; 3) A peak in ν_A is detected at a density $N_1 \simeq 40 \times 10^{20}$ cm^{-3} and second peak at a density $N_2 \simeq 110 \times 10^{20}$ cm^{-3}. The peak densities scale roughly as the energies of the first two accessible vibrationally excited states of O_2^- with v=4 and v=5.

EXPERIMENTAL METHOD

We used the same pulsed photoinjection technique [6] as in our previous work [1] . Two parallel plane electrodes separated by a distance d (1.0 or 0.4 cm) delimit the drift space. Electrons are emitted by shining the cathode with a short U.V. light pulse from a Xe flashlamp. The electrons drift toward the anode under an externally applied electric field. From the signal induced by the electron motion in an external RC circuit connected to the anode the time of flight τ_e as well as the attachment frequency can be measured.

The high pressure cell is mounted on the head of a cryogenerator by means of 4 copper cantilevers in order to decouple it mechanically from the head while ensuring a good thermal contact. The cell was thermoregulated within 0.03 K by means of standard techniques. The gas in the cell can be forced to circulate through an activated charcoal trap at 77 K and through an Oxisorb trap by means of a bellow circulator [7] in order to remove both O_2 and N_2. For the mobility measurements the gas was purified to a high degree. The typical range of ν_A was $0.5 < \nu_A < 10$ kHz. For the attachment frequency measurements we typically had $20 < \nu_A < 200$ kHz.

The amount of injected charge $Q_e = en_0$ is very tiny : $(4 < Q_e < 400) \times 10^{-15}$ C and the induced current is $2 \times 10^{-12} < i_e < 3 \times 10^{-8}$ A depending on the experimental conditions. In this situation it is necessary to integrate the current signal by means of large load resistors in order to minimize the signal-to-noise ratio. If the circuit time constant RC is much larger than τ_e (in our case RC $\simeq 50 \times 10^{-3}$ s, and $(\tau_e)_{max} \simeq 0.5 \times 10^{-3}$ s), the electron signal waveform is given by

$$v_e(t) = -v_T(1 - e^{-\nu_A t})/A \qquad \text{for } t \leq \tau_e \qquad (1)$$

where $v_T = en_0/C$ and $A = \nu_A \tau_e$. For $t > \tau_e$ with $\tau_e \ll$ RC, v_e takes on the constant value $v_e = v_S = -v_T(1 - \exp(-A))/A$. In absence of attachment, A=0 and the waveform is linear with slope $-v_T/\tau_e$. The time of flight can be easily determined from an extrapolation of the linearly changing voltage waveform to find its start and end points. For the electron mobility measurements we purified the gas thoroughly by circulating it through the traps until A $\ll$ 0.1. In the case of attachment, A $\neq$ 0, both ν_A and τ_e can be accurately determined by fitting relation (1) to the experimental data [8] .

In our experimental setup it is also possible to detect the motion of the O_2^- ions created by the electrons passage. Because the ions are much slower than the electrons, the ionic transit time is larger than the electronic one, $\tau_i \gg \tau_e$. To integrate the ionic signal we have used a 10^{11} Ω load resistor yielding a 5 s time constant. This was made possible by using an ultra-low input bias current ($< 10^{-13}$ A) op-amp (Burr-Brown OP A128). Because RC is not much larger than τ_i $((RC/\tau_i)_{min} \simeq 3)$, the ionic waveform is much more complicated than the electronic one and it is given (assuming $\tau_e \ll \tau_i$) by

$$v_i(t) = -v_T[I - Pe^{-t/RC} - Qe^{At/\tau_i}] \qquad \text{for } 0 \leq t \leq \tau_i \qquad (2)$$

and

$$v_i(t) = v_i(\tau_i)e^{-(t-\tau_i)/RC} \qquad \text{for } t \geq \tau_i \qquad (3)$$

where $I = RC/\tau_i$, $P = (\exp(-A) + (AI)^2 - 1)/((1+AI)A)$, and $Q = I\exp(-A)/(1 + AI)$. We have developed some numerical techniques [8] to fit relation (2) and (3) to the experimental data in order to measure the ionic time of flight and the attachment number $A = \nu_A \tau_e$.

The optimum conditions in order to maximize the accuracy of ν_A measurements are such that A $\simeq$ 1. In this case the amount of electrons arriving at the collector is of

28

the same order of magnitude as the ions. For the attachment frequency measurements we measured along isothermal paths. We filled the cell at the highest pressure required in that run at the desired temperature. We then circulated the gas until we reached the condition $A \simeq 1 - 2$. We waited until the O_2 impurity concentration was stationary as indicated by the constancy in time of the electronic signal amplitude. The lower densities were obtained then by successively spilling out of the cell the desired amount of gas.

EXPERIMENTAL RESULTS

Mobility measurements

In Figure 1 we show the experimental results of the zero-field mobility measurements as a function of the Ne gas density between 45 and 48 K. The gas density was calculated by means of the McCarty-Stewart equation of state [9]. We also plot here the ionic mobility measured at 46.5 K. It can be noted that the electron mobility decreases by nearly 6 orders of magnitude in the range $(5 < N < 180) \times 10^{20} \, \text{cm}^{-3}$, while the ionic mobility is well separated from the electronic one even at the highest densities and does not depend very much on N.

In order to describe the electron mobility we can divide the density range in two regions separated by the density $N^* \simeq 95 \times 10^{20} \, \text{cm}^{-3}$. For $N < N^*$ the mobility shows an upward concavity, while for $N > N^*$ the concavity is downwards. Moreover, for $N > N^*$, the μ vs. N curve slope becomes steeper. We believe that for $N < N^*$ the electrons are quasifree, while for $N > N^*$ there is coexistence of localized and extended electron states. As the density is increased the equilibrium between the two kinds of states shifts towards the localized ones. This point of view is supported by the behavior of the μ vs. electric field strength E curves. For $N < N^*$ μ decreases with increasing field, while for $N > N^*$ μ first increases with E, it goes through a maximum, and then it recovers the classical $E^{1/2}$ dependence at high field strengths. In Figure 1 we also show the classical prediction for the zero-field mobility (curve 1) given by

$$\mu = \frac{4}{3N} \left\{ \frac{e}{[2\pi m (k_B T)^5]^{1/2}} \right\} \int_0^\infty \frac{\epsilon}{\sigma_{MT}(\epsilon)} e^{-\epsilon/k_B T} d\epsilon \qquad (4)$$

where e is the elementary charge, m is the electron mass, k_B is the Boltzmann constant, T is the absolute temperature, and $\sigma_{MT}(\epsilon)$ is the energy dependent electron-atom momentum transfer scattering cross section [10]. Curve 2 is the prediction of the multiple scattering theories in the form proposed by O'Malley [11] for positive scattering length gases. The theory replaces the exponential in relation (4) with $\exp[-(\epsilon + \Gamma)/k_B T]$ where $\Gamma = (h/\pi)(2m)^{-1/2} N \sigma_T(\epsilon) \sqrt{\epsilon}$. σ_T is the total scattering cross section and ϵ is the energy. Because σ_T is very small particulary at low energy $(\sigma_T(0) = 0.161 \, \text{Å}^2)$ the multiple scattering correction is small and cannot account for the experimental data. Curve 3 is the result of the model proposed in our previous paper [1]. As discussed before, we assume that σ_{MT} in relation (4) must be evaluated at the shifted energy $\epsilon' = \epsilon + \epsilon_0$, where ϵ_0 can be calculated self-consistently with the Wigner-Seitz model using an effective radius $a = [\sigma_T(\epsilon_0)/4\pi]^{1/2}$. Moreover, the calculated mobility has been divided by the long wavelength part of the static structure factor $S(0)$ to take into account the correlations among the scatterers, as suggested by Lekner [12]. The agreement between this curve and the experimental data is satisfactory up to $N = N^*$. However, this way of calculating ϵ_0 is not theoretically justified and how to calculate it is still an open problem.

The high density branch of the mobility curve for $N > N^*$ shows, in our opinion, evidence of coexisting extended and localized electron states. Curve 4 has been calculated [13] according to a simple bubble model closely related to the approach of Hernandez [14]. The model assumes that the electron is confined in a partially empty

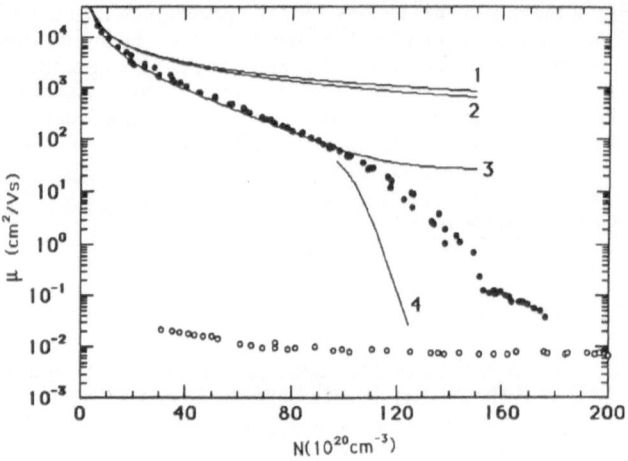

Fig. 1. Electronic mobility(closed points) at 45< T<49 K and ionic mobility (open points) at T=46.5 K (see text).

bubble that acts as a spherically symmetric quantum well. The polarizability of the atom has not been neglected and has been accounted for to first order in the density. The Helmholtz free energy of the system has been calculated by adding the mechanical work needed to create the bubble to the electron ground state energy in the well and by neglecting the configurational entropy contribution. Finally, the mobility is given by the weighted average of the quasifree electron mobility (calculated by extrapolating curve 3 for $N > N^*$) and of the localized state mobility (calculated according to the Stokes hydrodynamic formula). This model gives a threshold density value of $N \approx 95 \times 10^{20}$ cm^{-3} $\simeq N^*$ for the appearance of stable localized states in good agreement with the experimentally observed value. Nonetheless, there is disagreement between the calculated curve and the experimental mobility data, but we have to stress the fact that the actual problem is far more complicated than a simple 2-level problem and that the density of states should be crucial to this goal.

Attachment frequency measurements

At each density the attachment frequency ν_A was measured at several electric field strengths. ν_A turned out to depend weakly on the field in our range ($10^{-21} < E/N < 3 \times 10^{-20}$ V cm^2). The ν_A was thus extrapolated to zero field in order to get information at thermal energies. To rule out the dependence on the number of O_2 molecules, the ν_A data are divided by the Ne gas density. We show in Figure 2 the values of ν_A/N as a function of N at T=70 K. ν_A/N shows a peak at $N = N_1 \simeq 44 \times 10^{20}$ cm^{-3}. Assuming that the coefficient of stabilization per collision of O_2^{-*} in Ne saturates, or, at least, that it does not vary very rapidly with N, and that the capture cross section is narrow, then $\nu_A/N \propto f_M(\epsilon_R)$, the value of the energy distribution function of the electrons at the resonance energy ϵ_R. Assuming that ϵ_R is density independent, the distribution function is sampled at fixed energy, and the peak structure of ν_A/N is a consequence of the fact that the average electron energy is shifted to higher energies by increasing the gas density. In figure 3 we plot ν_A/N as a function of N at T=46.5 K, where the accessible density range is larger. Again, we clearly see the first peak at $N = N_1$, but we also see a second peak at the much larger density $N_2 \simeq 110 \times 10^{20}$ cm^{-3}. It is interesting to note that the value of the ratio N_1/N_2 is very close to that of the energies of the two vibrationally excited states with v=4 and v=5 of O_2^-. This fact seems to confirm that the optical model for the calculation of the electron self energy in a disordered medium should apply. Nonetheless, there are still some controversial points. Aside from the assumption of a nearly density independent stabilization coefficient,we point out that the peaks

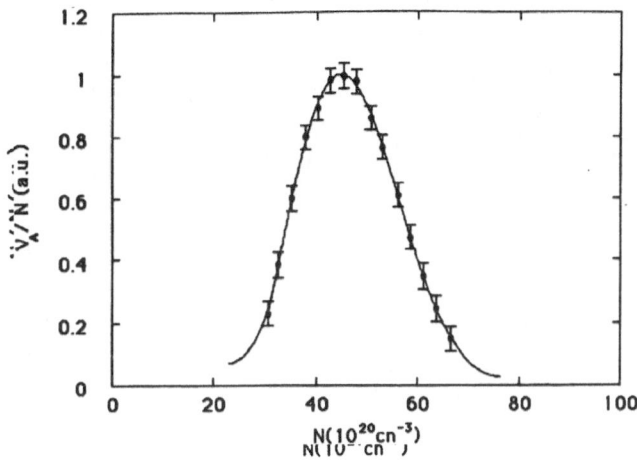

Fig. 2. The peak in ν_A/N at T=70 K. The solid line has no theoretical meaning.

appear at densities much lower than expected on the basis of the standard optical potential $\epsilon_0 = (2\pi\hbar^2/m)Na$. Indeed, if we take $a \simeq 0.113\,\text{Å}$, we get $\epsilon_0 \simeq 23\,\text{meV}$ at $N = N_1$, which is much lower than the expected $\epsilon_R \simeq 90\,\text{meV}$. Even using the self-consistent Wigner-Seitz potential we are off by a factor of 2 approximately. Moreover, the peaks appear much broader than expected on the basis of the results obtained in He gas.

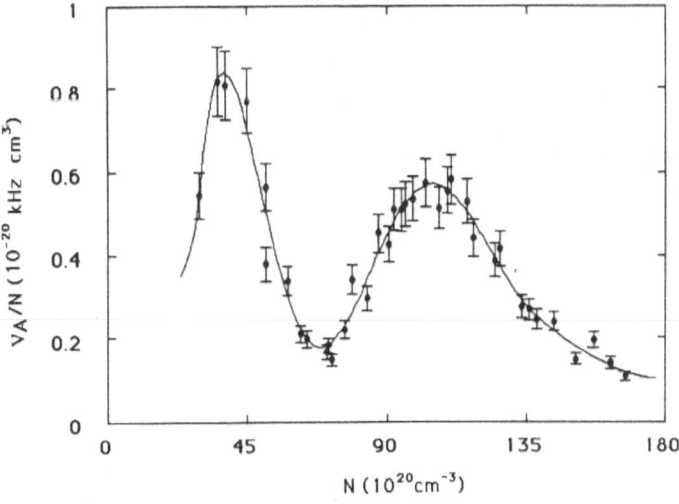

Fig. 3. First and second peak in ν_A/N at T$\simeq 46$ K. The solid line is only an eye guideline.

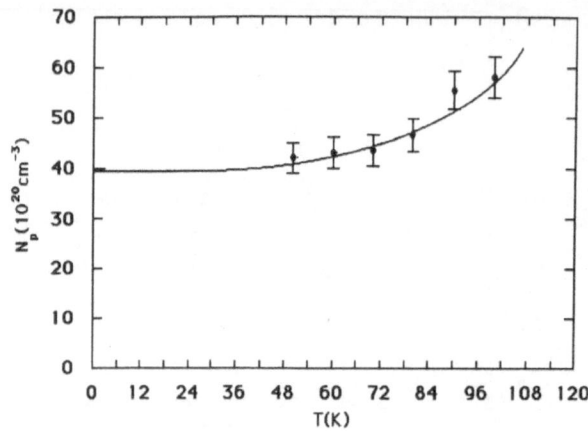

Fig. 4. The peak density plotted as a function of temperature. The solid line has no theoretical meaning.

Finally, as shown in figure 4, the density of the first peak increases unexpectedly with temperature.

REFERENCES

1. A.F.Borghesani, L.Bruschi, M.Santini, G.Torzo, Phys. Rev. A 37:4828 (1988).
2. L.L.Foldy, Phys. Rev. 67:107 (1945).
3. A.K.Bartels, Phys. Lett. 45 a:491 (1973).
4. L.Bruschi, M.Santini, G.Torzo, J.Phys. B 17:1137 (1984).
5. L.G.Christophorou, D.L.McCorkle, A.A.Christodoulides, in " Electron-Molecule Interactions and their Applications", L.G.Christophorou, ed., Academic Press, Orlando,(1984).
6. A.K.Bartels, Ph.D. Thesis, Hamburg University (1971).
7. G.Torzo, Rev. Sci. Instrum. 64:1162 (1990).
8. A.F.Borghesani , M.Santini, Meas. Sci. Technol., 1:939 (1990).
9. R.D.McCarty, R.B.Stewart , National Bureau of Standards Note # 8726 (1965) unpublished.
10. T.F.O'Malley, R.W.Crompton, J.Phys. B 13: 3451 (1980).
11. T.F.O'Malley, J.Phys. B 13:1491 (1980).
12. J.Lekner, Phil. Mag. 18:1281 (1968).
13. A.F.Borghesani, M.Santini, to be published.
14. J.P.Hernandez, Phys. Rev. B 11:1289 (1975).

DISCUSSION

L. G. CHRISTOPHOROU: These studies are rather basic for our efforts to relate breakdown and prebreakdown phenomena in gases to those of liquid dielectrics. For example the transition from electron delocalization to electron localization is equivalent to the transition from electronic to ionic conduction.

E. MARODE: (1) Is ϵ_0 related to the fact that we are no longer in a two—body collision controlled process? (2) Is bubble formation linked in some way to the very small amount of residual impurities?

A. F. BORGHESANI: (1) Yes, it does. The first experimental evidence of this energy shift was produced by the observation of the emission spectra of alkali atoms in buffer noble gases. The emission lines of highly excited electronic states of those atoms were shifted with respect to the same lines in absence of the host gas. The interpretation of this experimental fact was given by Fermi who realized that the wave function of the electron in the highly excited state spans over a rather large volume containing more than just one host gas atom. Therefore, the electron no longer undergoes a two—body collision process, but rather a multi—body one. The overall result of this scattering process, which can be calculated within the frame of multiple scattering theory, is that the electron energy is shifted by an amount, ϵ_0, which depends on the gas density and, of course, on the electron—gas atom interaction potential. (2) Bubble formation, or better, formation of localized electron states is not related to the amount of residual impurities (mainly O_2) present in the gas, because the electron energy distribution function is not significantly modified by impurities, at least when the impurity content is so low as in our case (it is estimated to be in the p.p.b. range). However, when the thermodynamical conditions of the gas allow only for localized electron states it turns out to be difficult from an experimental point of view to discriminate between a localized electron and an O_2^- ion just by measuring drift mobilities because these are of the same order of magnitude.

TEMPERATURE DEPENDENCE OF THE DISSOCIATIVE ELECTRON

ATTACHMENT TO CH_3Cl AND C_2H_5Cl.[1]

P. G. Datskos[2], L. G. Christophorou[2] and J. G. Carter
Atomic, Molecular, and High Voltage Physics Group
Health and Safety Research Division
Oak Ridge National Laboratory
Oak Ridge, Tennessee 37831 USA

ABSTRACT

The electron attachment rate constant k_a for CH_3Cl and C_2H_5Cl has been measured as a function of the gas temperature and the mean electron energy. At room temperature these molecules exhibit exceedingly small electron attachment ($k_a < 10^{-14} cm^3 s^{-1}$ for CH_3Cl and $k_a \sim 10^{-13} cm^3 s^{-1}$ for C_2H_5Cl) which increases greatly with small increases in T (i.e., molecular internal energy). In fact, it has been found that as T increases from 400 to 750 K the electron attachment rate constant over a wide range of mean electron energies below ~ 1 eV increased by over 4×10^3 times for CH_3Cl and by ~ 50 times for C_2H_5Cl. In this paper we report our findings on the temperature enhanced electron attachment to these two molecules and indicate their potential use for dielectric and pulsed power switching applications.

INTRODUCTION

It has been established that the breakdown strength of gases depends strongly on their electron attaching properties.[1-3] It is also known that electron attachment can be profoundly affected by changes in the internal energy of a molecule (effected, for example, by laser light[4] or heat[5]). Studies over the past decade have revealed that the electron attachment rate constant k_a (or electron attachment cross section σ_a) of electronegative gases can be a strong function of the temperature T.[6-8] In fact it has been established that for molecules which attach electrons dissociatively the effect of an increasing T is generally an increase in k_a (and σ_a).

The knowledge of how much the electron attaching properties of a molecule vary with increases in the molecular internal energy is crucial in controlling or modifying the insulation/conduction properties of a medium. With this in mind we undertook a study of the electron attaching properties of CH_3Cl and C_2H_5Cl as a function of the gas temperature. In this paper we report the results of the low—energy (< 1 eV) electron attachment to these two molecules and the dramatic enhancement in the electron attachment rate constant we observed as the gas temperature was increased from 300 to 750 K.

[1]Research sponsored by the Office of Health and Environmental Research, Department of Energy, under Contract No. DE–AC05–84OR21400 with Martin Marietta Energy Systems, Inc.

[2]Also, Department of Physics, The University of Tennessee, Knoxville, Tennessee 37996 USA.

EXPERIMENTAL METHOD

The high temperature electron–swarm (HTES) technique[7,9] was employed to study electron attachment to CH_3Cl for $T \geq 500$ K and to C_2H_5Cl for $T \geq 400$ K. The CH_3Cl was also studied using the modified–pulsed Townsend (MPT) technique[10] for $T \leq 400$ K. The HTES technique could not be used to study electron attachment to CH_3Cl below $\simeq 500$ K since at these temperatures $k_a(E/N,T)$ is exceedingly small and large amounts of the electron attaching gas were used (5 to 10 Torr in 700 to 3000 Torr of total gas pressure). The total gas pressure is the sum of the pressures of the electronegative gas and the buffer gas (N_2 in the present study). Such high concentrations of electron attaching gas in N_2 increased the number of the electrons in the swarms produced by each α–particle[9] and also caused large changes in the electron drift velocity. These changes resulted from changes in the electron energy distribution function $f(\epsilon,E/N,T)$ in the mixture as compared to that in pure N_2.

For $T \geq 400$ K, $k_a(E/N,T)$ are sufficiently large so that smaller concentrations of the electronegative gas were used for which we were able to correct the effect of CH_3Cl and C_2H_5Cl on $f(\epsilon,E/N,T)$. This we accomplished by measuring the electron attachment rate constant as a function of the attaching gas number density, N_a, for each E/N and then extrapolating k_a to $N_a \longrightarrow 0$ (Fig. 1). The values of $k_a(E/N,T)$ for $N_a \longrightarrow 0$ are the ones plotted in Fig. 2 (for CH_3Cl) and in Fig. 3 (for C_2H_5Cl). The methyl chloride and ethyl chloride samples used were purchased from Matheson Gas Products and had a stated purity of 99.9%. They were further purified with several vacuum distillation cycles. The N_2 buffer gas was also obtained from Matheson Gas Products and had a quoted minimum purity of 99.999%. In the present work the number density of N_2 was varied from 2.25 to 9.66 x 10^{19} molecules cm^{-3}. The number density, N_a, of CH_3Cl was varied from 1.46 to 10.43 x 10^{15} molecules cm^{-3} at 750 K and from 4.29 to 42.9 x 10^{16} molecules cm^{-3} at 400 K. The N_a for C_2H_5Cl was varied from 3.86 to 14.81 x 10^{15} molecules cm^{-3} at 700 K and from 1.77 to 8.21 x 10^{16} molecules cm^{-3} at 400 K.

The measured $k_a(E/N,T)$ for both CH_3Cl and C_2H_5Cl were independent of the total gas number density, as expected for a dissociative electron attachment process; low–energy electrons are captured dissociatively to these molecules and yield Cl^- ions.[11,12]

RESULTS AND DISCUSSION

The total electron attachment rate constant for CH_3Cl and C_2H_5Cl was measured as a function of E/N, T and N_a. The extrapolated values of $k_a(E/N,T)$ for $N_a \longrightarrow 0$ are listed as a function of the mean electron energy $\langle \epsilon \rangle$ in Table 1 for CH_3Cl and in Table 2 for C_2H_5Cl and are plotted in Figs. 2 and 3. The mean electron energy $\langle \epsilon \rangle$ was calculated at each E/N and T from the corresponding $f(\epsilon,E/N,T)$ for N_2 which was itself obtained by solving the Boltzmann transport equation using the code developed by Luft.[13]

The measurements in Figs. 2 and 3 show that as T is increased from 300 to 750 K the $k_a(\langle \epsilon \rangle,T)$ increases by 3 to 4 orders of magnitude for CH_3Cl and as T is increased from 400 to 700 K the $k_a(\langle \epsilon \rangle,T)$ increases by $\sim$ 50 times (depending on $\langle \epsilon \rangle$) for C_2H_5Cl. At each T at which $k_a(\langle \epsilon \rangle,T)$ was measured the total electron attachment cross section was determined for CH_3Cl and C_2H_5Cl using an iterative technique[14] from

$$k_a(\langle \epsilon \rangle,T) = \left[\frac{2}{m}\right]^{1/2} \int_0^\infty d\epsilon \; \epsilon^{1/2} \; \sigma_a(\epsilon,T) \; f(\epsilon,\langle \epsilon \rangle,T) \tag{1}$$

The $\sigma_a(\epsilon,T)$ for CH_3Cl are presented in Fig. 4 and the $\sigma_a(\epsilon,T)$ for C_2H_5Cl are shown in Fig. 5. From Fig. 4 it is clearly shown that there are two maxima (one at $\sim$ 0.03 eV and another at $\sim$ 0.75 eV) which indicate that the dissociative electron attachment to CH_3Cl proceeds via at least one low–lying negative ion state. It is also apparent

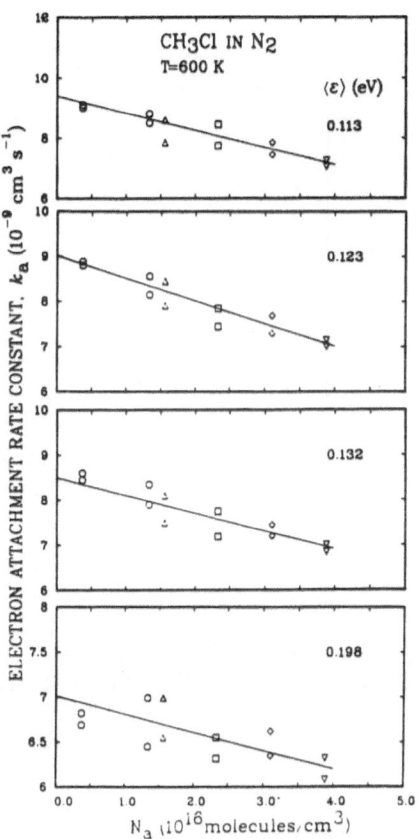

Fig. 1.   Total electron attachment rate constant $k_a(\langle\epsilon\rangle,T)$ for CH_3Cl in N_2 as a function of the CH_3Cl gas number density N_a (T = 600 K). The total gas number densities were: 2.25 x 10^{19} (o), 3.22 x 10^{19} (Δ), 4.82 x 10^{19} (□), 6.44 x 10^{19} (◇), and 9.66 x 10^{19} (▽) molecules cm^{-3} for a number of $\langle\epsilon\rangle$.

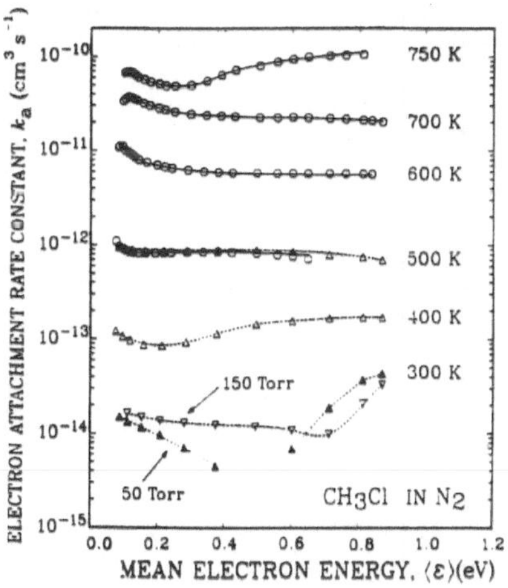

Fig. 2.   Total electron attachment rate constant $k_a(\langle\epsilon\rangle,T)$ (for $N_a \longrightarrow 0$; see text) as a function of the mean electron energy $\langle\epsilon\rangle$ for CH_3Cl in N_2 of T = 300, 400, 500, 600, 700, and 750 K. The broken lines are measurements made using the modified–pulsed Townsend technique and the solid lines using the high temperature electron swarm technique.

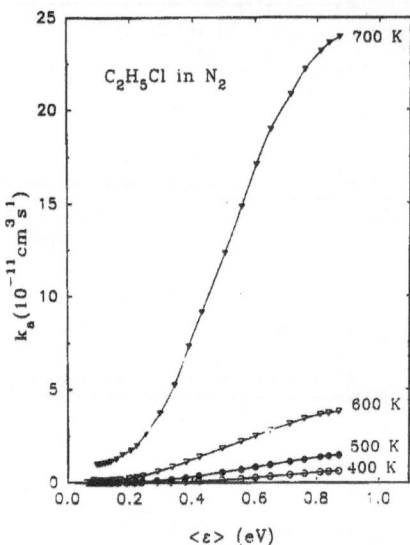

Fig. 3.   Total electron attachment rate constant $k_a(\langle\epsilon\rangle,T)$ as a function of the mean electron energy $\langle\epsilon\rangle$ for C_2H_5Cl in N_2 at temperatures T of 400 (o), 500 (•), 600 ($\triangledown$) and 700 ($\blacktriangledown$) K.

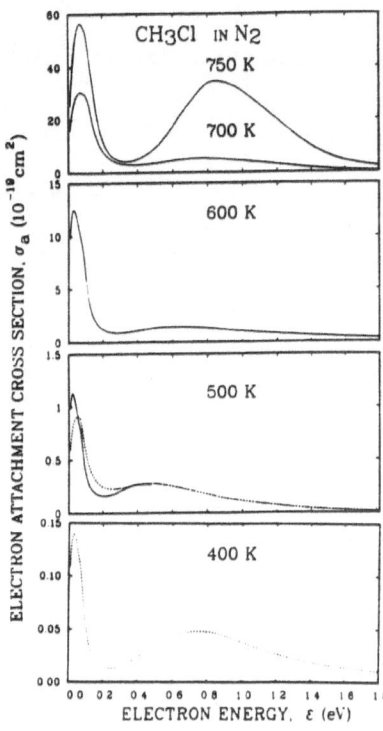

Fig. 4.   The swarm–unfolded total electron attachment cross section $\sigma_a(\epsilon,T)$ for CH_3Cl determined from the $k_a(\langle\epsilon\rangle,T)$ data in Fig. 2.

Table 1. Total electron attachment rate constants, k_a ($\langle\varepsilon\rangle$,T), for CH$_3$Cl in N$_2$, as a function of E/N and $\langle\varepsilon\rangle$, at various temperatures; $w(E/N,T)$ and $\langle\varepsilon\rangle$ $(E/N,T)$ for N$_2$.

E/N	400 K			500 K				600 K			700 K			750 K		
	$\langle\varepsilon\rangle$	ω	k_a^c	$\langle\varepsilon\rangle$	ω	k_a^a	k_a^b	$\langle\varepsilon\rangle$	ω	k_a^b	$\langle\varepsilon\rangle$	ω	k_a^b	$\langle\varepsilon\rangle$	ω	k_a^b
(10^{-17} V cm²)	(eV)	(10^6 cm/s)	(10^{-14} cm³/s)	(eV)	(10^6 cm/s)	(10^{-13} cm³/s)	(10^{-13} cm³/s)	(eV)	(10^6 cm/s)	(10^{-12} cm³/s)	(eV)	(10^6 cm/s)	(10^{-11} cm³/s)	(eV)	(10^6 cm/s)	(10^{-11} cm³/s)
0.062				0.070	1.32	11.02		0.063	1.20	10.81	0.092	1.08	3.35	0.097	0.96	5.21
0.093				0.076	1.77	9.60		0.086	1.60	11.12	0.097	1.49	3.54	0.102	1.34	6.72
0.124	0.075	2.26	12.20	0.085	2.10	9.12	9.87	0.096	1.92	11.10	0.104	1.81	3.73	0.108	1.65	6.86
0.155				0.095	2.31	8.77		0.104	2.16	10.00	0.112	2.06	3.65	0.116	1.91	6.91
0.186	0.096	2.66	10.64	0.104	2.48	8.46	9.42	0.113	2.36	9.49	0.121	2.27	3.52	0.124	2.17	6.72
0.217				0.115	2.65	8.27		0.123	2.52	9.01	0.130	2.41	3.41	0.133	2.29	6.33
0.248	0.118	2.88	9.73	0.125	2.77	8.20	9.14	0.132	2.67	8.49	0.139	2.61	3.18	0.142	2.44	6.01
0.310				0.144	3.00	8.20		0.144	2.91	7.98	0.157	2.88	2.99	0.160	2.71	5.68
0.373	0.159	3.22	8.69	0.164	3.20	8.20	8.96	0.168	3.11	7.49	0.176	3.12	2.81	0.178	2.94	5.36
0.457				0.193	3.44	8.25		0.198	3.39	7.02	0.202	3.41	2.70	0.204	3.21	5.11
0.528	0.213	3.57	8.53	0.217	3.59	8.30	8.98	0.221	3.57	6.73	0.221	3.58	2.56	0.223	3.41	4.89
0.621				0.238	3.75	8.40		0.242	3.75	6.50	0.249	3.81	2.42	0.250	3.66	4.86
0.776	0.263	4.10	9.31	0.286	4.12	8.43	9.06	0.289	4.17	6.21	0.296	4.20	2.36	0.297	4.02	4.99
0.931				0.333	4.37	8.50		0.336	4.36	5.98	0.343	4.44	2.33	0.344	4.35	5.52
1.087	0.376	4.62	11.46	0.378	4.62	8.35	9.24	0.381	4.70	5.88	0.388	4.70	2.29	0.389	4.66	6.39
1.240				0.421	4.92	8.17		0.423	5.00	5.81	0.429	5.00	2.24	0.431	4.95	7.22
1.550	0.494	5.42	14.36	0.495	5.44	7.83	9.25	0.496	5.50	5.80	0.503	5.57	2.23	0.504	5.51	8.02
1.860				0.556	5.99	7.60		0.556	6.02	5.77	0.556	6.12	2.25	0.556	6.07	8.83
2.170	0.601	6.50	15.50	0.602	6.54	7.10	8.83	0.602	6.58	5.72	0.602	6.70	2.22	0.602	6.61	9.39
2.480				0.647	7.07			0.647	7.12	5.73	0.647	7.28	2.19	0.647	7.15	9.96
3.110	0.711	8.05	16.66	0.711	8.10		8.20	0.711	8.18	5.70	0.711	8.40	2.15	0.711	8.20	10.21
3.720				0.759	9.39			0.759	9.46	5.69	0.759	9.66	2.08	0.759	9.22	10.38
4.660	0.812	10.93	16.94	0.812	11.10		7.77	0.812	11.18	5.63	0.812	11.31	2.05	0.812	10.64	10.50
5.280				0.839	11.77			0.839	11.91	5.59	0.839	12.05	1.99			
6.210	0.872	13.00	16.90	0.872	13.10		7.20	0.872	13.22		0.872	13.40				

^{a} Measured by the modified-pulsed Townsend technique.

^{b} Measured by the high-temperature electron-swarm technique.

Table 2. Total electron attachment rate constants, k_a $(\langle\varepsilon\rangle, T)$, for C_2H_5Cl in N_2, as a function of E/N and $\langle\varepsilon\rangle$, at various temperatures; $w(E/N, T)$ and $\langle\varepsilon\rangle(E/N, T)$ for N_2.

E/N	400 K			500 K			600 K			700 K		
$(10^{-17}V\,cm^2)$	$\langle\varepsilon\rangle$ (eV)	w $(10^5cm/s)$	k_a $(10^{-14}cm^3/s)$	$\langle\varepsilon\rangle$ (eV)	w $(10^5cm/s)$	k_a $(10^{-13}cm^3/s)$	$\langle\varepsilon\rangle$ (eV)	w $(10^5cm/s)$	k_a $(10^{-12}cm^3/s)$	$\langle\varepsilon\rangle$ (eV)	w $(10^5cm/s)$	k_a $(10^{-11}cm^3/s)$
0.062	0.058	1.53	0.148	0.070	1.32	0.231	0.083	1.20	1.490	0.092	1.08	0.990
0.093	0.056	1.97	0.120	0.076	1.77	0.152	0.086	1.60	1.400	0.097	1.49	0.980
0.124	0.075	2.26	0.070	0.085	2.10	0.103	0.096	1.92	1.230	0.104	1.81	1.000
0.155	0.085	2.49	0.035	0.095	2.31	0.080	0.104	2.16	1.190	0.112	2.06	1.030
0.186	0.096	2.66	0.020	0.104	2.48	0.068	0.113	2.36	1.210	0.121	2.27	1.060
0.217	0.107	2.77	0.013	0.115	2.65	0.062	0.123	2.52	1.290	0.130	2.41	1.090
0.248	0.118	2.88	0.010	0.125	2.77	0.064	0.132	2.67	1.380	0.139	2.61	1.160
0.310	0.139	3.05	0.012	0.144	3.00	0.091	0.144	2.91	1.590	0.157	2.88	1.280
0.373	0.159	3.22	0.018	0.164	3.20	0.133	0.168	3.11	1.860	0.176	3.12	1.510
0.457	0.189	3.46	0.048	0.193	3.44	0.281	0.198	3.39	2.510	0.202	3.41	1.730
0.528	0.213	3.57	0.078	0.217	3.59	0.441	0.221	3.57	3.000	0.221	3.58	1.990
0.621	0.235	3.77	0.110	0.238	3.75	0.622	0.242	3.75	3.780	0.249	3.81	2.610
0.776	0.283	4.10	0.210	0.286	4.12	1.170	0.289	4.17	6.080	0.296	4.20	3.730
0.931	0.330	4.38	0.330	0.333	4.37	1.950	0.336	4.36	8.650	0.343	4.44	5.230
1.087	0.376	4.62	0.551	0.378	4.62	2.790	0.381	4.70	11.33	0.388	4.70	7.320
1.240	0.419	4.92	0.832	0.421	4.92	3.760	0.423	5.00	13.98	0.429	5.00	9.130
1.550	0.494	5.42	1.420	0.495	5.44	5.610	0.496	5.50	18.39	0.503	5.57	12.31
1.860	0.555	5.93	1.990	0.556	5.99	7.120	0.556	6.02	22.10	0.556	6.12	14.83
2.170	0.601	6.50	2.680	0.602	6.54	8.380	0.602	6.58	24.99	0.602	6.70	17.08
2.480	0.647	7.05	3.330	0.647	7.07	9.550	0.647	7.12	27.97	0.647	7.28	18.96
3.110	0.711	8.05	4.150	0.711	8.10	11.21	0.711	8.18	31.48	0.711	8.40	20.83
3.720	0.759	9.29	4.810	0.759	9.39	12.42	0.759	9.46	34.28	0.759	9.66	22.20
4.660	0.812	10.93	5.590	0.812	11.10	13.71	0.812	11.18	36.53	0.812	11.31	23.19
5.280	0.839	11.58	5.940	0.839	11.77	14.28	0.839	11.91	37.26	0.839	12.05	26.33
6.210	0.872	13.00	6.240	0.872	13.10	14.69	0.872	13.22	38.20	0.872	13.40	23.95

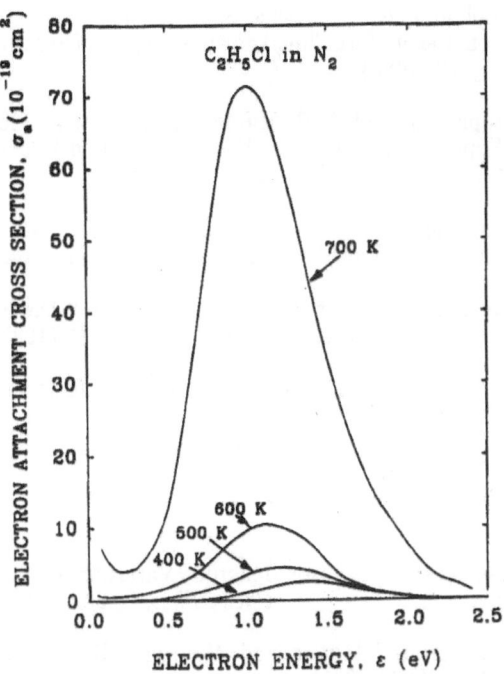

Fig. 5. The swarm–unfolded total electron attachment cross section $\sigma_a(\epsilon, T)$ for C_2H_5Cl determined from the $k_a(\langle \epsilon \rangle, T)$ data in Fig. 3.

that the $\sigma_a(\epsilon, T)$ for CH_3Cl is extremely sensitive to small changes in the gas temperature; the peak positioned at ~ 0.75 eV is enhanced by ~ 400 times as T is increased from 400 to 750 K. From Fig. 5 it is shown that the $\sigma_a(\epsilon, T)$ for C_2H_5Cl is also sensitive to the temperature; the peak value of $\sigma_a(\epsilon, T)$ increases in magnitude by ~ 30 times at T increases from 400 to 700 K. The electron energy at which the maximum in $\sigma_a(\epsilon, T)$ occurs shifts towards lower electron energies as T increases. This does not appear to be the case for CH_3Cl (see Fig. 4) for which both the peaks in $\sigma_a(\epsilon, T)$ increase with increasing T at about the same rate but their energy positions remain practically unchanged. The behavior of electron attachment to C_2H_5Cl is analogous to those observed earlier.[7,8]

CONCLUSIONS

In the temperature range (300 to 750 K) the $k_a(\langle \epsilon \rangle, T)$ for CH_3Cl and C_2H_5Cl increase with increasing T. The dielectric strength of these gases is expected to also increase with T. The profound changes in $k_a(\langle \epsilon \rangle, T)$ with small variations in T indicates a possible potential use of these gases in diffuse discharge switching applications.

REFERENCES

1.   L. G. Christophorou, R. A. Mathis, D. R. James, and D. L. McCorkle, On the role of electron attachment in the breakdown strength of gaseous dielectrics, *J. Phys. D: Appl. Phys.* 14:1889 (1981).

2.   L. G. Christophorou, R. A. Mathis, S. R. Hunter, and J. G. Carter, Effect of temperature on the uniform breakdown strength of electronegative gases, *J. Chem. Phys.* <u>63</u>:52 (1988).

3.   L. G. Christophorou, and S. R. Hunter, in: "Electron–Molecule Interactions and Their Applications," L. G. Christophorou (Ed.) Academic Press, New York, Vol. 2, Chapter 5 (1984).

4.   L. A. Pinnaduwage, L. G. Christophorou, and S. R. Hunter, Optically enhanced electron attachment to thiophenol, *J. Chem. Phys.* <u>90</u>:6275 (1989).

5.   P. G. Datskos and L. G. Christophorou, Variation of the electron attachment to $n-C_4F_{10}$ with temperature, *J. Chem. Phys.* <u>86</u>:1982 (1987).

6.   L. G. Christophorou, Negative ions of polyatomic molecules, *Environ. Health Perspectives* 36:3 (1980).

7.   S. M. Spyrou and L. G. Christophorou, Effect of temperature on the dissociative electron attachment to $CClF_3$ and C_2F_6, *J. Chem. Phys.* <u>83</u>:2620 (1985).

8.   P. G. Datskos and L. G. Christophorou, Variation with temperature of the electron attachment to SO_2F_2, *J. Chem. Phys.* <u>90</u>:2626 (1989).

9.   L. G. Christophorou, "Atomic and Molecular Radiation Physics," Wiley–Interscience, New York (1971).

10.   S. R. Hunter, J. G. Carter, and L. G. Christophorou, Electron transport measurements in methane using an improved pulsed Townsend technique, *J. Chem. Phys.* <u>60</u>:24 (1986).

11.   F. H. Dorman (1966), Negative fragment ions from resonance capture process, *J. Chem. Phys.* <u>44</u>:3856 (1966).

12.   A. A. Christodoulides, R. Schumacher, and R. N. Schindler, Studies by the electron cyclotron resonance technique. X. Interactions of thermal energy electrons with molecules of chlorine, hydrogen chloride, and methyl chloride, *J. Phys. Chem.* <u>79</u>:1904 (1975).

13.   P. E. Luft, Description of a backward prolongation program for computing transport coefficients, JILA Information Center Report No. 14, University of Colorado, Boulder (1975).

14.   L. G. Christophorou, D. L. McCorkle, and V. E. Anderson, Swarm–determined electron attachment cross sections as a function of electron energy, *J. Phys. B* <u>4</u>:1163 (1971).

MULTIBODY ELECTRON CAPTURE PROCESSES IN THE GAS PHASE

Iwona Szamrej and Mieczyslaw Forys

Department of Chemistry, Agricultural and Teachers University
08-110 Siedlce, Poland

INTRODUCTION

The well-established dissociative capture of electrons by isolated molecules of hydrogen chloride, hydrogen bromide, methyl bromide and hydrogen sulphide is endothermic and thus occurs at the electron energies high above the thermal region.[1] On the other hand, the attachment of the thermal energy electrons to these compounds has been observed in a number of the γ radiolysis experiments.[2-10] Moreover, it was established in these experiments that the kinetics of this process is of the third or even fourth order:

$$e^- + Sc + (1 \text{ or } 2)M \longrightarrow Products \qquad (1)$$

In this paper we summarize the results of a series of our experiments on these systems in which the electron swarm method has been applied.[11-14]

EXPERIMENTAL

The swarm chamber has been built based on the experience of authors from Christophorou's laboratory and the technical details are similar to those described in Ref.15. The experimental procedure has been described in detail in Ref.11. Here we make only some general remarks pertinent to the goal of this paper.

The swarm experiment in its classic form is limited to the use of only few buffer gases for which the electron energy distribution in the swarm is well known, like argon, nitrogen or ethylene. Of those, only ethylene has the range of E/N at which electrons have the thermal energy distribution, wide enough to be safely used for the thermal electrons experiment. However it is in turn reactive towards some radical species produced in our systems by α particles. In seeking the proper buffer gas we have found that carbon dioxide has a wide range of E/N in the thermal region and thus have chosen it for the experiment. As it will be seen from further consideration, carbon dioxide has also other advantages from the point of view of this work.

RESULTS

The rate constant, as defined for two-body electron capture in the swarm experiment, is equal to:

$$k = \alpha W \qquad\qquad (I)$$

where W is the electron drift velocity and α is the electron attachment coefficient:

$$\alpha = f/[Sc]d \qquad\qquad (II)$$

Here [Sc] denotes the concentration of the electron scavenger, d is the electron drift distance and f is found from experiment. If the reaction is of the second order, the k value is constant, i.e. does not depend on the pressure of any constituent of the investigated mixture. The concentration dependence of k indicates that more than one molecule takes part in the electron capture process. The electron drift velocity, W, depends on the composition of the mixture. As the literature lacks of such data we have determined W values using relative method which consists in measuring $\alpha(SF_6)$ in the given mixture and finding W from the relationship: $W = k/\alpha(SF_6)$, where k is the well established thermal electron capture rate constant for SF_6.

In this paper we present the results of measurement of the k values for the following systems: $HCl\text{-}CO_2$, $HBr\text{-}CO_2$, $HCl\text{-}HBr\text{-}CO_2$, $H_2S\text{-}CO_2$, $CH_3Br\text{-}CO_2$ and $H_2S\text{-}CH_3Br\text{-}CO_2$. The corresponding rate constants are shown in the first row of the Table 1. Two rate constants are placed when two parallel processes occur. The results for the $CH_3Br\text{-}C_2H_4$, $N_2O\text{-}C_2H_4$, $N_2O\text{-}CO_2$, $SF_6\text{-}CO_2$ and $SF_6\text{-}C_2H_4$ systems are also included for comparison. The value of the rate constant for SF_6 corresponds very well with the generally accepted data.[1] Also the two-body rate constant for CH_3Br falls within range of most data including the last ones obtained with FALP and CDE techniques.[16,17] Furthermore, the N_2O data are in remarkable agreement with those of Shimamori and Fessenden[18,19] which were obtained with the microwave conductivity technique. This comparison reveals that the method employed in this study is reliable both in the low and high additive concentration range.

Table 1. The rate constants for the thermal electron capture obtained by the electron swarm method.

Mixture investigated	k^a	vdW complex involved	K_{eq}^b	k_{vdW}^a
HCl-CO_2	1.6×10^{-31}	HCl-CO_2	5.0×10^{-22}	3.0×10^{-10}
HBr-CO_2	5.0×10^{-29}	HBr.HBr	1.9×10^{-22}	2.6×10^{-7}
	6.4×10^{-31}	HBr.CO_2	3.6×10^{-22}	1.5×10^{-9}
HCl-HBr-CO_2	5.0×10^{-29}	HCl-HBr	4.3×10^{-22}	1.2×10^{-7}
H_2S-CO_2	4.6×10^{-33}	H_2S-H_2S	3.1×10^{-22}	1.4×10^{-11}
CH_3Br-H_2S-CO_2	3.1×10^{-48}	CH_3Br-H_2S	7.2×10^{-22}	4.3×10^{-27}
	5.3×10^{-12}	-		
CH_3Br-CO_2	$5.3 \times 10\text{-}12$	-		
CH_3Br-C_2H_4	5.3×10^{-12}	-		
N_2O-CO_2	7.2×10^{-48}	N_2O-CO_2	4.0×10^{-22}	4.3×10^{-27}
	2.4×10^{-31}	-		
N_2O-C_2H_4	2.5×10^{-31}	-		
SF_6-CO_2	2.5×10^{-7}	-		
SF_6-C_2H_4	2.5×10^{-7}			

[a] The rate constants for second-order processes are in $cm^3 molec.^{-1} s^{-1}$, third-order - $cm^6 molec.^{-2} s^{-1}$ and fourth-order - $cm^9 molec.^{-3} s^{-1}$.
[b] In $cm^3 molec.^{-1}$

DISCUSSION

It is evident from Table 1 that the kinetics of electron capture is greater than second order for the first five systems. For the three-body process one could consider the simple Bloch-Bradbury (B-B) process with the second molecule acting as a stabilizing agent. However there are strong arguments against this mechanism in our systems:

1. It implies that the isolated molecules of HCl, HBr and H_2S accept thermal electrons to form transient excited negative ions which can be stabilized towards autoionization by a collision yielding either the stable parent or fragment negative ion. Both of these options should be excluded on the grounds that the lowest potential energy curves of the negative ion and parent neutral molecule cross high above the lowest vibrational levels of the molecule and thus far off the Franck-Condon range.[1,5-7,20-22]

2. Table 1 reveals that presumable stabilizing efficiency of the CO_2 molecule in the case of the HBr-CO_2 system would be by two orders less than that of the HBr one; and in the case of the H_2S-CO_2 system this molecule

does not participate in the process at all, while in the true B-B processes the stabilizing efficiency of the most of molecules is of the same order.[2] This behavior is only observed for the three-body process in the N_2O-CO_2 and N_2O-C_2H_4 systems (cf. Table 1).

If the simple B-B mechanism should be rejected, the alternative mechanism to account for the third and fourth order kinetics in these systems is to assume that thermal electrons are accepted by preexisting Van der Waals (vdW) complexes. The resulting negative ions are either stable towards the autoionization per se (third order kinetics) or should be collisionally stabilized (fourth order kinetics). This mechanism was postulated for hydrogen chloride as early as in 1970,[3] and then was developed for other systems in the γ radiolysis[5-10], microwave cavity[18,19,23] and electron swarm[11-14] experiments.

If we adopt this mechanism then the general reaction scheme for electron attachment to vdW complexes can be written as follows:

$$HX + B \xrightarrow{\quad K_{eq} \quad} (HX.B) \qquad (2)$$

$$e^- + (HX.B) \underset{k_{-3}}{\overset{k_3}{\rightleftharpoons}} (HX.B)^{-*} \qquad (3)$$

$$(HX.B)^{-*} + M \xrightarrow{\quad k_4 \quad} \text{Products} \qquad (4)$$

The measured "effective" rate constant $k_{eff} = \alpha W$ can be related to the corresponding rate constants in this scheme by the following expression:

$$k_{eff} = \frac{K_{eq}k_3k_4[B][M]}{k_{-3} + k_4[M]} \qquad (IV)$$

or

$$k_{eff}^{-1}K_{eq}[B] = k_3^{-1} + k_{vdW}^{-1}[M]^{-1} \qquad (V)$$

From Eq.V the three-body rate constant for the electron capture by (HX.B), $k_{vdW} = k_3k_4/k_{-3}$, can be found as well as the rate constant for formation of the excited negative ion, k_3, provided the equilibrium constant, K_{eq}, is known. If the vdW complex forms the stable negative ion ($k_{-3}=0$) then Eq.IV takes the simple form:

$$k_{eff} = K_{eq}k_3[B] \qquad (V)$$

from which the two-body rate constant for the electron capture by the vdW complex can be eluded. The equilibrium constants were calculated on the basis of second virial coefficient[24] and are shown in Table 1 along with corresponding rate constants.

Now the question arises why the vdW complexes are effective as electron acceptors while the constituent isolated molecules are not?

The first explanation given by Armstrong for the case of hydrogen halides was based on the short life-time of the transient negative ion HX^{-*}, of the order of 10^{-13} s. The distance the colliding molecule with the thermal velocity can traverse during this time is in fact of the atomic dimension (ca. 10^{-9} cm) which means that such a molecule already undergoes vdW interaction. However this explains only those cases when the individual molecule is principally able to form a negative ion.

It is instructive to compare the dipole moments of the vdW molecules with those of the constituent monomers: the dipole moments of $(HCl)_2$ and the HCl molecule are equal to 3.1 D[25] and 1.08 D[26], those of (H_2S) and the H_2S molecule – 2.3 D[25] and 0.97 D[26], correspondingly; and that of $(HCl.H_2S)$ – ca. 2 D.[12] The dipole moment increases also in the complexes with one unpolar molecule, as in the case of the $(HCl.CO_2)$ complex for which $\mu=1.45$ D.[27] While the data on the dipole moments of the vdW molecules are still scattered, nonetheless it is clear that dipole moments of these molecules are larger than it would follow from the simple vector sum of the constituent molecular dipoles.[28] This indicates that the interaction of the molecules in the complex is not confined to the weak vdW bond. All the complexes under consideration which accept electrons are hydrogen-bonded with nearly linear X-H-B configuration.[25,29-33] On the other hand, the $(H_2S.CO_2)$ complex, which is inactive towards electrons has a cyclic configuration and no hydrogen bond.[34] Also inactive the $(CH_3Br.CO_2)$ molecule obviously does not have a conventional hydrogen bond. Thus it seems, that the enhancement in the dipole moment of the vdW molecule should be caused by overlapping the lone pair of electrons on the electron donor (B) with the lowest unoccupied molecular orbital of the proton donor (HX). This, in turn, may result in a new molecular orbital energetically accessible for the thermal electron and thus activate the vdW complex as an electron acceptor.

Acknowledgement: This research was supported by CPBP 01.19 (01.01).

REFERENCES

1. For review see L.G.Christophorou,D.L.McCorkle and A.A.Christodoulides, in "Electron Molecule Interactions and their Applications", edited by L.G.Christophorou,Academic Press,New York,1984,v.1,p.477.
2. R.S.Davidow and D.A.Armstrong,J.Chem.Phys.,48:1235(1968).
3. G.R.A.Johnson and J.L.Redpath,Trans.Faraday Soc.,66:861(1970).
4. M.Forys, Radiat.Eff.,25:111(1975).
5. S.S.Nagra and D.A.Armstrong,Can.J.Chem.,54:3580(1976).
6. S.S.Nagra and D.A.Armstrong,J.Phys.Chem.,81:599(1977).
7. S.S.Nagra and D.A.Armstrong,Radiat.Chem.Phys.,11:305(1978).
8. I.Szamrej and M.Forys,Radiat.Eff.,77:237(1983).

9. I.Szamrej,I.Chrzascik and M.Forys,Radiat.Eff.,**83**:291(1984).

10.I.Szamrej and I.Chrzascik,Radiat.Phys.Chem.,**29**:149(1987).

11.I.Szamrej and M.Forys,Radiat.Phys.Chem.,**33**:393(1989).

12.I.Szamrej,H.Janicka,I.Chrzascik and M.Forys,Radiat.Phys.Chem.,**33**:387(1989).

13.I.Szamrej,H.Kosc,B.M.Zhytomirski and B.G.Dzantiev,Radiat.Phys.Chem. (submitted).

14.I.Szamrej,M.Forys and W.Tchorzewska,Radiat.Phys.Chem.(submitted).

15.L.G.Christophorou,"Atomic and Molecular Radiation Physics",Wiley-Interscience,New York,1971.

16.E.Alge,N.G.Adams and D.Smith,J.Phys.B,**17**:3827(1984).

17.Z.Lj.Petrovic and R.W.Crompton,J.Phys.B,**20**:5557(1987) and refs.therein.

18.H.Shimamori and R.W.Fessenden,J.Chem.Phys.,**69**:4732(1978).

19.H.Shimamori and R.W.Fessenden,J.Chem.Phys.,**71**:3009(1979).

20.D.Smith and N.G.Adams,J.Phys.B,**20**:4903(1987).

21.W.C.Wang and L.C.Lee,J.Appl.Phys.,**63**:4905(1988).

22.Z.Lj.Petrovic,W.C.Wang and L.C.Lee,J.Appl.Phys.,**64**:1625(1988).

23.Y.Hatano and H.Shimamori,in "Electron and Ion Swarms",edited by L.G.Christophorou,Pergamon Press,Oxford,p.103.

24.J.O.Hirschfelder,C.F.Curtis and R.B.Bird,"Molecular Theory of Gases and Liquids",Wiley,New York,p.111.

25.R.C.Kerns and L.C.Allen,J.Am.Chem.Soc.,**100**:6587(1978).

26."Handbook of Chemistry and Physics",56 ed.,CRC Press,1975.

27.R.S.Altman,M.D.Marshall and W.Klemperer,J.Chem.Phys.,**77**:4344(1982).

28.A.C.Legon and D.J.Millen,Chem.Rev.,**86**:635(1986).

29.M.J.Frisch,J.A.Pople and J.E.Del Bene,J.Phys.Chem.,**89**:3664(1985).

30.R.T.Arlinghaus and L.Andrews,J.Phys.Chem.,**88**:4032(1988).

31.M.M.Szczesniak and S.Scheiner,J.Chem.Phys.,**83**:1778(1985).

32.M.A.Spackman,J.Chem.Phys.,**85**:6587(1986).

33.E.J.Goodwin and A.C.Legon,J.Chem.Soc.,Faraday Trans.2,**80**:51(1983).

34.J.K.Rice,L.H.Coudert,K.Matsumura,R.D.Suenram,F.J.Lovas,W.Stahl,
D.J.Pauley and S.G.Kukolich,J.Chem.Phys.,**92**:6408(1990).

LEADER BREAKDOWN IN COMPRESSED SF$_6$: RECENT CONCEPTS

AND UNDERSTANDING

Lutz Niemeyer
ABB Research Center
Baden, Switzerland

ABSTRACT

A review is given of the present understanding of the breakdown of non–uniform gaps in compressed electronegative gases. Breakdown under these conditions occurs by the leader mechanism, the physical background of which is briefly discussed. It is shown how it can be quantified and how the temporal and spatial aspects of the breakdown process can be related to the parameters of the insulation system and the applied voltage. Some practical problems occurring in high voltage GIS will be discussed as examples such as particle induced impulse flashover, formative time lags under fast rising pulses, and flashover to ground in disconnector switching.

INTRODUCTION

Breakdown of compressed electronegative gases in non–uniform field gaps is known to be controlled by the stepped leader mechanism the physics of which is now understood sufficiently well to allow quantitative modelling of insulation systems. The scope of this paper is to summarize the present state of physical understanding to show how it can be used for breakdown modelling and to give a few examples for applications in GIS design.

BASIC PHYSICAL PROCESSES

The physical processes underlying leader breakdown in an electronegative gas are (in sequential order):

(1) the production of an initiating electron close to a high field electrode
(2) the development of a critical avalanche leading to streamer inception
(3) the development of a streamer corona by the propagation, multiplication, and branching of streamers
(4) an ionization/relaxation process which, after a time delay, transforms one or more of the streamer channels into conducting filaments thereby creating a leader step
(5) the repetition of the sequence (3) – (4) at the tip of the leader resulting in stepped leader propagation.

The major characteristics of these processes will be briefly reviewed with reference to publications in which more detailed information can be found.

Initiating electron and streamer inception

The mechanisms that provide avalanche initiating electrons in electronegative gases are polarity dependent. For negative polarity, field emission from the electrode is dominant. It is strongly dependent on the curvature radius, microstructure, and surface state of the electrode. For positive polarity, field detachment from negative ions in the gas is the dominant mechanism. It can be quantified by the modified volume – time law[1]. The initial electron triggers an avalanche which develops into a filamentary streamer if the streamer criterion is fulfilled[2]. Its evaluation will not be discussed here.

Streamer corona development

The first streamer triggers further ones by photoionization and leads to the development of a complex streamer pattern, the streamer corona. The growth dynamics and structure of the corona is controlled by the local space charge distorted field at each streamer tip and can be simulated by a three–dimensional random growth model[3]. Figure 1a shows a typical result of such a simulation. It is seen that the corona is a compact structure with approximately spherical envelope and with a maximal streamer length ℓ. Inside the corona the field is determined by the field in the streamer channels which is about the critical field E_{cr}[4]. With the corona, a space charge Q_c is injected into the gap (figs. 1b and c) which distorts the original background field $E_0(x)$ (broken curve) as indicated by the solid curve. The length ℓ of the longest streamer is controlled by the streamer channel field E_{cr} and scales as

$$\ell = f(U/E_{cr}) \approx C_\ell (U/E_{cr}) \qquad C_\ell \approx 0.5 \qquad (1)$$

where U is the applied voltage. The functional dependence is determined by the electrode geometry[5] and can be roughly approximated by a linear relationship with a proportionality factor C_ℓ. Note that due to the random character of the corona growth not all streamers are equally long; a few of them protrude out of the corona bulk (arrows in fig. 1a). These carry increased space charge and are exposed to a higher local field.

To simplify numerical calculations, an approximate corona model can be constructed as illustrated by fig. 1d. The space charge cloud containing the injected charge Q_c is replaced by an equivalent conducting sphere of radius r_Q carrying the same charge on its surface. This concept reduces field calculations from a Poisson problem to a Laplacian problem. It has been shown[6], that r_Q is approximately proportional to ℓ

$$r_Q \approx C_Q \cdot \ell \qquad C_Q \sim 0.3 \qquad (2)$$

Thus, with ℓ from eq (1) inserted into eq (2) the injected corona charge Q_c can be obtained by solving the Laplacian problem with the electrodes as boundary conditions and integrating the surface charge on the sphere r_Q:

$$Q_c = Q_c (U/E_{cr}, \text{gap geometry}) \qquad (3)$$

It can furthermore be shown by numerical field calculations that the maximal electric field $\hat{E}$ at the vertex of the equivalent sphere r_Q, the field E_ℓ at the corona periphery, and the average field $\bar{E}$ between the equivalent sphere and the corona periphery (fig. 1d) are approximately proportional to the injected charge Q_c:

$$\bar{E} \propto E_\ell \propto \hat{E} \propto Q_c \qquad (4)$$

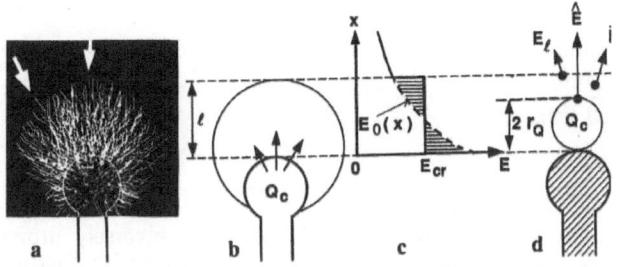

Fig. 1.   Streamer corona: random growth simulation (a), approximate structure
            (b), field distribution (c), and equivalent conducting sphere concept (d).

This porportionality is useful for simplifying leader propagation modelling. The time
scale of the corona development is controlled by the streamer velocity and is typically
in the ns range[5].

Leader inception

The transformation of a corona streamer into a leader step has been shown to
require that the gas in the channel be heated to above its dissociation temperature[7]
which is about 2500 K for SF_6. The required energy can be provided by various
mechanisms which depend on the parameters of the insulation gap and the applied
voltage stress. Hitherto, the following mechanisms have been identified:

— The **stem** mechanism (known from the breakdown of long air gaps) has been
  observed at negative polarity[8].
— The **precursor** mechanism[5,6] seems to be dominant under conditions encountered
  in HV–GIS.
— The **high frequency (HF) heating** mechanism has been found to control VFT
  breakdown at lower gas pressures.

Also, combinations of mechanisms have been observed. It is, at present,
neither clear if still further mechanisms exist nor have the criteria for the occurrence
of the above mechanisms been completely understood. In this paper, the discussion
will be restricted to the precursor mechanism as the most relevant for HV–GIS.

The precursor process can be briefly described as follows[5,6]: At the end of
corona growth, electrons are rapidly attached so that positive and negative ions
remain in the former streamer channels. These drift apart in the electric field and
create local space charge dipoles, an effect which is strongest at the corona periphery
and particularly at the protruding streamer tips (Fig. 1a) because of the locally
enhanced ion densities and fields. The space charges eventually create local field
enhancement zones at the corona periphery in which ionization is restarted. Thereby,
an ionization instability is triggered which propagates towards the electrode and,
upon reaching it, establishes an electronically conducting filament bridging the
corona. The outer end of this filament serves as an electrode from which a new
corona burst is launched. The current associated with it is channelled into the
filament and heats it to above the dissociation temperature.

The major characteristics of the precursor mechanism are the following:

(1) An **inception level** exists which can be expressed in the form of a corona charge criterion[5,6]: A precursor is activated if the injected corona charge Q_c exceeds a critical value Q_{cr}

$$Q_c \geq Q_{cr} = C_{cr}/p_0^2 \tag{5}$$

with a critical inception charge Q_{cr} which is inversely proportional to the square of the gas pressure p_0. The proportionality factor C_{cr} has been measured to be 40 AsPa² (160 AsPa²) for positive (negative) polarity.

(2) An **inception delay** exists between corona formation and the establishment of the leader channel which scales as [5,6]

$$\tau \approx C_\tau/p_0^2 U \tag{6}$$

with a proportionality factor C_τ that has been experimentally determined for SF_6 to be 4×10^8 VsPa²(80×10^8 VsP$_a$²) for positive (negative) polarity.

(3) The **parameter range** within which the precursor is the dominant leader formation mechanism is limited in two ways: At low pressures and overvolted gaps a transition to the streamer mechanism occurs[5]. For very rapidly varying applied voltages and lower pressures HF heating contributes to or dominates the leader formation process[9]. An onset criterion for the latter mechanism can be derived for the case of VFT stress. With a voltage of amplitude U_0 oscillating with a frequency f, HF heating can only occur if more than one oscillation period $T=1/f$ occurs during the inception delay τ, i.e., with reference to eq. (6), if

$$f \gg f_{lim} \approx p_0^2 \hat{U}_0/C_\tau \tag{7}$$

This criterion is only a rough approximation and will require a more precise formulation which will also have to account for the damping of the VFT waveform and the energetic aspects of the heating process.

Leader channel field

The heat input accompying the streamer–to–leader transition causes a pressure rise in the leader channel which leads to a gas dynamic expansion. This, together with the molecular dissociation, leads to a strong reduction of the critical field E_{cr} so that electronic conductivity becomes possible at low field values[7,11]. It has been shown[7] that the average leader channel field is approximately given by the analytic expression:

$$\bar{E}_\ell = a \ln(1 + b\, p_0 t)/t \tag{8}$$

Here, a and b are combinations of dielectric and thermodynamic gas properties, which for SF_6 have the values a = 0.5 (1.5) V s/m and b = 70 (20) Pa⁻¹s⁻¹ at positive (negative) prolarity. E_ℓ is seen to decrease with time t and to be only weakly dependent on the gas pressure p_0.

Leader propagation and breakdown

Once leader inception has taken place, the second corona developing from the first leader section may undergo another streamer to leader transition after another delay time and establish a second leader step. The repetition of this process results in stepped leader propagation and continues until the leader reaches the opposite electrode and causes breakdown or until the leader tip voltage U becomes too low to fulfill the leader formation criterion. A reduction of U may be caused either by a decrease of the applied voltage U_0 or by the increasing voltage drop along the elongating leader.

Spatial randomness of leader propagation

The tortuosity of leader trajectories is caused by the randomness with which the strongest streamers develop during the growth of the streamer corona. It has been shown[12,13] that the probability p of a leader step to occur in a specific spatial direction approximately obeys the proportionality

$$p \propto (\bar{E} - \bar{E}_0) \quad \text{for } \bar{E} > \bar{E}_0 \ , \ p = 0 \quad \text{for} \quad \bar{E} \leq \bar{E}_0 \tag{9}$$

where $\bar{E}$ is the average field component in this direction and $\bar{E}_0$ is a threshold field which is proportional to the critical charge Q_{cr} according to eq. (4). This probabilistic relation allows to introduce directional randomness into leader propagation modelling.

LEADER BREAKDOWN MODELLING

With the above elements, leader breakdown can be modelled as indicated by the flow diagram Fig. 2. Input parameters are the applied voltage waveform $U_0(t)$, the gap geometry, the gas pressure p_0, and the kind of gas. To determine **leader inception** (Fig. 2a), the field distribution E around the electrode is calculated (taking into account drifting ionic space charge that has remained from a previous discharge, if necessary). The streamer criterion is then evaluated and a stochastic model for the generation of an initiating electron is activated. If the streamer criterion is fulfilled and if a first electron is available, the streamer corona model is activated to provide the corona extension ℓ which determines the step length and the corona charge Q_c as input quantity for the leader inception criterion. If the latter is fulfilled, a delay module is activated to determine the inception delay τ.

The subsequent **leader propagation** (Fig. 2b) is modeled similarly to the inception except that first electron and streamer inception do not have to be verified because they are always fulfilled at the narrow, ionized leader tip. Additionally, the presence of the leader is taken into account for the field calculations as an electrode extension with a potential distribution according to the average channel field $\bar{E}_\ell$. In three dimensional modelling, additionally, the step orientation has to be randomly chosen according to eq. (9).

A key component of all leader models is the calculation of electric fields and charges. According to electrode geometry and dimensionality of the model different levels of approximation are used ranging from analytical expressions over charge simulation methods to 3–dimensional grid methods[14].

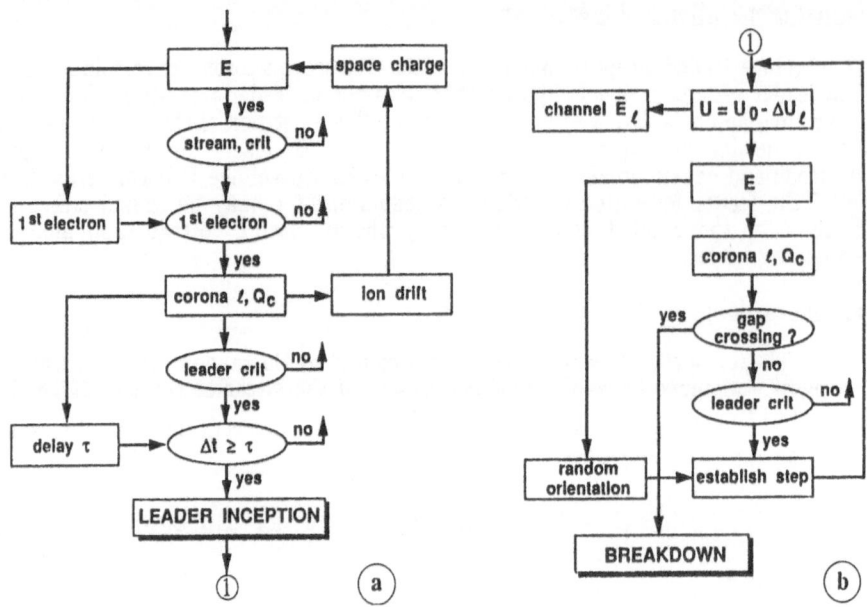

Fig. 2.   Flow diagram for modelling leader inception (a) and leader propagation (b).

EXAMPLES

Leaders modelling as described above has already been applied to various problems such as impulse breakdown of non–uniform field gaps[15], breakdown between rough electrode surfaces[16], and the determination of random spatial breakdown trajectories[6,12,13]. Here some further examples will be discussed which are of particular interest for GIS design.

Particle induced breakdown

We consider a weakly non–uniform ("as designed") gap in which a small oblong conducting particle is present as a contamination. The particle is assumed to be oriented parallel to the electric background field E_0 and is characterized by its effective tip curvature radius r_0 and its length ℓ_0. The length ℓ_0 is assumed to be small in the sense that E_0 does not vary appreciably along the particle. The field distribution around the particle then is determined by the background field E_0, the particle length ℓ_0, and the particle aspect ratio ℓ_0/r_0. Together with the gas pressure p_0 and the corresponding critical field E_{cr} all quantities controlling the discharge processes are then defined locally, i.e. independently of the shape of the electrodes and of the applied voltage by which the background field E_0 is produced. This allows to formulate the discharge criteria in terms of local parameters.

The **streamer inception** criterion can be expressed in the form[17]

$$E_0^{(\text{str})}/E_{cr} = f(p_0 \cdot \ell_0, \; \ell_0/r_0) \tag{10}$$

if the particle sits on an electrode. If the particle is uncharged and floats between the electrodes, only half of the real particle length has to be inserted for ℓ_0 as the equipotential surface through the particle center corresponds to the electrode surface as indicated in Fig. 3. The figure shows a plot of eq. (10) for 2 values of the aspect ratio ℓ_0/r_0 (broken curves).

For **leader inception**, numerical calculations indicate a scaling relation similar to eq. (10) with an additional direct pressure dependence at high $p_0 \cdot \ell_0$ values:

$$E_0^{\text{lead}}/E_{cr} = f(p_0 \cdot \ell_0, \ell_0/r_0, p_0) \tag{11}$$

Figure 3 shows this relation calculated for a typical GIS pressure of $p_0 = 450$ kPa as solid curves. The points are experimental data for fixed (full points) and floating (open points) particles in uniform field gaps stressed with step voltage pulses.

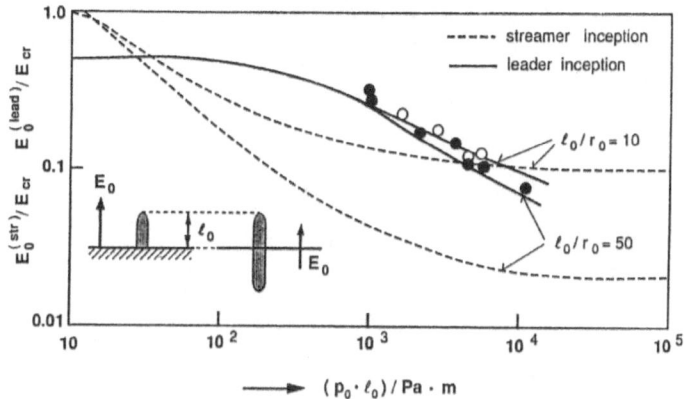

Fig. 3. Normalized background fields $E_0^{(\text{str})}/E_{cr}$ (broken curves) and $E_0^{(\text{lead})}/E_{cr}$ (solid curves) for streamer inception and positive leader inception ($p_0 = 450$ kPa). Measured data for particle on positive electrode (full points) and floating particles (open points). Parameter range: 100–500 kPa, $\ell_0 = 5$–15mm.

The following conclusions can be drawn from Fig. 3:

(1) Both streamer inception and leader inception fields decrease with increasing product $p_0 \cdot \ell_0$ and increasing aspect ratio ℓ_0/r_0. This means that the longer and the more pointed a particle and the higher the gas pressure are the lower the streamer and leader inception field levels become.

(2) Once a leader has started, it produces an effective elongation of the particle thereby further reducing the inception level for the next step. This inevitably leads to breakdown if the background field E_0 remains constant during leader propagation so that leader inception is equivalent to breakdown.

(3) The leader inception field is less sensitive to the aspect ratio ℓ_0/r_0 than the streamer inception field i.e. the details of the particle shape enter the breakdown field less sensitively than they enter the PD inception field.

(4)   The streamer and leader inception curves have a cross–over point at low $p_0 \cdot \ell_0$ values. For typical GIS pressures the range to the left of the cross–over point corresponds to particle lengths below some $100\mu m$, i.e. to electrode surface roughness structures. As in this range $E_0^{(str)} > E_0^{(lead)}$, streamer inception automatically entails leader breakdown so that the streamer criterion is sufficient as a breakdown criterion.

(5)   The $p_0 \cdot \ell_0$ range to the right of the cross–over point corresponds to particulate contamination. Here we have $E_0^{(str)} < E_0^{(lead)}$, i.e. particles produce partial discharges before causing breakdown. In this case it is the leader criterion which is relevant for breakdown.

Leader propagation and formative time lags

The knowledge of formative time lags, particularly under rapidly varying voltage waveforms, is important for insulation coordination. To determine them, a simplified model can be used[10] which assumes the leader to propagate along the shortest field line in the gap. This concept neglects the random spatial character of the leader trajectory and therefore gives a lower limit estimate of the time to breakdown. Figure 4 shows an example for lightning impulse breakdown of a uniform field gap with a particle on the negative electrode. The photomultiplier (PM) signal

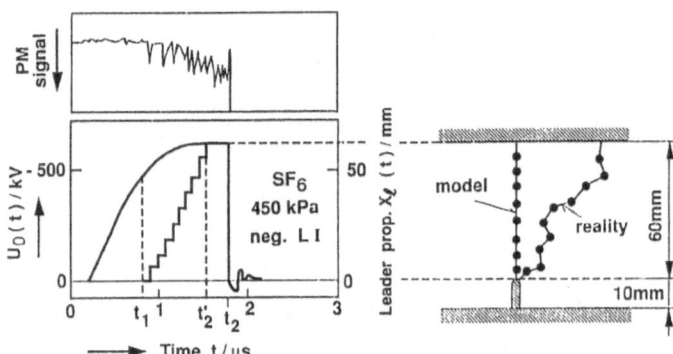

Fig. 4.   Leader propagation through a uniform field gap with a protrusion on the negative electrode under lightning impulse. Breakdown times t_2 (measured) and t_2' (calculated).

indicates the light emitted by the leader steps. The calculated leader propagation curve X (t) is shown as a stepped curve together with the applied voltage waveform $U_0(t)$. For the calculation it has been assumed that the first electron is produced at the time t_1 at which the first light signal is experimentally observed. It is seen that the model predicts gap crossing after $N_s = 9$ steps and a formative time lag of $t_f = t_2 - t_1 = 650$ ns whereas the measurement gives $N_s = 12$ and $t_f = 950$ ns. Both N_s and t_f are by about the same factor higher than the model prediction. This is due to the randomness of the leader propagation in space as schematically indicated in the figure.

Random aspects of leader propagation

As an example we discuss breakdown to the enclosure in a GIS disconnector switch which closes on an open bus section with a trapped charge. Here, the spatial randomness of the leader and its branching are critical for the insulation performance and a fully three–dimensional and random modelling is required. As indicated in Fig. 5a the mobile contact 1 connected to an AC power source moves towards the fixed contact 2 connected to a bus section with a trapped charge symbolized by a capacitance C charged to − 1pu. When the source polarity is opposite to the trapped charge polarity a leader starts from contact 1 and branches (thin lines). Eventually one of its branches reaches electrode 2 and establishes a breakdown channel (arrowed line) through which the open bus section is recharged to the source voltage + 1pu. The two electrodes and the branched leader pattern now become an equipotential structure at + 1pu against the grounded enclosure. For wrongly designed electrode geometry it then can happen that one of the leader side branches protrudes so much towards the enclosure that it starts to propagate to ground and establishes breakdown (dotted line). Figure 5a is the result of a simulation carried out for the shown geometry. Figure 5b shows a still photograph taken in an experimental device with similar parameters as those assumed in the simulation. Repetitions of both the experiment and the simulation show that the flashover to the enclosure is in fact a random event the probability of which is determined by the geometry of the device.

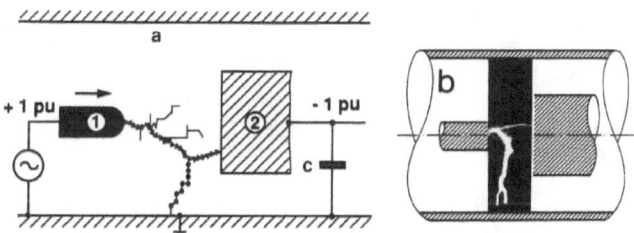

Fig. 5.   Breakdown to the enclosure is a GIS disconnector switch closing onto a bus section with a trapped charge. (a): Simulation with 3D random leader propagation model. (b): Still photograph of an experiment.

CONCLUSIONS

It has been shown that some major aspects of leader breakdown in compressed electronegative gas insulation have become accessible to physically based modelling and that this modelling can be used for practical insulation design by introducing appropriate simplifications. Other aspects of leader breakdown still require more detailed study. An important issue is leader inception in the presence of drifting ionic space charge under slowly varying waveforms (corona stabilization). Another issue is an improved understanding of other leader formation mechanisms such as the stem and the HF heating process and an establishment of precise criteria for their range of relevance.

REFERENCES

1. N. Wiegart, L. Niemeyer, F. Pinnekamp, W. Boeck, J. Kindersberger, R. Morrow, W. Zaengl, M. Zwicky, I. Gallimberti, and S. A. Boggs, Inhomogeneous Field Breakdown in GIS — The Prediction of Breakdown Probabilities and Voltages, IEEE Trans. on Power Deliv., 3, 923 (1988).

2. A. Pedersen, I. W. McAllister, G. C. Crichton, and S. Vibholm, Formulation of the Streamer Breakdown Criterion and its Application to Strongly Electronegative Gases and Gas Mixtures, Archiv für Elektrotechnik, 67, 395–402 (1984).

3. L. Niemeyer and H. J. Wiesmann, Structure of the Impulse Corona in Electronegative Gases, IXth Int. Conf. on Gas Discharges, Venice, 223–226 (1988).

4. I. Gallimberti, G. Marchesi, and R. Turri, Corona Formation and Propagation in Weakly or Strongly Attaching Gases, 8th Int. Conf. on Gas Discharges, Oxford, 167–170 (1985).

5. I. Gallimberti and N. Wiegart, Streamer and Leader Formation in SF_6 and SF_6 Mixtures Under Positive Impulse Conditions, Pts. I and II, J. Phys. D: Appl. Phys. 12, 2351–2379 (1986).

6. L. Niemeyer, L. Ullrich, and N. Wiegart, The Mechanism of Leader Breakdown in Electronegative Gases, IEEE Trans. El. Ins. Vol. EI–24, 309–324 (1989).

7. L. Niemeyer and F. Pinnekamp, Leader Discharges in SF_6, J. Phys. D: Appl. Phys. 16, 1031–1045 (1983).

8. I. Gallimberti, L. Ullrich, and N. Wiegart, Experimental Investigation of the Streamer to Leader Transition in SF_6 Under Negative Polarity, Gaseous Dielectrics V, L. G. Christophorou and D. W. Bouldin (Eds.), Pergamon, New York, 126–133 (1987).

9. H. Hiesinger, The Calculation of Leader Propagation in Point/Plane Gaps Under Very Fast Transient Stress, These Proceedings, p. 129.

10. T. Dunz, L. Niemeyer, and G. Riquel, The Effect of Leader Propagation on the V–t–Curves Under LI and VFT in GIS. These Proceedings, p.255.

11. L. Niemeyer, A Model of SF_6 Leader Channel Development, Proc. 8th Int. Conf. on Gas Discharges, Oxford, 223–226 (1985).

12. L. Niemeyer and H. J. Wiesmann, Modelling of Leader Branching in Electronegative Gases, Gaseous Dielectrics V, L. G. Christophorou and D. W. Bouldin (Eds.), Pergamon, New York, 134–139 (1987).

13. L. Niemeyer, A Stepped Leader Random Walk Model, J. Phys. D: Appl. Phys. 20, 897–906 (1987).

14. L. Niemeyer, L. Pietronero, and H. J. Wiesmann, Fractal Dimension of Dielectric Breakdown, Phys. Rev. Lett. 52, 1033–1036 (1984).

15.   N. Wiegart, A Semi—Empirical Leader Inception Model for SF_6, Proc. 8th Int. Conf. in Gas Discharges, Oxford, 227–230 (1985).

16.   T. Dunz, B. Fruth, L. Niemeyer, L. Ullrich, K. Diederich, and M. Hässig, Electrical Field on Rough Electrode Surfaces and Its Influence on the Statistical Properties of SF_6 Breakdown, 6th ISH, New Orleans, paper no. 23.04 (1989).

17.   C. Cooke, Ionization, Electrode Surfaces, and Discharges in SF_6 at EHV, IEEE–PAS **94**, 1518 (1975).

DISCUSSION

I. GALLIMBERTI: The complete discharge development to breakdown is related to the leader propagation in steps: in the statistical "random walk" model you need a criterion to continue or arrest the leader propagation at each step. Can you describe this criterion?

L. NIEMEYER: The random propagation probability law, equation (9), contains a threshold field $\bar{E}_0$ below which the probability of propagation $p = 0$, i.e. no further propagation is possible. The numerical value of $\bar{E}_0$ is determined by calculating the leader inception voltage U_{1i} and the first leader step length ℓ with the model given in Fig. 2a and by calculating the average field $\bar{E}$ along ℓ with a numerical field calculation method. This value is taken as the threshold value $\bar{E}_0$ and is found to be proportional to and of the order of the critical field E_s.

N. G. TRINH: The propagation of the leader discharge will modify the field distribution in the gap. Can you elaborate on how this aspect is covered in the simulation program?

L. NIEMEYER: In the random leader propagation model the three–dimensional potential distribution in the gap is calculated at the lattice points by which the volume has been discretized by a fast iterative Laplace solution (Ref. 14). This calculation is carried out anew after each propagation step and takes into account the average voltage drop along all leader branches (Ref. 12).

R. T. WATERS: The paper shows significant progress in the modeling of the spatio–temporal characteristics of leader growth in SF_6. To what extent are transient currents and electric field data available to augment the image converter and photomultiplier records in order to test more quantitatively the simulations?

L. NIEMEYER: Experimentally, fast measurements of the true current with sub–nanosecond risetime have been made (see Refs. 5 and 6) with a shielded probe. They have shown that the current pulses associated with the leader step have durations in the ns range and correspond roughly to typical streamer propagation velocities and corona extensions.

R. J. VAN BRUNT: Could you comment on the possibilities of making meaningful comparisons between results of your simulations and actual discharges, e.g., in prediction of probability of breakdown or leader propagation in a given direction.

L. NIEMEYER: A comparison of simulated and experimental leader patterns and breakdown trajectories is given in Ref. 12. It shows satisfactory agreement. A limitation of the random lattice model is the limited number of propagation directions possible on a cubic lattice. This does not allow us to reproduce completely the smoothness of the real trajectories.

ELECTRICAL BREAKDOWN IN THE SPACE ENVIRONMENT

E. E. Kunhardt, S. Barone, J. Bentson and S. Popovic
Weber Research Institute
Polytechnic University
Route 110, Farmingdale, NY 11735

ABSTRACT

The domain of parameters for breakdown in the space environment involves extreme conditions such as non-uniform field, $Nd/(Nd)_m \ll 1$ (where N is the density of the background gas, d is the electrode separation, and $(Nd)_m$ is the value of Nd at the Paschen minimum), and magnetic field, and in addition, the influence of a weakly ionized background gas. We have decomposed the problem of breakdown in the space environment into two: 1) breakdown in partially enclosed structures, where the effect of the "ground" electrode (i.e., the enclosing structure) needs to be taken into account, and 2) "sheath" breakdown, where the "ground" electrode is assumed to be at infinity. The first of these problems is further decomposed into two: a) when the size of the structure (or openings in the enclosure) is such that the effect of the space-plasma is mainly to eliminate the statistical time lag for breakdown, and b) where the effect of the space-plasma needs to be taken into account. In this paper, the problem of breakdown in partially enclosed structures neglecting the effect of the space-plasma is considered.

INTRODUCTION

Recent interest in the behavior of higher power systems in the space environment has renewed interest in breakdown under extreme conditions. In general, this refers to the situation where the applied field, $\underset{\sim}{E}_a$, is non-uniform (i.e., $\underset{\sim}{E}_a = \underset{\sim}{E}_a(r)$), and/or $Nd/(Nd)_m \ll 1$ or $Nd/(Nd)_m \gg 1$ (where N is the density of the background gas, d is the electrode separation, and $(Nd)_m$ is the value of Nd at the Pashen minimum) (Meek and Craggs, 1978). The problem of breakdown in the space environment is characterized by having $Nd/(Nd)_m \ll 1$, with the additional complications of a weakly ionized background gas and the presence of a crossed magnetic field, B. In the absence of these complications, "gas" breakdown (in contrast to "vacuum" breakdown) can occur in this regime for geometries in which electrons can be trapped in orbit, whose lengths in the field direction, are at least of the order of the effective free path for ionization, α_{eff}^{-1}. In this case, ionization growth can occur before the primary electrons are collected, and a partial glow-like discharge can be sustained near the high field electrode. These discharges are observed, for example, in geometries where the anode area is sufficiently small to create geometrically trapped orbits (Alexeff, 1987).

In the presence of the magnetic field, the orbits become more complex and the dynamics of the transport is complicated by trapping due to both geometry and field, and by the fact that the energy distribution of the electrons at a point $\underset{\sim}{r}$ is highly non-local in space due to the nearly ballistic transport of electrons.

Significant contributions to the formulation of the breakdown characteristics in this regime have been made by Redhead (1958) and Hobson and Redhead (1958) in connection with the development of vacuum gauges. He assumed that breakdown develops through the avalanche growth of trapped electrons and obtained an approximate expression for the primary ionization coefficient (the effective pressure or field concept is not applicable in this regime (Blevin and Haydon, 1958). Presently, the physics of the breakdown process for breakdown under extreme conditions in the presence of a magnetic field is yet to be clarified. Further, for the altitudes of interest (200-400 km), the background gas can be considered to be highly ionized since electron and ion densities approach 1% of the neutral densities. Thus, the application of a voltage to an electrode in the space plasma causes the development of a plasma sheath, which, under some circumstances, may radically alter the breakdown characteristics.

A general formulation of a breakdown criterion requires the introduction of parameters that can be used to calculate ionization growth in strongly non-local conditions, including the effect of trapped orbits and space charge. An interesting aspect of the particle dynamics, relevant to the determination of these parameters, is the difference in the influence of elastic collisions on the energy gain of the electrons for the two cases with and without (crossed) magnetic field. In the absence of a magnetic field, elastic collisions with the background gas contribute to the energy loss mechanism of the electrons. In the presence of a magnetic field, the energy gain is limited by the maximum potential in a given orbit; however, elastic collisions allow access to regions with higher potential, and, therefore, contribute to the energy gain mechanism.

We have decomposed the problem of breakdown in the space environment into two: 1) breakdown in partially enclosed structures, where the effect of the "ground" electrode (i.e., the enclosing structure) needs to be taken into account, and 2) "sheath" breakdown, there the "ground" electrode is assumed to be at infinity. The first of these problems is further decomposed into two: a) when the size of the structure (or openings in the enclosure) is such that the effect of the space-plasma is mainly to eliminate the statistical time lag for breakdown, and b) where the effect of the vacuum plasma needs to be taken into account. Further consideration need to be given to the orientation of the electrodes with respect to the magnetic field.

In this paper, we discuss breakdown in partially enclosed structures in which the inter-electrode separation is assumed to be smaller than the sheath dimension at breakdown, so that initially, the contribution from the background space-charge to the local field is negligible in comparison to the applied field. Moreover, we will restrict our discussion to a generic geometry consisting of a spherical anode of radius r_a, inside a cylindrical cathode of radius, r_k, with $r_k \gg r_a$. As discussed below, the Townsend coefficients, α and ω, can be used in this case to arrive at a breakdown criterion. Although orbit theory is not sufficient to determine these parameters, careful consideration of the electron orbits is necessary in order to develop a physical understanding of the breakdown characteristics. In the next section, the properties of these orbits are discussed for the case of an isolated spherical anode (i.e., $1/r$ potential). Some of the orbits have been found to be extremely sensitive to initial conditions (i.e., chaotic). Following this discussion, the formulation of a breakdown criterion for the generic geometry is discussed. In the final section, results from breakdown experiments in the generic geometry are presented.

SINGLE ELECTRON TRAJECTORIES IN A $1/r$ POTENTIAL IN THE PRESENCE OF A MAGNETIC FIELD

Trajectories have been computed numerically for electrons starting from rest for a variety of initial positions relative to the spherical anode in order to study the effect of initial condition on the subsequent orbit (Bentson et al., 1990). The equations of motion for an electron (charge q, mass m) in a uniform magnetic field, B, in the vicinity of a spherical anode (radius a) biased to a potential $V_0 (> 0)$ have been written in dimensionless form:

$$\frac{d\hat{v}_x}{d\tau} = -\frac{\hat{x}}{\hat{r}^3} - \hat{v}_y \tag{1a}$$

$$\frac{d\hat{v}_y}{d\tau} = -\frac{\hat{y}}{\hat{r}^3} - \hat{v}_x \tag{1b}$$

$$\frac{d\hat{v}_z}{d\tau} = -\frac{\hat{z}}{\hat{r}^3} \tag{1c}$$

where $\tau = \omega_c t$, $\hat{r} = r/\lambda$, $\hat{v} = d\hat{r}/d\tau$, ω_c (the cyclotron frequency) $= -qB/m$, and λ (a characteristic length) $= (-qV_o a/(\omega_c^2))^{1/3}$. Two of the constants of motion for this problem are the (dimensionless) total (kinetic + potential) energy, $\hat{E}$, and the component of the total angular (orbital + field) momentum parallel to $\underline{B}, L_z$. Specifically,

$$\hat{E} = \frac{1}{2}(\hat{v}_x^2 + \hat{v}_y^2 + \hat{v}_2^2) - \frac{1}{\hat{r}} \tag{2a}$$

and

$$\hat{L}_z = \hat{x}\hat{v}_y - \hat{y}\hat{v}_x - \frac{1}{2}(\hat{x}^2 + \hat{y}^2) \tag{2b}$$

where the dimensionless energy and angular momentum are defined by

$$\hat{L}_z = \frac{L_z}{m\omega_c\lambda^2} \qquad \hat{E} = \frac{E}{m\omega_c^2\lambda^2}$$

Notice that the non-dimensional differential equations of motion are independent of parameters. Thus, the properties of an orbit are entirely determined by the dimensionless initial position and velocity of the electron in the six-dimensional phase-space of the charged particle.

These equations are also the governing equations for the orbits of an electron in a hydrogen atom in the presence of a magnetic field (Delos et al., 1984; Friedrich and Wintgen, 1989). We have numerically integrated the equations of motion, Eq. (1), and made Poincare maps by recording the radial position and radial velocity of the electron as it crosses the plane of symmetry perpendicular to the magnetic field (i.e., z=0). Both "chaotic" and "regular" orbits have been observed (Bentson et al., 1990).

Calculations were performed using the non-dimensional equations and the results were then converted to dimensional values for the specific case of a 10 cm radius sphere at 10 kV potential. The magnetic field was taken to be 0.4 Gauss. This configuration corresponds to a geometry representative of the SPEAR 1 rocket experiment and leads to normalization constants of $\lambda = 1.5206$m and $\omega_c = 7.035\times10^6$rad/sec. The initial velocity was assumed to be zero in the majority of the runs. Additional computations were performed with initial velocities corresponding to both electron thermal velocity at 160K altitude (2×10^5 m/sec) and to values representative of secondary electrons produced by electron-neutral collision ($\sim 10^6$ m/sec). There was no qualitative difference between the Poincare maps obtained with these values of the initial velocity compared to those corresponding to zero initial velocity.

The x and z initial positions were varied from z = 0.05m to z - 10m and from x = 1.3m to x = 10m. The upper value was chosen to correspond to typical plasma sheath dimensions at 160 km altitude for the given potential. The lower value of x corresponds to a value slightly above the electron capture limit for zero initial velocity and the given sphere radius. It was verified numerically that any trajectory started from rest inside this line eventually hits the sphere and is captured. After the regions of initial conditions leading to chaos were delineated from these runs, the actual boundary between chaos and regular trajectories were refined by fixing z and then decreasing the spacing in x.

Approximately 700 Poincare maps were computed and examined corresponding to both cases with both zero and low ($\leq 10^6$ m/sec) initial velocity. Figure 1 shows a set of representative maps at z=0 for initial conditions of x = 1.5 and 4.5 meters, and z =

0.5 and 6.0 meters, all starting from rest. The solid curves shown are the permissible limit boundaries for the given energy and angular momentum associated with the initial conditions.

Electrons which start near the plane $z=0$ are always regular and tend to have values of the map parameters ρ, $\dot{\rho}$ close to the permissible limit boundary curve. This is due to the fact that the range of z values for these trajectories are small and thus $\dot{z}$ is also small as the electron crosses the plane $z=0$. As a result the orbit is close to the limiting boundary on which $\dot{z} = 0$.

For electrons that start close to the electron capture region ($x \leq 1.3m$) the maps start as closed curves for initial positions near the sphere. As the initial value of z is increased, these maps evolve into islands of closed curves as we approach the boundary of chaos at which point the random distribution of points in the map appears. We have continued these calculations for initial values of z as far as $z=40m$ and still found the same chaotic nature of the resulting maps for ρ near the electron capture limit.

Figure 2 shows the region of initial conditions leading to chaos in dimensional space for the specific case discussed above and includes the electron capture limit. All trajectories used to construct the figure were started from rest but the results are also valid for low initial velocities as described previously.

The results obtained for trajectories started near the $z=0$ plane and can be used to formulate a one-dimensional breakdown criterion for the sphere-cylinder problem. This is presented in the next section. Moreover, the properties of the chaotic orbits can be used to justify the criterion for current collection of the electrode in the presence of a background ionized gas based on the conservation laws (Parker and Murphy, 1967).

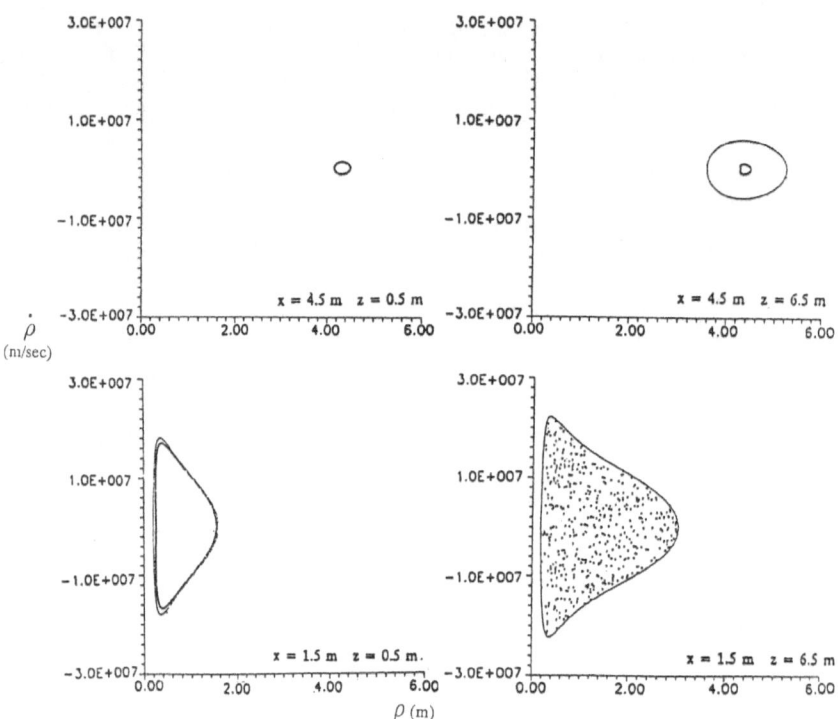

Fig. 1.   Poincare maps for electrons starting from initial positions shown in the figure.

BREAKDOWN IN PARTIALLY ENCLOSED STRUCTURES

In the generic geometry of a spherical anode inside a cylindrical cathode with a uniform magnetic field parallel to the axis of the cylinder, the trajectories presented in the previous section show that the breakdown problem is at least two-dimensional since secondary electrons and ions may be generated outside the plane symmetry. These ions maintain the generation of electrons at the cathode in the region outside of the symmetry plane. Since the electrode separation is much larger than the radial component of the ionization free path, and the electron trajectories parallel to $\underline{B}$ are unaffected by $\underline{B}$ (so that avalanching along $\underline{B}$ is small), we can assume that the threshold characteristics for breakdown are determined by conditions existing in the plane perpendicular to $\underline{B}$ and passing through the origin. In this plane, the avalanche gain is the largest. Thus, we can treat this problem as being one-dimensional in the symmetry plane perpendicular to $\underline{B}$. For 1-D problems, the electron trajectories are regular and the perigee and apogee of the trajectory depend only on the coordinate perpendicular to $\underline{B}$ and are determined by the phase-space boundaries discussed in the previous section. It is then possible to define approximate space-dependent primary (α) and secondary (ω) ionization coefficients that can be used to obtain the steady-state growth in current, given the electron current at the cathode (Kunhardt et al., 1988). From this (as done in the Townsend theory for breakdown, we can formulate a breakdown criterion. For the values of $\underline{B}$ of interest, the ion trajectories are assumed to be unaffected by the $\underline{B}$ field. Thus for $Nd/(Nd)_m \ll 1$, the ions have ballistic trajectories between the electrodes. In general, the breakdown characteristics in non-uniform fields can be expressed in the form

$$\int_{r_l}^{r_u} \alpha_{\mathrm{eff}}(r')dr' = K \tag{3}$$

where, α_{eff} is the effective Townsend ionization coefficient along the field direction, K is a constant that is determined by the feedback mechanism and depends on surface conditions, geometry, density, and $\underline{B}$ field, and r_l and r_u are the lower and upper radius defining the region over which $\alpha_{\mathrm{eff}} > 0$. For the conditions of interest (Nd < < 1), breakdown develops by the Townsend mechanism; namely, avalanche regeneration at the cathode (ω processes).

(a) The limits r_l and r_u (see Fig. 3): r_l is the radius where $\phi(r) = \eta\phi_i$, where ϕ_i is the ionization potential and η the ratio of total cross section to ionization cross section, and r_u is the radius where the secondary electrons produced by ionization no longer have trapped orbits, and are immediately collected by the anode,

$$r_u = \left(1 + \left[e(\phi_a - \phi(r_u))/(m\omega_c^2\, r_a^2) \right]^{1/2} \right)^{1/2} r_a \tag{4}$$

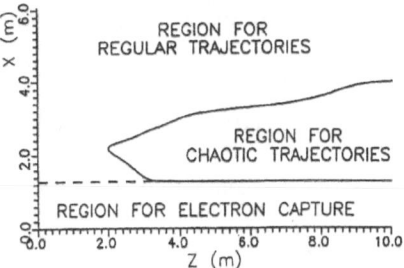

Fig. 2. Properties of trajectories starting from low velocity. Dimensional results corresponding to B=.4G, V=10Kv, a=10 cm.

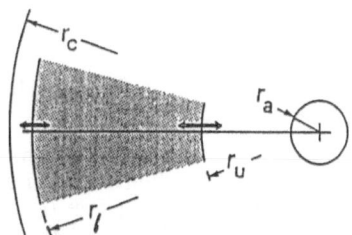

Fig. 3. Boundaries r_1 and r_u of the region over which avalanching occurs ($\alpha_{\mathrm{eff}} > 0$).

This equation is based on the condition that at the anode surface, $\partial_t r(r_a) = 0$.

(b) The effective ionization coefficient: To find the effective ionization coefficient along the direction of the electric field (i.e., in the r direction), knowledge of the energy distribution function as a function of r is necessary. Since this distribution is not known at this time, we have pursued an approach that makes use of the most probable trajectories. For α_{eff} we obtain (Kunhardt, et al., 1988):

$$\alpha_{eff} = \frac{\pi}{4} \frac{\overline{Q}_i}{\overline{Q}} \frac{1}{\overline{x}} \tag{5}$$

where $\overline{x}$ is the mean free path in the r direction, and represents the most probable position within the radial excursion limits of an electron at radius r, for a collision to occur; $\overline{Q}_i$ and $\overline{Q}$ are (average) ionization and total collision cross sections, respectively. (c) The constant K: The constant K is determined by the feedback mechanisms which give rise to new electron avalanches. For the conditions discussed here $(Nd \ll (Nd)_m, B \neq 0)$ we will only have to consider secondary electron emission from the cathode through ion bombardment (Γ process, where Γ is the secondary emission coefficient). Many materials have significant yields for secondary electron emission ($\Gamma > 1$) at these energies. Some electron, however, may be recaptured when it collides again with the cathode surface after one or more gyration radii. In general,

$$K = \ln\left[(GP_s\Gamma + 1)/GP_s\Gamma\right] \tag{6}$$

where G is the geometric factor that accounts for the fact that secondary electrons produced at the cathode by the Γ process make little contribution to breakdown unless they come from regions of the cathode that are perpendicular to the B field; and P_s is the probability that an electron, once ejected from the cathode, enters the avalanche process:

$$P_s \simeq R^s \tag{7}$$

where R is the reflection coefficient of electrons at cathode surface, and $S = \lambda_c' / \pi \delta(r_c)$ is the number of collisions than an emitted electron makes with the surface of the cathode. The mean free path $\lambda_c' \ll \lambda_c$ is based on the collision cross-section for small angle scattering. Thus, using Eqs. (4)-(7) in Eq. (3), the breakdown characteristics become:

$$\frac{\pi}{4} \frac{\overline{Q}_i}{\overline{Q}} \int_{r_1}^{r_u} \frac{1}{\overline{x}} dr = \ln\left[\frac{R_e^s G\Gamma + 1}{R_e^s G\Gamma}\right] \tag{8}$$

Calculated breakdown characteristics obtained by numerical solutions of Eq. (8) are shown in Fig. 4. These characteristics are observed to consists of two branches as in the cylinder-cylinder geometry (Readhead, 1958). The value B_{crit} corresponds to the lowest B field for which breakdown can occur in a chamber of radius r_k and a gas density, N.

EXPERIMENTAL MEASUREMENTS OF THE BREAKDOWN CHARACTERISTICS

The DC breakdown characteristics in the sphere-cylinder configuration have been measured as a function of pressure, magnetic field, and anode radius. The results are shown in Fig. 4 for a sphere of 20 cm diameter in a cylindrical chamber of 304 cm diameter and 7m length. The experimental results are in general agreement with theory. The lower branch, however, has been found to consist of a series of toroidal discharges whose fluorescence is confined to a region around the meridian plane of the electrode system and whose radial extend increases with threshold voltage. One of these modes has been observed by Greaves et al. (1990). This series of discharges terminates with the establishment of a glow-like discharge that fills the cylindrical chamber. In the upper branch, only the glow-like discharge is observed.

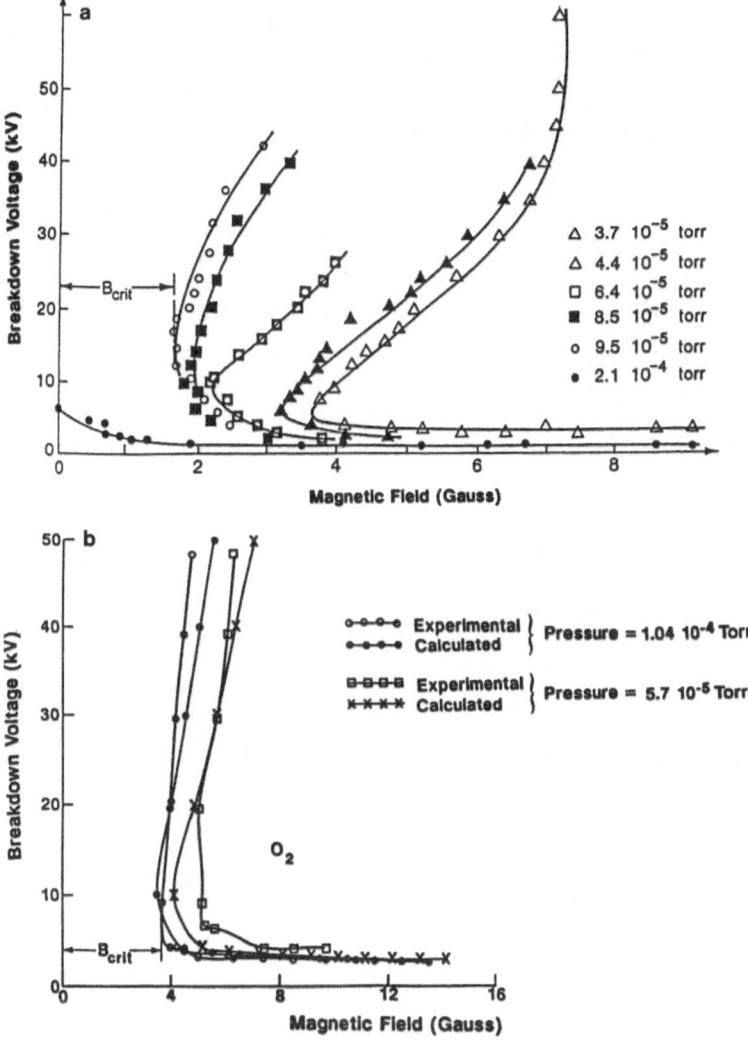

Fig. 4. Measured and calculated breakdown voltage vs. magnetic field for the sphere-cylinder configuration (r_a = 5.08 cm, r_c = 117 cm).

Four types of experiments have been performed to measure the two branches of the breakdown characteristics and classify the phenomenology of the intermediate stages that lead to the formation of the glow discharge in the lower branch. In the first type of experiments, positive potential is gradually applied to the anode from zero to the breakdown value, while the axial magnetic field is kept constant. The lower branch of the breakdown characteristics is approached from below and the possibility of getting a distorted picture of the phenomenon due to overvoltage is avoided. Formative time lags for breakdown are of the order of seconds or more, which indicate that overvoltage is practically negligible. Experiments with a pulsed voltage applied to the anode have also been performed. This type of experiment allows the application of an overvoltage relative to the DC characteristics. In the cases where rigorous measurements of the anode current of a particular discharge are to be made, the axial magnetic field is applied in a stepwise manner until the desired discharge is obtained, while the anode voltage was preset and kept constant. Toroidal discharge modes are associated with low anode currents. These currents are measured with a current sensing resistor, having a resistance comparable to the impedance of the power supply. In the third type of experiments, both anode voltage and magnetic field are kept constant and the gas pressure in the chamber is slowly reduced. The anode current indicates the changes in discharge mode. In the last type of experiment, the characteristics of the upper branch have been measured by first biasing the anode at a high voltage and then increasing the magnetic field from ambient until breakdown is obtained. No toroidal discharges have been observed for this branch.

The I-V characteristics associated with the lower branch of the breakdown characteristics is shown in Fig. 5. The sequence of toroidal discharges preceding the formation of the glow-like discharge are clearly evident. However, the structure of the curve is less clear at higher pressure, consisting of the first and third mode only, the later also being rather unstable. The transition between the discharge modes is marked by sharp increases in anode current, by more than an order of magnitude in some cases. The high current mode develops into the glow-like mode, which is characterized by a much lower sustaining voltage and high current. The development of anode current as a function of time for an applied voltage above the self-sustaining voltage of the highest toroidal mode is shown in Fig. 6. The transition between modes is denoted by the sharp changes in the current trace.

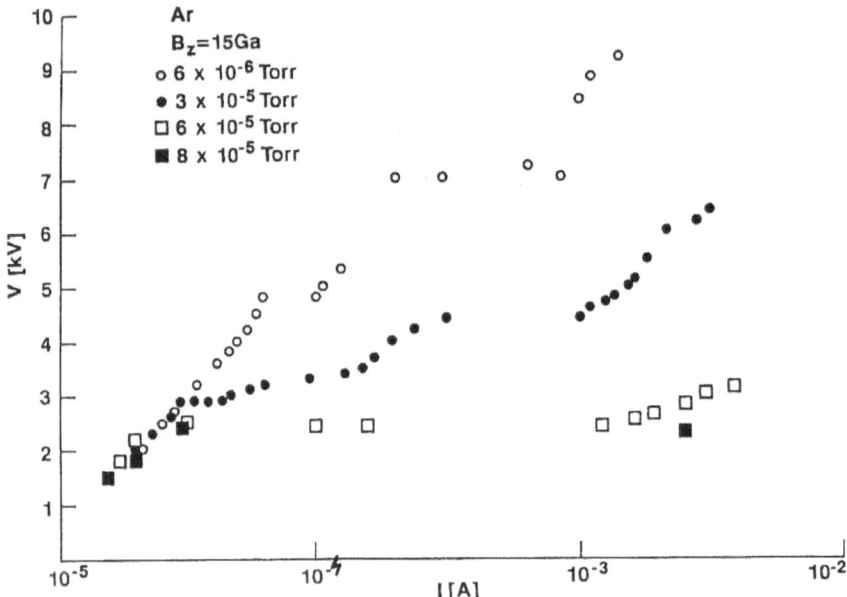

Fig. 5. V-I characteristics of toroidal discharges in A_r at axial magnetic field strength B_z = 15 Ga. Anode-to-cathode ratio was 4.19×10^{-2}.

The shape of the V-I characteristics is substantially different from its plane-electrode analogue. The sequence of toroidal discharge modes ending with the glow-like discharge present a different evolution than that found in plane geometry. These V-I characteristics and the turbulent nature of the transitions between the modes, indicated that breakdown is developing via ionization growth in the ExB direction in the meridian plane, to form a toroidal discharge. Therefore, the main current in the toroidal discharge flows in the θ direction. The radial current due to electron drift in the electric field is smaller by roughly 10^3. This is the current that is collected at the anode. The toroidal discharge builds up a zone with large space charge between the

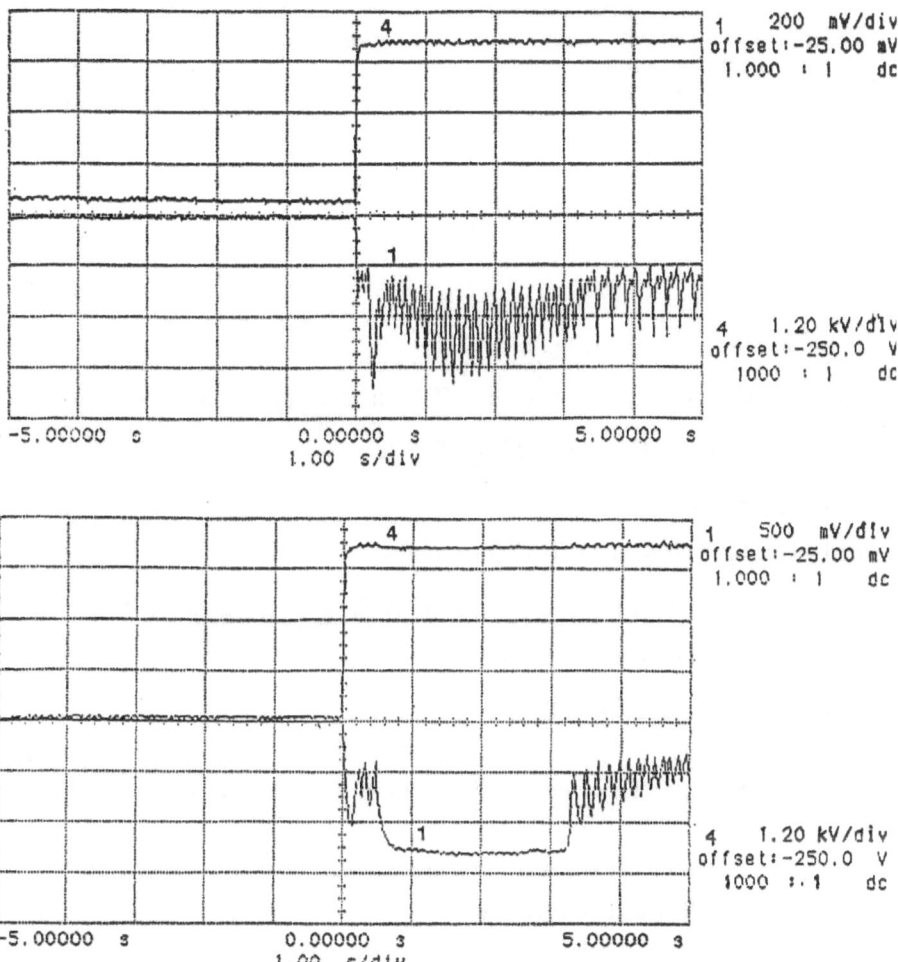

Fig. 6. Current [1] and voltage [4] traces of toroidal discharges indicating transition between second and third discharge modes.

anode and cathode that ultimately develops into high current radial breakdown leading to voltage collapse and a current pulse of ~ $100\,\mu s$ duration. Similar voltampere characteristics as shown in Fig. 5 have been observed in the pressure range between 1×10^{-6} Torr and 1×10^{-4} Torr and for magnetic fields higher than B_{cr}. At fields near B_{cr}, the toroidal modes merge and transitions between modes have not been observed.

The high voltage branch of the breakdown characteristics does not exhibit the same development as the lower branch. In particular the toroidal discharges have not been observed. To understand this observation, we recall how the two branches of the breakdown characteristics have been obtained. The lower branch is obtained first by raising the magnetic field, which implies that the collection radius starts at the anode (r_a) and increases to $\sim 2r_A$ as the anode voltage is subsequently increased. The inner radii of the first toroidal discharges mode is larger than and not in correlation with collection radius. Consequently, a large space charge is formed between anode and cathode. In contrast, the upper branch is obtained by first setting the anode voltage and subsequently increasing the magnetic field from ambient. In this case, the collection radius starts from infinity and as the field increases it decreases to a value smaller than the cathode radius. When this happens, and since the electric field is large, the ionization growth in the region between cathode and collection radius results in radial breakdown.

ACKNOWLEDGMENT

This work is supported by SDIO/IST.

REFERENCES

Alexeff, I., 1987, Phys. Rev. Lett., 53:1423.

Bentson, J., Kunhardt, E. E., Barone, S., 1990, submitted for publication.

Blevin, H., Haydon, S. C., 1958, Z. Phys., 151:340.

Delos, J. B., Knudson, S. K., Noid, D. W., 1984, Phys. Rev. A, 30:1208.

Friedrich, H. and Wintgen, D., 1989, Phys. Repts., 183:37.

Greaves, R. G., Antonaides, J. A., Boyd, J. A., 1990, Phys. Rev. Lett., 64:886.

Hobson, J. P., Redhead, P. A., 1958, Can. J. Phys. 36:271.

Kunhardt, E. E., Lederman, S., Levi, E., Schaefer, G., Nunnaly, W. W., Dillon, W. E., Smith, C. V., 1988, Proc. Int. Discharges Electr. Insul. Vac. 13th, 247.

Meek, J. M., Craggs, J. D., editors, 1978, "Electrical Breakdown of Gases,: Wiley & Sons, New York.

Parker, L.W., Murphy, B.L., 1967, J. Geophys. Res., 72:1631.

Redhead, P.A., 1958, Can J. Phys., 36:255.

DISCUSSION

I. D. CHALMERS: My compliments to the authors for a most interesting presentation. In the experimental facility which you described can you give some indication of the range of plasma density produced and briefly indicate the method of production?

E. E. KUNHARDT: A hollow cathode discharge is used and the plasma is injected through a nozzle and a conical section that is 2.5 m long and expands to 3 m, the diameter of our chamber. At this time, we have no good measurements of the electron density in the test section.

R. L. CHAMPION: What is the connection between the chaotic behavior observed for your macroscopic system and that observed for its microscopic analog, viz., a hydrogen atom, large principal quantum number in an external magnetic field?

E. E. KUNHARDT: The Hamiltonian of both is the same. In fact, we have found scaling parameters such that no characteristic parameters appear in the Hamiltonian. However, physically, the magnitude of the field angular momentum is higher than the mechanical angular momentum for our conditions.

POSITIVE SYNERGISM AND TIME-RESOLVED SWARM EXPERIMENTS

J.M. Wetzer and C. Wen[*]

High-Voltage Group
Eindhoven University of Technology
P.O. Box 513 5600 MB Eindhoven, The Netherlands

ABSTRACT

Positive synergism in binary gas mixtures is studied by means of fast time-resolved current measurements with a time resolution of 1.4 ns. In this paper we present experiments in pure $1\text{-}C_3F_6$, in pure $c\text{-}C_4F_8$ and in an optimized mixture of 40% $1\text{-}C_3F_6$ and 60% $c\text{-}C_4F_8$. We have experimentally verified that the mixture of $1\text{-}C_3F_6$ and $c\text{-}C_4F_8$ has a higher rate of conversion from unstable negative ions to stable ones than either of the two constituent gases. This enhanced stabilization rate is responsible for the observed synergetic effect. The conclusions are supported by an extended avalanche model which includes electron detachment and stabilizing ion conversion processes, next to the commonly assumed ionization and attachment processes.

INTRODUCTION

For properly chosen mixing ratios, mixtures of SF_6 with C_3F_8, SO_2 and OCS, and mixtures of $1\text{-}C_3F_6$ with SO_2 and $c\text{-}C_4F_8$ have a higher breakdown strength than either of the constituent gases[1]. Hunter and Christophorou[1] have proposed requirements for the observation of this positive synergism, and explain it by the enhanced production rate of stable negative ions. No experimental verification was presented up to now because it was not possible to distinguish from experiments what processes are responsible. In this paper we present fast time-resolved swarm measurements in pure $1\text{-}C_3F_6$, pure $c\text{-}C_4F_8$, and in an optimized mixture of 40% $1\text{-}C_3F_6$ and 60% $c\text{-}C_4F_8$. These measurements confirm that the positive synergism can be explained by the enhanced rate at which unstable negative ions, formed by attachment, are stabilized.

---

[*] present address: Claymount Assemblies, Marconistraat 25-27
                          6942 PX Didam, The Netherlands

Gaseous Dielectrics VI, Edited by L.G. Christophorou and
I. Sauers, Plenum Press, New York, 1991

AVALANCHE MODEL

The avalanche model used in this paper is presented elsewhere[2,3,4]. Here we confine ourselves to a brief outline. Apart from neutral molecules, four species are involved: electrons (index e), positive ions (p), unstable negative ions (nu) and stable negative ions (ns). The processes considered are ionization (coefficient α), attachment (η), detachment (δ), ion conversion (β) and (electron and ion) drift. Secondary emission and diffusion can be incorporated, if necessary[4]. It is assumed that all negative ions formed via attachment are unstable: they can either release their electron (detachment), or can be converted into stable ones (ion conversion). More detailed studies show that this assumption is not critical for the resulting waveforms. Of course the values of the coefficients depend on the type of initially produced negative ions.

The coefficients α and η are defined, as usual, as the mean number of ionization or attachment events in the time that an electron travels a unit length in the field direction. In contrast to conventional definitions, δ and β are defined as the mean number of detachment or conversion events per unstable negative ion in the time that an *electron* travels a unit length in the field direction. The conventional definitions relate δ and β to the *ion* drift velocity which is not convenient for the study of avalanche growth.

Conventional avalanche models do not involve electron detachment and ion conversion. In combination with ionization and attachment processes, electron detachment causes:
-   an apparent reduction of the electron drift velocity as a result of temporary electron trapping[5], and
-   an apparent increase of the (effective) ionization coefficient as a result of the secondary electrons produced by detachment[3].

Both mechanism are counteracted by ion conversion processes which transform the unstable negative ions to stable ones[3]. The neglect of electron detachment and ion conversion processes may lead to "apparent" values for the electron drift velocity and for the effective ionization coefficient which seem to disobey scaling laws[3,4,5].

Based on this model and on waveforms observed, three types of avalanches are distinguished[4]: electron avalanches without delayed electrons (N_2, SF_6), electron avalanches with delayed electrons (air, $1\text{-}C_3F_6$), and ion-dominated avalanches ($c\text{-}C_4F_8$). Only the first type of avalanche can be described by means of the conventional model.

DIELECTRIC PROPERTIES OF $1\text{-}C_3F_6$ AND $c\text{-}C_4F_8$

In contrast to many other gases, $1\text{-}C_3F_6$ does not obey Paschen's law: the breakdown voltage increases more rapidly with pressure than with electrode separation, and the limiting E/p value increases substantially with pressure[4,6]. Only at pressures higher than 4 bar this increase can be ascribed to compressibility[6]. Moreover, the electron drift velocity and the (pressure reduced) effective ionization coefficient seemingly disobey scaling laws: at constant E/p, v_e and $(\alpha\text{-}\eta)/p$ seem to decrease with increasing pressure[4,7,8]. The observed waveforms show electron avalanches with delayed electrons[4]. The observed behavior can be explained with the extended avalanche model mentioned before. From our work we have derived a simplified reaction scheme[4,5].

Also $c\text{-}C_4F_8$ does not obey Paschen's law: again, the breakdown voltage increases more rapidly with pressure than with electrode separation, and the limiting E/p increases substantially with pressure[9]. As in $1\text{-}C_3F_6$, the (pressure reduced) effective ionization coefficient seems to decrease with increasing pressure at constant E/p[4,9]. Other interesting features are the high dielectric strength and the high figure of merit[9]. The measured waveforms are those of ion-dominated avalanches[4,10]. Also for such avalanches the extended model presented is valid.

POSITIVE SYNERGISM IN THE MIXTURE OF $1\text{-}C_3F_6$ AND $c\text{-}C_4F_8$

At a pressure of 500 Torr the uniform field breakdown strength relative to SF_6 is 1.08 for $1\text{-}C_3F_6$, 1.22 for $c\text{-}C_4F_8$, and 1.38 for an optimized mixture of 60% $c\text{-}C_4F_8$ and 40% $1\text{-}C_3F_6$[1]. Hunter and Christophorou[1] have stated that the observation of positive synergism requires that at least one of the two gases should have an attachment coefficient with a pressure dependence other than $\eta = p.f(E/p)$. This requirement is fulfilled for both gases under study. Note that the attachment coefficient referred to is an "apparent" coefficient, which is related to the "real" attachment coefficient according to[3,4]:

$$\eta_{app} = \eta_{real} - \eta_{real}.\delta/(\delta+\beta)$$

Here all coefficients are related to the *electron* drift velocity (see Avalanche Model).

The present study serves to verify that, indeed, the stabilization rate of both constituents is enhanced by addition of the other constituent.

SWARM EXPERIMENTS

The method which we used for the avalanche current measurement is the so-called time-resolved swarm method. This method is based on the detection of the total displacement current due to the drift of electrons and ions across a parallel-plate gap under the influence of the applied electric field. The time-resolution of the whole measuring system is 1.4 ns. A detailed description of the measuring technique and the experimental setup can be found elsewhere[4,11,12]. Before the (high purity) gas under investigation is admitted, the vessel is evacuated down to 10^{-5} Pa. All measured pressures are reduced to 20°C. In all experiments the electrode separation was kept at 1 cm.

The observation and evaluation of waveforms in $1\text{-}C_3F_6$ and $c\text{-}C_4F_8$ have been extensively reported in previous work[4,5,10]. In this paper we present and compare measured waveforms in $1\text{-}C_3F_6$, in $c\text{-}C_4F_8$ and in a mixture of 40% $1\text{-}C_3F_6$ and 60% $c\text{-}C_4F_8$. Note that this is the optimum mixing ratio at a pressure of 500 Torr. The optimum may be different at other pressures. In this work, however, the mixing ratio is kept constant.

Avalanches in $1\text{-}C_3F_6$ are electron avalanches (with delayed electrons), with a duration in the order of 100 ns. Avalanches in $c\text{-}C_4F_8$ are ion-dominated avalanches: a distinct electron component is observed only at very low pressure (below 10 Torr). The time duration is in the order of micro-seconds. A comparison therefore involves both timescales.

Figure 1 shows a number of waveforms measured at a total pressure of 5 Torr for different values of E/p, on a timescale of 200 ns. In $1\text{-}C_3F_6$ the current grows to breakdown for E/p values above 210 V/cm.Torr. This is the result of detachment from unstable negative ions formed during the electron transit from cathode to anode. In the mixture a second peak is observed after the electron transit time. The growth is, however, limited by the conversion of unstable ions, and no breakdown occurs.

Also in $c\text{-}C_4F_8$ a second peak is observed. A comparison with the mixture shows that the height of the second peak, relative to the heigth of the first one, is reduced by the addition of $1\text{-}C_3F_6$. Again, this is caused by a higher ion conversion rate by which unstable ions are stabilized. Apparently, the constituent gases mutually enhance the production of stable negative ions. Note that both in $c\text{-}C_4F_8$ and in the mixture the first peak is predominantly caused by electrons, the second peak by ions.

On a longer timescale only $c\text{-}C_4F_8$ and the mixture of $c\text{-}C_4F_8$ and $1\text{-}C_3F_6$ are relevant. Figure 2 shows a number of waveforms measured at a pressure of 1 Torr and an E/p-value of 700 V/cm.Torr. Again the pressure is low enough for an electron component to show up in the current waveform. The ion-dominated current that flows after the electron transit time is, however, very pronounced. The waveforms are shown on three different timescales. At short times (below 200 ns) the waveforms are comparable, and the addition of $1\text{-}C_3F_6$ seems to result only in a slight enhancement of the ion conversion rate. On a long time scale (up to 20 μs) this enhancement is very pronounced. As a result of the addition of $1\text{-}C_3F_6$, the second peak in the current waveform is strongly reduced with respect to the first one.

At pressures above 10 Torr the current waveform does not show a distinct electron component, and the effect of the admixture of $1\text{-}C_3F_6$ to $c\text{-}C_4F_8$ can only be observed on a μs-timescale.

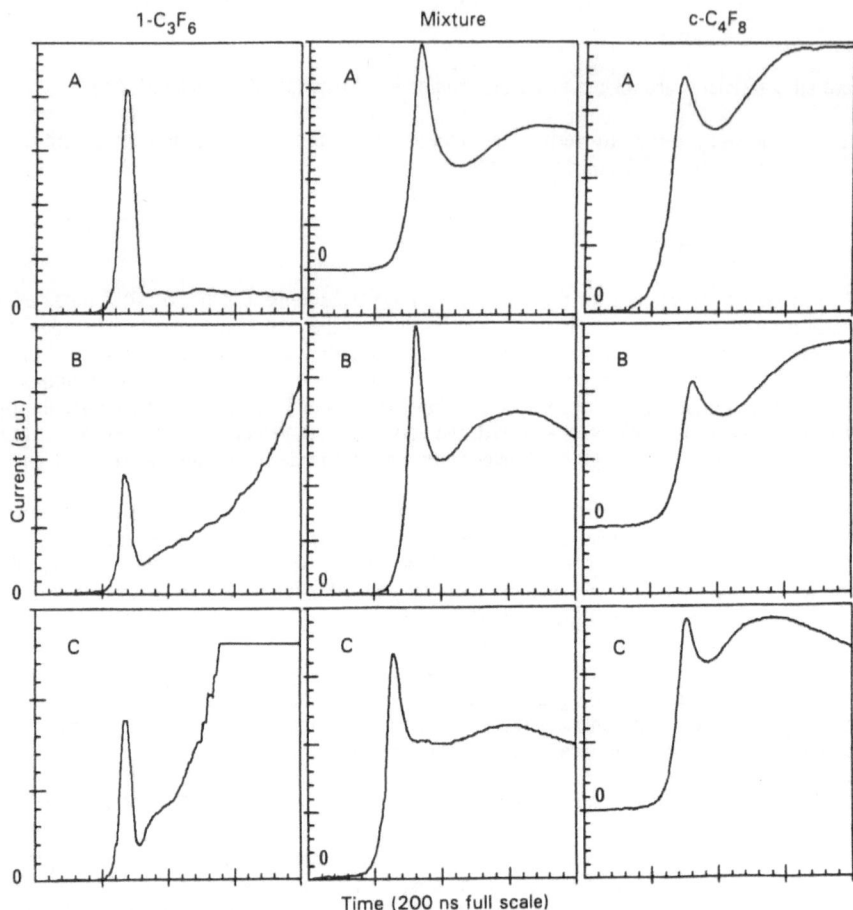

Fig. 1.   Avalanche current waveforms measured in 1-C_3F_6, c-C_4F_8 and in an optimized
mixture of 40% 1-C_3F_6 and 60% c-C_4F_8, for different E/p values
A: E/p = 210 V/cm.Torr (1.575 V/cm.Pa)
B: E/p = 220 V/cm.Torr (1.650 V/cm.Pa)
C: E/p = 230 V/cm.Torr (1.725 V/cm.Pa)
Total pressure: 5 Torr (0.67 kPa).

76

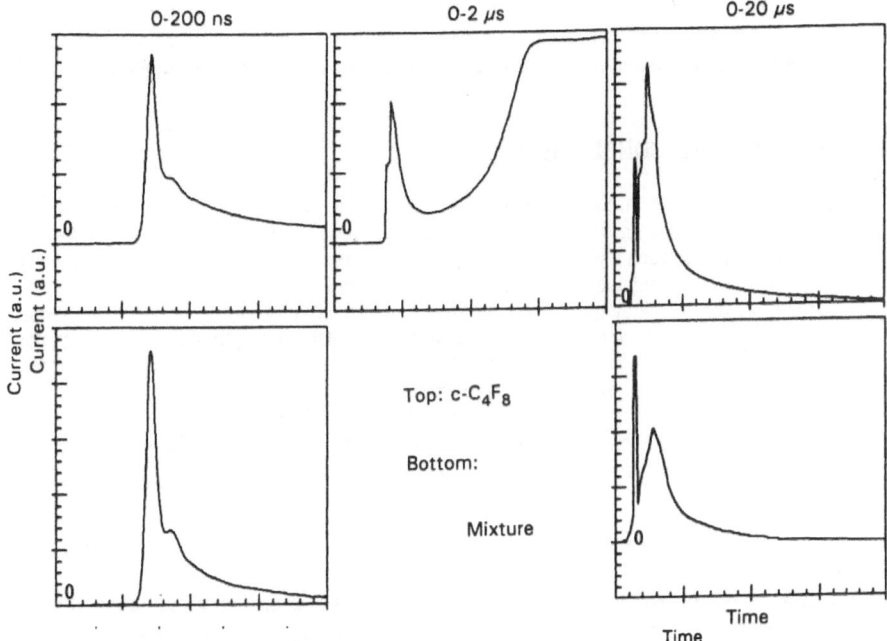

Fig. 2. Avalanche current waveforms measured in c-C_4F_8 and in an optimized mixture of 40% 1-C_3F_6 and 60% c-C_4F_8, at three different time-scales.
Pressure 1 Torr (0.133 kPa); E/p: 700 V/cm.Torr (5.25 V/cm.Pa).

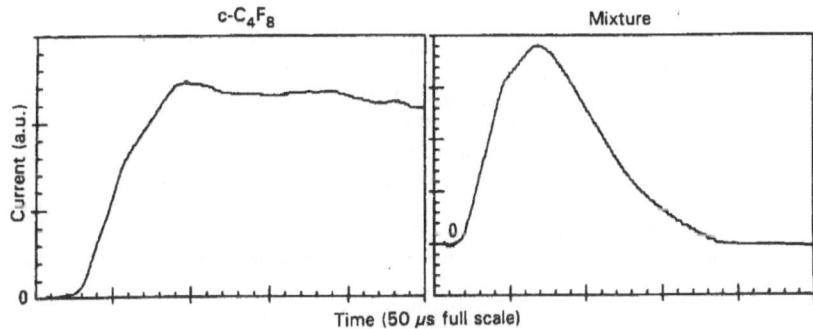

Fig. 3. Avalanche current waveforms measured in c-C_4F_8 and in an optimized mixture of 40% 1-C_3F_6 and 60% c-C_4F_8.
Pressure: 30 Torr (4 kPa); E/p: 157 V/cm.Torr (1.1775 V/cm.Pa).

Measurements performed at pressures of 30 and 50 Torr confirm the enhanced stabilization rate in the mixture as compared to the constituent gases. Figure 3 shows an example of measurements at 30 Torr.

DISCUSSION AND CONCLUSION

Hunter and Christophorou have shown that the synergetic effect in binary gas mixtures is related to the pressure dependence of the (apparent) attachment coefficient[1]. At least one of the constituent gases should possess such a pressure dependence. They further stated that the synergetic effect can be explained by the enhanced production rate of stable negative ions. From the experiments presented, we can conclude that, compared to either of the two constituents, the mixture of 1-C_3F_6 and c-C_4F_8 has indeed a higher rate of conversion from unstable negative ions to stable ones. This conclusion is supported by the extended avalanche model which includes electron detachment and stabilizing ion conversion processes, next to ionization and (real) attachment processes. Among the responsible candidate processes are several charge exchange and dissociative attachment processes. It is further concluded that swarm experiments with a good time-resolution provide a tool to recognize what type of processes are responsible for synergism in mixtures.

ACKNOWLEDGEMENT

The authors thank Prof.dr. P.C.T. van der Laan for his contribution and support, Mr. A.J. Aldenhoven for his technical assistance during the experiments, and R.W.A. Hinz who participated in this work as a student.

REFERENCES

1. S.R. Hunter and L.G. Christophorou, Pressure dependent electron attachment and breakdown strength of unitary gases and synergism of binary gas mixtures: a relationhip, J.Appl.Phys.,57, 4377-4385 (1985).
2. H.F.A. Verhaart, Avalanches in insulating gases, Ph.D.thesis, Eindhoven University of Technology (1982).
3. C. Wen and J.M. Wetzer, Electron avalanches influenced by detachment and conversion processes, IEEE Trans.on El.Ins., 23, 999-1008 (1988).
4. C. Wen, Time-resolved swarm studies in gases with emphasis on electron detachment and ion conversion, Ph.D. thesis, Eindhoven University of Technology (1989).
5. C. Wen and J.M. Wetzer, Swarm parameters in hexafluoropropene 1-C_3F_6, Proc.IEEE Int.Symp.on El.Ins., Boston, pp.108-111 (1988).
6. G.Biasiutti, C. Amman, E. Engler and W.S. Zaengl, Electric strength of hexafluoropropylene (C_3F_6) and its mixtures with SF_6 at practical pressures, Proc.4th.Int.Symp.on High Voltage Eng., Athens, paper 33.02 (1983).
7. Th. Aschwanden, H. Böttcher, D. Hansen, H. Jungblut and W.F. Schmidt, Mobility and recombination of ions and effective ionization coefficient in hexafluoropropylene (C_3F_6), Gaseous Dielectrics III (Ed. L.G. Christophorou), Pergamon Press, pp.23-33 (1982).
8. H.F.A. Verhaart and P.C.T. van der Laan, Fast avalanche measurements in C_3F_6, Proc.4th.Int.Symp.on High Voltage Eng., Athens, paper 33.12 (1983).
9. J. Berril, J.M. Christensen and I.W. McAllister, Measurement of the figure of merit for several perfluorocarbon gases, Gaseous Dielectrics V (Eds. L.G. Christophorou and D.W. Bouldin), Pergamon Press, pp.304-310 (1987).
10. C. Wen and J.M. Wetzer, Time-resolved avalanche current waveforms in octafluorocyclobutane, IEEE Trans.on El.Ins., 24, 143-149 (1989).
11. H.F.A. Verhaart and P.C.T. van der Laan, Fast current measurements for avalanche studies, J.Appl.Phys., 53, 1430-1436 (1982).
12. J.M. Wetzer, C. Wen and P.C.T. van der Laan, Bandwidth limitations of gap current measurements, Proc.1988 Int.Symp.on El.Ins., Boston, pp.355-358 (1988).

DISCUSSION

I. GALLIMBERTI: The stable ion formation process in the mixture has to involve (in order to explain the positive synergism) collision processes that do not occur in the parent gases, or occur at a much lower rate; for example, dissociative charge exchange reactions involving both species. Can you comment on these possible processes and on their effect?

J. M. WETZER: Apart from the processes that occur in each of the constituent gases, positive synergism indeed requires additional, parent processes. Hunter and Christophorou have proposed a reaction scheme with charge exchange and dissociative attachment processes involving parent negative ion formation. As to the effect of these processes, the time—resolved swarm experiments clearly show an enhanced production of stable negative ions. The present experiments, however, do not resolve which one of the candidate processes is responsible for the observed stable anion formation.

DISCHARGE DEVELOPMENT IN AIR

AT VERY HIGH HUMIDITY LEVELS

A. J. Davies, J. Dutton, M. Matallah and R. T. Waters*

Department of Physics, University College of Swansea
Singleton Park, Swansea, SA2 8PP, Great Britain
* School of Electrical, Electronic and Systems Engineering
University of Wales College of Cardiff, P.O. Box 904
Cardiff, CF1 3YH, Great Britain

INTRODUCTION AND EXPERIMENTAL ARRANGEMENT

In many practical situations, especially in hot humid climates, absolute humidity levels up to 30 gm^{-3} frequently occur, and thus it is highly desirable to investigate the various phases of discharge development and breakdown in air gaps with the aim of elucidating the different discharge processes.

Previous papers in this series (Davies *et al.* 1982, 1984, 1987) have described tests in air for absolute humidities in the range 0–15 gm^{-3}, and for pressures from 0.8 to 1.4 bar. A schematic diagram of the experimental arrangement is shown in Fig. 1 and, as in the previous tests, a sealed ionization chamber (1.2 m internal diameter and 3 m internal height) was used. The humidity of the air was controlled by a closed-loop system and measured by a precision dew-point meter. In the new work, however, higher humidity was achieved by thermally insulating the chamber and heating it to 33°C. Accurate temperature control was attained by winding Raychem trace heating tape around the body of the chamber and then insulating the whole of the chamber with 10 cm of fibreglass covered with a reflecting coating. The heating tape has the special property of increasing rapidly in resistance at a critical temperature and, by carefully balancing the length of heating tape used (and thus the heat input) against the heat lost through the insulation, very stable temperatures could be attained.

The gap geometry and diagnostics were those developed in the earlier programme. Positive polarity switching impulses (100/2500 µs) were applied to a 50 mm diameter sphere located 200 mm above a plane cathode 1 m in diameter. Measurements were made of the applied voltage, apparent charge injection at the anode, light output, and both the field variation and charge arrival at the centre of the cathode.

The earlier experiments had already established that the enhancement of the electric field at the cathode by the corona space charge could be measured and that the spatial distribution of charge could be simulated. Furthermore, the conduction of charge to the plane when the corona streamers interacted with the cathode could be discriminated by means of a composite field/charge probe (Davies and Turri, 1990) which was capable of measuring both the surface electric field and charge conducted to the probe with a bandwidth of 0.001 Hz to 3 MHz. The upper frequency limit was a factor almost one thousand higher than the high-speed fluxmeter probe previously used. The impulse corona was shown to consist of a primary charge-injection sequence,

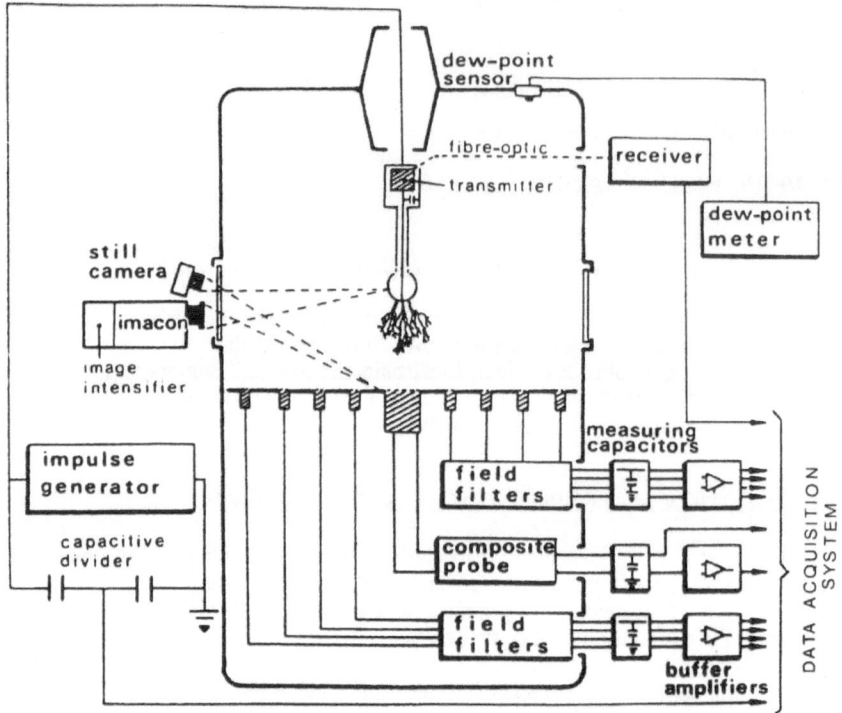

Fig. 1. Schematic diagram of experimental arrangement.

sometimes followed by series of smaller fast secondary pulses which could lead to breakdown.

Within the range of humidity of that work, the withstand voltage was deduced to be much less dependent upon absolute humidity than the voltage V_{50} corresponding to 50% probability of breakdown, which showed the expected 1% increase per gm^{-3} increase in absolute humidity. It was also found that the probability of breakdown was closely related to the average charge injected during the primary corona events, which strongly decreased with increasing humidity for a given crest voltage. The breakdown probability distribution was found to satisfy the cumulative normal distribution with parameters V_{50} and σ.

In the present paper, the results are extended to report the variation of the probability of breakdown, V_{50}, σ, and the injected charge, for absolute humidity values up to 30 gm^{-3}. These refer to an air pressure of 1 bar. Further experiments are in progress at other pressures.

BREAKDOWN STATISTICS

The present series of measurements were all carried out at 1 bar pressure and for absolute humidities in the range 2 to 30 gm^{-3}.

Fig. 2 shows the breakdown probability as a function of crest voltage plotted on normal probability paper. We note that at all humidities less than 21.5 gm^{-3} the breakdown data all satisfy the cumulative normal distribution but at the highest humidity levels studied there is evidence of a cut-off at low probability levels suggesting that a Weibull distribution may be more appropriate. Further data is being taken in order to clarify this point.

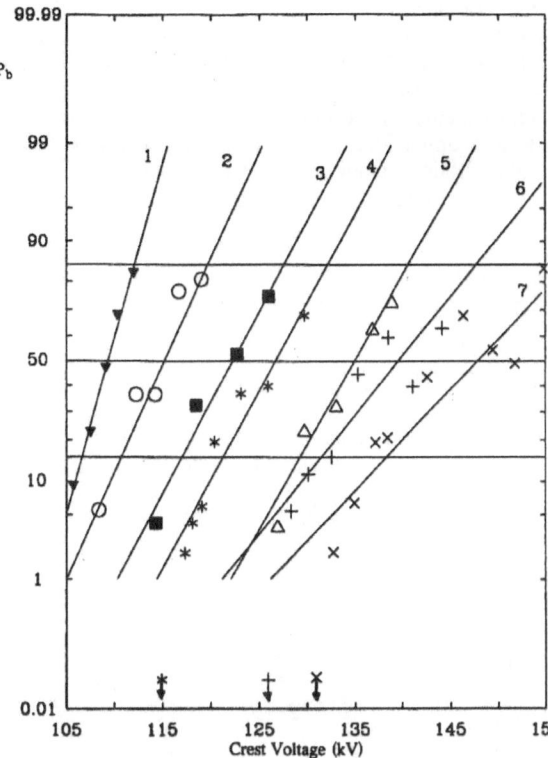

Fig. 2. Breakdown probability curves at different absolute humidities.
1 ▼ 2.5; 2 o 7.0; 3 ■ 11.7; 4 * 15.2; 5 △ 21.5; 6 + 25.3; 7 × 28.7 gm⁻³

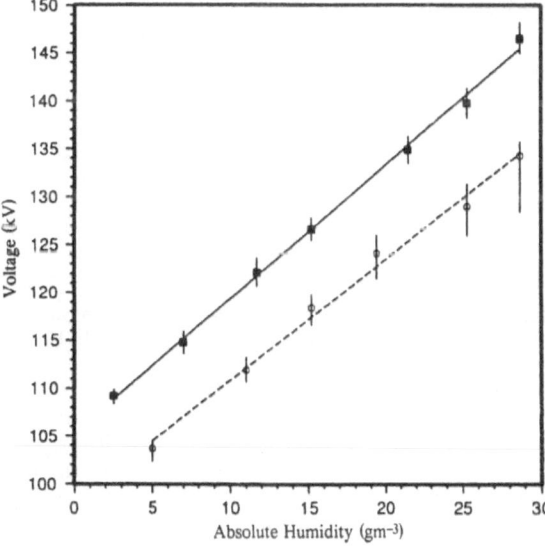

Fig. 3. Voltages V_{50} (——) and V_5 (– – –) corresponding to 50% and 5% breakdown probabilities as function of absolute humidity. The vertical bars represent the 90% confidence intervals.

Fig. 3 shows the voltages V_{50} and V_5 corresponding to fifty and five percent probabilities of breakdown plotted as a function of absolute humidity. It can be seen that V_{50} continues to increase at 1% per gm^{-3} over the whole range of humidities studied and that the 90% confidence intervals associated with V_{50} are very small due to the closely controlled experimental conditions. All the breakdown parameters and confidence intervals were estimated using the Maximum Likelihood technique (Davies *et al*, 1988). The V_5 data, obtained from extended up and down tests, also gave narrow 90% confidence intervals and increased with humidity at a slightly lower rate than the V_{50} data.

The present experiments at high humidity confirm the tendency for σ to increase linearly from about 2.5% at 2 gm^{-3} to about 6% at 29 gm^{-3}.

CHARGE AND FIELD MEASUREMENTS

Fig. 4 shows the average total charge injected during primary corona events as a function of crest voltage at different humidities. Note that there is a sharp rise in $\overline{Q}_i$ with crest voltage at a given humidity especially at the lower humidity values where σ is also small. The curves suggest that extrapolation of the lower values may yield more or less the same minimum $\overline{Q}_i$. Note also that at a constant value of the crest voltage there is a sharp decrease in $\overline{Q}_i$ with increasing humidity.

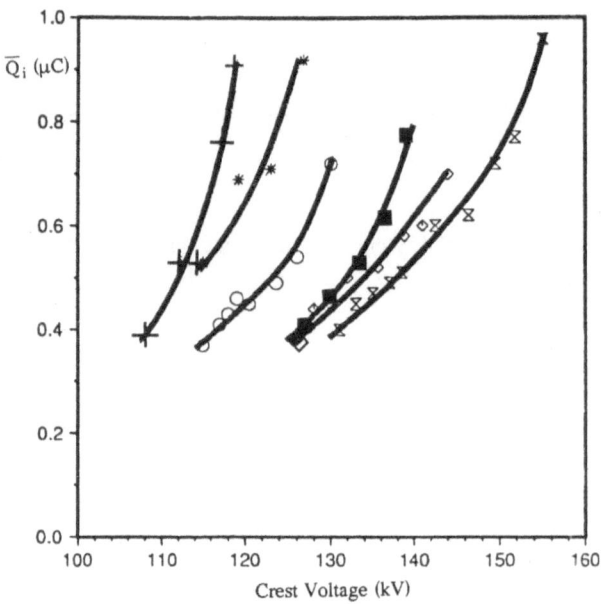

Fig. 4. Average total charge injection during primary corona events (as a function of crest voltage) at different humidities. + 7.0; * 11.7; o 15.2; ■ 21.5; ◇ 25.3; ⊠ 28.7 gm^{-3}.

In Fig. 5 the probability of breakdown for a given average total primary corona charge is plotted for different humidities on normal probability paper. It can be seen that there is a unique relationship between the breakdown probability and the charge injection which is independent of absolute humidity. This relationship does not appear to be normal and, if the experimental points are compared with the cumulative normal distribution derived from the data (full line), a minimum level of $\overline{Q}_i$ is again indicated.

Fig. 6 shows typical experimental records obtained at low (6.0 gm^{-3}) and high (28.7 gm^{-3}) absolute humidities at crest voltages corresponding to the same probability of breakdown (5%) and the same primary corona charge injection. In Fig. 6a the signals

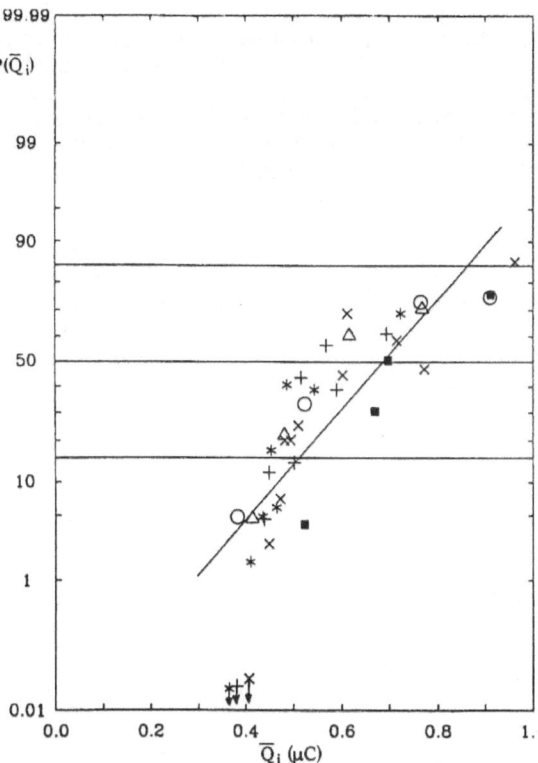

Fig. 5. Probability $P(\overline{Q}_i)$ of breakdown for a given $\overline{Q}_i$ at different humidities. o 7.0; ∎ 11.7; * 15.2; △ 21.5; + 25.3; × 28.7 gm^{-3}

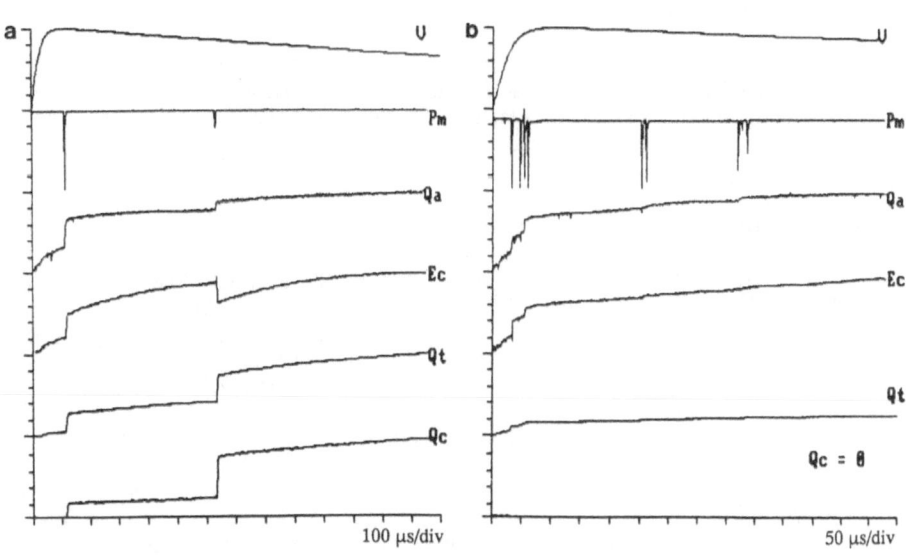

Fig. 6. Typical records at (a) 6.0 gm^{-3}, V_{cr} = 108.8 kV; (b) 28.7 gm^{-3}, V_{cr} = 139 kV. P_b = 5% for both records. Scales / div:– V = 22 kV, Q_a = 0.15 μC, E_c = 0.69 kV, Q_t ≡ 2.64 kVcm^{-1}, Q_c = 6.4 nC, P_m arbitrary.

show that two primary corona occur. In the first event, which occurs near the crest of the applied voltage, there are sharp increases in the apparent charge injection at the anode, Q_a, the central cathode field, E_c, and the total induced charge, Q_t, on the composite probe. There is considerable interaction with the cathode leading to conducted charge, Q_c, arriving at the surface of the probe. The second corona event also gives increases in Q_a, Q_t, and Q_c, but there is a <u>decrease</u> in the electric field, E_c, at the surface of the cathode. The corona events are accompanied by single bursts of light emission as shown on the photomultiplier signal P_m.

In Fig. 6b, on the other hand, the primary corona events are accompanied by multiple bursts of light probably associated with reilluminations of the corona stem. There is no cathode interaction ($Q_c = 0$) and the step changes in E_c are generally much smaller. The long slow increase in E_c is due to the drift towards the cathode of the injected space charge. These measurements clearly indicate that different breakdown mechanisms are operative at low and high humidities.

CONCLUSIONS

The present series of measurements at high humidity levels indicate that, whereas the breakdown probability satisfies the cumulative normal distribution at low humidity levels, there is evidence that there is a departure from this distribution for humidity levels from 21.5 gm^{-3} upwards. As a consequence of the closely controlled experimental conditions the values of V_{50}, estimated using the Maximum Likelihood technique, have narrow confidence intervals and show an increase of 1% per gm^{-3} over the whole range of absolute humidity up to 29 gm^{-3}.

Previous work demonstrated that the probability of breakdown was closely related to $\bar{Q}_i$ and the present measurements confirm that this result still holds when higher humidity levels are included. This is clearly useful in developing predictive models of breakdown. The present results also demonstrate that different breakdown mechanisms are operative at low and high humidity.

Futher work is being undertaken at other pressures in the range 0.8 - 1.4 bar.

ACKNOWLEDGEMENTS

This work was carried with the aid of Research Grants GR/E 66507 and GR/E 51824 jointly funded by the UK Science and Engineering Research Council and The National Grid Company P.L.C. One of the authors (M.M) is grateful to the Ministry of Higher Education, Algeria, for the award of a student research grant.

REFERENCES

1. A. J. Davies, J. Dutton, E. O. Selim, and R. T. Waters, Positive Switching Impulse Breakdown of Sphere–Plane Gaps in Dry and Moist Air, Gaseous Dielectrics III, L. G. Christophorou (Ed.), Pergamon, New York, 242–243 (1982).
2. A. J. Davies, J. Dutton, J. Jarvis, A. Robledo–Martinez, and R. T. Waters, Corona Inception, Breakdown and Charge–Injection Measurement in Sphere–Plane Gaps Subjected to Impulse Voltages, Gaseous Dielectrics IV, L. G. Christophorou and M. O. Pace (Eds.), Pergamon, New York, 128–136 (1984).
3. A. J. Davies, J. Dutton, R. Turri, and R. T. Waters, The Effect of Humidity on Space–Charge Growth in Short Air Gaps, Gaseous Dielectrics V, L. G. Christophorou and D. W. Bouldin (Eds.), Pergamon, New York, 249–254 (1987).
4. A. J. Davies, A. Rowlands, R. Turri, and R. T. Waters, The Statistical Analysis of Flashover Data Using a Generalized Likelihood Method, <u>Proc. IEE</u>, A135:79–87 (1988).
5. A. J. Davies and R. Turri, A Composite Probe for Simultaneous Electric Field and Conducted Charge Measurements, <u>J. Phys. E.</u>, 1990 (in press).

DISCUSSION

L. NIEMEYER: Can you calculate the streamer inception voltage by the streamer criterion using the known effective ionization coefficients for humid air? This would allow us to determine theoretically the 0% cutoff in the voltage dependent probability distribution of corona discharge.

A. J. DAVIES: It is certainly possible to use the streamer criterion to calculate a streamer inception voltage. Due to the statistical nature of the ionization processes, however, it is questionable whether this voltage would represent a true 0% cutoff in the breakdown probability distribution.

I. GALLIMBERTI: Should the time scale for charge neutralization be 100 ns rather than $100\mu s$ as shown on your slide?

A. J. DAVIES: This is the correct time scale. Remember, we have a band width of 5 MHz. I don't know if it's charge neutralization or something more complicated.

D. KÖNIG: Your studies on humid air insulation are of high interest for high–voltage test engineers, who apply "correction factors" according to IEC–Publ. 71. Your studies enable an improvement and extention of these factors. I hope you will transfer your knowledge to the relevant IEC Committee TC 42. According to my knowledge only a few electrode configurations are dealt with in IEC–Publ. 71. Do you intend to restrict your studies to spheres or to extend them to other configurations, which are applied in practice, but where correction factors are still missing? I would like to encourage you to do so! Could you comment on this?

A. J. DAVIES: We intend to extend our tests to other electrode geometries and applied voltage shapes. In the first instance we will be studying a hemispherically–tipped rod high–voltage electrode situated at various distances from the existing plane cathode. In addition to switching surges applied voltage shapes with faster fronts, in particular lightning $(1/50~\mu s)$ impulses, will be used for both sphere/plane and rod/plane geometries. Measured humidity correction factors will be published and communicated to IEC Committee TC42 as they become available.

E. MARODE: Do you think that the difference in discharge behaviour between low and high humidity rate may be related to the formation of a leader stage. In a small discharge cell, 1 cm gap, 1 mm positive electrode radius, in SF_6, experiments performed in Gif–sur–Yvette, showed evidence that for low humidity rate a so called "direct breakdown" prevails, i.e., a streamer which crosses the gap and is followed by a spark without a leader phase, while at high humidity rate a leader stage develops before the first streamer reaches the cathode, and the final spark is consequently delayed.

A. J. DAVIES: The phenomena we have observed appear to be very similar. At low humidity streamer interaction with the cathode is nearly always observed corresponding to your case of "direct" breakdown. At high humidity the initial corona events are restricted to a small region near the sphere. Repeated reilluminations of the corona stem occur and a leader–like channel propagates towards the cathode before the final breakdown phase where the channel bridges the gap.

R. J. VAN BRUNT: Could you comment on relative effects of initiating electron release versus amount of charge injection in governing the breakdown probability?

A. J. DAVIES: In order to have any corona activity, possibly leading to breakdown, it is of course necessary to have an initiatory electron. We find that the spread in corona inception times decreases with humidity indicating that initiatory electrons are more plentiful at high humidities due to the greater density of residual negative ions in the discharge space at the time of impulse application. Given that corona activity is present, however, the probability of breakdown in any given case then appears to be related to the total charge injection during the primary corona. Initiatory electron release thus determines the time at which corona events occur while the subsequent breakdown probability is governed by the charge injection.

NEGATIVE $\overline{\eta}$ SYNERGISM IN $CF_2Cl_2-N_2$ AND $CF_2Cl_2-CO_2$ GAS MIXTURES

Y. Qiu, X. Ren, Z.Y. Liu and M.C. Zhang

High Voltage Division
Xi'an Jiaotong University
Xi'an 710049, China

INTRODUCTION

Though difluorodichloromethane(CF_2Cl_2) is of about the same dielectric strength as that of SF_6(Qiu and Weng, 1987), it has been shown that the dielectric strength of the $CF_2Cl_2-CO_2$ gas mixture is apparently different from that of the SF_6-CO_2 mixture with the same CO_2 concentration in the mixture (Qiu et al., 1989). Our recent work(Qiu et al., 1990) has also indicated that in contrast to the SF_6-N_2 mixture, the dielectric strength of the $CF_2Cl_2-N_2$ mixture as a function of CF_2Cl_2 concentration is quite close to a straight line similar to that of $CF_2Cl_2-CO_2$.

As difluorodichloromethane is one of the five chlorofluorocarbons whose production should be restricted in order to protect the ozone layer in the stratosphere(Joyce and MacKenzie, 1987), it is unlikely to be extensively used as a constituent gas of insulating gas mixtures. However, an analysis of the synergistic effect in the above-mentioned CF_2Cl_2 gas mixtures might be of some value to those who are involved in basic research of insulating gas mixtures.

IONIZATION AND ATTACHMENT COEFFICIENTS

The Townsend first ionization coefficient α and the electron attachment coefficient η were measured in $CF_2Cl_2-N_2$ and $CF_2Cl_2-CO_2$ gas mixtures using the steady-state Townsend method over the range $40 \leqslant E/p \leqslant 140$, where E is the electric field in Vcm^{-1}, and p the gas pressure in Torr reduced to 20 °C. The stainless steel ionization chamber and its accessories used in the present work were the same as those described previously(Qiu and Weng, 1987).

The effective ionization coefficients $\overline{\alpha}$ ($\overline{\alpha}=\alpha-\eta$) measured in $CF_2Cl_2-N_2$

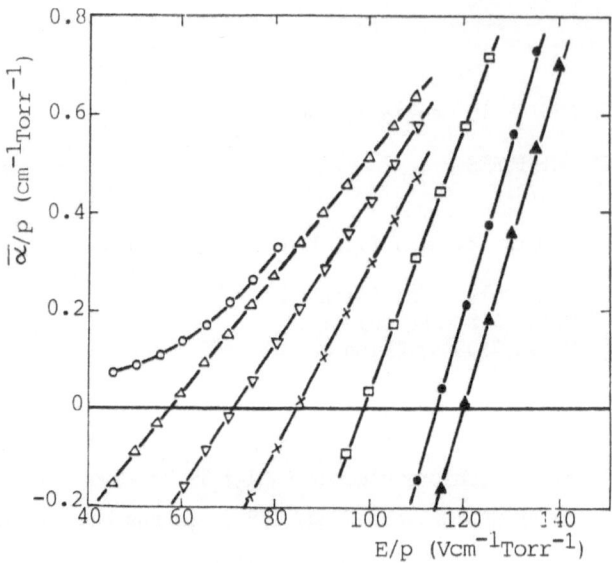

Fig. 1. $\bar{\alpha}/p = f(E/p)$ measured in CF_2Cl_2/N_2 of different
mixing ratios: o, 0/100; Δ, 10/90; ▽, 25/75;
x, 50/50; □, 75/25; ●, 90/10; ▲, 100/0 .

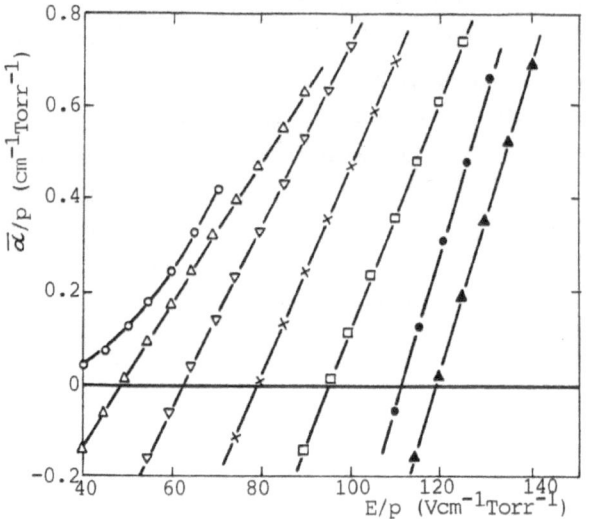

Fig. 2. $\bar{\alpha}/p = f(E/p)$ measured in CF_2Cl_2/CO_2 of different
mixing ratios: o, 0/100; Δ, 10/90; ▽, 25/75;
x, 50/50; □, 75/25; ●, 90/10; ▲, 100/0 .

and CF_2Cl_2-CO_2 gas mixtures are given in Fig. 1 and Fig. 2, respectively, as a function of the pressure-reduced electric field. It can be seen in both figures that $\bar{\alpha}/p = f(E/p)$ is a good straight line when the percentage concentration of CF_2Cl_2, k, is equal to or greater than 10.

RELATIVE ELECTRIC STRENGTH OF THE MIXTURES

Figure 3 shows the relative electric strength(RES) of CF_2Cl_2-N_2 gas mixtures compared with CF_2Cl_2 gas. It can be seen that data derived from the present pre-breakdown current growth measurements are in agreement with the static breakdown experiments both performed in this work, and reported by Somerville et al.(1980). Static breakdown results presented by Berril et al. (1986) are also in agreement with the present work, but not shown in Fig. 3 for avoiding being too crowded. The swarm experiments reported by Fréchette and Novak(1987) are in agreement with the present work when CF_2Cl_2 concentration in the mixture is equal to or greater than 50%. The relative electric strength of the CF_2Cl_2-N_2 mixture derived from the swarm parameters given by Siddagangappa et al.(1983) is, however, much too high compared with other data sources, and hence not included in Fig. 3.

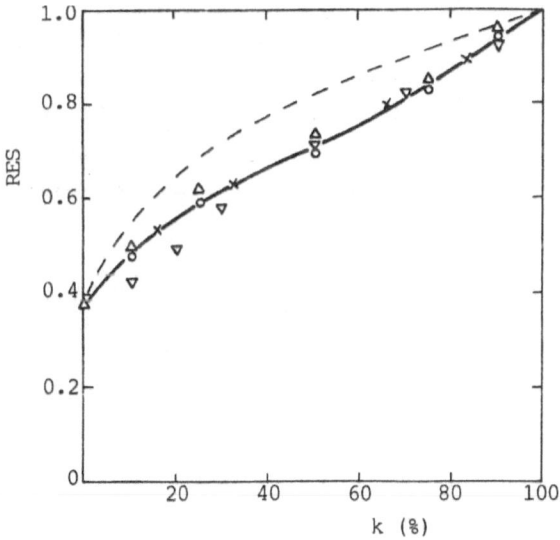

Fig. 3. Relative electric strength of CF_2Cl_2-N_2
o: from Fig. 1; ∇: Fréchette & Novak(1987)
Δ: breakdown data; x: Somerville et al.(1980)
---: Wieland's approximation .

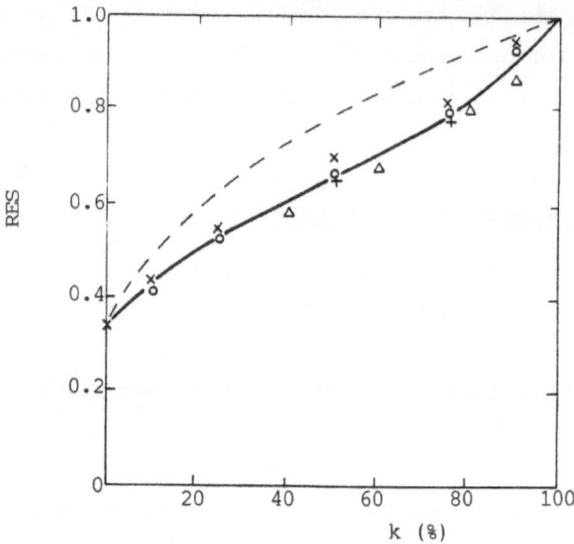

Fig. 4. Relative electric strength of CF_2Cl_2-CO_2
    o: from Fig. 2; Δ: Dincer & Govinda Raju(1985)
    +: Qiu et al.(1988); x: breakdown data
    ---: Wieland's approximation.

It is quite interesting to note that although the effective ionization coefficient in CF_2Cl_2 is close to that measured in SF_6(Qiu and Weng, 1987), the RES curve pattern for the CF_2Cl_2-N_2 mixture is significantly different from that for the SF_6-N_2 mixture, which rises rapidly at first and then slowly with the increasing concentration of the electronegative gas(Qiu and Liu, 1987).

The RES of the CF_2Cl_2-CO_2 gas mixture is given in Fig. 4, where the static breakdown results and data derived from the present pre-breakdown measurements are in agreement with our previous swarm measurements(Qiu et al., 1988), and those reported by Dincer and Govinda Raju(1985). The RES curve pattern for CF_2Cl_2-CO_2 is quite similar to that for CF_2Cl_2-N_2, indicating that CF_2Cl_2 is the gas component responsible for the unusual curve pattern for both gas mixtures.

DISCUSSION ON THE SYNERGISTIC EFFECT

Synergism is defined as the simultaneous action of separate agencies which, together, have greater total effect than the sum of their individual effects. By "effect" most investigators mean the breakdown strength(Chris-

92

tophorou et al., 1983), and therefore synergism is said to be present even in gas mixtures like SF_6-N_2. As the breakdown strength of the SF_6-N_2 gas mixture can be predicted by the Wieland approximation(Wieland, 1973) based on the assumption that no interaction takes place between constituent gas molecules, the concept of synergism with respect to breakdown strength seems to be misleading in this case. In this paper the concept of $\overline{\eta}$ synergism proposed by Chantry and Wootton(1981) is used by comparison of experimental data with the prediction of the Wieland approximation. The Wieland approximation predicts the effective attachment coefficient $\overline{\eta}$ ($\overline{\eta}=\eta-\alpha$) in the gas mixture by assuming that it consists of the sum of the partial-pressure-weighted effective attachment coefficients of the component gases, and therefore this approximation represents a criterion to judge whether there exists $\overline{\eta}$ synergism, and whether the $\overline{\eta}$ synergism is positive or negative.

Figures 3 and 4 show that the experimentally determined E/p values for both CF_2Cl_2 mixtures are apparently below their respective Wieland approximation, indicating that negative $\overline{\eta}$ synergism exists in both CF_2Cl_2 gas mixtures.

CONCLUSION

Unlike SF_6-N_2 and SF_6-CO_2 gas mixtures whose dielectric strength can be predicted by using the Wieland approximation suggesting no interaction taking place between molecules of different constituent gases, the dielectric strength of CF_2Cl_2-N_2 and CF_2Cl_2-CO_2 gas mixtures is, however, apparently lower than that predicted by the Wieland approximation, indicating that negative $\overline{\eta}$ synergism exists in both CF_2Cl_2-N_2 and CF_2Cl_2-CO_2 gas mixtures.

REFERENCES

Berril, J., Christensen, J.M., and McAllister, I.W., 1986, Measurement of the figure of merit M for several strongly electronegative gases, in: "Conference Record of the 1986 IEEE International Symposium on Electrical Insulation," IEEE Publishing Services, New York.

Chantry, P.J., and Wootton, R.E., 1981, A critique of methods for calculating the dielectric strength of gas mixtures, J. Appl. Phys., 52:2731.

Christophorou, L.G., Sauers, I., James, D.R., Rodrigo, H., Pace, M.O., Carter, J.G., and Hunter, S.R., 1984, Recent advances in gaseous dielectrics at Oak Ridge National Laboratory, IEEE Trans. Elec. Insul., EI-19:550.

Dincer, M.S., and Govinda Raju, G.R., 1985, Ionization and attachment coefficients in Freon-12 and CO_2 gas mixtures, IEEE Trans. Elec. Insul., EI-20:595.

Fréchette, M.F., and Novak, J.P., 1987, Limit-field behavior of various gas mixtures discussed in the framework of a Boltzmann-equation analysis, IEEE Trans. Elec. Insul., EI-22:691.

Joyce, C., and MacKenzie, D., 1987, Hot air threatens ozone in Montreal, New Scientist, 1578:30.

Qiu, Y., and Weng, X., 1987, Measurement of ionisation and attachment coefficients in difluorodichloromethane and SF_6 gas mixtures, J. Phys. D: Appl. Phys., 20:1203.

Qiu, Y., and Liu, Y.F., 1987, A new approach to measurement of the figure-of-merit for strongly electronegative gases and gas mixtures, IEEE Trans. Elec. Insul., EI-22:831.

Qiu, Y., Ren, X., Liu, Z.Y., and Zhang, M.C., 1989, Measurement of ionisation and attachment coefficients in gas mixtures of difluorodichloromethane and carbon dioxide, J. Phys. D: Appl. Phys., 22:1553.

Qiu, Y., Ren, X., and Liu, Z.Y., 1990, Ionisation and attachment coefficients measured in nitrogen and difluorodichloromethane gas mixtures, J. Phys. D: Appl. Phys., (to be published).

Siddagangappa, M.C., Lakshminarasimha, C.S., and Naidu, M.S., 1983, Ionisation and attachment in binary mixtures of SF_6-N_2 and $CCl_2F_2-N_2$, J. Phys. D: Appl. Phys., 16:763.

Somerville, I.C., Tedford, D.J., and Thurogood, A., 1980, Automatic measurement of spark breakdown potential in gas mixtures, in: "Sixth International Conference on Gas Discharges and their Applications, part 2," The Institution of Electrical Engineers, London.

Wieland, A., 1973, Gasdurchschlagmechanismen in elektronegativen Gasen(SF_6) und in Gasgemischen, ETZ-A, 94:370.

Wootton, R.E., Dale, S.J., and Zimmerman, N.J., 1980, Electric strength of some gases and gas mixtures, in: "Gaseous Dielectrics II," L.G. Christophorou, ed., Pergamon Press, New York.

EXCITATION AND IONISATION IN THE TRANSIENT
STATE DISCHARGE IN PURE NITROGEN

Hans Korge, Urmas Kuusk, Matti Laan and Jaan Susi

Gas Discharge Laboratory

Tartu University, ESTONIA

INTRODUCTION

In pure nitrogen at the atmospheric pressure there exist two stationary discharge forms in the case of a negative point in the point-plane discharge gap[1]. These forms are the weak one (current $\leqslant 10^{-9}$ A) and the strong one (current $\geqslant 10^{-4}$ A). The transition from the weak discharge to the strong one can be spontaneous or initiated, e.g. by the additional voltage pulse. The duration of the transient state is about 10^{-8} s in both cases of transition.

In air, the other experimental conditions being similar, the discharge exists only in the form of Trichel pulses. The current pulse shape and the time-spatial distribution of the Trichel pulse radiation[2] resemble those of the transient state discharge in pure nitrogen.

The possibility of existence of the two stationary discharge forms with current differences more than five orders at the same voltage indicates that the change of the mechanisms of excitation and ionisation must take place during the transient state. In the present study the population dynamics of some $C^3\Pi_u$ and $B^3\Pi_g$ state vibrational levels are measured for the transient discharge in nitrogen at atmospheric pressure. The solution of the balance equations for excitation of these states enables us to speculate on the possible mechanisms of excitation and ionisation.

EXPERIMENTAL CONDITIONS AND METHODS OF REGISTRATION

The experiments were carried out in the point-plane 40 mm

Gaseous Dielectrics VI, Edited by L.G. Christophorou and
I. Sauers, Plenum Press, New York, 1991

discharge gap. The point was a hemispherically tipped platinum rod of 0.5 mm radius. The gap was placed in a vacuum chamber which was evacuated to 8×10^{-8} Torr and then filled with nitrogen of 99.996 % purity to 770 Torr.

The steady voltage of 5.6 – 5.8 kV was supplied to the point electrode to raise the weak discharge. Then the additional voltage pulse (first front $\approx$ 10 ns, duration 100 ns, and amplitude 4 kV) was applied to initiate and strictly fix the starting of the transient state. After the strong discharge had been established, it was interrupted by shorting the gap with a thyratron. The repetition rate of discharge pulses was 50 s^{-1}. The spectrum of the discharge was registered by photon counting with the 1 ns time resolution. The diameter of discharge was 0.1 mm in the investigated area at the height of 0.2 mm above the point tip.

DETERMINATION OF THE POPULATION OF $C^3\Pi_u$ AND $B^3\Pi_g$ STATES

To determine the absolute population of a certain vibrational level of the $C^3\Pi_u$ state one has to know the absolute intensity of the corresponding band of the 2$^+$ system. In our experiments the band head intensity was measured with the spectral width $\Delta\lambda$ and the absolute intensity of the band $v' \rightarrow v''$ is

$$I_{v'v''}(\frac{1}{cm^3 \cdot s}) = B(\lambda) \cdot R(\lambda,t) \cdot D(t) \cdot F(\Delta\lambda, T_r) \cdot [I^m_{v'v''}(t) - I^b], \quad (1)$$

where v' and v'' are the vibrational level numbers respectively of the $C^3\Pi_g$ and $B^3\Pi_g$ states, λ the band head wavelength, $B(\lambda)$ the coefficient of the absolute calibration, $R(\lambda,t)$ the correction factor of self-absorption, $D(t)$ the correction factor of multiphoton counting, $F_{v'v''}(\Delta\lambda, T_r)$ the weight function of the band intensity for spectral width $\Delta\lambda$ and rotational temperature T_r, $I^m_{v'v''}(t)$ the photon counting rate or the measured intensity, I^b the background intensity of neighbour bands.

The coefficient $B(\lambda)$ was determined by the calibration of the registration set-up with the tungsten ribbon lamp. The self-absorption factor $R(\lambda,t)$ and the absorption coefficient $\mathscr{x}_0$ were determined from the measurements of relative absorption of the respective band radiation by the single mirror method[3]. The factor $D(t)$ was determined from the photon counting rate supposing the Poisson distribution[4] and it did not exceed 1.1.

The weight functions $F_{v'v''}(\Delta\lambda, T_r)$ were calculated by G. Hartmann[4]. Knowing their dependence on the rotational tempera-

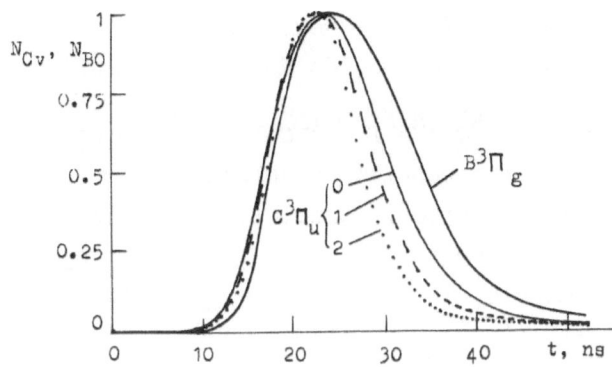

Fig. 1. Population of vibrational levels in transient discharge for $C^3\Pi_u$ ($v' = 0$, 1, and 2) and $B^3\Pi_g$ ($v'' = 0$) states.

ture T_r, one can easily determine T_r. The band intensity $I_{v'v''}$ is related to the level v' population $N_{v'}$ as

$$I_{v'v''} = A_{v'v''} \cdot N_{v'}, \qquad (2)$$

where $A_{v'v''}$ is the absolute transition probability[5]. From Eqs. (1) and (2) at the same $\Delta\lambda = 2.5$ Å the bands $0 \to 0$ and $0 \to 3$ yield the rotational temperature $T_r = (300 \pm 100)$ K for the whole transient stage of discharge.

Having determined all correction factors of Eq. (1), the absolute intensities of the 2^+ system bands $0 \to 0$, $1 \to 0$, and $2 \to 1$ as well as the absolute populations $N_{Cv'}$ of the $C^3\Pi_u$ state vibrational levels $v' = 0$, 1, and 2 were obtained. The absorption coefficient æ_0 enables us to determine the population of the level $v'' = 0$ for $B^3\Pi_g$

$$N_{B0}(t) = 8.1 \times 10^{32} \, \Delta\lambda_D \, \text{æ}_0(t) \, / \, A_{00}\lambda_{00}^4, \qquad (3)$$

where $\Delta\lambda_D$ is the Doppler's contour half-width. These populations are presented in Fig. 1 and are normed to the unity in maxima, which are the following: $N_{B0}^{max} = 4 \times 10^{14}$, $N_{C0}^{max} = 8 \times 10^{13}$, $N_{C1}^{max} = 1.5 \times 10^{13}$, and $N_{C2}^{max} = 4 \times 10^{12}$ cm^{-3}.

These measured variations of $N_{Cv'}(t)$ and $N_{B0}(t)$ serve as a basis for discussing possible ionisation and excitation mechanisms in the transient state discharge.

DISCUSSION

As the first approximation, we suppose, that the excitation is effected by the electron impact with the nitrogen molecules in the ground state $X^1\Sigma_g^+$ and the deexcitation is caused by the quenching with neutral molecules and the spontaneous radiation:

$$dN_{Cv'}/dt = k_{Cv'}^e . n_e N_X - (k_{Cv'}^q . N_X + \sum_i A_{v' v_i''}) N_{Cv'}, \qquad (4)$$

where $dN_{Cv'}/dt$ is the population velocity from the measured $N_{Cv'}(t)$, $k_{Cv'}^e$ the electronic excitation coefficient of level v', n_e the electron concentration, $k_{Cv'}^q$ the quenching coefficient of level v' by neutrals, N_X the molecule concentration. The different $k_{C0}^q = 1.2 \times 10^{-11}$, $k_{C1}^q = 2.4 \times 10^{-11}$ and $k_{C1}^q = 8 \times 10^{-11}$ cm^3 . s^{-1} [6] cause difference in decay of different levels (Fig. 1), since the total transition probabilities $\sum A_{v' v_i''}$ of these levels are nearly equal ($\approx 2.25 \times 10^7$ s^{-1}) [5].

Within experimental errors the population distribution of the $C^3\Pi_u$ levels $v' = 0, 1, 2$ can be characterized by the vibrational temperature T_{vC}. But in our conditions T_{vC} does not reflect the excitation since the vibrational population is strongly influenced by quenching which depends on the level number. For that reason it is expedient to investigate the fictitious $C^3\Pi_u$ vibrational population which would arise by the exclusion of quenching. Such populations N_0, N_1 and N_2 can be obtained by integrating Eq. (5) over $N_{Cv'}$ with the condition $k_{Cv'}^q = 0$. The new population distribution being non-Boltzmannian, it is suitable, instead of T_{vC}, to introduce the "temperatures" T_{01} and T_{12}:

$$T_{01} = (k/\Delta\varepsilon_{01}) \ln (N_0/N_1), \qquad T_{12} = (k/\Delta\varepsilon_{12}) \ln (N_1/N_2) \qquad (5)$$

where k is the Boltzmann constant, $\Delta\varepsilon_{01}$ and $\Delta\varepsilon_{12}$ are the energy differences of the neighbour vibrational levels.

In case of low pressure glow discharge T_{vC} is coupled with the temperature T_{vX} of the ground state[7]. Replacing T_{vC} with T_{01} and supposing the latter to be analogously coupled with T_{vX}, one can evaluate the energy pooling into the vibrational reservoir. Thus the discharge current must attain 500 mA, which is about two orders greater than the experimental value. The estimations of T_{vX} growth, based on the measured discharge current, show that it does not exceed 100 K while the growth of T_{01} is about 2000 K.

Thus we are having a situation, where the energy pooling from the spectroscopic measurements is in contradiction with the discharge current. To clarify this discrepancy, we must determine the variation of the electric field to gas density ratio E/N and

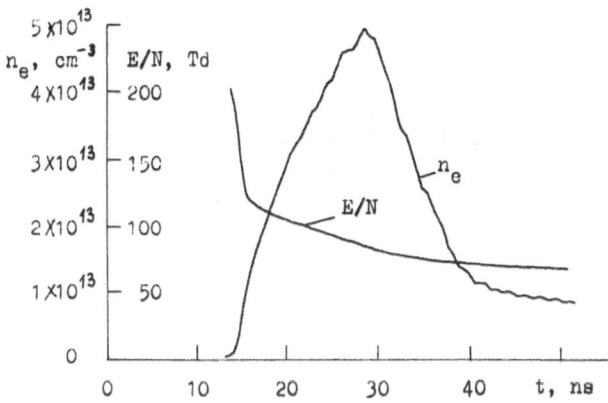

Fig. 2. Variation of electric field to gas density ratio E/N
and electron concentration n_e at the transient stage.

the electron concentration n_e. For this purpose we compose the
balance equation of population for the $B^3\Pi_u$ level $\upsilon''' = 0$, which
is analogous to Eq. (4)

$$dN_{BO}/dt = k^q_{BO}n_eN_X + \sum_i A_{\upsilon_i'0}N_{C\upsilon_i} - (k^q_{BO}N_X + \sum_i A_{0\upsilon_i'''}) N_{BO}. \quad (6)$$

The levels υ_i''' are those of the $A^3\Sigma_u^+$ state.

From Eqs. (4) and (6) one obtains the excitation velocity of
the $C^3\Pi_u$ and $B^3\Pi_g$ state levels $\upsilon = 0$, i.e. the temporal variation
of terms $k^e_{CO}n_eN_X$ and $k^e_{BO}n_eN_X$, and their ratio k^e_{CO}/k^e_{BO} being exclusively determined by E/N. On the basis of data for $k^e_C(E/N)$ and
$k^e_B(E/N)$ from[8], the variation of $E/N(t)$ and $n_e(t)$ is calculated
(Fig. 2). The maximum discharge current is about 5 mA relying on
these E/N and n_e, what is an acceptable result.

Further, to estimate the reliability of E/N and n_e calculations, the balance of electrons was examined. This examination
shows that up to 20 ns the direct ionisation is really the pre-
valent ionisation mechanism, but later it becomes insufficient to
compensate the electron losses by recombination. To evaluate the
influence exerted by the metastable states on excitation, the
growth of the $A^3\Sigma_u^+$ population was estimated on the basis of calculated E/N and n_e, and it reaches 10^{15} cm^{-3} by $t \geqslant 30$ ns. In
these conditions the following reactions

$$e + N_2(A^3\Sigma_u^+) \rightarrow N_2(C^3\Pi_u) + e, \quad 2 N_2(A^3\Sigma_u^+) \rightarrow N_2(C^3\Pi_u) + N_2(X^1\Sigma_g^+),$$

become effective and as a result the population of higher vibrational levels of the $C^3\Pi_u$ state is favoured and the coupling between the temperatures T_{vX} and T_{vC} is violated.

As regards for ionisation, evidently the associative ionisation by collision of the metastable $a'\Sigma_g^-$ (8.5 eV) or $a^1\Pi_g$ (8.66 eV) molecules will be influential, as their total amount reaches 10^{14} cm^{-3}. In this case the ionisation rate by collision

$$2\ N_2(a'\Sigma_u^-) \rightarrow N_2^+ + N_2(X^1\Sigma_g^+) + e$$

becomes comparable to that of direct ionisation. Thus the variation of E/N and n_e with time, as calculated on the assumption of the direct electron excitation mechanism, is true in the very beginning stage of the transition.

CONCLUSION

During the transient state the accumulation of metastable states occurs, but by the initiation with the voltage pulse up to the moment $t \leqslant 20$ ns, the field strength is sufficiently high and the main ionisation mechanism is the direct electron impact. The influence of metastables is revealed in vibrational population of the $C^3\Pi_u$ state, i.e. in the alteration of the coupling between the vibrational temperatures T_{vC} and T_{vX}. Later, $t \geqslant 20$ ns, because of the diminishing of field strength, the direct ionisation becomes ineffective and the associative ionisation may prevail.

REFERENCES

1. H. Korge, K. Kudu, and M. Laan, The discharge in pure nitrogen at atmospheric pressure in point-to-plane discharge gap, 3rd ISHVE, Paper 31.04, Milan (1979).
2. H. Korge, K. Kudu, and M. Laan, Development of d.c. corona pulses at atmospheric pressure, Proc. XIIIth ICPIG, Contr. papers, 451, Berlin (1977).
3. S. E. Frish, Determination of concentration of normal and excited atoms and oscillator strengths by methods of light emission and absorption, in: "Spektroskopiya gazorazryadnoy plazmy," S. E. Frish, ed., Nauka, Leningrad (1970).
4. G. Hartmann, Spectroscopie de la décharge couronne: étude des mécanismes de collisions dans le dard (streamer), Thèse, Université Paris-sud, Orsay (1977).
5. D. E. Shemansky and A. L. Broadfoot, Excitation of N_2 and N_2^+ systems by electrons I. Absolute transition probabilities, J. Quant. Spectrosc. Radiat. Transfer, 11:1385 (1971).
6. V. V. Urošević, J. V. Božin, and Z. Lj. Petrović, Excitation of $C^3\Pi_u$ state of N_2 by an electron swarm, Z. Phys. A - Atoms and Nuclei, 309:293 (1983).
7. V. N. Ochkin, Vibrational temperature in molecular discharge plasmas, XIth ICPIG, Contr. papers, 12, Prague (1973).
8. H. Brunet, P. Vincent, J. Rocca Serra, Ionization mechanism in a nitrogen glow discharge, J. Appl. Phys., 54:4951 (1983)

DISCUSSION OF SOME SPACE-CHARGE EFFECTS IN THE CONTEXT OF

LOW-TEMPERATURE AND HIGH-FIELD STEADY-STATE GAS DISCHARGES

M.F. Fréchette and Danny Roberge

Câbles et Isolants, Vice-Présidence Recherche (IREQ)
Hydro-Québec, Varennes, Québec, Canada J0L 2P0

ABSTRACT

Simplified analytical description of current growth in a parallel-plate steady-state discharge is achieved with the aim of discussing the breakdown behavior of a gas insulating medium under low-temperature environment and high-field conditions. The formulation of the model is believed to be applicable for weakly electronegative gases in the presence of space charge. In the event of a heavy-ion mobility dependence on temperature, it is observed that breakdown behavior in pure N_2 remains generally unaffected unless space charge forms. Relevance of the undervoltage-breakdown situation to that of field-emission assisted gas breakdown is discussed for SF_6/N_2 -type mixtures.

INTRODUCTION

Usually implicit in conducting breakdown experiments and reporting associated data is the assumption that the functionality of the measured characteristics has a unique dependence upon the thermodynamic variables. This has considerable advantages allowing for instance to choose a reference set of macroscopic variables (pressure p and temperature T) which in turn permits the comparison of data taken at different temperatures (p_T, T varied) or direct application of room-temperature data for low-temperature applications.

Great costs and technical complexities associated with performing low-temperature tests have resulted in an almost ad hoc general acceptance of the above described methodology: there is an obvious lack of experimental low-temperature breakdown data pertaining to electrotechnical gases and dielectric testing of high voltage outdoor equipments at low temperature is rarely done. Recently, Fréchette (1990) raised interrogations on the reasoning behind the above described approach for SF_6. Initiated from a review of earlier works, the analytical discussion of some of the data that followed, has outstressed a temperature-dependence of SF_6 breakdown under the experimental conditions prevailing for outdoor insulation (a pressure of a few bar and gas temperature down to -50 ^{0}C). The potential causes for this observed temperature dependence were qualified to be rather systematic or circumstantial in nature.

In the present contribution, the breakdown behavior of outdoor gas-insulated systems is adressed in a more basic framework, an attempt at relating some of the discharge parameters relevant to such a situation: low temperature, space charge and high-field conditions. A simple model, first-order approximation and semi-empirical in essence, that allows for a crude representation of space charge formation, is used at first to investigate the possible effect on breakdown of a heavy-ion mobility variation due to a temperature drop. Results are presented for pure N_2. As breakdown of gas medium insulating HV equipments often arises under very high-field conditions, it is likely that the manifestation of a gas breakdown behavior affected by lowering the experimentation temperature will intervene concurrently with that associated with an active role of field emission (under negative polarity). Bearing this in mind, the present model is further used to discuss, in the context of an undervoltage-breakdown situation, some discharge characteristics such as emission currents and breakdown departures from Paschen curve for weakly electronegative SF_6/N_2 -type mixtures.

FORMULATION

The steady-state discharge model is based on the Townsend mechanism of breakdown with the electric field satisfying the Poisson equation; the collisional processes taken into account are exclusively ionization (α) and attachment (η). The present derivation follows in essence that of Crowe et al. (1954). The one-dimensional formulation assumes a reciprocal ratio of the positive-ion mobility dominant compared to that of other species and leaves the secondary ionization coeficient (γ, independent of field) as a fitting parameter. Negative-ion densities contributing to the total current are taken to be small yet non-negligible, a model assumption that limits the applicability to high-E/p discharges in weakly electronegative gases.

For the parallel-plate discharge with the cathode at x=0 and for a specified electrode spacing (d), the following expression relating the electric field and the current growth equation, is obtained (SI units):

$$(\alpha-\eta).E.dE \ = \ \frac{i_0/(\varepsilon_0\mu_+) \cdot [\exp(u) - \exp(v)].du}{[\ 1 - \gamma.(\exp(v)-1)]} \tag{1}$$

where u and v are defined such as:

$$u = \int_0^x (\alpha-\eta).dx' \quad \text{and} \quad v = \int_0^d (\alpha-\eta).dx \ ,$$

in which expressions, E stands for the electric field, i_0 is the initial current density; ε_0 and μ_+ are respectively the permittivity of free space and positive-ion mobility. Assuming further that the net ionization coefficient may be expressed in the form of an exponential inversely proportional to a linear function of the electric field:

$$\frac{(\alpha-\eta)}{p} \ = \ A \cdot \exp \left[\frac{-B}{(E/p - c)} \right] , \tag{2}$$

where the parameters A, B and c are determined from numerical fitting to relevant experimental data, the thus obtained system equation , from Eq.(2) into Eq. (1), becomes analytically integrable, taking the form:

$$F(z) \ = \ F(z_0) \ - \ \frac{i_0/(\beta\varepsilon_0\mu_+)[(u.\exp(v) - \exp(u)+1)]}{[1 - \gamma.(\exp(v)-1)]} \tag{3}$$

where $z = Bp/E(x)$, $z_0 = Bp/E(0)$, $\beta = Ap(Bp)^2$ and $2F(z) = e^{-z}/z^2 + G \cdot [E_1(z) - e^{-z}/z]$

in which: $G = 1 - 2c/B$ and $E_1(z) = \int_z^\infty e^{-t}/t \cdot dt$ with z on the positive real axis.

The final solution is sought from a simple numerical procedure, in the course of which the gap voltage is obtained from the integration of the field distribution over the total gap length (d). In particular, the electric field is calculated through the variable z, an operation that necessitates the inversion of the function F; the current density in the gap is given by:

$$i(x) = \frac{i_0 \cdot \exp(u)}{[\,1 - \gamma.(\exp(v)-1)\,]} \quad , \text{ assessed at d.} \tag{4}$$

RESULTS AND DISCUSSION

The selection of the relevant discharge data used in the course of simulation was the result of a balanced choice between exactness (when known), representative nature (trends and orders of magnitude) and, above all, tractableness, as the present performed exercise is basically one of illustration.

Values of the discharge parameters for N_2 are essentially those retained by Crowe et al. Figure 1 (curve 1) shows the approximated variation of the ionization coefficient with the reduced field (Eq.(2),with $A = 52$ cm^{-1}kPa^{-1}, $B = 1950.6$ Vcm^{-1} kPa^{-1} and $c = 0$). Calculations for N_2 were performed for a unique value of pressure, i.e. 93.3 kPa, and at a gap length of 1 cm. At this pressure, the values of the positive-ion mobility and secondary ionization coefficient were taken to be respectively 3.26 cm^{2}v^{-1}s^{-1} and 3.19 x 10^{-4}.

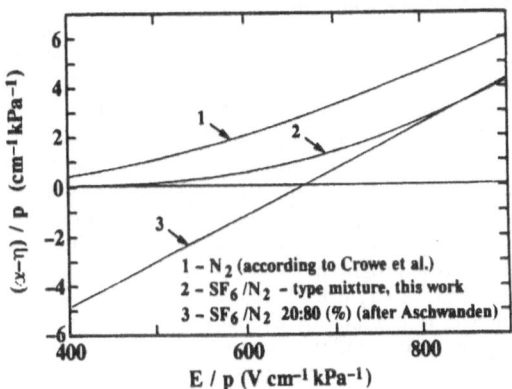

Fig. 1. Effective ionization coefficient as a function of reduced field.

The model gas featuring a weakly electronegative SF_6/N_2 mixture has an effective ionization coefficient (Fig. 1, curve 2) that lies between that of pure N_2 and the coefficient associated with a mixture having 20% SF_6 in content (curve 3). The analytical form of the coefficient is given by Eq.(2) with $A = 88.4$ cm^{-1}kPa^{-1}, $B = 2256.9$ Vcm^{-1}kPa^{-1} and $c = 160$ Vcm^{-1}kPa^{-1}. The secondary ionization coefficient was kept constant throughout, unless otherwise stated, the value retained being close in magnitude to that of pure N_2,i.e. 5 x 10^{-4}. An expression giving an estimate of the positive-ion mobility for the type of mixture considered was worked out as follows. Mobilities for SF_6 ions were averaged for reduced fields between approximatively 525 and 1230 Vcm^{-1}kPa^{-1} using the expression reported by Morrow (1986). Assumed independent of field, this particular value and that corresponding to pure N_2 were proportionnally added according to the ratio of a SF_6/N_2 20:80 (%) mixture. The final expression reads: $\mu_+ = 257.7/p[kPa]$ cm^{2}V^{-1}s^{-1}.

Context: Low-Temperature Conditions

Validity of the Paschen's law for N_2 was reported to be verified for a considerable range of experimental conditions. Under non-uniform field conditions, however, Hara et al.

(1989) found that the discharge characteristics were dependent on the gas temperature. Experimentation carried on using a uniform-field geometry does not imply necessarily a discharge growth developping under a space-charge free environment; field-emission assisted gas breakdown is such an exemple.

In this section, results from a simulation featuring discharge growth under space-charge and low temperature, are presented. The low-temperature condition comes about through an assumed heavy-ion mobility variation with temperature. There is some supporting evidence of such a dependence which was provided by Kovar et al. (1957). In particular, mobility of the positive ions in N_2 was observed to increase considerably with lowering the gas temperature in the region of reduced fields ranging from 300 to 375 Vcm^{-1} kPa^{-1}. This variation was identified with an ion-conversion reaction scheme dependent on temperature, resulting in the time-dependent presence of several ion species having different mobilities.

To make our point in an illustrative way, current-voltage characteristics were generated using several initial current densities. Starting the series with a current magnitude (Fig. 2, 6.36 x 10^{-11} Acm^{-2}) insuring a discharge growth undisturbed by space charge, the consecutive i-V curves were calculated with the initial current densities: 1, 2, 4, 10 in unit of 10^{-8} Acm^{-2}. By doing so, the thus obtained breakdown voltage was reduced consequently on the prescribed increase in initial current, the maximum reduction from V_p (on Fig. 2) being of the order of 5%. While the absolute magnitude of an i-V curve is dependent on the initial current [see Eq.(4)], the field distribution and, therefore the space-charge effect, are a function of the ratio i_o/μ_+ [see Eq.(1)]. Thus, current-normalization aside, a unique shape of current-voltage characteristic could be the result of calculations done with several couples (mobility , initial current) as long as the ratio of the current over mobility remains unchanged. This artifice was utilized in the foregoing discussion to transform conceptually a breakdown reduction due to an increase in initial current into that resulting from a decrease of the mobility value. Figure 2 summarizes the results. Normalization of the current-voltage curves was done with reference to the highest initial currrent (10^{-7} Acm^{-2}), the reference curve giving the evolution of the current as a function of gap voltage for an experiment conducted at room temperature (T_0). As the gas temperature is decreased below room temperature (from T_0 to T_3 , gas molecular density kept constant), the corresponding increase in magnitude of the ion mobility affects the shape of current-voltage characteristics and leads to a progressive yet slight increase in the breakdown voltage. Starting from room temperature (T_0), each consecutive specified temperature at which the i-V characteristic is shown, represents the almost geometrical doubling (1, 2.5, 5, 10) of the mobility value, a progression unlikely to reveal itself in any compound nor physical situation (except maybe when phase transition occurs).

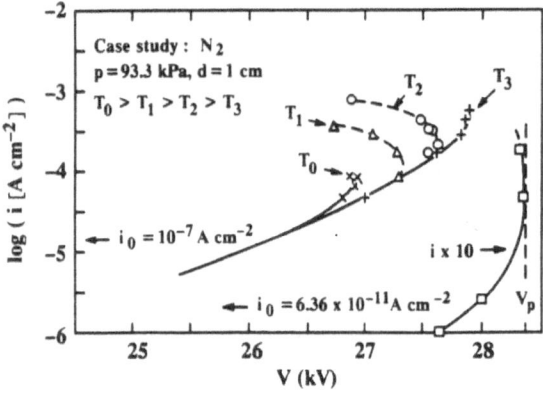

Fig. 2. Illustration of the space charge effect on the current density-voltage curve when gas temperature is varied.

Context: High-Field Conditions

In this section, field-emission assisted gas breakdown and associated deviations from values given by the Paschen curve are discussed in a framework connecting into that of an undervoltage-breakdown situation. All calculations reported here were done for a fixed gap length of 0.2 cm and pertain to the behavior of SF_6/N_2 -type mixtures. A mixture of interest in the context since threshold for the occurence of field emission in electronegative gases usually appears at lower field-values.

In the foregoing, it is proposed to picture the occurence of a field-emission current as a triggering event that would class as an initial electron current to be amplified through a gas under the influence of an applied electric field. Since electron emission current due to field effect is an electrode-related phenomenon, the field enhancement remains closely localized at the cathode surface: the field gradient does not extend through the gap. Thus, once passed the stage of current generation, it is contended that the final observational phenomenology of the physical situation can be ascribed to that of a discharge growth under space charge. Indeed, magnitudes of electron emission current currently observed at threshold are generally above those for which the applied field remains undisturbed; identification of such field-emission current with that usually photoelectrically produced at the cathode will bring about an undervoltage-breakdown situation. For the sake of simplicity, the nature of these initial currents was considered continuous. Limited by the underlying assumptions of the model, assessment of the undervoltage-breakdown values needed to be done in a pressure region far from the expected breakdown-departure threshold. For the present simulation, the pressures in the range of 2 to 4 bar were selected. Given a reasonable set of values for the initial current density, secondary ionization coefficient and pressure, a defined deviation from the Paschen curve can be generated. The initial current used to get a particular deviation was taken to be the corresponding electron field-emission current at breakdown. Extrapolation from the deviated breakdown curves was subsequently used to evaluate the magnitude of the emission currents at the voltage threshold (and associated pressure) at which the current emission starts.

Figure 3 shows typical current-voltage characteristics obtained for a pressure of 266.6 kPa. On such a graph, the breakdown value corresponds to the voltage at which there is an inflexion point of the relation i-V . The breakdown voltage ascribed to that of the Paschen curve is calculated using the lowest current density (10^{-9} to 10^{-10} Acm^{-2} throughout). It is found that an increase of the initial current produces a decrease of the breakdown voltage; here, at most 10% for the highest current. Breakdown decrease is caused by field distortion resulting from the presence of space charge, the field being enhanced at the cathode as illustrated in Fig. 4. The field distribution along the discharge axis is calculated for several voltages, including that corresponding to breakdown, for one of the cases shown on the previous graph ($i_o = 10^{-4}$ Acm^{-2}).

From the observation of the trends displayed by the i-V curves (Fig. 3), that is a reduction of the breakdown voltage resulting from the increase of the initial current, comes to mind the possibility of generating particular deviated breakdown curves from that of Paschen. Weakness of the nonuniformity factors encountered, at maximum 1.2 for the present calculations, gives however some indication of the limited potential of the model in producing large breakdown deviations. Yet, the corresponding physical situation could be compared with that of breakdown occuring between somewhat well polished electrodes. Two specific illustrations (simulations a and b) are presented in Fig. 5. Simulation a (curve 2) which gives the smallest deviations from the Paschen curve (numbered 1) results from calculations performed with the initial current densities: 3.5 x 10^{-7}, 1.8 x 10^{-6} and 4 x 10^{-5} Acm^{-2}. The corresponding percent deviations are: 1.3, 2.7 and 7 at respective pressures: 200, 266.6 and 400 kPa. Simulation b (curve 3) brings the following characteristics: deviations of 3.6, 4.8 and 10.9 (%) for the same respective pressures. The current densities used take respectively the values: 3.5 x 10^{-6}, 10^{-5} and 4 x 10^{-4} Acm^{-2}. Quoted values of initial current densities associated with the largest deviations in both simulations were coupled with $\gamma = 10^{-2}$ in order to keep current magnitudes on a realistic basis. If γ is left unchanged (5 x 10^{-4}) but initial densities doubled, comparable deviations are obtained. Figure 6 summarizes the results concerning the variation of the field-emission currents (initial current interpreted as) as a function of pressure for both simulations.

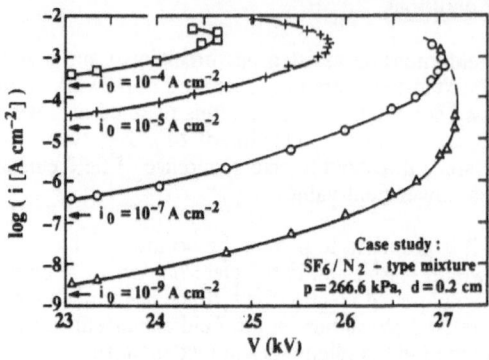

Fig. 3. Variation of the current density as a function of voltage for several given initial current densities.

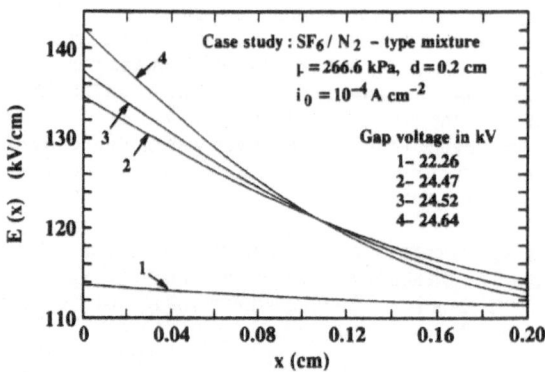

Fig. 4. Electric field distribution along the discharge axis for several applied voltages.

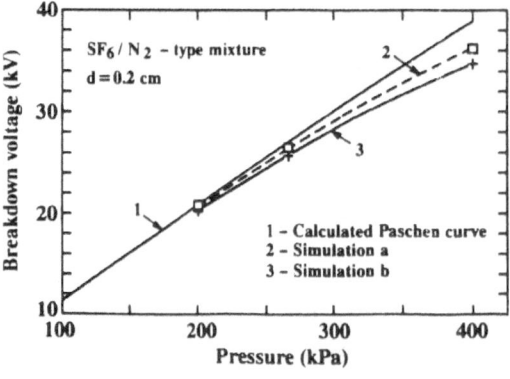

Fig. 5. Breakdown voltage as a function of pressure.

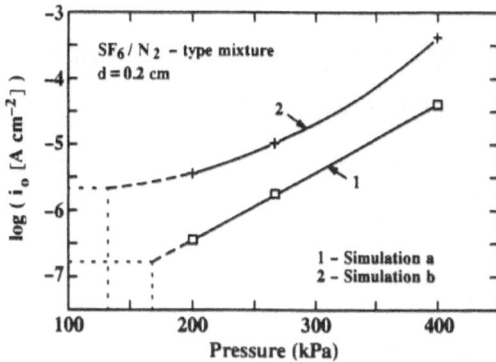

Fig. 6. Initial current densities as a function of pressure for selected deviations from Paschen curve.

General trends are in agreement with common experimental observations, e.g. - the larger the deviation from Paschen, the higher the emission current, - current increases with applied field. In particular, the magnitudes of the emission currents at threshold compare consistently with those obtain from a critical-current analysis (Latham 1986). Finally, the found relationship illustrates well an argument developped by Fréchette (1990). Assume that curve 2 (Fig. 6) corresponds to the initial emission characteristics of a cathode; voltage and emission at threshold are respectively: 74 kVcm^{-1} and 2 x 10^{-6} Acm^{-2}. Following an assumed deterioration of the emission characteristics (featured by curve 1) due to surface-state variations, the resulting least deviated breakdown curve from Paschen (Fig. 5, curve 2) would arise from a lower value of emission current (1.6 x 10^{-7} Acm^{-2}) appearing at a higher voltage threshold (89 kVcm^{-1}).

SUMMARY

Results from a theoretical investigation of some space-charge effects in the context of low-temperature and high-field steady-state gas discharges were reported. If assumed that the positive ion-mobility varies with temperature, it can be stated that the breakdown voltage will be affected when under the influence of space charge. Yet, the magnitude of the resulting effect is considered to be negligible for pratical situations. Gas breakdown in the presence of electron field-emission was discussed in the framework of an undervoltage-breakdown situation. Results from the simulation, in particular the estimation of the magnitude of the field-emission currents at threshold and resulting effects on breakdown behavior, show consistency with current phenomenology and provide some grounds for the understanding of foreseeable effects due to the variation of the field-emission characteristics proceeding from some modification of the surface state of the emitting cathode.

ACKNOWLEDGEMENT

This work was sponsored in part by Etudes et Normalisations, Equipement de Transport, Vice-Présidence exécutive Equipement, Hydro-Québec. Danny Roberge is with Ecole Polytechnique de Montréal and beneficiates from a graduate student fellowship from Hydro-Québec.

107

REFERENCES

Aschwanden, Th., 1985, Die ermittlung physikalischer entladungsparameter in isoliergasen und isoliergasgemischen mit einer verbesserten swarm-methode, Doctorate dissertation, ETH, Zurich, Switzerland, unpublished.

Crowe, R.W, Bragg, J.K,. and Thomas, V.G., 1954, Space charge formation and the Townsend mechanism of spark breakdown in gases, Phys. Rev., 96:10.

Hara, M., Suehiro, J., Matsumoto, H., and Kaneko, T.,1989, Breakdown characteristics of cryogenic gaseous nitrogen and estimation of its electrical insulation properties, IEEE Trans. Electr. Insul., EI-24: 609.

Fréchette, M.F., 1990, SF_6 gas breakdown at low temperature, Proc. 1990 IEEE Int. Symp. on electrical insulation, Toronto, Canada, June 3-6 1990, pp. 229-235.

Kovar, F.R., Beaty, E.C. and Varney, R.N., 1957, quoted in "Collision phenomena in ionized gases", E.W. McDaniel, John Wiley and sons, NY, 1964, p. 475.

Latham, R.V., Bayliss, K.H., and Cox, B.M., 1986, Spatially correlated breakdown events initiated by field electron emission in vacuum and high pressure SF_6, J. Phys. D, 19:219.

Morrow, R., 1986, A survey of the electron and ion transport properties of SF_6, IEEE Trans. Plasma Sci., PS-14:234.

CHAPTER 3: MODELING

NON-EQUILIBRIUM EFFECTS IN THE INITIATION OF

PSEUDOSPARK DISCHARGES

L.C. Pitchford and J.-P. Boeuf

Centre de Physique Atomique de Toulouse
CNRS, Unité de Recherche Associée 277
Université Paul Sabatier, 118, route de Narbonne
31062 Toulouse, Cedex, France

INTRODUCTION

We have recently developed a model of the initiation phase of pseudospark discharges[1] which has elucidated the sequence of physical events leading to the onset of this unusual discharge mode[2]. The main conclusion of that work was that the existence of a transient hollow cathode configuration during the initiation is essential for the rapid current rise characteristic of pseudospark discharges. The large and localized source of ionization which is a consequence of the transient hollow cathode is inherently a nonequilibrium effect. The term "nonequilibrium" is used here to denote conditions for which the electron transport and rate coefficients are non-local; that is, these coefficients cannot be described simply by "swarm" parameters[3] which are functions of the local reduced field, $E(r,t)/N$, $E(r,t)$ being the electric field strength at a position r and a time t and N being the neutral density. The purpose of this article is to describe more fully the nonequilibrium effects in pseudospark discharges.

PSEUDOSPARK DISCHARGES

Pseudospark discharges are distinguished from other discharge modes by three main features which are interesting from a scientific as well as a technical point of view; e.g., the *extremely rapid* (and triggerable) *current rise* a short time after application of a voltage; the *generation of electron and ion beams* during the current rise and the final high current phase; and the *diffuse and relatively long-lived final, high current phase*. Pseudospark discharges have been observed in a variety of gases including the rare gases, nitrogen, oxygen and hydrogen. Their properties are more dependent on the discharge geometry and gas pressure than on the exact composition of the gas, although the influence of the gas composition on the electron and ion beam formation seems to be important[4].

The simplest configuration in which pseudospark discharges occur consists of a planar anode separated from the parallel face of a cathode by a distance such that the product of this distance and the gas pressure is on the left hand side of the Paschen curve. The cathode face has a central hole whose diameter is much less than the diameter of the electrode face leading to a hollow backspace. This geometry and the representative dimensions which we have used for our model calculations are illustrated in fig. 1. More complicated, multi-gap geometries have also been investigated and show behavior qualitatively similar to the single gap discharges.

The remarkable properties of pseudospark discharges can be seen in the following examples of their operating characteristics reported from the literature. (For a recent survey of the state-of-the-art, see reference 5.)

Rapid and Triggerable Current Rise

Upon application of a sufficient voltage and after a brief delay which depends on the voltage and trigger (if any), the discharge current begins to increase very rapidly. The pd product (pressure x distance) in these discharges is such that the breakdown voltages are high (10's of kV), and increasing hold-off voltage is achieved by decreasing pd. In either a self-break or in a trigger mode, the transition to the highly conducting state (several 10's of kA) occurs in a very short time (several 10's or 100's of nanoseconds.). For example, in a simple, single gap configuration, a current rise of 5.0×10^{11} A/sec was reported[6] for a current and an applied voltage of up to 20 kV and 10 kA, respectively, with a repetition rate of 100 kHz.

Various schemes for introducing charge into the hollow cathode backspace have been demonstrated as triggers for pseudospark discharges[6-9]. A distinction has been made in the literature between electrical triggering of pseudospark discharges and the optically triggered version which has been referred to as a back-lit thyratron (BLT) [8-9].

Generation of Charged Particle Beams

The transition to the high-current discharge is often accompanied by the generation of an intense electron beam as observed through a central hole in the anode. Reported electron beam properties appear to be highly dependent on the particular conditions, but they can reach up to 10 - 20% of the total discharge current for 10's of nanoseconds during the current rise, with beam energies up to the applied voltage[10]. In addition to the electron beams, positive ion beams[4,11] and x-rays[11] have also been observed in pseudospark discharges, although quantitative experimental results are few.

Super-Dense Glow

The term "pseudospark" was originally coined to describe the high-current density discharge phase[2]. In contrast to the arc-like filaments usually seen at such high current densities, pseudospark discharges appear to be diffuse, and the electrodes seem to be free of the damage (erosion, pitting, etc.) associated with arcs[12]. The duration of the high-current phase or "super-dense glow" phase, as it has been termed[13], can be microseconds and longer[14]. Field enhanced, thermionic electron emission from the cathode has been invoked to explain the observed high current densities in these diffuse discharges[13].

Applications

Given these interesting and unusual properties, a number of potential applications of pseudospark discharges have been discussed in the literature. Considerable effort has been devoted to the exploitation of pseudosparks for switching applications[15]. The diffuse nature of the discharge and the fact that the cathodes are not externally heated (in contrast to thyratrons) suggest that switches based on triggered pseudospark discharges may have lifetimes far greater than thyratrons for comparable electrical characteristics. The beam generation capabilities of pseudosparks are potentially interesting from the point of view of materials processing[10]. There may also be interesting applications to be discovered for the "super dense glow" since such a volume of high plasma density is difficult to generate in other discharge devices.

PHYSICAL MODEL AND NUMERICAL MODELS

Our purpose in this work has been to develop a numerical model which can be used to identify the physical processes controlling the initiation of the pseudospark discharge mode. In this article we emphasize the role of nonequilibrium effects, and in particular, the nonlocal ionization rate, in the sequence of events leading to the rapid rise of current characteristic of pseudospark discharges. The numerical method for the resolution of the equations described below is similar to that used in the context of electron transport in semiconductor devices as discussed in references 1 and 16.

Physical Model

The models of the electrical behavior of gas discharges developed in Toulouse over the past few years[17] are the basis of the studies in pseudospark discharges reported here. This pseudospark model is a two-dimensional, time-dependent model in cylindrical coordinates. Results from the model include the space and time dependent electric field and charged particle densities, and hence, the components of the current density as functions of time and space.

The physical model is based on a hybrid fluid - particle (Monte Carlo) description of the electron and ion transport coupled with Poisson's equation for the electric field. The ideal model of a low pressure discharge would consist of Boltzmann equations for the electrons and the ions coupled to Poisson's equation for the electric field. Such a procedure is impractical in most situations, and it is usually necessary to seek some simplified description of the electron and ion transport to couple to Poisson's equation.

Lawler[18] has shown that, except within a mean free path of the electrode surfaces, the ion energy distribution function is well described by the local value of the reduced electric field, $E(\mathbf{r},t)/p$, where E is the magnitude of the electric field and p is the gas pressure. Thus, in our model, ion continuity and momentum transfer equations with a drift velocity and diffusion coefficient given as functions of the local field are used to represent mathematically the ion transport.

In contrast to the ions, a description of the electron transport is precisely the aspect which requires considerable care in a model of pseudospark discharges as well as in other low pressure discharge devices. The bulk electrons are important in pseudospark discharges primarily through their effects on the space charge electric field and on the discharge conductivity. The high energy electrons, on the other hand, contribute little to the overall electron density (and hence the space charge field) or mean energy. The importance of these electrons lies in their ability to ionize the gas atoms or molecules; i.e., to provide a source of electrons and ions. At low pressures and because of the energy dependence of the electron scattering cross sections, the mean free path for electron impact ionization may be comparable to the dimensions of interest while the energy or momentum exchange distance, reflecting primarily the low energy electrons, can be much shorter. Thus, a knowledge of the distribution of the low energy electrons does not yield the distribution for the high energy tail, and descriptions of both the bulk and the high energy electrons over a wide range of discharge conditions are essential if we are to understand the physics of pseudospark devices.

The low energy electrons which form the bulk of the distribution can generally be characterized by their mean properties (density, mean velocity, mean energy). A model where the charged particle transport properties are characterized by their mean properties is termed a **fluid model** and is based on the solution of the first two or three of the infinite series of coupled velocity moments of the Boltzmann equation. The system of moment equations is truncated after two (the equations of continuity and momentum conservation) or three (continuity and momentum and energy conservation) with some choice of closure relations, and, hence, fluid representations contain inherent assumptions as to the form of the energy distribution function of the bulk electrons. Various ways of closing the system of moment equations after the first few have been described in previous work. For example, an approximation which is often used in discharge modeling is the local field approximation (which we mentioned above in the context of the ion transport). In this approximation one assumes that the ionization frequency and the transport coefficients at a given position and time $(\mathbf{r}, t)$ in the discharge depend only on the value of the local reduced electric field $E(\mathbf{r}, t)/p$. This is equivalent to assuming that the electron distribution function at $(\mathbf{r}, t)$ is that which would exist in a uniform reduced field of magnitude $E(\mathbf{r}, t)/p$, the condition in swarm experiments. We use this local field or equilibrium approximation in our fluid model for the electron transport due to mobility and diffusion because these transport effects are primarily controlled by the bulk electrons. We will return to this point below.

Since the ionization source term is an extremely sensitive function of the high energy tail of the electron distribution and since the electron and ion densities increase exponentially with this source term, it is of paramount importance to accurately represent the high energy part of the distribution. Although the local field approximation might be a reasonable assumption to use in the calculation of the mean properties of the bulk electrons, it is completely unrealistic for the tail especially under conditions where the ionization mean free path is of the order or larger than the characteristic length of variations of the electric field (e.g. cathode sheath). It is impossible, for example, to describe the electron "pendulum" effect in a hollow cathode[19] using the local field approximation. In fact, the hollow cathode "pendulum" effect is inherently a nonequilibrium effect as we will discuss.

In our model, the high energy part of the electron distribution function (above the ionization threshold) is obtained from a **particle model** (Monte Carlo simulation). From this distribution one can deduce the ionization rate which is used as a source term of the continuity equation of the fluid model for bulk electrons and ions.

Fluid Model

The fundamental variables, the electron density, n_e, the positive ion density, n_p, and the electric potential, V, are found from the solution of the following equations,

112

the electron continuity equation,

$$\frac{\partial n_e}{\partial t} + \nabla \cdot n_e \mathbf{v_e} = S \qquad (1)$$

the ion continuity equation,

$$\frac{\partial n_p}{\partial t} + \nabla \cdot n_p \mathbf{v_p} = S \qquad (2)$$

and Poisson's equation,

$$\Delta V = - \frac{|e|}{\varepsilon_0} (n_p - n_e) \qquad (3)$$

where $\mathbf{v_{e(p)}}$ is the electron (positive ion) mean velocity, S represents the ionization source term, e is the electronic charge, and ε_0 is the permittivity of free space.

The charged particle mean velocities appearing in the continuity equations are obtained from the second moment of the Boltzmann equation; i.e., from the momentum transfer equation. Neglecting the electron and ion inertia, this equation takes a simple form leading to the representation of charged particle flux as the sum of a drift term and a diffusion term,

$$\phi_{e(p)} = n_{e(p)} \mathbf{v_{e(p)}} = n_{e(p)} \mu_{e(p)} \mathbf{E} - \nabla (n_{e(p)} D_{e(p)}) \qquad (4)$$

where $\mu_{e(p)}$ is the electron (ion) mobility and $D_{e(p)}$ is the electron (ion) free diffusion coefficient where these two quantities are assumed to be functions of only $E(\mathbf{r},t)/p$.

Monte Carlo Simulation of Ionization Source Term

We use a Monte Carlo calculation[20] for the source term assuming the electric field distribution given by the solution of eqs. 1-3 at a previous time. The Monte Carlo simulation follows the trajectories of the initial electron density and the cathode emitted electrons and of their progeny until it becomes energetically impossible for them to ionize further before being absorbed by the anode. We therefore use the term "high energy electrons" for those electrons whose kinetic energy plus potential energy at a given point $(\mathbf{r},t)$ is larger than the ionization threshold (the potential energy being defined as $e(V_a - V(\mathbf{r},t))$ where V_a and $V(\mathbf{r},t)$ are respectively the anode potential and the potential at $(\mathbf{r},t)$).

RESULTS

The geometry we have used in our model of pseudospark discharges is shown in figs. 1. The cathode and anode diameters are 2 cm; the cathode hole diameter is 0.6 cm; the hollow cathode length is 0.6 cm; and the anode-cathode gap length is 0.7 cm. The results presented in this section have been obtained for helium at a pressure of 0.5 torr and for an anode voltage of 2 kV. The secondary emission on the cathode is supposed to be due only to ion bombardment and is equal to 0.3 inside the hollow cathode. Secondary emission is zero on the part of the cathode facing the anode. The calculations we describe start with

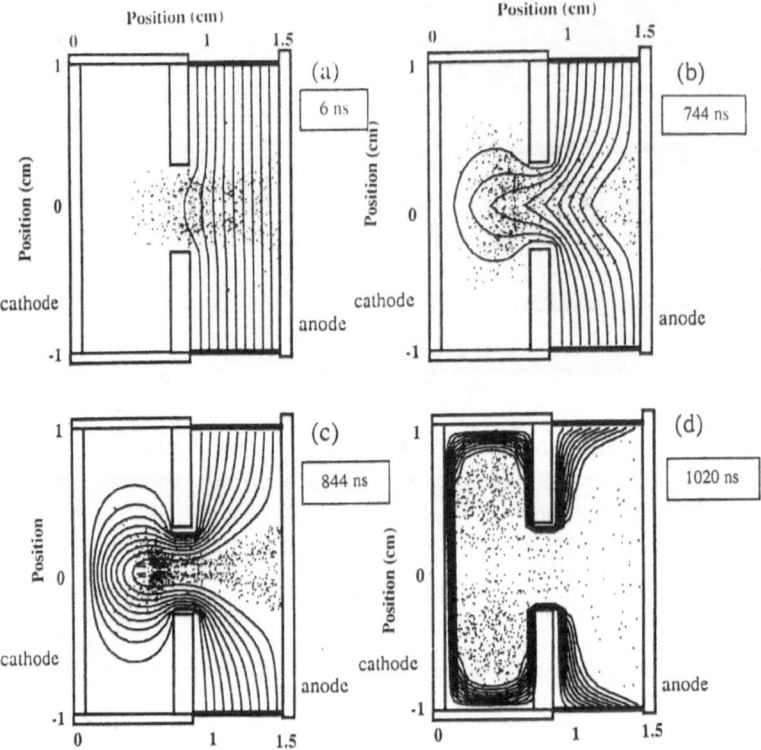

Fig. 1.   Equipotential curves from the hybrid fluid-particle model
          for the conditions described in the text.   The four figures
          correspond to (a) t=6ns, (b) t=744ns, (c) t=844ns, and (d)
          t=1020ns.

uniform initial electron and ion densities of 10^9 cm^{-3} inside the hollow
cathode to simulate the effect of a trigger.

Figure 1 shows the potential distribution in the device at
different times of the discharge development in the form of contours of
constant potential.   The electric field and electron and ion densities
on the axis of the discharge are shown in figs. 2 at six different
times.   It is convenient to separate the discharge evolution seen in
these figures into four phases as described below.   Figure 3 shows the
discharge current as a function of time, and the four phases through
which the discharge passes during its evolution are indicated in the
figure.

Phase 1: Townsend Discharge

At t=6 ns (figs. 1a and 2a), the potential is almost uniform in
the cathode-anode gap and the penetration of the electric field in the
hollow cathode is weak due to the small dimension of the cathode hole.
Free electrons inside the hollow cathode which are created by the
trigger are, however, extracted from the cavity by this field,
undergoing a a small number of ionizing collisions in this region and in
the main gap.   At this time the electron and ion densities are too small
to induce a significant space charge field, and the potential and field

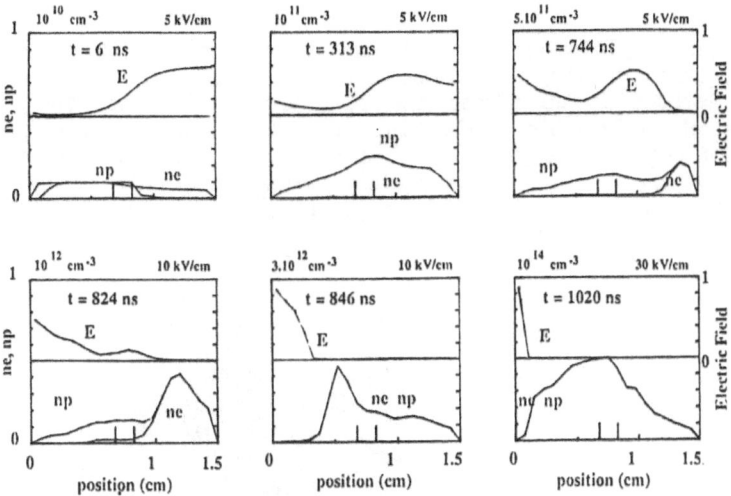

Fig. 2.   Electric field strength and electron and ion densities on the
axis of the discharge at (a) t=6ns, (b) t=313ns, (c) t=744ns, (d)
t=824ns, (e) t=844ns, and (f) t=1020ns.

distributions of figs. 1a and 2a in the main gap are mostly due to the
geometry of the device. The first phase of the discharge development is
similar to a Townsend discharge where successive avalanches which are
initiated by secondary electrons emitted from the inside surfaces of the
hollow cathode by ion impact, contribute to the build up of the ion
density in the cavity and in the main gap. If, however, the initial
electron density produced by an external trigger is high enough, the
duration of this phase can be considerably reduced.

Phase 2: Plasma formation

      The increase in the ion space charge leads progressively (see
figs. 1 and 2) to a distortion of the electric field both in the hollow
cathode region and in the main gap. The changes in the potential and
field distributions are such that the electron multiplication increases.
As in any low pressure discharge, this is because the build-up of the
ion density in the gap leads to a decrease of the electric field on the
anode side and an increase of the field on the cathode side; i.e., in
the hollow cathode, This transient field configuration is the precursor
to the formation of ion sheaths at the cathode.

      After t=744 ns (figs. 1b and 2c), the ion density has grown to the
point that the ion-induced, space charge field is of the same order as
the geometric field. This time corresponds to the beginning of a second
phase of the discharge development characterized by the formation of a
plasma. Because the ion space charge field leads to a decrease in the
total electric field in the main gap, the electrons no longer speed from
cathode to anode, but rather are slowed down in the region of high ion
space charge density. Hence, a plasma begins to form. The consequence
of the plasma formation in the main gap is an increase of the potential
close to the cathode hole. The field penetration in the hollow cathode
is then substantially enhanced as is the multiplication in this region.
During this second phase of the discharge development the anode
potential approaches progressively the cathode aperture.

Phase 3: Plasma entering the hollow cathode

When the plasma enters the hollow cathode region (figs. 1c and 2e) the potential on the axis close to the cathode hole and inside the hollow cathode is at almost the full anode potential. The electron multiplication in this region increases again due to the confining effect of the cathode geometry. This increase in the multiplication due to the hollow cathode geometry characterizes the third phase of the discharge ignition. As the plasma enters and expands in the hollow cathode region, the ion sheaths along the cathode surface contract.

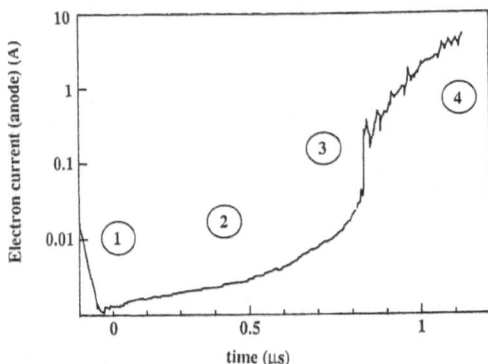

Fig. 3.   Total discharge current as a function of time from the hybrid fluid-particle model calculations.   The four phases of the discharge evolution are indicated by the numbers in the figure.

The efficiency of a hollow cathode configuration in increasing the discharge current is related to two distinct mechanisms[19].   The first one, mentioned above, is simply due to the confining effect of this geometry. The electrons emitted and accelerated in the sheath on one side of the cathode can be reflected by the opposite sheath. The electron can undergo a number of successive reflections before being extracted through the cathode hole and reaching the anode.   Since the time spent by the electron inside the hollow cathode is substantially increased by the confining effect, the probability that the electron undergoes an ionizing collision in this region is increased accordingly. A large part of the electron energy which otherwise would have been lost on the anode can therefore be released in the hollow cathode in the form of excitation and ionization.

Another effect, related to the sheath length in the hollow cathode can further enhance the electron multiplication[19].   In figs. 1d and 2f the field configuration is such that the sheath regions between the zero potential contour and the contour representing the full anode potential are compressed near the inner walls of the hollow cathode.   There is thus an extended region of near zero field inside the hollow cathode, and secondary electrons are created in ionization events predominately in the low field regions. In contrast, in figs. 1c and 2e, the high field sheath regions fill the hollow cathode. It is this latter configuration which favors a high electron multiplication because the secondary electrons created by ionization events are themselves born and accelerated in these high field regions. Thus, each secondary electron will produce in its turn a large number of secondary electrons, and a maximum in the multiplication rate occurs for the field configuration in figs. 1c and 2e.

Phase 4: Sheath contraction

Once the plasma enters the hollow cathode and the electron multiplication increases, there is nothing to prevent the plasma from growing to fill the volume of the hollow cathode. This expansion of the plasma leads to a field configuration where the sheaths are compressed more and more against the inner walls of the hollow cathode cavity. In figs. 1f and 2d, the contraction of the sheaths is evident. The electron multiplication passes through a maximum just as the plasma enters the cavity and then decreases to a minimum value (about 10 in our conditions) which is larger than the multiplication needed to sustain the discharge. Since the electron multiplication is greater than that for a the self-sustained discharge, the plasma continues to expand in the hollow cathode, and the sheaths contract without limit in our model. The subsequent evolution of the discharge cannot be predicted by our model because eventually field emission or field enhanced thermionic emission will begin to play a role as the sheath thickness continues to decrease. These effects not included in our model are needed to explain the observed rates of current increase. Alternate electon emission mechanisms from the cathode of pseudospark discharges have been discussed in detail in by Hartmann et al[13].

DISCUSSION AND CONCLUSIONS

In order to quantify in more detail the importance of the nonequilibrium ionization rate, we show in fig. 4 the electron multiplication calculated from the Monte Carlo simulation as a function of time. Also shown for comparison is the electron multiplication calculated assuming a local field value of the ionization rate coefficient for the potential distributions from the hybrid fluid-particle model. Even with the potential distributions from the hybrid model, the local field multiplication considerably underestimates the total electron multiplication. In fact, a steady-state is reached when the multiplication is equal to $(1 + \gamma)/\gamma$ where γ is the secondary electron emission coefficient. Thus, the pseudospark discharge mode cannot be predicted with a local field value of the ionization rate coefficient.

We have not included in the model phenomena which undoubtedly begin to play a role in the plasma expansion phase above. In particular, it has been proposed that field enhanced thermionic emission becomes the operative electron emission mechanism from the cathode in the final, high current phase[13]. Our model predicts a current rise of up to 10^9 A/sec during the initiation of the pseudospark mode, a figure lower than those observed experimentally. In agreement with Hartmann et al[13] , we conclude that alternate cathode emission mechanisms are needed to explain the high rate of current rise observed experimentally.

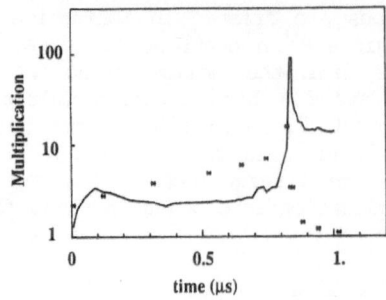

Fig. 4.   Electron multiplication (ratio of electron current collected
at the anode to that leaving the cathode) as a function of time
calculated from the fluid model (solid line) and calculated
assuming a local field ionization rate coefficient but with the
potential distributions from the hybrid fluid-particle model
(symbols).

Although additional physical phenomena must be considered in the
model in order to follow the evolution of the pseudospark mode through
the initiation and into the high current phase, the model described
above does provide a framework for understanding the onset of the
pseudospark mode.

REFERENCES

1. J.P. Boeuf and L.C. Pitchford, "Pseudosparks via Computer
    Simulations", to appear in IEEE Trans. Plasma Sci., special issue
    on computer modeling (1991).
2. J. Christiansen and C. Schultheiss, "Production of High Current
    Particle Beams by Low Pressure Spark Discharges", Z. Phys. A290:35
    (1979).
3. L.G.H. Huxley and R.W. Crompton, "The Diffusion and Drift of
    Electrons in Gases", Wiley-Interscience, New York (1974).
4. D. Bloess, I. Kamber, H. Reige, G. Bittner, V. Brueckner, J.
    Christiansen, K. Frank, W. Hartmann, N. Leiser, C. Schultheiss, R.
    Seeboeck, and W. Steudtner, "The Triggered Pseudo-Spark Chamber as
    a Fast Switch and as a High-Intensity Beam Source", Nucl. Instrum.
    Methods 205:73 (1983).
5. M.A. Gundersen and G. Schaefer, eds., "The Physics and Applications
    of Pseudosparks", Plenum Press, New York (1990).

6. G. Mechtersheimer; R. Kohler, T. Lasser, and R. Meyer, "High Repetition Rate, Fast Current Rise, Pseudospark Switch", J. Phys. E 19:466 (1986).

7. K. Frank, E. Boggasch, J. Christiansen, A. Goertler, W. Hartmann, C. Kozlik, G. Kirkman, C. Braun, V. Dominic, M.A. Gundersen, H. Riege, and G. Mechtersheimer, "High Power Pseudospark and BLT Switches", IEEE Trans. Plasma Sci. 16:317 (1988).

8. G. Kirkman and M. Gundersen, "A Low Pressure, Light Initiated, Glow discharges Switch for High Power Applications", Appl. Phys. Lett. 49:494 (1986)

9. G. Kirkman, W. Hartmann, and M.A. Gundersen, "Flashlamp Triggered High Power Thyratron Type Switch", Appl. Phys. Lett. 52:613 (1988).

10. W. Benker, J. Christiansen, K. Frank, H. Gundel, W. Hartmann, T. Redel and M. Stetter, "Generation of Intense Pulsed Electron Beams by Psekduospark Discharges", IEEE Trans. Plasma Sci. 17:754 (1989).

11. M. Bauer, A. Brandelik, A. Citron, H. Ehrler, K. Mittag, A. Rogner, W. Schimassek, and Chr. Schultheiss, "A 20 GW-Pinch as an Ion and X-Ray Source", in: Proceedings of the 6th Pulsed Power Conference, Arlington, VA. (1987).

12. H. Riege and E.P.Boggasch, "High-Power, High-Current Pseudospark Switches", IEEE Trans. Plasma Sci. 17:775 (1989).

13. W. Hartmann, V. Dominic, G. Kirkman and M.A. Gundersen, "Evidence for Large-Area Superemission into a High Current Glow Discharge", Appl. Phys. Lett. 53:1699 (1988).

14. W. Hartmann, "Cathode-Related Processes in High Current Density, Low Pressure Glow discharges", in: "The Physics and Applications of Pseudosparks", M.A. Gundersen and G. Schaefer, eds, Plenum Press, New York (1990).

15. J. Christiansen and R. Tkotz, "The Pseudospark and its Applicationa as a High Power Switch", in: Proceedings of the XIX International Conference on Phenomena in Ionized Gases, invited talks, V.J. Zigman, ed., Studio Plus, Belgrade (1989).

16. M. Kurata, "Numerical Analysis for Semiconductor Devices", Heath, Lexington (1982).

17. J.P. Boeuf, "Numerical Model of rf Glow Discharges", Phys. Rev. A36:2782 (1987); J.P. Boeuf, A Two-dimensional Model of dc Glow Discharges", J. Appl. Phys. 63:1342 (1988); J.P. Boeuf, "Self-consistent Models of DC and Transient Glow Discharges", in: The Physics and Applications of Pseudosparks", M.A. Gundersen and G. Schaefer, eds, Plenum Press, New York (1990).

18. J.E. Lawler, "Equilibration Distance of Ions in the Cathode Fall", Phys. Rev. A32:2977 (1985).

19. H.-B. Valentini, "Electron Kinetics in the Cathode Region of a Glow Discharge and in Hollow Cathodes", Contrib. Plasma Phys. 27:331 (1987).

20. J.P. Boeuf and E. Marode, "A Monte Carlo Analysis of an Electron Swarm in a Non-uniform Field: the Cathode Region of a Glow Discharge in Helium", J. Phys. D 15:2169 (1982).

DISCUSSION

I. GALLIMBERTI: I would like to know a bit more on the coupling of the fluid equations with the Monte Carlo (MC) calculation of the source term: how the MC is run, how you keep "memory" of the electron energy distribution in space and time, etc.

L. C. PITCHFORD: Since we use the Monte Carlo simulation to calculate the ionization rate, we only need to keep track of those electrons which are energetically capable of producing further ionization. Thus, the time–consuming tracking of the many low energy electrons in low field regions is not included in our simulations. The MC simulation is performed after every 50 or so integration time steps (the integration time steps vary according to how rapidly the potential is changing). Between successive calls to the MC routine, the source terms are updated by the ratio of the electron flux leaving the cathode surface (calculated from the fluid equations) to that at the time of the last call to the MC routine. By doing this we are including the effect of changes in the flux distribution in time but not changes in the potential distribution.

P. F. WILLIAMS: Can you estimate the heating rate of the cathode around the hole, and if so, where is this rate a maximum? How does this correlate with Gundersen's results showing melting of the cathode near the hole?

L. C. PITCHFORD: We have not yet looked in detail at the heating of the cathode surfaces. From our calculations of the electric field, however, we can infer that the position of the maximum of the cathode heating is time–dependent with a maximum in the region of the hole at the time when the electron multiplication is a maximum. Later on, after the plasma has expanded to fill the cathode backspace, the heating should be more uniform over the cathode surface. This is qualitatively consistent with experimental observations.

E. E. KUNHARDT: The characteristics of the final state of conduction you described may be related to the glow–to–arc transition in parallel–plane geometry as follows. The final state of conduction in such configuration is filamentary in nature because, for a glow discharge of finite radius, the radial mode with largest growth is maximum at (and concentrated near) the origin. If a hole is now made in one electrode, this mode is suppressed and higher order modes which are more "diffuse" become dominant.

S. M. MAHAJAN: It appears that the "inside" contour of a hollow cathode will have an enormous effect on the plasma density. Would you please comment on such an effect?

L. C. PITCHFORD: In pseudospark discharges, our calculations would suggest that it is the size of the cathode hole more than the inside contour which controls the plasma density. The electron multiplication after the plasma has entered the cathode backspace will increase with decreasing hole diameter, but if the hole is too small, the discharge cannot be initiated.

DISSOCIATION PROCESSES IN PLASMA CHEMISTRY
AND GASEOUS DIELECTRICS

L. E. Kline
Westinghouse STC
Pittsburgh, PA 15235

INTRODUCTION

Plasma processing is widely used in integrated circuit manufacturing. In a typical plasma process a chemically nonreactive feed gas flows into a discharge reactor where the discharge electrons dissociate the feed gas and produce chemically reactive products. The discharge dissociation products react at the surface of the Si, GaAs or other wafer being processed and either remove material in an etching process or deposit a thin film. Some of the gases which are used in these etching and deposition processes are listed in Table 1. Our present understanding of these plasma processes is limited, in part, by the limited availability of fundamental information about the microscopic physical and chemical processes which lead to the observed macroscopic etching and deposition processes. The important processes include electron impact ionization and dissociation, gas phase chemical reactions, surface chemical reactions and physical processes at the surfaces. Electron impact dissociation can also change the chemical composition of gaseous dielectrics. Therefore, dissociation cross section data is needed to understand and model both plasma processing reactors and the behavior of gaseous dielectrics. Dissociative ionization and dissociative excitation cross sections are available for many gases. However, for gases where the dissociation channels are known, a significant fraction of the total dissociation is due to processes which produce neutral fragments in their electronic ground state. Cross sections for these neutral dissociation processes are difficult to measure because the products are difficult to detect. Computer modeling of plasma processing discharges and the available cross section and chemical kinetic data have recently been reviewed.[1]

This paper describes recent progress toward the determination of electron cross section data for some of the gases which are used in plasma processing and for gases which have been proposed as new dielectric gases. The electron impact processes which take place in plasma processing and in the electrical breakdown of gaseous dielectrics are elastic scattering, vibrational excitation, electronic excitation, ionization and dissociation. Ionization is important in plasma processing because the electrons and ions produced in ionizing collisions are needed to sustain a plasma processing discharge, i.e. to replace electrons and ions which are lost by diffusion to the electrode surfaces and chamber walls. This replacement condition also sets the threshold voltage for electrical breakdown. Ions are also important in plasma processing because they are more chemically active than their parent gases and because electron impact ionization is a dissociative process for most of the molecules used in plasma processing. Dissociation of the feedstock gas produces neutral fragments which are chemically active. The resulting change in the gas composition may raise or lower the threshold voltage for electrical breakdown. The neutral dissociation channel is typically as important or more important than the dissociative ionization channel in producing the chemically active species which drive the plasma processes. These neutral dissociation processes are the main topic of this paper.

Many gases are used in plasma processing. A representative list of plasma processing gases is given in Table 1. The gases are grouped into etching gases and deposition gases. Note that some gases appear in both groups. Some of the gases listed in Table 1 are used or have been proposed as gaseous dielectrics. For example, SF_6 and N_2 are widely used as dielectrics. Some members of the C_xF_y family of gases have been proposed as new dielectrics or as components of dielectric gas mixtures.

Table 1. Gases used in plasma etching and deposition.

Etching Gases (for Si, GaAs, Si Oxide, Si Nitride, Metals)

CF_4, C_xF_y, SF_6, SiF_4, NF_3, CCl_xF_y, Cl_2, CCl_4, BCl_3 and HCl,

PCl_3, $SiCl_4$, CH_4, H_2, O_2, CF_3H, Br_2, CBr_xF_y, Rare Gases

Plus numerous stable and intermediate product species

Deposition Gases (For C, Si, SiC, Ti Oxides, Si,B Nitride and Metals)

CH_4, SiH_4, Si_2H_6, TEOS (tetraethoxysilane), O_2, N_2O, CO_2, NH_3,

B_2H_6, BCl_3, BBr_3, $TiCl_4$, WF_6, MoF_6, $MoCl_5$, H_2, Rare Gases

Plus numerous stable and intermediate product species

CROSS SECTION DETERMINATION FROM BEAM EXPERIMENTS

The cross sections which are needed to understand the microscopic processes in plasma processing and gas breakdown are usually measured in electron beam experiments. In a beam experiment, a monoenergetic beam of electrons is reacted with a volume filled with gas or with a beam of gas molecules or atoms. Conditions are adjusted so that the probability of more than one collision is negligible. The methods which are used to measure the various types of cross sections listed above are reviewed in Ref. 2.

Total cross sections are usually measured by measuring the attenuation of the electron beam. Total cross section measurements are difficult at low energies because it is difficult to produce a monoenergetic electron beam at low energies. Thus, total cross sections are often not available from beam experiments at low energies.

Vibrational excitation processes play a role in determining the shape of the electron energy distribution and the resulting excitation, neutral dissociation, ionization and dissociative ionization rate coefficients. Vibrational cross sections are usually measured by detecting electrons which have lost one or more vibrational quanta of energy. Detection of electrons which have lost a specific amount of energy is more difficult as the energy loss becomes smaller. In addition, vibrational excitation energies decrease as the size of the molecule increases. Consequently, vibrational cross section measurements become more difficult as the size of the molecule increases. Cross sections for vibrational excitation have been measured for most of the small molecules listed in Table 1 but only for a few of the larger molecules.

The fact that charged products are produced, and can easily be detected, facilitates the measurement of ionization and attachment cross sections. As a result ionization cross sections have been measured for almost all of the gases listed in Table 1 and attachment cross sections have been measured for most of the gases which form negative ions. Since ionization cross sections are usually measured by detecting the positive ion products, the differential cross section data for ionization is limited. The dominant positive and negative ion products for the polyatomic gases listed in Table 1 are fragment ions rather than the parent ion. In fact, the parent ions are observed only for the rare gases, the diatomic molecules and the tri-atomic molecules which are listed in Table 1. All of the product ions are fragments for the larger molecules. The product ions are identified in positive ionization and attachment cross section measurements by using a mass spectrometer as the ion detector.

NEUTRAL DISSOCIATION CROSS SECTIONS

Cross sections for electronic excitation have been measured for the subset of the molecules listed in Table 1 which have bound electronically excited states. However, many of the larger molecules listed in Table 1 do not have bound, electronically excited states. Therefore, the cross sections needed for these molecules are dissociation cross sections. Dissociation cross sections where one (or more) of the dissociation fragments is in an elec-tronically excited radiating or metastable state have been measured for many molecules. Becker[3] has reviewed some of the recent measurements of these dissociative excitation cross sections for plasma processing gases. Dissociative excitation cross section data for processes where an excited fragment is produced is available for many gases. However, the data for gases where a complete cross section set is available suggests that many dissociative excita-tion processes produce two (or more) fragments which are all in their electronic ground states. The cross sections for these neutral dissociation processes are difficult to measure because the products are difficult to detect. In fact, direct detection of neutral dissociation products has been performed for only a few molecules.[4,5]

Another approach to measuring neutral dissociation cross sections is to measure the cross sections for 1) total dissociation and 2) dissociative ionization. The neutral dissociation cross section is then the difference between the total dissociation cross section and the disso-ciative ionization cross section. Total dissociation cross sections have been measured for a number of plasma processing gases by Winters.[6]

CROSS SECTION DETERMINATION FROM SWARM EXPERIMENTS

Low energy total cross sections can be determined from swarm experiments and this approach has been used to study some of the gases listed in Table 1. Comparison of pre-dicted and measured swarm data can be used to check the consistency of a cross section set and to determine unknown cross sections. This approach to cross section determination is discussed by Phelps[7] and Crompton.[8] In this approach a "complete" set of electron cross sections is assembled by estimating the missing cross sections. Then the consistency of the cross section set is checked by using the cross section to predict swarm data including the electron drift velocity and diffusion coefficients and ionization and attachment coefficients. Comparison of the predicted swarm data with measured swarm data provides an indication of the consistency of the cross section set and can also be used to make adjustments to the cross section set. In fact, this approach can be used to determine an unknown cross section in some cases where information about all other cross sections is available. This procedure is shown schematically in Figure 1. Recently, Hayashi[9,10] has used this approach to develop cross section sets for many plasma processing gases.

CROSS SECTION DETERMINATION FROM DISCHARGE DISSOCIATION DATA

Since the data for neutral dissociation cross sections is very limited, it is useful to consider other approaches to determining rate dissociation coefficients and/or cross sections. One approach is to measure the products which are formed by discharge dissociation of a plasma processing gas. Ryan and Plumb have used this approach to study several gases including CF_4, $CF_4 + O_2$ mixtures, SF_6 and $SF_6 + O_2$ mixtures. They used mass spec-trometric composition measurements in conjunction with a chemical kinetics model to deter-mine rate coefficients. In principle, it is possible to "work backwards" from this kind of experimental result to determine the neutral dissociation cross section. This approach can be used where all of the other cross sections are known. This approach is shown schematically in Figure 2. The needed data are known, at least approximately, for many of the gases listed in Table 1. In these cases, the shape of the neutral dissociation cross section can be estimated and used, together with the other cross sections, to calculate electron energy distributions (EED's) and rate coefficients. For example, the shape of the neutral dissociation cross section can be estimated by using the ionization cross section shape for the same gas and shifting

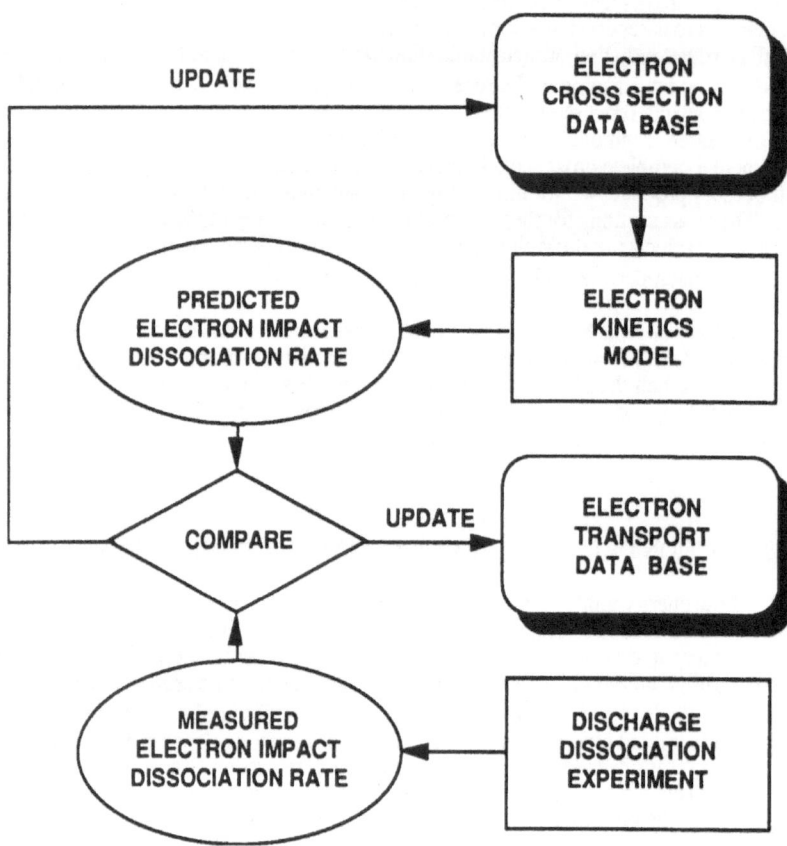

Fig. 1. Procedure for determining cross sections from swarm data.

the threshold or analogy with known neutral dissociation cross section shapes for similar gases. Since dissociative ionization cross sections must be known to use this approach, the ionization contribution to dissociation will be one of the outputs of the EED calculations. Therefore, the magnitude of the neutral dissociation cross section can be estimated by comparing the predicted and measured neutral dissociation rates. Although this approach is approximate, is should allow plasma process modeling over a much wider range of discharge conditions than would be possible by simply using the rate coefficients determined for one specific set of discharge conditions.

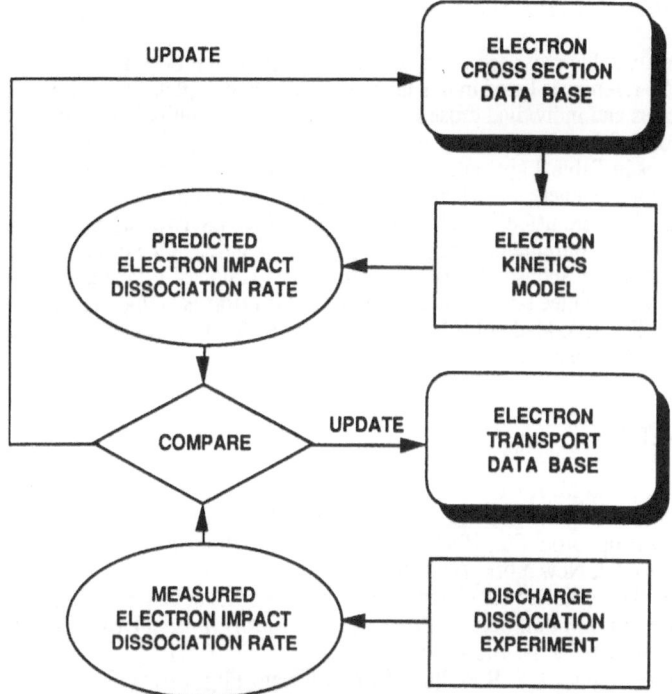

Fig. 2. Procedure for neutral dissociation cross section determination from discharge experiments. The conventional procedure, shown in Fig. 1, is used to determine all of the other cross sections, including the dissociative ionization cross section.

Table 2. Gases of Interest to Plasma Processing For Which Comprehensive Electron Impact Cross Sections are Available.

Etching Systems:

CF_4	[6,9,11,12,13,14]	C_2F_6	[9,14]
CCl_2F_2	[9,15,16,17]	CCl_4	[9,18]
F_2	[19,20,21]	HCl	[9,22]
Cl_2	[21,23]	SF_6	[20,24,25]

Deposition Systems:

SiH_4	[9,,26,27,28,29]	Si_2H_6	[9,27,29]
CH_4	[9,29,30,31]	C_2H_6	[10,29]
N_2O	[10]		

Diluent Gases:

He	[8]	Ar	[32,33]

CROSS SECTION DATA SOURCES

The available cross section data for various plasma processing gases is summarized in Table 2. The references listed in this table include both complete cross section compilations for these gases and individual cross sections. Complete compilations of cross sections or cross section references are given in Refs. 8, 12, 15, 16, 19, 22-26, 28 and 30. The remaining references in Table 2 give either individual cross sections or cross section sets which are useful only for low energies. There are many other sources of specific cross sections for these gases which are listed in Ref. 1. In addition, there are two ongoing projects to develop cross section data compilations for plasma processing gases. One of these is being performed under the sponsorship of the International Union of Pure and Applied Chemistry (IUPAC) Subcommittee on Plasma Chemistry. This effort is being chaired by L. E. Kline. The second effort is being carried out by W. L. Morgan of the Joint Institute for Laboratory Astrophysics in Boulder, Colorado.

REFERENCES

1. L. E. Kline and M. J. Kushner, 1989, "Computer Simulation of Materials Processing Plasma Discharges," CRC Critical Reviews of Solid State and Materials Sciences, 16:1.
2. L. G. Christophorou, Ed., 1984, Electron-Molecule Interactions and Their Applications, Vols. 1 and 2, New York, Academic.
3. K. H. Becker, "Electron Molecule Collision Cross Sections for Etching Gases," to be published in Non Equilibrium Processes in Partially Ionized Gases, NATO ASI Series, M. Capitelli and J. N. Bardsley, Eds.
4. C. E. Melton and P. S. Rudolph, 1967, J. Chem. Phys., 47:1771.
5. P. C. Cosby and H. Helm, 1989, Bull. Am. Phys. Soc., 34:325.
6. H. F. Winters and M. Inokuti, 1982, Phys. Rev. A, 25:1420.
7. A. V. Phelps, 1968, Rev. Mod. Phys., 40:399.
8. R. W. Crompton, 1983, Proc. 16th International Conf. on Phenomena in Ionized Gases, Duesseldorf, Invited Papers, p. 58.
9. M. Hayashi, 1985, "Electron Collision Cross Sections for Molecules Determined From Beam and Swarm Data", in Swarm Studies and Inelastic Electron Molecule Collisions, L. C. Pitchford, B. V. McKoy, A. Chutjian, and S. Trajmar,Springer-Verlag, New York, pp. 167-187.
10. M. Hayashi, 1985, "Electron collision cross sections for C_2F_6 and N_2O," in Gaseous Dielectrics V, L. Christophorou and D. Bouldin, Eds., Pergamon, New York.
11. K. Stephan, H. Deustch, and T. D. Mark, 1985, J. Chem. Phys., 83:5712.
12. K. Masek, L. Laska, R d'Agostino and F. Cramarossa, 1987, Contrib. Plasma Phys., 27:15.
13. M. G. Curtis, I. C. Walker and K. J. Mathieson, 1988, J. Phys. D, 21:1271.
14. S. M. Spyrou, I. Sauers, and L. G. Christophorou, 1983, J. Chem. Phys., 78:7200.
15. J. P. Novak and M. F. Frechette, 1985, J. Appl. Phys., 57:4368.
16. S. Okabe and T. Kuono, 1986, Japanese J. Appl. Phys., 24:1335.
17. K. Leiter and T. D. Mark, 1985, "Electron Ionization Cross Sections for CF_2Cl_2," in Proc. 7th Int. Symp. Plasma Chem., Eindhoven.
18. K. Leiter, K. Stephan, E. Mark, and T. D. Mark, 1984, Plasma Chem.and Plasma Proc., 4:235.
19. M. Hayashi and T. Nimura, 1983, J. Appl. Phys., 54:4879.
20. H. Deutsch, P. Scheier, and T. D. Mark, 1986, Int. J. Mass Spect., 74:81.
21. F. A. Stevie and M. J. Vasile, 1981, J. Chem. Phys., 74: 5106.
22. D. K. Davies, 1982, "Measurements of swarm parameters in chlorine-bearing molecules," Report No. AFWAL-TR-82-2083, Air Force Wright Aeronautical Lab., Wright-Patterson Air Force Base, Ohio.
23. G. L. Rogoff, J. M. Kramer, and R. B. Piejak, 1986, IEEE Trans. Plasma Sci., PS-14:103.
24. A. V. Phelps and R. J. VanBrunt, 1988, J. Appl. Phys., 64:4269.
25. M.Hayashi and T. Nimura, 1984, J. Phys. D, 17:2215.
26. Y. Ohmori, M. Shimozuma and H. Tagashira, 1986, J. Phys. D, 19:1029.

27. J. Perrin, J. P. M. Schmitt, G. De Rosny, B. Drevillon, J. Huc, and A. Lloreyt, 1982, Chem. Phys., 73:383.
28. A. Garscadden, G. L. Duke and W. F. Bailey, Appl. Phys. Lett., 43:1012.
29. H. Chatham, D. Hils, R. Robertson and A. Gallagher, J. Chem. Phys., 81:1770.
30. Y. Ohmori, K. Kitamori, M. Shimozuma and H. Tagashira, 1986, J. Phys. D, 19:437.
31. D. K. Davies, L. E. Kline and W. E. Bies, 1989, J. Appl. Phys., 65:3311.
32. K. Tachibana, 1986, Phys. Rev. A, 34:1007.
33. Y. Nakamura and M. Kurachi, 1988, J. Phys. D, 21:718.

DISCUSSION

R. J. VAN BRUNT: Do you have the E/N dependences of the rate coefficients for electron–impact dissociation of N_2O into the various channels that work best in your model in accounting for observations of product distributions?

L. E. KLINE: The E/N value used in the model is 240 Td. This E/N value is the time– and space–averaged value of E/N given by $(2/\pi)V_p/Nd$ and where V_p is the peak value of the applied 13.56 MHz RF voltage, $N = 1.6 \times 10^{16}$ cm^{-3} is the gas density and $d(= 2$ cm$)$ is the electrode spacing. The E/N dependence of the rate coefficient for neutral dissociation is similar to the E/N dependence of the ionization rate, but with a threshold at a lower E/N value.

A. GARSCADDEN: What about surface effects?

L. E. KLINE: Gas phase recombination is not very important. Surface recombination is not important except in maintaining the N_2O balance.

I. GALLIMBERTI: Which is the fraction of the total energy given to the gas which goes into dissociation? Or, if you prefer, how many eV do I have to expend for each dissociation event?

L. E. KLINE: The threshold for the N_2O neutral dissociation process described in the talk is 5 eV, based on measured energy–loss spectra. About half of the discharge input energy is lost in this neutral dissociation process at the time– and space–averaged value from the experiments which is 240 Td. The details of the N_2O RF discharge experiments and modeling will be published in the IEEE Transactions on Plasma Science in April, 1991.

THE CALCULATION OF LEADER PROPAGATION IN POINT/PLANE GAPS

UNDER VERY FAST TRANSIENT STRESS

Heinrich Hiesinger

High Voltage Institute
Technical University Munich
Federal Republic of Germany

ABSTRACT

The discharge development in SF_6 in case of strong inhomogeneous fields under Very Fast Transient (VFT) stress was investigated. In addition to voltage and current signals the light emission and visual appearance of the breakdown was recorded by means of different photomultipliers and an image converter camera with a digital image processing system. Based on these data and with consideration to existing models a description for the VFT breakdown mechanism was found. A program was developed which allows the calculation of leader propagation from inception to the final breakdown. The main steps of the program are described. Computed and measured data are compared for different experimental conditions.

INTRODUCTION

Gas insulated switchgears have shown good reliability in long time operation. The quality requirements are essential influenced by special problems, which arise in the presence of strong imperfections and moving particles in case of disconnector induced VFT. For pressures in the range from 0.07 MPa to 0.3 MPa the lowest breakdown values were obtained under VFT conditions (Hiesinger and Witzmann, 1988; Hiesinger, 1989). Nearly equal voltage levels for VFT and lightning impulse (LI) stress were measured for 0.45 MPa (Riquel et al., 1989). These investigations made the main parameters and their influence on the insulation withstand level evident. In addition with optical methods the details of discharge development can be observed. The experimental results obtained with a high sensitive image converter camera and different photomultiplier tubes (PMT) will be reported.

A detailed description of the leader mechanism in SF_6 for standard impulse voltages was given by Niemeyer (1985), Niemeyer et al. (1989). The physical data were collected by Chalmers et al. (1987). The fundamental knowledge of these work in addition with new experimental data for VFT breakdown enables the calculation of discharge development under VFT conditions. Measured VFT waveforms and the gap geometry are the necessary input data for the modeling. Essential parts of the program will be described. The result may be either the detailed leader development step by step from inception to breakdown or a complete voltage-time characteristic. Computed and measured data are compared for different gap geometries and pressures.

Gaseous Dielectrics VI, Edited by L.G. Christophorou and
I. Sauers, Plenum Press, New York, 1991

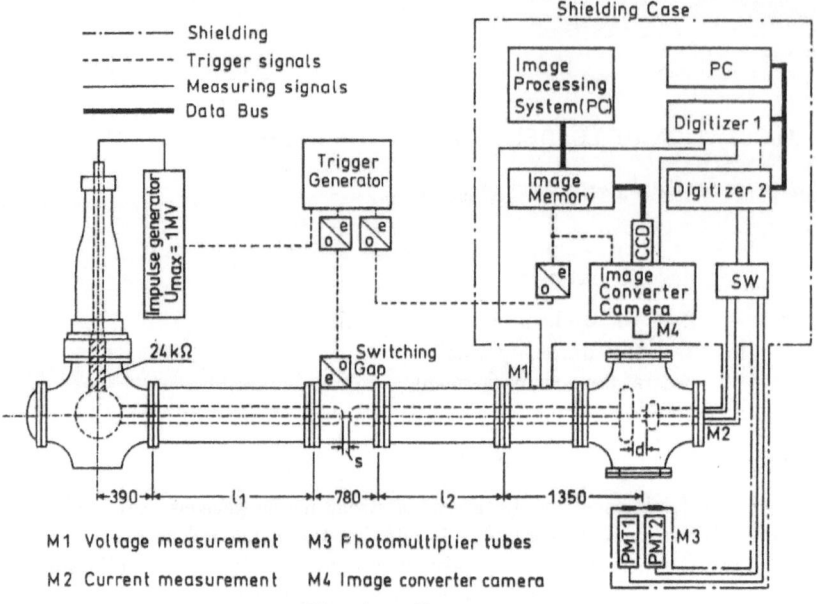

─··─··─	Shielding
───────	Trigger signals
─────────	Measuring signals
━━━━━━	Data Bus

Fig. 1. Test setup.

M1 Voltage measurement   M3 Photomultiplier tubes
M2 Current measurement   M4 Image converter camera

EXPERIMENTS

The test setup consists of 420 kV GIS components (Fig. 1). A variable spark gap separates the bus duct into a source and load side. The triggerable switching gap initiates the VFT inside the GIS. The length of source and load side is variable. Therefore frequencies of the resulting VFT main component from 10 to 45 MHz can be generated. At the open end of the bus duct a rod/plane gap is installed with the plane on negative high potential. Several measuring points (M1-M4) enable the recording of voltage, current signals and the light emission of the discharge. The current measuring system includes a compensation circuit for the high displacement current (Hiesinger, 1989). The photomultiplier tubes (PMT) have different spectral sensitivities. PMT1 is sensitive in the UV range from 180-300 nm. PMT2 was used with a 680-880 nm transmission filter or a 694 nm interference filter. A time-resolved picture of discharge development could be obtained with the image converter camera including an image intensifier. The system is able to record the first corona onset in technical pure SF_6 without additions of nitrogen. The high dynamic of the light emission from corona onset to the bright leader development is well handled by the 12 bit resolution per pixel of the cooled CCD-chip and the subsequent image processing system.

An example for a streak record and the corresponding PMT-signals is given in Fig. 2. The UV-PMT clearly detects the corona onset. Later on each light emission produces significant signal pulses with an amplitude depending on the brightness of the event. Thus the UV signal gives a complete information about the light intensity over time. A separation into different discharge steps is not possible. The time between two leader reilluminations was about 6 ns for a leader length of 10 mm. As a consequence the information content of the UV signal is very high and unsuitable for an analysis of the leader development. Although the IR-PMT is insensitive for the first corona pulses, the advantage of the IR signal is the different behavior for leader steps and leader reilluminations. In case of a new leader step (marked with s in Fig. 2) the IR-signal shows a significant pulse with high amplitude due to the strong fluorine lines in the range from 685-691 nm for dissociated SF_6. The later leader reilluminations are also detected but their amplitude is much lower. Therefore the IR-PMT is a valuable tool to separate the high frequent reilluminations from the stepwise leader propagation.

The criterion for a new leader step is always a bright event at the streamer head of the leader (Fig. 3), which occurs in majority in the steep rising front of the VFT pulses. These characteris-

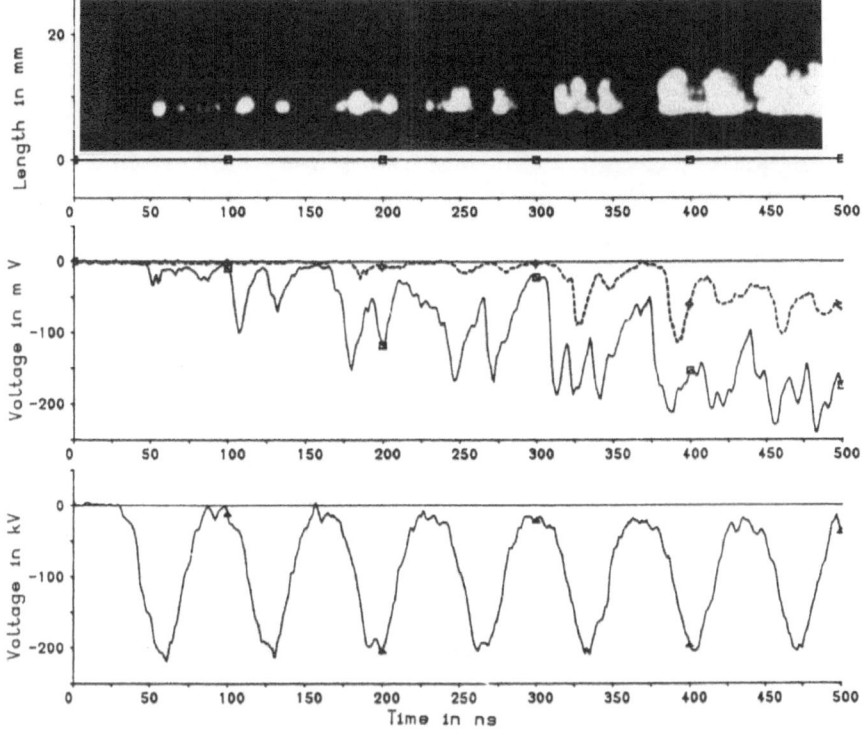

Fig. 2. Streak record with UV (solid line), IR signals (dotted line) and VFT waveshape.

tic yields the fundamental point for the modeling. The following reilluminations have the same channel length. An especially bright reillumination is visible in the steep falling part of the VFT waveshape. The leader stepping occurs very regularly. The time between two steps corresponds with the fundamental frequency of the VFT. For extreme high voltage amplitudes more than one leader step can be observed for each VFT peak. The leader grows about 2-3 mm each step with higher values for the final steps. Therefore a lot of steps are necessary to bridge the whole gap. For a gap distance d = 85 mm about 25-40 steps gives a typical number.

MODELING OF LEADER PROPAGATION

The discharge development starts if the applied voltage exceeds the critical voltage. The corona onset at the tip of the needle is influenced by the first electron statistic, which is included in the model. The corona diameter $D(t)$ and the corresponding capacitance $C(t)$ are time-dependent as a result of the fast oscillating VFT (Fig. 4a). The corona region is characterized by a nearly constant field strength of about 89 kV / (mmMPa) inside it. The corona diameter can

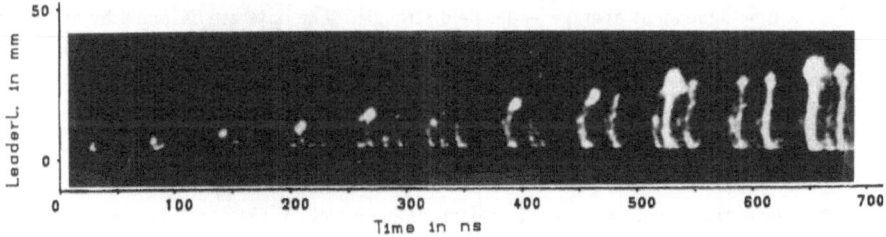

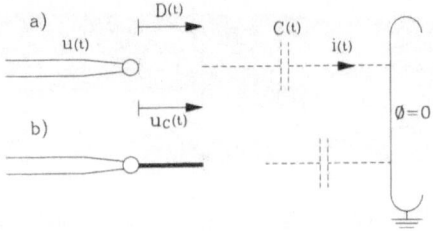

Fig. 4. Schematic representation of discharge development in SF_6.

be estimated by calculating the intersection of internal and external potential curve for each time step. The capacitance $C(t)$ is gained by a set of field calculations using the Charge Simulation Method. The fast oscillating VFT $u(t)$ and time-dependent capacitance $C(t)$ produce a high displacement current $i(t)$.

$$i(t) = C(t) \cdot \frac{du(t)}{dt} + u(t) \cdot \frac{dC(t)}{dt}$$

Both displacement current $i(t)$ and voltage drop $u_C(t)$ lead to a significant energy input into the corona region.

$$W(t) = \int |(u_C(t) \cdot i(t))| dt$$

If the enthalpy rise Δh exceeds $10 \times 10^6 \, J/kg$ the gas is completely dissociated and heated up to about 2100 K (Niemeyer and Pinnekamp, 1983). The corona region is now bridged by a leader channel with a much smaller internal field strength.

The modified field arrangement with the actual leader is included in a new field calculation. The program needs the leader radius at its tip as input data, which is not available. Only data about the leader channel radius near the electrode are measured, obtained with the Schlieren technique (Chalmers et al., 1987). The leader radius at the tip can be estimated from two limiting cases. The initial radius should not be smaller than the positive streamer radius, which is about $50 \, \mu m$ for 0.1 MPa (Niemeyer et al., 1989). As an upper boundary the channel radius for a fully developed leader can be used. For 0.1 MPa Chalmers et al. (1987) gave a value of $250 \, \mu m$ after $3 \, \mu s$ propagation time. In our case the time between two leader steps was not longer than 100 ns, therefore we choose a leader radius in the range from $100 - 170 \, \mu m$ for the calculation depending on pressure. Streamer development starts again at the tip of the leader and creates spherical ionized regions (Fig. 4b). The same threshold energy is necessary to trigger the next leader step. A stepwise leader propagation can be observed.

For an accurate modeling of leader propagation the time-dependent leader channel radius must be considered. An increasing leader channel radius leads to a reduced current density inside the channel. As a consequence the voltage drop along the leader decreases. This behavior can be modeled by a time-dependent average leader field strength. The data are obtained by fitting the experimental with the calculated results for different configurations (Fig. 5). For higher pressures the channel radius becomes smaller and the average leader field strength reaches higher values which must be considered in the model.

COMPARISON OF MEASURED AND CALCULATED V-T CURVES

The accuracy of the model was tested by calculating complete V-t curves. The necessary input data are VFT waveshapes, obtained by measurements or calculation, and the geometry of the rod/plane gap for the field calculation. The oscillation frequency of the VFT main component

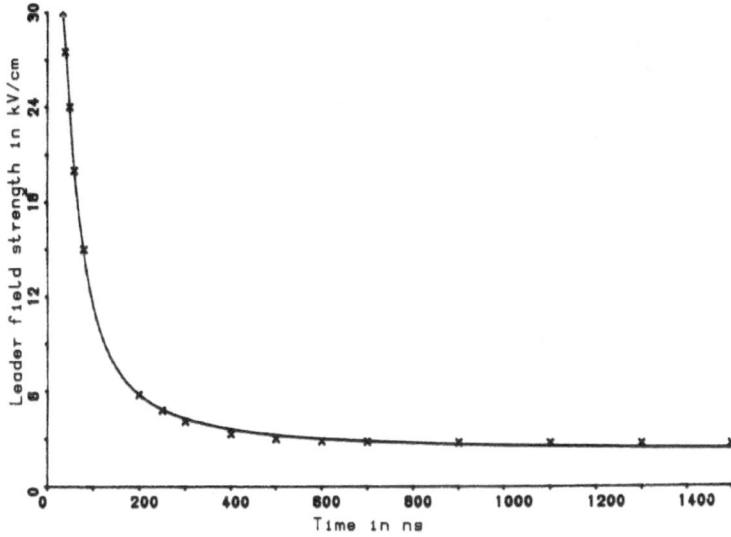

Fig. 5. Average leader field strength versus time for 0.1 MPa.

was 15 MHz. The remaining gap distance is given by (100 mm - rod length l). Fig. 6 shows the results for two different protrusion lengths. The field enhancement and faster leader propagation for the longer rod which results in lower breakdown voltages is well treated by the included field calculation. A reduced leader channel radius (Chalmers, 1986) and an increased average leader field strength must be considered for higher pressures. In this case the pressure dependence of the breakdown voltages is in good agreement with measured values (Fig. 7).

CONCLUSION

The breakdown behavior of SF_6 in case of VFT stress is mainly influenced by the protrusion geometry, gas pressure and the VFT waveshape, especially its steepness, oscillation amplitude

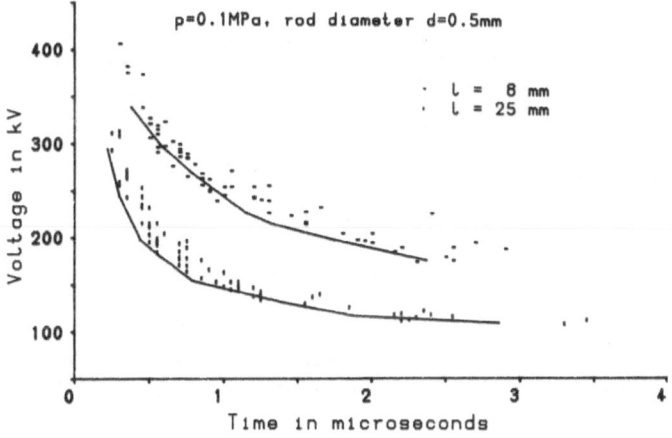

Fig. 6. V-t curves for different rod lengths (calculation: solid lines).

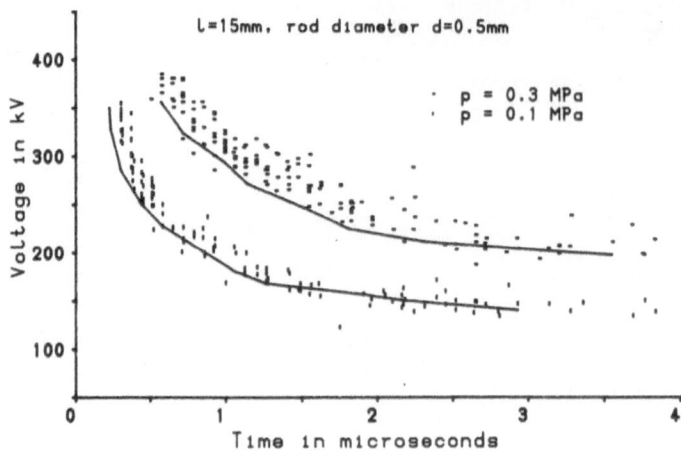

Fig. 7. V-t curves for different pressures (calculation: solid lines).

and frequency content. For pressures from 0.1-0.3 MPa the breakdown values under VFT conditions are lower than the level obtained with lightning impulse testing. Riquel at al. (1989) present LI as the most severe voltage shape, although they mentioned equal insulation withstand level for LI and VFT in case of extremely small rod diameter. The different experimental results can be explained by the different VFT waveshapes. The VFT in the second case is the result of a LC-circuit, which is smooth compared with real VFT. The maximum oscillation amplitude of the VFT is only 70% of the peak voltage which prevents time-to-breakdown longer than 1 μs. In Fig 6 7 it is clearly demonstrated that a further reduction of breakdown voltages can be observed for longer times and weakly damped VFT.

The application of an extensive field calculation is necessary for an acceptable estimation of breakdown levels. An energy criterion in the streamer head combined with a time-dependent average leader field strength was used in the model. Calculated and measured data are in good agreement for various electrode arrangements and different pressures.

REFERENCES

Chalmers, I.D., Gallimberti, I., Gibert, A. and Farish, O., 1986, The development of electrical leader discharges in a point-plane gap in SF$_6$. Proc. R. Soc. London A 412, p. 285-308.
Hiesinger, H. and Witzmann, R., 1988, Very Fast Transient Breakdown at a Needle Shaped Protrusion. 9. Int. Conf. on Gas Discharges and their Applications, Venedig, p. 323–326.
Hiesinger, H., 1989, Statistical time lag in case of very fast transient breakdown.
6. ISH, New Orleans, No. 32.23
Niemeyer, L., Pinnekamp, F., 1982, Leader discharges in SF$_6$.
J. Physics D: Appl. Physics, 16, p. 1031-1045.
Niemeyer, L., 1985, A Model of SF$_6$ Leader Channel Development. 8th Int. Conf. on Gas Discharges and their Applications, Oxford, England, p. 223-226.
Niemeyer, L., Ullrich, L. and Wiegart, N., 1989, The Mechanism of Leader Breakdown in Electronegative Gases. IEEE Trans., Vol.24, No.2, p. 309-324.
Riquel, G., Ren, Z.Y. and Lefrancios, L., 1989, Comparison between VFT and lightning impulse breakdown voltages for GIS insulation in presence of defects on live-conductors.
6. ISH, New Orleans, No. 23.09.

DISCUSSION

I. GALLIMBERTI: In your streak photographs it is possible to observe in some cases (e.g. between the first corona and the first leader step) bright spots which could be associated with the formation of a "precursor" channel. Do you have evidence of the competition between the two possible leader mechanisms (displacement current heating and precursor formation and development)?

H. HIESINGER: I agree with you that the weak reilluminations between the first corona and the first leader step are similar to a precursor channel, but only the first leader step is influenced by this mechanism. For the second and later steps these phenomena could not be observed. If you take into consideration that many steps are necessary to cross the gap, the delay of the first step is not evident for the total time to breakdown. This statement was proven for VFT frequencies between 8 and 45 MHz. Up to now I can not say what will happen for frequencies from 2–5 MHz. I think we will observe a mixture of the different mechanisms.

I. GALLIMBERTI: The fact that the stepping times appear to be related to the oscillation frequency of the applied voltage seems to indicate that the displacement heating of the corona filaments is the dominant leader formation mechanism; in fact the precursor formation is essentially related to ion drift within the corona filaments, and therefore is not directly related to the times of the voltage oscillations.

C. M. COOKE: In your modeling you have included a current component from a time varying calculated capacitance. Could you give typical values for that calculated capacitance and its associated current component? Because the system is modelled from a distributed structure there may also be an influence of the channel inductance. Have you included such a series inductance in your model and if so under what conditions does the inductance influence the model results?

H. HIESINGER: For a gap distance $d = 85$ mm and a rod radius $r = 0.25$ mm a typical value for C is 0.04 pF depending on the corona diameter. The current amplitudes are in the range from 1–3 A. A leader channel inductance is not included in the program.

L. NIEMEYER: Which is the volume of the corona into which the displacement current power imput is considered to be deposited in the theoretical model?

H. HIESINGER: The volume for the energy input are the dotted spheres drawn in Fig. 4 (Schematic model). The corona sphere is growing and decreasing according to the applied VFT voltage.

FRACTAL DESCRIPTION OF ELECTRICAL DISCHARGES

L. Egiziano, N. Femia

Inst. of Electronic Eng.
Univ. of Salerno
via Allende,
Baronissi (SA), Italy

G. Lupó, V. Tucci

Dept. of Electrical Eng.
Univ. of Naples
via Claudio, Napoli, Italy

INTRODUCTION

The discharge propagation at the interface between gaseous and solid dielectrics exhibits different complex mechanisms depending not only on the physical properties of the investigated materials but also on the characteristics of the applied electrical stress. A full understanding of the discharge mechanism has not been achieved due to the not yet complete knowledge of the interface physics. A rather accurate description of the physical processes taking place during a surface discharge would, in fact, require the solution of a complicated coupled system composed of Poisson and transport equations together with suitable "constitutive" relations.

Several approaches have been proposed to investigate the discharges occurring along the surface of solid insulating materials (the well–known *Lichtenberg Figures*); among them the dust figure method, the photographic plate method and the use of the image–converter can be considered. So far the experimental researches in this field have only put in evidence the differences between patterns relative to positive and negative discharges and their topological characteristics as a function of the gas pressure and some semi–empirical relationships between global averaged quantities (like the maximum length of the Lichtenberg figures and the characteristics of the applied electrical stress). Through such relationships some effective, although not refined, diagnostics, which are useful in operating conditions, can be obtained.

However the results of a recent research by Kawashima et al.[1] have put in evidence that the photographed discharge patterns do not show the real discharge structure since the sensitivity and processing of the film and the photographic techniques may reduce the image resolution level. In particular, the effective area involved in the discharge is much greater than that appearing on photographic records and can be suitably reconstructed.

On the other hand, the discharge patterns resulting from experiments carried out under different conditions show some common structural characteristics which recently have stimulated studies aimed at characterizing this aspect of the phenomenon. Such studies have profited from the concepts of *fractal geometry* firstly introduced by Mandelbrot.[2] In fact, the topology of the filamentary "self–similar" structures, which are typical of stepped electrical discharges, can be suitably described by means of fractal parameters.

In this paper are reported results of a numerical simulation procedure whose output are surface discharge patterns in a point–plane geometry exhibiting fractal characteristics. The growth process of the discharge is obtained by iteratively solving the Laplace equation in the considered region with moving boundary conditions and on the basis of two laws of probability depending respectively on the position and the electrical field intensity.

Niemeyer, Pietronero and Weismann[3] (NPW) in 1984 have for the first time proposed such a *fractal model for dielectric breakdown*. They have shown that such approach leads to fractal structures which can be characterized in terms of *fractal dimension D*.

Although our work shares with the NPW model the same basic procedure for generating the discharge pattern it presents the following distinctive aspects:

a) the surface discharge pattern is obtained by solving a three–dimensional field problem instead of a two–dimensional one;

b) two stochastic (power) laws are introduced in order to obtain a better simulation of the discharge patterns pertaining to different experimental conditions;

c) a filtering procedure is considered in order to put in evidence the effect of the photographic processing on the discharge pattern;

d) another characteristic parameter in addition to the fractal dimension is introduced in order to take into account the transition of the discharge from a *homogeneous* structure to a *dendritic* pattern.

The structures obtained from our simulation compare very well with physical realizations of discharges of positive (point) polarity. Moreover, some relationships between parameters pertaining to the simulation and relevant physical characteristics are put in evidence. The dependence of the topologies obtained from the numerical simulations in relation with the exponents of the growth laws governing the structure dynamics are also discussed.

Moreover, the flexibility of the adopted algorithm allows also the study of other physical processes such as, for example, the treeing in polymers.

DESCRIPTION OF THE METHOD

In Fig. 1 the point–plane configuration (according to Toepler[4]) considered for the numerical simulation is reported. The electrical field distribution in the different regions is obtained by solving the Laplace equation with the appropriate boundary conditions through the finite difference method (FDM). The discharge pattern is obtained by adding at each step of the procedure a certain number of branches connecting two nodes of the adopted mesh. It is supposed that the branches which are dynamically added to the discharge pattern switch from a zero to an infinite conductivity. As mentioned in the previous section, according to the setup depicted in Fig. 1, a 3D field problem with moving boundary conditions is considered.

The propagation of the discharge starts from the point electrode (hv electrode) and is modulated by two probabilistic laws. During the n–th step of the procedure at each candidate point P, belonging to the (moving) boundary, from which a new branch can have origin, a probability value is assigned as follows:

$$p_1(P) = [r(P)/R_{n-1}]^{\alpha} \qquad (1)$$

where R_{n-1} is the maximum distance from the point electrode reached by the discharge at the (n–1)–th step and $r(P)$ is the distance of P from the point electrode; the exponent α is a real positive number whose influence upon the discharge pattern development will be discussed in the next section. Only those points whose values p_1

satisfy the relation $p^* < p_1$, where p^* is a random variable, are considered for the successive elaborations.

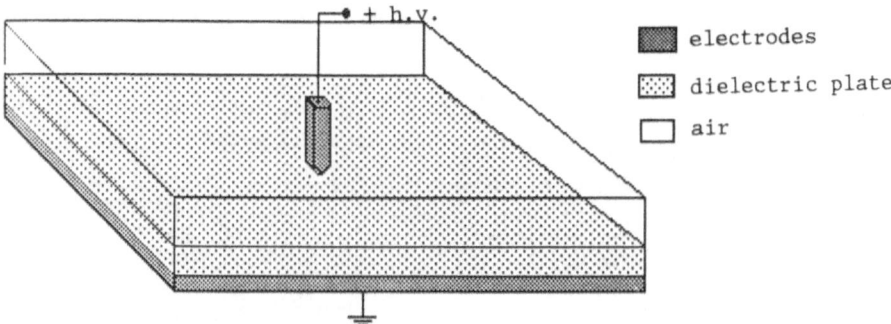

Fig. 1. Configuration considered for the numerical simulation.

Moreover, it is checked if among all possible candidate bonds starting from P there exists one (P,P′) for which the electrical field component along it satisfies the following condition:

$$E(P,P') > E_{crit} \qquad (2)$$

where E_{crit} is a critical value of the electric field. Among all points that fulfill the previous conditions, the growth direction is obtained on the basis of a probabilistic law depending on the electrical field:

$$p_2(P,P') = [E(P,P') / \Sigma\, E(P,P')]^{\beta} \qquad (3)$$

where the sum is performed on all points that can be reached from P through a single bond step (also diagonally) and β is a real positive number modulating the discharge pattern development. It can be noted that this procedure takes into account both the growth dependence on the electrical field and on the position.

RESULTS AND DISCUSSION

In Fig. 2a a typical output of the numerical procedure obtained with $\alpha=\beta=0.5$ is presented. It shows a very good resemblance with physical realizations of surface discharge, as proposed by Kawashima et al.[1] using image processing. In analogy with the experimental results, in the simulated pattern a distinction in *main* (thicker) and *secondary* (thinner) branches can be done. The thickness of a branch starting from the node P is determined by the number of the nodes pertaining to the descending sub–tree starting from P. Moreover, in Fig. 2b a filtered version of the Fig. 2a is reported. Such image, obtained by eliminating the secondary branches (i.e., those with a number of the nodes of the descending sub–tree lower than a suitable value), reproduces with good accuracy the well–known photographic records of surface discharges (Lichtenberg figures). A process conceptually inverse to the image reconstruction can be performed in this way; its accuracy is obviously dependent on the dimensions of the chosen grid.

By acting on the values of the exponents α and β of the two probabilistic laws, different patterns can be obtained, simulating diverse physical realizations: Figs. 3 and 4 show some typical results. From the analysis of such figures the different influence of α and β can be recognized. In particular, Fig. 3 shows that α determines the *global pattern aspect* in terms of the number of main branches; for $\alpha < 1.0$ the patterns exhibit a *radial symmetry*, whereas for $\alpha > 1.0$ only a few main branches can be noted. The exponent β, on the other side, shows its influence only for $\alpha > 1.0$; under this condition, as shown in Fig. 4, β determines the *side growth level* around the main branches. A correspondence between such influence of α and β on the

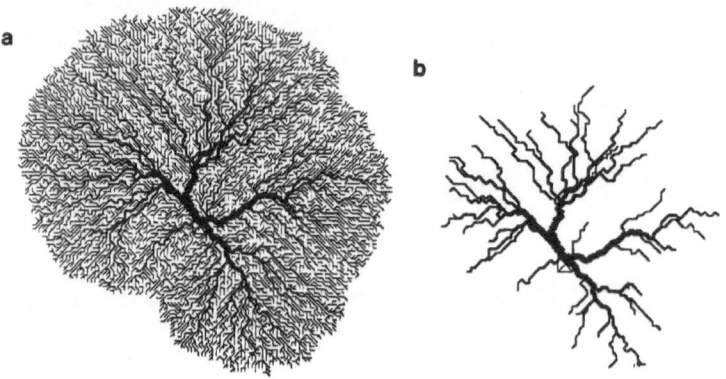

Fig. 2.   Simulated discharge obtained with $\alpha=\beta=0.5$:   a) complete pattern; b) filtered pattern.

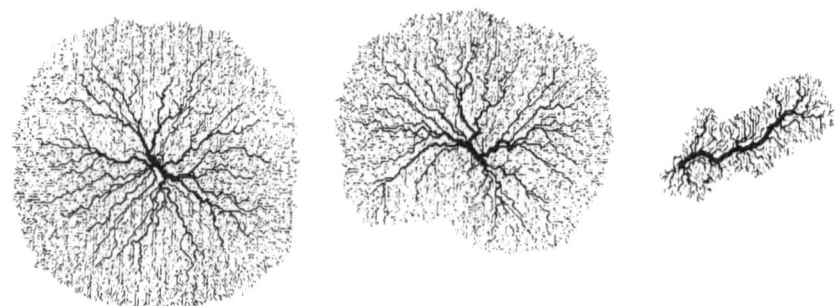

Fig. 3.   Topological variations in the discharge pattern for $\beta=1.4$ and (from left to right) $\alpha=0.1, 1.0, 2.0$.

topological structure of the simulated patterns and that of some physical parameters on the development of real discharges can be attempted. In fact, α can be related to the electrical stress characteristics (i.e., peak value or dV/dt of the applied impulse), whereas β may take into account the physical environmental conditions (such as gas pressure, surface contamination, etc.).

For the obtained patterns a set of significant parameters have been calculated. In Table 1 the ratio between the *maximum radius* R reached by the discharge and the *number of branches* N is reported for several values of α and β. Such ratio can be interpreted as the *advancement speed* of the simulated discharge. Appreciable variations are evidenced by such ratio with respect to α according to the preceding conjecture.

140

Table 2 shows the values of the ratio L/R, where L is the *global pattern length*, representing the *ramification level*; in fact, the greater its value, the higher arborescence of the structure. Moreover, in order to individuate the fractal parameters of the pattern an analysis of the scaling properties has been conducted. In Fig. 5 the plots of the total length L(r) of the pattern branches, contained within a square of side 2r, vs r are reported for two pairs of α and β values. They represent the two typical behaviors that can be obtained for $\alpha < 1.0$ (Fig. 5a) and for $\alpha > 1.0$ (Fig. 5b). In Fig. 5a a single interpolating line can be determined, whereas the experimental points of Fig. 5b are interpolated by two lines with different slopes. The slope of the line m represents the fractal dimension D_c of the structure, typical of a whole *class* of phenomena (*diffusion limited aggregation*) i.e., characterized by the same topological structure. The slope of the line n gives a second relevant topological parameter, D_s,

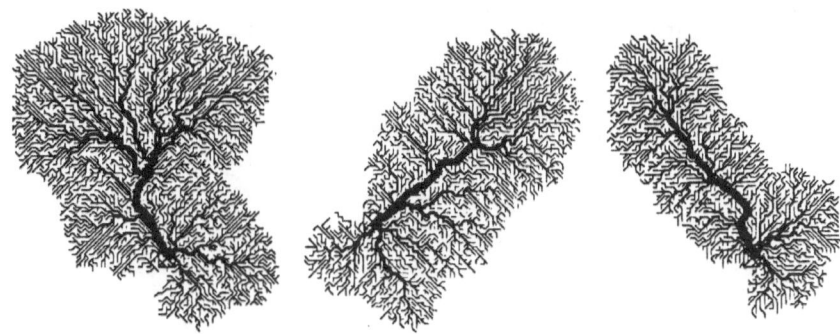

Fig. 4.   Topological variations in the discharge pattern for $\alpha=1.4$ and (from left to right) $\beta=0.1, 1.0, 2.0$.

which takes into account the ramification level of the pattern, typical of the *species* of the phenomenon. It can be observed (see Fig. 6a) that D_c strongly depends on the values of α rather than on those of β; furthermore, for $\alpha > 1.0$ the variations are much more appreciable than for $\alpha < 1.0$, and so in accordance with the transition from *uniform* to *dendritic* structures. In Fig. 6b the strongest variations of D_s are exhibited for $1.0 < \alpha < 1.6$ and for $0.1 < \beta < 0.5$. In general, D_s shows percent variations with respect to β higher than those of D_c. The preceding results seem to enable a significant correlation between the global topology (D_c) of the patterns and the exponent α (i.e., the characteristics of the applied electrical stress) from one side, and between the local topological aspect (D_s) and the exponent β (i.e., the environmental physical parameters).

Table 1. *The discharge advancement speed* R/N (x10^{-3} mm) for several values of α and β.

α \ β	0.10	0.50	1.00	1.40	2.00
0.10	1.66	1.62	1.58	1.58	1.55
0.50	1.72	1.96	1.74	1.98	1.66
1.00	2.68	2.88	2.63	2.49	2.77
1.40	5.25	6.00	6.26	6.05	6.22
2.00	11.19	10.87	9.90	10.87	10.92

Table 2. The *ramification level* L/R for several values of α and β.

α \ β	0.10	0.50	1.00	1.40	2.00
0.10	219	223	227	224	225
0.50	212	186	205	180	209
1.00	136	125	137	142	126
1.40	71	60	57	65	59
2.00	37	34	39	32	31

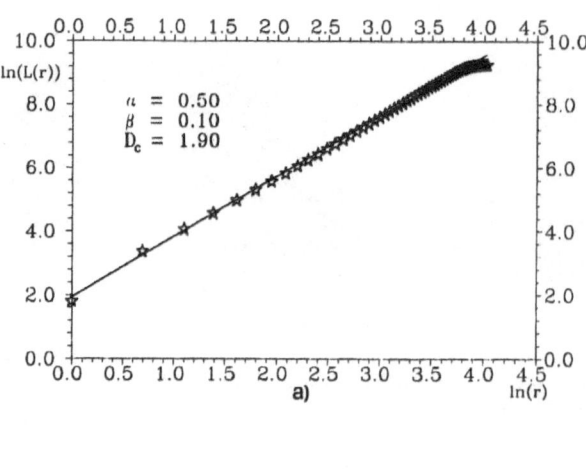

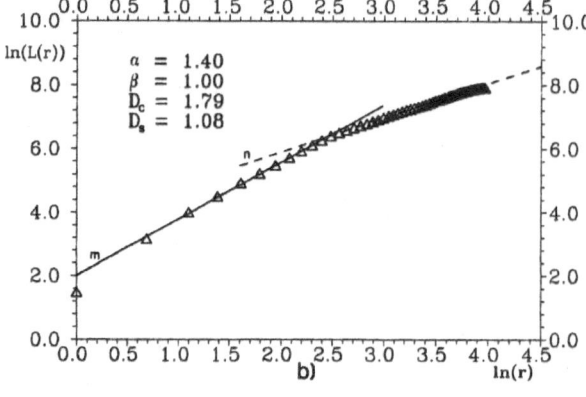

Fig. 5. Scaling characteristics of pattern length: a) homogeneous structure; b) dendritic structure.

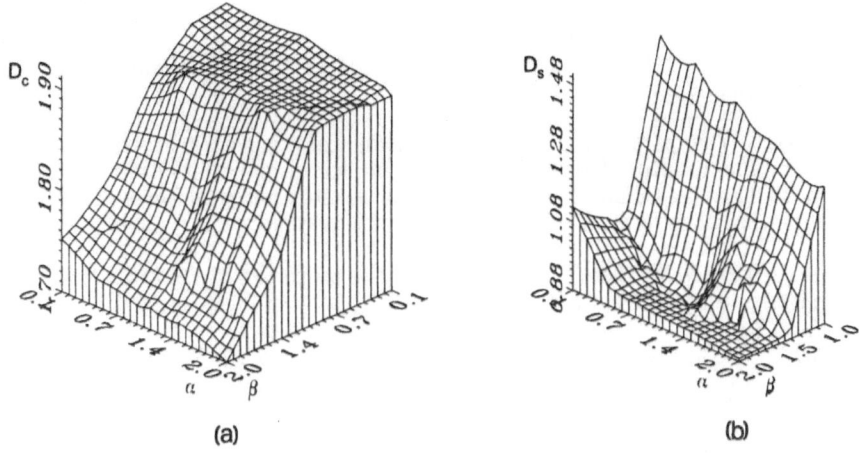

Fig. 6.   Dependence on the exponents α and β of D_c (a) and D_s (b).

CONCLUSIONS

A numerical simulation of surface discharges in a point–plane geometry has been presented in this paper. The discharge patterns obtained show a good accordance with physical realizations. Based on such patterns similarities a correlation study between the exponents of the probabilistic laws adopted in our simulations and some relevant physical parameters such as the electrical stress characteristics, the environmental operating conditions, etc., has been performed. This topological approach allows to obtain, also without a deep knowledge of the underlying physical mechanisms, the general properties of the experimental realizations. The algorithm adopted is characterized by a good flexibility and a relatively low execution time.

REFERENCES

1. A. Kawashima, S. Shoda, T. Momma, K. Kubota, Image Analyzing of Surface Discharges, VI Int. Symp. on High Voltage Eng., New Orleans, LA, paper 22–17 (1989).

2. B. Mandelbrot, "The Fractal Geometry of Nature," W. H. Freeman, New York, (1982).

3. L. Niemeyer, L. Pietronero, H. J. Weismann, Fractal Dimension of Dielectric Breakdown, Phys. Rev. Lett., 52:1033 (1984).

4. M. Toepler, Uber die Physikalischen Grundgesetze der in der Isolatorentechnik Auftretenden Elektrischen Gleiterscheinungen, Archiv. für Elektrot., 5/6:157 (1921).

DISCUSSION

L. NIEMEYER: Has the parameter α been introduced to limit the radial extension of the simulation, i.e. to create a finite pattern?

N. FEMIA: The parameter α has been introduced to obtain several discharge topologies. The resulting figures show that α influences the radial extension of the simulations for a fixed number of iterations, but the stop has been imposed to all the discharges when they reached a fixed maximum observation radius.

A. GARSCADDEN: Is there a way to bring in the properties of the gas? How can I distinguish between different gases?

N. FEMIA: We are trying to do laboratory experiments to compare with different β parameters which may take into account the nature of the gas or the corona parameters. Also an exponent could be introduced to take into account the different physical parameters.

A TWO–DIMENSIONAL SIMULATION OF

A HOLLOW–CATHODE DISCHARGE

W. Niessen and A. J. Davies

Department of Physics, University College of Swansea
Singleton Park, Swansea SA2 8PP, Great Britain

INTRODUCTION

In recent years hollow–cathode discharges in the E/n–range above 500Td (E is the electric field strength and n the gas number density) have gained increasing interest in several fields of research. They have been used as sources of high intensity electron–beams[1], which can carry currents of several kA's, as well as transmitters of X–rays[2], covering a wide spectrum. Their role as ion–beam sources has also been investigated[1]. They have found application as a mechanism for fast switching of high currents at high voltages in pulsed power technology[3] and another possible application is their use as a medium for sputter processes.

The rotationally symmetric diodes in which such discharges are produced, often called pseudospark diodes or thyratrons, consist of an anode–cathode gap of several millimeters in length and a hollow–cathode chamber which is connected with the gap by a hole in the cathode of typically 5mm diameter. Some versions have guard–rings placed between anode and cathode. The working gas is usually a noble gas, hydrogen or nitrogen. Due to the geometry the more or less homogeneous electric field in the anode–cathode gap is much higher than the non–homogeneous field in the hollow–cathode. If the diode is being operated at high voltage (>1000V) and low pressure (<5Torr) most of the electrons become run–away electrons after entering the gap; i.e. they have too much energy to have an appreciable probability of undergoing ionization collisions and the discharge mainly takes place in the hollow–cathode.

Despite the widespread recognition of pseudospark phenomena a thorough understanding, based on a comprehensive and quantitative theory, is still lacking. Due to the relatively complex geometry and the variety of physical processes involved, even quite simplified mathematical models cannot be solved analytically and numerical methods have to be employed. Assuming local equilibrium between charged particles and the electric field, Mittag[4] simulated the spatio–temporal development of a pseudospark in the hollow–cathode region. This approach implies that swarm parameters such as ionization coefficient, drift velocity and diffusion coefficient are uniquely defined by the local value of the electric field and the density of the working gas. Using a Monte Carlo technique, Niessen[5] investigated the non–equilibrium behavior of ions, electrons and fast neutrals at a very high reduced field ($E/n = 3.8 \times 10^6$ Td) in the region between anode and cathode. At a lower E/n (1.5×10^5 Td) in the high–field region the equilibrium approach has been used successfully by Pak and Kushner[6]. They simulated the switching performance

Gaseous Dielectrics VI, Edited by L.G. Christophorou and
I. Sauers, Plenum Press, New York, 1991

of a hydrogen–filled thyratron by applying the equilibrium model over the whole diode region.

This path is also pursued in the present work. The model employed is described in the next section and is applied to the simulation of a nitrogen–filled diode. After a discussion of different breakdown mechanisms, the development of an electron swarm is presented and the focusing effect of the diode geometry is clearly demonstrated.

THE SIMULATION MODEL

Electrons and positive ions are described by their continuity equations:

$$\frac{\partial n_e}{\partial t} + \nabla j_e = \alpha\, j_e \tag{1}$$

$$\frac{\partial n_i}{\partial t} + \nabla j_i = \alpha\, j_e, \tag{2}$$

where n_e and n_i are the number densities, j_e and j_i current densities and α the ionization coefficient. The spatial particle transport is a combined flux of drift, caused by the electric field, and diffusion:

$$j_e = n_e\, u_e - D_e \nabla n_e \tag{3}$$

$$j_i = n_i\, u_i - D_i \nabla n_i, \tag{4}$$

where u_e and u_i are the drift velocities whose direction is parallel to the field. D_e and D_i are the diffusion coefficients. Isotropic diffusion is assumed, i.e. the lateral and longitudinal diffusion coefficients are equal. As already mentioned, α, u_e, u_i, D_e and D_i are functions of the local value of the field and the working gas density only.

The boundary condition for the electrons at the cathode surface is defined by the flux of secondary electrons released by ions and photons hitting the cathode:

$$j_{e,c} = \gamma_i\, j_{i,c} + \gamma_{ph} \int_V \alpha j_e\, g(\mathbf{x}, \mathbf{x}_c)\, dV, \tag{5}$$

where γ_i and γ_{ph} are the secondary electron emission probabilities for ions and photons. αj_e is the source term for electron–ion pair production within the gas and it is also used as source term for photon production. $g(\mathbf{x}, \mathbf{x}_c)$ is a geometric factor, which accounts for the special shape of the diode. It represents the fraction of photons, produced in a unit volume around the point $\mathbf{x}$, falling on a unit area around the cathode surface point $\mathbf{x}_c$. The direction of $j_{e,c}$ is perpendicular to the surface. The whole boundary, including anode, cathode and insulator walls, are deemed to be perfectly absorbing for electrons as well as ions.

The electric field satisfies the Poisson equation

$$\nabla^2 \phi = -\frac{e}{\varepsilon_0}\, (n_i - n_e) \tag{6}$$

with
$$\mathbf{E} = -\nabla \phi, \tag{7}$$

where ϕ is the potential and $\mathbf{E}$ the field. The potential at the electrodes is assumed to be constant in time, i.e. the applied voltage does not change.

Since the diode studied has rotational symmetry (Fig. 1), the two coordinates z and r were used for the specific formulation of Eqns. (1) to (7). They were discretized on a rectangular mesh containing 100×50 points and in order to achieve best resolution the point density increased exponentially towards the axis of rotation and towards the back wall of the hollow-cathode. A simple second-order finite difference scheme was employed to solve Poisson equation (6) with the resulting set of linear finite difference equations being solved by a library routine. The hyperbolic continuity equations (1) and (2) were solved by a version of the so called Upstream-centered Scheme for Conservation Laws[7] (MUSCL).

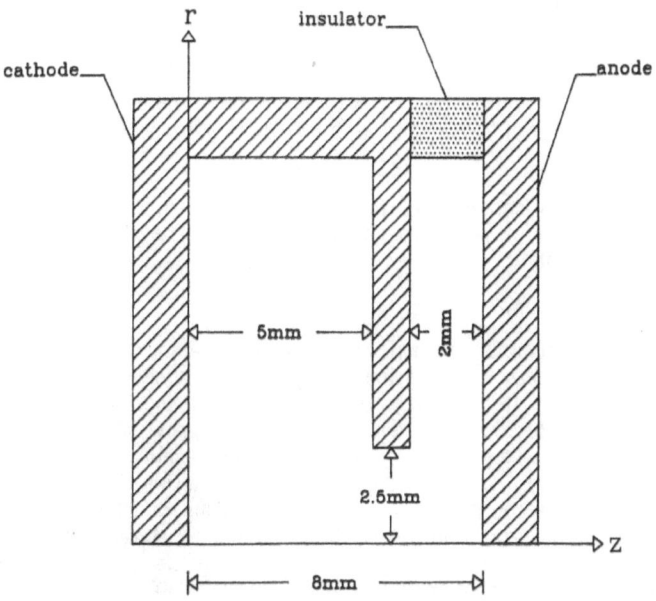

Fig. 1.   Cross section of half of the diode in the z–r plane.

RESULTS

The assumed initial distribution of charged particles consisted of 1500 electrons, released at the back of the hollow-cathode. The lateral and longitudinal spread was Gaussian with a half-width of 0.5mm. The working gas was nitrogen and the applied voltage 4000V. Values of the swarm parameters were taken from publications by Davies and Evans[8], Yousfi et al.[9] and Phelps et al.[10].

Fig. 2 shows the amplification f_a of the avalanche solely due to ionization within the gas at various pressures. The values were obtained by switching off the γ_i and γ_{ph} processes and neglecting the space charges. It can clearly be seen that f_a increases exponentially with pressure. From Fig. 2 one can obtain the pressure above which breakdown will occur, providing the total secondary coefficient γ is known. The general breakdown criterion is then given by

$$\gamma \, (f_a - 1) > 1. \tag{8}$$

This means that f_a must be greater than $1 + 1/\gamma$. Using typical values for γ_i and γ_{ph}[8], which are of the order of 0.05 and 0.002 respectively (i.e. $\gamma = 0.052$), an amplification of at least 20.23 is needed in order to get breakdown. According to Fig. 2 this is achieved at pressures above 160Pa. The corresponding threshold for fast breakdown (built up by photo-electrons released at the cathode) is around 330Pa.

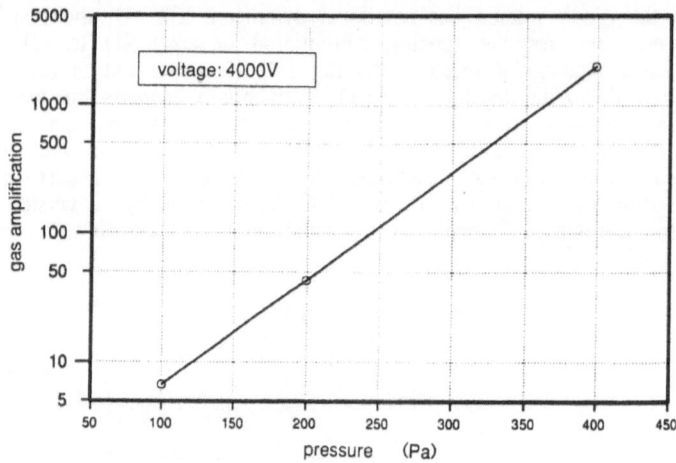

Fig. 2. Amplification f_a of the avalanche due to ionization within the gas as function of pressure; voltage = 4000V. The circles indicate simulated points.

Electron distributions on axis at several instants are shown in Fig. 3. in which secondary processes and space charges are now accounted for. The pressure was 400Pa (≈3Torr). After a decrease of the peak of the initial distribution due to diffusion (from 0ns to 2ns), the densities begin to grow (4ns and 7ns) just inside the hollow cathode where α is greatest and fall again (10ns) as the avalanche passes into the anode. Although the breakdown criterion is met, no signs of the development of the secondary avalanche can be seen at 10ns. This is because the photo–electrons come from the whole hollow–cathode surface so that not only do most of them have to travel a great distance but in addition they are slower because the field is lower in regions remote from the axis.

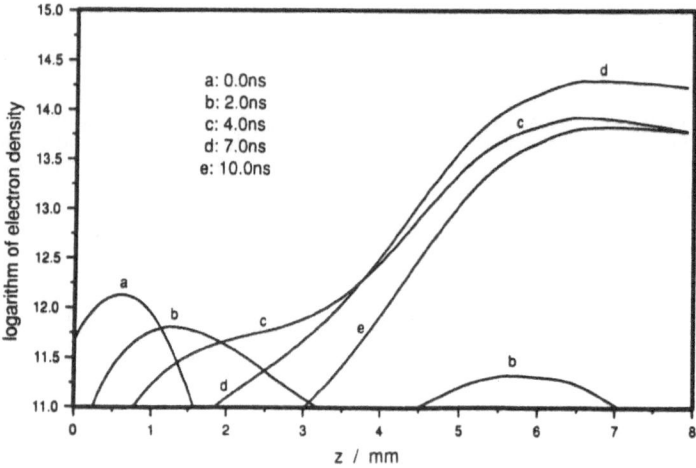

Fig. 3. Logarithm of electron number density (in m^{-3}) on axis at various times. Voltage = 4000V, pressure = 400Pa (≈3Torr).

A two–dimensional view of the swarm development is shown by the contour plots of electron densities in Figs. 4a–4d. The windows cover the whole length of the diode from its cathode back wall to its anode while the radial extension goes to 0.9mm. The times are the same as in Fig. 3. The focusing effect of the cathode hole upon the electrons can be seen if Fig. 4a is compared with 4b–4d. Even though the effect of diffusion is appreciable the relative broad lateral extension of the distribution at 2ns evolves into a narrower one of about half the radius when entering the main gap. Subsequent avalanches are slightly wider because their seed electrons come from all over the hollow–cathode surface and the increasing negative space charge on axis tends to broaden the beam as well. At the stages shown in Figs. 3 and 4a–4d, however, the internal field does not yet have a significant influence. If the voltage is decreased to 1000V the focusing effect is less, but still dominates over diffusion as shown in Fig. 5.

CONCLUSION

The fluid model described in the present work, although not incorporating the effects of non–equilibrium, gives a good qualitative description of a pseudospark discharge. In the hollow–cathode the values of E/n are low enough (< 2000Td) for equilibrium to be justified and in the main gap, where E/n is much higher, the ionization is in any case negligible. Information is given of the spatio–temporal development of the charged particle densities including secondary effects at the cathode surface and space–charge distortion. Further results will be presented in the near future.

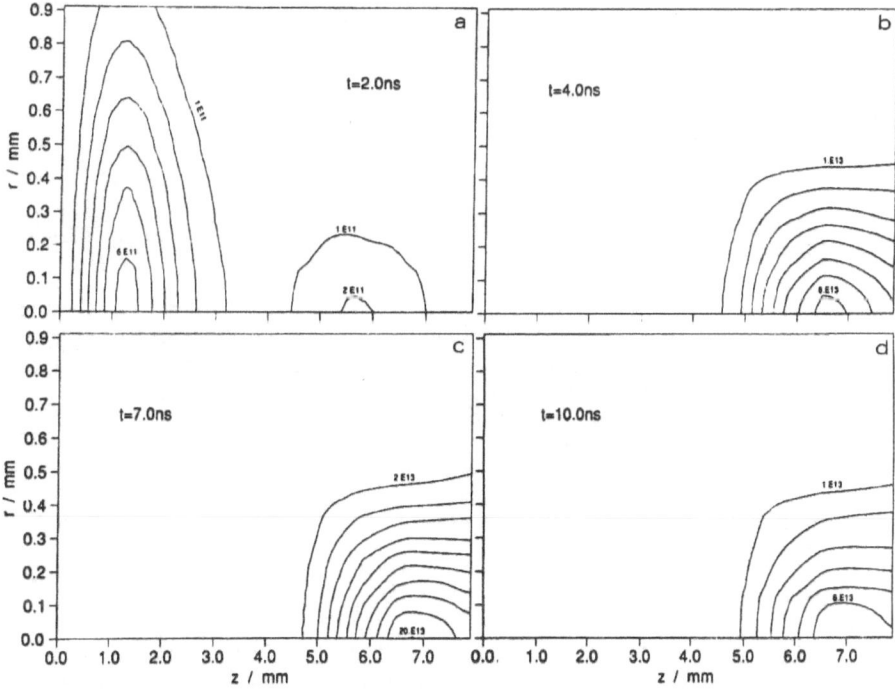

Figs. 4. Contour plots of the electron density at various times. Conditions as in Fig. 3. The differences between two successive contours are 10^{11}m^{-3} in (a), 10^{13}m^{-3} in (b) and (d) and 2×10^{13}m^{-3} in (c).

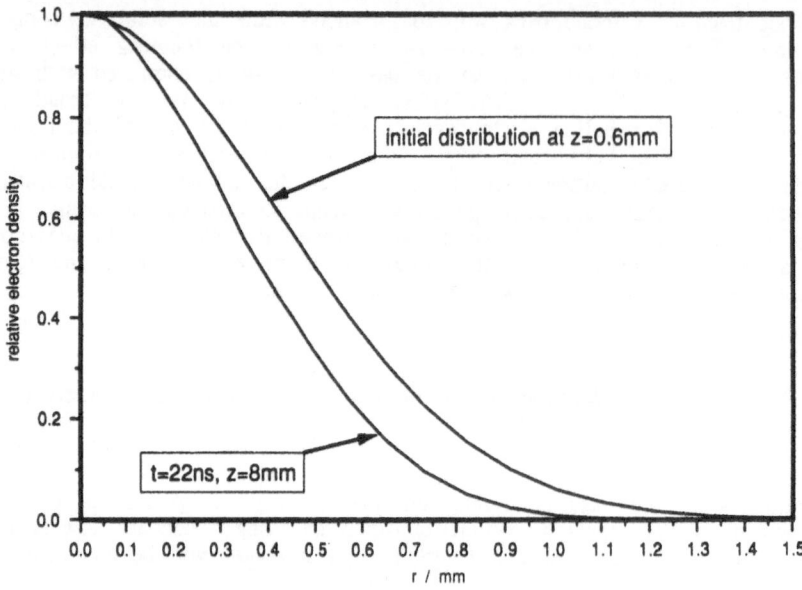

Fig. 5. Relative radial density profiles at initial instant and 22ns. Voltage = 1000V, pressure = 400Pa.

ACKNOWLEDGEMENT

This work was supported by the EC Grant SCI–0055 'Computer modeling of the fundamental mechanisms involved in the operation of pseudospark devices'.

REFERENCES

1.   D. Bloess, I. Kamber, H. Riege, G. Bittner, V. Brueckner, J. Christiansen, K. Frank, W. Hartmann, N. Lieser, C. Schultheiss, R. Seebock, and W. Steudtner, Nuclear Instruments & Methods 205:173 (1983).

2.   G. Jung, A. Kitamura, A. Rogner and c. Schultheiss, IX International Conference on Gas Discharges and Their Applications, page 681, Venice, September 1988.

3.   K. Frank, E. Boggasch, J. Christiansen, A. Goertler, W. Hartmann, C. Kozlik, G. Kirkman, C. Braun, V. Dominic, M. A. Gundersen, H. Riege, and G. Mechtersheimer, IEEE Transactions on Plasma Science 16:317 (1988).

4.   K. Mittag, IX International Conference on Gas Discharges and Their Applications, page 673, Venice, September 1988.

5.   W. Niessen, IX International Conference on Gas Discharges and Their Applications, page 670, Venice, September 1988.

6.   H. Pak and M. J. Kushner, Journal of Applied Physics 66:2325 (1989).

7.   B. Einfeldt and C.–D. Munz, Karlsruhe Nuclear Research Center Report No. 4191 (1987).

8.   A. J. Davies and C. J. Evans, "The Theory Of Ionization Growth in Gases under Pulsed and Static Fields", CERN Yellow Report 73-10, Geneva (1973).

9.   M. Yousfi, N. Azzi, P. Segur, I. Gallimberti, and S. Stangherlin, "Electron–Molecule Collision Cross Sections and Electron Swarm Parameters in Some Atmospheric Gases", European Group on Gas Discharges (1987).

10.   A. V. Phelps, B. M. Jelenkovic, and L. C. Pitchford, Phys. Rev. A 36:5327 (1987).

CHAPTER 4: GAS BREAKDOWN AND ITS RELATION TO VACUUM AND LIQUID BREAKDOWN

SIMILARITIES BETWEEN HIGH ELECTRIC FIELD ELECTRON EMISSION AND CONSEQUENT BREAKDOWN PROCESSES IN COMPRESSED GASES AND VACUO

Albert E. D. Heylen

Department of Electronic and Electrical Engineering
The University
LEEDS, LS2 9JT
England

INTRODUCTION

It is exactly 40 years ago, in 1950, that Trump and co-workers at M.I.T. showed that the electric strength of a 2 cm gap at a gas pressure of almost 30 atmospheres (3 MPa) could be increased from about 600 kV to 900 kV by using a stainless steel cathode instead of an aluminium one. This cathode dependence of the sparking voltage was unexpected at that time as most workers had accepted that near to atmospheric pressure, the sparking voltage becomes less and less influenced by the cathode with increase in gas pressure. Trump writes: 'the cathode is contributing exponentially to the sparkover of the gap by photo-electric, secondary and high field emission and the latter mechanism must be regarded as a probable one even at electric gradients of 20 MV/m since the average field is increased by localised projections and since the surface is necessarily contaminated by relatively low work function material.' Previously in 1928-29, Fowler-Nordheim had established that for electron field emission to occur, a value of 1000 MV/m was required; so a field enhancement factor of $50 \times$ was accepted.

At about the same time, in 1953, the startling experiments by Llewellyn-Jones and co-workers revealed that electron emission of some 10^5 e /s could be achieved with a field as low as 4 MV/m; however these workers sparked their electodes at the rate of 50 per second so that the surface became rough and pitted.

Although Trump et al·observed a field dependence of the sparking voltage in their measurements, later work showed more clearly, notably the graphs issued by GIGRE in 1977, that deviations from Paschen's law occur at an electric field of 30 MV/m for small gaps (1 mm) to 15 MV/m for bigger gaps (50 mm) and this is exactly the field region where our later work of 1985 has shown high electric field electron emission (HEFEE) to commence. It should be pointed out that our work involves deducing emission rates by counting giant pre-breakdown avalanches in the absence of sparking.

Also at that time it came to our attention that conduction leading breakdown occurs in vaccum in that electric field range of 10 to 20 MV/m and so this sets the scene for a comparison between electron emission leading to breakdown in compressed gases and vacuo.

SIMILARITY BETWEEN MOGAM AND MOVAM STRUCTURES IN MORE DETAIL

A crucial matter is that the electric field at which electron emission starts in compressed gases ($\simeq 10^6 Pa$) and vacuo is very similar ($\simeq 10 MV/m$) and that in both cases the high electric field electron emission (HEFEE) follows almost a Fowler/Nordheim law. A significant difference is that in a vacuum, currents switch on to reach μA's before insulation failure occurs, whilst in compressed gases, this can be due to a single electron emanating from the cathode; this can be measured statistically.

In a MOVAM structure which is an acronym for Metal-Oxide-Vacuum Media, for virgin electrodes with a good vacuum of 10^{-6} Pa, the current will suddenly rise from $\leq 10^{-11}$ A to between 10^{-7} to 10^{-5} A, apparently due to a switch on phenomenon at some threshold of about 10 MV/m. Hard materials such as T_i, M_o and SS, having unspecified surface properties, emit the least electron current and makes them suitable for vacuum insulation. Oxidation of the surface tends to suppress electron emission and a general requirement is cleanliness. It has been shown (Latham,1988) that current, glow, gas and spark conditioning all lead to enhanced vacuum insulation by destroying or suppressing dominant electron emission sites.

In a MOGAM structure, which stands for Metal-Oxide-Gas Media, for virgin, unsparked cathodes, single electron emission of about 10 e/s only commences for hard materials at 14 MV/m, and for soft materials, like Al, at 20 MV/m, the surfaces having been carefully polished. The electric field range covered for Cu (a fairly hard material) is 17 to 25 MV/m before a spark occurs. Contrary to expectation and to vacuum, the harder the material cathode, the more the electron emission; also the thicker the oxide, the more the emission. The latest discovery is that an insulating layer (like enamel) reduces the electron emission somewhat. It is an established fact that spark conditioning of a MOGAM structure can raise the sparking voltage by almost a factor of two (Cookson, 1981).

ELECTRON MISSION PROCESSES LEADING TO BREAKDOWN

Using complex apparatus, it has been shown (Latham,1981), that in MOVAM structures, vacuum emission sites are broadly of the point type or the extended one. The point sites, of μm dimension, hardly act as electron emitters; they invariably consist either of the base metal or Al, Ag or C and may be electrically insulated from the cathode surface. Extended sites, of about 10 μm size, generally are found to involve carbon, situated along grain boundaries or cracks in the cathode. Using electron image studies, Bayliss and Latham (1985) have established that sites consist of a group of independent subsites that individually switch on and off, giving rise to current fluctuations, the frequency of which increases with increase in background gas pressure. The electron emission is accompanied by photons and is of a non-metallic nature, perhaps pointing towards an avalanching process which indeed involves hot (exciting and ionising) electrons. The general property of emission sites appears to be independent of the basic substrate electrode material, but is strongly influenced by electrode surface treatment.

For MOGAM structures, it has long been surmised (Trump, Cloud, Mann and Hanson 1950) that HEFEE, in the broadest sense of the word, plays a leading role in compressed gas sparking characteristics and is responsible for departures from Paschen's law. The difficulty has usually been to separate the initiating process of HEFEE from the primary ionisation mechanism leading to avalanche formation and the secondary process of electron regeneration at the cathode by photon radiation from excitation gas molecules. By using a gas mixture of 10% ethane added to another gas (Heylen, Hitchcok and Guile, 1978), typically N_2, the secondary ionisation process is completely suppressed and the initiating electrons, giving rise to giant avalanches ($e^{\alpha d} = 10^{+8}$) can be readily observed on an oscilloscope and counted by a Pulse Height Analyser (Salim, Williams, Heylen and Guile,1985). For this gas mixture, it was invariably found that those cathodes giving least electron emission (typically Al) yield sparking characteristics which lay above those for cathodes giving high HEFEE such as mild steel or stainless steel. However in pure N_2, where secondary ionisation takes place, the difference in sparking characteristics is reduced and in air, where secondary ionisation is presumably predominant, it was found that at the highest gas pressure used (about 1.4 MPa), the sparking characteristic for the aluminum cathode falls below that for the stainless steel one (Salim, Heylen and Guile,1988), in agreement with the M.I.T. workers. Very close to the sparking voltage, switching mechanisms in the cathode surface take place which result in large current (10^{-11}A) activity; it is difficult to separate this switching process from the breakdown one. Cookson (1981) has reviewed the situation for MOGAM structures and Latham for MOVAM structures in 1988.

PRESENT DAY POSITION AND FUTURE OUTLOOK

M. J. Morant (1955) interpreted the sparking gap experiment of Llewellyn-Jones (1953) in terms of the Schottky-Richardson thermal emission equation, obtaining reasonable work function values and emitting area the size of Llewellyn-Jones' broad-area electrodes. Llewellyn-Jones (1953), though obtaining reasonable work functions at field intensification factors of about 100 ×, obtained emitting areas of atomic dimensions, whilst the S.G.F. (1929) interpretation of our 1984 results revealed emitting areas of electronic size; the S.G.F. equation was really developed for Na on tungsten tips and gave good agreement with Gosling's experiments.

The S.G.F. law (see Heylen, Guile and Morgan,1984) gives, for idealised HEFEE, the functional relationship between the number of electrons emitted per second, N_o, and the applied electric field of voltage/gap, E, in $V\ cm^{-1}$ as

$$N_o = C E^2 e^{-D/E} \quad ; \qquad (1)$$

thus, a logarithmic plot of N_o/E^2 versus $1/E$ should yield a straight line with negative slop (gradient) equal to D and an intercept lnC. This relationship was found to hold for all our broad-area electrodes as shown in Fig. 1. It is observed that the slope increases strongly in going from the 2.5-4.0 nm oxide thick copper surface to the 25-30 one, remaining fixed thereafter for a 380 nm thick oxide layer.

According to the original F/N (1928) interpretation of the constants, D is given by

$$D = \frac{6.8 \times 10^7\ \phi^{3/2}}{M} \qquad (2)$$

in which ϕ is the work function (in volts) and M the field intensification factor. Thus, for a given slope D, the work function (ϕ) increases with increases in the field intensification factor, as shown in Fig. 2. The intercept constant C is given in terms of the Fermi level ς (volts), the surface area $A(cm^2)$, ϕ and M as follows:-

$$C = \frac{38.5 \times 10^{12}\ A\ \varsigma^{1/2}\ M^2}{(\varsigma + \phi)\ \phi^{1/2}} \qquad (3)$$

so that the emitting area A decreases with increase in M for a given C as shown in Fig. 3, using $\varsigma = 7$ volts for Cu; also for a given M, the area A increases with increase in work function.

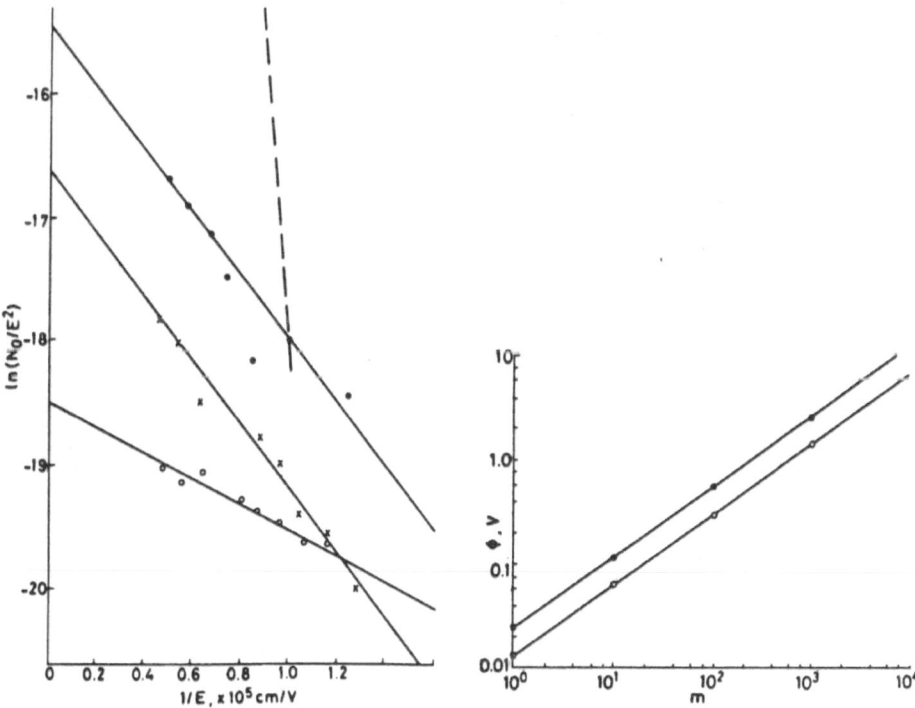

Fig. 1. Fowler-Nordheim type of plot for three oxides. 0, 2.5 - 4 nm; ×, 25-30; • 380 nm.

Fig. 2. Work function as a function of field-enhancement factor. 0, 2.5-4.0 nm oxide film; • 25-30 and 380 nm.

The complete S.G.F. equation won't be gone into here because, for this presentation, it is too complicated (sophisticated), but it is noted that, using $\phi_1 = 4.5$ volts for copper and taking the natural oxide layer of 2.5-4.0 nm, the derived area was in line with that used experimentally, though the area found for the two other oxides differs by a factor greater than 10^{100}.

Using the M.J. Morant (1955) equation, based on the Schottky-Richardson law, he obtained good agreement, plotting ln I versus $E^{1/2}$, which yielded a straight line relationship; however our 1984 results do not yield a straight line on this basis.

For a MOVAM structure, a typical F/N plot is shown in Fig. 4. This was obtained by Bayliss (unpublished) for a single site. Whilst the slope does not change for the oxidised site ($\simeq 59$ nm thick), there is a marked change in the HEFEE region of the characteristic and in the spectral characteristic. Similarity with MOGAM structures is that an oxide layer increases the HEFEE for Cu. The interpretation by Bayliss and Latham (1986) is that the oxide gives rise to a bulk-limited rather than a contact-limited HEFEE mechanism.

Bayliss and Latham (1986) have analysed their results using as a model a massive oxide particle (800 nm thick) on the metal base and obtain good agreement with their sophisticated measurements. In fact they (BL) can account for the saturation affect at HEFEE's and can, by their treatment, straighten the F/N plot as shown in Fig. 5. They use electric field intensification factors of only $15 \times$ in the particle and $80 \times$ at the boundary of the particle/vacuum. Whereas the traditional F/N theory relies on electrons escaping from the base metal (cathode) surface by quantum mechanical tunnelling through the surface potential barrier, the BL model proposes HEF penetration of a MOVAM structure where electrons are heated up internally in the 'insulator' through conducting channels, so that they can be emitted over the surface potential barrier. Using the BL method applied to our results in a MOGAM structure, we find that for aluminum a minimum thickness of 'insulator' of 140 nm is required and then a singularity occurs in the calculations. Reasonable work-function values ensue and the emitting area is about $10^{-33}\ m^2$ (classical cross section of electron: $2.5 \times 10^{-29}\ m^2$).

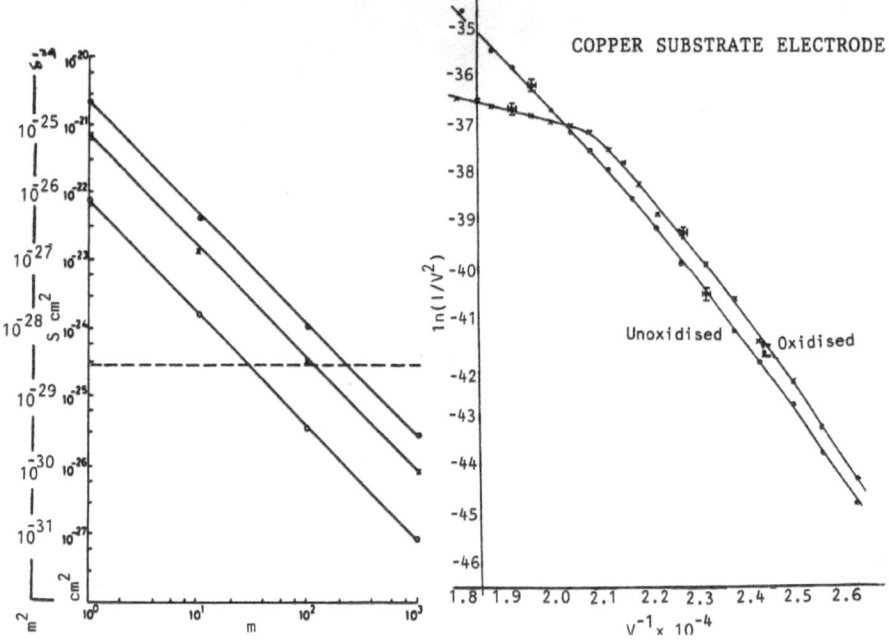

Fig. 3. Emission areas as a function of field-enhancement factor. ○, 2.5-4.0 nm, × 25-30 nm: ● 380 nm oxide film. ---- cross sectional area of electron.

Fig. 4. F/N plot for oxided and unoxided copper in vacuum.

CONCLUSIONS

Work in recent years has shown an increasing number of similarities between HEFEE and its role in breakdown in compressed gases and vacuo. These are: significant HEFEE current starts to flow between electrodes in the electric field range of 10 - 30 MV/m and this is where departure from Paschen's law occurs; in both cases, the current/electric field characteristics follow Fowler/Nordheim type plots, in the broadest sense; cathode material and the nature of the surface greatly affect HEFEE; switching phenomena take place; the sparking voltage is prone to electrode conditioning; in sparking, followed by arcing, a few parts per million of foreign molecules can affect HEFEE; finally, spatial correlation has been established

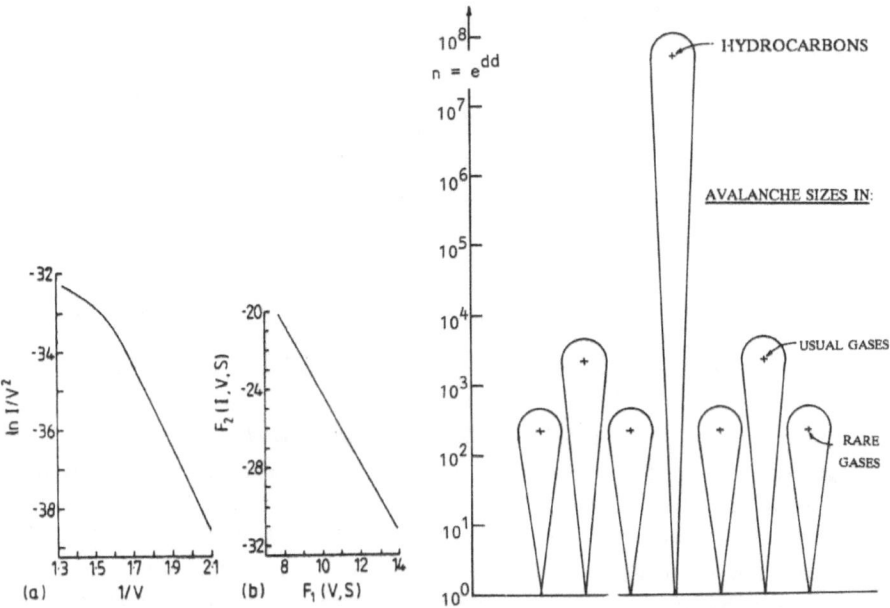

Fig. 5. F/N plot (a) conventional (b) recent analytical model (BL).

Fig. 6. Avalanches in gases.

I. ALUMINUM OXIDES

	(a)	(b)	(c)
$1/E = 6 \times 10^{-8}$ m/V	2–3nm (d_1)	50–60 nm (d_2)	160 nm (d_3)
$\ln\left[\frac{I}{E^2}\right]_{ordinate}$	−33.43	−33.16	−32.94
Thickness	d_1	$d_2 = d_1 + \Delta_1$	$d_3 = d_1 + \Delta_2$
Avalanche Size	$e^{\alpha d_1}$	$e^{\alpha d_1} e^{\alpha \Delta_1}$	$e^{\alpha d_1} e^{\alpha \Delta_2}$
Ratio $\frac{b \cdot c}{a \cdot a}$		$e^{\alpha \Delta_1} = 0.27$	$e^{\alpha \Delta_2} = 0.49$
α		5.14×10^8/m	3.11×10^8/m
Multiplication	1.01x	1.25x	1.93x
$1/E = 3 \times 10^{-8}$ m/V			
Multiplication	1.024x	1.69x	4.57x

II. NIOBIUM OXIDES

$1/E = 6 \times 10^{-8}$ m/V	2–3 nm	50–60 nm	200 nm
Multiplication	1.095x	7.32x	1,394x
$1/E = 3 \times 10^{-8}$ m/V			
Multiplication	1.074x	4.82x	305x

Fig. 7. Avalanching in oxide layers.

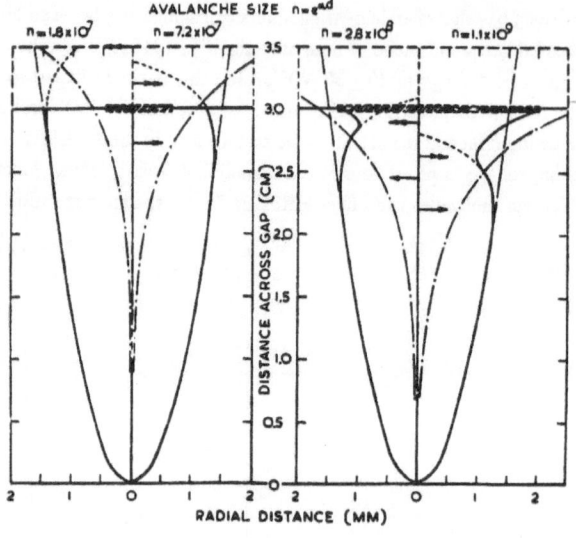

AVALANCHE SIZE n=e^{ad}

The spatial development of single avalanches.

— — diffusion radius,
· · · · constriction radius.
· - — repulsion radius.
—— resultant avalanche envelope,

/ / / / / / / / / / / experimental radii /4/,
→ diffusive = constrictive velocity.
→ > constrictive = repulsive velocity.

Fig. 8. Spatial development of single avalanche.

(Latham, Bayliss and Cox 1986) between vacuum and the first compressed gas spark/arc on the same cathode provided a convenient delay of at least 18 hours is allowed to elapse between these two processes, when it is thought that charge transfer occurs in the cathode surface during this interval.

Major differences are that:- in compressed gas, sparking may ensue from one initiatory electron, while in vacuum, the current has to reach at least $\mu A \mathring{s}$; the conduction current in vacuum is presumably purely electronic, while in compresses gas it is ionic, i.e., electrons and ions flow; the measuring analytical techniques used in one area can at present not be applied to the other area of research or vice-versa; other contradictions also exist like hard materials provide good insulation in vacuum, whilst in compressed gas they provide poor insulation, at least up to a gas pressure of about 1.4 MPa and a gap distance of 1/2 mm, requiring about 20 kV.

Finally, giant avalanches occur in hydrocarbons, as shown in Fig. 6. We postulate that avalanching, which is a cosmological phenomenon in all areas, (animal, vegetable and mineral) under HEF conditions, occurs in MOVAM structures and this is possibly a rival theory to BL's, Fig. 7. The influence of space charge on the spatial developement of an electron avalanche (Heylen,1966) is shown in Fig. 8.

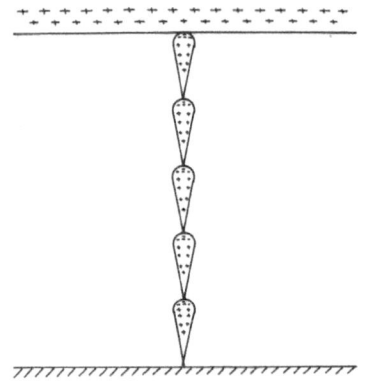

Fig. 9. Bead lightning.

Bead lightning, which occurs in the Midi in France, is in the form of a 'chapelet' as shown in Fig. 9 (Heylen,1971).

ACKNOWLEDGEMENT

The author thanks Professor Michael J. Howes for providing the facilities of the Department, Professor Alan E. Guile and Rodney V. Latham for encouragement and Moyra Culbert and Judith Watson for processing this text.

REFERENCES

Bayliss, K. H., and Latham, R. V., 1985, The spatial distribution and spectral characteristics of field-induced electron emission sites on broad-area high voltage electrodes, Vacuum, 35: 211.

Bayliss, K. H., and Latham, R. V., 1986, An analysis of field-induced hot-electron emission from metal-insulator microstructures on broad area high voltage electrodes, Proc. Roy. Soc., A 40 3: 285.

Cookson, A. H., 1981, Review of high voltage gas breakdown and insulators in compressed gas, IEE Proc. A128 (4): 303.

Fowler, R. H., and Nordheim, L., 1928, Electron Emission in Intense Electron Fields, Proc. Royal Soc. A119: 173.

Guile, A. E., Latham, R. V., and Heylen, A. E. D., 1986, Similarities between electron emission and consequent breakdown processes in high-pressure gases and in vacuum, IEE Proc. A 133: 280.

Heylen, A. E. D., 1966, The influence of space charge on the spatial development of an electron avalanche, Proc. VII Int. Conf. on Phenomena in Ionised Gases, I Beograd, 563.

Heylen, A. E. D., 1971, Influence of baffle plates on the uniform electric field strength of air, Proc. X Int. Conf. on Ionisation Phenomena in Gases, Oxford, 162.

Heylen, A. E. D., Hitchcock, A. H., and Guile, A. E., 1978, A novel non-destructive method of measuring electron field emission, IEE Conf. Proc. 165: 308.

Latham, R. V., 1981, High voltage vacuum insulation - the physical basis, Academic Press, London.

Latham, R. V., Bayliss, K. H., and Cox, B. M., 1986, Spatially correlated breakdown events initiated by field electron emission in vacuum and high pressure SF_6, J. Phys. D. 19 (2): 219.

Latham, R. V., 1988, High voltage vacuum insulation - New horizons,IEEE Trans. Electrical Insulation 23 (5): 881.

Llewellyn-Jones, F., and De La Perrelle, E.T., 1953, Field emission of electrons in gas discharges, Proc. Roy. Soc., A216: 267.

Llewellyn-Jones, F., and Morgan, C. G., 1953, Surface films and field emission of electrons, Proc. Roy. Soc., A218: 88.

Morant, M. J., 1955, Interpretation of experiments on electron emission in spark gaps, Proc. Phys. Soc. B68: 513.

Salim, M. A., Williams, L., Heylen, A. E. D., and Guile, A. E., 1985, High field electron emission for various materials with varying thicknesses of oxide film, Proc. VIII Int. Conf. Gas Discharges and Appl.: Oxford, 303.

Salim, M. A., Heylen, A. E. D., and Guile, A. E., 1988, Correlation between HEFEE and sparking characteristics in compressed gases, IX Int. Conf. on Gas Discharges and Their Applications, Venice, 339.

Stern, T. E., Gosling, R. S., and Fowler, R. H., 1929, Further studies in emission of electrons from cold metals, Proc. Roy. Soc. A 124: 699.

Trump, J. G., Cloud, R. W., Mann, J. G., and Hanson, E. P., 1950, Influence of electrodes on d.c. breakdown in gases at high pressure, Electrical Engineering, 65: 961.

DISCUSSION

E. E. KUNHARDT: How is it possible that electron/ion avalanching takes place in the oxide layer as the band energy gap of an insulator is about 8 eV?

A. E. D. HEYLEN: Bayliss and Latham (BL) showed for vacuum insulation that an insulator of 800 nm sits on the surface in their proposed model. Taking an electric field of 20 MV/m for the breakdown field, a field enhancement factor in the insulator of 16 (BL) and a dielectric constant of 4 (BL), the energy due to the voltage across the insulating particle is 64 eV ; thus at least eight avalanches in series can be produced to provide filamentary conducting channels through the particle.

M. GOLDMAN: On Al we have been able at atmospheric pressure to start electron emission with light of $\sim$ 1 eV. Have you studied the surface layers of Al and the cracks which can increase with surface thickness? If so, they could be emission sites where exoelectrons can be emitted as in our experiments.

A. E. D. HEYLEN: The standard work function accepted for Al is some 3–4 eV; yet we also observed in 1955 electron emission from Al, giving rise to giant avalanches ($e^{\alpha d} = 10^8$) in hydrocarbon gases, simply due to illumination by an ordinary tungsten light bulb. As Al is a soft metal, we would expect fewer cracks than in a hard metal like stainless steel; this agrees with lower electric–field electron–emission having been observed in a compressed gas in the latter case.

LIQUID BREAKDOWN AND ITS RELATION TO GAS BREAKDOWN

Robert Tobazéon

Laboratoire d'Electrostatique et de
Matériaux Diélectriques
Centre National de la Recherche Scientifique
166 X, 38042 Grenoble Cedex, France

INTRODUCTION

Breakdown of dielectric media occurs via a fast succession of irreversible processes which produce a high conductivity between electrically stressed conductors when the voltage exceeds a certain value. Breakdown can also be achieved in gases, liquids and solids through the action of an intense beam of light produced by a laser (laser-induced breakdown). Depending on the particular situation, breakdown takes place in times ranging from ps to ms. The breakdown voltage depends not only on the dielectric itself, its purity and the way it is handled, but also on the geometry of the gap, the nature and finish of the electrodes, pressure, temperature and on the shape and duration of the applied voltage (DC, AC, impulse) . Another point of importance is that the vast majority of high voltage equipment is insulated with a combination of at least two of the generic media, breakdown of the whole insulation issuing from complex interactions between each other.

As a matter of fact, in most applications, liquids are used in association with solid insulators. Their main role, as known (and used) since the end of the last century, is to replace air, in order to avoid partial discharges in cavities possibly subjected to voltages above the Paschen threshold. The wide variety of applications of liquid impregnants has led to the use of either natural or synthetic products, polar or non polar, viscous or non viscous, etc .... In addition to their dielectric properties, technical liquids have to ensure thermal transfer and to satisfy various criteria, such as compatibility with solids, thermal stability, flammability, toxicity and other economics. Mineral oil has been, and still is, the most widely used (mainly in power transformers, bushings). In the field of power capacitors, the ban of polychlobiphenyls during this decade has resulted in the development of replacement liquids possessing much better properties than their predecessors, especially concerning their "gassing" properties: this qualifies the way a gaseous cavity trapped or created within the liquid, will expand or shrink (which is obviously desirable) when subjected to discharges. In practice, when "medium" or "large" interstices of free liquids (mm to tens of cm) are used (oil ducts, bushings), the field in service is only a few kV/mm. In capacitors, the liquid layers do not exceed a few microns; the field is typically 60 Vr.m.s/μm , reaching eventually 200 V/μm (2 MV/cm) at the edges of the thin aluminium electrodes. Although, for certain applications, liquids offer the advantage of a high permittivity, on the one hand their breakdown strength is often comparable to that of gases under moderate pressures (a few bars), and on the other hand, their conductivity (and the corresponding AC losses) is not negligible (as for gases), and thermal breakdown of the insulation may occur.

Gaseous Dielectrics VI, Edited by L.G. Christophorou and
I. Sauers, Plenum Press, New York, 1991

Conduction, even in well purified non-self dissociated liquids and up to several hundreds of kV/cm, results from the presence of tiny amounts of impurities and is mainly ionic in nature: ions may preexist (dissociation of electrolytes and/or of the liquid itself) or may be created either in the bulk (field-aided dissociation) or at the electrodes via electrochemical reactions of oxydo-reduction of the liquid or of neutral species (unipolar or bipolar injection). Electrical forces will be able to induce electrohydrodynamic motions, analogous to the electric wind in gases, but their contribution is, by far, more important to the transport of charges and to the space charge distribution. Electronic conduction can only be observed in thoroughly purified liquids or liquefied gases, requiring special techniques of injection of excess charge carriers, the resulting species being generally short lived. Compared to gases, breakdown in liquids shows many phenomenological similarities, but if there is no doubt that electronic processes are at work, many other more or less correlated phenomena take place, thermal, hydrodynamical, mechanical. Although the nature and the chronology of various events happening in prebreakdown phenomena are not fully understood, recent progress allows us to answer, at least partly, questions such as : how does the energy deposit in a liquid proceed? What are the nature, origin and mobilities of charge carriers? Is there an electronic multiplication by avalanches ? Is a gaseous phase generated, and is it ionized? Are there streamers or leaders, and how are they generated and propagated? Is it possible to predict breakdown voltages ? Most of these questions have received an answer in gases, and this paper is an attempt to relate breakdown mechanisms in gases and liquids.

GAS AND LIQUID BREAKDOWN : SOME GENERAL FEATURES AND ORDERS OF MAGNITUDE

Similarities And Differences With Regard To Phenomenology

Several features are common to gas and liquid breakdown :
- breakdown is localized, the spark or arc being preceded by luminous tree-like patterns ;
- the actual strength depends on the shape and the duration of the applied voltage, generally being much higher for impulses of short duration ;
- scale effects are important : the lower the gap, or the area, or the volume subjected to the field, the higher the strength ;
- conducting particles (either isolated in chains), can reduce the strength to 10 % or less of the uncontaminated value, by locally enhancing the field or by triggering microdischarges when approaching an electrode surface ;
- polarity effects are often very important in divergent fields : the breakdown voltage is generally much lower when the acute electrode is positive, especially under impulse voltage ; this situation, considered the most "dangerous" has been the most widely studied ;
- the surface state of the electrodes (roughness, presence of oxides, adsorbed layers, etc ...) may result in conditioning processes, and contribute to give, especially in uniform fields, a more or less large dispersion in rigidity measurements ;
- breakdown is facilitated along solid spacers.

There are several other mechanisms specific to liquids which have been shown to be at the origin of their breakdown :
- moisture content, especially in uniform field, when it approaches and exceeds the saturation level (reductions of a factor of 3 have been reported) ;
- gas which comes out from the bulk (dissolved gas), from the electrodes or insulating solids (occluded gas), or gas generated by electrochemical processes, thermal heating, or, as we shall see later, by more complicated electronic processes.

However, if very careful attention is given to chemical and physical purification of the samples, control of the finish of electrodes, etc ... to reduce the influence of the "extrinsic" mechanisms encountered in practice, one can measure high or very high strengths using short pulses, more or less characteristic of the liquid itself, allowing comparisons between liquids tested under the same conditions. Nevertheless, the "intrinsic strength" of any liquid (or liquefied gas) has still not been measured. This considerably differs from the situation in gases where breakdown thresholds are well-known for many pure gases and gas mixtures. Widely different is also the dependence on pressure : in uniform field, an increase of less than a factor 2 results in applying pressures up to 25 bars, subatmospheric pressures having little effect ; in divergent field, although there is evidence that breakdown strengths usually are much increased by increasing pressure in the negative polarity than in the positive, studies are scarce and nothing similar to corona stabilization, critical pressures, etc ... thoroughly studied in electronegative gases, has been described. This is indeed direct experimental evidence (among others, as shown later on) that a gaseous phase, to a greater or lesser extent, plays a significant role in liquid breakdown.

Exemplifying Liquid Breakdown Versus Gas Breakdown.

In the early thirties[1], it was reported that liquids were able to sustain very high fields under impulse voltage and that breakdown was completed in a very short time (1.2 MV/cm in transformer oil with a 0.5 mm gap subjected to a 5 ns impulse) ; the decrease in strength when the gap was increased, the beneficial action of hydrostatic pressure favoured an electron multiplication theory of breakdown analogous to the Townsend mechanism in gases. Other similarities with gas discharges were observed in point-plane geometry, such as the different shapes of patterns (supported by Lichtenberg figures) when the polarity was changed, and the general trend of voltage breakdown dependency on electrode distance under impulse and AC up to large distances (0.6 m in transformer oil)[2]. It must be said that, over the years, many other interpretations than the collisional ionization model have been offered, as reviewed in detail by Sharbaugh et al.[3]. A few selected results are presented in figures 1 and 2 in order to illustrate the behaviour of clean transformer oils (of comparable composition) subjected to different voltage waves in the widest range of gaps available from the literature, both in uniform and non-uniform fields[4-9]. We notice that exceptionally high values of strength are attainable for small sphere gaps and short pulses (6

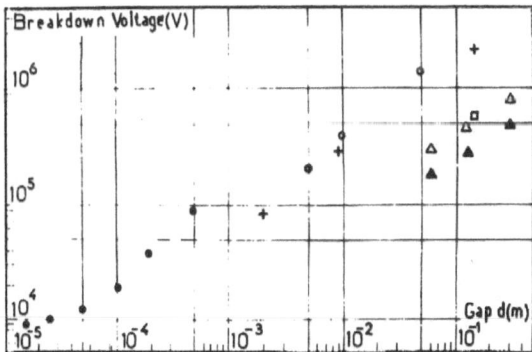

Fig. 1. Illustrating the variation of breakdown voltages of transformer oil vs gap spacing in uniform and quasi uniform fields. Spheres: (●) [4] 1.3 cm in diam. Impulses 1/3 μs. (o) [5] 25 cm in diam. 3 kV/s AC rising voltage. Plane electrodes (+) [6] Lightning impulses (plane diam.: 6 cm, 22 cm, 200 cm respectively for d: 0.2 cm, 0.9 cm, 14.6 cm). (□) [6] AC, max. value 1 min. withstand). (Δ) [7] DC, 5 s. withstand. (▲) [7] DC, 30 s withstand (plane diam.: 120 cm).

MV/cm), whereas with large plane gaps it is reduced by around 2 orders of magnitude. Moreover, the overall tendency is that the increase in the breakdown voltage V_B, versus the electrode separation d, is lower than in air or SF_6. Unfortunately, we do not possess comparable data in such a wide range of experimental conditions in other liquids. However, using spheres and microsecond pulses, one could measure a strength depending on the molecular constitution : for example, increasing the chain length (and then the density) in the homologous series of alkanes, resulted in an increase of strength, whereas in the alkylbenzene homologs (with nearly the same density) the reverse was observed[3], so density alone, as for gases, was not the dominant factor. The strength of CCl_4 was found to be much higher (4.8 MV/cm) than those of $CHCl_3$ or CH_2Cl_2 (2.5 MV/cm, comparable to that of n-hexane)[10] : although being an isolated result, this bears some similarity to the behavior of electronegative gases. Nanosecond pulses applied to spherical gaps (50-500 µm) revealed still higher, and surprisingly comparable strengths in transformer oil and water (8 MV/cm)[11]. Interestingly enough, these high strengths are comparable to the laser-induced breakdown values[12]. In practice, AC tests with mm. spherical gaps, although showing significant differences between technical liquids (up to a factor of 2), are not very sensitive to differences between similar liquids, such as mineral oils. Highly divergent field tests are much more sensitive and much easier to work up at large gaps :

- the negative impulse breakdown voltage can be much higher than the positive (fig. 3), in certain liquids[13] reaching values higher than a factor of 7 ; it is very sensitive to oil composition, being lowered in oils containing aromatic compounds ; the adverse effect is especially marked by addition of polyaromatic hydrocarbons (such as pyrene) to a white naphtenic oil[13] ;

- although the positive impulse is less sensitive to changes in composition in the usual range of gaps, the order of rating for negative and positive polarity may differ for different liquids, or be reversed at large distances for the same liquid (fig 2. and 3) ;

- the times to positive impulse breakdown are usually shorter than for the negative polarity.

As a conclusion, although liquids are able to sustain several MV/cm (to be compared to 89 kV/cm. bar for SF_6, both under uniform and divergent field the adverse effect of electrode separation appears to be more marked than in gases.

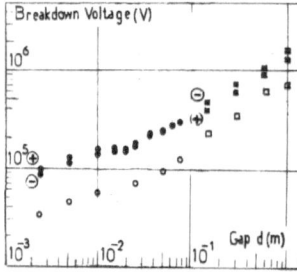

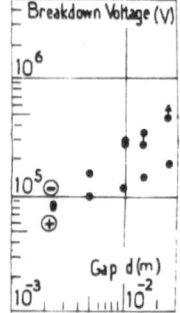

Fig. 2. Rod-plane breakdown voltages of transformer oils versus gap spacing. Rounded rod 1.4 mm in diam [8] : (●) Lightning impulses (o) AC, max. value,1 min. withstand. Square cut rod 12.5 mm [9] : (■) Lightning impulses (□) AC, max. value.

Fig. 3. Rod-plane lightning impulse breakdown voltages of a naphtenic white oil. Rounded rod 1.4 mm in diam. [8].

There is still no classification of liquids, as known for gases (according to their density and to their electronic affinity) ; no synergetic effect has been reported for breakdown in uniform field ; aromatic additives play an important role in divergent field.

PREBREAKDOWN PHENOMENA IN LIQUIDS

Breakdown Time Lag And Successive Steps Leading To Breakdown

The statistical study of breakdown time-lags in liquids has shown, as in gases, that it is possible to separate an "initiation time" (its statistical nature being controversial[14]) from a "formative time-lag" much better characterized in most of the experimental conditions. It is now generally accepted that the total time-lag corresponds to the following successive steps :
- an "initiation phase", which corresponds to the onset and development of a precursor event (or succession of events) able to give rise to the next phase ;
- a "propagation phase" of figures identified as more or less ramified, luminous and conducting tree-like patterns issuing from a highly stressed region (generally called "streamers") able to cross part or the whole gap ;
- an "arcing phase", where the "main stroke" is established, in a very short time (1 ns or less) ; this final stage has been the least studied.

This separation into distinct phases is questionable, especially when breakdown takes place within a very short time (ns) : breakdown could then take place via means somewhat different from the so-called streamer mechanism.

Streamers In Liquids

The term "streamer" has been used to name the luminous and ramified figures preceding the arc in gases, solids or liquids. In gases, a terminology is well established, and the processes leading to breakdown by the streamer or the leader mechanisms are well-documented and fairly understood[15-17]. Although named "streamers" the figures propagating along solid surfaces immersed in oil[2], the luminous channels propagating across large divergent oil gaps (5-20 cm) were called "initiating streamers"[18], whereas others[19] claimed that "leaders" developed ; in general, almost any detectable tree-like event is named "streamer". With divergent fields, the development of streamers is much less limited before breakdown than in uniform field and, whatever the liquid (non polar, polar, liquefied gas), one can select some of their general characteristics[20]:
- their optical index is different from that of the liquid ;
- their velocity varied within a very wide range[21], being much lower than the accoustic velocity (around 1 km/s) for negative streamers (originating from the cathode) in hydrocarbons (100 m/s), and much higher in halogenated compounds (10-80 km/s) for both polarities ; for a given liquid, positive streamers are faster than the negative ; the velocity can go through a minimum or be essentially constant accross the gap ; it generally increases with the voltage, but is sometimes independent of it (as in mineral oils, except at large overvoltages) ;
- their shapes can be markedly different, but "slow" subsonic figures are bush-like (fig. 4-5), while "fast" supersonic are filamentary (fig. 6) ;
- the addition of electron scavengers in .small concentration to a liquid with no electronic affinity, surprisingly increases (by one order of magnitude) the velocity of slow negative streamers[22] ; the increase in the velocity of positive streamers by additives is much less marked ; no slowing down of streamers has been reported ;
- streamers produce a current composed of either discrete pulses or of a dominant continuous component, and also emit light ; current and emitted light have similar shapes ;
- fast streamers are the seat of energetic processes, a large amount of decomposition products being produced[20] ;
- shock waves are associated with the propagation of streamers ;
- the streamer is arrested if the amplitude or the duration of the voltage are not

large enough, producing a string of microbubbles ; an increase in the hydrostatic pressure impedes the propagation of slow streamers, the fast form being much less sensitive.

Therefore, several facts support the view that streamers are constituted, at least partly of a gaseous phase (optical index, shock waves, microbubbles, influence of pressure), and that electronic processes are operating (emitted light, influence of electron scavengers, presence of decomposition products).

<u>Streamers And Breakdown</u>

Breakdown is controlled either by initiation or by propagation of streamers. Their voltage propagation length can be widely different according to the nature of the liquid, the polarity, the shape of the voltage wave. Thus, not only the velocity of streamers, but mainly their ability to propagate is the factor controlling breakdown.

<u>Impulse voltage</u>. Breakdown is usually lower for a positive point, which exemplifies the "danger" of easily propagated streamers. Slow negative streamers cannot propagate far away, so the negative lightning impulse breakdown values are high (they are lowered by addition of electron scavengers which increase the velocity of negative streamers). In uniform fields, the conditions of total propagation are generally fulfilled, and breakdown is controlled by initiation either in the bulk of the liquid or at the electrodes (particles, asperities).

<u>A.C voltage</u>. The streamer appearance is randomly distributed in time, the mean appearance frequency increasing sharply with the voltage[25]. In mineral oil, two extreme situations are possible : 1) at low divergence and high mean fields ($\geq$ 80 kV/cm), the propagation of streamers is possible, breakdown is controlled by streamer initiation : since generally negative streamers are produced at lower voltages, breakdown happens when the point is positive ; 2) at high divergence and low mean fields (< 30 kV/cm), positive or negative streamers can be generated but the positive being more easily propagated, breakdown takes place when the point is positive.

<u>D.C. voltage</u>. It has still been almost impossible in liquids to find evidence in uniform fields of breakdown mechanisms resulting either from a Townsend-

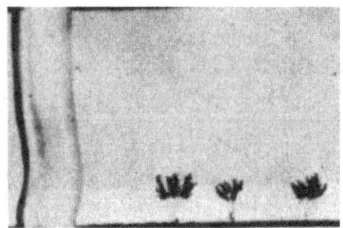

Fig. 5. Negative streamers and spark channel in n-hexane. Gap : 0.55 cm ; crest voltage : 225 kV[24].

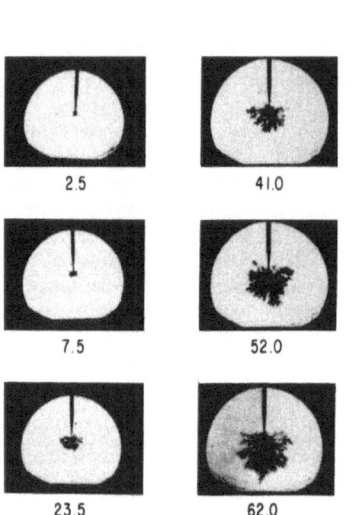

2.5   41.0

7.5   52.0

23.5   62.0

Fig. 4. Negative streamer growth in a white naphtenic oil. Gap : 1.27 cm ; voltage step : 185 kV ; time in microseconds [21].

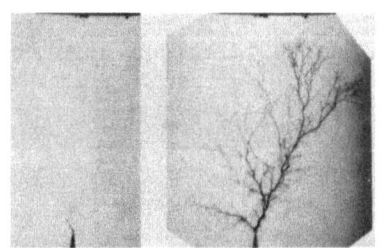

Fig. 6. Positive filamentary streamers in silicone oil [23].

process, or, indeed, from a streamer process (as described by Raether or Meek and Loeb). On the contrary, with the point to plane geometry, the study of high field D.C. conduction has highlighted many basic processes implied in prebreakdown phenomena.

BASIC PROCESSES AND POSSIBLE MECHANISMS INVOLVED IN LIQUID BREAKDOWN

Electronic And Ionic Processes

Depending on the interactions between electrons and the liquid molecules[26-28], electronic mobilities vary in a wide range between the two extreme cases : 1) the electron is quasi-free, the mobility is high for quasi-spherical molecules (90 cm^2/V.s. in tetramethylsilane) ; 2) the electron is trapped on localized states, its mobility is low (0.4 cm^2/V.s. in cyclohexane, 0.08 cm^2/V.s. in n-hexane), its velocity is expected to increase with the field above say 1 MV/cm (no experimental data are available). Hole conduction is still not well-documented, the mobilities known being lower than 0.01 cm^2/V.s.

The life-time of excess electrons can exceed 1 ms in ultra-purified liquids ; they readily form negative ions by attachment to electron scavengers (as an example 10^{-4} mole/liter of CCl_4 in hexane reduces the life-time to 10^{-8} s).

Ionic mobilities are generally low (~ 10^{-4} cm^2/V.s.) ; conduction, even at high field is mainly governed by ions produced in the bulk or at the electrodes. Electrohydrodynamic turbulence can develop even in less than microseconds[20] ; the so-called electrohydrodynamic mobility $K_H = (\varepsilon/\rho)^{1/2}$ (ε : permittivity, ρ : mass per unit volume) is an upper limit for the apparent mobility corresponding to the convective transport of charges ($K_H \sim 1.5 \times 10^{-3}$ cm^2 /V.s. in non-polar liquids).

Let us point out that, in contrast with gases, natural radiation or U.V. irradiation of liquid is not an abundant source of ("seed") electrons ; moreover, it has not been reported that radiations produced even by the luminous fast streamers produce any photoionization far ahead in the liquid (whereas in gases the photoionization is very prolific).

D.C. Conduction At Very High Fields

The study of highly purified hydrocarbons in point-plane geometry[29,30] has clarified the conditions in which well-identified mechanisms of generation multiplication and transport of charge carriers could be observed ; typical threshold fields in these liquids have been evaluated, the radius of the point r_p strongly influencing the observed phenomena. For $r_p < 0.2$ μm, the current-voltage curves can be interpreted in terms of field-emission (point negative) or field ionization[26] (point positive), noticeable currents being produced when the field E_s calculated at the point surface exceeds 20 MV/cm. For $r_p > 0.5$ μm, the phenomena strongly depend on the polarity. With the point negative, above a threshold field (7 MV/cm), there is a regular pulse regime, quite similar to the Trichel regime ; this shows that electron multiplication by avalanches in liquids does exist, and the characteristic discharge parameters have been evaluated[30] : a (first Townsend coefficient) vs E, x_i (ionization length), E_i (field at $x = x_i$, limit above which electron multiplication is negligible). A remarkable result is that $x_i \sim r_p$, and that $E_i \sim 2.5$ MV/cm (independent of r_p) ; moreover, the calculation shows that no avalanches can be produced if $r_p < 0.6$ μm (which is confirmed by the studies with very sharp points). With the point positive, phenomena are more erratic, only a few pulses of current being detected before breakdown.

In liquids of lesser purity, there are erratic impulses of current for both polarities ; but the conditions can be more critical (higher E_s, energy input in the liquid more localized) and the transition to a streamer regime is rapidly observed.

Generation, Growth and Collapse of Bubbles

It has now been well-established that in point-plane geometry current impulses systematically produce bubbles : the current is therefore the cause of the gas generation, and not the consequence (current being produced by discharges in a gaseous pocket initially in the vicinity of the point). The sequence of generation, growth and collapse of the bubbles has been made evident either in industrial liquids under A.C. [25], or in ultrapurified liquids [29-31] (fig. 7). Electron avalanches inject in a very brief time (ns) a large localized energy (10^{-10} - 10^{-8} J) in the liquid which is raised to a supercritical state : a bubble appears (essentially the liquid vapor in hydrocarbons), provided that the pressure is lower than the critical pressure. The dynamic of phenomena, limited by the inertia of the liquid, is well illustrated by the Rayleigh model [31]. Notice that the maximum radius of bubbles is a few μm, and their life-time a few μs. When the point was positive, the generation of bubbles was occasionally observed

Transition To The Streamer

This is illustrated by Fig. 8 : a train of bubbles is produced at a negative point by successive current pulses, the new growing bubbles being generated ahead of the preceding ones[31]. The shape of these figures, their propagation velocity (~ 100 m/s), the presence of peaks in the instantaneous current, closely correspond to "bush-like" streamers. These streamers, at least in the initial phase, are constituted of gaseous pockets connected to the point, in which gaseous discharges can then take place ; another proof of their gaseous nature is that they vanish when a low hydrostatic pressure is applied. As a general rule, with the point positive, the streamer regime is immediately established [25,29].

Streamer Propagation

To get orders of magnitude of the energy needed to generate a unit volume of the expanding streamer, Félici [32] expresses each term corresponding to the various possible processes involved as $(\varepsilon E_p^2)/2$ (E_p field at the tip of the

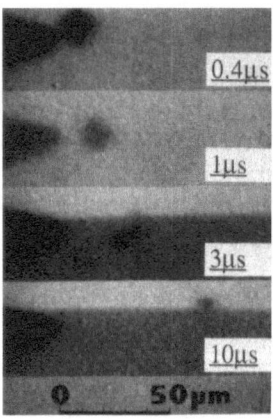

Fig. 7. Bubble growth, motion and disappearance into transformer oil as a function of time (AC voltage, point negative)[25].

Fig. 8. Train of bubbles produced at a negative point in cyclohexane by successive current pulses : the new growing bubbles are generated ahead of the preceding ones (200 ns delay after the first pulse ; current : 100 mA/div ; 100 ns/div)[31].

conducting streamer ; taking $\epsilon/\epsilon_0=2$, he found that vaporization needs 3.5 MV/cm, decomposition or ionization 10 MV/cm, whereas mechanical energy needed to repell the liquid becomes dominant at supersonic velocities. Since in the case of slow negative streamers, discharges in a gaseous phase are likely to occur [32], energy conversion is due to free electrons from the gas, which could explain why electron scavengers suppress the slow mode [32]. As regards the initial phase of streamer expansion, by combining electrostatic and hydrodynamic concepts, equations for cavity growth, instability growth rate and streamer velocity have been derived, in reasonable agreement with experimental data on negative streamers [33].

As concerns fast streamers, many experimental data show that their conductivity is high (transient currents, emitted light, field plots by Kerr effect [34]). The field at the tip is high enough to ionize liquid molecules, the channel being loaded with highly excited and ionized species, reminiscent of "leaders" in gases. Spectroscopic studies combined with other methods promise to give more information about the nature and the conductivity of these channels.

CONCLUSION

In many ways, liquid breakdown is related to gas breakdown. Basic mechanisms, as electron multiplication, are similar, at least in the initiation stages of breakdown. However, the following steps involve more specific processes such as phase change, electrohydrodynamic phenomena, resulting in the generation and propagation of subsonic or supersonic streamers, likely constituted of a more or less ionized gaseous phase.

REFERENCES

1. A. Nikuradse, "Das Flüssige Dielektrikum", J. Springer, Berlin (1934).
2. B. L. Goodlet, F. S. Edwards and F. R. Perry, Dielectric phenomena at high voltages, Journal IEE 69 : 695 (1931).
3. A. H. Sharbaugh, J. C. Devins and S. J. Rzad, Progress in the field of electric breakdown in dielectric liquids, IEEE Trans. on Elec. Insul. 13 : 242 (1978).
4. M. E. Zein El-Dine and H. Tropper, The electric strength of transformer oil, Proc. Inst. Elect. Engrs. 103, Pt. C : 35 (1955).
5. Y. Kawaguchi, H. Murata and M. Ikeda, Breakdown of transformer oil, IEEE Trans. on Pow. App. and Systems 91 : 9 (1972).
6. N. Giao Trinh, C. Vincent and J. Régis, Statistical dielectric degradation of large-volume oil-insulation, IEEE Trans. on Pow. App. and Systems 101 : 3712 (1982).
7. I. Ohshima, S. Motegi, M. Honda, T. Yanari and Y. Ebisawa, HVDC breakdown of transformer oil and the effect of space charge on it, IEEE Trans. on Pow. App. and Systems 102 : 2208 (1983).
8. EPRI Report, "Uniform and non-uniform field electrical breakdown of naphtenic and paraffinic transformer oils", RP 562-1, Contract Nr CCR-78-07, prepared by the General Electric Co (1978).
9. Y. Kamata and Y. Kako, Flashover characteristics of extremely long gaps in transformer oil under non-uniform field conditions, IEEE Trans. on Elec. Insul. 15 : 18 (1980).
10. W. D. Edwards, High values for the electrical breakdown strength of liquids, J. Chem. Phys. 20 : 753 (1952).
11. N. S. Rudenko and V. I. Tsvetkov, An investigation of the electric strength of some liquid dielectrics subject to nanosecond voltage pulses, Sov. Phys. - Tech. Phys. 10 : 1417 (1966).
12. S. Sakamoto and H. Yamada, Optical study of conduction and breakdown in dielectric liquids, IEEE Trans. on Elec. Insul. 15 : 171 (1980).
13. K. N. Mathes and T. O. Rouse, Influence of aromatic compounds in oil on Pirelli gassing and impulse surge breakdown, Ann. Rep. Conf. on Elec. Insul. and Dielec. Phenomena, NAS-NRC : 129 (1975).

14. T. H. Gallagher, "Simple dielectric liquids", Oxford University Press, London (1975).

15. "Electrical breakdown of gases", J. M. Meek and J. D. Craggs,eds , J. Wiley and Sons, New York, (1978).

16. N. Wiegart, L. Niemeyer, F. Pinnekamp, W. Boeck, J. Kindersberger, R. Morrow, W. Zaengl, M. Zwicky, I. Gallimberti and S. A. Boggs, Inhomogeneous field breakdown in GIS-The prediction of breakdown probabilities, IEEE Trans. on Pow. Delivery 3 : 923, 931, 939 (1988).

17. L. G. Christophorou and L. A. Pinnaduwage, Basic physics of gaseous dielectrics, IEEE Trans. on Elec. Insul. 25 : 55 (1990).

18. T. W. Liao and J. G. Anderson, Propagation mechanism of impulse corona and breakdown in oil, Trans. Amer. Instn. Electr. Engrs. 72, PtI : 641 (1953).

19. I. S. Stekol'nikov and V. Ya. Ushakov, Discharge phenomena in liquids, Sov. Phys. - Tech. Phys. 10 : 1307 (1966).

20. R. Tobazéon, Streamers in liquids, in : "The Liquid State and its Electrical Properties", E. E. Kunhardt, L. G. Christophorou and L. H. Luessen, eds. , Plenum Press, New York (1988).

21. K. Yoshino, Dependence of dielectric breakdown of liquids on molecular structure, IEEE Trans. on Elec. Insul. 15 : 186 (1980).

22. J. D. Devins, S. J. Rzad and R. J. Schwabe, Breakdown and prebreakdown phenomena in liquids, J. Appl. Phys. 52 : 4531.

23. W. G. Chadband, On variations in the propagation of positive discharges between transformer oil and silicone fluids, J. Phys. D : Appl. Phys. 13 : 1299 (1980).

24. E. O. Forster and P. Wong, High speed laser schlieren studies of electrical breakdown in liquid hydrocarbons, IEEE Trans. on Elec. Insul. 12 : 435 (1977).

25. O. Lesaint and R. Tobazéon, Streamer generation and propagation in transformer oil under A. C. divergent field conditions, IEEE Trans. on Elec. Insul. 23 : 941 (1988).

26. W. F. Schmidt, Electronic conduction processes in dielectric liquids, IEEE Trans. on Elec. Insul. 19 : 389 (1984).

27. L. G. Christophorou and K. Simos, Interphase physics : linking knowledge on electron-molecule interactions in gases to knowledge of such processes in condensed matter, in : "Electron- molecule interactions and their applications", L. G. Christophorou, ed. , Academic Press, New York, Vol. 2, (1984).

28. T. J. Lewis, Electronic processes in dielectric liquids under incipient breakdown stress, IEEE Trans. on Elec. Insul. 20 : 123 (1985).

29. A. Denat, J. P. Gosse and B. Gosse, Electrical conduction in purified cyclohexane in a divergent electric field, IEEE Trans. on Elec. Insul. 23 : 545 (1988).

30. M. Haidara, N. Bonifaci and A. Denat, Corona discharges in liquid and gaseous hydrocarbons: The influence of pressure, this conference.

31. R. Kattan, A. Denat and O. Lesaint, Generation, growth, and collapse of vapor bubbles in hydrocarbon liquids under a high divergent electric field, J. Appl. Phys. , 66 : 4062 (1989).

32. N. J. Félici, Blazing a fiery trail with the hounds, IEEE Trans. on Elec. Insul. 23 : 497 (1988).

33. P. K. Watson and W. G. Chadband, The dynamics of pre-breakdown cavities in viscous silicone fluids in negative point- plane gaps, IEEE Trans. on Elec. Insul. 23 : 729 (1988).

34. E. F. Kelley and R. E. Hebner, The electric field distribution associated with prebreakdown phenomena in nitrobenzene, J. Appl. Phys. 52 : 191 (1981).

DISCUSSION

E. E. KUNHARDT: Can you explain the microscopic origin of impact ionization in a liquid?

R. TOBAZEON: The next paper will show how the α and η coefficients are measured. If you make a calculation of the order of magnitude at the tip of the streamer you have fields in excess of a few MV/cm so that there is sufficient energy for ionization and decomposition processes.

E. MARODE: Is the bubble formation due to an electrical discharge effect within the liquid phase? What is the spatial extent of the electrical discharge before bubble formation?

R. TOBAZEON: Yes, the bubble generation is the consequence of current pulses due to avalanches in the <u>liquid</u> phase. The spatial extent of electrical discharges, before bubble generation, is a few tens of microns in "slow" liquids in which electrons are trapped, and their mobilities are low, such as hydrocarbons — whereas it can reach several hundreds of microns in TMS (a "fast" liquid, in which electrons are quasifree, and their mobilities are very high). This has been clearly demonstrated by the work of Dr. A. Denat and co-workers (see page 171 of these proceedings).

H. WANG: Would you please comment on the effect of Electrohydrodynamics (EHD) on breakdown characteristics or mechanisms of dielectric liquids in uniform field or quasi–nonuniform field under AC or DC voltage?

R. TOBAZEON: In uniform or quasi–uniform fields, under AC or DC voltages, especially long–term breakdown takes place at low or very low mean fields. We may suppose that there is somewhere — in the bulk, or at the electrodes — a field–enhancement (by particles(s) or protrusion(s))able to create the conditions of streamer generation. If this occurs at the negative electrode, the streamer expansion is likely to be described by the EHD model of P. K. Watson. Another possibility of intervention of EHD on breakdown could be the generation of a gas pocket, if the pressure is reduced below the vapor pressure of the liquid, either by the EHD liquid motion induced by ionic space charge, or in the wake of charged particles.

L. G. CHRISTOPHOROU: To what extent are the initial phenomena which lead to breakddown in liquids, at least in the initial stages, related to gas–phase processes in the liquid?

R. TOBAZEON: Clearly, when the so–called slow "bush–like" streamers develop — as in the case of a negative point in hydorcarbons — the processes are controlled by discharges in a gas–phase: the first bubble is followed by a trail of bubbles (following successive peaks of current); each further current pulse contributes to the increase of the streamer volume (and its propagation). On the contrary, the propagation of "fast" supersonic streamers is unlikely to be interpreted in terms of a bubble mechanism, but is rather related to ionization in the liquid phase.

CORONA DISCHARGES IN LIQUID AND GASEOUS HYDROCARBONS:

THE INFLUENCE OF PRESSURE

M. Haidara, N. Bonifaci and A. Denat

L.E.M.D.-C.N.R.S.
25, Avenue des Martyrs, BP 166 X
38042 Grenoble (France)

INTRODUCTION

Previous studies[1,2] in a point-plane electrode assembly of prebreakdown phenomena in very pure liquid cyclohexane have shown that the results depended greatly on the radius, r_p, of the point. For sharp points of r_p of the order of 0.1 μm, the observed phenomena were found in good agreement with a field emission (point cathode) or field ionization (point anode) regime. In this case, an electrical field strength, E_s, at the surface of the emitter of about 20 MV/cm is required for current onset. For larger point radii ($r_p > 0.5$ μm) the phenomena were completely different. For negative points, above a threshold electrical field of about 7 MV/cm (a field much lower than required for field emission) a highly regular pulse regime, very similar to the Trichel one found in air[3] can be observed. However, for positive points in the same range of point radius and voltage, no current instability and pulse regime can be detected. Extended studies in other pure liquid hydrocarbons (propane and n-pentane) and in compressed gases (air, propane and methane),[2,4] later showed that the same phenomena were observed in all these media. However, in liquids, contrary to results in gases, the characteristics of the Trichel-like pulse regime displayed no pressure dependence. All these results led us to consider a mechanism of charge multiplication in purified liquid hydrocarbons, although the existence in these liquids of hot electrons and avalanches is questionable.

The purpose of the present work was to investigate the processes of corona discharges in gaseous and liquid hydrocarbons as a function of point radius, pressure and properties of liquids and gases. Experiments were undertaken using ultrafast techniques of electrical and optical phenomena recording and spectroscopic analysis of emitted light. These experimental studies were carried out with liquid cyclohexane, propane, n-pentane and n-decane and with cyclohexane vapor and gaseous methane and propane. However, as these liquids have all electronic mobility K_e lower than 10^{-4} $m^2V^{-1}s^{-1}$ (electrons in such media are localized for low field), a study of conduction phenomena in liquid tetramethylsilane ($K_e = 9 \times 10^{-3}$ $m^2V^{-1}s^{-1}$; quasi-free electrons at low electrical field), isooctane ($K_e = 7 \times 10^{-4}$ $m^2V^{-1}s^{-1}$) and neohexane ($K_e = 1.2 \times 10^{-3}$ $m^2V^{-1}s^{-1}$) with elevated electronic mobility was carried out.
From these results, an estimation of multiplication coefficients and ionization lengths in liquids are deduced.

EXPERIMENTAL TECHNIQUES

The liquids were first purified according to classical methods, and then transferred into the cell by vacuum distillation. Coaxial test cells with a well-defined wave impedence of 50 Ω were used in order to measure significant current pulses in the nanosecond range without reflections or oscillations. The cells used for liquids could also be pressurized up to 12 MPa using an argon-operated piston. For gases, the maximum pressure that cells could sustain was 6 MPa. Each cell was equipped with a point electrode and a stainless steel plane electrode and stressed with dc voltages. The tungsten or steel needles had a point radius between 0.5 to 20 μm. The electrode gap d was varied from 0.5 to 3 mm in liquids and from 5 to 15 mm in gases. Stabilized voltage sources and sensitive current meters (Keithley 610C) were used. Time-resolved measurements of the conduction current and of the light intensity were made via an oscilloscope (Tektronix 7834). The spectral analysis of emitted light was effected with a spectrograph and an optical multichannel analyser. The useful spectral range extends from 200 to 800 nm. A charge storage technique[5] was used in experiments where current pulses could be observed; it allows us to measure simultaneously the pulse charge Q_i and the elapsed time Δt between two successive pulses. The measured values were transferred and processed in a microcomputer to obtain histograms of Q_i and Δt and to research a correlation between the two quantities. Numerical results are obtained via the hyperboloidal approximation in which the needle is considered as a hyperboloid of revolution.

RESULTS

Our investigations of electrical conduction of purified liquid hydrocarbons show that phenomena similar to those described in cyclohexane[1] are observed. Indeed, the radius r_p of the point is the main parameter which determines different conduction regimes. However, the results presented here refer mainly to point radii larger than 0.5 μm where corona discharges are detected. In this case, for equivalent densities, there is a great similarity of the discharge regimes in these liquids and in gaseous methane and propane and in air. In particular, in all these dense media, the discharge characteristics depend on the voltage polarity applied to the point.

The main results of these studies are summarized here.

Point cathode

-The pulse regime is always observed when the voltage is above a threshold value, V_s, depending on the density and the tip radius of the point. The corresponding threshold field at the tip, E_s, being independent of r_p. E_s is about 7 MV/cm in liquids of low electronic mobility (LM liquids), approximately 4 MV/cm in tetramethylsilane and around 5 MV/cm in isooctane and in neohexane.

-This pulse regime is stable in air (electronegative gas) and in LM liquids, but unstable in liquids with high electronic mobility (HM liquids) and in gaseous hydrocarbons.

-For a given point radius in air and LM liquids, the repetition frequency, F, of the pulses is proportional to the mean current, I_m ($I_m = QF$); in TMS, there is no correlation between F and I_m.

-Pulses of emitted light are correlated to the current pulses. In liquids, the area of light emission is centred around the point on a distance function of K_e. For example, about 20 μm in cyclohexane and up to 300 μm in TMS. The emitted light spectrum in LM liquids and in gaseous methane and propane at high pressure is composed of a continuum from 350 to 800 nm and of some peaks at 430, 470 and 516 nm emerging very weakly in the continuum. In liquids, the same spectrum was obtained between 1 and 12 MPa of hydrostatic pressure.

The discharge characteristics, as a function of applied voltage and tip radius, show the same variation laws, both in liquids and gases, for a given pressure, P, of the medium ($\equiv$ gas density).

-In liquids, each current pulse induces the generation of a bubble near the point electrode if the hydrostatic pressure is below the critical pressure of the liquid;[6]

-Direct measurements of the charge quantity, Q_i, per pulse and of the rest time, Δt, between two successive pulses, show that: (i) stable and regular current pulses occur, (ii) there is a correlation between $Q_i(n)$ and $\Delta t(n-1)$ (fig. 1) and (iii) the ratio, Q_i/Q, is lower than 0.5, both in LM liquids and air .

-While the different characteristics ($I_m(V)$, $I_m(F)$, $V_s(P)$, $Q_i(V)$, etc.) are pressure-dependent in gases, these are no longer influenced by hydrostatic applied pressure in liquids, at least in the range of our experimental values (0.1 to 12 MPa).

<u>Point anode</u>

Unlike when the point polarity is positive, the pulse regime is observed only for low density values, N, of the medium with $N < N_s$, where N_s is a threshold density value depending on the medium. N_s is decreasing when the atom number of the molecule increases. For instance, $N_s = 7.5 \times 10^{20}$, 3.8×10^{20} and 1.3×10^{20} mol./cm^3 in air, methane and propane respectively.

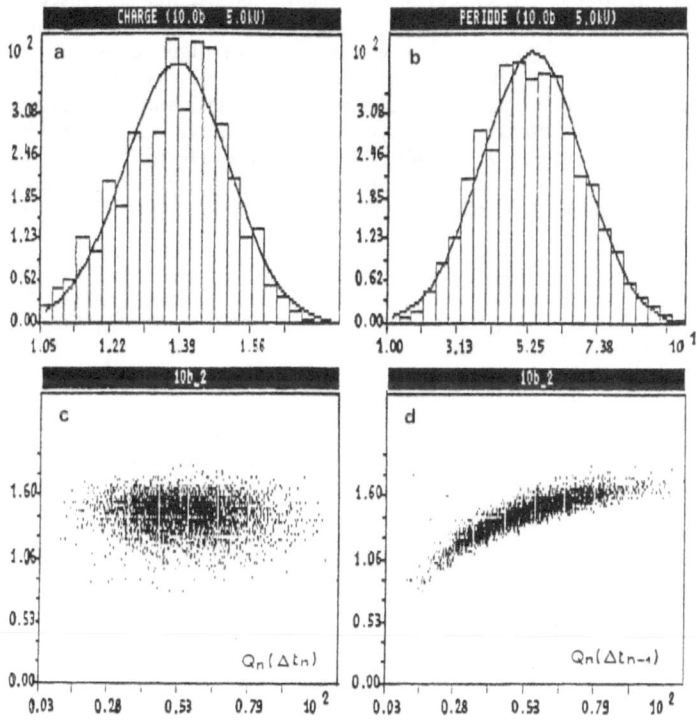

Fig. 1. Histograms of the pulse charge Q_i (a) and of the elapsed time Δt between two successive current pulses (b). Solid line: normal distribution function. $Q_i(\Delta t)$ distributions with: (c) Δt being the following time interval of the pulse; (d) Δt being the preceding time interval of the pulse. Liquid n-decane, P=1 MPa and V=-5 kV.

N_s being lower than the liquid phase density, no regular pulses are observed in liquids. In this case, and for $N>N_s$ in the gaseous phase, only a few erratic pulses are detected. The pulse regime observed for $N<N_s$ in gases, displays the same characteristics as for the negative polarity of the point electrode.

DISCUSSION

These experimental results which show a relationship between K_e and the characteristics of conduction regimes indicate the great influence of hot electron mechanisms in purified liquid hydrocarbons.

The identity of the discharge characteristics in LM liquids and in air justifies the idea of an identity of the mechanisms involved. The pulse regime in air is explained very well by the mechanism suggested by Loeb which involves an electron multiplication process (bipolar phenomena) in the high field region, near the tip, and a stabilization effect of the space charge built beyond this multiplication region (unipolar space charge of the point polarity). In LM liquids whose results indicate that $Q_i/Q<0.5$, Q and Q_i are proportional to r_p, $Q_i(n)$ is a function of $\Delta t(n-1)$, $I_m=QF$, the pulse period $T=1/F$ is only a small fraction of the ion transit time (except at $V=Vs$),etc.., these processes are well-illustrated. Furthermore, as the corona characteristics are independent of the applied hydrostatic pressure and as visual observations of the tip have established that the current pulse effectively preceeds the bubble generation, we may conclude that the corona discharges in liquid hydrocarbons occur in the liquid phase and not in a bubble, as we may be tempted to think.

In TMS, the results can be explained by the same mechanism of electron avalanche in the liquid phase, but without self stabilization by space charge. This implies that the capture of the electrons is done far from the point and this is demonstrated by the large extension of the area of light emission. The different values of E_s in liquids is justified by the difference in electron properties in these liquids. In fact, in weak field, the electrons are considered as quasifree in TMS and localized in LM liquids (e.g. cyclohexane) and in an intermediate state in isooctane and neohexane. In liquids in which the electron is in the quasifree state it must therefore be easier to increase the energy of electrons above the thermal energy, under the action of the electrical field to ionize and dissociate molecules in the liquid.

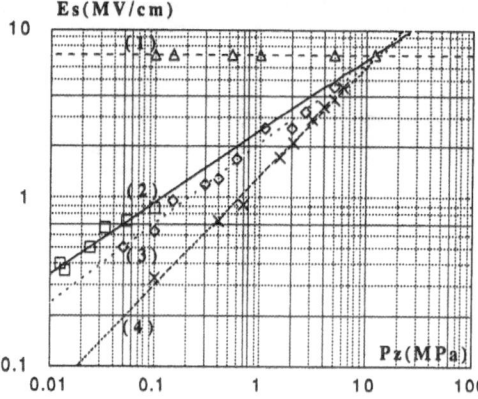

Fig. 2. Threshold electrical field E_s vs corrected pressure P_z in LM liquids (e.g. cyclohexane) (1), cyclohexane vapor (2), gaseous methane (3) and compressed air (4).

The spectral analysis reveals the same continuum in liquids and compressed methane and propane and the emission of the C_2 Swan band at 470 and 516 nm. The only hypothesis which can explain these results is a process of dissociation and fragmentation of the liquid molecules and the recombination of these molecular fragments. Some of these fragments can have a positive electron affinity (e.g. C_2, CH, H, ..) and form negative ions constituting the space charge. Indeed, direct electron attachment process to hydrocarbonous molecules is unlikely.

Evaluation of the multiplication coefficient

The electron avalanche in gases is characterized by a primary ionization coefficient, α, depending on the field strength, E, and on the gas density, N, via the pressure, P. To evaluate the multiplication coefficient in liquids, we examine the plots (fig. 2) of the threshold field, E_s, at the point for LM liquids and gaseous methane, cyclohexane and air vs the corrected pressure, P_z, ($P_z=P/z$ where z is the compressibility factor, so P_z is proportional to N). E_s for liquids is not pressure-dependent (about 7 MV/cm) while for gases it increases with P_z approximately to $P_z^{0.5}$. The two lines cross at $P_z{}^c=10$ MPa for cyclohexane and 12 MPa for propane. At these pressures, the ratio of the gas number density to the liquid number density is about 0.44. With the same value of the threshold field, we can suppose that the electrical discharge in the liquid phase has the same characteristics as in its gas phase at a pressure $P_z{}^c$. Thus, taking equation $\alpha/N=A\exp(-BN/E)$ with A and B measured for cyclohexane vapor[7] and gaseous propane,[8] $\alpha(E)$ in these gases calculated for $P=P_z{}^c$ can be identified as $\alpha(E)$ in these liquids (fig. 3, curves 1 and 3). A similar result has already been found by Derenzo et al[9] in liquid xenon, but with a density ratio of 0.05. This can be explained by the very different electronic properties of these liquids.

The regular Trichel pulse regime observed in LM liquids implies that electron attachment is the dominant process at low field values.

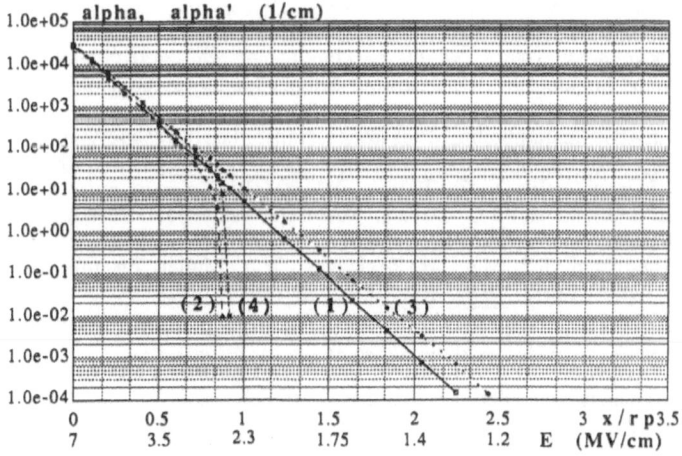

Fig. 3. Multiplication coefficient α and α' ($=\alpha-\eta$) vs non-dimensionalized distance x/r_p from the tip [$\equiv E(x/r_p)$] in liquid cyclohexane (1 and 2) and liquid propane (3 and 4).

From an estimation of the attachment coefficient, η, for fragments,[10] the variation of $\alpha'=\alpha-\eta$ as a function of E or of x/r_p where x is the distance from the point can be calculated (fig. 3, curves 2 and 4). Indeed, if $r_p<<d$, $E=E_sf(x/r_p)$. The length of the multiplication region, x_i, in the liquid phase defined by $\alpha=\eta$ is of the order of the point radius (fig. 3). In compressed air at 13 MPa where $E_s=7$ MV/cm, a value of $0.6r_p$ is obtained. The critical field E_i at $x=x_i$ is about 2.5 MV/cm. We can note that the coefficient α' has elsewhere been evaluated in n-hexane by Arii et al,[11] but with markedly higher values. So, for E=1MV/cm, they still observe the multiplication phenomenon ($\alpha'=80$ cm^{-1}), while we find that this does not exist for E<2.3 MV/cm (fig.3).

CONCLUSION

For large point radii ($r_p>0.5$ μm), the regular pulse regime observed in LM liquids and conduction phenomena observed in HM liquids are shown to correspond to hot electron mechanisms in the liquid phase. A rough estimation of the multiplication coefficients, α and α', in the liquid phase can be made from studies in compressed hydrocarbonous gases and liquids. The length of the ionization region and the corresponding critical electrical field can be deduced. It is then shown that fields, E>2.5 MV/cm, are required for electronic ionization in liquid cyclohexane and propane.

REFERENCES

1. A. Denat, J.P. Gosse and B. Gosse, Electrical conduction of purified cyclohexane in a divergent electric field, IEEE Trans. Electr. Insul. 23:545 (1988).
2. M. Haidara, "Impulsions de Trichel dans le cyclohexane liquide et les gaz comprimés," Doctorat de l'Université J. Fourier, Grenoble, France (21/12/1988).
3. W.L. Lama and C.F. Gallo, Systematic study of the electrical characteristics of the Trichel current pulses from negative needle-to-plane coronas, J. Appl. Phys. 45:103 (1974).
4. M. Haidara and A. Denat, Influence of pressure on electrical discharges in liquid and gaseous hydrocarbons, 9th Int. Conf. on Gas Discharges and their Applications, Venezia (19-23 sept. 1988).
5. C. Marteau, M. Haidara and A. Denat, Dispositif d'étude des décharges électriques, Revue Phys. Appl. 24:597 (1989).
6. R. Kattan, A. Denat and O. Lesaint, Generation, growth and collapse of vapor bubbles in hydrocarbon liquids under a high divergent electric field, J. Appl. Phys. 66:4062 (1989).
7. Von H. Schlumbohm, Elektronen-stoßionisierungskoeffizient (α) für organische dämpfe und sauerstoff (aus der trägerstatistik von elektronenlawinen), Z. Physik 4:156 (1959).
8. O.H. LeBlanc and J.C. Devins, Townsend ionization constants in n-alkanes, Nature 189:219 (1960).
9. S.E. Derenzo, T.S. Mast, H. Zaklad and R.A. Muller, Electron avalanche in liquid Xenon, Phys. Rev. A9:2582 (1974).
10. K. Arii, I. Kitani and M. Kawamura, Avalanche breakdown in n-hexane, J. Phys D: Appl. Phys. 12:787 (1979).
11. D.K. Davies, L.E. Kline and W.E. Bies, Measurements of swarm parameters and derived electron collision cross sections in methane, J. Appl. Phys. 65:3311 (1989).

DISCUSSION

F. SCHWIRZKE: Concerning the liquid phase pressure dependence, over what range did you change the pressure?

A. DENAT: From 0.1 MPa to 12 MPa, i.e. a pressure 3 to 6 times higher than the critical pressure of the studied liquids.

E. KUNHARDT: (1) Is there local ordering around the tip to affect field emission and local heating? (2) Also, are the electrons considered quasi—free with respect to ionizing molecules in the liquid?

A. DENAT: (1) The electric field at the tip ($\sim$ 7 MV/cm) leads to an electrostrictive pressure of 3 MPa, so the total pressure can go to 15 MPa. This is certainly too low to cause local ordering. (2) At this high field, the electrons can be considered as quasi—free whatever the liquid, so that they can lead to avalanches.

ELECTRON DRIFT VELOCITIES IN FAST DIELECTRIC LIQUIDS AND THEIR VAPORS[1]

H. Faidas, L. G. Christophorou, D. L. McCorkle and J. G. Carter

Atomic, Molecular and High Voltage Physics Group
Health and Safety Research Division
Oak Ridge National Laboratory
Oak Ridge, Tennessee 37831 USA

and

Department of Physics
The University of Tennessee
Knoxville, Tennessee 37996 USA

ABSTRACT

Excess electron drift velocities, w, as a function of the applied electric field, E, have been measured in liquid 2,2–dimethylpropane (TMC), tetramethylsilane (TMS), tetramethylgermanium (TMG), tetramethyltin (TMT), 2,2,4,4–tetramethlypentane (TMP), mixtures of TMS with TMP and n–pentane and in TMC, TMS, and TMP vapors. Our findings on the electron transport and the electrical properties of these liquids/vapors are discussed.

INTRODUCTION

Fast room temperature liquids (say, those with electron drift velocity $w > 10^6$ cm s^{-1} at an applied electric field E of $\sim 5 \times 10^4$ V cm^{-1}) are finding wide applications and are being considered for use in liquid–filled radiation detectors[1] and pulsed power switches.[2] Similarly, the vapors of these liquids in the pure form or in mixtures with gases (e.g., Ar) can be used in gas–filled (e.g., muon) detectors.[3] The molecules of such liquids have "spherical" shapes usually containing a central atom and four methyl groups; they include: 2,2–dimethylpropane (TMC), tetramethyl– silane (TMS), tetramethylgermanium (TMG), tetramethyltin (TMT), 2,2,4,4– tetramethylpentane (TMP) and their mixtures. The fundamental electrical properties of these liquids that make them desirable for many applications and determine their usefulness are their high dielectric strength (generally $> 10^5$ V cm^{-1}) and their ability to transport excess electrons at speeds approaching 10^7 cm s^{-1}.

In this paper we report w versus E measurements in liquid TMC, TMS, TMG, TMT, TMP, and mixtures of TMS with TMP (1.31:1 mole ratio) and n–pentane

---

[1]Research sponsored by the Office of Naval Research and the U.S. Department of Energy under respectively, Contract Nos. N00014–89–J–1990 and DE–AS05–76–ER03956 with the University of Tennessee, and by the Office of Health and Environmental Research, U.S. Department of Energy, under Contract No. DE–AC05–84OR21400 with Martin Marietta Energy Systems, Inc.

(102:1, 17:1, and 5.6:1 mole ratios), and w versus E/N, where N is the number density, in TMC, TMS, and TMP vapors. We report also on our findings concerning the electrical properties of these liquids that are relevant for applications.

EXPERIMENTAL

Unlike earlier measurements of w or of thermal electron mobility, μ_{th}, in room temperature liquids where bulk ionization techniques were employed,[4] in the present measurements a fast photoinjection technique was employed.[5,6] Briefly, a laser pulse (308 nm, ~15 ns or 337 nm, ~600 ps) struck a metal cathode and injected electrons into the medium that drifted across the cathode–anode distance, d, under the influence of an applied uniform electric field E = (V/d), where V is the applied voltage. The signal was measured as a voltage (Fig. 1a) or current (Fig. 1b) waveform. From these waveforms the drift time, τ, and w = d/τ and μ = w/E = $d^2/\tau V$ were determined. Numerical differentiation of the current waveforms gave the rate of charge injection at the cathode and charge collection of the anode (Fig. 1c).

Analysis of the charge injection waveforms (curves in Fig. 1d) as a function of E and correlation of the injection and collection pulses can give information on the longitudinal electron diffusion coefficient, D_L. When an ultrashort pulse is used (e.g., 600 ps), then the broadening of the collection pulse due to the laser pulse duration can be ignored and the width of the collection pulse (2 δt), can be used to calculate[7] directly D_L: $D_L = d^2 (\delta t)^2/4\tau^3$.

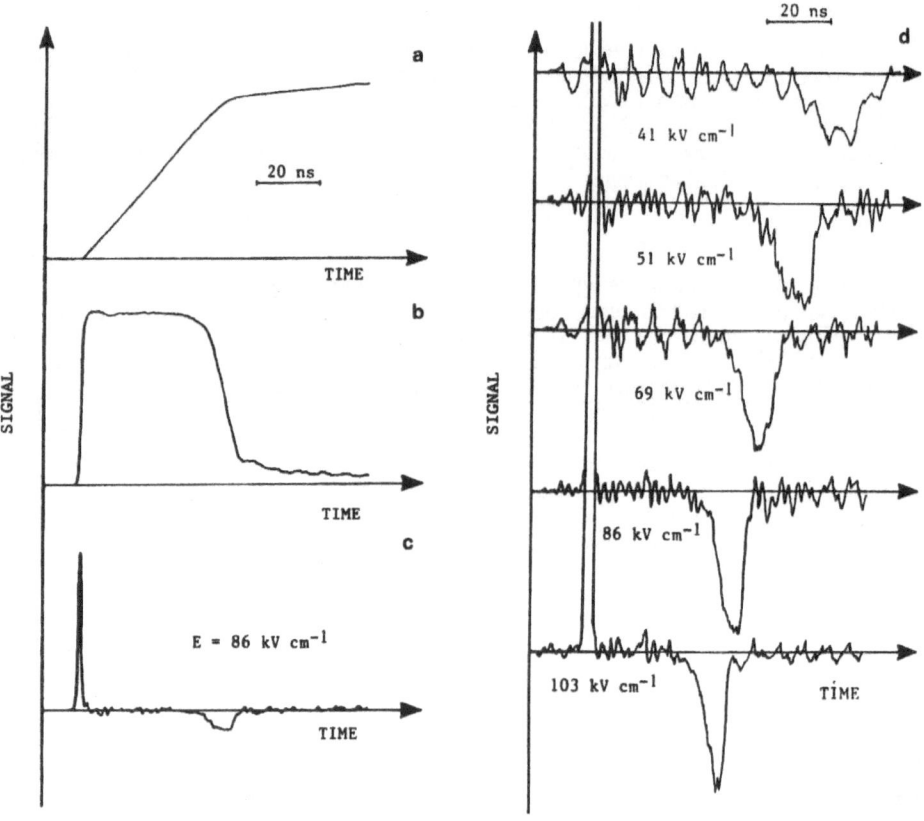

Fig. 1. Voltage (a), current (b), and charge injection rate (c, d) waveforms in TMG with a 600 ps laser pulse.

To check the accuracy of the technique we measured w and D_L for gaseous Ar and compared our values with those in literature. The w values were within 1–2%, while the D_L values were within the experimental error (5% to 20% depending on E/N) which was mainly due to the uncertainty in determining δt.

RESULTS AND DISCUSSION

Liquids

Figures 2a,b show w(E) and μ(E) for neat TMC, TMS, TMG, TMT, and TMP, while Figs. 2c,d show the same data for the mixtures. For all liquids there is a critical field value, E_c, such that for $E < E_c$, $\mu = w/E = \mu_{th}$ = constant. For $E > E_c$, the w versus E dependence becomes sublinear. At higher E–values, w(E) tends toward saturation.[8] In our measurements, however, saturation did not occur up to the maximum field applied, E_{max}. An interesting feature in Figs. 2a, b is that because the E_c–values for TMT and TMG are significantly larger than that for TMS, the w(E) curves of TMT and TMG cross over those of TMS for $E > 60$ kV cm⁻¹ and 25 kV cm⁻¹ respectively. Table 1 lists the mole ratio M, E_c, μ_{th}, E_{max}, and the corresponding maximum electron drift velocity measured, w_{max}, for the liquids/mixtures studied. The w(E) and μ(E) data for the mixtures indicate that: (i) small (~1 to 2%) amounts of nonelectronegative impurities do not significantly affect w and (ii) these impurities may actually help improve the dielectric strength.

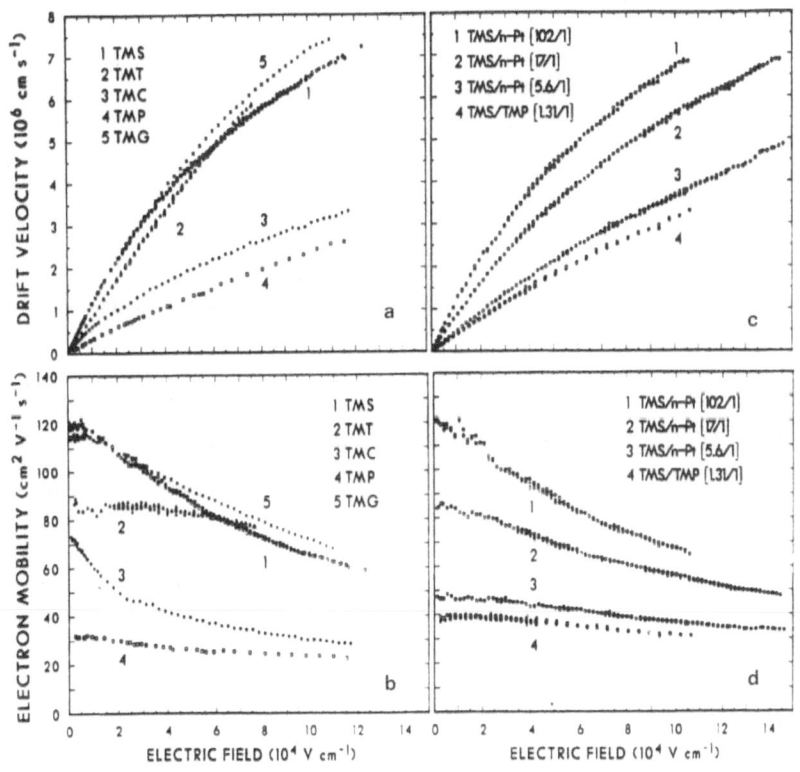

Fig. 2.   Electron drift velocities and electron mobilities in pure fast liquids and mixtures.

Table 1: Mole Ratio M, Critical Electric Field E_c, Thermal Electron Mobility μ_{th}, Maximum Electric Field E_{max}, and Maximum Electron Drift Velocity w_{max}, for the Fast Neat Liquids and Mixtures Studied.

Liquid/Mixture	M	E_c (kV cm^{-1})	μ_{th} (cm^2 s^{-1} V^{-1})	E_{max} (kV cm^{-1})	w_{max} (10^6 cm s^{-1})
TMC	Neat	3.5	71.5	116[a]	3.3
TMS	Neat	7	119.3	125	7.2
TMG	Neat	15	114.7	109[b]	7.4
TMT	Neat	30	85.7	75	6.0
TMP	Neat	15	31.8	115	2.6
TMS/TMP	1.31/1	18	39.1	105	3.2
TMS/n–pentane	102/1	7	118	105[a]	6.8
TMS/n–pentane	17/1	8	85	145[a]	6.8
TMS/n–pentane	5.6/1	15	47.6	145	4.9

[a]E_{max} was not limited by electrical breakdown.
[b]Electrical breakdown occurred at 120 kV cm^{-1}.

Table 2: Vapor Pressure, Number Density N, Thermal Electron Mobility, μ_{th}, Density-Normalized Electron Mobility μ_n, and the Ratio, R, for Liquid and Gaseous TMC, TMS, and TMP.

	TMC		TMS		TMP	
	Liquid	Gas	Liquid	Gas	Liquid	Gas
Vapor Pressure [Torr]		600		680		21.15
N [Molecules cm^{-3}]	5.121×10^{21}	1.963×10^{19}	4.423×10^{21}	2.224×10^{19}	3.378×10^{21}	6.804×10^{17}
μ_{th} [cm^2 s^{-1} V^{-1}]	71.5	1,338	119.3	354.7	31.8	31,959
μ_n [s^{-1} V^{-1} cm^{-1}]	3.6×10^{23}	2.84×10^{22}	5.97×10^{23}	7.89×10^{21}	1.05×10^{23}	9.5×10^{21}
$R = (\mu_n)_L/(\mu_n)_G$		13.7		66.8		11.0

182

Figure 3 shows the w(E/N) measurements of gaseous and liquid TMC, TMS, and TMP. The measurements were taken at room temperature ($\sim$ 295 K) and at vapor pressures of 500 and 680 Torr for TMS, 600 Torr for TMC and 21.15 Torr for TMP. In Table 2 are listed the gas pressures, the gas and the corresponding liquid number densities N, μ_{th}, the density normalized electron mobility $\mu_n = \mu_{th}$ N and the ratio R = $(\mu_n)_L/(\mu_n)_G$ for TMC, TMS, and TMP. The gas has a much lower μ_n value than the corresponding liquid; R is 66.8, 13.7 and 11 for TMS, TMC, and TMP, respectively. This would indicate that at thermal and near thermal energies the electron scattering cross sections in these liquids are smaller than in the corresponding gas. Such changes are known to occur especially for molecules that exhibit a Ramsauer–Townsend minimum at low energies.[9] Another difference between the gas and the liquid is that the former exhibits a supralinear w–E dependence at intermediate E/N values before it reaches a saturation value for w at $\sim$ 5 x 10^6 cm s^{-1} while the liquid shows a sublinear w–E dependence.

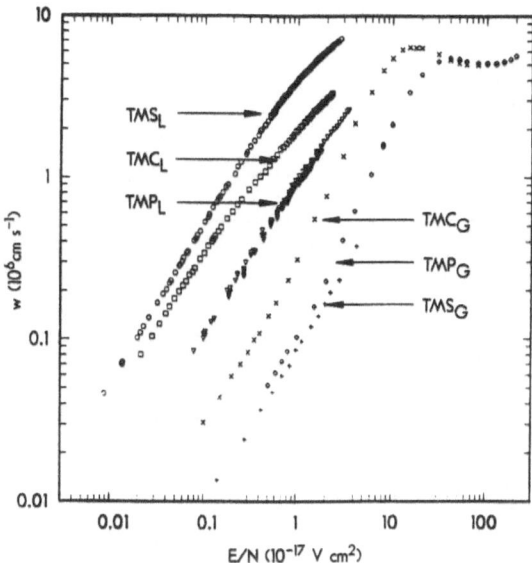

Fig. 3.   Electron drift velocities in liquid (TMS$_L$) and gaseous (TMS$_G$) TMS, liquid (TMG$_L$) and gaseous (TMC$_G$) TMC, and liquid (TMP$_L$) and gaseous (TMP$_G$) TMP.

ELECTRICAL PROPERTIES: FINDINGS RELEVANT TO APPLICATIONS

Table 3 shows the comparative properties of TMC, TMS, TMG, TMT, and TMP. In all liquids except TMT electrical breakdown occurred at E > 10^5 V cm^{-1}. The TMS/n–pentane mixtures appear to have relatively higher dielectric strengths. All liquids underwent various degrees of decomposition after an electrical breakdown. In TMS decomposition was minimal and did not affect noticeably the subsequent performance of the system or the measurements. In TMG and (more so) in TMT extensive decomposition occurred, the cell surfaces were covered with heavy layers of Ge and Sn, respectively, and the dielectric strength of the liquid was lowered to $\sim$ 20 to 30 kV cm^{-1}. In TMP and TMG electronegative impurities were formed that reduced the electron lifetime to less than $\sim$ 100 ns and 10 ns respectively. From our studies it appears that unless electrical breakdown can be totally avoided, TMS is the most promising liquid for applications.

Table 3: Comparative "Practical" Properties of TMC, TMS, TMG, TMT, and TMP

	Purification	Drift Velocity	Dielectric Strength	Effects of Electrical Breakdown
TMC	Easy/Quick	Inter-mediate	High	
TMS	Easy/Quick	Fast	High	No apparent liquid deterioration Damage to electronics/pitting of electrodes
TMG	Easy/Quick	Fast	High	Extensive liquid decomposition rendered system unusable. Electronegative impurity(ies) formed. Damage to electronics/pitting of electrodes
TMT	Difficult reacts with H_2SO_4 and NaK	Fast	Low	Same as TMG
TMP	Slow	Slow	High	Electronegative impurity(ies) formed

REFERENCES

1. M. G. Albrow and collaborates, Performance of a uranium/tetramethylpentane electromagnetic calorimeter, Nucl. Instr. and Meth. A265:303 (1988).

2. L. G. Christophorou and H. Faidas, Dielectric liquids for possible use in pulsed power switches, Appl. Phys. Lett. 55:948 (1989).

3. J. Va'vra, High resolution drift chambers, Nucl. Instr. and Meth. A244:391 (1986).

4. W. F. Schmidt, Electronic conduction processes in dielectric liquids, IEEE Trans. Electr. Insul. EI–19:389 (1984).

5. H. Faidas, L. G. Christophorou, and D. L. McCorkle, Drift velocities of excess electrons in TMP and TMS: A fast drift technique, Chem. Phys. Lett. 163:495 (1989).

6. H. Faidas, L. G. Christophorou, D. L. McCorkle, and J. G. Carter, Electron drift velocities and electron mobilities in fast room–temperature dielectric liquids and their corresponding vapors, Nucl. Instr. and Meth. A294:575 (1990).

7. E. B. Wagner, F. J. Davies, and G. S. Hurst, Time–of–flight investigations of electron transport in some atomic and molecular gases, J. Chem. Phys. 47:3138 (1967).

8. W. Döldissen and W. F. Schmidt, Saturation of electron drift velocity in liquid TMS, Chem. Phys. Lett. 68:527 (1979).

9. L. G. Christophorou, Gas/liquid transition: Interphase physics, in "The Liquid State and Its Electrical Properties," E. E. Kunhardt, L. G. Christophorou, and L. H. Luessen, Eds, Plenum Press, 283–316, New York (1988).

DISCUSSION

E. KUNHARDT: (1) Could you clarify what you mean by the lowering of the cross section in a liquid relative to a gas? (2) Will you be able to carry out similar experiments in liquid argon?

H. FAIDAS: (1) For the liquids/vapors studied (TMS, TMC, TMP) the electron scattering cross section in the liquid, i.e. the cross section due to scattering by an ensemble of molecules (nearest neighbors or more extended, depending on the electron energy) appears to be smaller than the cross section in the gas due to isolated electron–molecule collisions. Evidence for this is the lowering of w from the liquid to the gas for the same E/N. (2) We have constructed the experimental apparatus and are planning to measure w, μ and other transport parameters in liquid Ar using our laser photoinjection technique.

R. TOBAZEON: How is the breakdown voltage connected to the transit time of electrons? Can you comment on the breakdown voltage of the studied liquids?

H. FAIDAS: We did not find any correlation between breakdown voltage and the drift velocity, w (the transit time of the electrons) for the liquids we studied. The breakdown voltage of these liquids is relatively low compared to other dielectric liquids (n–hexane for example) The low breakdown voltage value could be due either to the presence of micro particles in the liquid, coming from the cell walls, or to the possibility that high mobility liquids have low breakdown voltage.

GAS BREAKDOWN AND ITS RELATION TO VACUUM AND LIQUID BREAK–DOWN UNDER THE INFLUENCE OF ELECTRIC AND MAGNETIC FIELD

J. C. PAUL

Department of Electrical Engineering
Tripura Engineering College
Tripura 799 055 India

ABSTRACT

Gas, vacuum and liquid breakdown have been the topics of considerable study and research for a long time. From Townsend onwards a considerable study has been made and several theories have been put forward by different researchers. The effects of magnetic field on the breakdown mechanism of gases, vacuum and liquid dielectrics have been studied earlier, but their relationships have not yet been established so far in the available literature. In this paper gas breakdown and its relation to vacuum and liquid breakdown under the influence of electric and magnetic fields are presented. It is evident from the analysis that gas phase exists in the liquid breakdown and the breakdown voltage increases under the influence of cross magnetic field (c.m.f.). A generalized current density equation is also presented in this paper.

INTRODUCTION

Pre–breakdown and breakdown phenomena of dielectrics whether, solid, liquid, gas or vacuum, play a significant role in the application of these materials in science, engineering and technology. The mechanism of conduction of these dielectrics leading to their ultimate electrical breakdown has drawn the attention of a large number of investigators. Research work on dielectric materials has in most cases been related to the study of the mechanism of ionization, conductivity and breakdown under the influence of electric field only. The performance characteristics of a gaseous medium subjected to the simultaneous application of electric and magnetic fields have been reported by some researchers.[1-3] Investigations on gases, vacuum, liquid and solid dielectrics under the influence of both electric and magnetic fields have been studied and reported by the author.[4-10]

GASEOUS BREAKDOWN

It is known that when a magnetic field, whose direction is at right angle to that of an electric field is applied to a gas Townsend discharge, the charged particles instead of moving in straight lines between collisions are deflected to a cycloidal path. Such a deflection increases the probability of collision in the direction of the electric field resulting in an effective reduction of the electron mean free path (m.f.p.). This reduction of m.f.p. by the magnetic field may be interpreted as an apparent increase of pressure which gives rise to the equivalent pressure concepts studied[3] extensively in the past. Under uniform field geometry equivalent pressure in presence of both electric and magnetic fields has been developed[7] in terms of m.f.p. (λ), collision frequency (ν), magnetic field(T) and gap distance between the electrodes (d) which may be expressed as:

$$P' = P \left\{ 1 + \left[\frac{\lambda \nu Td}{VP} \right]^2 \right\}^{\frac{1}{2}} \qquad (1)$$

The sparking potential (V) with c.m.f. may be written[7] as:

$$V = \frac{BPd \left\{ 1 + \left(\frac{\lambda \nu Td}{VP} \right)^2 \right\}^{\frac{1}{2}}}{\log \left[\frac{APd \left\{ 1 + \left(\frac{\lambda \nu Td}{VP} \right)^2 \right\}^{\frac{1}{2}}}{\log \left(1 + \frac{1}{\gamma} \right)} \right]} \qquad (2)$$

where A and B are constants and γ is the secondary Townsend ionization coefficient.

LIQUID DIELECTRIC

The conduction mechanism of liquid dielectrics is a complex phenomenon and there is no generally accepted theory which may be applied to analyze the behavior of liquid dielectrics under an electric field. However, a generalized current density equation has been developed and is presented here. Neglecting fringing the conduction current density (J) in perpendicular electric and magnetic field may be expressed as:

$$J = n \, q \, \mu (E + v \times T) + \sigma (E + v \times T) \qquad (3)$$

where n, q, and μ are number density, ionic charge and mobility respectively. The first part of equation (3) is due to the ionic contribution while the second part is due to the electronic contribution. The applied electric and magnetic fields are perpendicular to each other and in the above equation v and T are cross product. From equation (3) the generalized current density equation may be expressed as shown below (v = velocity):

$$J = \left[\frac{nq\mu_p \nu_p}{\sqrt{\left(\frac{qT}{m_p} \right)^2 + (\nu_p)^2}} + \frac{n_n q\mu_n \nu_n}{\sqrt{\left(\frac{qT}{m_n} \right)^2 + (\nu_n)^2}} + \frac{n_e e\mu_e \nu_e}{\sqrt{\left(\frac{eT}{m_e} \right)^2 + (\nu_e)^2}} \right] E \qquad (4)$$

First, second and third part of equation (4) are the contributions of conduction for positive ions, negative ions, and electrons respectively.

VACUUM BREAKDOWN WITH SOLID DIELECTRIC

Solid insulators in vacuum have been studied in the presence of electric and c.m.f. and the results are presented here.

EXPERIMENTAL SET UP

The experimental set up is shown in Figure 1 which may be used for the studies of gases and vacuum and in case of liquid dielectric the bottom portion of the system gas/vacuum is to be disconnected.

Figure 2 shows the gaseous breakdown under the influence of c.m.f. Figures 3 and 4 represent the behavior of the conduction current and photograph during the time of discharge with and without magnetic field. Figures 5 and 6 represent the vacuum breakdown behavior of solid dielectrics.

188

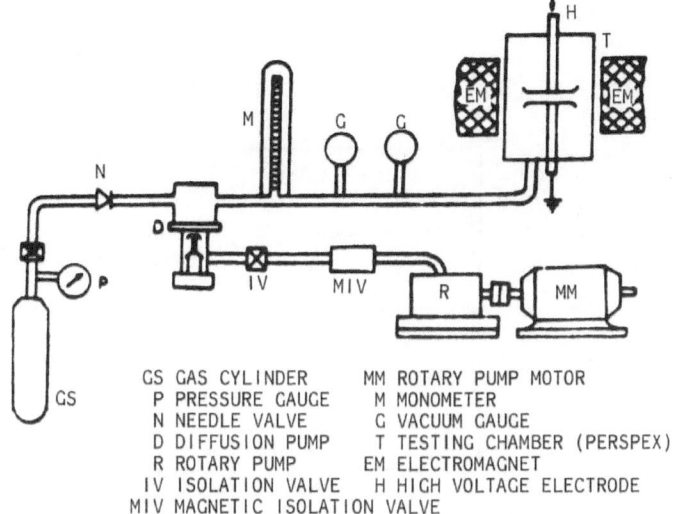

GS GAS CYLINDER	MM ROTARY PUMP MOTOR
P PRESSURE GAUGE	M MONOMETER
N NEEDLE VALVE	G VACUUM GAUGE
D DIFFUSION PUMP	T TESTING CHAMBER (PERSPEX)
R ROTARY PUMP	EM ELECTROMAGNET
IV ISOLATION VALVE	H HIGH VOLTAGE ELECTRODE
MIV MAGNETIC ISOLATION VALVE	

Fig. 1.   Experimental set up for gaseous and vacuum breakdown studies.   For
liquid breakdown studies the bottom portion to be disconnected.

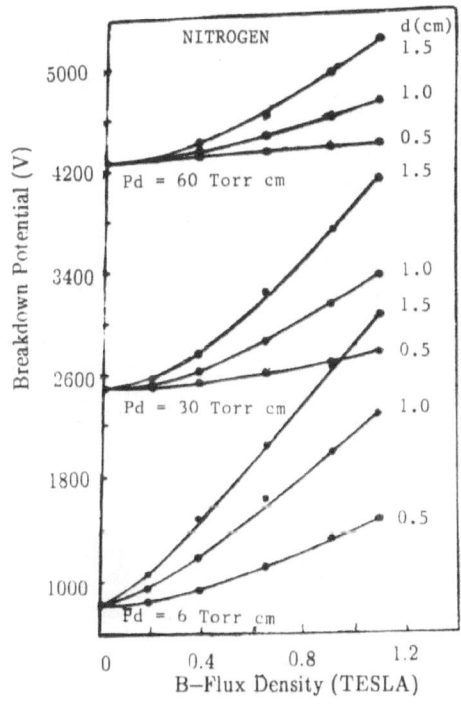

Fig. 2.   Relation between the breakdown potential and flux density for nitrogen
gas at different values of gap distance and gas pressure.

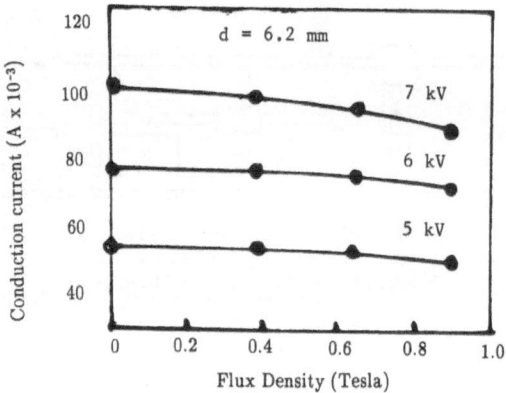

Fig. 3. Relation between the conduction current and flux density at different values of applied voltages for paraffin oil.

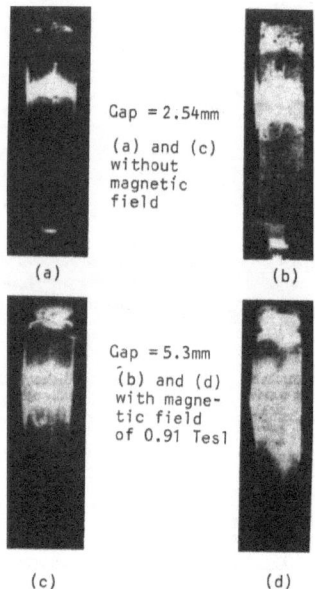

Gap = 2.54mm

(a) and (c) without magnetic field

(a)   (b)

Gap = 5.3mm

(b) and (d) with magnetic field of 0.91 Tesl

(c)   (d)

Fig. 4. Photographic observations during the time of discharge for paraffin oil with and without magnetic field.

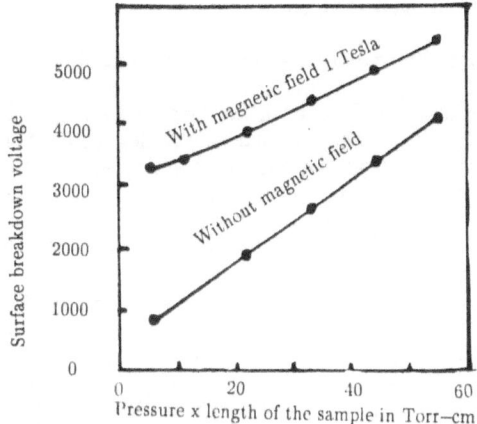

Fig. 5 · Surface breakdown voltage of solid insulator at different values of Torr–cm with and without magnetic field.

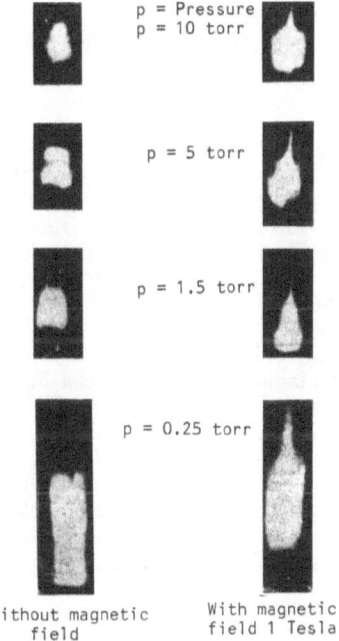

Fig. 6 · Photographic observations of solid insulator in vacuum at 0.25 Torr to 10 Torr with and without magnetic field.

191

DISCUSSION AND CONCLUSIONS

According to Paschen's law the sparking potential will remain constant if the product of pressure and gap distance remain constant. But this is not true when cross magnetic field is applied as it will be evident from Fig. 2. In the case of liquid dielectric it is generally accepted that molecules acquire the additional energy required for reaction known as the energy of activation, as a result of interchanging occurring in collision. It may be shown that with c.m.f. the conductivity of a liquid dielectric decreases due to the increase of activiation energy. This in turn reduces the conduction current and increases the breakdown voltage. In the case of insulators in vacuum it may be seen from Figure 6 that the discharge paths are channelized to a particular path under c.m.f. Free electrons emitted at the cathode insulator junction bombard the anode or insulator near the anode and release secondary electrons. Thus the surface of the insulators near the anode becomes positively charged to a potential approaching that of the anode. The charged area gradually moves closer to the cathode so that the stress at the cathode eventually increases to a value sufficient to cause breakdown. However, the following general conclusions may be drawn for the gases, vacuum and liquid dielectrics under c.m.f.

1. There is an increase in collision frequency.
2. There will be reduction in mean free path.
3. There will be increase of equivalent pressure.
4. There may be change in the dipole moment of molecules.
5. Total polarization of molecules likely to be changed.
6. There may be orientation of the spins of molecules.
7. There will be increase of activation energy.
8. The conductivity will decrease.
9. The net result will be the increase of breakdown voltage.

REFERENCES

1. E. Mayer, <u>Ann. Phys.</u> 58:297 (1919).
2. H. A. Belvin and S. C. Haydon, <u>Aust. J. Phys.</u> 11:18 (1958).
3. S. C. Haydon, <u>5th Int. Conf. on Ionization Phenomena, Munich</u> (1961).
4. J. C. Paul, Effective Field Calculation of Dielectric Under The Influence of Electric and Magnetic Fields, <u>Int. Sym. on H.V. Eng.</u>, Zurich (1975).
5. J. C. Paul, Conduction and Breakdown Mechanism of Some Liquid Dielectrics Considering Their Chemical Structures, <u>CEIDP</u>, Gaithersburg (1975).
6. J. C. Paul, Effect of Magnetic Field on the Breakdown Voltage of Liquid Dielectric, <u>Ind. J. Tech.</u> Vol. 16 (1978).
7. J. C. Paul, Effect of Magnetic Field on Mean Free Path and Viscosity and Its Influence at Low Pressure Gaseous Discharge, <u>Ind. J. Phys.</u> 96 (1979).
8. J. C. Paul, Surface Breakdown of Solid Insulator in Vacuum Under the Influence of Electric and Magnetic Fields, <u>J. Inst. Engrs. (Ind.)</u>, EL 1,65 (1984).
9. J. C. Paul, Behavior of Ferroelectric Materials Under the Influence of Magnetic Field, 5th Int. Sym. on Electrets, Heidelberg (1985).
10. J. C. Paul, Polarization Phenomena of Biological Dielectric Under The Influence of Electric and Magnetic Fields, <u>6th Int. Sym. on Electrets</u>, Oxford (1988).

MEASUREMENT OF THE FIGURE OF MERIT M FOR

SEVERAL SF_6-BASED BINARY GAS MIXTURES

J. Berril and I. W. McAllister

Physics Laboratory II, Building 309B
The Technical University of Denmark
DK-2800 Lyngby, Denmark

INTRODUCTION

The figure of merit M for a strongly electronegative gas was introduced by Pedersen (1980) as a means of quantifying the sensitivity of such gases to the effects of electrode surface roughness: the smaller M is, the more sensitive is the gas to the microscopic field perturbations produced by the inherent roughness of electrode surfaces. Subsequently Pedersen and colleagues (1984) have shown that this parameter has a wider significance as it is possible to evaluate the discharge onset characteristics of a strongly electronegative gas from a knowledge of M and the pressure-reduced limiting electric field strength $(E/p)_{lim}$.

In the present paper, high precision measurements of these parameters for several SF_6-based binary gas mixtures are reported. Such mixtures, which enable the $M, (E/p)_{lim}$ relationship to be investigated over an extended range of $(E/p)_{lim}$ values, are obtained by diluting SF_6 with several dielectrically-weaker gases. The gases in question are $CHClF_2$ (chlorodifluoromethane, Freon 22), CHF_3 (trifluoromethane, fluoroform, Freon 23), $1,1,1,2-C_2H_2F_4$ (1,1,1,2-tetrafluoroethane, CF_3CH_2F, Arcton 134a) and dry air.

For $(E/p)_{lim} < (E/p)_{lim}(SF_6)$, the present results supplement the existing $M, (E/p)_{lim}$ data, see Berril and McAllister (1989), and confirm that, in this range of $(E/p)_{lim}$, the value of M is not uniquely related to $(E/p)_{lim}$ although a well-defined lower limit exists.

FIGURE OF MERIT FOR A STRONGLY ELECTRONEGATIVE GAS

As the basis of the figure of merit concept and its determination are described in the literature (Pedersen et al., 1984), only a brief resumé will be provided in order to indicate the manner in which the experimental data are analysed.

Although the Paschen curve as a whole is non-linear and exhibits a min-

Table 1. Paschen curve data for $CHClF_2/SF_6$ mixtures.

% SF_6	$\dfrac{(E/p)_{lim}}{kV(mm\ bar)^{-1}}$	$\dfrac{U_0}{kV}$	$\dfrac{M}{bar\ mm}$	r^2
100.0	8.86	0.35	0.040	0.9_6
90.3	8.60	0.36	0.041	0.9_5
80.2	8.43	0.38	0.045	0.9_5
70.3	8.29	0.39	0.047	0.9_5
59.9	8.13	0.43	0.052	0.9_5
50.2	7.95	0.46	0.058	0.9_5
40.2	7.71	0.52	0.067	0.9_5
30.1	7.38	0.58	0.079	0.9_5
20.1	6.88	0.77	0.112	0.9_5
15.0	6.57	0.79	0.120	0.9_5
10.1	6.12	1.01	0.165	0.9_5
5.0	5.44	1.35	0.248	0.9_5

Note: $0.9_5 \equiv 0.99999$

imum value, it is, for a strongly electronegative gas, essentially linear over the range of interest to practical applications. This relationship may be represented as

$$U_s = U_0 + (E/p)_{lim}\ pd, \tag{1}$$

where U_s is the breakdown voltage, p the gas pressure, d the gap length and U_0 a constant voltage. $(E/p)_{lim}$ is the value of (E/p) for which the value of the effective coefficient of ionisation $\bar{\alpha}$ is zero. From the definition of the figure of merit M, it can be shown that M is expressible in terms of Paschen curve parameters, viz.

$$M = \frac{U_0}{(E/p)_{lim}}. \tag{2}$$

Consequently the measurement of the linear part of the Paschen curve for a strongly electronegative gas provides data on both $(E/p)_{lim}$ and M.

RESULTS AND DISCUSSION

The experimental system and the basic measuring procedure have been fully described (Berril et al., 1982, 1986). Measurements are performed on a uniform-field gap using a highly stabilized DC voltage. The gap is ir-radiated by a radioactive source located axially within the high voltage electrode. In the present study. all the breakdown measurements were made at a nominal total pressure p_t of 0.7 bar at 20 °C.

With the present experimental apparatus, the measured onset breakdown voltage is observed to be an exceptionally well-defined quantity, such that

Table 2. Paschen curve data for 1,1,1,2-$C_2H_2F_4$/SF_6 mixtures.

% SF_6	$\dfrac{(E/p)_{lim}}{kV(mm\ bar)^{-1}}$	$\dfrac{U_0}{kV}$	$\dfrac{M}{bar\ mm}$	r^2
100.0	8.86	0.35	0.040	0.9_6
90.0	8.63	0.40	0.046	0.9_6
80.0	8.42	0.43	0.051	0.9_6
70.1	8.20	0.40	0.049	0.9_6
60.0	7.97	0.44	0.055	0.9_6
50.1	7.72	0.46	0.060	0.9_6
40.0	7.40	0.60	0.080	0.9_5
29.9	7.05	0.63	0.089	0.9_5
25.0	6.85	0.62	0.090	0.9_5
19.9	6.57	0.76	0.116	0.9_6
14.9	6.26	0.76	0.122	0.9_6
10.7	5.92	0.85	0.143	0.9_5
4.8	5.24	1.00	0.190	0.9_5

Table 3. Paschen curve data for dry air/SF_6 mixtures.

% SF_6	$\dfrac{(E/p)_{lim}}{kV(mm\ bar)^{-1}}$	$\dfrac{U_0}{kV}$	$\dfrac{M}{bar\ mm}$	r^2
100.0	8.86	0.35	0.040	0.9_6
90.0	8.68	0.42	0.049	0.9_5
80.1	8.48	0.41	0.048	0.9_5
69.8	8.20	0.47	0.057	0.9_5
60.0	7.90	0.48	0.060	0.9_5
50.1	7.52	0.62	0.083	0.9_5
40.1	7.11	0.62	0.088	0.9_5
30.1	6.60	0.70	0.105	0.9_6
25.1	6.30	0.80	0.126	0.9_5
20.0	5.98	0.78	0.130	0.9_5
15.0	5.62	0.84	0.150	0.9_5
9.4	5.11	0.93	0.181	0.9_5
5.0	4.57	1.01	0.222	0.9_5

the standard deviation $s(U)$ associated with the determination of U_s is ~10 V.
The breakdown voltage measurements were analysed in terms of (1) and the
results are summarized in the tables. From the values of the coefficient of

Table 4. Paschen curve data for CHF_3/SF_6 mixtures.

% SF_6	$\dfrac{(E/p)_{lim}}{kV(mm\ bar)^{-1}}$	$\dfrac{U_0}{kV}$	$\dfrac{M}{bar\ mm}$	r^2
100.0	8.86	0.35	0.040	0.9_6
88.9	8.42	0.37	0.044	0.9_5
80.2	8.08	0.27	0.033	0.9_5
70.3	7.69	0.27	0.035	0.9_4
60.1	7.25	0.35	0.049	0.9_5
50.1	6.82	0.34	0.050	0.9_5
40.0	6.34	0.38	0.060	0.9_5
29.9	5.83	0.40	0.069	0.9_5
20.0	5.24	0.47	0.090	0.9_5
15.1	4.89	0.54	0.110	0.9_5
10.0	4.47	0.64	0.144	0.9_5
5.0	3.91	0.75	0.193	0.9_6

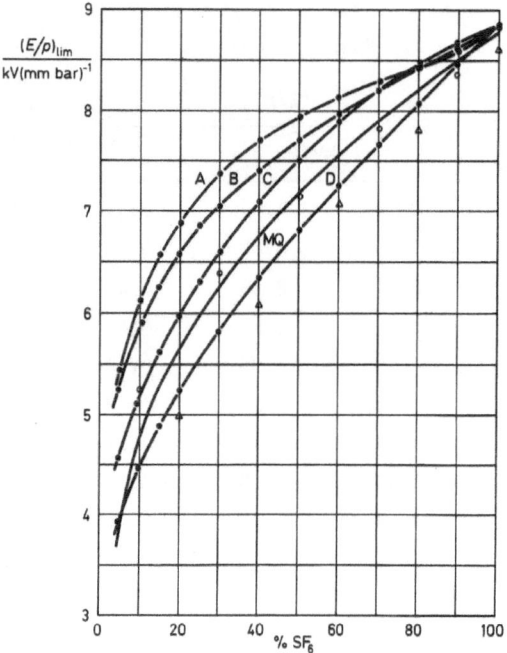

Fig.1. Variation of $(E/p)_{lim}$ with % SF_6 in mixture.
A – $CHClF_2/SF_6$ B – $1,1,1,2-C_2H_2F_4/SF_6$
C – dry air/SF_6 D – CHF_3/SF_6
MQ Malik and Qureshi (1980)
o Lee (1983), and Chatterton et al. (1985), Fig.4
Δ Christophorou et al. (1981), Fig.7

determination r^2, it is clear that the variation of the measured values of U_s with pd fit almost uniquely to a straight line. The Paschen curve data for the four SF_6-based mixtures are listed in Tables 1 – 4, and the variations of $(E/p)_{lim}$ for these mixtures are shown in Fig.1. For comparison, previously reported $(E/p)_{lim}$ values are included.

On the basis of the partial-pressure ratios of the mixture components, Malik and Qureshi (1980) assumed an analytical expression for the pressure-reduced effective coefficient of ionisation $\bar{\alpha}/p$ of dry air/SF_6 mixtures and thereafter deduced values for the corresponding $(E/p)_{lim}$. This variation of $(E/p)_{lim}$ is shown in Fig.1. In addition, these authors reported values of $(E/p)_{lim}$ obtained from breakdown voltage measurements at $0.1 \leq p/bar \leq 0.2$, see Fig.5 of Malik and Qureshi (1980). However as these experimentally-derived data are in reasonable agreement with the analytically-derived data, only the latter have been included in Fig.1. Lee (1983) and Chatterton, Moruzzi & Lee (1985) have also reported values of $(E/p)_{lim}$ for dry air/SF_6 mixtures. As these data, which are obtained from breakdown voltage measurements and current measurements, both steady-state and pulse, agree to within 3 %, only average values are indicated in Fig.1. In general, the $(E/p)_{lim}$ values of Malik and Qureshi are ~ 0.3 kV/(mm bar) less than those obtained in the present study, whereas the $(E/p)_{lim}$ values of Lee and colleagues straddle both sets of results.

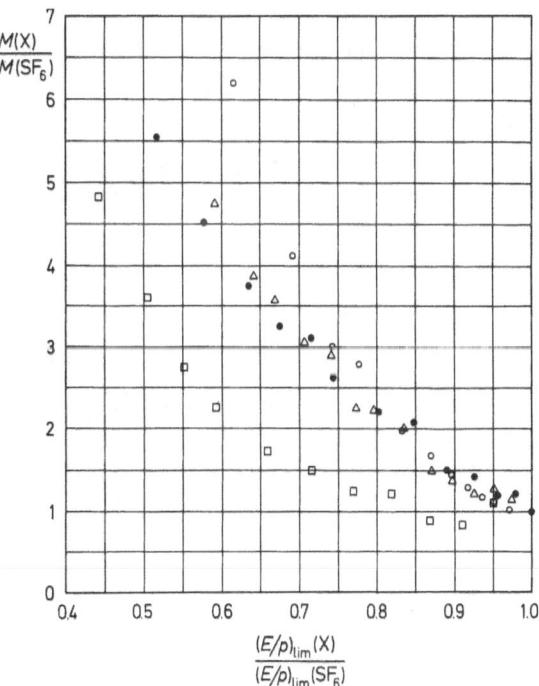

Fig.2. Variation of M with $(E/p)_{lim}$.

o $CHClF_2/SF_6$,　● dry air/SF_6

Δ $1,1,1,2-C_2H_2F_4/SF_6$,　□ CHF_3/SF_6

Christophorou and colleagues (1981) measured the breakdown voltages for several CHF_3/SF_6 mixtures, and the corresponding $(E/p)_{lim}$ values are shown in Fig.1. The agreement with the present results is good.

The present measurements represent the first study of the dielectric strength of $1,1,1,2-C_2H_2F_4/SF_6$ mixtures. This situation has arisen as $1,1,1,2-C_2H_2F_4$ is one of the new compounds developed to replace the widely used CFC, CCl_2F_2 (Steven & Lindley, 1990). The dielectric strength of $1,1,1,2-C_2H_2F_4$ itself is discussed in McAllister (1989). No directly comparable $(E/p)_{lim}$ data have been found in the literature for $CHClF_2/SF_6$ mixtures.

The variation of M with $(E/p)_{lim}$ for the mixtures examined in this study is illustrated in Fig.2. The present results confirm the trend in the M, $(E/p)_{lim}$ relationship observed earlier for $(E/p)_{lim} < (E/p)_{lim}(SF_6)$ (Berril and McAllister, 1989). This trend, which is associated with a bounded spread in the M values, implies that the value of M is not determined uniquely by the $(E/p)_{lim}$ value. Nevertheless, there are clear indications that, for any value of $(E/p)_{lim}$, there exists a well-defined lower limit to the range of possible M values. This is significant as one is particularly interested in worse case situations in electrical insulation.

CONCLUSION

With the present results, there now exists sufficient data in the literature to enable a viable analysis of the $M, (E/p)_{lim}$ relationship to be made. Such an analysis will be the subject of a future study.

ACKNOWLEDGMENTS

This work was funded in part by EPRI through project RP2669-1. The authors wish to thank ICI Chemicals & Polymers Ltd. for the Arcton 134a.

REFERENCES

Berril, J., Christensen, J. M., and Pedersen, A., 1982, Seventh International Conference on Gas Discharges and their Applications, London, 266.
Berril, J., Christensen, J. M., and McAllister, I. W., 1986, Conference Record of the 1986 IEEE International Symposium on Electrical Insulation IEEE Press, New York, Publication 86CH2196-4-DEI, 251.
Berril, J. and McAllister, I. W., 1989, Sixth International Symposium on High Voltage Engineering, New Orleans, Paper No.32.08.
Chatterton, P. A., Moruzzi, J. L., and Lee, Z. Y., 1985, Eighth International Conference on Gas Discharges & Their Applications, Oxford, 307.
Christophorou, L. G., James, D. R., and Mathis, R. A., 1981, J. Phys. D: Appl. Phys., 14:675.
Lee, Z. Y., 1983, Fourth International Symposium on High Voltage Engineering, Athens, Paper No.31.01.
Malik, N. H. and Qureshi, A. H., 1980, IEEE Trans. Elect. Insul., EI-15:413.
McAllister, I. W., 1989, J. Phys. D: Appl. Phys., 22:1783.
Pedersen, A., 1980, in: "Gaseous Dielectrics II," L. G. Christophorou, ed., Pergamon, New York, 201.
Pedersen, A., McAllister, I. W., Crichton, G. C., and Vibholm, S., 1984 Arch. Elektrotech., 67:395.
Steven, H. and Lindley; A., 1990, New Scientist, 126(1721):48.

DISCUSSION

J. CASTONGUAY: How can the "summarized" results of your work presented in Fig. 2 be used to select an optimum binary gas mixture with SF_6: best gas partner and best composition?

I. MCALLISTER: As the sensitivity to surface roughness is less the greater the value of M, the results illustrated in Fig. 2 suggest that, for a given $(E/p)_{lim}$, the mixtures studied may be rated as $CHClF_2/SF_6$, 1,1,1,2–$C_2H_2F_4/SF_6$, dry air/SF_6, and CHF_3/SF_6. However to properly evaluate the insulation capabilities of the different mixtures, it is necessary to undertake onset calculations to determine which mixture composition exhibits the optimum combination of M and $(E/p)_{lim}$.

MODE TRANSITIONS IN HOLLOW-CATHODE DISHARGES

G.A. Gerdin, K.H. Schoenbach, L.L. Vahala and T. Tessnow

Department of Electrical and Computer Engineering
Old Dominion University
Norfolk, VA 23529

INTRODUCTION

Hollow-Cathode Discharges (HCD) have been studied for many years.[1] A major cause for the interest in these discharges is the relatively high current densities that can be conducted through the HCD; currents of over 100 kA with rates of rise of 2×10^{12} A/s have been observed.[2] This ability to carry high currents, plus the observation of sharp impedance changes with pressure,[3-7] seem to make the HCD a candidate for a high-power switch. Furthermore, the observation that an axial magnetic field of sufficient strength B_z, can be used to suppress the transition from the high-impedance mode to its low-impedance mode,[7] could indicate a means of switch control. To put these possibilities on a firmer footing, it is desirable to simulate the HCD computationally, to see if these HCD phenomena can indeed be explained on the basis of first principles. Once this is achieved, it is much easier to understand the underlying physical mechanism causing these mode transitions. The nature of this mechanism could then suggest confirming experimental tests, and suggest a means of control of these impedance transitions, and hence lead the way toward the development of a high-power switch. In this paper, substantial improvements in our ability to simulate the HCD are reported, and the results of some preliminary calculations are compared with experimental results.

Previous Theoretical Work

Since the pressure and gap length at which these HCD mode transitions occur is on the left-hand side of the Paschen curve,[3-7] the mean-free path for electron elastic collisions λ_{ec}, is not small with respect to the dimensions of the system, so models using the local field approximation (LFA),[8,9] where the electron drift speed is directly proportional to the local electric field times a mobility, can not describe the entire electron population. Some high-speed electrons can oscillate back and forth between the surface of the cathode hole and the axis of the hole several times before escaping, and cause increased ionization over that predicted by an LFA model; these electrons are called "pendel" electrons. In the LFA models, the effect of these pendel electrons can be included in terms of a modified ionization coefficient,[8] but one expects these models to only give qualitative results in general, although certain features of the electrical breakdown in the HCD can be modeled reasonably well.[9]

To account for the pendel and other non-equilibrium electrons (i.e., those electrons not treated by the LFA models), methods involving Monte-Carlo calculations, solutions to the Boltzmann equation, and the convective scheme[10] could be employed. These techniques have been applied to glow discharges, and the results of the first two methods have been reviewed in the literature.[11,12]

PRESENT APPROACH

To model the HCD plasma and mode transitions, a one-dimensional Monte-Carlo computational technique has been implemented. The single dimension represents the radial dimension of the cathode-hole plasma, which is assumed to have both axial and azimuthal symmetry to a first approximation. The motion of the electrons is in the (r,ϕ) plane, which is represented in Figure 1 by a dashed line labeled S. This is justified in treating the HCD mode transitions, because it is experimentally observed,[3-8] that the pressure at which these transitions occur, is independent of the distance between the cathode and the anode, and is dependent only on the applied voltage[7] and the diameter of the hole.[3]

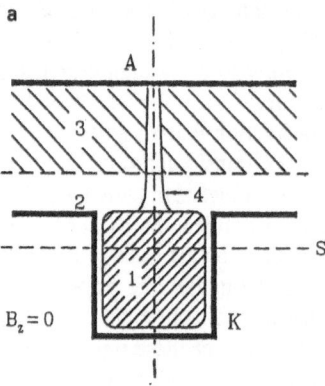

Fig. 1a. A schematic of the model for the high-impedance HCD mode (HIM) with zero or weak axial magnetic fields. Area 1) is the cathode-hole plasma (CHP); area 2) is the cathode-fall region (CFR); area 3) is the positive-column negative-glow region (PCNGR); and area 4) is the region through which the collimated electron beam passes.

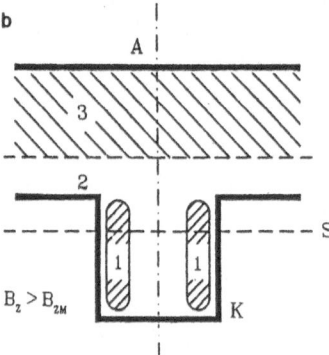

Fig. 1b. A schematic of the magnetic-insulation effect showing the constriction of the CHP (again area 1)) if the axial magnetic field B_z is greater than some threshold value B_{zm}. In both a) and b) A, K, and S label the anode, cathode, and (r,ϕ) plane where the electron motion is followed, respectively.

The overall scheme to be employed in the present simulations is a hybrid approach,[13] where the electron kinetics is modeled by means of a Monte-Carlo technique, and the results (e.g., electron drift velocity and the ionization rate coefficients) are used in a fluid model, to determine the charge density and current distributions of the ions and electrons between the inner wall of the cathode hole and its axis. This is an iterative method,[13] which starts with an initial guess of the electric field distribution between the cathode wall and the device axis, with the subsequent results of the Monte-Carlo and fluid calculations (i.e., the spacial distribution of the electron and ion charge densities) inserted into Poisson's equation which yields a new electric field distribution.

The results reported here represent the initial stage of the iteration scheme. Here the initial field distribution is imposed, and the electron distributions, in space, and in velocity are computed from our Monte-Carlo code for various neutral gas pressures. The subsequent fluid and iterative calculations are beyond the scope of this work. Thus, the results of these calculations will be selected to illustrate the ability of this approach to treat non-equilibrium electrons and their predicted motion under experimental conditions. Even these initial calculations could be helpful in revealing other phenomena associated with these electrons, which may have been previously overlooked. In addition, a substantial improvement has been made in the Monte-Carlo code, which now treats the fully circular nature of the cathode hole (see Figure 1), instead of the two-parallel-plate cathode configuration previously considered.[14]

Physical Model of the Mode Transition.

As an example of the utility of such a model, calculations were performed to simulate the transitions from the high-impedance mode (HIM) to the low-impedance mode (LIM), using experimental parameters. To model this transition, a model for the HIM is needed which is compatible with the experimental observations. It is assumed that the HIM consists of three regions (see Figure 1a): the positive column-negative glow region (PCNGR), the cathode-fall region (CFR), and a cathode-hole plasma (CHP). In this model, the plasma densities of the CHP and the PCNGR are comparable in the HIM so that each represents a source of ions which bombard the inner surface of the hole, giving rise to secondary electrons which carry the current through the PCNGR and CFR. Since the magnitude of the component of ion flux from each region is proportional to the plasma density in that region, the device current is determined by roughly equal ion fluxes from each. In the LIM, the density of the CHP becomes several orders of magnitude higher than that in the PCNGR, so that the CHP dominates as a source for ions which ultimately generate the device current. Thus, if increasing the pressure by a small percentage caused a sharp increase in CHP density in the model, this would be viewed as a transition from the HIM to the LIM.

Intuitively, one might expect that for a given applied voltage V_0, an azimuthally symmetric potential difference should exist between the wall of the cathode hole and its axis, with the ionization mean-free path λ_i being greater than the hole radius R_0 for the discharge to be in the HIM. Increasing the gas pressure should shorten λ_i and hence cause a transition to the LIM near the point where $\lambda_i \sim R_0$. Since λ_i is inversely proportional to the gas pressure for such discharges with low percentage ionization, then the transition pressure should only be a function of V_0 and R_0 as is consistent with experiment.[3-8] Thus, to a first approximation, the effect of the anode is only to establish a boundary condition for the electrical potential on the device axis, and that the dominant mechanism in determining the mode transition, is the radial (and azimuthal) motion, production, and loss of the charges in the CHP. In the one-dimensional model, axial motion, will be approximated by a loss time τ.

This three region model for the HIM is consistent with experimental observations involving the length of the anode-cathode gap, the energy and spatial distribution of the electron beam generated in this mode, and the appearance of the HIM. First, the position of the anode should not be important as long as the length of the anode-cathode gap is greater than the thickness of the CFR.[3-8] Second, the PCNGR and the CFR constitute the basic configuration of the high-voltage glow discharges between parallel plates,[15-17] which also produced electron beams collimated along the device axis.[15] For the high-voltage parallel-plate discharges, all but about 10 V of V_0 appear across the CFR, and the energy spread of electron beams produced were within 10% of eV_0.[15] For high-voltage hollow-cathode configurations, Rocca et al.,[18] observed a dark space in the region corresponding to the CFR at the mouth of the gun and determined from the energy distribution of the electron beam generated, that over 90% of V_0 appeared across this region; it was estimated that less than 10% of the remainder was responsible for the generation of the hole plasma.[18] Third, the electron beam produced by the HCD in the HIM, was found to be strongly collimated along the device axis;[18] presumably the filamentary appearance of the discharge in this mode[7] is due to the excitation and ionization of the gas caused by the passage of the electron beam (see Figure 1a).

To complete the model, one needs to predict how τ will scale with pressure and to "guess" the radial potential profile for the CHP. Since $\lambda_{ec} \geq R_0$ near the transition pressure, $\tau \propto h/v_m$, where h is the depth of the hole and v_m is the mean speed of the electrons in the CHP. Thus to a first approximation, τ should be independent of gas pressure. For the radial potential distribution of the CHP, $\Phi_v(r)$, the vacuum potential distribution at the open end of the hole[19] is used as the initial "guess"; here $\Phi_v(r)$ is parabolic, where the potential difference between the central potential and the cathode potential $\Delta\Phi = 0.2 * V_0$. This is a region which is believed to have the strongest influence on the transition.[20] However, since the three region model for the HIM appears to represent a considerable distortion of the potential distribution from the vacuum case, results will also be presented with $\Delta\Phi = 0.1 * V_0$, which may be more compatible with experimental conditions.[18] The answer as to whether either of these "guesses" is close to being correct, awaits the results of the iterative scheme discussed above, which is beyond the scope of the results reported here.

The Monte-Carlo Model

In this model, the planar motion (r, ϕ) is calculated for primary electrons originally released from the cathode at a random initial energy W_{e0} (where $0 < W_{e0} < 10$ eV), and random initial planar direction. This motion is in the plane represented by the dashed line labeled S in Figure 1. Each primary particle is followed in the model for a constant length of time τ, which represents a mean loss time due axial motion. The secondary electrons are produced by the ionization events, and the energy of the to outgoing electrons is assumed to be shared equally. The planar motion of the secondary electrons is also followed in either of two possible models: 1) that the lifetime of the secondary electron is a full τ and that of the primary is that what is left of its original τ, say τ - t, or 2) that since both electrons have the same energy after ionization event, they are indistinguishable, and hence both have a lifetime of τ - t. Both models have been used and they give qualitatively similar results. However, calculations using the first model take considerably longer to perform, since each of the secondaries have a higher probability to create additional secondaries, because every particle created has the same lifetime τ. Results of the second model will be presented here for the purposes of illustration.

Scattering, excitation, and ionization events encountered by all these electrons, are treated through the use of the total cross sections for these processes in helium gas.[21] The scattering process is assumed to be isotropic; the total number of primary electrons is held fixed at one hundred particles.

RESULTS

In the present results M, the primary electron multiplication factor (i.e. the number of secondary electrons produced for each primary electron released from the cathode) will be calculated as a function of gas pressure p_0 for $B_z = 0$, to model the mode transitions.[7] If the potential distribution were fixed, and p_0 is varied, one might expect M to increase rapidly in the vicinity of p_0 where a radial breakdown occurs. This behavior could explain the mechanism for the HCD impedance transitions, if these transitions are related to radial breakdown. As can be imagined, M is roughly inversely proportional to the length of time (or lifetime) τ, that the electrons are followed in the program; τ is taken to be independent of the pressure at fixed V_0 as consistent with the model. τ was previously taken to be 80 ns, to simulate the net electron drift velocity in the axial direction,[14] in this study, τ was varied about 80 ns until an increase in M appears at about the same pressure as the mode transition occurred in the experiment. In the model, this occurs when $\tau \approx 40$ ns, as can be seen in Figure 2. In that figure, the variation of M with p_0 is presented for $\tau = 40$ ns, $V_0 = 15$ kV, $R_0 = 0.5$ cm, and $\Delta\Phi = 0.2*V_0$. As can be seen there is a significant increase in M for 100 mTorr $\leq p_0 \leq 200$ mTorr, whereas for similar V_0 and R_0, the experimental transition pressure $p_{0t} = 165$ mTorr.[7] This value for τ (40 ns) was then be used in the subsequent calculations so that those results may be more consistent with experiment. Thus the results of the model show at least the qualitative trend of a mode transition, even though τ itself can only be treated as a parameter at present. However, the existance of this trend in M with p_0 is very encouraging.

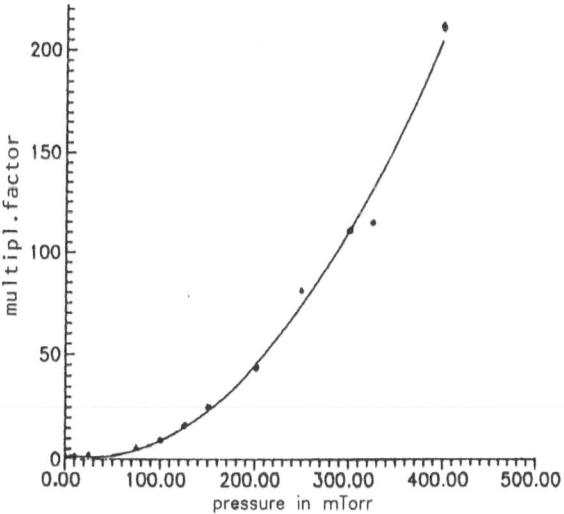

Fig. 2. The variation in the primary electron multiplication factor M with gas pressure p_0 in mTorr for a primary lifetime of $\tau = 40$ ns, a cathode-hole radius $R_0 = 0.5$ cm, an applied potential of $V_0 = 15$ kV, and $\Delta\Phi = 0.2*V_0$. The corresponding transition gas pressure for this R_0 and V_0 was $p_{0t} = 165$ mTorr.

To consider the effect of an axial magnetic field, the radial distributions of electron number density $n_e(r)$, the ionization rate coefficient $k_i(r)$, and the electron energy distribution $f_e(E,r)$ were calculated for various B_z between 0 and 750 mT at $p_0 = 50$ and 200 mTorr. Here the B_z required to suppress the transition into the LIM would be evidenced by a "hole" appearing in the $n_e(r)$ and $k_i(r)$ distributions indicating the discharge can no longer be sustained. This is a more quantitative version of the magnetron model discussed earlier[7], where the inability of the primary electrons to reach the center was taken as the criterion for the discharge to stay in the HIM. In this latter model $\Delta\Phi = V_0$ which would represent the case where the PCNGR extends down the axis of the CHP; this seems to be somewhat extreme for the initial guess for $\Delta\Phi$.

The results for $n_e(r)$ for various B_z are shown in Figure 3; here $\Delta\Phi = 3$ kV (corresponding to $V_0 = 15$ kV for $\Phi_v(r)$), $p_0 = 200$ mTorr, and $R_0 = 0.5$ cm (fixed throughout this work). As can be seen in the Figure, a depression appears in the center of the hole for $B_z = 250$ mT, whereas a peak in $n_e(r)$ occurs at the center for $B_z = 0$. This result could be interpreted as the B_z necessary to prevent the transition into the LIM.[7] This is also reflected in the corresponding $k_i(r)$, which is peaked on the axis for small B_z, but also is hollow for $B_z = 250$ mT. While the experimental value for the minimum B_z for LIM suppression B_{zm} was 150 mT, this may be due to the uncertainty in $\Delta\Phi$.

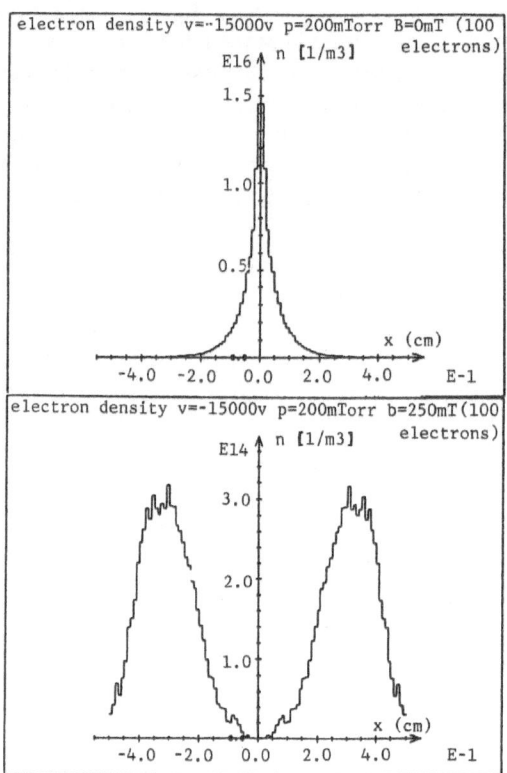

Fig . 3.   The radial electron density profiles in the plane labeled S in Figure 1, for an applied $\Delta\Phi = 3$ kV, a gas pressure $p_0 = 200$ mTorr, and a cathode hole radius $R_0 = 0.5$ cm. The vertical center line is the center axis of the hole, and horizontal distances are in mm. In the top figure, the axial magnetic field $B_z = 0$, and the vertical scale is linear in units of $10^{10}/cm^3$. In the bottom figure, the axial magnetic field $B_z = 250$ mT, and the vertical scale is linear in units of $10^8/cm^3$.

It should be noted that the absolute drop in the electron density to zero inside a certain radius illustrates one of the limitations of the model. That is, the diffusion processes that would at least partially fill the interior regions, take longer than the fixed lifetime τ, and so steep density gradients are predicted. Thus these results exaggerate the underlying trends predicted by the model and seen in experiment. However the model does indicate that through a lack of energetic particles in this interior region, the reaction rate coefficient is considerably reduced, and hence that the discharge cannot be sustained in this region. Such a "hollowing" of the electron density profile, has in fact been observed in HCDs, when an axial magnetic field was applied.[22] This process appears to be a magnetic insulation effect (see Figure 1b), which can be utilized as an opening effect for gas discharge switches.

CONCLUSIONS

By modification of our Monte-Carlo code to treat electrons moving in polar coordinates and in an axial magnetic field B_z, it can be used to model certain features of the mode transitions observed in the hollow-cathode discharge, HCD. These features include the rapid increase in primary electron multiplication factor M with increasing pressure and fixed applied potential, and the "hollowing" of the electron density distribution when a sufficiently strong B_z is applied to the discharge. The former could indicate the transition from the high-impedance HCD mode, HIM, to the low-impedance HCD mode, LIM, observed experimentally,[7] although the absolute value of the pressure where the increase in M occurs cannot be obtained from the model self-consistently at present. That is, the transition pressure can be changed by adjustment of τ the primary-particle loss time, which is used to treat the effect of axial motion of these electrons. However, when the same value of τ, needed to obtain the experimental transition pressure from the code, is used to treat the case which includes the effect of an axial magnetic field B_z, a "hollowing" of the radial electron density profile is predicted when the magnitude of B_z is sufficiently high. This hollowing has been observed experimentally,[22] and could be intrepreted as the B_z required to suppress the transition into the LIM which would normally occur at this pressure. Then, the magnitude of B_z required to hollow out the discharge predicted by the code is within a factor of two, of that observed neccessary to suppress the mode transition.[7]

ACKNOWLEDGEMENTS

The authors would like to thank Drs. V.K. Lakdawala and J.J. Rocca for helpful discussions. The authors also gratefully acknowledge the support of ONR/SDIO through Contract # N00014-85-K-0602; the program monitor is Dr. G. Roy.

REFERENCES

1. P.F. Little and A. von Engel, "The hollow cathode effect and the theory of glow discharges," Proc. Roy. Soc. (London) A 224:209 (1954).
2. K. Frank, E. Boggasch, J. Christiansen, A. Goertler, W. Hartmann, C. Kozlik, C.G. Braun, V. Dominic, M.A. Gundersen, H. Riege, and G. Mechtersheimer, "High power pseudospark and BLT switches," IEEE Trans. Plasma Sci. 16:317 (1988).
3. Z. Yu, J. Rocca, J. Meyer, and G. Collins, "Transverse electron guns for plasma excitation," J. Appl. Phys. 53:4704 (1982).
4. K. Frank and J. Christiansen, "The fundamentals of the pseudospark and its applications," IEEE Trans. Plasma Sci. 17:748 (1989).

5. P. Choi, H. Chuaqui, J. Lunney, R. Reichle, A.J. Davies, and K. Mittag, "Plasma formation in a pseudospark discharge," IEEE Trans. Plasma Sci. 17:770 (1989).

6. G. Kirkman-Amemiya, H. Bauer, and M.A. Gundersen, "Analysis of the high current glow discharge occurring in the BLT and pseudospark switch," Bult. Amer. Phys. Soc. 31:2131 (1989).

7. M.T. Ngo, K.H. Schoenbach, G.A. Gerdin, and J.H. Lee, "The temporal development of hollow cathode discharges," Trans. Plasma Sci. 18:669 (1990).

8. H. Helm, F. Howorka, and M. Pahl, "Ueber den Fallraum in einer zylindrischen Hohlkathode," Z. Naturforsch. 27a:1417 (1972).

9. H. Pak and M.J. Kushner, "Simulation of the switching performance of an optically triggered pseudo-spark thyratron," J. Appl. Phys. 66:2325 (1989).

10. T.J. Sommerer, W.N.G. Hitchon, and J.E. Lawler, "Self-consistent kinetic model of the cathode fall of a glow discharge," Phys. Rev. A 39:6365 (1989).

11. A.J. Davies, "Discharge simulation," IEE Proc. 133:217 (1986).

12. P. Segur, M. Yousfi, J.P. Boeuf, E. Marode, A.J. Davies, and J.G. Evans, in: "Electrical Breakdown and Discharges in Gases," E.E. Kunhardt and L.H. Luessen, ed., Vol. 89a of NATO ASI Series B, Plenum, New York (1983).

13. K.H. Schoenbach, H. Chen, and G. Schaefer, "A model of dc glow discharges with abnormal cathode fall," J. Appl. Phys. 67:154 (1990).

14. K.H. Schoenbach, L.L. Vahala, G.A. Gerdin, N. Homayoun, F. Loke, and G. Schaefer, "The effect of pendel electrons on breakdown and sustainment of a hollow cathode discharge," in: "Proceedings of the NATO Workshop on the Physics and Applications of Hollow Cathode Switches," Lillehammar, Norway, 17-21 July 1989," M.A. Gundersen, ed., Plenum, New York (1990).

15. G.W. McClure, "High-voltage glow discharges in D_2 gas. I. Diagnostic measurements," Phys. Rev. 124:969 (1961).

16. B.B. O'Brien, Jr., "Characteristics of a cold cathode plasma electron gun," Appl. Phys. Lett. 22:503 (1973).

17. G.G. Isaacs, D.L. Jordan, and P.J. Dooley, "A cold-cathode glow discharge electron gun for high-pressure CO_2 laser ionization," J. Phys. E: Sci. Instrum. 12:115 (1979).

18. J.J. Rocca, J. Meyer, and G.J. Collins, "Hollow cathode electron gun for the excitation of cw lasers," Phys. Lett. 87A:237 (1982).

19. T. Tessnow, private communication.

20. W. Hartmann and M.A. Gundersen, "Origin of anomalous emission in superdense glow discharge," Phys. Rev. Lett. 60:2371 (1988).

21. M. Hayashi, "Recommended values of transport cross sections for elastic collisions and total cross sections for electrons in atomic and molecular gases," Inst. Plasma Phys., Nagoya University, Report IPPJ-AM-19 (Nov. 1981).

22. J.J. Rocca, G.J. Fetzer, and G.J. Collins, "The effect of an axial magnetic field on the Spontaneous emission from an argon hollow cathode discharge," Phys. Lett. 84A:118 (1981).

FORMATION OF CATHODE SPOTS BY UNIPOLAR ARCING

Fred Schwirzke

Department of Physics
Naval Postgraduate School
Monterey, CA 93943

ABSTRACT

Despite the fundamental importance of cathode spots for the breakdown process and the formation of a discharge, the complicated processes, the structure of the cathode spot and the source for the high current density were not fully understood. Experiments show that cathode spots are formed by unipolar arcing. The localized build–up of plasma above an electron emitting spot naturally leads to a pressure gradient and electric field distribution which causes unipolar arcing. The high current density of an arc provides explosive plasma formation of a cathode spot.

INTRODUCTION

Many discharges form small cathode spots which provide such a high current density that the cathode material "explodes" into a dense plasma within a very short time. The concept of explosive electron emission is well established in the literature.[1-4] However, estimates of current density j for a cathode spot vary by orders of magnitude. Breakdown in a vacuum diode is initiated by field emission of electrons from a whisker or other emitting micropoints on the cathode. The field emission current density, j_{FE}, is assumed to become large enough to explode the whisker by joule heating, forming a dense plasma. The plasma surface then acts as a virtual cathode and the diode current density becomes space charge limited by Child–Langmuir's law at a value j_{CL}, which represents an upper limit, i.e., $j_{FE} < j_{CL}$. The whisker explosion model requires $j \gtrsim 10^8$ A/cm² to produce a plasma by resistive heating within nanoseconds. However, for the approximate parameters of this experiment, a diode voltage pulse of 25 ns of 0.7 MV over a gap of d = 2 cm, assuming a uniform E, $j_{CL} = 3.8 \times 10^2$ A/cm². The field emitted electron current becomes space charge limited at a value which is insufficient for the explosive like transition of the whisker into a dense plasma. The problem then is, how can the relatively small j_{FE} and j_{CL} increase to $j \approx 10^8 - 10^9$ A/cm²? Such values of j are typical for arc discharges.

FORMATION OF UNIPOLAR ARCS

Unipolar arcing represents a discharge which easily causes explosive plasma formation. The current density and hence the power dissipation for an arc are considerably higher than those of field emitted or space charge limited current flow. Using a laser produced plasma[5] it has been demonstrated that unipolar arcs ignite and burn on a nanosecond time scale. Fig. 1 shows the damage on a stainless steel surface once breakdown occurs. The craters result from electrical microarcs burning between the plasma and the surface. Without any external voltage applied the arcs are driven by sheath electric fields and plasma pressure gradients. The hole at the center is the electron emitting cathode of the arc, the surrounding ring with crater rim is the electron receiving anode area. Since both "electrodes" are located on the same metal surface the arc is called "unipolar". Similar unipolar arc craters have now been observed[6] on the cathode surface of a pulsed vacuum diode with an externally applied voltage of 0.7 − 1.1 MV, Fig. 2.

Electrode surfaces are usually far from ultra high vacuum clean. The electric field, the emission of electrons, and the impact of ions will stimulate desorption[7-8]. Suddenly desorbed layers form an expanding neutral gas cloud of varying density n_0. The mean free path length of field emitted electrons for ionizing neutrals $\lambda = 1/(n_0\sigma_0)$ depends on the ionization cross section σ_0 which for many gases has a maximum value of about $\sigma_0 \approx 10^{-16}$ cm^2 for ≈ 100 eV electrons. In an electric field of 5×10^7 V/m a field emitted electron has gained 100 eV at the distance of 2 μm. For a diode the onset of breakdown is typically delayed by 1 to 10 ns which corresponds to the expansion time of the desorbed neutrals until they reach the maximum of the ionizing zone, i.e., where the electrons have about 100 eV. Ions produced there are accelerated back towards the electron emitting spot. The ions hit the spot with about 100 eV and recombine. The localized energy deposition leads to surface heating and further release of adsorbed gases[8]. The build−up of positive space charge enhances E and thus strongly j_{FE} which will further increase the ionization rate and so on, Figs. 3a and 3b. As the positive space charge increases, a double layer forms between the ions moving from the ionization zone to the cathode and the electrons moving to the anode. This reduces the externally applied E_{ext} until the electron flow becomes space charge limited at j_{CL}. A plasma has formed which now screens E_{ext} from reaching the cathode. However, the plasma in "contact" with the cathode still forms a positive space charge sheath of width $d_s = (\epsilon_0 V_p/n_e e^2)^{1/2}$. V_p is the plasma potential. The sheath electric field $E_s \approx V_p/d_s$ controls now the electron emission. Essentially the same situation exists for a laser produced plasma in contact with a metal surface. As the plasma density above the electron emitting spot increases, so does the surface heating of the spot by ion bombardment. However, for the considered one dimensional geometry the electron flow to the anode is limited by j_{CL}.

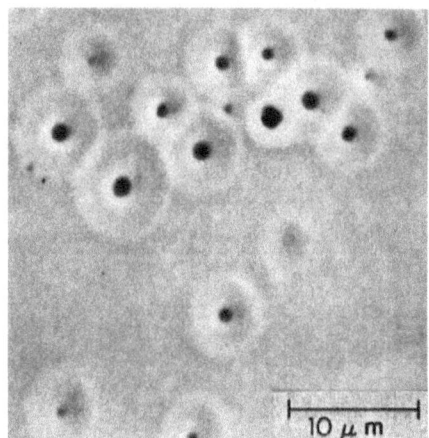

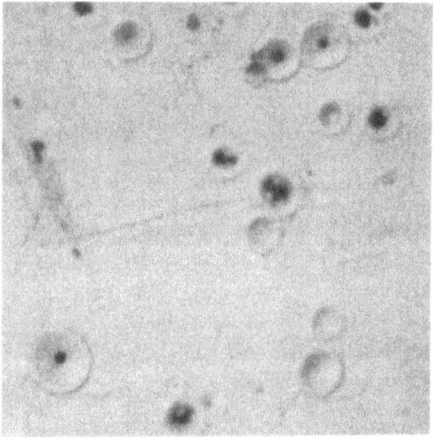

Fig. 1. Laser produced unipolar arc craters on stainless steel, voltage V=0.

Fig. 2. Crater formation on stainless steel cathode of diode, V=0.7 MV.

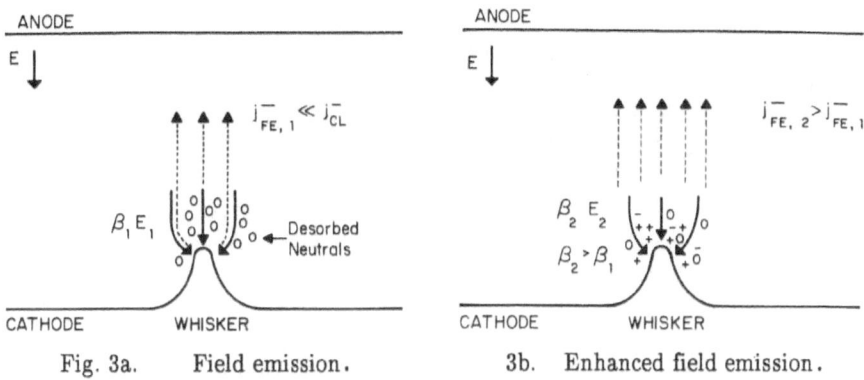

Fig. 3a.   Field emission.   3b.   Enhanced field emission.

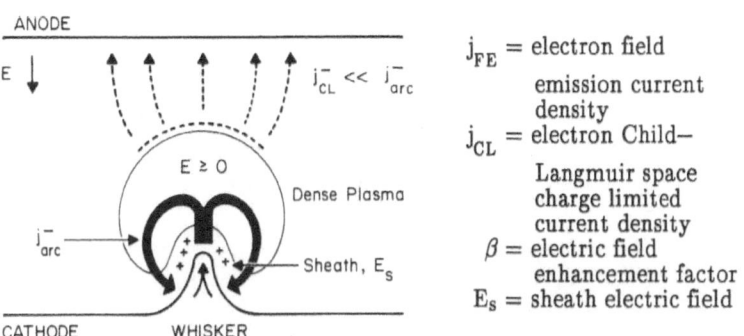

j_{FE} = electron field emission current density

j_{CL} = electron Child– Langmuir space charge limited current density

β = electric field enhancement factor

E_s = sheath electric field

Fig. 3c.   Sequence of events leading to the formation of a cathode spot by unipolar arcing.

In reality, since E_{ext} is screened from the plasma, the electrons can flow sideways and even back to the cathode surface, Fig. 3c, as explained by the unipolar arc model.

Electron emission from a spot on the surface and desorption and ionization of adsorbates lead to the formation of a small dense plasma blob above the electron emitting spot, Fig. 4. A sheath forms as the radially expanding plasma sweeps over the metal surface. The plasma pressure gradient leads to an electric field E_r in radial direction, tangential to the surface. Without any current flowing this field would be the ambipolar electric field $\vec{E}_{amb} = -\nabla P_e/en_e$. Associated with this field the plasma potential decreases in radial direction. Consequently, the plasma sheath potential V_s also decreases in a ring–like area A around the cathode spot. The distribution of V_s will be such that the quasineutrality of the plasma is assured. At some radial distance r_f from the cathode spot the sheath potential will be equal to the floating potential, $V_f = (kT_e/2e)\ln(M_i/2\pi m_e)$, providing equal ion and electron flow rates to the surface at this location, i.e., the net current through the sheath is zero,

$$i_s = +en_i A \left[\frac{kT_e}{M_i}\right]^{1/2} - \frac{1}{4}\,|e|\,n_e\bar{v}\,A \left\{\exp\left[-\frac{|e|\,V_f}{kT_e}\right]\right\} = 0 \qquad (1)$$

$\bar{v}$ is the average speed of the electrons. The first term corresponds to the ion saturation current which remains essentially constant for $V_s > 0$. The second term gives the electron flow through the sheath. At distances $r \lessgtr r_f$ the plasma potential, and thus $V_s \gtrless V_f$, and the electron flow will change exponentially. The net wall current density is $j_s > 0$ for $V_s > V_f$, and $j_s < 0$ for $V_s < V_f$, while $j_s = 0$ for $V_s = V_f$.

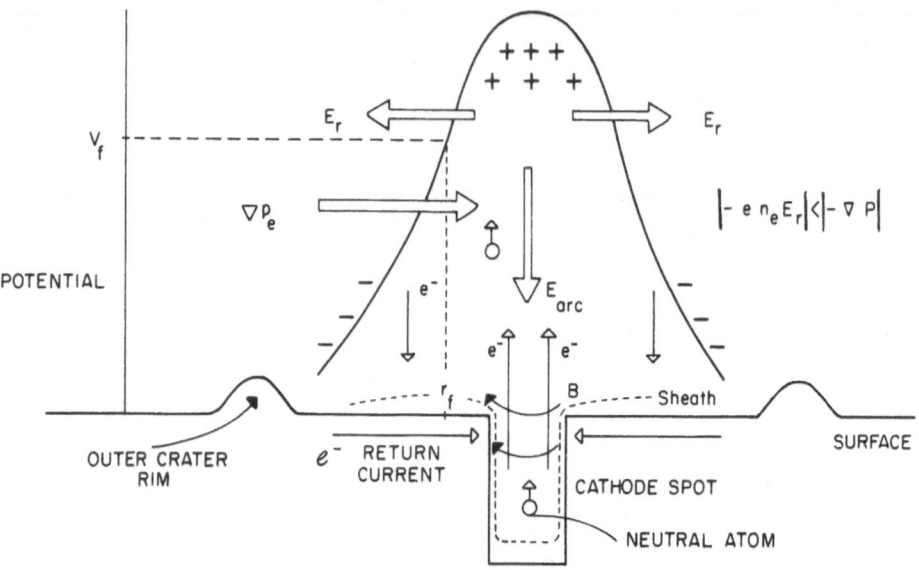

Fig. 4. Unipolar Arc Model.

$$j_s = -\frac{1}{4}\,|e|\,n_e\bar{v}\left\{\exp\left[-\frac{|e|\,V_s}{kT_e}\right] - \exp\left[-\frac{|e|\,V_f}{kT_e}\right]\right\} \tag{2}$$

The ion saturation current has been expressed by Eq. 1. $j_s > 0$ implies that more ions than electrons leave the plasma and the current is in direction of the sheath electric field $\vec{E}_s$, from the positive plasma potential to the wall at zero potential. $j_s < 0$ means that more electrons than ions reach the wall. The increasing flow of electrons to the surface at $r > r_f$ due to $V_s < V_f$ implies a reduction of the negative space charge. This results in a reduced radial electric field $E_r(r) = -\Delta V/\Delta r < E_{amb}$. Consequently, the radially outward directed electron pressure gradient force becomes larger than the electric field force which holds the electrons back $|-\nabla P_e| > |-en_e\vec{E}_r|$. A net force, $\vec{F}_{net}$, acting on the electron fluid is pointing outward in radial direction

$$\vec{F}_{net} = -\nabla P_e - |e|\,n_e\vec{E}_r \tag{3}$$

This is the driving force of the unipolar arc. The current density j of the arc follows from the equation of motion for the electron fluid $-|e|\,n\vec{E} - \nabla P_e + |e|\,n\vec{j}/\sigma = 0$. σ is the electrical conductivity. Solving for $\vec{j}$ and assuming a constant kT_e, we find

$$\vec{j} = \sigma\left[\frac{\nabla P_e}{|e|\,n_e} - \frac{dV}{dr}\right] = \sigma\left[\frac{kT_e}{|e|\,n_e}\frac{dn_e}{dr} - \frac{dV}{dr}\right] \tag{4}$$

Again, $j = 0$ if the expression in the parentheses is zero, i.e., $-dV/dr = E_{amb}$. It is essentially the mass difference between electrons and ions which causes a difference between the two terms on the right hand side. The plasma density profile is determined by the dynamics of the slow ions, $dn_e/dr = dn_i/dr$, while dV/dr is

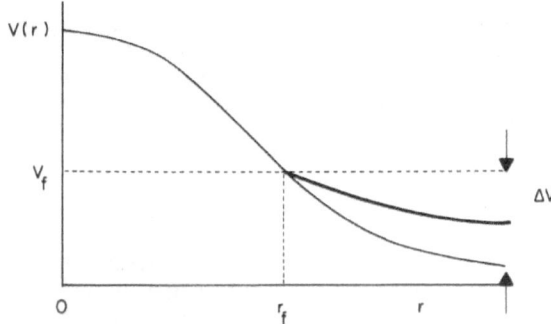

Fig. 5. Plasma sheath potential as function of radial plasma dimension. Light curve is without electron losses to the surface. Heavy curve is with electron losses. The electron emitting spot is at $r = 0$. r_f gives the location of the floating potential V_f.

determined by the more mobile electrons. From Poisson's law, $(d/dr)(dV/dr) = (|e|/\epsilon_0)(n_e - n_i)$, follows that for a strictly one dimensional density profile the curvature of the associated potential profile is proportional to $(n_e - n_i)$. It is concave downwards if $n_i > n_e$, Fig. 5. At the location r_f where $n_i = n_e$, $(d^2V/dr^2) = 0$. The curvature is concave upwards for $r > r_f$, i.e., $n_e > n_i$. In the two–dimensional geometry of the unipolar arc, the loss of electrons to the wall due to the reduced sheath potential $V_s < V_f$ in the region $r > r_f$ leads to a slight decrease of n_e by Δn_e. Consequently the curvature will be reduced in this region.

$$\frac{d^2V}{dr^2} = \frac{|e|}{\epsilon_0}\left[(n_e - \Delta n_e) - n_i\right] > 0$$

still $|n_e - \Delta n_e| > n_i$. The more pronounced curve in Fig. 5 schematically shows this change of the potential profile. As more electrons flow from the plasma to the surface, $\Delta V(r)$ is decreased and the second term on the right hand side of Eq. 4 $|dV/dr| < |(kT_e/en_e)(dn_e/dr)|$. Since $(dn/dr) < 0$ a $\vec{j} < 0$ results, i.e., $\vec{j}$ has the same sign as the current through the sheath $j_s < 0$ in the region $r > r_f$: $\Delta V(r)$ and the sheath potential $V_s(r)$ will adjust in the region $r > r_f$ to values which provide a continuous, increasing current flow as more neutrals become ionized thus increasing the driving ∇P_e term. The increasing temperature of the surface of the electron emitting spot by ion bombardment and ohmic heating leads then to thermionic electron emission and a large arc current begins to circulate. This is the mechanism by which a cathode spot forms.

CONCLUSIONS

Unipolar arcing can supply the necessary current density for the explosive formation of cathode spots in a nanosecond time frame. Unipolar arcing does occur on the cathode surface of a high voltage vacuum diode. It is identical to the arcing observed between a laser produced plasma and a metal target surface when no external voltage was applied.

Surface breakdown is initiated by ionization of desorbed contaminants by field emitted electrons. Since this requires energy deposition only within a few monolayers at a time instead of an entire whisker volume, and since the neutral contaminants are only loosely bound to the surface, the onset of surface breakdown by this mechanism requires much less current and energy than vaporization and ionization of the entire cathode whisker by joule heating.

The localized build–up of plasma above an electron emitting spot naturally leads to pressure and electric field distributions which cause unipolar arcing.

Since the external electric field is screened from the cathode surface, the whisker current is no longer determined by the space charge limited current. Hence, the unipolar arc current density can be many orders of magnitude greater than the Child–Langmuir space charge limited diode current density.

This work was sponsored by the Naval Research Laboratory and the Naval Postgraduate School.

REFERENCES

1. E. A. Litvinov, G. A. Mesyats, and D. I. Proskurovski, Sov. Phys. Usp. 26, 138 (1983).
2. G. N. Fursey, IEEE Transactions on Electrical Insulation, EI–20, 659 (1985).
3. V. I. Rakhovsky, IEEE Transactions on Plasma Science, PS–15, 481 (1987).
4. B. Jüttner, IEEE Transactions on Plasma Science, PS–15, 474 (1987).
5. F. Schwirzke, Laser Induced Unipolar Arcing, in: "Laser Interaction and Related Plasma Phenomena", Vol. 6, pp. 335–352, H. Hora and G. H. Miley, eds., Plenum Publishing Corporation, New York (1984).
6. F. Schwirzke, X. K. Maruyama and S. A. Minnick, Bull. Am. Phys. Soc. 34, 2103 (1989).
7. J. Halbritter, IEEE Transactions on Electrical Insulation, EI–20, 671 (1985).
8. F. Schwirzke, H. Brinkschulte and M. Hashmi, J. Appl. Phys. 46, 4891 (1975).

DETACHMENT OF ELECTRONS FROM NEGATIVE IONS IN ELECTRICAL DISCHARGES

Timm H. Teich

High Voltage Engineering Group
Swiss Federal Institute of Technology
CH - 8092 Zürich

ABSTRACT

Electron detachment from negative ions may considerably influence
the spatial charge distribution and the temporal development of
discharges, as all unstable negative ions generated become potential
electron sources. Explicit solutions of the continuity equations for all
charge carriers involved permit fast simulations of charge carrier
densities which are used here to illustrate - for a hypothetical gas
modelled on oxygen - the differences between plane discharges with and
without detachment. Most obvious is the delayed production of electrons
and the consequent additional ionization on the anode side of the place
of detachment which may permit electrical breakdown well below
critical E/N.

INTRODUCTION

In electrical discharges, detachment of electrons from negative ions
can be important as the negative ions involved form a potential source
of electrons which can be liberated investing considerably less energy
than would be required , for instance, for the direct ionization of
neutral gas molecules. *Photoionization* of neutral molecules would require
high photon energies in a spectral region for which the absorption of
radiation is very high; *photodetachment*, on the other hand, requires in
many cases very modest photon energies at which there is little absorption
in the neutral gas, so that effects might be found more distant from the
origin of the radiation. Photodetachment has been proposed as a maintaining
mechanism for positive point corona (Boylett,Edwards and Williams 1970;
Beattie, 1975). It has also been used as a diagnostic tool in the
determination of negative ion densities (Taillet 1969; Teich and Morris
1987). - *Collisional detachment* may be quite dependent on electrical
field strength - thus potential electron sources may be resident in
discharge space and be triggered by the occurrence/arrival of an adequate
potential gradient, be it due to an applied pulse voltage or an approaching
streamer head. The process is considered of major importance in the
development of positive coronas (Sigmond 1978; Nelson 1985) in
electronegative gases and as a streamer trigger in fast breakdown
(Gallimberti and Wiegart 1985).

Electron detachment from negative ions has been ably reviewed (Massey 1976; Christophorou 1987). In connection with homogeneous field discharges, collisional detachment made sporadic appearances in textbooks (Raether 1964; Llewellyn-Jones 1967; Dutton in Meek and Craggs 1978) and was subject of a modest number of experimental and theoretical investigations, among these the work of Frommhold (1964) for oxygen and air, of Pack and Phelps (1966) for oxygen at low E/N, of Ryzko and Åström (1967) for dry air, of Moruzzi, Ekin and Phelps (1968) and Price, Lucas and Moruzzi (1972) for oxygen with hydrogen admixture, of O'Neill and Craggs(1973), Price, Lucas and Moruzzi(1973) and Frommhold, Corbin and Goodson(1973) once more for oxygen. There was also evidence pointing towards electron detachment in SF_6 (Teich and Branston 1974) which could be readily simulated (Teich, Jabbar and Branston 1975), but the density dependence of coefficients thus arrived at defied ready interpretation (Vidhyashankar 1979) which had to wait for the more profound investigations of recent times (Wang and colleagues 1989, Champion 1990). Further investigations of charge carrier swarm development in homogeneous fields in air, oxygen, c-C_4F_8 and other gases showing electron detachment have been carried out in the last decade by P.C.T. van der Laan and his colleagues at Eindhoven (Verhaart 1982; Wen and Wetzer 1988; Wen 1989). The relative scarcity of reliable quantitative data may have its origin in the difficulties in the mathematical treatment of detachment and connected processes (see below) and the associated tedium of the evaluation of measurements; it is, as far as investigations of swarm current measurements are concerned, also due to the difficulty of *separating* the contributions to the discharge current of the different charge carriers which forms the basis of conventional attachment and ionization rate coefficient determination. For such parameter determination, this separation may be essential and thus there have been efforts in this direction *a)* by using selective electron and negative ion filters in the determination of local charge carrier densities (Pack and Phelps 1966), *b)* by measuring relative local electron density in transient discharges in an electronegative gas by way of detection of radiation (Brennan and Teich 1988) or *c)* by determination of local negative ion density by way of photodetachment (Teich and Morris 1987; Gallimberti and colleagues 1988). The separation of the contributions of *negative* and *positive* ions has also been brought about by measurement of ion arrival rates at the anode and the cathode of a discharge gap by means of screened electrodes (Davies 1987).

With some gases and for a limited range of conditions, one may get away with ignoring detachment at very low gas densities and thus determine, besides charge carrier drift velocities, two-body attachment and ionization rate coefficients which may remain applicable at higher densities when detachment must be taken into account. For instance, with oxygen near critical E/N (ca. 112 Td) and at gas densities not above 3.3×10^{23} m^{-3} (10 torr), the discharge currents are hardly modified by detachment.

APPROACH TO SIMULATION OF DETACHMENT

The basis of all simulations are the continuity equations for all charge carriers involved and attempts at their solution. Linear small-step approximations (Jabbar 1974) or more sophisticated numerical solutions (M. dalla Francesca 1986) can be very universal as regards processes involved but their use may also be very time-consuming. For a simple reaction scheme such as that of Fig. 1b analytic solutions of the coupled continuity equations for electrons and primary negative ions have been given by Frommhold (1964) without derivation; a derivation for electrons based on Laplace transforms can be found in the literature (Dutton 1978) and a revised and improved version has obligingly been provided by Colin Evans (1989) at the request of the author. A solution for electrons which takes diffusion into account has been given by Wen (1989). A solution for primary negative ions can also be arrived at using Laplace transforms. The densities of secondary negative and of positive ions refused to be approached in this way. (An approximate analytic solution (via Laplace transforms) for a situation *without detachment* but *with ion conversion* has been derived by de Urquijo-Carmona (1980)). Therefore the derivation of the densities (and currents) of all charge carriers involved was reworked on the basis of the solution method of Cauchy (Courant and Hilbert 1968) and the method of characteristics (Garabedian 1964) by W. Blumer (Electromagnetic Ambients, ETH Zürich) and will be subject of a separate publication (Blumer and Teich, 1990/91). The resulting expressions have the advantage that they no longer require the full numerical solution of a set of differential equations so that the calculation of the densities of the charge carriers becomes straightforward and rapid, including the convolution for the time-dependent electron release from the cathode.

Equations for the densities of charge carriers involved in a reaction scheme like that of Fig. 1b are presented below. For these, release of a δ pulse of n_0 electrons per unit area at $(x,t) = (0,0)$ is assumed. The reaction rate coefficients k_α, k_η, k_δ and k_χ stand for ionization, attachment, detachment and conversion respectively; the drift velocities $v_e = v_1$, $v_{n1} = v_2$, $v_{n2} = v_3$ and $v_p = v_4$ stand for electrons, primary negative ions, secondary negative ions and positive ions. I_0 and I_1 are modified Bessel functions.

Electrons (adapted from Frommhold 1964): (1)

$$N_e(x,t) = \frac{n_0}{v_e - v_{n1}} \, \exp\left\{ \frac{N(k_\alpha - k_\eta)(x - v_{n1}t) + N(k_\delta + k_\chi)(x - v_e t)}{v_e - v_{n1}} \right\} \left\{ \delta\left(\frac{v_e t - x}{v_e - v_{n1}} \right) + \sqrt{\frac{N^2 k_\delta k_\eta (x - v_{n1}t)}{v_e t - x}} \cdot I_1\left[\frac{2N}{v_e - v_{n1}} \sqrt{k_\eta k_\delta (v_e t - x)(x - v_{n1}t)} \right] \right\}$$

Primary negative ions (adapted from Frommhold 1964):

$$N_{n1}(x,t) = \frac{n_0}{v_e - v_{n1}} \cdot N k_\eta \cdot \exp\left\{ \frac{N(k_\alpha - k_\eta)(x - v_{n1}t) + N(k_\delta + k_\chi)(x - v_e t)}{v_e - v_{n1}} \right\} I_0\left[\frac{2N}{v_e - v_{n1}} \sqrt{k_\eta k_\delta (v_e t - x)(x - v_{n1}t)} \right]$$ (2)

The solution for the densities of positive and secondary negative ions is rather involved and has to be dealt with in the separate paper mentioned above. It must suffice here to indicate the general structure of these solutions, with c_m ($m = 1,..4$) containing only the drift velocities and the reaction coefficient combinations $(k_\alpha - k_\eta)$ and $(k_\delta + k_\chi)$.

Secondary negative ions: $N_{n2}(x,t) = c_1 \cdot n_0 k_\eta k_\chi \exp\left\{ c_2(x - v_{n2}t) \right\} \sum_{\kappa=0}^{\infty} A_\kappa \big]_{\kappa\kappa}$ (3)

Positive ions: $N_p(x,t) = \frac{n_0 N k_\alpha}{v_e + v_p} \exp\left\{ N(k_\alpha - k_\eta) \frac{x + v_p t}{v_e + v_p} \right\} \left\{ F + \sum_{n=0}^{\infty} B_n \big]_{n,n+1} \right\}$ (4)

where

$$A_\kappa = \frac{(c_3 k_\eta k_\delta)^\kappa}{(\kappa!)^2} \quad , \quad B_n = \frac{(c_4 k_\delta k_\eta)^{n+1}}{n!(n+1)!}$$

$$F = \begin{cases} 1 \text{ for } x \leq d - v_p(t - T_e) \\ 0 \text{ otherwise} \end{cases}$$

$$J_{\kappa n} = \int_{s_{j b}}^{s_{j e}} s_j^n (s_j + C_j)^\kappa e^s \, ds$$

with integration variables s_j, integration limits s_{js} and s_{je} and terms C_j each proportional to neutral gas density N and specific to the ion species $j = 3, 4$ considered. The terms in the sums originate from the term-wise integration of the modified Bessel functions. For a discharge without detachment, all B_n and all A_κ except A_o disappear and (4) is reduced to the well-known solution for positive ions. - The terms brought into the equations (1), (2) and (4) by detachment are generally proportional to N_2 or higher powers of N. When the coefficients k_δ and k_χ for onward reactions are not too large and only two-body reactions are significant, we may thus have a chance to apply *initially* the simpler model with no detachment to measurements made at low gas density and modest electrode spacing, thus determining k_α, k_η and $(k_\alpha - k_\eta)$ along with the drift velocities of the charge carriers observed in the conventional way from discharge current measurement (Raether 1964), utilizing of course the potential of short-pulse lasers to generate primary electron swarms of practically any desired size (Branston and Teich 1973). The ab-initio knowledge of the parameters mentioned may then greatly facilitate the simulation of the detachment situation with a view towards determining the remaining parameters k_δ and k_χ or at least their combination.

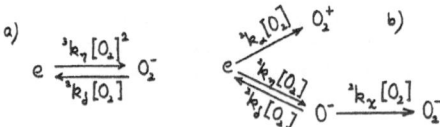

Fig. 1. a) Reaction scheme of Pack and Phelps (1966) for O_2 at low E/N. b) Reaction scheme for "model O_2" to apply near critical E/N.

SIMULATION EXAMPLE : EXPERIMENT OF PACK AND PHELPS (1966)

In this "classical" experiment, the authors measured the density of electrons and negative ions at a particular position (x = 25 mm) in their homogeneous field discharge gap. Under the conditions of the authors (O_2, for instance, E/N = 0.31 Td and 3.1 Td) there was neither ionization nor ion conversion to be considered, Fig.1a. Using the rate coefficients as determined by these authors and extrapolations of mobility data in the literature, the simulation of the measurements was straightforward.

From their "high pressure experiments", Pack and Phelps concluded that there is - under the conditions studied - a joint motion of electrons and negative ions subject to repeated attachment-detachment cycles; the existence and motion of this joint swarm is illustrated in Fig.2 . The drift velocity of this combined swarm is $v_s = (v_e N_e + v_{n1} N_{n1}) / (N_e + N_{n1})$ and its determination will, via the ratio N_e/N_{n1}, also yield the ratio $^2k_\delta/^3k_\eta$ of the detachment and attachment rate coefficients. The density dependence of the joint swarm drift velocity [$(N_e/N_{n1}) \sim 1/N$] is shown as simulation in Fig.3 together with the original measurement plot of Pack and Phelps.

In their "low pressure experiments" these authors could separate the incremental current contributions of electrons from the joint contribution of electrons and negative ions; the simulated densities of the two charge carrier species are, however, given here separately. Assuming perfect

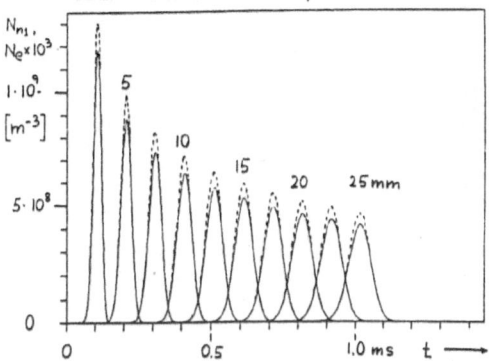

Fig. 2. Joint motion of the negative ion and the electron swarm: Densities of negative ions and of electrons (x1000, dashed curve) in dependence upon time and position in the gap, indicated in mm from the cathode; O_2, 0.31 Td, $N = 6 \times 10^{24}$ m^{-3}.

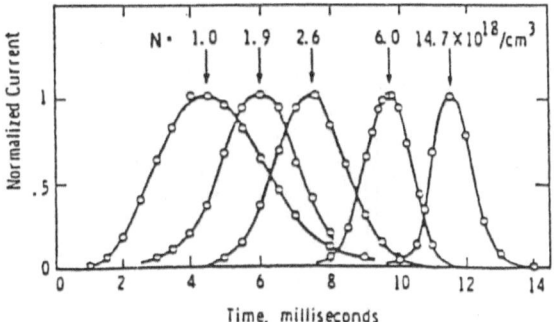

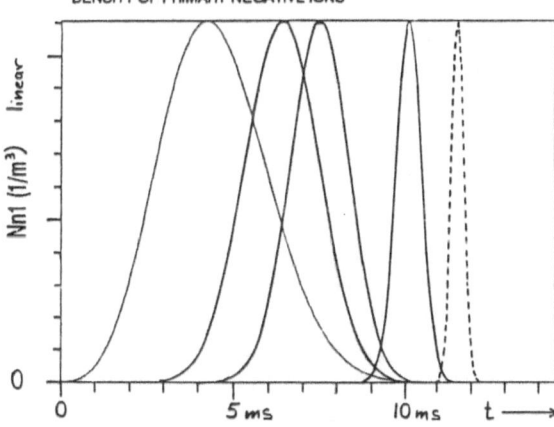

Fig. 3. Charge carrier density in dependence upon gas density N and time at a plane 25 mm from the cathode, E/N = 0.31 Td : Measurements of Pack and Phelps (top) and simulation (below) for the same gas densities except for the dotted curve for which the gas density is only 13.5×10^{24} m^{-3}. For the simulations, the plot of joint swarm mobility vs. N^{-1} produces a perfect straight line (see text). The plot of the measurements has been reproduced from the Journal of Chemical Physics **44**, 1876(1966).

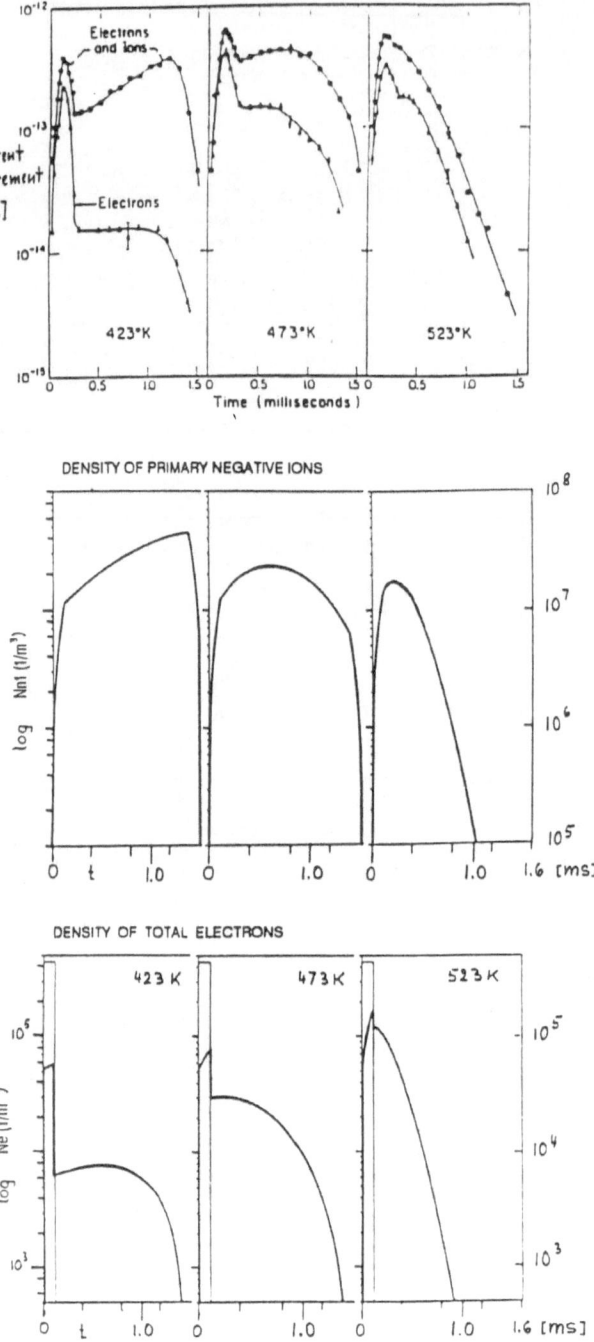

Fig. 4. Simulations of primary negative ion and electron densities
at a plane 25 mm from the cathode, shown as logarithmic plots
together with the measurements of Pack and Phelps (top) which
are reproduced here from the Journal of Chemical Physics **44**,
1873 (1966). Oxygen, 3.1 Td, $N = 10^{24}$ m^{-3}.

220

shutter action, the electron density would have to be multiplied by v_e/v_{n1} before being added to the negative ion density. The simulation plots presented here in Fig.4 give the instantaneous values of the charge carrier densities at the plane of the shutter, x = 25 mm; the averaging brought about by the long shutter times (120 μs) of the authors has not been included in this simulation. The square pulse section on the left margin of the electron density plot represents the electron density at the cathode of the discharge gap.

SIMULATION WITH IONIZATION, ATTACHMENT, DETACHMENT AND ION CONVERSION

It is intended to demonstrate here the basic differences between homogeneous field discharges with and without detachment. For most of the examples, data for a "hypothetical oxygen" following the hypothetical reaction scheme of Fig.1b have been chosen; the E/N values most widely used here are just above critical, say 113 Td, to illustrate some ionization growth of the electron numbers even in the absence of detachment. The initial reaction rate coefficients k_s and k_z are taken from the work of Gallimberti and colleagues (1988) where they gave a reasonable fit to observed current pulse shape as well as optogalvanically measured negative ion density. as both the total currents and the time-dependent spatial distributions of at least some of the carge carrier species are accessible to measurement, examples for both will be presented. As the 3-body conversion from O^- to O_3^- will not, under the conditions chosen, significantly alter what is being shown here (assessment data taken from de Urquijo-Carmona, 1980), this process is ignored here for the sake of clarity. We may also assume that cathodic feedback would not play a noticeable role under the conditions considered.

The most immediately obvious consequence of electron detachment is the continued existence of an electron population in the discharge gap after the passage of the primary electron swarm (here assumed to develop from the release of electrons from the cathode in a short pulse) which may generate further charge carriers - while without detachment the production of charge carriers would have ceased with the arrival of the primary electron swarm at the anode, that is, after one electron transit time T_e. Most readily recorded is the charge carrier current of a swarm, Fig.5 a)(with detachment) and b)(without). The simulation gives us also

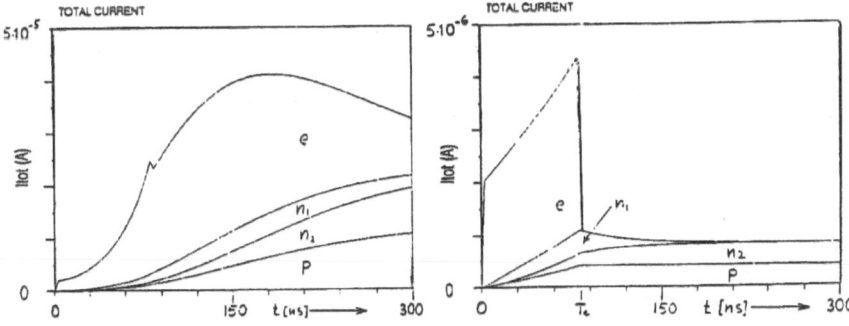

Fig. 5. Early part of a simulated current pulse ("O_2", 113 Td, N = 15.8X10²⁴ m⁻³, 15 mm gap) with detachment (left) and without (right), all other parameters equal. Note current scale for left plot one order of magnitude larger than for the right. For both cases the shares of the current contributed by each species of charge carrier are indicated (see text).

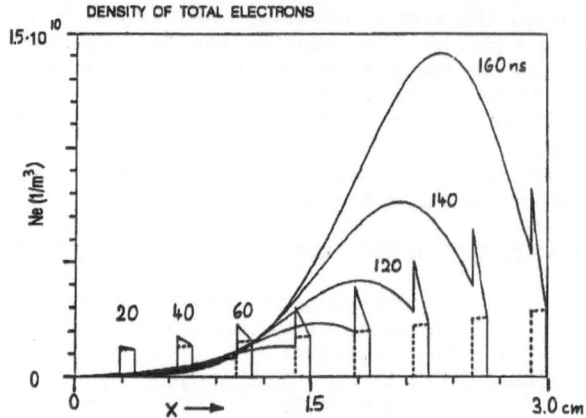

Fig. 6. Spatial distribution of electron density in a 30 mm homogeneous field discharge gap in dependence upon time (every 20ns up to 160ns) with detachment, solid curves; the broken lines give the electron density when no detachment occurs. 10^6 electrons/m^2 released by a 5ns square light pulse. The plot shows the growth of the effects of detachment with distance from the cathode. ("O_2", 113 Td)

the current share contributed by each species of charge carrier (from bottom: positive ions p, secondary negative ions n_2, primary negative ions n_1, electrons e) which simple measurements unfortunately do not so directly provide us with. For the discharge with detachment we note the following: 1) The arrival of (what is left of) the primary electron swarm remains only just discernible; here we assumed primary electron generation by a square light pulse of 3 ns duration; with the more usual near-Gaussian shape and/or longer pulse duration the arrival information (and thus the chance to determine electron drift velocity) may well be lost. 2) The production of ions continues after the primary electron swarm has reached the anode (transit time T_e = 81 ns). 3) At the chosen gas density (15.8×10^{24} m^{-3}) the final ion current level is one order of magnitude greater with detachment - Fig.6 shows the electron density in dependence upon time and position in a 3 cm gap. We see 4) that the influence of detachment increases considerably with gap spacing. When one has to limit the detachment effects in discharge parameter determination, one may have to work with small gap spacing though this may reduce the timing accuracy.

In the solutions for the charge carrier densities we saw N^2 or higher powers of neutral density N in the terms taking detachment into account while there are only dependences on N^1 in the absence of detachment. Not surprisingly, we find 5) significant differences in the density dependences of discharges with and without detachment; this is illustrated in Fig.7. Right down to a gas density of 2×10^{24} m^{-3} (60 torr) the dominant current share after the primary electron swarm transit (81ns) and during the entire time range depicted remains that of electrons; without detachment - see also Fig.7, right - there is of course no electron current remaining after T_e. The more than linear increase of total current with density will probably not hold ad infinitum : Three body processes will become more important and reduce the number of unstable negative ions ready to have an electron detached; this can be illustrated for N_2O

222

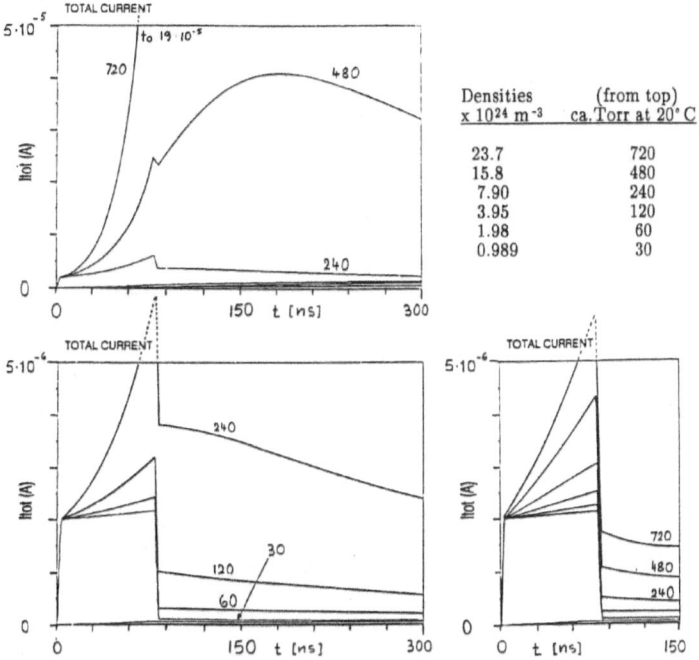

Fig. 7.   Total current in dependence upon time and gas density with
          detachment (left plots) and without (right plot); the two
          lower diagrams have the same current scale; for the top
          diagram the current scale had to be increased by a factor
          10 to accommodate the drastic increase in electron numbers
          with density when detachment is important. The downward step
          in the current at 81ns is due to the arrival of the primary
          electron swarm at the anode. ("O_2", ca.113 Td)

at quite modest densities (see Fig.12) and would also be expected with
O_2 at high densities by way of O_3^- formation.

   We should still demonstrate a further property of discharges with
electron detachment: 6) There can be overall growth of the electron swarm
below critical E/N, that means for $\alpha < \eta$. For Fig.8, k_η was chosen to be
7.5% greater than k_α; after the expected rapid fall in electron current,
there is a remarkable recovery producing additional ionization,
particularly towards the anode, and thus a higher ion current level,
Fig.9; it is also interesting that the maximum value of the electron
current is only 20% higher than the eventual ion current, as against an
order of magnitude for the case without detachment.

   One may be tempted to describe a discharge with reasonably fast
detachment by means of an augmented "apparent" ionization rate
coefficient. We shall show that such a description is inadequate under
conditions considered here. Let us take the eventually attained positive
ion current as a yardstick. The ionization rate coefficient for a discharge

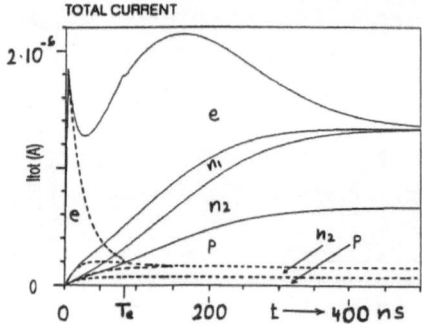

Fig. 8. In the presence of detachment, the electron population may well increase even below critical E/N when $k_\alpha < k_\eta$, here $k_\eta = 1.075 \, k_\alpha$ and $N = 15.8 \times 10^{24} \, m^{-3}$. The dashed curves represent the case without detachment; there is then a monotonous fall in the electron population which disappears totally at T_e. Primary electron swarm arrival at the anode is only just discernible in both cases. The contributions of the different charge carrier species to the current are also indicated (see also Fig.12).

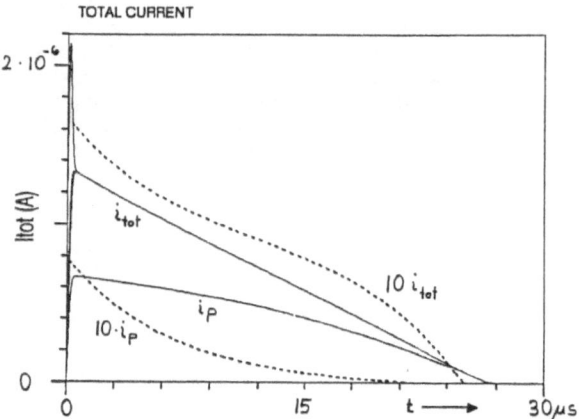

Fig. 9. Case of Fig.8 with the time scale extended to cover the entire ion transit; the (dashed) curves for the discharge without detachment have been scaled up by a factor 10. The time-dependent currents of positive ions (lower) and secondary negative ions (difference to upper curve) are very different in shape in the two cases; ion arrival rate measurement at both electrodes (Davies 1987) should facilitate parameter determination; to try this from conventional total current measurement looks less encouraging.

224

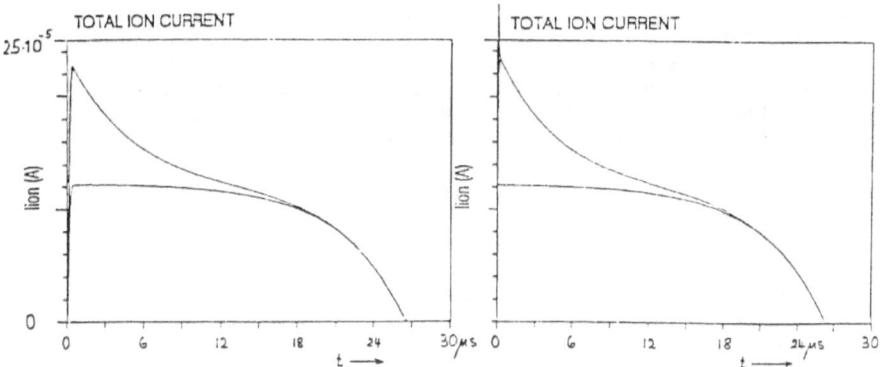

Fig. 10. Ion current with detachment ($k_\alpha = 0.4070 \times 10^{-16}$ m³/s, E/N = 113 Td), left graph, and without detachment with the ionization rate coefficient raised ($k_\alpha = 0.4443 \times 10^{-16}$ m³/s) to obtain the same positive ion current in both cases ("O_2", N = 15.8×10^{24} m⁻³). From 500 ns onwards, the two plots are practically identical. Lower curve: positive ion current.

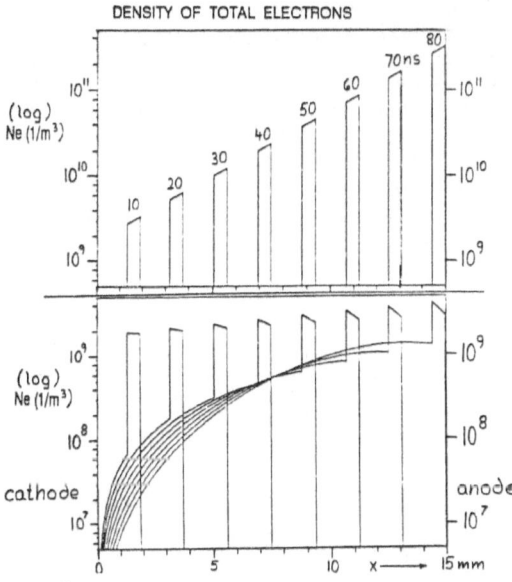

Fig. 11. Logarithmic plots of electron density (top: without detachment, bottom: with detachment) across a 15 mm gap, in 10ns time steps, for the conditions of Fig.10. The two cases, though resulting in the same positive ion current, have rather different spatial distributions and considerable differences in electron density near the anode (note different logarithmic density scales).Primary electron release at the cathode (x = 0) by a 3 ns square pulse. – Detachment produces an altogether "softer" distribution with lower gradients (see text).

225

without detachment was increased until the positive ion current matches that of the equivalent discharge with fast detachment (primary negative ion decay rate, for instance, $3 \times 10^7 s^{-1}$ or higher); under conditions chosen here this implies an increase in the ionization rate coefficient of about 10%. Resulting ion current pulse shapes are illustrated in Fig.10 and seem to show a perfect match from 500ns onwards to the end of the positive ion transit. The spatial development of electron density shows a rather different picture, Fig.11. Though eventually producing the same number of positive ions, to do so the discharge without detachment attains electron densities two orders of magnitude higher than the discharge with detachment which also leaves delayed electrons distributed throughout much of the gap during the electron transit time T_e and well beyond. One asks oneself about the consequences of this behaviour and suspects that any simple criterion of streamer initiation based on electron concentration within one diffusion radius would have to be considerably modified to take detachment properly into account.

A SIMULATION FOR NITROUS OXIDE

The reaction scheme for N_2O (Dutton, Harris and Hughes 1975) is more complicated than that shown in Fig.1b), so that the explicit solutions the simulations presented hitherto were based on cannot be directly applied. The full facilities of the Padova program (mentioned above) had therefore to be used. Fig. 12 shows as an example the time-dependent current at 185 Td (well below critical E/N) for different values of gas density. The pulse shapes have so far only roughly been matched to our measurements, but the illustration will suffice to demonstrate the eventual reduction in delayed electron population with increasing gas density, due to unstable ion conversion in *3-body* processes (O^- to $N_2O_2^-$, NO^- to $N_3O_2^-$). Nevertheless, the detachment remains so strong that the breakdown of N_2O at the higher densities tends to occur well below critical E/N. On the other hand, even at densities as low as $3.3 \cdot 10^{22}$ m^{-3} (1 torr) the current after the first electron transit is dominated by delayed electrons for at least ten electron transit times; this may have to be taken into account when attempting to determine discharge parameters at low gas density (see above).

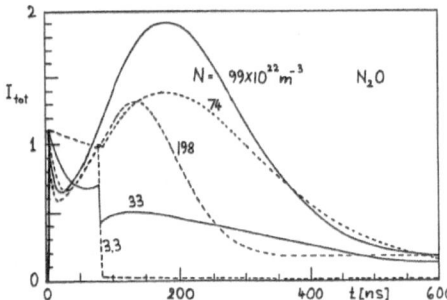

Fig. 12. Current in a 15 mm plane gap in N_2O at 185 Td (ca. 10% below critical E/N) for different gas densities from 3.3 to 198×10^{22} m^{-3} (1 to 60 torr). At densities above 75×10^{22} arrival of the primary electron swarm at the anode is no longer discernible. Note the eventual reduction in electron swarm size at the highest shown gas density due to 3-body reactions.

CONCLUSIONS

Experimental evidence of detachment in homogeneous field discharges can be usefully elucidated by fairly comprehensive simulations based on explicit solutions of the continuity equations. The effects of detachment increase with gap spacing and gas density, although eventually three-body reactions may stop this increase by reducing the population of *unstable* negative ions. Due to detachment, growth of charge carrier swarms is possible well below critical E/N. For a given eventually attained positive ion population, the local electron densities in a discharge without detachment are orders of magnitude higher than with detachment. Detachment produces relatively smooth spatial electron distributions. Simple streamer criteria based on electron number may have to be reconsidered taking detachment into account.

ACKNOWLEDGEMENTS

The author wishes to express his gratitude to Walter Blumer for preparation of the simulation program and to Kurt Lehmann for valuable assistence with the preparation of the text. Colleagues at Padova provided the simulations used for Fig. 12.

REFERENCES

Battie, J. (1975). *The positive glow corona discharge*. Ph.D.Thesis, University of Waterloo, Ontario, p.44 seq.

Boylett, F.D.A.; Edwards, H.G.J. and Williams, B.G.(1970). Impulse breakdown of positive point-plane gaps in air containing charged particles. *J.Phys.D,3,* 1219-1225.

Branston, D.W. and Teich, T.H. (1973). Observation of electron swarms produced by laser light. *Nature,* **244,** No.5417, 504-505.

Brennan, M.J. and Teich, T.H. (1988). Swarm parameters in N_2-O_2-mixtures – using the photon flux technique. *Proc. 9th Intern. Conf. Gas Discharges and Their Applications*. Benetton Editore, Padova, 343-346.

Champion, R.L. (1990). Collisional electron detachment in dielectric gases. in Christophorou, L.G. and Sauers, I. (editors), *Gaseous Dielectrics VI*, Plenum Press, New York, p. 1.

Christophorou, L.G. (1987). Electron attachment and detachment processes in electronegative gases. *Contrib. Plasma Phys.,* **27,** 237-281.

Courant, R. and Hilbert, D. (1968). *Methoden der mathematischen Physik*. 2nd Ed., Springer, Berlin, Chapter 5.

Davies, D.K. (1987). Measurement of swarm parameters in dry and humid air. *Proc. XVIII Int. Conf. on Phenomena in Ionized Gases*, Swansea; Adam Hilger, Bristol, U.K., 1, 2-3.

Dutton, J. (1978). *Spark breakdown in uniform fields*. Ch.3 in Meek, J.M. and Craggs, J.D. (Eds.), *Electrical Breakdown of Gases*. John Wiley, Chichester U.K.

Dutton, J., Harris, F.M. and Hughes, D.B. (1975). Ionization, electron attachment and negative-ion reactions in N_2O. *J. Phys. B,* **8,** 313-324.

Evans, C. (1989). Solution of the continuity equations for electrons and primary negative ions; derivation of the analytic solution for electrons. *Private communication* conveyed by J. Dutton, U.C. Swansea, U.K.

dalla Francesca, Maria (1986). *Modelli matematici delle scariche elettriche nei gas*. Doctoral Thesis, University of Padova, Italy.

Frommhold, L. (1964). Über verzögerte Elektronen in Elektronenlawinen, insbesondere in Sauerstoff und Luft, durch Bildung und Zerfall negativer Ionen (O^-). *Fortschr. Phys.,*12 , 597-642.

Frommhold, L., Corbin, R.J. and Goodson, D.W. (1973). Electron avalanches in oxygen : Theory. *Phys. Rev. A,* **8,** 1403-1411.

Gallimberti, I., Poli, E., Stangherlin, S. and Teich, T.H. (1988). Photodetachment probing of negative oxygen ion populations. *Proc. 9th Intern. Conf. Gas Discharges and Their Applications.* Benetton Editore, Padova, 351-354.

Gallimberti, I. and Wiegart, N.J. (1985). Impulse corona and streamer to leader transition in SF$_6$. *Proc. 8th Intern. Conf. Gas Discharges and Their Applications,* Oxford, 219-222.

Garabedian, P.R. (1964). *Partial differential equations.* John Wiley, New York.

Jabbar, M.A.A. (1974). *Simulation of discharge processes in electronegative gases.* Ph.D. Thesis, University of Manchester, U.K.

Llewellyn-Jones, F. (1967). *Ionization avalanches and breakdown.* Methuen, London

Massey, Sir Harrie (1976). *Negative ions.* 3rd ed., Cambridge University Press, Cambridge, U.K., Chapters 11-13.

Moruzzi, J.L., Ekin jr.,J.W. and Phelps, A.V. (1968). Electron production by asssociative detachment of O$^-$ ions with NO, CO and H$_2$. *J. Chem. Phys.,* **48**, 3070-3076.

Nelson, J.K. (1985). Positive corona processes in electronegative gaseous dielectrics. *IEEE Trans. El. Insul.,* **EI-20**, 601-607.

O'Neill, B.C. and Craggs, J.D. (1973). Collisional detachment of electrons and ion-molecule reactions in oxygen. *J. Phys. B,* **6**, 2625-2633.

Pack, J.L. and Phelps, A.V. (1966). Electron attachment and detachment. I. Pure O$_2$ at low energy. *J. Chem. Phys.,* **44**, 1870-1883.

Price, D.A., Lucas, J. and Moruzzi, J.L. (1972). Ionization in oxygen-hydrogen mixtures. *J. Phys. D,* **5**, 1249-1259.

Price, D.A., Lucas, J. and Moruzzi, J.L. (1973). Current growth in oxygen. *J. Phys. D,* **6**, 1514-1524.

Raether, H. (1964). *Electron avalanches and breakdown in gases.* Butterworths, London . Section 7.7 .

Ryzko, H. and Åström, E. (1967). Electron attachment-detachment processes in dry air. *J. Appl. Phys.,* **38**, 328-331.

Sigmond, R.S. (1978). *Corona discharges.* Ch.4 in Meek, J.M. and Craggs, J.D. (Eds.), *Electrical brekdown of gases,* John Wiley, Chichester U.K.

Taillet, J. (1969). Détermination des concentrations en ions négatifs par photodétachement-eclair. *C.R. Acad. Sci. Paris,* **269 B**, 52-54.

Teich, T.H. and Branston, D.W. (1974). Time resolved observation of ionisation and electron detachment in SF$_6$. *Proc. 3rd Intern. Conf. on Gas Discharges, IEE Conf Publ.* **118**, 109-113.

Teich, T.H., Jabbar, M.A.A. and Branston, D.W. (1975). Observation and simulation of discharge development in electronegative gases. *Proc. Intern. High Voltage Symposium,* Zürich, **2**, 390-394.

Teich, T.H. and Morris, E.C.A. (1987). Galvanooptical measurement of photodetachment from negative ion swarms in oxygen. in Christophorou, L.G. and Bouldin, D.W. (Eds.), *Gaseous Dielectrics V,* Pergamon, New York, 18-26.

de Urquijo-Carmona, J. (1980). *Determination of discharge parameters in sulphur hexafluoride and oxygen by observation of laser light initiated electron swarms.* Ph.D. Thesis, University of Manchester, U.K.

Verhaart, H.F.A. (1982). *Avalanches in insulating gases.* Doctoral Thesis, Eindhoven University of Technology, The Netherlands.

Vidhyashankar, N.S. (1979) *Electron detachment and other ion reactions in sulphur hexafluoride.* M.Sc. Dissertation, Univ. Manchester, U.K.

Wang, Yicheng, Champion, R.L., Doverspike, L.D., Olthoff, J.K. and Van Brunt, R,J. (1989). Collisional electron detachment and decomposition cross sections for SF$_6^-$, SF$_5^-$ and F$^-$ on SF$_6$ and rare gas targets. *J. Chem. Phys.,* **91**, 2254-2260.

Wen, Chuan (1989) *Time-resolved swarm studies in gases with emphasis on electron detachment and ion conversion.* Doctoral Thesis, Eindhoven University of Technology, The Netherlands.

Wen, C. and Wetzer, J.M. (1988). Electron avalanches influenced by detachment and conversion processes. *IEEE Trans. on Electr. Insul.,* **23**, 999-1008.

228

DISCUSSION

J. M. WETZER: Your presentation agrees well with work performed at Eindhoven. With regard to your statement on avalanche growth below E/N (where $\bar{\alpha} = 0$) it is important to note the difference between the "real" $\bar{\alpha}$ and the "apparent" $\bar{\alpha}$ which is derived from steady state measurements. In fact, from our model we have derived a relation between the real and apparent values. Your statement refers to the $(E/N)_{crit}$ derived from the real $\bar{\alpha}$. It should not be directly applied to measured $(E/N)_{crit}$ values. Can you comment on this?

T. H. TEICH: The relation stated by Wen and Wetzer (1988) merely compares total electron numbers as produced by the models with and without detachment. We measure and consider only the "real" values of $k_\alpha - k_\eta$ as these determine, together with the detachment and conversion rate coefficients and the electron drift velocity, the "real" spatial distribution of charge carriers responsible for streamer and other forms of space–charge–controlled breakdown; accordingly, we define $(E/N)_{crit}$ by $k_\alpha - k_\eta = 0$. Admittedly, the criteria of Townsend breakdown refer only to total numbers of *positive* ions formed per electron released from the cathode, and one could attempt to introduce a simpler model which includes detachment in ionization (see Fig. 10) and define some critical E/N which would now become a function of gas density. However, the simpler model would produce an unrealistic spatial electron density distribution (see Fig. 11) and is thus not considered very useful.

E. MARODE: Does your result mean that collisional detachment is negligible at high pressures, since at high pressure all negative ions are already collisionally stabilized?

T. H. TEICH: It need not mean that at all. The high energy required for detachment from SF_6^- in its ground state as shown by Dr. Champion is probably specific to those ions. When stabilization and detachment are both 2–body processes, there can still be drastic growth of the total charge carrier population with increasing gas density at a given E/N, say, $(E/N)_{crit}$, as is shown in Fig. 7. When, however, stabilization in 3–body processes becomes important, the growth (with increasing density) of ionization due to detachment may eventually be halted and even reversed as is illustrated in Fig. 12 which applies to N_2O below the critical E/N. Even then detachment may retain a significant influence: at twice atmospheric density, breakdown in N_2O occurs at 172 Td, more than 15% below the critical E/N (corresponding to $k_\alpha = k_\eta$).

L. G. CHRISTOPHOROU: Is "spontaneous" detachment you refer to the same as "autodetatchment"?

T. H. TEICH: Yes, I intended to refer to any detachment process not brought about by collisions as "spontaneous".

TIME LAGS AND OPTICAL INVESTIGATIONS OF PRE-DISCHARGE IN SF₆/N₂-MIXTURES AT VERY FAST TRANSIENT VOLTAGES

TIME LAGS AND OPTICAL INVESTIGATIONS OF PRE-DISCHARGE IN
SF_6/N_2-MIXTURES AT VERY FAST TRANSIENT VOLTAGES

W. Pfeiffer, V. Zimmer and P. Zipfl
Technische Hochschule Darmstadt, Landgraf-Georg-Str. 4
6100 Darmstadt, FRG

ABSTRACT

This paper deals with the statistical and formative time lags in SF_6 and SF_6/N_2-mixtures under strongly non-uniform field distribution, stressed by very fast transient voltages (VFT).
In most of our experiments, the dielectric strength of the insulating gas, stressed with the highest investigated frequency of 20 MHz, was equal or less compared to a 8 MHz VFT. Nevertheless it was found that with certain parameters a 20 MHz VFT may provide a higher dielectric strength than a 8 MHz VFT.
In pure N_2 the time lags are distributed over the whole investigated range between 10 ns and 100 ms. By increasing the portion of SF_6 the events in the time range between 100 nanoseconds and some microseconds become rarer and have a minimum for a mixture ratio of 50 to 75 % SF_6/N_2, depending on the other parameters.
Breakdown often occurs at a time when the damped VFT surges are already vanished and the DC voltage remains at the test gap. Obviously the preceeding charge injection has greatly reduced the dielectric strength of the insulating gas. Even though the predischarge current hardly exceeds the noise level of approximately 3 A, weak luminosity could be observed in the gap during the first periods of the VFT. This results in a reduced formation time of the discharge channel, which may happen hundreds of microseconds later.

INTRODUCTION

During disconnector operation in GIS steep fronted travelling waves are generated [1, 2]. These waves are reflected at each irregularity of characteristic impedance inside the GIS or at the connection with other devices and are superimposed to the applied voltage. Though there is a frequency mixture existing in real switchgear [3], a monofrequent voltage shape has been used in order to evaluate the effects of the fundamental frequency upon the insulating behaviour of the test gap.
Optical investigations were made in order to resolve the pre-breakdown developement. With a high-speed camera and an optical delay line luminous phenomena could be detected approximately 45 ns before the inception of the high current discharge.

Gaseous Dielectrics VI, Edited by L.G. Christophorou and
I. Sauers, Plenum Press, New York, 1991

EXPERIMENTAL SETUP

A detailed description of our experimental setup is given in [5]. The investigations were performed using a high frequency sinusoidal voltage superimposed to a voltage step. This voltage shape is generated by excitation of a damped series resonant circuit. The highest frequency that could be obtained is 20 MHz. The maximum peak voltage is approximately 160 kV. Figure 1 shows a typical VFT-voltage shape and the discharge current.

Measuring Setup

The waveforms of the applied voltages and the currents were measured with capacitive dividers with a high frequency cutoff at 175 MHz. The low frequency cutoff is 700 kHz for the voltage divider and 7 MHz for the current probe. The equivalent tail time constants are 500 ns for the voltage – and 50 ns for the current – time curves. This has to be taken into account in order to evaluate the actual voltage and current values from fig. 1.

Both signals were recorded with a two channel digital storage oscilloscope with a sampling rate of 200 megasamples per channel. The analog input bandwidth is 50 MHz.

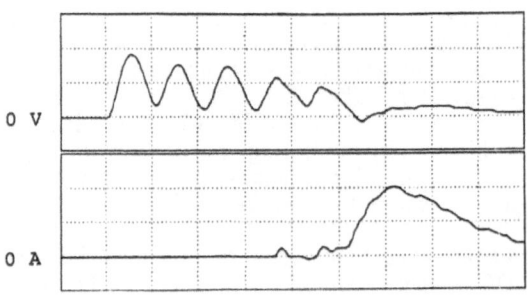

Fig. 1. Characteristical voltage – and current curves for a 20 MHz VFT; hor.: 50 ns/div; vert.: upper trace: 51 kV/div, lower trace: 350 A/div.

Test Gap

In all experiments the cathode, which is made of brass, is hemispherical with a diameter of 33 mm. The hemispheric top contained a steel needle which has a length of 2 mm and a diameter of about 200 μm.

An point – shaped UV – radiation source was used to pre – ionize the discharge volume around the cathode. The UV – source was a mercury vapour lamp which has its maximum emission at a wavelength of 253.7 nm. The radiation was precisely focused to the fairly small area on the top of the needle.

Setup For Optical Measurements

Figure 2 shows the optical measuring setup. The luminous phenomena in the gap is focused by a parabolic mirror and a UV – lens on the high – speed camera. The mirror is placed in a distance of 10 m from the gap in order to obtain an optical delay line of 20 m length.

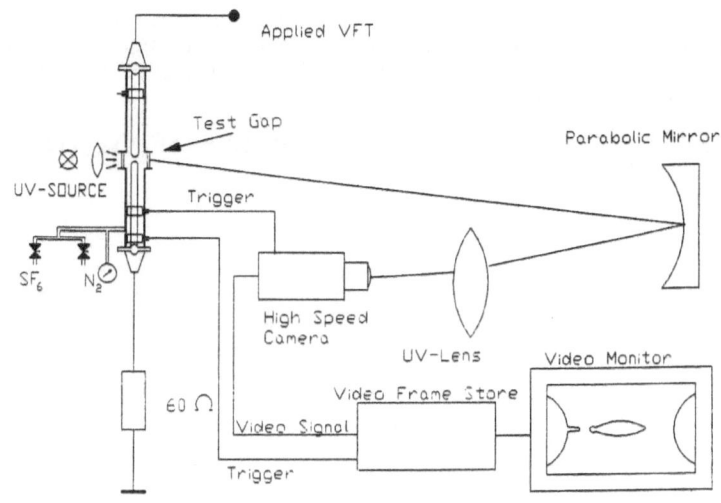

Fig. 2. Experimental setup for optical measurements.

The camera is triggered by the current signal. The minimum total delay time of the camera is only 15 ns. This allows to take pictures of the prebreakdown phenomena approximately 45 ns before the current signal triggers the camera.

The high-speed camera contains an ITT F4144 image intensifier tube with dual microchannel plate and has a minimum exposure time of 700 ps. Gating the photo-cathode with an electric pulse determines the exposure time, which was set to 2 ns in all our experiments because of the weak luminousity of the prebreakdown phenomena.

MEASUREMENTS

Dielectric Strength

It was found out that the dielectric strength of the insulating gas stressed with a 20 MHz VFT was less compared to the 8 MHz VFT for most of the

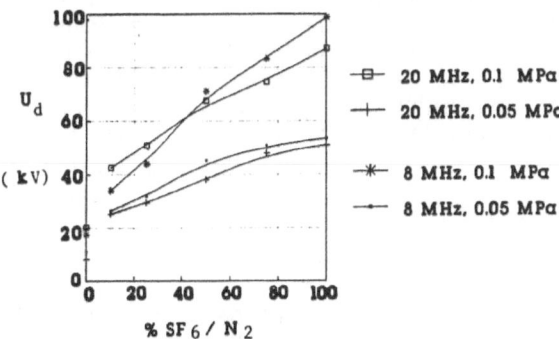

Fig. 3. Minimum breakdown voltage as a function of mixture ratio, gas pressure and applied VFT; gap distance d = 10 mm.

investigated values of the gap distance and the gas pressure [4]. But for a pressure of 0.1 MPa, a gap distance of 10 mm and a portion of less than 40 % SF_6 to N_2 a 20 MHz VFT provides a better dielectric strength than the 8 MHz VFT (Figure 3).

In figure 4 the time lags for pure N_2 and 50 % SF_6/N_2 are shown for example.

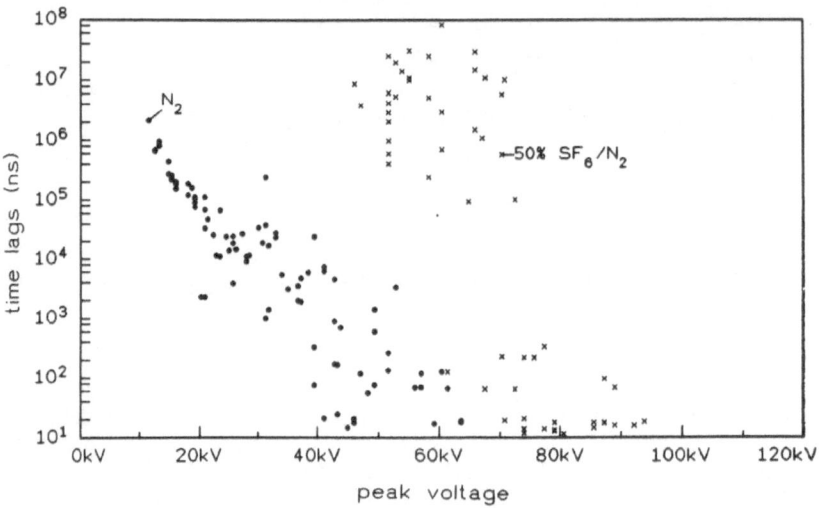

Fig. 4. Time lags as a function of the peak voltage for pure N_2 and 50 % SF_6/N_2, f_{VFT} = 20 MHz, p = 0.05 MPa, d = 15 mm.

For 50 % SF_6/N_2 there are no discharge events in the time range between 300 ns and 100 µs. For pure N_2 events still happen in this time region. Even for an admixture of 10 % SF_6 there is a small region where no breakdown occurs between 200 ns and 2 µs. For pure SF_6 this region becomes smaller again compared to 50 % SF_6/N_2.
For gases including portions of SF_6 events of discharge are increasing again after the area of no events for further hundred milliseconds. In pure N_2 time lags of that largeness were never observed.
It has to be taken into account that 1 µs after the VFT was applied the magnitude of the peak value of the oscillation was damped to 15 % of the magnitude of the peak of the first period and after 2 µs no oscillation could be measured. After this time a constant voltage remained across the test gap.
Discharges that occur within the first two µs are due to the fast transient overvoltages during this time. After that time different conditions are given because the leading VFT surges weaked the insulating gas and a DC voltage remains across the gap.
It was expected that two different mechanisms of discharge take place in the two time areas discribed above.
Optical investigations were made to find out the different discharge developments.
The camera triggered by the main discharge current resolved the predischarge development at different moments. Using the constant optical and the variable electrical delay line the following pictures were taken from different discharges by sampling technique.

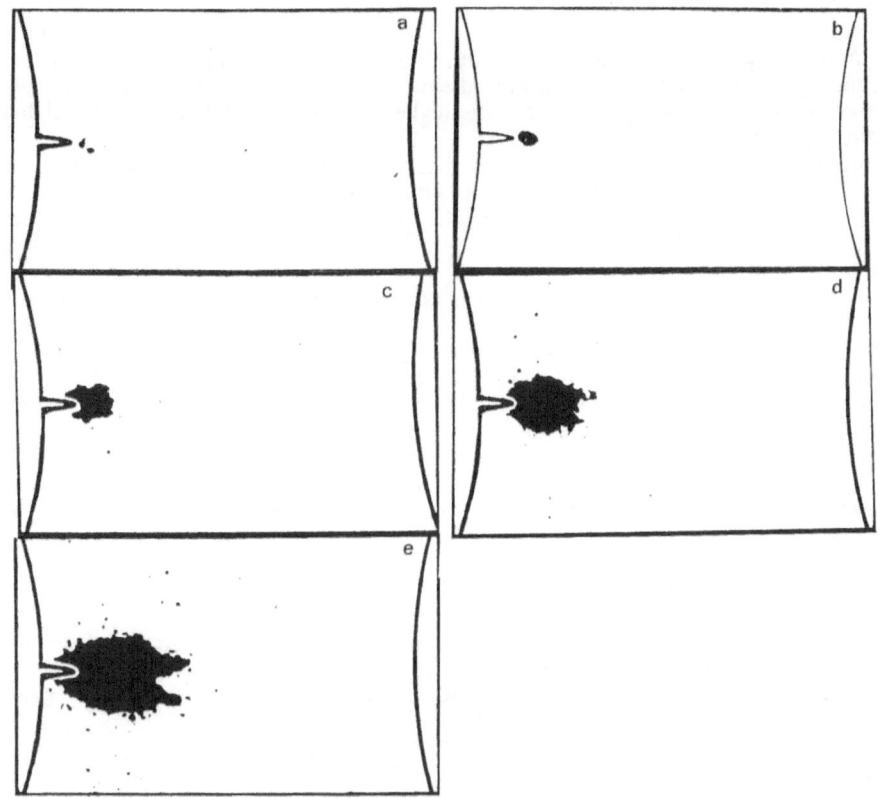

Pictures 1a–e: Predischarge development, early phenomena; f_{VFT} = 20 MHz, d = 15 mm, p = 0.05 MPa, 50 % SF_6/N_2.

Pictures 1a–e were taken during the first few VFT surges (early phenomena). The predischarge starts at the needle shaped cathode. A corona develops in front of the needle and initiates a streamer discharge. This behaviour was found out to be typical for a discharge caused by the sinusoidal overvoltage.

Pictures 2a–b show the predischarge development for a discharge occuring when the oscillation was already damped (late phenomena).

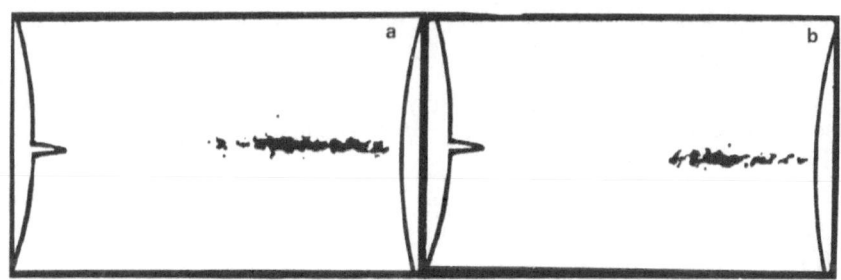

Pictures 2a–b: Predischarge development; late phenomena f_{VFT} = 20 MHz, d = 15 mm, p = 0.05 MPa, 50 % SF_6/N_2.

Compared to pictures 1a–e the predischarge starts at the front of the opposite electrode, i. e. the anode. Obviously the discharge mechanisms are different.

CONCLUSIONS

In most of our experiments it was found out that the application of a higher frequency VFT results in a lower dielectric strength of the insulating gas. Nevertheless for some parameters the higher frequency VFT show the higher dielectric strength.

Different discharge mechanisms have been observed dependent whether the discharge takes place during the first sinusoidal voltage shapes or when the oscillation is already damped to a DC voltage. An early discharge will be initiated by a corona starting at the needle shaped cathode, a late discharge often starts at the anode.

The short statistical time – lags for early discharges are due to electron multiplication processes being typical for streamer discharges. According to the long statistical time lags for the late discharges it is assumed, that the discharge process is based on ion movement. Negative SF_6 – ions are drifting to the anode were they detache positive metal – ions from the electrode. These positive ions are drifting in the oposite direction to the cathode were they generate free electrons which are attached by the electronegative gas. The above described process starts again. The ion movements through the gas increases its thermal energy because of the collision processes until a conductive plasma will be generated.

FURTHER INVESTIGATIONS

Further investigations will bee made to increase the confidence level of the statistical evaluations. The measurement for every set of parameters shown in this paper is based on merely 100 events. A fully automatic test setup is going to be built to meet these requirements.

To resolve the propagation velocity of the discharge a new camera system is planned which allows to take at least four frames of each predischarge.

ACKNOWLEDGEMENTS

This research program is sponsored by the German Research Foundation (DFG).

REFERENCES

[1] G. Luxa, E. Kynast, W. Boeck, H. Hiesinger, A. Pigini, A. Bargigia, D. Schlicht, N. Wiegart, Recent research activity on the dielectric performance of SF_6, with special reference to very fast transients , 1988 CIGRE session, paper 15 – 06.

[2] R. Witzmann, Schnelle transiente Spannungen in metallgekapselten SF_6 – isolierten Schaltanlagen, VDI Fortschritt Berichte, Reihe 21, Nr. 55, Düsseldorf VDI – Verlag 89·

[3] H. Hiesinger, R. Witzmann, Very fast transient breakdown at a needle shaped protrusion, 9th Int. Conf. on Gaseous Diel. and their Appl., Venezia 88, pp. 323 – 326·

[4] W. Pfeiffer, V. Zimmer, P. Zipfl, Insulating Characteristics of SF_6 and SF_6/N_2 – Mixtures for Very Fast Transient Voltages (VFT), 6th Int. Symp. on High Volt. Eng. 89, paper 49.06·

[5] W. Pfeiffer, V. Zimmer, P. Zipfl, Insulating Characteristics of SF_6 and SF_6/N_2 – Mixtures for Very Fast Transient Voltages, 1990 IEEE Int. Symp. on Electrical Insulation, pp. 240 – 243·

DISCUSSION

Y. QIU: Have you checked the effectiveness of the artificial irradiation you used? According to my experience, UV irradiation is often insufficient for gaps in SF_6 due to photon absorption in pressurized SF_6 gas.

W. PFEIFFER: In principle I agree that the efficiency of UV radiation in SF_6 is less than in other gases. However, the wavelength of the source being used is very short (254 nm). Comparative measurements without irridiation showed that especially the spatial discharge development becomes more irregular without irradiation.

E. MARODE: Have you any indication of the spectral composition of the recorded light by the high—speed camera?

W. PFEIFFER: No spectrally resolved measurements were performed within these experiments. However, from other work on DC breakdown of SF_6 it was found that the spectrum of the predischarge phenomena is mainly identical with that of N_2 which is usually present in the gas in tiny fractions. A typical SF_6 spectrum could not be obtained from those experiments.

O. FARISH: I just want to comment that we have observed similar time distributions with a similar pulse applied to a point—plane gap.

Y. DOIN: What is the influence of pressure on the results presented in the paper?

W. PFEIFFER: The frequency dependence of the breakdown voltage for VFT stress increased with gas pressure. Therefore it would have been of great interest to further increase the pressure above 100 kPa. However, at present the maximum amplitude of the test voltage, which is approximately 160 kV, does not allow that.

NON-UNIFORM FIELD FLASHOVER CHARACTERISTICS IN SF$_6$ GAS

UNDER NEGATIVE STEEP-FRONT AND OSCILLATING IMPULSE VOLTAGES

T. Ishii*, M. Hanamura*
S. Matsumoto**, H. Aoyagi**, H. Murase**, M. Hanai** and I. Ohshima**

* The Tokyo Electric Power Company, Inc. — Tokyo, Japan
** Toshiba Corporation — Kawasaki, Japan

ABSTRACT

The predischarge phenomena in compressed SF$_6$ gas have been investigated to clarify the flashover mechanism under the very fast transient overvoltages. Experiments were carried out for non-uniform electric field gaps with negative non-oscillating and oscillating impulse voltages. Predischarge current measurements and high-speed streak photographs showed that the back discharges under 2.2MHz oscillating impulse voltage occurred in the space charges, and leader propagation velocity was slower than that under non-oscillating impulse voltages. Flashover voltages under the oscillating impulse voltages were higher than those under the non-oscillating impulse voltages. As a result, the volt-time characteristic was affected by leader propagation velocity. The test results show that the lightning impulse test for GIS is sensitive to failure detection.

INTRODUCTION

The gas-insulated switchgears (GIS's) have made high-reliability, high-safety, compact substations possible by using SF$_6$ gas. They are likely to be further developed toward higher voltages and larger capacities, along with greater compactness and lower cost. Although SF$_6$ gas has excellent insulation performance, flashover voltages are reduced by the non-uniform electric field gap. Thus, extensive studies[1-5] have been made on the flashover characteristics of SF$_6$ gas in the non-uniform electric field. In 1986 Boeck et al.[6] reported that when steep-front surges caused by the switching of the GIS disconnector were applied to a non-uniform electric field gap, the breakdown voltage fell a great deal. This phenomenon has been attracting much attention.[7-13] However, the discharge characteristics of SF$_6$ gas for non-uniform electric field at steep-front surging have not been explained satisfactorily and are still being studied.[8-13]

Previously, we reported the flashover characteristics of SF$_6$ gas on positive-polarity voltages.[11] Recently, we conducted experiments on the negative-polarity voltages, discovering precursor discharges in negative-polarity discharges, differences in V-t characteristics between positive and negative polarities, and differences in discharge propagation due to steep-front and oscillating impulses. This is a report on the details.

Gaseous Dielectrics VI, Edited by L.G. Christophorou and
I. Sauers, Plenum Press, New York, 1991

EXPERIMENTAL APPARATUS

We have already reported the details of the experimental apparatus[11]; therefore, in this section we describe the major items only. The electrode used in the experiment was a coaxial aluminum cylinder: 42mm I.D., 150mm O.D., and 400mm long. The needle electrode for forming non-uniform electric fields was a nichrome wire, 0.26mm in diameter and 10mm long, set erect on the surface of the central electrode. To measure predischarge currents, the needle electrode was insulated from the central conductor, and a 10Ω resistor was set between them. The voltage between both ends was measured with an oscilloscope. Optical measurements were carried out with a high-speed streak camera (IMACON790) equipped with an image intensifier (I.I.). Gas pressures ranged from 0.1 to 0.45 MPa. The applied voltages were steep-front and oscillating impulses. In the experiments, fast-oscillating and non-oscillating impulse voltages were used. Waveforms of oscillating impulse voltages were three types. Frequencies, voltage rise times, and attenuation times up to $1/e$ were (1) 2.47MHz, $0.18\mu s$, $1.58\mu s$, (2) 289kHz, $1.44\mu s$, $58.7\mu s$, and (3) 140kHz, $2.85\mu s$, $128\mu s$ respectively. As to waveforms of non-osicllating impulse voltages, virtual front times were 0.32, 2.3, 8.3, 33, 59, 186, 250, $306\mu s$, and virtual times to half value were longer than $1400\mu s$.

EXPERIMENTAL RESULTS

V-t Characteristics

First, to compare the flashover voltages with positive and negative polarities, we investigated V-t characteristics. Figure 1 shows the experimental results when gas pressure was 0.45 MPa. From these results we conclude the following:

1) Flashover voltages are higher with negative impulse voltages than with positive impulse voltages. This trend is prominent in short-time ranges (below $10\mu s$) and smaller in long-time ranges.

2) In negative polarity, flashover voltages with an oscillating impulse (frequency: 2.2 MHz) are higher than flashover voltages with a steep-front impulse voltage. In positive polarity, there is little difference between them.

3) The time domain where the minimum flashover voltage appears shifts to a longer-time domain in negative polarity than in positive polarity.

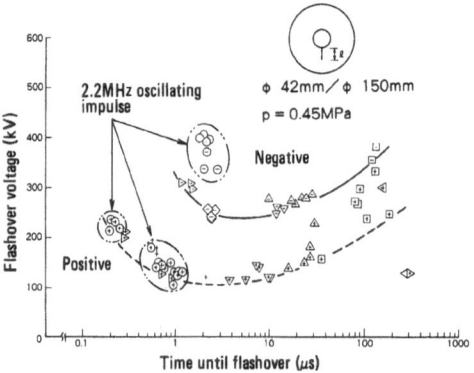

Fig. 1. Volt-time characteristics of SF_6 gas
($\ell = 10mm$, $p = 0.45MPa$, $\phi 42mm/\phi 150mm$).

4) Time lags are generally longer in negative polarity than in positive polarity. When a voltage with a rise time of less than 1µs, for example, is applied, there is a consistent time lag of about 2µs. This time lag remains essentially unchanged when the voltage is raised by about 1 σ (the standard deviation of discharge scatter).

V-p Characteristics

The movement of electrons and ions depends on gas pressure. Thus, the relation between flashover voltage and gas pressure (V-p characteristics) is important in studying flashover mechanisms.

Figure 2 shows V-p characteristics obtained when a steep-front impulse voltage with a rise time of 0.24µs was applied. With negative polarity, as the gas pressure rises, the flashover voltage also rises with some saturation trend. With positive polarity, on the other hand, the flashover voltage shows a prominent saturation trend with respect to the gas pressure, hardly changing at all as the gas pressure rises from 0.1 to 0.45 MPa. Thus, with negative polarity, within the gas pressure range used in our experiments, the corona stability effect occurs consistently with steep-front surges. We assume that under any gas pressure this effect probably occurs.

Figure 3 shows the V-p characteristics obtained when an oscillating impulse voltage with a frequency of 2.2 MHz was applied. With positive polarity, the flashover voltage is nearly the same as that shown in Fig. 2, though the data with the steep-front impulse is a little lower. With negative polarity, on the other hand, under any gas pressure, the flashover voltage is definitely higher with the oscillating impulse than with the steep-front impulse. Pressure characteristics, for both polarities, are similar to those shown in Fig. 2.

Optical Observation of Discharge Propagation and Current Measurement

Figure 4 shows the results of a current measurement and a streak photograph with gas pressure at 0.45 MPa. Leader discharges propagate in a step-wise fashion, and pulsive-currents flow. The discharge propagates nearly straight in the streak photograph; its speed is assumed to be constant. Also, a precursor discharge due to the movements of positive and negative ions[5, 9] can be seen at the head of the leader.

Figure 5 shows the result of applying an oscillating impulse with a frequency of 2.2 MHz under a gas pressure of 0.45 MPa. In this case, a bipolar current flows in accordance with voltage variations. The streak photograph shows that illumination and extinction repeat in

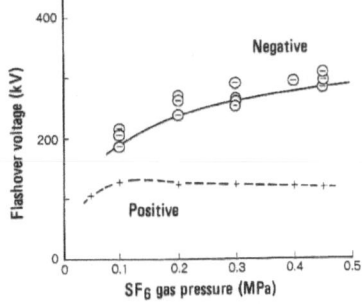

Fig. 2. Pressure dependence of flashover voltages with steep-front impulse voltages (±0.24/2560µs, ϕ42mm/ϕ150mm, ℓ = 10mm).

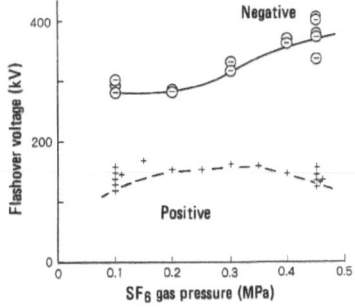

Fig. 3. Pressure dependence of flashover voltages with 2.2MHz oscillating impulse voltages (ϕ42mm/ϕ150mm, ℓ = 10mm).

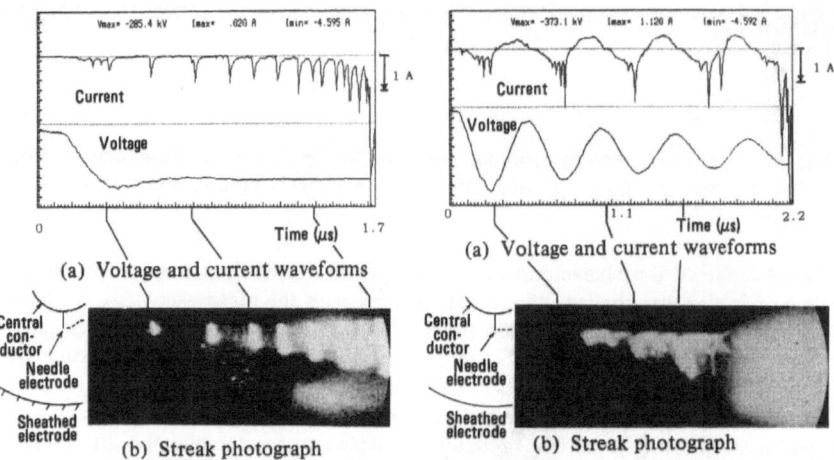

Fig. 4. Voltage-current records and high-speed streak photograph with steep-front impulse voltages (p = 0.45MPa, ϕ42mm/ϕ150mm, ℓ = 10mm).

Fig. 5. Voltage-current records and high-speed streak photograph with 2.2MHz oscillating impulse voltages (p = 0.45 MPa, ϕ42mm/ϕ150mm, ℓ = 10mm).

accordance with the bipolar current flow; while a negative current flows, the leader propagates, and while a positive current flows, the leader does not propagate. In addition, the current is continuous when it turns from negative to positive, but it is nearly interrupted when it turns from positive to negative. As these phenomena are repeated, the discharge propagates. These phenomena mean that back discharges occur in the space charges.

Discharge Propagation Velocity and Intervals of Discharge Steps

Applying a steep-front impulse, we measured leader propagation velocities and discharge step intervals.

Figure 6 shows leader propagation velocities for both polarities, obtained from streak photographs. Depending on voltage and gas pressure, discharge propagation velocity with negative polarity is 3-5 X 10^4 m/s, which is close to the value—3 cm/μs (3 X 10^4 m/s)—reported by Takuma et al.[1] Discharge propagation velocity with positive polarity is 4-10 X 10^4 m/s. With the gas pressure varied, the value changes less than it does with positive polarity.

Figure 7 shows the intervals of discharge steps. With negative polarity, the interval changes less than with positive polarity, as the voltage and the gas pressure vary. Under the same pressure, the interval is longer with negative polarity than with positive polarity; under higher pressures, the difference between polarities is even larger.

DISCUSSION

On the basis of the experimental results described above, let us consider the breakdown mechanisms.

These are the factors which have an effect on flashover characteristics:
1) Corona stabilizing effect on negative corona discharges
2) Back discharges (neutralizing discharges)
3) Leader propagations

One of the factors that has an effect on flashover characteristics is the corona stabilizing

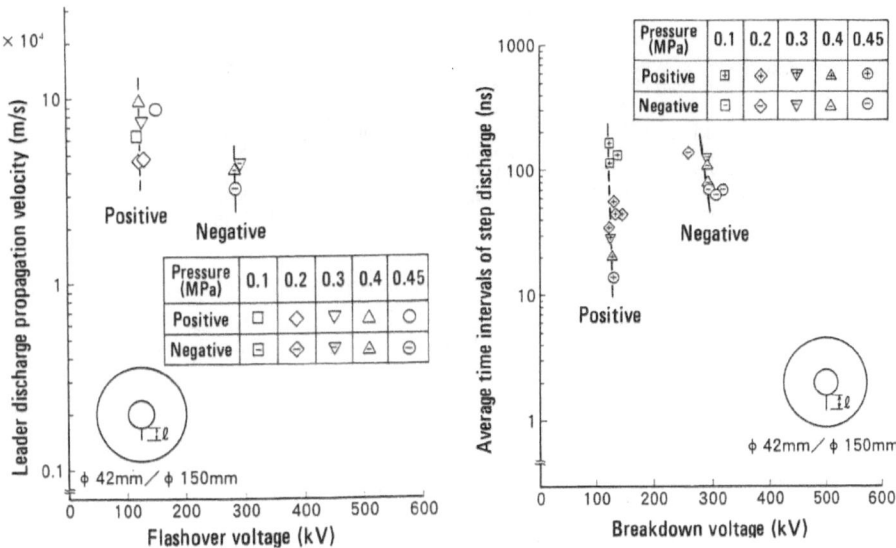

Fig. 6. Leader propagation velocity.

Fig. 7. Time interval of stepwise leader propagation.

effect. With positive polarity, the corona stabilizing effect is caused by positive ions, and with negative polarity, it is caused by negative ions. From Ref. 14, the drift velocities of positive ions (SF_6^+) and negative ions (SF_6^-) are estimated to be nearly equal, about 10^{-4} m/μs, with the value for positive ions a little higher than that for negative ions; the velocity of electrons is about 0.1 m/μs, which is 1000 times larger than the velocity of ions. Electrons can move into the gaseous space away from the needle electrode and form negative ions there. This movement may be the cause of the differences in the corona stabilizing effect.

Another factor having an effect on flashover characteristics is the back discharge. The neutralization of the negative space charge promoted by the back discharge is assumed to be caused by the negative charge being allowed into the needle electrode by the needle electrode potential, which is positive toward the space charge. Possible negative charges may be the movement of electrons, electrons from negative ions, and negative ions. In this case the movement of electrons can follow voltage variations completely, and this may be the reason why the positive current only, free of pulse components, flows continuously, shown in Fig. 5. Also, remaining ions in the corona move so slowly that the field relaxing effect may be sustained even for oscillating surges. This causes the flashover voltage with oscillating surges to be higher than when a steep-front impulse is applied.

The third factor having an effect on flashover characteristics is leader propagation. References 4 and 9 explain that a leader propagation occurs as the injected electric energy turns into thermal gas energy, causing fluorine atoms to be generated. (For details see Refs. 4 and 9.) The electric energy per step is estimated from the test results.

Table 1 shows the averages of injected electric energy per step and measured charges. From this table we conclude the following:

(1) With a constant gas pressure, the charge with negative polarity is only a few times larger than that with positive polarity; the injected electric energy with negative polarity is nearly 100 times larger than that with positive polarity.

(2) Under gas pressures above 0.2 MPa with positive polarity, electric energy injected per step is nearly constant at 2×10^{-4} J. Under gas pressures above 0.3 MPa with negative polarity, it is in the range of $1\text{-}2 \times 10^{-2}$ J.

243

Table 1. Leader propagation characteristics

Gas pressure (MPa)	Positive		Negative	
	Charge (nC)	Electric energy per step (J)	Charge (nC)	Electric energy per step (J)
0.1	30	5×10^{-4}	96	7.3×10^{-2}
0.2	12	2×10^{-4}	82	7.1×10^{-2}
0.3	9.5	2×10^{-4}	28	1.7×10^{-2}
0.4	10	1.9×10^{-4}	21	1.3×10^{-2}
0.45	9.6	2×10^{-4}	13	1.1×10^{-2}

According to (1) and (2), the gas space involved with discharge propagation is larger with negative polarity than with positive polarity; thus, the charge and the injected electric energy may also be larger with negative polarity. Discharge step intervals may be determined by the injection rate of the electric energy supplied from the power supply and the time constant for gas heating, and the extent of leader propagation is determined by the range of gas heating. Precursor discharges and illuminations may also make some contribution to this process.

CONCLUSIONS

To study the flashover characteristics of SF_6 gas in a steep-front, non-uniform electric field, we investigated discharge characteristics under a negative impulse voltage. This is a summary of the results:

1) With negative polarity, oscillating surges have a significant effect on flashover voltages. This may be because leader propagation velocity is lower with negative polarity than with positive polarity and because back discharges have a strong influence on flashover voltages.

2) Flashover voltages are minimum toward long-time ranges with negative polarity, and flashover voltages are higher with negative polarity than with positive polarity.

ACKNOWLEDGMENT

We express our thanks to Prof. T. Kawamura and Ass. Prof. M. Ishii at the Institute of Industrial Science, The University of Tokyo; Prof. T. Kouno and Ass. Prof. K. Hidaka of the Electrical Engineering Department, Engineering Faculty, The University of Tokyo; Prof. K. Gosho of the Electrical Engineering Department, Nippon Institute of Technology; and Prof. T. Takuma at the Kyushu University, with whom we had discussions regarding our study.

REFERENCES

1. T. Takuma, T. Watanabe, K. Kita, and Y. Aoshima, Discharge Development of Long Gaps in SF_6 Gas, Int. Symp. on High Voltage Engineering, München, pp. 386-390 (1972).

2. A. H. Cookson and O. Farish, Particle-initiated Breakdown between Coaxial Electrodes in Compressed SF_6, IEEE Trans., Vol. PAS-92, pp. 871-876 (May/June 1973).

3. T. Hattori, M. Honda, H. Aoyagi, N. Kobayashi, and K. Terasaka, A Study on Effects of Conducting Particles in SF_6 Gas and Test Methods for GIS, IEEE Trans. on Power Delivery, Vol. 3, No. 1, pp. 197-204 (Jan. 1988).

4. F. Pinnekamp and L. Niemeyer, Qualitative Model of Breakdown in SF_6 in Inhomogeneous Gaps, J. Phys. D: Appl. Phys., Vol. 16, pp. 1293-1302 (1983).

5. L. Niemeyer and F. Pinnekamp, Leader Discharge in SF_6, J. Phys. D: Appl. Phys., Vol. 16, pp. 1031-1045 (1983).

6. W. Boeck, W. Taschner, J. Gorablenkow, G. F. Luxa, and L. Menton, Insulating Behaviour of SF_6 with and without Solid Insulation in Case of Fast Transients, CIGRE 15-07 (Aug. 27, 1986).

7. I. Gallimberti and N. Wiegart, Streamer and Leader Formation in SF_6 and SF_6 Mixtures under Positive Impulse Conditions: Part 2 Streamer to Leader Transition, J. Phys. D: Appl. Phys. 19, pp. 2363-2379 (1986).

8. S. Kobayashi, Y. Yamagata, S. Nishiwaki, H. Okubo, Y. Kawaguchi, Y. Murakami, and S. Yanabu, Particle-initiated Flashover Caused by Disconnector Restriking Surge in GIS, 5th Int. Symp. on High Voltage Engineering, No. 12.03 (1987).

9. N. Wiegart, L. Niemeyer, F. Pinnekamp, W. Boeck, J. Kindersberger, R. Morrow, W. Zaengel, M. Zwicky, I. Gallimberti, and S. A. Boggs, Inhomogeneous Field Breakdown in GIS. The Prediction of Breakdown Probabilities and Voltages Part 1: Overview of a Theory for Inhomogeneous Field Breakdown in SF_6, IEEE Trans. on Power Delivery, Vol. 3, No. 3, pp. 923-930 (July 1988).

10. L. Ullrich, The Influence of Very Fast Transients on the Discharge Development in SF_6, Gas Discharge and their Applications 88GD, Venezia, pp. 335-338 (1988).

11. S. Matsumoto, H. Okubo, H. Aoyagi, and S. Yanabu, Non-uniform Flashover Mechanism in SF_6 under Oscillating Fast Transient Overvoltage, 6th Int. Symp. on High Voltage Engineering, No. 32.16 (1989).

12. S. Kobayashi, T. Kobayashi, H. Aoyagi, S. Matsumoto, K. Uehara, H. Okubo, and S. Yanabu, Particle-initiated Flashover Characteristics under Disconnector Surge Voltages in SF_6 Gas-insulated Bus, 6th Int. Symp. on High Voltage Engineering, No. 32.15 (1989).

13. K. Möller and A. Stepken, Leader Formation in SF_6 for Oscillating Impulse Voltages, Int. Symp. on High Voltage Engineering, No. 32.02 (1989).

14. D. T. A. Blair, B. H. Crichton, F. J. Al-Kindi, and T. L. Sharma, Drift Velocities of Positive Ions and Negative Ions in Cylinder SF_6, J. Phys. D: Appl. Phys. 22, pp. 755-758 (1989).

DISCUSSION

I. GALLIMBERTI: Under steep—front voltages you have indicated in your streak photographs the presence of "precursor" channels in the leader stepped propagation. Can you say the same thing for oscillating voltages (at the frequency of 2.2 MHz) or do you have evidence that the leader heating due to a backward neutralizing current is the major propagation mechanism?

M. HANAI: First, I think there are or must be precursor channels under oscillating voltage with negative polarity. Because the edge of the leader channel is thin and vague, I could not take a photograph of the precursor. With positive polarity, I have photographs of the precursor under oscillating voltage. Second, from the photograph of the leader, the leader with negative polarity is thinner than that with positive polarity, and the injected electric energy with negative polarity is also larger than with the positive polarity. So I think the leader heating is one of the most important propagation mechanisms.

BREAKDOWN CHARACTERISTICS OF NON-UNIFORM GAPS IN SF$_6$
UNDER FAST OSCILLATING IMPULSE VOLTAGES

H. Fujinami[*] and E. Kuffel[**]

[*]Central Research Institute of Electric Power Industry
Tokyo, Japan
[**]University of Manitoba, Winnipeg, Canada

ABSTRACT

The paper presents breakdown characteristics of non-uniform field gaps in SF$_6$ gas under voltages simulating very fast transient overvoltages (VFTO) generated by the operation of a disconnect switch in GIS. The results presented show that the breakdown voltages under the VFTO remain higher than the corresponding values obtained under standard lightning impulse voltages. However, the voltage-time characteristics for the VFTO fall below the lightning impulse V-t characteristics especially at the higher gas pressure. The effect of corona discharge on the breakdown characteristics for VFTO superimposed upon dc voltage is also reported.

INTRODUCTION

The operation of GIS disconnect switches may give rise to very fast transient overvoltages (VFTO) occasionally reaching values up to 2.5pu. Under normal operating conditions in the absence of localized field stresses caused by e.g. protrusions or the presence of metallic particles, such surges will not affect the dielectric strength of the insulation. Recent studies showed (Hiesinger and Witzmann, 1988; Matsumoto and colleagues, 1989; Riquel and colleagues, 1989), however, that localized highly non-uniform fields caused by the presence of metallic particles may reduce drastically the dielectric strength of the SF$_6$ insulation under VFTO, especially in the sub-microsecond time region.

In the present studies, we used the following parameters. The gap consisted of sphere-plane electrodes with a needle fixed at the center of the sphere simulating a presence of a metallic particle. The gas pressure ranged from 100 to 500kPa. The voltages used included standard lightning impulses, oscillating impulses of 6 MHz frequency and direct voltages with superimposed oscillating voltages.

EXPERIMENTAL SETUP AND PROCEDURE

The gap arrangement used is shown in Fig. 1. It consisted of a plane electrode 220mm in diameter and a sphere 100mm in diameter. In the center of the latter a needle (1mm diameter) was fixed, whose length (l) could be

varied.   The studies covered two gap length (g) of 50mm and 100mm
respectively, while the length of the needle was varied from 2mm to 10mm.
The test chamber consisted of a fiberglass reinforced cylinder 82.5cm high
and 35cm diameter.   Both ends of the chamber were sealed with brass
plates.

The circuit generating the test voltage is shown in Fig. 2.   The upper
circuit represents the generator for the steep fronted oscillatory
transient voltages, while the lower circuit shows the dc generator.   When
only lightning or oscillatory transient voltages were used the lower
electrode was grounded, while for the combined voltage studies the dc
voltages of the appropriate polarity were applied to the lower electrode.

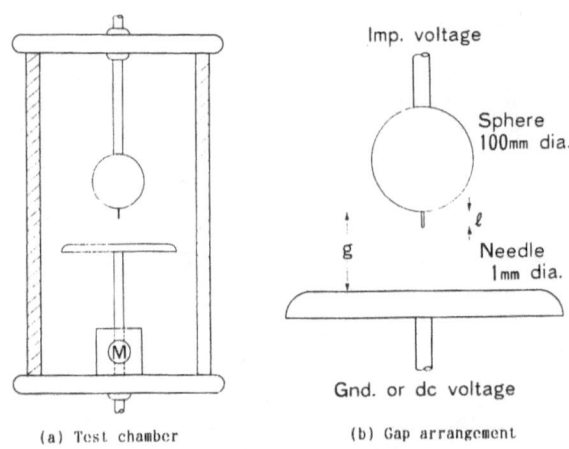

(a) Test chamber   (b) Gap arrangement

Fig. 1. Test chamber and Gap arrangement.

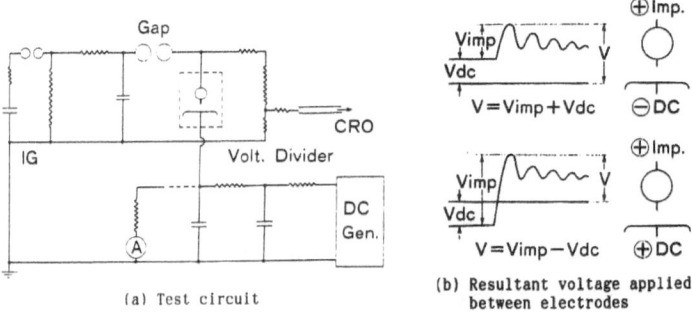

(a) Test circuit   (b) Resultant voltage applied
between electrodes

Fig. 2. Test circuit for generating VFTO superimposed upon dc voltage.

Thus the resultant voltage appearing between the electrodes equaled to either the sum of the transient voltage and the direct voltage or the difference between the transient and direct voltage depending upon the polarity of the dc voltage as shown in Fig. 2(b). The standard lightning impulse voltages were 1.3/40 μsec and were derived from a 12 stage impulse generator. For measuring the fast transient oscillatory voltages a special compact low resistance divider (as shown in Fig. 2(a)), with response time approximately 7 nsec, was used.

EXPERIMENTAL RESULTS AND DISCUSSION

Figure 3 shows the effect of needle length on the positive impulse breakdown characteristics for a 50mm gap and a pressure range 100-500kPa. Three lengths of needle 2mm, 5mm and 10mm were investigated. Each point represents the average value of 20-30 readings and the bars indicate the standard deviation σ. For the shortest length needle (l=2mm) the results become highly scattered, especially at the lower pressure when the lowest breakdown values noted equaled to those corresponding to highly non-uniform field gaps. The curve for the 10mm needle remained nearly flat over the entire pressure range investigated, which is a typical characteristic for highly non-uniform field gaps.

Figure 4 compares the breakdown voltage-pressure characteristics for lightning impulses under positive and negative polarities with those obtained for positive oscillating transient voltages for a pressure range 100 to 500kPa. The gap length (g) was 50mm and the needle length (l) was 5mm. The breakdown voltages for negative lightning impulses shown in Fig. 4 are included here as an example indicating that the presence of the

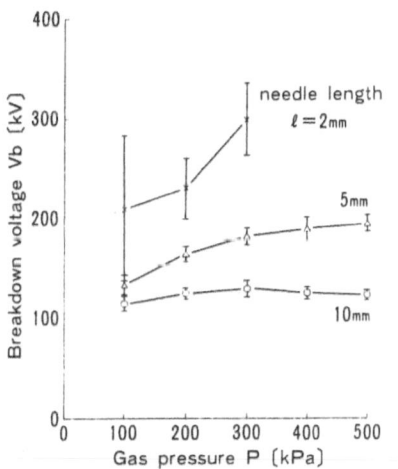 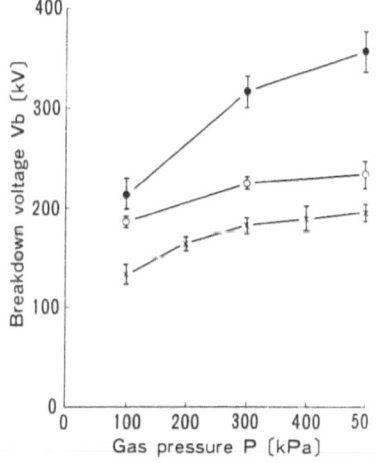

Fig.3. Influence of needle length on positive impulse breakdown characteristics.
[gap length (g)=50mm]

Fig.4. Breakdown voltage-pressure characteristics.
[gap length (g)=50mm, needle length (l)=5mm]
Standard lightning impulse:
X - positive polarity
● - negative polarity
Fast oscillating impulse:
O - positive polarity

needle reduces the breakdown voltage by much less than in the case of positive impulses. This trend was observed for all gap lengths and needle lengths studied. Consequently only the breakdown characteristics for positive polarity will be reported here.

When the needle length was increased to 10mm, the positive breakdown characteristics for both type of transients followed the pattern observed for highly non-uniform field gaps, as shown in Fig. 5. The figure gives the breakdown voltage-pressure relationship for the two types of surges over a pressure range from 100 to 500kPa. It is noted that in both cases the breakdown voltage remains mostly independent of gas pressure, with the oscillatory transient voltages giving approximately 30% higher values than the standard impulse voltages. Figure 5(b) shows the characteristics for a 100mm gap and a needle length 10mm. Under the oscillatory transient voltages the characteristic remained flat over the entire pressure range studied, while under the standard lightning impulses the breakdown voltage at first increased with increasing the gas pressure reaching maximum value and then decreased slightly at a pressure of 500kPa.

When the breakdown characteristics for oscillatory transient voltages and standard lightning impulses obtained at various gas pressures were compared in terms of breakdown voltage-time characteristics it was noted that in the time range below 1 μsec the values for the oscillatory transient voltages were lower than the corresponding standard impulse breakdown values. The effect increased with the gas pressure. Figure 6 compares the two sets of V-t characteristics for a pressure of 500kPa, gap g=50mm and needle length l=10mm.

Figure 7 presents the breakdown characteristics for combined dc voltage plus oscillatory transient voltage and dc voltage plus standard positive impulse voltages, for a range of dc voltages extending from -100kV to +100kV at a pressure of 100kPa. Included are also the measured dc-corona currents. The results demonstrate the large difference of breakdown characteristics between dc + standard impulse and dc + oscillatory transient (VFTO). In case of (neg.)dc + (pos.)impulse or VFTO, the breakdown voltages increased with increasing dc voltage in the region of 40kV and above, where the corona discharge took place before the

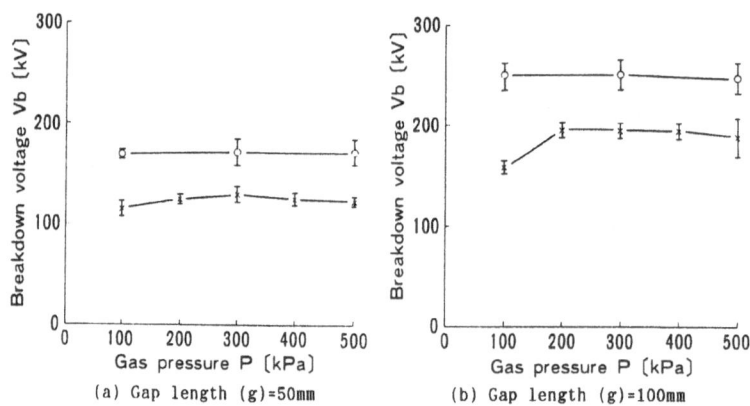

(a) Gap length (g)=50mm   (b) Gap length (g)=100mm

Fig. 5. Positive breakdown voltage - pressure characteristics
for needle length (l)=10mm.
X : Standard lightning impulse
O : Fast oscillating impulse

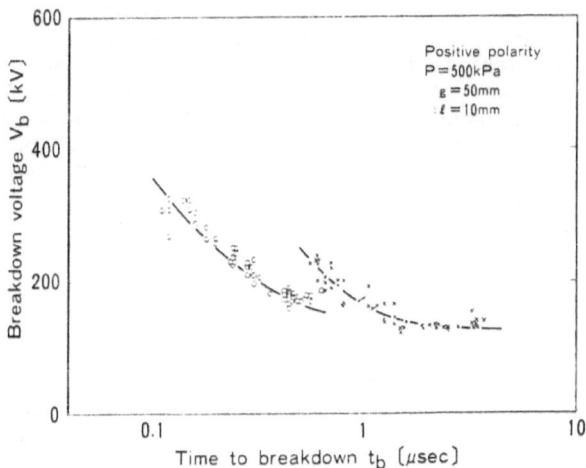

Fig. 6. Breakdown voltage - time characteristics for gas pressure 500kPa.
[gap length (g)=50mm, needle length (l)=10mm]
X : Standard lightning impulse
O : Fast oscillating impulse

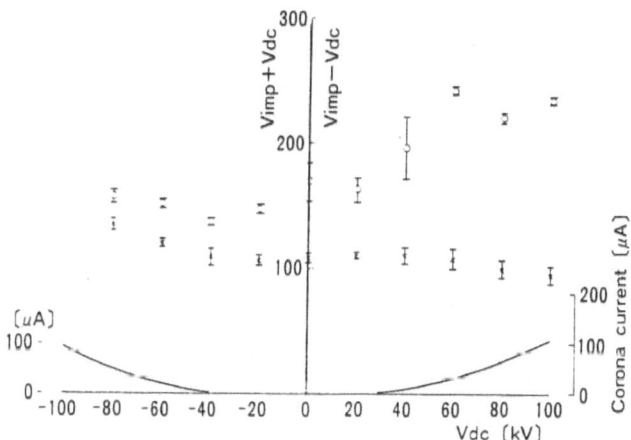

Fig. 7. Comparison of breakdown characteristics of superimposed lightning
impulses upon dc voltage with the corresponding VFTO superimposed
upon dc voltage in the presence of dc corona discharge.
X : DC + (pos.)Standard lightning impulse
O : DC + (pos.)Fast oscillating impulse

application of impulses or VFTO to the gap. As the polarity of the corona discharge at the needle tip was the same as the impulses or VFTO, corona stabilization is thought to be the reason for the increase of breakdown voltages in this condition. In case of (pos.)dc + (pos.)impulse, the breakdown voltage decreased slightly with increasing dc corona current, where corona discharge was of opposite polarity to the applied impulse. For (pos.)dc + (pos.)VFTO, on the other hand, the breakdown voltage increased with increasing dc corona current. For more quantitative description further investigations are necessary, especially in the higher gas pressure.

CONCLUSION

The investigation of the breakdown characteristics of non-uniform field gap under fast oscillating impulse voltages have shown that;

(1) The minimum breakdown voltages for the fast oscillating impulses (VFTO) are higher than those for the standard lightning impulses.
(2) The V-t curves of the fast oscillating impulses in the shorter time region (below 1 µsec) fall below those of the standard lightning impulses especially at the higher gas pressures.
(3) The effect of dc corona on the VFTO breakdown voltages differed in the polarity of the corona discharge.

ACKNOWLEDGMENT

The authors wish to thank Mr. J. Kendall, of the University of Manitoba, for his technical assistance and achievements.

REFERENCES

Hiesinger, H. and Witzmann, R., 1988, Very fast transient breakdown at a needle shaped protrusion., 9th International Conference on Gas Discharges and their Application, Venezia, pp.323-326.

Matsumoto, S., Okubo, H., Aoyagi, H. and Yanabu, S., 1989, Non-uniform flashover mechanism in SF_6 gas under fast-oscillating and non-oscillating impulse voltages., Sixth International Symposium on High Voltage Engineering, New Orleans, Paper 32.16.

Riquel, G., Ren, Z.Y. and Lefrancois, L., 1989, Comparison between V.F.T. and lightning impulse breakdown voltages for GIS insulation in presence of defects on line-conductors., Sixth International Symposium on High Voltage Engineering, New Orleans, Paper 23.09.

DISCUSSION

E. MARODE: If, in the case of an oscillating potential, one takes the mean oscillating potential value V_b' instead of the maximum potential $V_b = V_{imp} \pm V_{dc}$, would the standard lightning case V_b be similar to V_b'?

H. FUJINAMI: For the oscillating impulse voltage used in these experiments the ratio of the peak value to the mean value is about 1.4 and the breakdown voltages for oscillating impulses give approximately 30% higher values than the standard lightning impulse voltages as shown in Fig. 5. So breakdown voltages, using the mean values of oscillating impulses, are almost the same or slightly lower than standard lightning impulse breakdown voltages.

H. HIESINGER: First, how long did you apply the DC voltage in advance? Second, in cases when you superimpose the DC and impulse voltage and exceed the corona inception level so you have a charging of the gap, did you, then, find an influence of the DC application time on the breakdown level?

H. FUJINAMI: For the combined voltage studies, we applied the DC voltage for about 30 seconds before the application of the impulse voltage. Concerning the effect of the DC application time, we did not investigate breakdown characteristics with varying DC application time. I think the influence of the DC application time is negligible in the time range of over 30 seconds.

S. W. ROWE: The effect of the mean value of the oscillating wave can be shown to be of only secondary importance by studying highly damped waveforms where the oscillation is completely damped after the first peak. The work we carried out at Merlin Gerin showed that under these circumstances the breakdown voltage rises considerably. This is because the leader channel can only propagate during this first pulsed period. It then dies out. We clearly see this by fast photographic techniques.

J. OZAWA: I think the characteristics of surface flashover time along a spacer with a particle is also very important for GIS. What surface flashover time characteristics do you expect from your results?

H. FUJINAMI: Although I think the breakdown characteristics of the spacer surface are also important, our study is focussed on the gas gap.

R. T. WATERS: Do space charge effects have any role in increasing time to breakdown for lightning impulses compared with VFTO? Do you plan any experiments to relate time—to—breakdown with rate—of—voltage rise?

H. FUJINAMI: Space charge effects on time—to—breakdown are affected by the gap conditions, especially gas pressure and applied voltage waveforms. Space charge effects in fast oscillating impulse became less effective at higher gas pressure, resulting in the fact that the breakdown voltages for fast oscillating impulses are lower than the corresponding lightning impulse breakdown values as shown in Fig. 6. Yes, I plan for further investigations with different voltage waveshapes.

L. NIEMEYER: The comparison of LI and VFT may suggest that the VFT curves are systematically higher than the LI curves. In fact, they are essentially shifted not in the voltage scale direction but in the time scale direction. One major reason is that the VFT reaches the first peak much earlier than the LI and therefore can start the discharge much earlier leading to the earlier breakdown.

H. FUJINAMI: In our experimental results, V–t curves for fast oscillating impulses were shifted in the time scale at higher gas pressure. This tendency was restricted to the condition of higher gas pressure, such as 500 kPa.

THE EFFECT OF LEADER PROPAGATION ON THE V-T CURVES
UNDER LI AND VFT IN GIS

Th. Dunz, L.Niemeyer and G. Riquel*)

ABB Research Center
Baden, Switzerland
*) EDF Direction des Etudes et Recherches Les Renardières, France

ABSTRACT

A simple stepped leader propagation model is presented with which the formative time lag of breakdown in compressed gas insulation can be calculated. The model is based on physical principles and accounts for the insulation parameters gas, gas pressure, gap geometry, and voltage waveform. It is applied to particle induced breakdown of a uniform field gap under lightning impulse (LI) and very fast transient (VFT) stress.

INTRODUCTION

Breakdown in compressed gas insulation systems is controlled by the stepped leader mechanism which has become accessible to modeling from first principles [1,2]. In this paper we study the temporal aspects of the breakdown process, i.e. the formative time lag. Together with the statistical time lag, it determines the voltage time (V-t) curve of an insulation system which is important for insulation coordination considerations.

We present a simple numerical model with which the stepped propagation of a leader through the insulation gap can be simulated and the formative time lag can be determined for a given gap geometry and applied voltage waveform. The model will be applied to particle initiated breakdown in a uniform field gap under lightning impulse (LI) and very fast transient (VFT) stress. The results will be compared to measured data and the limitations of the model will be discussed.

EXPERIMENTAL

A detailed description of the experimental set-up is given elsewhere[3]. A 510 kV rms AC transformer energized by a capacitor bank is connected to a capacitor which is switched by a spark-gap to a 3 m long GIS bus-duct including either an oscillation inductor unit for VFT or a resistor for LI. Fig. 1 shows the VFT used for the present study.

Diagnostics include voltage measurements by a 300 MHz field probe, a broad band photomultiplier and a coaxial shunt for measuring the pre-discharge current. For VFT the displacement current is compensated by the signal from a discharge-free reference electrode.

Gaseous Dielectrics VI, Edited by L.G. Christophorou and
I. Sauers, Plenum Press, New York, 1991

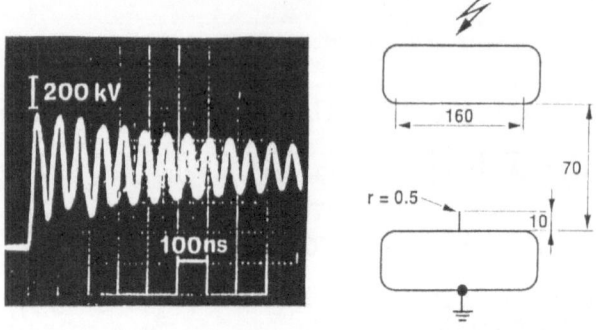

Fig. 1. VFT voltage signal (13 MHz) and gap configuration.

LEADER PROPAGATION MODEL

The dynamics of leader propagation in compressed gas insulation systems is controlled by the voltage U at the electrode or at the tip of the propagating leader, the kind of gas, its pressure p_o, and the geometry of the insulation gap. For a detailed description of the underlying physical processes and their modelling the reader is referred to [1,2].

The voltage U at the leader tip is determined by the voltage U_0 applied to the gap and the voltage drop ΔU_ℓ along the leader channel and can be approximated by

$$U = U_o - \Delta U_\ell \quad \text{with} \quad \Delta U_\ell = \overline{E}_\ell \cdot X_\ell \tag{1}$$

Here X_ℓ is the leader length and $\overline{E}_\ell$ the average leader field that can be calculated according to reference[2] eq. (8).

The leader step length ℓ can be approximated[2] by

$$\ell \approx \frac{C_\ell}{(E/p)_{cr}} \cdot \frac{U}{p_o} \quad \text{with} \quad \begin{array}{l} C_\ell \approx 0.5 \\ (E/p)_{cr} = 89 V / mPa \end{array} \tag{2}$$

The corona growth injects a charge pulse Q_c into the gap which can be determined from the voltage U, the step length ℓ, and the gap geometry by a numerical field/charge calculation which includes the leader as a conductor with a voltage drop $\overline{E}_\ell \cdot X_\ell$

$$Q_c = Q_c \left(U, \ell, \overline{E}_\ell, X_\ell \text{ gap geom.} \right) \tag{3}$$

Further, we assume that the leader propagates along the shortest field line in the gap and thus neglect its spatially random motion. This will result in systematically too low gap crossing times. With the corona charge Q_c the leader formation criterion

$$Q_c \geq Q_{cr} = C_{cr} / p_o^2 \tag{4}$$

is checked, where Q_{cr} is a pressure and polarity dependent critical charge. For SF_6 $C_{cr} = 50$ (250) $AsPa^2$ for positive (negative) polarity. If eq. (4) is fulfilled the leader step is established with a time delay τ given by

$$\tau = C_\tau / p_o^2 U \tag{5}$$

with $C_\tau = 4 \times 10^8 \ (80 \times 10^8)$ VsPa^2 for positive (negative) polarity in SF$_6$. This equation is valid if the voltage U remains constant during precursor development. For the case of rapid voltage variation, we generalize it by introducing a formal average leader propagation velocity

$$\bar{v}_\ell = \ell/\tau = C_v U^2 p_o \qquad \text{with} \qquad C_v = C_\ell / \left[(E/p)_{cr} C_\tau \right] \qquad (6)$$

which is an equivalent of the precursor development velocity. We now can state that the precursor development is completed if the precursor length x(t) has reached the step length ℓ as determined by eq. (2) i.e. if the condition

$$x(t) = \int_0^{\tau_s} \bar{v}_\ell [U(t)] dt \geq \ell \qquad (7)$$

is fulfilled.

Eqs. (1) - (8) allow to construct the inception of the leader at the electrode and its step-by-step propagation through the gap for a prescribed time varying voltage $U_o(t)$ applied to the gap, a given gas and its pressure p_o, and a given gap geometry.

It has to be remembered that this model is restricted to cases in which leader step formation occurs by the precursor mechanism. If other leader mechanisms[2] are involved, both the formation criterium eq. (4) and the delay time scaling eq. (5) may have to be modified or changed. As an example, in the case of VFT stress, displacement current heating[3] may contribute to or replace the precursor mechanism. An order-of-magnitude estimate of the parameters for which such an effect remains negligible, is given by

$$f < f_{max} \approx p_o^2 \hat{U}_o / C_\tau \qquad (8)$$

where f is the VFT frequency and $\hat{U}_o$ its amplitude. The maximum frequency, up to which the precursor model can be applied, is thus seen to strongly increase with pressure p_o and $\hat{U}_o$.

RESULTS AND DISCUSSION

We have studied leader propagation through an uniform field gap with a protrusion on one electrode according to fig. 1 under different gas pressures p_o, applied voltage amplitudes $\hat{U}_o$, both polarities, and for standard LI and 13 MHz VFT (Fig. 1). In order to exclude the scatter associated with the statistical time lag we have carried out direct comparisons of the model with single experiments taking the first light monitored by the photomultiplier to define the start of the calculation (time t_1).

As first example Fig. 2 shows the results obtained for negative LI and VFT. The figure shows the voltage waveforms $U_o(t)$ up to the instant t_b of breakdown and the PM signal on which the leader steps can be indentified. The results of the model calculation are represented by the stepped curves $X_\ell(t)$ and the calculated breakdown times t_2, and the number of steps to breakdown N. It is recognized that the experimental formative time lags $t_b - t_1$ and step number N are about 30 % above the model prediction. Note that in the VFT case the precursor applicability criterium eq. (8) is just at its limit with f = 13 MHz $\approx f_{max} = 10$ MHz. This is a consequence of the model simplification which assumes the leader to propagate exactly along the gap axis whereas, in reality, it follows a longer spatial random trajectory. This is consistent with the results of a previous study of the spatial randomness of leader propagation[5].

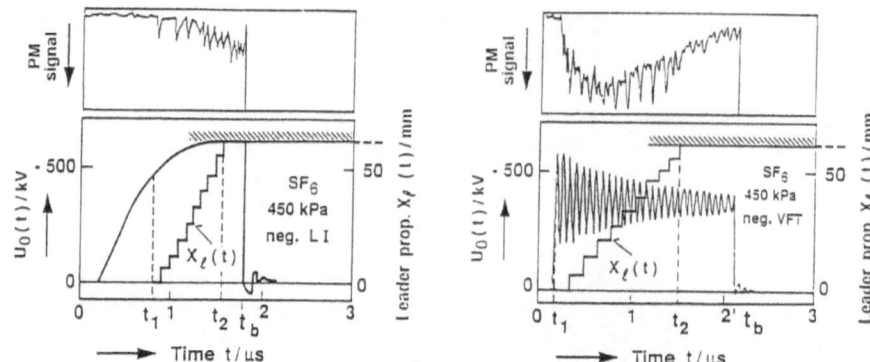

Fig. 2. Breakdown of the gap according to fig. 1 with $p_0 = 450$ kPa under negative LI and VFT waveforms. Experimental: $U_0(t)$, PM signal, discharge inception time t_1 and breakdown time t_b. Theoretical: Leader propagation curve $X_\ell(t)$ and breakdown time t_2.

The step sequence is no more resolved by the PM because of the much shorter step time. Fig. 3 shows an example for leader propagation under positive VFT at a gas pressure $p_0 = 270$ kPa. Again the experimental formative time lag $t_f = t_b - t_1$ is about 30 % higher than the calculated one. It is seen that leader stepping only occurs in bursts when the voltage is high whereas it stops during low voltage periods. The applicability criterium eq. (8) results positive with $f = 13$ MHz $<< f_{max} \approx 100$ MHz.

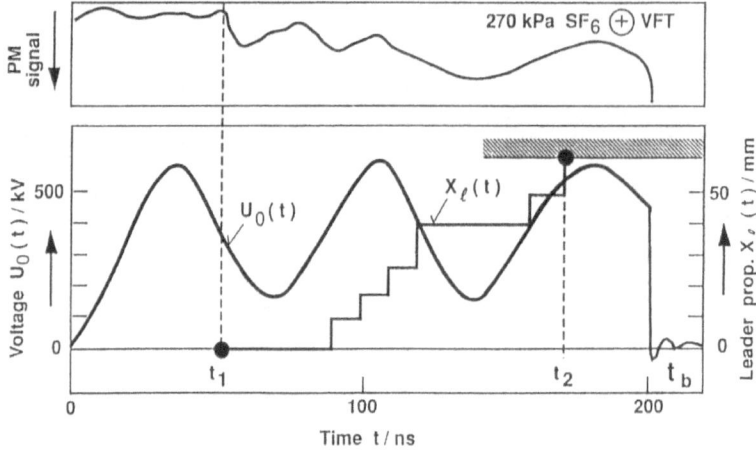

Fig. 3. Breakdown under positive VFT stress at $p_0 = 270$ kPa. Same symbols as in fig. 2.

In order to demonstrate, how the model can be used for the prediction of V-t-curves the latter have been measured for the gap of fig. 1 at 450 kPa for both porarities and for LI and 13 MHz VFT. The results are plotted as points in Fig. 4 according to the IEC convention. For the calculations it has been assumed that an initiatory electron is available immediately after the minimum leader inception voltage $U_{\ell i}$ has been exceeded. This implies that statistical time lags are not accounted for so that the calculations predict zero probability lower limits of the V-t-curves.

At long times the LI curves asymptotically approach the minimum voltage $U_{\ell i}$ which is about 50 % higher for negative polarity than for positive polarity. The negative VFT curves do not approach $U_{\ell i}$ because the voltage plotted according to IEC is the first peak value of the VFT and not the actual breakdown voltage. Under negative polarity the experimental data are rather close to but above the theoretical curves indicating low statistical time lags for both waveforms. Under positive polarity the measured breakdown times are substantially larger than the theoretical by predicted ones for all times. This is a consequence of the initatory electron generation mechanism, namely, field detachment from negative ions. The efficiency of this process strongly increases with the field and therefore with the applied voltage so that the statistical time lags become short for high voltages/short times and long for low voltages/long times. The experimental data thus are systematically shifted towards longer times (not towards higher voltages!).

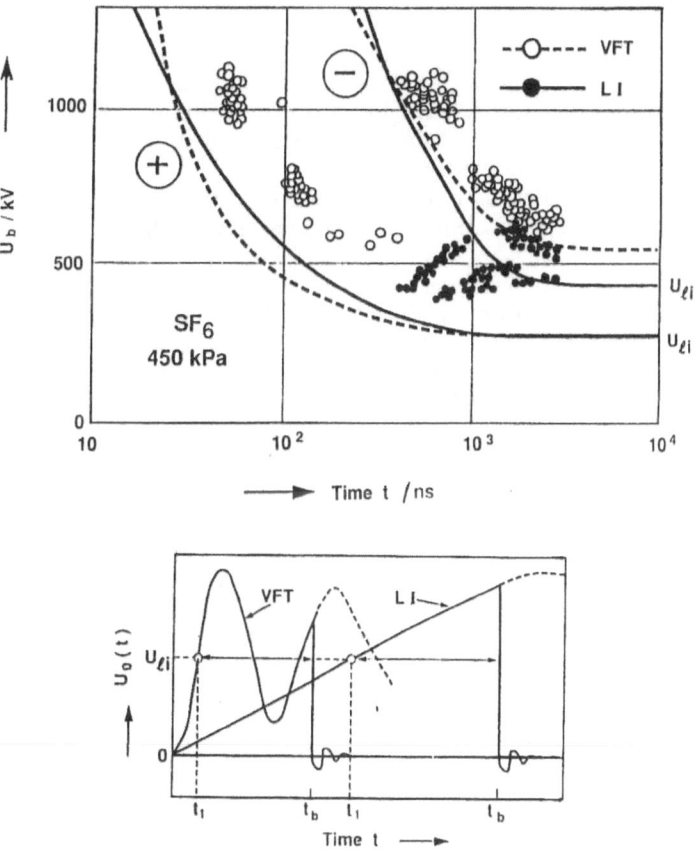

Fig. 4. V-t-curves for the gap according to fig. 1 with po = 450 kPa for both polarities and for LI and 13 MHz VFT waveforms. Points are measured data. Theoretical curves are lower bounds calculated with assumption of first electron occurence as schematically indicated on the right side of the figure.

CONCLUSIONS

The simple model presented for the simulation of stepped leader propagation through electronegative gases has the following main features:

- It is based on the precursor mechanism for leader formation.
- It assumes one-dimensional leader propagation along the shortest field line in the gap and does not account for the 3-dimensional random propagation trajectory. The latter can, however, be accounted for by a correction factor.
- It accounts for all design parameters of the insulation gap like gas composition, pressure and gap geometry including particulate contamination.
- It is applicable for rapidly varying waveshapes within parameter limits which can be verified by a simple criterium.

The model has been demonstrated to correctly predict formative time lags and lower bounds to V-t-curves for a particle contaminated uniform field gap for both polarities and for LI and VFT voltage waveforms. For typical SF_6 pressures used in high voltage GIS (350 - 700 kPa) the precursor mechanism controls the breakdown process and there is no substantial difference between the V-t-curves and minimum breakdown levels for LI and VFT stress. It is only beyond the validity limit of the precursor mechanism, e.g. at lower pressures and/or very high frequencies that displacement current heating contributes to the leader formation and reduces the insulation level under VFT with respect to LI stress.

REFERENCES

1. L. Niemeyer, L. Ullrich, N. Wiegart. The Mechanism of Leader Breakdown in Electronegative Gases. IEEE EI 24 No. 2 (1989), pp. 309-323.
2. L. Niemeyer. Leader Breakdown in Compressed SF_6: Recent Concepts and Understanding. These proceedings, p. 49.
3. G. Riquel, Z.Y. Ren. Insulation Behaviour of SF_6 Gaps Subjected to Fast Oscillatory Overvoltages. IXth Int. Conf. on Gas Disch. Venice, pp. 331-334 (1988).
4. H. Hiesinger, R. Witzmann. Very Fast Transient Breakdown at a Needle Shaped Protrusion. IXth Int. Conf. on Gas Disch. Venice, pp. 323-326 (1988).
5. L. Niemeyer, H.J. Wiesmann. Modelling of Leader Branching in Electronegative Gases. Gaseous Dielectrics V, L.G. Christophorou and D.B. Bouldin (Eds.), Pergamon, New York (1987), pp. 134-139.

INSULATING CHARACTERISTICS OF SMALL SF$_6$ GAS GAPS

UNDER LIGHTNING IMPULSE VOLTAGES

T.Ishii[*],M.Hanamura[*]
M.Hanai[**],K.Toda[**],T.Teranishi[**],H.Murase[**] and S.Yanabu[**]

*The Tokyo Electric Power Company,Inc. - Tokyo,Japan
**Toshiba Corporation - Kawasaki,Japan

ABSTRACT

SF$_6$ gas is very sensitive to electric fields and local electric field concentrations such that triple junctions cause partial discharges. Such partial discharges can cause fatal dielectric breakdowns in   power equipment composed of solid-gas composite insulations. Thus, research   on partial discharges in small gaps is important for the developing SF$_6$-gas-insulated equipment.

This paper deals with dielectric breakdowns at quasi-uniform SF$_6$ gas gaps. At a gas gap of below 2mm, with fine finished bare electrodes,under 0.1 MPa, a dielectric breakdown occurred at 2 or 3 times higher the voltage than Paschen's law voltage,under a lightning impulse. Under 0.4 MPa,a dielectric breakdown occurred in accordance with Paschen's law. If the electrodes are covered with PET film, however,a SF$_6$ gas breakdown voltage under 0.4 MPa is about twice as the voltage indicated by Paschen's law.

These phenomena were demonstrated clearly by a model which shows initial electrons being emitted from ions in the gas or from the electrode surfaces.

Also,the breakdown voltages expected from the simulation on the assumption that ion density in gas under 0.1MPa pressure is in the range of 10 to 100 ions/cm^3 agree well with experimental results with gas gaps 5mm or less.

INTRODUCTION

Making a high-voltage power equipment using SF$_6$ gas more compact and more reliable requires research on insulating characteristics of solid-gas composite systems. Insulation depending on small gas gaps,which are likely to occur between solid insulations such as triple junctions, is becoming very important.

Using bare and covered electrodes, we investigated dielectric breakdown characteristics with lightning impulse voltages at small SF$_6$ gas gaps under pressures of 0.1MPa and 0.4MPa. The results indicated that gas pressure and electrode-surface condition determine whether the initial electrons which caused a breakdown were emitted from ions in gas gaps or from the electrodes.

EXPERIMENTAL APPARATUS AND METHOD

As shown in Fig.1,two electrode structures were used in the experiment. Electrode structure (a) was composed of a spherical electrode and a plane one. Electrode structure (b) was composed of two spherical electrodes covered with PET(polyethylene terephthalate) film, 25μm thick (dielectric constant 3.2). For the spherical electrodes, steel ball bearings of diameter 40mm were used. The plane electrode was made of stainless steel. The ball bearing's surface was fine finished to less than 0.15μm roughness. The plane electrode was also fine finished to less than 0.1μm roughness. To adjust gaps, a micrometer with a minimum increment of 1μm was used.

Before installation, the electrodes were washed with ethyl alcohol. The gas inlet and outlet were provided with 0.1μm filters. After setting up, a vacuum pump got a vacuum within the tank to below 0.1 torr, and then the tank was filled with SF_6 gas from a commercial cylinder (purity 99.998%, less than 2ppm humidity) to the specified pressure.

All experiments were carried out at room temperature. The tank was kept dark to prevent initial electrons from being generated by external light. A small mercury lamp was installed in the tank. When this lamp is lit, light irradiates the gas gap.

The applied voltage was lightning impulse (1.2/55μs). The peak voltage was increased by a step of 2.5%, starting from 50% of the theoretical breakdown voltage, obtained from the Paschen's curve (George and Richards, 1969), and the voltage was applied at intervals of 30 seconds to obtain the initial breakdown voltage.

MECHANISM OF GENERATION OF INITIAL ELECTRONS

There are two main causes of the generation of initial electrons.

a) emission of electrons from ions in the gas due to the electric field
b) emission of electrons from the electrode surface due to the electric field.

Other possible causes include electron emission due to impurities, such as moisture in the gas and oxidation or deposition of electrode surface, and excitation by photons. However, since the electrodes are cleaned before use and gas is high purity, the effect of these other causes may be smaller than the effect of the two causes.

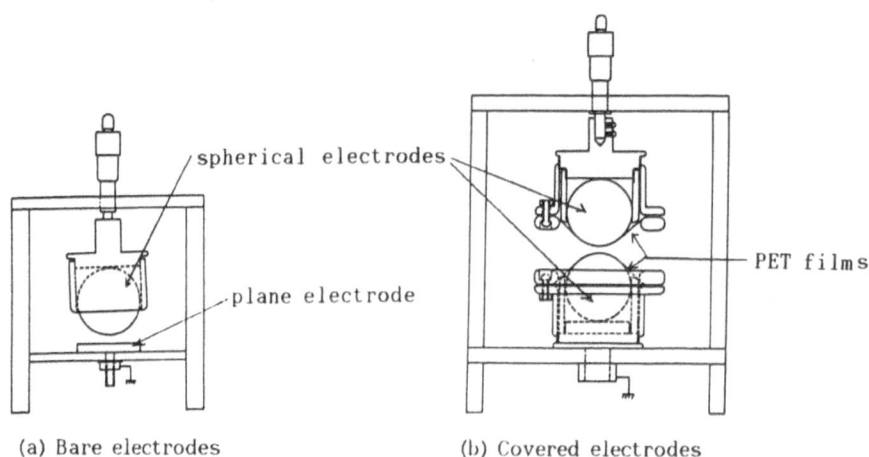

(a) Bare electrodes   (b) Covered electrodes

Fig.1. Electrode systems

ION DENSITY IN GAS GAPS

As for the production of ion pairs by cosmic rays and recombination and diffusion of ions, the number of ions in a gas gap is represented as a function of location and time:

$$\frac{dn^+}{dt} = No - Kr\, n^+ n^- - div\,(\mu^+ E\, n^+ - D^+\, grad\, n^+)$$

$$\frac{dn^-}{dt} = No - Kr\, n^+ n^- - div\,(\mu^- E\, n^- - D^-\, grad\, n^-) \qquad\qquad (1)$$

where n^+ and n^- are the densities of positive and negative ions respectively, No is the rate of ion production by cosmic rays, Kr is the coefficient of ion recombination , μ^+ and μ^- are the mobilities of positive and negative ions respectively, E is the electric field, D^+ and D^- are the diffusion coefficients of positive and negative ions respectively (Wiegart et al.,1987).

If ions in a gas gap are in equilibrium state, without an electric field, ion density is solved under a boundary condition of zero ion/cm^3 on the electrode surface. The maximum ion density in the gap increases in proportion to the square of the gap size; its upper limit of 0.1MPa is about 4000 ions/cm^3. For example, also at 0.1MPa, a gap length 5mm gives the ion density of about 90 ions/cm^3.

CALCULATION OF PROBABILITY OF INITIAL ELECTRON GENERATION BY IONS

On the assumption that initial electrons are generated by the electric field from ions in gas, it is reported that the distribution of statistical time lags of dielectric breakdown of SF$_6$ gas agrees well with measured values (Kindersberger and Taschner, 1985; Wiegart et al., 1987).

First, let's assume certain ion density (ion/cm^3) and divide the gas gap between electrodes into small volumes. Then, let's divide the application times of lightning impulse voltages and obtain total probability of electron emission P by numerically double-integrating the electron emission probability Po varying with the electric field (Teich and Zaengl,1978; Wiegart et al.,1987).

$$P = 1 - \exp\,[\int_{ts1}^{ts2} \int_V n\, Po\, dV\, dt] \qquad\qquad (2)$$

where it is assumed that the application voltages in the time-integration range between ts1 and ts2 exceed the breakdown voltage of Paschen's curve; Po is the probability of electron emission of one ion in a unit of time, and V is the total volume involved with electric fields above the Paschen's curve.

If it is assumed that breakdowns occur simultaneously with the generation of the initial electrons, the probability of electron emission from ions is the probability of breakdowns in gas gaps.

ELECTRON EMISSION FROM ELECTRODE SURFACE BY ELECTRIC FIELD

In a small gap with a small area, the macroscopic electric fields of electron emission from the electrode vary from 20 to 60 kV/mm with the surface roughness of the electrode. It is recognized that if the electrode surface is fine finished, in particular, electron emission occurs in the electric fields ranging from 40 to 60 kV/mm, regardless of pressure(Ozawa,1987).

MODEL OF INITIAL ELECTRON GENERATION

If all the electrodes have the same surface roughness, electric fields of electron emission from the electrode surfaces are constant, regardless of gas pressure. On the other hand, as gas pressure rises, the critical electric field

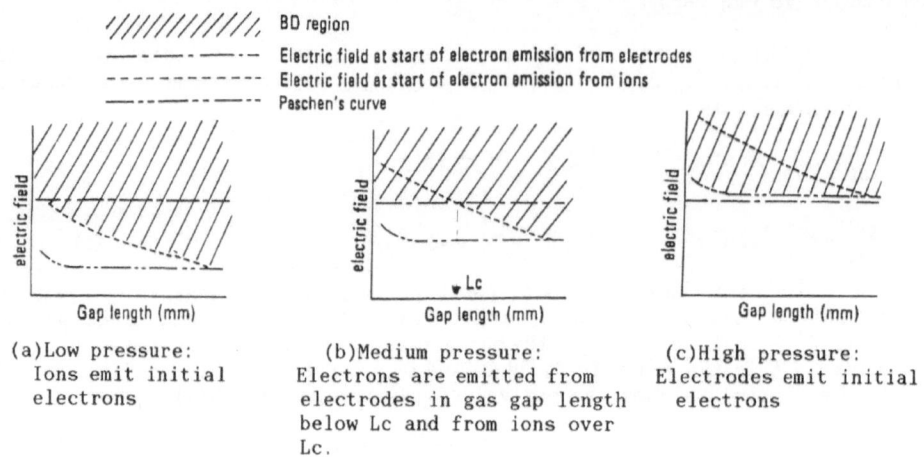

(a)Low pressure:
Ions emit initial
electrons

(b)Medium pressure:
Electrons are emitted from
electrodes in gas gap length
below Lc and from ions over
Lc.

(c)High pressure:
Electrodes emit initial
electrons

Fig.2.Breakdown models for small gas gaps under lightning impulse.

of electron emission from ions rises. However, as gas pressure rises, the number
of ions increases, and the rise of the avalanche electric field is relatively
small.

Thus, as shown in Fig.2 (a), if gas pressure is low, the electric field of
electron emission from the electrode is higher than that of electrons emission
from ions; the breakdown voltage becomes higher than Paschen's curve in a small
gap because it is determined by the generation of initial electrons from ions
(Hanai et al.1989).

Under a medium pressure,the electric field of electron emission from ions
becomes higher than that of electron emission from the electrode below a certain
gap length Lc. So breakdowns occur with a certain constant electric field in gap
length below Lc, and are caused by the generation of initial electrons from ions
in longer gap length, as shown in Fig.2 (b). If, under a high pressure, the
electric field of electron emission from the electrode is below the Paschen's
curve, breakdown voltages agree with the Paschen's curve, as shown in Fig.2 (c).

Thus, the precedence of the generation of initial electron either from ions
in gas or from the electrode surface can be considered to be determined by gas
pressure and electrode surface roughness.

DIELECTRIC BREAKDOWN CHARACTERISTICS WITH BARE ELECTRODES

As shown in.Fig.3, in a dark space under 0.1MPa, breakdown voltages with
gap length 2mm or less were about 3 times as high as those of the Paschen's
curve. This trend is more prominent with smaller gaps. When the gas space between
electrodes was irradiated with ultraviolet rays, breakdowns occurred at voltages
close to the Paschen's curve (Hanai et al.,1989).

Assuming that initial electrons under 0.1MPa were supplied from ions in the
gas, as shown in Fig.2 (a), breakdown voltages were calculated from eq.(2). An
initial breakdown happens when total breakdown probability (Ptotal) comes to be

$$\text{Ptotal} = 1 - \prod_{i=1}^{N} (1 - Pi) > 0.5 \tag{3}$$

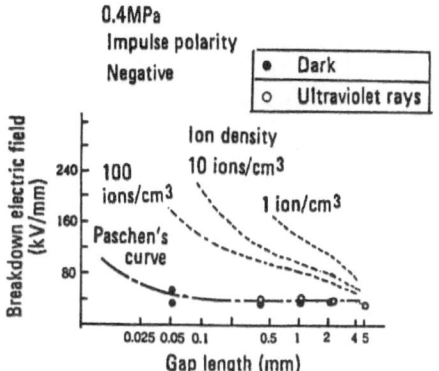

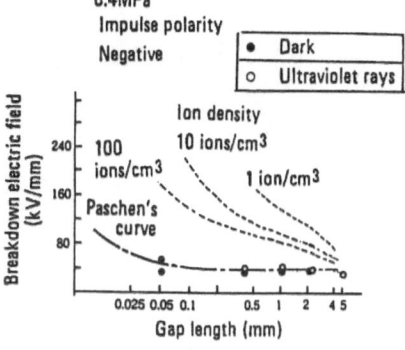

Fig. 3. Relation between the gap length and the breakdown electric field (0.1MPa).

Fig. 4. Relation between the gap length and the breakdown electric field (0.4MPa).

with lightning impulse voltages applied, until the initial breakdown occurs (N times) for ion densities in gas gaps: 1, 10, 100 ions/cm^3 (Pi is the probability of breakdown by i-th application). The calculated breakdown voltages are shown in relation with gap length by dotted lines in Fig.3. Evidently, almost all measured values for the dark space stand between the curves for ion densities 10 and 100 ions/cm^3 curves. This suggests that under 0.1MPa, breakdown voltages in the dark, higher than Paschen's curve, are the result of ion densities in gas gaps which generated initial electrons in the range of about 10 - 100 ions/cm^3.

Under 0.4MPa, on the other hand, in the gap length of 50μm to 5mm, breakdown voltages were not observed to rise even in the dark, but stayed consist with the Paschen's curve, as observed when ultraviolet rays were irradiated. This may be because electron emission from the fine finished electrode surfaces occurred at about 40 kV/mm, causing initial electrons to be generated from the electrode, before an electric field with initial electrons emitted from ions in the gas was formed, as shown in Fig.2(c).

INSULATION CHARACTERISTICS WITH COVERED ELECTRODES

Under a pressure of 0.4MPa, electron emission from the electrode surfaces preceded that from ions. This implies that insulating characteristics, shown in Fig.2(b), can be obtained by controlling electron emission from the electrode surfaces.

Using the electrodes covered with 25μm PET film, shown in Fig.1(b), we obtained breakdown characteristics of gas. As shown in Fig.5, under a pressure of 0.4MPa, breakdowns occurred with electric field 1.5 to 2.5 times higher than the electric field indicated by the Paschen's curve in the gas gap ranges below about 2mm. Also, these breakdowns were constant with the electric field of the Paschen's curve with a gap length of 5mm.

These results show that covering the electrodes with PET film controls the emission of initial electrons from the electrode and raises breakdown voltages.

These are two possible explanations of rising breakdown voltages: an initial electron occurs with 60 to 100 kV/mm from the PET film which covered the electrodes: the PET film covering changed the boundary conditions of ion density on the electrodes, resulting in an ion density are about 4000 ions/cm3 in the gas gaps. So, now we are investigating which one is the main cause.

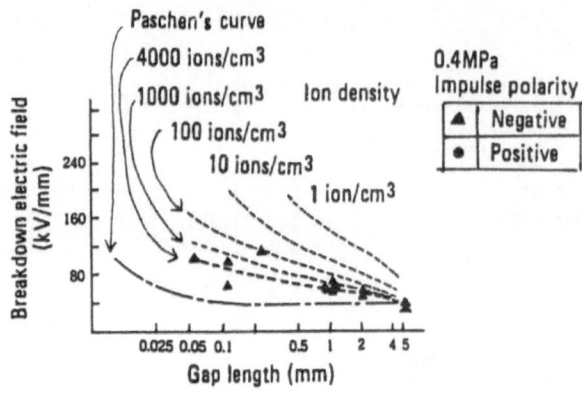

Fig. 5. Relation between the gap length and
the breakdown electric field
(0.4MPa, covered electrode).

CONCLUSION

Using fine finished electrodes, we investigated insulating characteristics
of SF_6 gas of small gaps from 25μm to 5mm under lightning impulse voltages.
These are the findings:

(1) Under 0.1MPa , breakdowns occur at voltages   about 2 to 3 times higher than
the Paschen's curve voltages with the gap length bellow 2mm.

(2) Under 0.4MPa, breakdowns occur at voltages consistent with the Paschen's
curve voltages.

(3) If electrodes are covered with PET film, breakdowns occur at voltages twice
as high as the Paschen's curve voltages, even under 0.4MPa.

(4) Breakdown characteristics depend largely on the generation of initial
electrons.This can be explained by a model in which initial electrons are
emitted from ions in the gas gap under 0.1MPa and from the electrode surfaces
under 0.4MPa.

(5) On the assumption that ion density is 10 - 100 ions/cm^3 in a dark space
under 0.1MPa with bare electrodes, the calculated probability of breakdowns
is consistent with experimental results.

REFERENCES

George,D.W. and Richards,P.H.,1969,Electrical Field Breakdown in Sulphur
Hexafluoride,Brit.J.Appl.Phys.(J.Phys.D.),Vol.2,No.2.
Hanai,M.,Teranishi,T.,Okubo,H. and Yanabu,S.,1989,Insulating Characteristics of
Small Gas Gaps against Lightning Impulse Voltages,Electrical Engineering
in Japan ,Vol.109-A,No.6,pp.255-262.
Kindersberger,J. and Taschner, W.,1985,Impulse Voltage Breakdown in SF6 with
AC and DC Bias Voltages,Proc.Int.Symp. on Gas Insulated Substation,
Pergamon,p.225.
Ozawa,J.,1987,Voltage-Time Characteristics of Impulse Breakdown in SF$_6$ Gas,
Gaseous Dielectrics V, L. G. Christophorou and D. W. Bouldin (Eds.), Pergamon, New York (1987), 438.
Teich,T.H. and Zaengl,W.S.,1978, The Dielectric Strength of an SF6, Brown
boveri Symp.Curr.Interruption High Voltage Network,pp.269-297.
Wiegart,N., Niemeyer,K., Pinnekarp,F., Boeck,W., Kindersberger,J., Morrow,R.,
Zaengl,W., Zwicky,M.,Gallimberit,L. and Boggs,S.A.,1987,Inhomogeneous
Field Breakdown in GIS-The Prediction of Breakdown Probabilities and
Voltage, IEEE/PES,WM191-0,1 .

LOW AND HIGH VOLTAGE DISTRIBUTION ALONG A CAP AND PIN INSULATOR STRING

SUBJECTED TO AC AND IMPULSE VOLTAGES

A. E. D. Heylen, S. E. Hartles, A. Noltsis and Derek Dring

High Voltage Laboratory
Department of Electronic and Electrical Engineering
The University of Leeds
Leeds, LS2 9JT, U.K.

INTRODUCTION

It is well known that under working ac voltage, the voltage distribution along a cap and pin insulator string is very uneven due to the presence of stray capacitances from the caps to the line conductor and especially to the earthed supporting tower. The ends of the string, particularly that near the line end, are subjected to voltages well above average and this leads to possible interference with radio and television signals especially under wet or polluted conditions, and results in considerable energy loss due to flowing corona currents. The standard way of alleviating this is to fit grading rings at both ends which considerably evens out the voltage distribution so that each cap and pin can be stressed to a higher voltage.

In the present paper, the contrasting voltage distribution with and without a grading ring at the working level and particularly near to flashover is examined under ac and impulse voltages of the lightning stroke and switching surge kind and not surprisingly it is found that the voltage distribution is evened out in the first and last case when corona currents flow over the sheds. This suggests that, as far as the ultimate flashover is concerned, grading rings could in general be dispensed with, thus contributing to an overall saving in installation costs. The results are backed up by calculations based on measured stray capacitance values.

EXPERIMENTAL PROCEDURE AND RESULTS

A representative string of five cap and pin insulators was chosen so that the ultimate flashover was within reach of a seven stage, 100 kV/stage, impulse generator and a 300 kV RMS ac supply. The string was suspended from the laboratory ceiling and the top end of the string was earthed by means of a cross-arm and stand which simulated the supporting pylon. The line end consisted of a 5 cm diameter tube of 3 metres length with loops at the ends to reduce corona effects. The classical way of fitting in turn a spark gap to each cap and pin insulator was used to measure the voltage distribution.

AC Voltage Distribution

The ac working voltage, V_w, was determined from

$$V_w = \frac{66}{\sqrt{3}} + \frac{1}{4} \left(\frac{132 - 66}{\sqrt{3}} \right) = 47.6 \text{ kV RMS}$$

in which 66 and 132 kV are the nominal three phase working voltages for a string of four and eight insulators (BSS 137:1960). The voltage distribution with and without one grading ring at the line end is shown in Figure 1.

For the high voltage distribution near to the flashover value (285 kV RMS), the low voltage pin electrodes of the spark gap were replaced by parallel plate electrodes 3.5 cm in diameter and domed according to a profile equivalent to 5 cm radius spheres so as to keep the sparking distance reasonably short. The derived voltage distribution is also shown in Figure 1 and it was noted by audible and visible means that corona currents started to flow only above 140 kV RMS.

Gaseous Dielectrics VI, Edited by L.G. Christophorou and
I. Sauers, Plenum Press, New York, 1991

Lightning Stroke Voltage Distribution

As no British Standard is available for determining the working voltage under impulse conditions, this was determined from the ac value according to $47.6 \times \sqrt{2} \times 1.1 = 74$ kV peak where 1.1 is the accepted impulse ratio. As the spark gap voltage was approximately $74/5 = 15$ kV corresponding to only a half cm. gap, the overall voltage was increased to 150 kV peak under which condition still no corona currents flow (see preceding section). The applied impulse was a 0.8/49.5 wave as measured oscilloscopically. The measured voltage distribution with and without one grading ring at the line end is shown in Figure 2.

At high impulse voltage, the 50% flashover voltage for negative polarity under a lightning stroke and switching surge occurred at 500 kV peak, but the positive polarity one for the lightning stroke was at 450 kV peak so that the positive polarity was used for voltage distribution measurement at the 10% flashover value. As the distribution varied widely (see Figure 2), the value of applied peak voltage was kept constant and the gap distance of the spark gap was varied to d_1, d_2, d_3, d_4, and d_5. As $d = d_1 + d_2 + d_3 + d_4 + d_5$, the fractions d_1/d, d_2/d, d_3/d, d_4/d and d_5/d were normalised to the sum of unity giving the required distribution as decimal fractions. The voltage distribution with and without one grading ring is shown in Figure 2.

Switching Surge Voltage Distribution

For the switching surge the 50% positive polarity flashover value occurred at 408 kV (near to the ac peak of 403 kV) and as this is the lowest, this polarity was used for determining the voltage distribution as it has applications in insulation co-ordination where the system has to be guarded against the lowest flashover voltage. The switching surge used was a 120/3500 wave as Hughes and Roberts[1] have shown that for spark gaps the lowest breakdown voltage occurs near to a wave front of 100 μsec. The working voltage distribution is shown in Figure 3.

At high voltage, the 10% flashover voltage was applied and the variable gap distance of the spark gap was used as in the preceeding section. The voltage distribution obtained is shown in Figure 3.

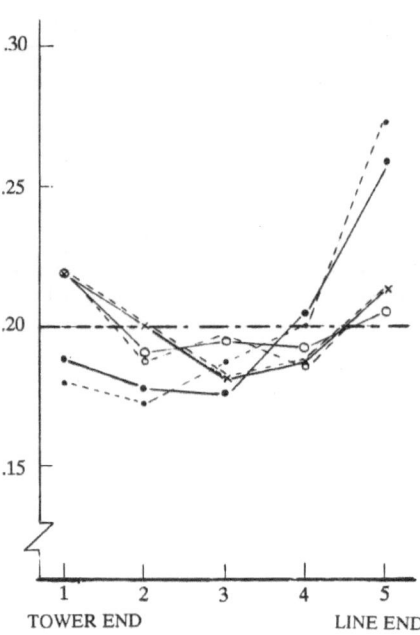

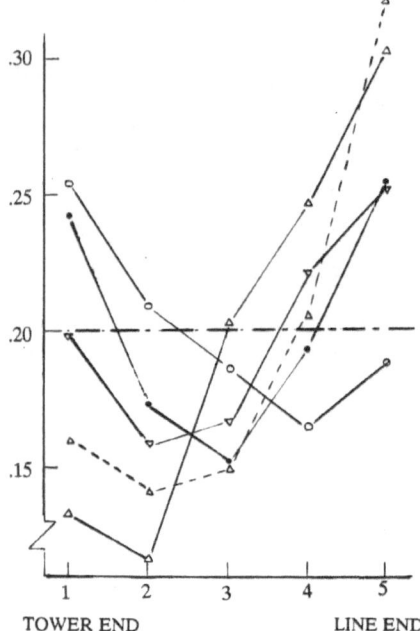

Fig. 1. A C voltage distribution for five cap and pin insulators. Solid Lines:- •, without grading ring at working voltages; ○, with grading ring at working voltage; x, near to flashover. Dotted lines:- theory. Chain line:- even distribution.

Fig. 2. Lightning stroke voltage distribution for five cap and pin insulators. Solid lines:- •, without grading ring at working voltage; ○, with grading ring at working voltage; △, near flashover, no grading ring; ∇, near to flashover, with grading ring. Dotted line:- theory. Chain line:- even distribution.

THEORETICAL MODEL

The uneven voltage distribution along a cap and pin insulators string is due to stray capacitances between the caps and the line and especially the earthed tower. Text book treatment (see reference 2 for instance) usually assumes these capacitances to be constant, but present measurements show that the stray capacitances increase gradually towards the ends and that even the self- capacitance of apparently identical cap and pin insulators is not constant, as given in Table 1.

TABLE 1 - Capacitance values

1. Insulator capacitance (in pF), starting from earthed to tower end.

C_1	C_2	C_3	C_4	C_5
33.0	35.2	33.5	33.05	34.0

2. Spark gap capacitance (in pF)

    low voltage electrodes:- 4.1
    high voltage electrodes:- 1.3

3. Stray capacitances (in pF); see Figure 5

without spark gap	C_2^E	C_3^E	C_4^E	C_5^E
	5.0	4.3	3.8	3.8
	C_2^L	C_3^L	C_4^L	C_5^L
	1.2	1.5	1.9	3.9
with spark gap [*]	C_2^E	C_3^E	C_4^E	C_5^E
	6.6	5.8	5.4	5.1
	C_2^L	C_3^L	C_4^L	C_5^L
	1.6	2.0	2.3	3.6

[*] values chosen for calculations

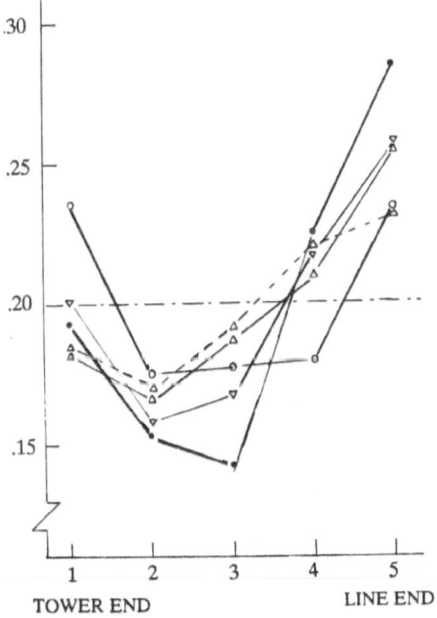

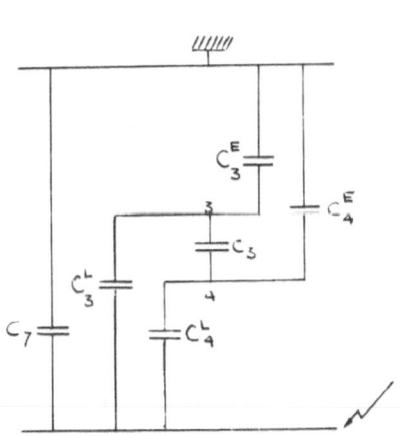

Fig. 3.Switching surge voltage distribution for five cap and pin insulators. Solid lines:- ● without grading ring at working voltage; ○, with grading ring at working voltage; △ near to flashover, no grading ring; ▽ near to flashover, with grading ring. Dotted line:- theory. Chain line:- even distribution.

Fig. 4.Equivalent circuit of one suspended cap and pin insulator in position No. 3.

269

To measure the stray capacitances, one insulator at a time was suspended in its line position by means of a rope, thus reducing the equivalent circuit to that shown in Figure 4 for position 3. In this circuit, the line to earth capacitances, C_7, comprising stray and equipment capacitance, is included for, although it does not affect the voltage distribution, it does have a large bearing on capacitance measurements which were made using a three terminal method, the instrument having long coaxial leads.

At normal working voltage, the insulator string may be represented by the electrical circuit shown in Figure 5 with the resistances equal to infinity. Taking currents and voltages as indicated, let

$$V_{12} = v \text{ then}$$

$$V_{23} = I_{23}/j\omega C_2 = \left[\left(C_1 + C_2^E + C_2^L\right)v - C_2^L V\right]/C_2 \quad;$$

similarly

$$V_{34} = I_{34}/j\omega C_3 = \left[\left(C_2 + C_3^E + C_3^L\right)V_{23} + \left(C_3^E + C_3^L\right)v - C_3^L V\right]/C_3 \quad;$$

also

$$V_{45} = I_{45}/j\omega C_4 = \left[\left(C_3 + C_4^E + C_4^L\right)V_{34} + \left(C_4^E + C_4^L\right)\left(V_{23} + v\right) - C_4^L V\right]/C_4$$

and

$$V_{56} = I_{56}/j\omega C_5 = \left[\left(C_4 + C_5^E + C_5^L\right)V_{45} + \left(C_5^E + C_5^L\right)\left(V_{34} + V_{23} + v\right) - C_5^L V\right]/C_5$$

where

$$V = V_{12} + V_{23} + V_{34} + V_{45} + V_{56}.$$

This enables one to determine V and by substitution the values of V_{12}, V_{23} etc. If one wishes to determine the voltage distribution in terms of decimal fractions, then V is put equal to unity. These equations were solved on the university computer and the distributions found are given in Figures 1, 2 and 3.

At high voltage, corona currents flow and these can be represented by the respective voltages divided by resistances as shown in Figure 5 which can be analysed as follows:

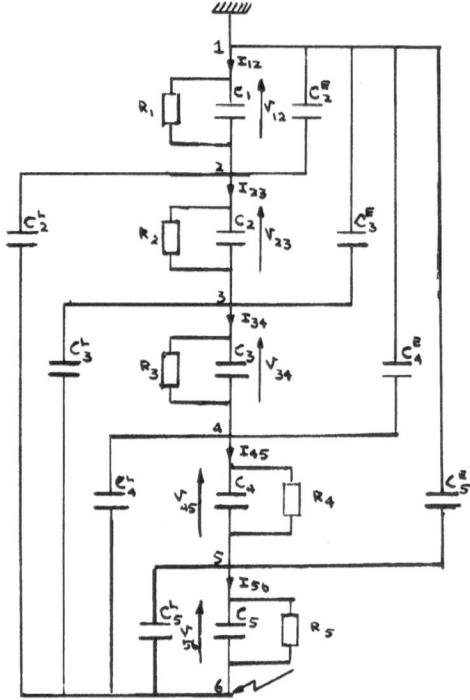

Fig. 5. Equivalent circuit for a string of five cap and pin insulators.

270

Let $V_{12} = v$ then

$$I_{12} = \left(\frac{1}{R_1} + j\omega C_1 \right) v$$

and

$$V_{23} = \left(\frac{R_2}{1 + j\omega C_2 R_2} \right) \times$$

$$\left[\frac{1}{R_1} + j\omega \left(C_1 + C_2^E + C_2^L \right) v - j\omega C_2^L V \right]$$

which can be modified to

$$V_{23} = \frac{R_2}{\left[1 + (\omega R_2 C_2)^2 \right]} \times$$

$$\left[\frac{1}{R_1} + \omega^2 C_2 R_2 \left(C_1 + C_2^E + C_2^L \right) + j\omega \left(C_1 + C_2^E + C_2^L - C_2 \frac{R_2}{R_1} \right) v \right.$$

$$\left. - \left(\omega^2 C_2 R_2 C_2^L - j\omega C_2^L \right) V \right]$$

etc.

These equations were put on a computer programme and the solutions are shown graphically in Figures 1, 2 and 3. With corona current flowing, the resistance values are R_1 = 219 $k\Omega$, R_2 = 199 $k\Omega$, R_3 = 180$k\Omega$, R_4 = 187 $k\Omega$, and R_5 = 213$K\Omega$ under a c conditions and agreement with experiment was complete. With switching surges at high voltage, the resistance values were R_1 = 1.85 $M\Omega$, R_2 = 1.62 $M\Omega$, R_3 = 1.86 $M\Omega$ R_4 = 2.09 $M\Omega$ and R_5 = 2.54 $M\Omega$. The computer programme was run for ac voltages and from a Fourier analysis, the fundamental frequency of the lightning stroke was 23.58 KHz and for the switching surge 2.464 KHz. These were the frequencies used in the computations.

DISCUSSION OF RESULTS

For the a/c distribution, it is seen from Figure 1 that at working voltage, the distribution is very uneven with about 26% of the total voltage occurring across the line-end insulator. As expected the grading ring evens out the voltage distribution considerably with only 21% of the voltage now appearing at the line end which is very close to the ideal of 20%. The behaviour of the string without grading ring near to flashover is interesting in that considerable corona currents flow which tend once more to even out the voltage distribution, the maximum deviation from a uniform distribution being only 10%. The calculated values are very close to the measured ones. This sets the scene for the examination of the voltage distribution under impulse conditions.

Under lightning strokes, the distribution at working voltages is again uneven, as seen from Figure 2, the line end insulator again having 26% of the voltage across it; the grading ring at the line end reduces this value to 19%. Near to flashover, the voltage across the line end insulator rises to 30% and a grading ring reduces this to 25%. It should be clear from this that with such a fast wave front, corona currents have no time to establish themselves and play no role in evening out the voltage distribution. The calculated value agrees reasonably well with the experimental one.

Under switching surge conditions, at the working voltage with no grading ring, the line and insulator has a voltage across it of 26%; a grading ring reduces this to 23%. What is interesting is that near to flashover the corona currents have time to establish themselves with the result that the voltage across the line insulator drops to 24% and the grading ring has virtually no effect. The calculated value agrees well with experiment.

CONCLUSION

The voltage distribution near to flashover evens out considerably under a/c and positive polarity switching surge conditions and the ultimate flashover of a five cap and pin insulator string is virtually the same. As this voltage is considerably below that under lightning stroke conditions, it can be concluded that as far as the ultimate flashover voltage for clean cap and pin insulators is concerned, no grading rings would be required thus saving in the cost of initial installation, although grading rings (stirrups) are still required to guard the porcelain sheds from the follow through arc.

ACKNOWLEDGEMENTS

Thanks are due to Professor Alan E. Guile for encouragement, to Mr G. L. Bibby for advice on the measurement of the stray capacitances, to Paul Clark for help in computer programming and to Doulton Insulators Ltd., Staffordshire, for their gift of cap and pin insulators. John Christoforou and M. D. Khairi contributed to the project measurements.

REFERENCES

1. R. C. Hughes, and W. J. Roberts, Application of flashover
   characteristics of air gaps to insulation co-ordination,
   Proc. IEE, 112, 198, (1965).

2. A. E. Guile, and W. Paterson, Electrical Power Systems,
   Vol. 1, 139-142, Oliver and Boyd: Edinburgh, 1962.

STARTING PROCESSES AND THE INFLUENCE OF THE RADIOACTIVITY OF

THORIUM ON THE STATISTICAL LAG TIME IN METAL HALIDE LAMPS

Scott R. Hunter[*] and A. Bowman Budinger[**]

[*]GTE Electrical Products
100 Endicott Street, Danvers, MA 01923
[**]GTE Laboratories, Inc.
40 Sylvan Road, Waltham, MA 02254

INTRODUCTION

High pressure metal halide discharge lamps are widely used as high efficiency lighting sources in a wide variety of applications from high wattage (175 to 1500 W) outdoor and stadium lighting to the more recent lower wattage (75 to 150 W) lamps used in retail store and display lighting[1,2]. Both low and high wattage lamps usually contain thorium; either in the form of ThI_4 as an additive to the chemical fill in the lamp, or as ThO_2 in the pressed, sintered and drawn tungsten wire used as the electrode material in the lamp. The presence of thorium in the discharge improves the colour rendering properties of the light from the lamp due to the many emission lines in the radiative spectrum of atomic thorium[3]. The presence of a few monolayers of elemental thorium on the surface of the tungsten electrodes also aids in the starting of the lamp and lowers the operating temperature of the thermionically emitting electrode by lowering the work function of W from $\phi = 4.6$ eV to $\phi \simeq 2.7$ eV[4]. The stages involved in the starting of these lamps are complex and are briefly outlined in the next section.

Recently, it has been suggested that a second mechanism exists in which thorium aids in the initial ignition of the discharge within the arc tube of these lamps. Thorium is weakly radioactive and the ionizing radiation produced in the radioactive decay of thorium is a source of free electrons in the arc tube and may aid in reducing the statistical lag time of the gas breakdown process after the application of the high voltage[5]. The influence of the thorium on the breakdown process has been studied by calculating the production rates for the intermediate daughter radionuclides and the various forms of ionizing radiation that are produced during the complete decay cycle of thorium as a function of time after the initial purification of the thorium. The electron production rates in the arc tube derived from these calculations are compared with those produced by natural background radiation sources. Finally, experimental measurements of the starting time for lamps containing tungsten electrodes with varying percentages of thoria (0% to 2%) are presented which confirm the dependence of the statistical lag time on the percentage of thorium in the electrodes of these lamps.

METAL HALIDE LAMP STARTING

The ignition or starting of a metal halide lamp consists of several stages which can be characterized as follows: prebreakdown and gas breakdown; glow discharge; glow-to-arc transition; low pressure thermionic arc; and finally, the high pressure arc. The issues involved in lamp starting and particularly in the glow-to-arc transition have been addressed in several

recent studies[5-10]. The initial stages after the application of the high
voltage and the actual breakdown of the gas can be divided into two
processes, namely the time required for the initial electron or electrons to
appear in the discharge gap – called the statistical time lag and varies from
milliseconds to 100's seconds for these lamps – and the time for the
subsequent growth in the electron number density to achieve a self-sustained
low current glow discharge – called the formative time lag and is usually in
the sub microsecond range.

One of the basic requirements for gas breakdown to occur within the
lamp is that the voltage applied across the electrodes must be of a
sufficient amplitude and duration to ensure that electron growth by electron
impact ionization will produce an avalanche of critical size (i.e. $\simeq 10^8$
electrons) such that breakdown will occur. The DC breakdown voltages of the
lamps used in this study (100 watt metal halide lamps containing $\simeq 100$ torr
of Ar, rod-rod electrodes 0.4 mm in diameter and electrode separation 14 mm)
are in the range 1.1 to 1.2 kV. The applied voltage waveform is shown in
Figure 1a for several cycles and on an expanded scale in Figure 1b showing
the details of one of the high voltage pulses imposed on the 60 Hz AC input
voltage. Usually 3 to 8 high voltage pulses (peak voltage $\simeq 3$ kV, half width
$\simeq 3$ µs) are imposed on each half cycle of the applied voltage (voltage $\simeq 300$
V RMS) until the gas breaks down and a glow discharge is established within
the lamp[5].

A second requirement for gas breakdown is that there must be at least
one free electron within the discharge gap; it is the sources of the free
electrons that form the basis of the present paper. In lamps that do not
contain sources of UV radiation (UV sources are sometimes included to provide
electrons by photoelectric emission[9]), initiatory electrons are provided by
background radiation sources (primarily natural radioactivity and cosmic
radiation), by charges stored within the walls of the arc tube from previous
gas discharges, and by radiation sources deliberately added to the lamp.

The glow discharge phase, which follows the initial breakdown of the
gas and may last from $\simeq 100$ ms to several seconds, depending on the electrode
materials, design and degree of coating with Hg, is characterized by a high
discharge impedance. Electrode heating, primarily by positive ion bombardment[6-8], occurs during this phase until the point where thermionic emission
from the electrodes becomes significant and the discharge experiences a glow-
to-thermionic arc transition. During this phase, the impedance of the
discharge drops and the discharge current increases. Lamps made with pure
tungsten electrodes tend to 'hang up' in the glow discharge causing electrode
sputtering and poor lumen maintenance. The presence of ThO_2 in the electrodes
reduces the the glow time by lowering the temperature at which the electrodes
become thermionic emitters[4]. The thermionic arc phase is initially characterized by low gas pressures (primarily Ar) but as the arc tube heats, the
vapour pressures of Hg and finally the metal halide salts increase to 3–20
atmospheres and 3–20 torr respectively. Voltage and current waveforms and
waveform envelopes obtained during this phase of lamp operation are shown in

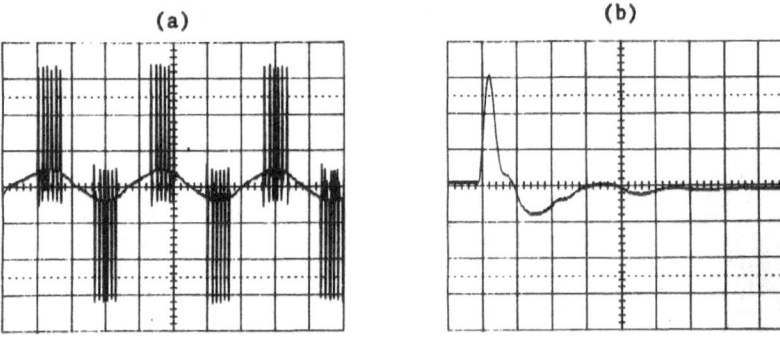

(a)   (b)

Fig. 1.   Lamp starting voltage waveforms. (a) 5 ms/div., 1 kV/div.
          (b) Expanded scale; 10 µs/div., 1 kV/div.

(a)　　　　　　　　　　　　　　　　(b)

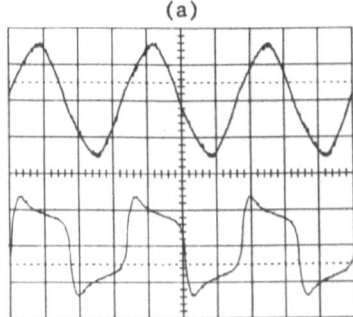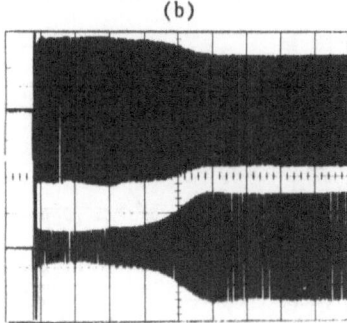

Fig. 2. Voltage (lower) and current (upper) waveforms during the
thermionic arc. (a) 5 ms/div.; (upper) 1 A/div., (lower)
100 V/div. (b) 10 s/div.; (upper) 1 A/div., (lower) 100 V/div.

Figure 2a and 2b respectively. Several minutes are sometimes required to
achieve full equilibrium within the discharge, during which time the light
output from the discharge also reaches a maximum value.

CALCULATED ELECTRON PRODUCTION

An estimate of the effect that the presence of thorium has on the
statistical time lag can be obtained by calculating the electron production
rate attributable to thorium, and comparing this rate to that estimated from
background radiation sources. Electron-ion pairs are formed in the Ar gas
within the arc tube by the energy decay of the α particles produced by the
radioactive decay of thorium and its daughter products. Electron production
from the β and γ radiation produced in the decay of thorium and its daughters
is assumed to be negligible in comparison with that from the α particles.
Thorium, in the form of ^{232}Th, is a naturally occurring radionuclide
with a half-life of 1.39×10^{10} years. The processes involved in the
radioactive decay of thorium are quite complex as it is the first member of a
long chain of successive radionuclides[11]. There are ten intermediate steps,
including one branched decay sequence, in the complete decay cycle to form a
stable end product (^{208}Pb) and involves the production of six α and four β
particles, along with several X and γ rays of varying energies. The half-
lives of the various daughter nuclides vary over many orders of magnitude
from 1.39×10^{10} years for ^{232}Th to 0.3 μs for ^{212}Po. To estimate the
electron production rate from the energy decay of the α particles, the
production rates of the various intermediate daughter nuclides must be estimated. In a naturally occurring sample of thorium, the production and decay
of the intermediate daughters are in secular equilibrium (i.e. approximately
constant over the time scale of interest), and the overall α particle
production rate is constant. However, during refining of the thorium, the
daughters are removed, and the α particle production rate is no longer
constant and will vary with time until the production rates of the daughters
are again in equilibrium. For the thorium series, the equilibrium time is $\simeq$
30 years. The calculation of the production and decay rates for the daughter
nuclides is not straightforward, as the amount of the parent material for the
succeeding decay cycles is not initially constant. In this situation the
amount of any daughter nuclide $N_i(t)$ produced at time t after refining is

$$N_i(t) = \int_0^t \int_0^t N_{i-1}(s)(1 - \exp[-\lambda_{i-1}\{t - s\}])ds\,dt \ .$$

This equation may be solved numerically from a knowledge of the amount of the
parent or intermediate daughter nuclide (i-1) in the sample. Once the

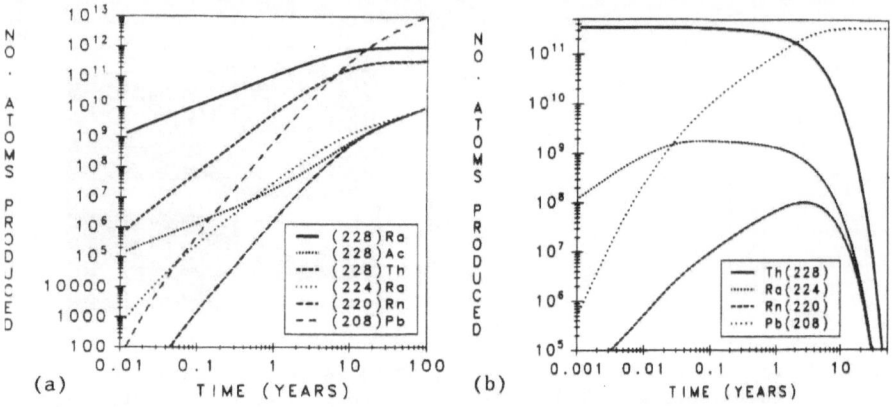

Fig. 3. The amounts of daughter radionuclides produced in the decay of
(a) 1 gm ^{232}Th, and (b) an equilibrium amount of ^{228}Th impurity.

amounts of the daughters are known for all the times of interest, then the
rates for the various ionizing particle production rates can be calculated.

The actual amounts of several of the daughter nuclides and the stable
^{208}Pb produced from the decay of ^{232}Th over the time period from a few hours
after purification to ≃ 100 years are shown in Figure 3a. These results show
that the relatively long lived daughters ^{228}Ra (half-life = 5.75 yr) and
^{228}Th (half-life = 1.91 yr) form secular equilibrium concentrations in the
thorium ≃ 30 years after purification. The shorter lived radionuclides shown
in Figure 3a (i.e. ^{228}Ac, ^{224}Ra and ^{220}Rn) as well as the other short lived
nuclides also form equilibrium concentrations on a similar time scale but do
not exhibit this behaviour in Figure 3a due to an artifact of the present
calculations. This does not effect the overall electron production rate.

During the chemical refining of thorium, the ^{228}Th isotope will remain
in the purified thorium at its equilibrium concentration (≃ 1.5 parts in
10^{10}). The decay of this isotope (half-life = 1.91 yr) and its subsequent
radioactive daughter nuclides will initially contribute to the total ionizing
particle production rates at short times (< 10 years) after the refining of
the thorium. Calculations have also been made of the daughter and ionizing
particle fluxes produced by the decay of this residual amount of ^{228}Th. The
amount of stable ^{208}Pb along with the final amounts of the transient radionuclides ^{224}Ra and ^{220}Rn and the amount of the remaining impurity ^{228}Th are

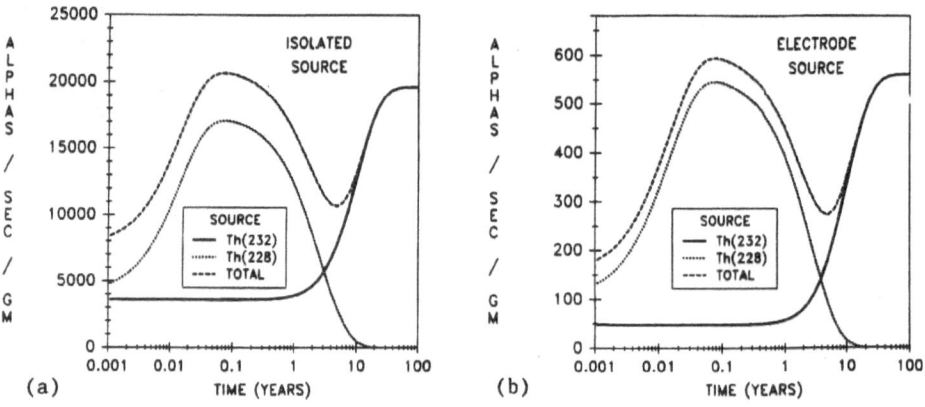

Fig. 4. α particle flux from a 1 gm ^{232}Th source (solid curve), an
equilibrium ^{228}Th source (dotted curve) and the total flux (dashed
curve) for an (a) isolated and (b) electrode source.

given in Figure 3b as a function of time after refining. These calculations show that virtually all the ^{228}Th impurity initially present in the refined thorium sample has decayed into lead after $\simeq$ 10 years and no longer contributes to the ionizing particle flux in the thorium sample.

The α particle production rate from a pure ^{232}Th sample (solid curve) is given in Figure 4a along with the production rate from the decay of the ^{228}Th remaining in a 1 gm chemically refined sample of thorium (dotted curve) and the total α particle flux for both of these sources (dashed curve). The thorium in the tungsten electrode is initially distributed uniformly throughout the electrode, and only α particles produced near the surface of the electrode have a high probability of escaping. The α particle production rates, corrected for attenuation through the tungsten electrode, have also been calculated and are given in Figure 4b. These calculations show that only $\simeq$ 3% of the α particles actually escape from the electrode to produce electrons in the arc tube. These calculations show that for a 100 watt metal halide lamp containing 1% thoria doped electrodes ($\simeq$ 0.2 mg of ThO_2), the α particle production rate in the arc tube is 1 every 9 seconds. The electron production rate in the arc tube can be calculated from this data and a knowledge of the energies of the α particles escaping from the electrodes. The average α particle energies as a function of time after refining are shown in Figure 5a for an isolated source and a source located within the electrode, while the electron production rate is given in Figure 5b. These calculations indicate that 10 to 100 times as many electrons are produced in the arc tube by the radioactivity of the thorium as are produced by natural sources[11], and consequently, the presence of thoria in the arc tube should have a significant effect on the statistical lag time in these lamps.

LAMP STARTING MEASUREMENTS

Statistical lag time measurements have been performed on lamps using a pulsed constant voltage technique where the risetime of the voltage pulse (< 10μs) is fast and the pulse duration ($\simeq$ 1 sec.) is slow in comparison with the lag time of the lamps under test. To remove uncertainties in knowing whether the electrodes were coated with Hg and the metal halide salts (and thus inhibiting the release of the α particles from the electrodes into the arc tube), the lamps were filled only with 100 torr of Ar. An external resistor limited the discharge current when a gas breakdown occured, and the time between successive measurements was long enough (typically 5 sec.) such that the starting times were independent of the preceeding gas breakdowns in the lamp. An average of 250 individual breakdowns were recorded for each measurement and stored on a computer for later analysis. For the present measurements, only the average time of all the breakdowns was recorded.

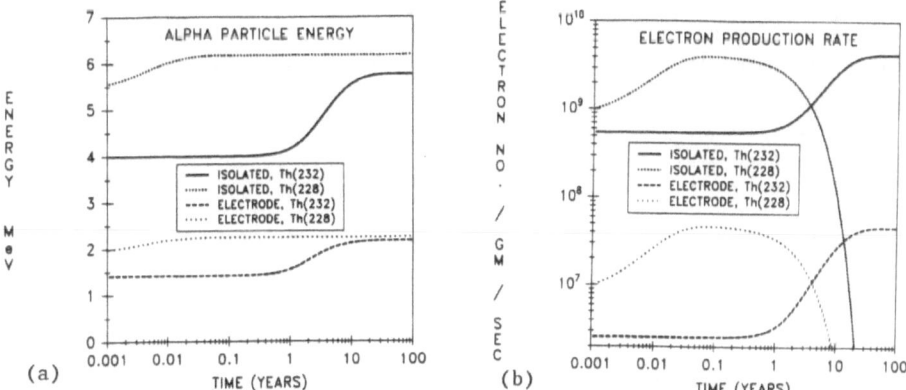

Fig. 5. (a) Average α particle energy and (b) electron production rate for ^{232}Th and the ^{228}Th impurity for an isolated source and a source located within the tungsten electrode.

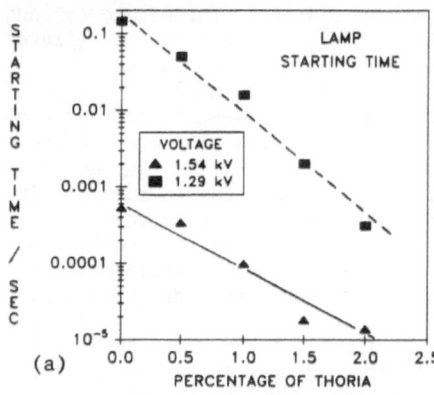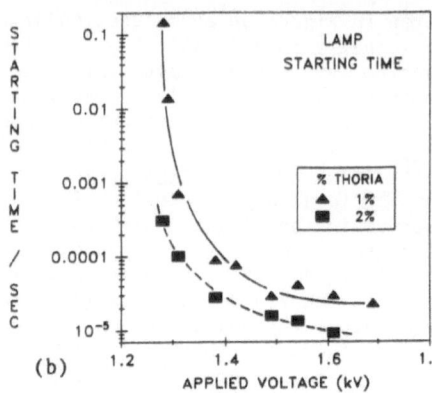

(a) (b)

Fig. 6. Lamp starting time as a function of (a) the applied voltage for lamps containing varying percentages of ThO_2, and (b) percentage of ThO_2 in the electrode for a low (1.29 kV)and a high (1.54 kV) overvoltage.

The measurements were performed on several groups of lamps containing electrodes with varying percentages of thoria (from 0% to 2% in 0.5% increments) as a function of overvoltage (breakdown voltages for these lamps were in the range 1.1 to 1.2 kV). The starting times of two groups of these lamps (containing 0% and 2% ThO_2) as a function of the applied voltage are shown in Figure 6a. As expected, the starting time increases dramatically at low overvoltages and decreases with increasing thoria content at the same overvoltage. This can be seen more clearly in Figure 6b which shows the starting time of the lamps as a function of the percentage of thoria in the electrodes at a low overvoltage (1.29 kV) and a high overvoltage (1.54 kV). These measurements tend to confirm the conclusion of the analysis given above that the presence of the thoria in the electrodes can reduce the lamp starting times by 1 to 2 orders of magnitude.

REFERENCES

1. W. M. Keeffe and Z. K. Krasko, A New Low Wattage Metal Halide Lamp, **Lighting Des. Appl.**, 15:48 (Nov.) (1985).
2. W. M. Keeffe, Z. K. Krasko, J. C. Morris, and P. J. White, Improved Low Wattage Metal Halide Lamp, J. Illum. Eng. Soc., 16:40 (Summer) (1988).
3. J. F. Waymouth, "Electric Discharge Lamps," M.I.T. Press, Cambridge (1971).
4. W. H. Kohl, "Handbook of Materials and Techniques for Vacuum Devices," Reinhold, New York (1967).
5. G. Zaslavsky, S. Cohen, and W. M. Keeffe, Improved Starting of the 100 Watt Metal Halide Lamp, **J. Illum. Eng. Soc.**, 17:50 (Summer) (1990).
6. J. F. Waymouth, The Glow-To-Thermionic-Arc Transition, **J. Illum. Eng. Soc.**. 16:166 (Summer) (1987).
7. G. M. J. F. Luijks and J. A. J. M. van Vliet, Glow-To-Arc Transitions in Gas Discharge Lamps, **Lighting Res. Technol.**, 20:87 (1988).
8. W. W. Byszewski, A. B. Budinger, and Y. M. Li, Transition to the Thermionic Arc in High Pressure Discharges, Proc. XIX Int. Conf. Phen. Ionized Gases, Belgrade, 10-14 July, (1989), pg. 700.
9. W. W. Byszewski and A. B. Budinger, UV Enhanced Starting of HID Lamps, 1989 Ann. Illum. Eng. Soc. North America Conf., Orlando, 6-10 August, (1989).
10. W. W. Byszewski, A. B. Budinger, and Y. M. Li, HID Starting: Glow Discharge and Transition to the Thermionic Arc, 1990 Ann. Illum. Eng. Soc. North America Conf., Baltimore, 29 July - 2 August, (1990).
11. K. Liden and E. Holm, Measurement and Dosimetry of Radioactivity in the Environment, in: "The Dosimetry of Ionizing radiation," K. Kase, B. Bjarngard, and F. Attix, eds., Academic, Orlando (1985).

BREAKDOWN OF AIR GAPS UNDER

OSCILLATORY SIMULATED SWITCHING SURGES

M.S. Abu-Seada, S.N. Salem and *H. Anis

Cairo University, Egypt
*Kuwait University, Kuwait

INTRODUCTION

The breakdown of air insulation under high frequency fields was treated by many workers[1,2,3]. Nasser reported[1] that below a certain frequency the alternating voltage will not have enough time to reverse the direction of field once the breakdown process has been initiated. That frequency was said to be inversely proportional to gap spacing. A critical frequency was determined[1] below which the breakdown mechanism is slightly modified by the longer presence of some positive ions resulting in a slight reduction in breakdown voltage. Beyond the critical frequency the ion space charge oscillates and new avalanches will then cause it to grow until instability and breakdown occurs under yet smaller voltages.

Breakdown was said to occur[2] in quasi-uniform field gaps if positive space charges accumulate and if it, in turn, augments the external field. This meant that the distance crossed by positive ions during the quarter cycle must be less than half the width of the ionization zone. The latter being bounded by points in space at which the net ionization coefficient variables (in air, it is where the field is equal to 24 kV/cm atm). An expression for the breakdown frequency was given for a given gap and applied voltage. Similar criterion was adopted[3] yet with more accurate representation of spatial field distribution using charge simulation[4]. Accumulation of positive ions was said to take place if during a half cycle positive ions do not cross the ionization boundary corresponding to the peak value of the applied voltage.

BREAKDOWN UNDER OSCILLATORY SWITCHING SURGES

In many cases in a power system a switching surge is basically a power frequency voltage wave superimposed by a high frequency damped oscillation. The period over which breakdown develops is much shorter than a power frequency cycle. The surge can thus be looked upon as a constant (dc) voltage superimposed by the high frequency oscillation as seen in Fig.1 which has the form

$$V(t) = V_{dc} + V_{ac} \sin(\omega t + \theta) \tag{1}$$

Gaseous Dielectrics VI, Edited by L.G. Christophorou and
I. Sauers, Plenum Press, New York, 1991

279

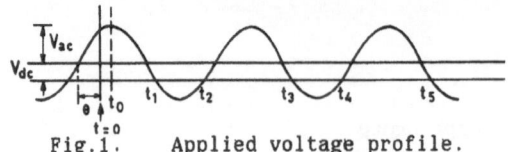
Fig.1. Applied voltage profile.

The times at which the voltage applied to an insulation reverses its sign are

$$t_i = [i\pi + \sin^{-1} (V_{dc}/V_{ac}) - \theta]/\omega \quad i \geq 1 \tag{2}$$

Under the given voltage profile positive ions may accumulate. For a given voltage magnitude the breakdown frequency is that under which positive ion space charges accumulate. Fig.2 displays the reciprocating movement of positive ions under the voltage profile of Fig.1. The distance r_1 is covered over the period t_2-t_3 ; r_2 is a retreat over the period t_3-t_4 followed by another advance r_3 over t_4-t_5, and so on. The distances R_i represent the resultant charge location at successive polarity reversals.

The effect of space charge field on ionic movement is ignored during the first cycle of ionic oscillation. The present criterion stipulates that if ionic accumulation begins during that cycle, i.e. prior to space charge formation, it will continue and a breakdown is bound to occur.

For a given gap arrangement the electric field $E(r,t)$ is expressed as a function of distance r along the ionizing trajectory while it varies with time according to eq.(1). The positive ion drift velocity is expressed by

$$\frac{dr}{dt} = K^+ E(r,t) \tag{3}$$

where K^+ is the positive ion mobility which is equal to 2.6 cm²/V sec at NTP[1]. The breakdown criterion is set by integrating for r between the high field electrode surface and the ionization zone limit (at which the net ionization coefficient vanishes) of 24 kV/cm (Fig.3) while the time is integrated between t_2 and t_3. This procedure establishes a relation between the applied voltage magnitude $V_{dc}+V_{ac}$ (for a given ratio of these two components) and the critical frequency of breakdown f_0.

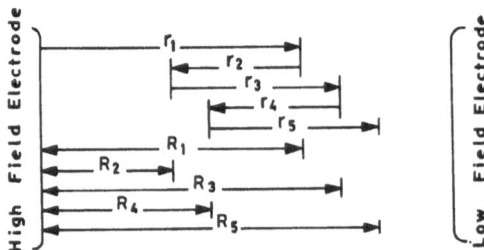
Fig.2 . Oscillatory movement of positive ions.

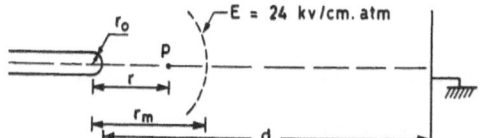

Fig.3 .   Ionization boundary.

Cylinder-to-plane Gap

Following the above treatment the critical frequency could be obtained by numerically solving two simultaneous equations

$$[(r_0^3 - r_1^3) + 3s(r_1^2 - r_0^2) + 3(y^2 - 2ys)(r_1 - r_0)]/(s-y)$$

$$= \frac{2K^+}{\ell_n A} (V_{dc}(\pi + 2\theta_1) + 2V_{ac} \cos\theta_1)/2\pi f_0 \qquad (4)$$

and

$$24000 = (V_{ac} + V_{dc})(\frac{1}{r_1 - y} + \frac{1}{2s - y - r_1})/\ell_n A \qquad (5)$$

where, r_0 is the cylinder radius, s height of cylinder axis, y displacement of cylinder line charge off axis and $A = (2s-y-r)/(r_0-y)$.

Sphere-to-plane

The critical frequency of breakdown could be calculated, based on the same procedure, by simultaneously solving the following derived two equations

$$r_1 - r_0 = \frac{k^+}{2\pi f_0 (r_1 - r_0)} \{[V_{dc}(\pi + 2\theta_1) + 2V_{ac} \cos\theta_1][\frac{1}{r_0 - y} - \frac{1}{r_1 - y}]$$

$$+ (\frac{1}{2s - y - r_1} - \frac{1}{2s - y - r_0})\} \qquad (6)$$

and

$$24000 = \frac{V_{dc} + V_{ac}}{M}[\frac{1}{(r_1 - y)^2} + \frac{1}{(2s - y - r_1)^2}] \qquad (7)$$

where, all dimensions are defined in the same manner as the cylinder-to-plane case.

$$M = (\frac{1}{r_0 - y} - \frac{1}{2s - y - r_0}) \qquad (8)$$

when all of the above cases were run on computer the initial travel of positive ions R_1 was found to be nearly equal to the ionization boundary limit. The subsequent reverse movement was insignificant a fact which justifies relying only on the first cycle travel for accumulation indication.

Point-to-Plane

For better evaluation of spatial field distribution modeling based on the method of charge simulation is resorted to. The field at a point on the gap axis at a distance r away from the center of the hemispherically capped rod is[3]

$$E(r) \cong 0.9 \ E(r_0) \ (\frac{r_0}{r})^2 \ e^{0.1r/r_0} \tag{9}$$

Also, the applied voltage can be related to field and gap length d by

$$V = r_0 \ E(r_0)/[0.465+0.155 \ e^{-.001 \ d/r_0}] \tag{10}$$

where, r_0 is the radius of the rod. Using an analytical procedure similar to that used in the preceding cases it can be found that the
following equation relates the frequency to the breakdown voltage

$$E_b(r_0) = 26.85 \ r_0 f_0 \{200 - e^{-0.1 \ r_1/r_0} \ [(\frac{r_1}{r_0})^2 + 20(\frac{r_1}{r_0}) + 200]\}$$

$$\{(1+C)/(\pi+2\theta_1 + 2C \ \cos\theta_1)\}$$

$$= 26.67 \times 10^3 \ (\frac{r_1}{r_0})^2 \ e^{-0.1(r_1/r_0)} \tag{11}$$

where, $C = V_{ac}/V_{dc}$, $E_b(r_0)$ is the field at the rod tip at the breakdown
threshold. Equation (11) should be read as follows. For a given rod r_0
and a given stress frequency f_0, eq.(11) yields the value of r_1 which is,
for breakdown, equal to the ionization boundary. That boundary can then
be used by means of eqs. (9) and (10) to evaluate the applied breakdown
voltage.

In Fig.4, generalized results are given where - as seen in eq.(11) -
the combined parameter $r_0 f_0$ is related to the breakdown field for
various ac to dc voltage component ratios. The case of pure ac voltage
($c \rightarrow \infty$) given by ref.3 is seen to fit the derived pattern.

Case of Smaller AC to DC Ratio

The preceding point-to-plane modeling is reasonably suitable for a
diminishing dc component. Under that assumption it was acceptable to
assume that relevant positive ion migration begins only under the first

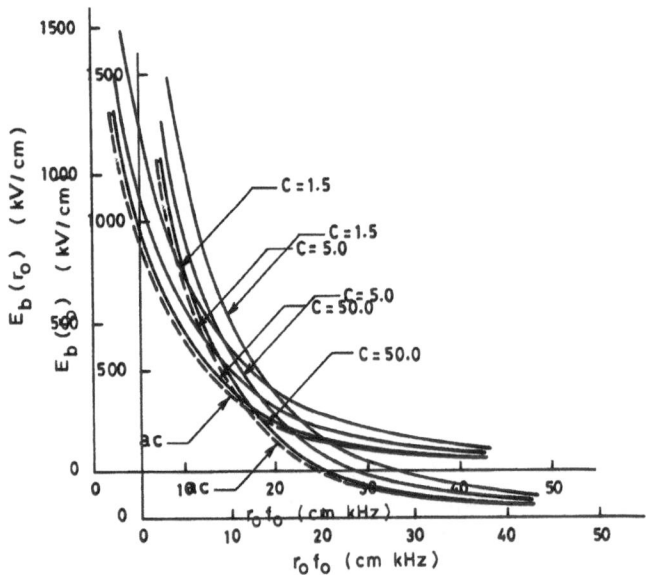

Fig.4 .   Generalized breakdown relation in a point-to-plane air gap.

"complete" positive cycle [(t_2-t_3) in Fig.1]. With substantial dc component the travel during the second half of the first incomplete positive portion [t_0-t_1 in Fig.1] can no longer be ignored. An avalanche develops at t_0 and advances towards positive electrode. During the negative portion t_1-t_2 the avalanche-produced positive ions retreat towards the rod and are partially absorbed. Under the full positive portion (t_2-t_3) a new avalanche is produced which is intensified by the field augmented by the presence of the remaining positive ions; the ionization boundary (Fig.3) extends further into the gap. This continues until breakdown.

The modified breakdown criterion is then such that positive ions produced at the first voltage peak remain within ionization boundaries for a complete voltage cycle.

Equation (3) is integrated between r_0 and r_1 over the period between t_0 and t_1. This is followed by integrating between r_1 and (r_1-r_2) for the time between t_1 and t_2 to evaluate the first retreat of positive ions. Finally, integration is done between (r_1-r_2) and $(r_1-r_2+r_3)$ for the period between t_2 and t_f to account for the second advance. If the final location is within the ionization boundary a breakdown is said to materialize.

EXPERIMENTAL RESULTS

The circuit shown in Fig.5 was built to generate high frequency high ac voltages (LHS of circuit) superimposed on a dc component (RHS of circuit). The values of the main components of the circuit were c_1=0.12 µF, c_2=0.27 µF, c_5=0.1 µF. The tapped inductor had a maximum value of about 0.1 mH. A capacitor and a resistive voltage divider, each operating at a ratio of 500:1, simultaneously monitored the output. The frequency of oscillation was varied by changing the taps on an air-cored inductor L which in turn oscillates with capacitors c_1 and c_2. A damping factor (R/2L) better that 5 sec^{-1} was achieved. The output was produced at frequencies 133, 160, 200 and 250 kHz. The ac peak voltage component V_{ac} was maintained at 5 kV above the dc component V_{dc}. With this constant difference the two components were raised simultaneously until flashover took place. The flashover voltage is said to be $V_{ac}+V_{dc}$.

A large number of experiments was performed to verify the breakdown model. Because of limited space only the results of testing two different air gap configurations are shown in Fig.6. On the figures are also plotted the computed flashover voltages using the present model. The calculated results of a point-to-plane gap are based on the latter model where migration of positive charges is accounted for at the first positive voltage peak t_0 (Fig.3). The agreement between the experimental and calculated results is reasonable.

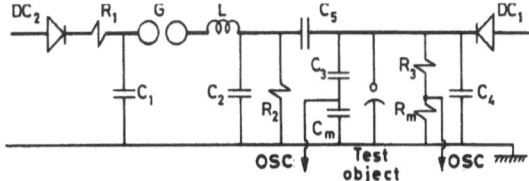

Fig.5.   Experimental set up.

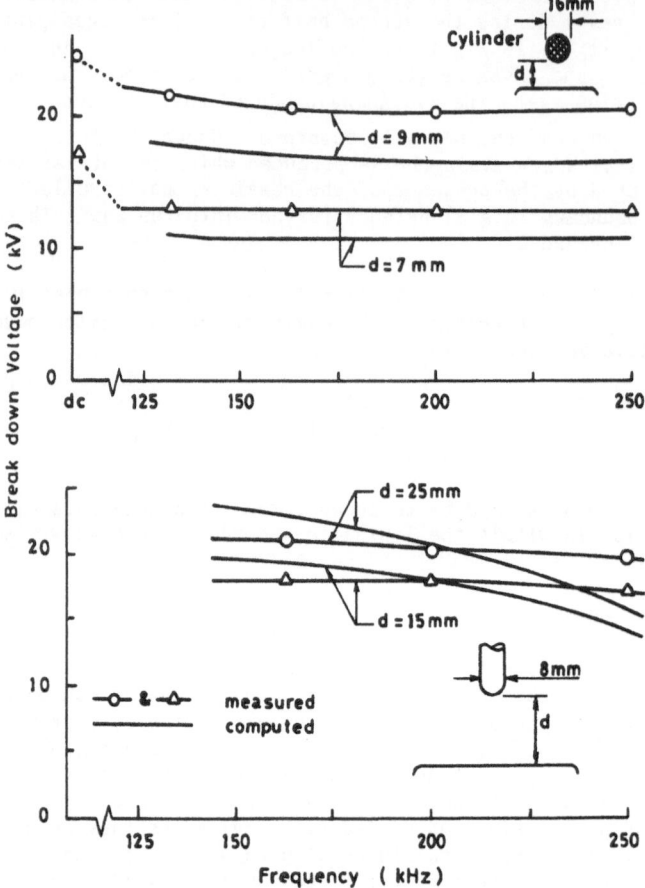

Fig.6.   Measured versus computed results.

REFERENCES

1.   E. Nasser, "Fundamentals of Gaseous Ionization and Plasma
     Electronics", John Wiley, New York (1971).

2.   T.N. Tarasova, The conditions of positive space charge accumulation
     at high frequency voltages. Proc. of 12th Int. Conf. on Phenomena
     in Ionized Gases, p.97 (1975).

3.   M.S. Abu-Seada, Calculation of high frequency breakdown voltages of
     point-to-plane airgaps. IEEE Trans. on Industry Applications,
     IA-20:1627 (1985).

4.   M.S. Abu-Seada, E. Nasser, Digital computer calculation of the
     electric potential and field of a rod gap. Proc. IEEE, 56:813
     (1968).

GAS-SOLID INTERFACE EMISSIONS DETERMINED BY

THE ESAW CHARGE DETECTION METHOD

C. M. Cooke and E. Gollin

High Voltage Research Lab, LEES
Massachusetts Institute of Technology
Cambridge, MA 01239

INTRODUCTION

The application of high voltages to highly insulating materials such as
employed in compressed gas insulated equipment can result in small, but
finite, charge transport. The high electric fields, can also stimulate the
release of charge carriers from gas/solid interfaces. Such small currents
over extended periods of normal operation, or under repeated over-voltages,
can cause significant distortions of the internal electric fields. Such
charging is significant because gas insulation is sensitive to breakdown
triggered by even modest amounts of over stress in local or broad area
regions[1,2]. To better control such charging and therefore prevent premature
failures caused by these overstresses a means to study interface charging
effects is needed.

In this paper the focus is on the interfaces in gas insulation. There is
little fundamental information concerning the response of interfaces to
electric stress, although is is well established that the interface regions are
where failure and breakdown can be triggered.[1-3] One reason for a lack of
information about interfaces has been the lack of methods to observe the
interactions which take place. This paper reports on a new non-destructive
means to measure electric charges at interfaces, called the Electrically
Stimulated Acoustic Wave method (ESAW).[4,5] Some results on PMMA and on
cast filled-epoxy solid-to-gas interfaces obtained with the new method are
also presented.

INTERFACE ELECTRIC FIELDS

Electric fields in gas insulated apparatus can be normal to the surfaces
and in the case of solid dielectrics may be tangential or have a tangential
component, Figure 1. At high gas pressures normal or tangential component
field values can be as high as 1MV/cm before breakdown develops. Yet often
surface effects occur at much lower stresses. Usual operating surface
stresses often are closer to 30kV/cm; and the higher the operating stress, the
more sensitive the insulation is to small perturbations which trigger
breakdown. In practice, fields vary in direction and magnitude along a
surface; this is illustrated in the capacitive equipotential plot for a post-type
insulator with inserts shown in Figure 2.

Gaseous Dielectrics VI, Edited by L.G. Christophorou and
I. Sauers, Plenum Press, New York, 1991

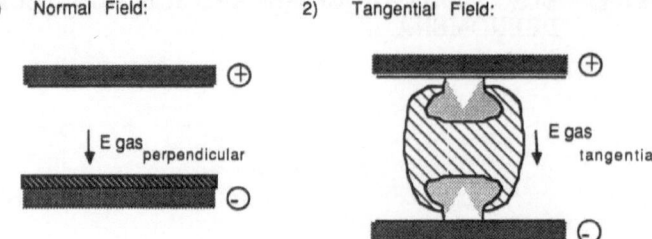

Fig. 1. Perpendicular and tangential electric field orientations at dielectric interfaces. Typical field magnitudes can attain 1MV/cm in compressed gas insulated apparatus

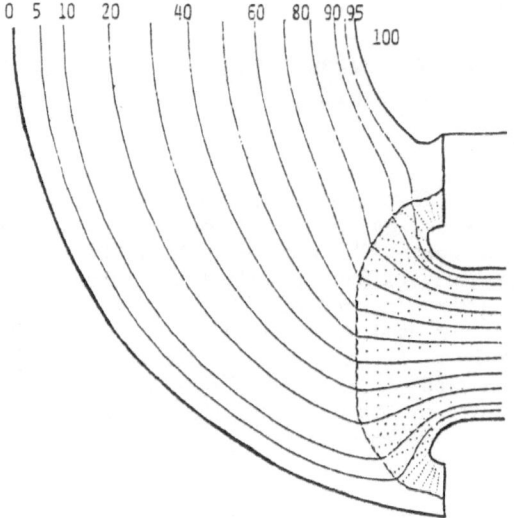

Fig. 2. Equipotential plot surrounding post-type epoxy insulator showing variation of electric field magnitude and direction.

CHARGE ACCUMULATION AT AN INTERFACE

To ascertain the basic response of a solid dielectric-to-gas interface a planar configuration composed of two regions, a bilayer structure, is used. Such an interface between two materials or media is represented in Figure 3.

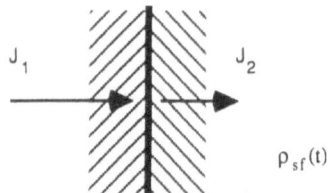

Fig. 3. Charge accumulation at interface from discontinuous current density, J.

286

where J_1 is in region 1 and J_2 is in region 2. If the two current densities differ, a resultant "free" surface charge accumulation, ρ_{sf}, will develop at the interface, where

$$\int_0^t (J_1 - J_2)\, dt = \rho_{sf}(t)$$

that is the surface charge increases/decreases with time as an integral of the difference. Or in differential form, the divergence in current density at the surface determines the rate of surface charge accumulation:

$$\nabla \cdot \vec{J}\big|_{surface} = -\frac{\partial \rho_{sf}(t)}{\partial t}$$

When the surface is a gas-solid interface this relation is the same, except that the current density J_2 is a more restricted form, J_{gas}, and is dependent on carrier mobility and diffusion in the gas. If the '1' media is metal then J_{gas} measurement is often based on currents from the metal or to a collecting electrode. But in the case where the '1' media is a dielectric, measurement can be by collected currents or by accumulated charges at a dielectric interface. Note that charge accumulation on a dielectric is by integration over time and hence very small currents may be detected when integration time is large.

Interface Emissions Measurement by Charge Accumulation:

The measurement of current emissions from an interface can be obtained from the accumulation of charge at a dielectric surface, $\Delta\rho_{sf}$. These accumulations may be on the emitting interface itself or on a collecting interface. The case of charges on the emitting interface is illustrated in Figure 4a, and the case of emission from a metal surface, where the measurement is the ρ_{sf} that collects on an opposite dielectric surface, is depicted in Figure 4b. Charge collection on a dielectric can also be used for emission measurement from a dielectric.

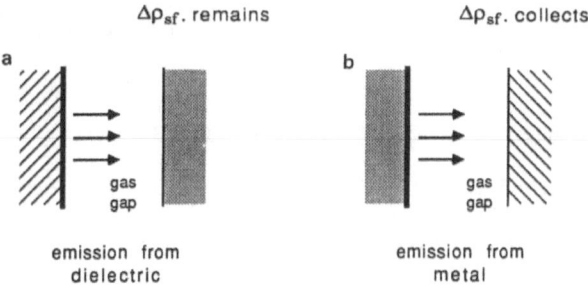

Fig. 4. Interface emission measurement by surface charge collection at dielectric, emission from (a) solid dielectric, and (b) metal.

Bilayer Theory

The charges that accumulate at a bilayer interface and at the electrodes which bound the bilayer can be calculated. Figure 5a and 5b show the charges present in two monolayer systems at the same applied voltage and distinguishes the case of a gas with dielectric constant ε_0, 5a, from the case where the gap is filled with a solid dielectric, 5b (of dielectric constant $k\varepsilon_0$). The charges on the conducting electrode and those adjacent to the surface caused by polarization within the dielectric are also identified, where ρ_{se} is the surface charge on the conducting electrodes for the gas gap.

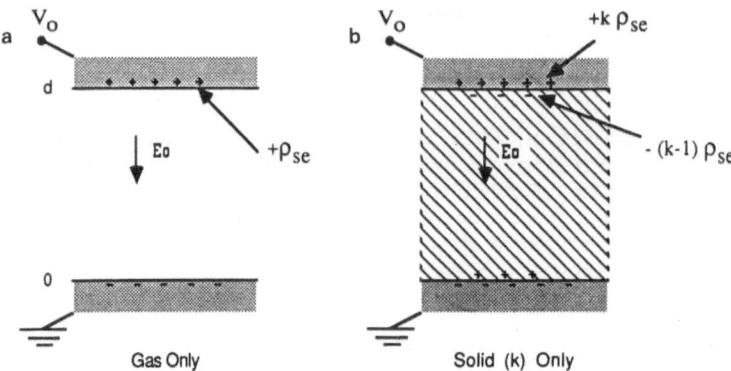

Fig. 5. Monolayer systems with electrode interface charges from applied voltage, V_o, case (a) gas gap only, and case (b) gap filled with solid dielectric material.

When the gap is half-filled with gas and half-filled with a solid dielectric the charges present are depicted in Figures 6a and 6b. Case '6a' has no free charges at the gas-dielectric interface and case '6b' includes these free charges.

Thus a measurement of the total charge at a bilayer interface will necessarily include two components, one associated with polarization and one with free charges at the interface. Furthermore, the charges at the electrodes can be distinguished to include components from induced image charges and from polarization charges.

SURFACE CHARGE MEASUREMENT IN-SITU BY THE ESAW METHOD

A new method to determine surface charges has been used with a bilayer structure. The measurement is non-destructive and can be active during high voltage excitation. This has enabled the detection of charge accumulations at the interface and hence determination of charges arriving via the gas gap.

The test system employed a planar configuration with a series two dielectric structure (bilayer) of metal : dielectric #1 (solid) : dielectric #2 (gas) : metal. A bilayer, two dielectrics in series, is sometimes referred to as a 'Maxwell capacitor' structure, in this case one of the dielectrics is a gas.

The charge measurement method is based upon pulsed acoustic wave measurements and is called the ESAW method (Electrically Stimulated Acoustic Wave method). A description of its principle and its application to bulk charge measurement has been published previously.[4, 5]

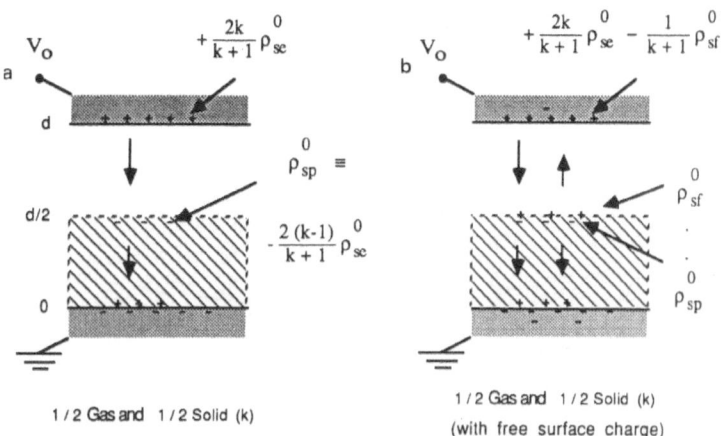

Fig. 6. Bilayer systems, half-filled with solid dielectric material showing interface charges from applied voltage, V_O; (a) applied voltage only, and (b) with added free charges at interface.

Briefly, in the ESAW method a small 'stimulation' pulse voltage is applied to the overall test sample. This pulsed voltage creates an internal pulsed electric field. This applied pulse field creates a pulsed Lorentz force on charges at the interface which causes a proportional pulsed acoustic wave that travels through the solid dielectric toward the surface electrode where it is detected by a fast response acoustic transducer. Figure 7 depicts the arrangement for a monolayer dielectric. Figure 8 shows an example of an acoustic wave response for a bilayer comprised of two layers of the same material, but with a charge at the interface between the two layers. The central part of the wave corresponds to the charges at the center of the structure, while the two exterior parts of the wave correspond to the two image charges at the electrodes.

The response of the ESAW method to a distributed volume bulk charge has been shown to yield good correspondence to theory for the case of e-beam implanted charges in plastics.[5]

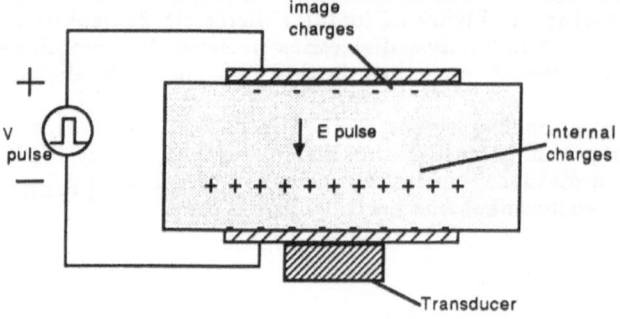

Fig. 7. Electrically stimulated acoustic wave (ESAW) measurement configuration, with monolayer sample.

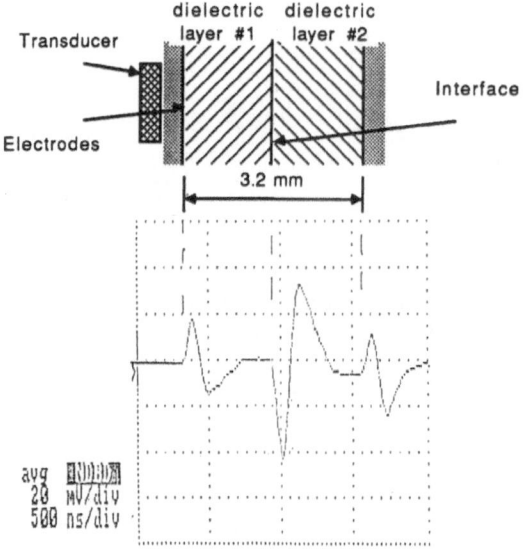

Fig. 8. ESAW acoustic response for bilayer dielectric of same material with charged central interface.

BILAYER TEST RESULTS

The measurement of interface charges in a bilayer structure with one layer a gas has been studied in three stages. The first demonstrates detection of charges (at the bilayer gas-solid interface) created by application of a DC voltage across the solid or by corona. The second concerns tests with an applied DC voltage across the full bilayer, and the third concerns charges that reside at the solid-gas interface after the applied DC voltage is set to zero.

Charge measurement with a solid-gas bilayer system was comprised of a solid dielectric layer of cast filled-epoxy or PMMA, and an air gap, Figure 9. The ESAW signal was obtained from a transducer attached to the exterior, left-side electrode. Interface charge detection was demonstrated by using a thin vacuum-metallized aluminum electrode and a dc voltage applied to this metallized electrode relative to the sample exterior ground. The result was equal surface charges on both sides of the solid. The ESAW charge measurement with this configuration for a 3.0mm PMMA solid layer is shown in Figure 10. The left-hand peak in Figure 10 corresponds to charges at the electrode side, negative polarity, while the right-hand peak corresponds to the charges at the interface, positive polarity in this case.

Fig. 9. ESAW configuration for solid-gas interface charge measurements.

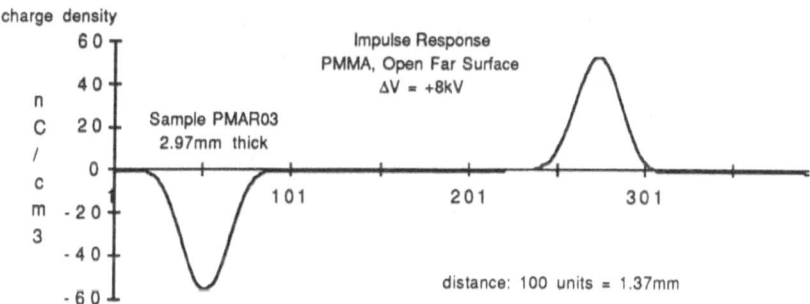

Fig. 10. ESAW measured surface charges on PMMA sample 2.97mm thick with voltage ΔV = 8kV across PMMA, air gap on right side.

The change in acoustic impedance at the solid-gas interface causes an almost doubling of the signal size for the charges at the interface. The open interface signal in the solid-gas configuration is therefore always halved in value for all plots in this paper. In the above case, the applied voltage of 8kV means the charges correspond to a signal level of 2.4nC/cm2. Integration of the charge density yields these values. Note too that the distance or position, ie. thickness, can also be quantified from the acoustic wave measurement.

A similar result is obtained when the solid layer is a cast filled-epoxy layer. Figure 11 shows the result for 5.3mm thick filled-epoxy sample. There is some spreading of the center charge in this epoxy sample which is the result of the sample sides not being parallel and so the acoustic wave arrives over a slightly spread time interval. Effectively, in this case the area under the curve should be used to represent the surface charge density.

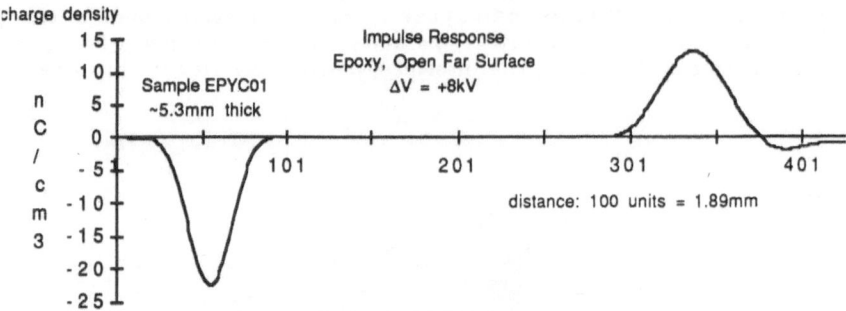

Fig. 11. ESAW measured surface charges on filled-epoxy sample 5.3mm thick with voltage ΔV = 8kV across epoxy, air gap on right side.

Bilayer Demonstration Tests

Tests with a bilayer of equal distances of solid and gas were made to demonstrate the detection of 'free' charges at the interface. In this case the free charges were delivered to the interface via corona spray; the solid was PMMA 3mm thick and the air gap 3.2mm. The solid-gas interface was aluminum covered. Figure 12 shows the same two-peaked type result as in Figure 9, because there are again charges at the interface and image charges at the electrode adjacent the acoustic wave transducer.

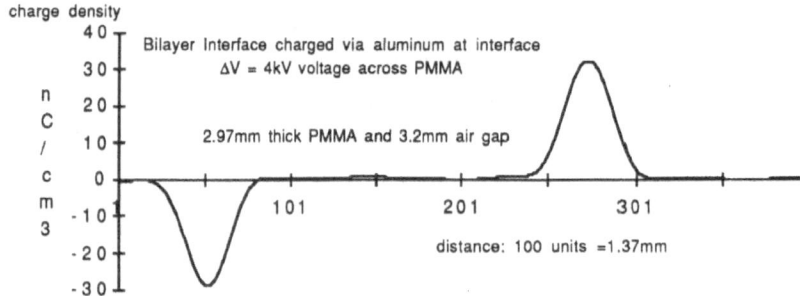

Fig. 12. ESAW measured surface charges on PMMA sample 2.97mm thick with voltage ΔV = 4kV across PMMA, 3.2mm air gap on right side.

When this structure is used with an applied DC voltage across the gas gap, so that no voltage is across the solid portion of the bilayer, a different charge distribution is expected. Figure 13 shows the charge at the interface and very importantly, that there is <u>no charge</u> at the electrode adjacent the transducer. This further demonstrates that the ESAW is accurately detecting charges at the interface.

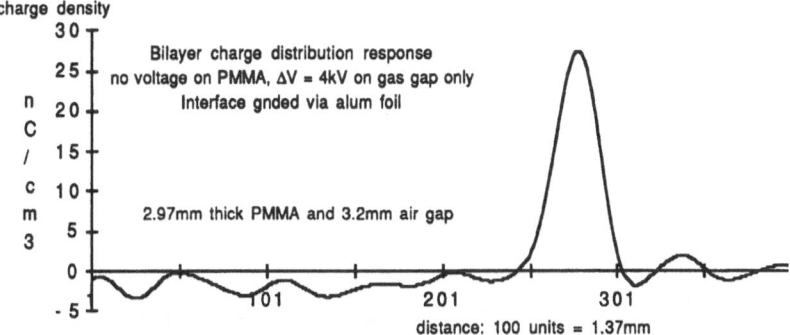

Fig. 13. ESAW measured interface charges on bilayer PMMA sample 2.97mm thick with voltage $\Delta V = 4$kV across 3.2mm gas gap only.

A final demonstration of charge detection at the interface was made with the PMMA-air gap configuration, but this time the aluminum electrode at the gas-solid interface was not applied. A surface charge was introduced by negative corona. This time the ESAW charge measurement is reversed in polarity and shows the interface charge and the image charge at the electrode, Figure 14. Because there is no electrode the presence of this charge, and it magnitude are only known by the ESAW result. The polarity correctly corresponds to the corona condition used to apply it, ie. negative.

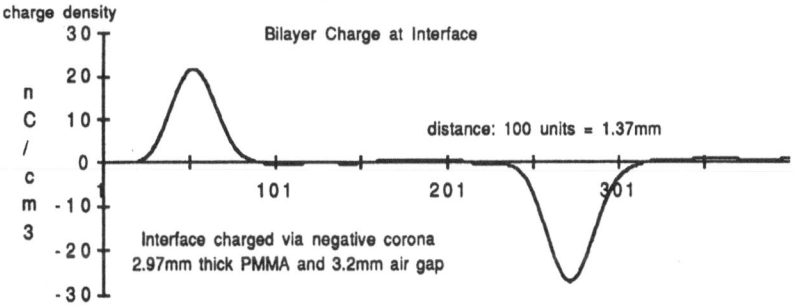

Fig. 14. ESAW measured surface charges on PMMA sample 2.97mm thick with corona charged interface, negative, 3.2mm air gap on right side.

Gas-Solid Bilayer Interface Charge Decay Tests

The variation in time of charge levels at the interface can be determined on a single sample by repeated non-destructive measurements with the ESAW. Such a repeated measurement is depicted in Figure 15. Here the charge distribution is the <u>change in charge</u>. That is, the plotted distribution is the difference between the charge present on the sample 20 hours after decay, referenced to the charge initially at the surface when the sample was corona charged negative. Since it is a reduction in the amount of negative charge, this difference corresponds to a positive <u>change</u> at the interface, as seen in Figure 15. The amount of change corresponds to almost 50% of the initial charge over the 20 hour decay period.

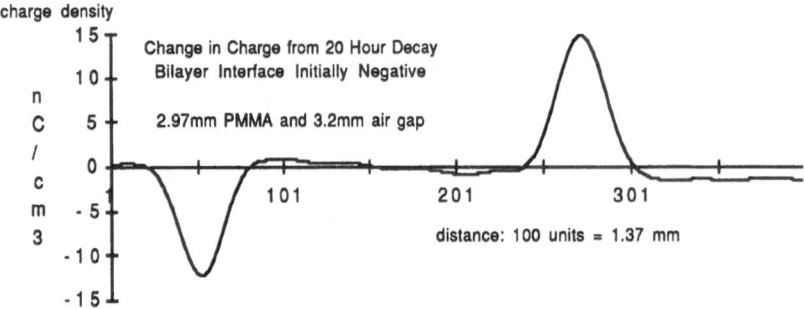

Fig. 15. ESAW measured charge decay, change in surface charges, after initial negative charge to interface on PMMA sample 2.97mm thick with 3.2mm air gap on right side.

Gas-Solid Bilayer Charge Accumulation Under DC Voltage

A cast filled-epoxy with gas gap bilayer structure was exposed to an applied DC voltage and the interface charge monitored. The epoxy was 5.3mm thick and the gas gap was a 3.2mm air gap. The result after 27 hours at voltage is shown in Figure 16. The interface has charged positive, indicating charge transport across the gas gap. The applied voltage caused an electric stress in the gas gap of close to 50% of the expected breakdown stress. This means that even at stresses well below breakdown levels, significant charge transport via the gas occurred indicating that substantial charge accumulations can take place at epoxy-gas interfaces.

CONCLUSIONS

The result of these studies has been to demonstrate that solid dielectric interfaces can accumulate charge and hence 'float' in an electrical sense. As a result significant distortions in electric stress can occur and thereby cause uncertain electrical performance of a high voltage insulation system. In the experimental case of a solid to gas gap interface charge transport across the gas as well as through the solid needs to be considered.

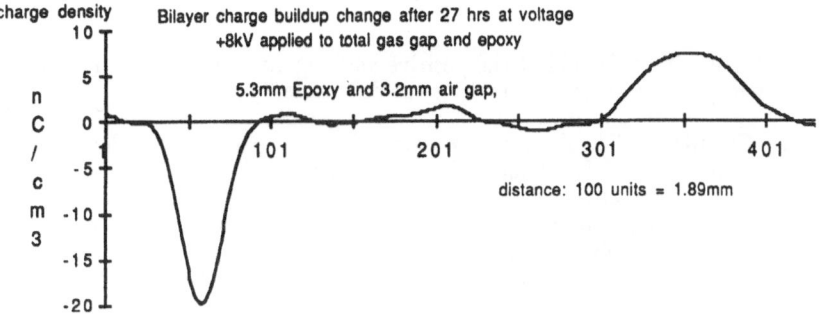

Fig. 16. ESAW measured surface charges on filled-epoxy sample 5.3mm thick after 27 hours at +8kV across bilayer, 3.2mm air gap on right side. Note interface charge is positive polarity.

For idealized bilayer configurations the initial charge distribution and resultant electric fields can be calculated. Furthermore, if the current densities in the two media of the bilayer can be modelled then the time evolution of charge and stress can also be modelled.

The results of measurement with bilayer systems clearly show that the ESAW method provides a useful means to determine the in-situ surface charges, even during excitation. It is non-destructive and can be used to quantify important interface factors such as, dielectric interface emissions and the influence of electrode material and surface condition.

The initial results show that filled-epoxy to gas interfaces readily charge and sustain substantial charge, even in atmospheric air, for extended periods of time. Applied gas gap stresses of only 50% of breakdown caused significant charging of a test interface.

REFERENCES

1. C.W. Mangelsdorf, C.M. Cooke, Static Charge Accumulated by Epoxy Post Insulation Stress at High DC Voltages. IEEE Annual Report CEIDP, p.220 (1978).

2. C.M. Cooke, Charging of Insulator Surfaces by Ionization and Charge Transport, IEEE Trans. on Electrical Insulation, EIS-17, p.172 (1982).

3. C.M. Cooke, Surface Flashover of Gas/Solid Interfaces; Gaseous Dielectrics III , Pergamon Press, ed. L.G.Christophorou, p.337, (1982).

4. T. Maeno., Futami, Kushibe, Takade, and Cooke, Measurement of Spatial Charge Distributions in Thick Dielectrics Using the Pulse Electroacoustic Method, IEEE Trans EI-23, p.433, (1988).

5. C. Cooke, Wright, Takasu, Bernstein, and Golin, Calibration of Volume Charge Measurements by Use of Electron Beam Implantation, Proc. IEEE Annual Report CEIDP, p.435, (1989).

DISCUSSION

M. KRISTIANSEN: Shouldn't the positive and negative signals have identical areas under them and how do you account for the differences?

C. M. COOKE: Thank you for reminding me to discuss the quantitative characteristics of the ESAW method. With an applied voltage equal positive and negative charges are observed across a solid dielectric as shown in the first example figure. When there is a bi–layer with gas gap an additional factor associated with the change in acoustic impedance must be included. The charges at the solid–gas interface generate a direct and a reflected wave from the impedance change. When this acoustic effect is included, there is good equality and the positive and negative charges are measured to be the same.

I. D. CHALMERS: In work involving dielectric charge phenomena there is always a problem in differentiating between surface charge, which is fairly easy to remove, and bulk charge, which can reside for long periods of time. Can the ESAW method be used to detect the presence of bulk charges where there may be charges of opposite polarity in close physical proximity (i.e. an effective dipole situation) or does the method only recognize the net charge?

C. M. COOKE: Yes, it is very easy to separate. If the positive and negative charges are in close proximity then there is a dipole response, while the bulk response is zero. The dipole response has a slightly different signature and is much smaller. We have more studies of metal–solid interfaces and observe the injections of charge that propagate into the solid very readily.

M. GOLDMAN: I have to congratulate you for your interesting work and talk. As the interface problems are important it is obvious that competitive methods are developed in parallel. We have been working for several years on a technique to measure thin layers of contamination on surfaces or the growth of oxide layers on metals. The principle is the change in the work function or the charge decay. It is important to compare the different methods we have achieved in order to measure from monolayer to the micron range. Could you tell us (1) your resolution and (2) if your method can also determine the chemical nature of the layers?

C. M. COOKE: The resolution is determined by the bandwidth of the system. This system, operating at 10–20 MHz has a spatial resolution of 100 microns. It is difficult to get resolution below 1 micron. In terms of sensitivity it is easy to measure charge layers of $0.1 \ nC/cm^2$.

L. NIEMEYER: What are the pros and cons of the ESAW and the pressure pulse methods? What can be achieved with the ESAW method that cannot be achieved with the pressure pulse method? Can the application of a voltage pulse disturb the charge distribution to be measured, e.g. by charge injection from an electrode?

C. M. COOKE: The voltage applied is 100 V. You can use 10–20 V so that there is no excess stress here. This method is very reproducible. The difficulty might be in applying the electric stress uniformly. In the pressure–pulse method it is not always certain how well the pressure is propagated.

CONTRIBUTION OF A SOLID INSULATOR TO AN

ELECTRON AVALANCHE IN NITROGEN GAS

S. M. Mahajan and K. W. Lam

Electrical Engineering Department
Tennessee Technological University
Cookeville, TN 38505 U.S.A.

ABSTRACT

A one dimensional simulation of electron avalanches has been performed near a solid insulator in nitrogen gas at 101.3 kPa. The solid insulator is assumed to modify the growth of electron avalanche in the nitrogen gas by (i) causing a non-uniform field and (ii) photoemission of electrons. Three different profiles of photoemissive contribution have been assumed. An electron avalanche simulated by assuming an exponentially decreasing contribution of photoemissive electrons appears very similar to the one obtained by an experiment.

INTRODUCTION

A solid insulating support is an integral part of most gaseous dielectric systems. A gas-solid dielectric interface is usually the weakest link in such a system. In order to fully exploit the dielectric strength of a certain gas, it is essential to overcome the limitations caused by flashover at the gas-solid interface. The exact mechanisms behind a surface flashover are still not known [1]. However, it is imperative to include the basic parameters of a solid dielectric in any model of a surface flashover. In a model of a discharge along a solid dielectric surface, Tanaka included photoemissive contribution from the solid surface [2, 3]. Studies of Yumoto indicated that such a photoemission from a solid dielectric is indeed possible [4].

In the present paper, electron avalanches have been simulated near a solid dielectric in nitrogen gas at a pressure of 101.3 kPa. Growth of primary and secondary avalanches has been formulated according to the Townsend mechanism. Three different profiles of photoemission from a solid insulator have been assumed and electron avalanches simulated. Comparison of simulated electron avalanches with the experimentally obtained avalanche supports the assumed concept of photoemission.

SIMULATION METHOD

Role of a solid insulating surface was included in the one dimensional simulation by considering (i) the electric field modification and (ii) photoemission. The electric field modification was based upon positive charging of the solid insulator [5]. As a

Gaseous Dielectrics VI, Edited by L.G. Christophorou and
I. Sauers, Plenum Press, New York, 1991

result of charging, an enhancement of electric field near the cathode triple junction and a corresponding reduction of electric field near the anode triple junction could be observed. The exact electric field profile used in the present simulation is illustrated in Figure 1. The mathematical equations for similar profiles were discussed in details in our previous paper [6]. An interelectrode spacing of 1 cm was divided into 500 equal segments and the growth of primary and secondary electrons was calculated. Number of primary electrons in an ' i 'th subsection is given by

$$N_i(t) = N_{i-1}(t-1) \cdot (1 + \alpha_i \cdot \Delta x) \tag{1}$$

where

$$\Delta x = (\text{gap spacing})/500$$

$$\alpha_i = 760 \cdot A \cdot \exp\left(-\frac{B \cdot 760}{E_i}\right) \tag{2}$$

$$A = 0.275 \ 1/(\text{cm·torr})$$

$$B = 144 \ V/(\text{cm·torr})$$

$$E_i = \text{Electric field in the ' i 'th subsection}$$

For the calculation of temporal growth of electrons, drift velocity was assumed to vary according to [7]

$$V_e = 482.45 \ E_i \tag{3}$$

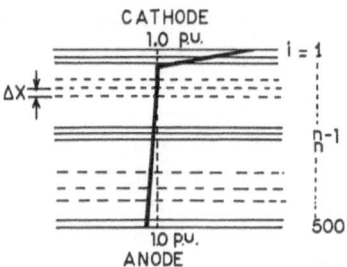

Fig. 1. An Electric Field Profile at the Gas-Solid Interface.

The number of secondary electrons generated at the cathode as a result of the growth of N_i electrons in the ' i 'th subsection is given by

$$S_i(t) = N_{i-1}(t-1) \cdot \alpha_i \cdot \Delta x \cdot \gamma \tag{4}$$

where

$$\gamma = \text{Townsend's secondary ionization coefficient in nitrogen}$$
$$\text{gas due to photoemission at the cathode [8]}$$
$$\doteq 1.67 \times 10^{-3}.$$

The number of secondary electrons generated at the 'i'th subsection as a result of the growth of N_i primary electrons in the 'i'th subsection is given by

$$SS_i(t) = N_{i-1}(t-1) \cdot \alpha_i \cdot \Delta x \cdot \gamma_s \qquad (5)$$

where

$$\gamma_s = \text{Townsend's secondary ionization coefficient due}$$
$$\text{to photoemission from the solid insulator.}$$

The basic difference between equations (4) and (5) is the location of secondary electrons.

The total number of electrons at any time 't' is then given by

$$N(t) = \sum_{i=1}^{n} \left(N_i(t) + S_i(t) + SS_i(t) \right) \qquad (6)$$

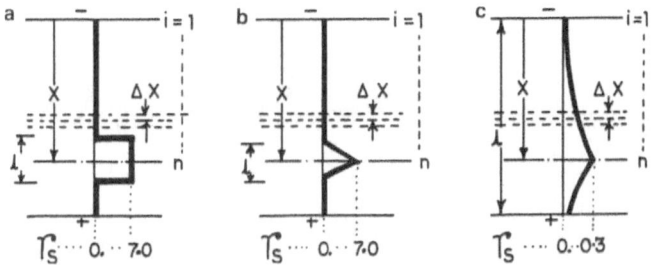

Fig. 2. Photoemission Profiles
(a) Constant γ_s, (b) Linearly decreasing γ_s, and
(c) Exponentially decreasing γ_s .

PROFILES FOR PHOTOEMISSION FROM A SOLID SURFACE

The photoemissive contribution of electrons, as described in equation (5), requires the knowledge of γ_s. In addition, photoemission from a solid insulator will be maximum near the tip of an electron avalanche. However, in the absence of precise data on photoemission, three different profiles of γ_s as a function of the active length of the insulator (1) are assumed. These profiles are illustrated in Figure 2.

It should be noted that x is a variable and therefore these are spatially moving profiles. For a constant and a linearly decreasing profile of γ_s, only 10% of the length of the solid insulator is photoemissive at any time. In an exponentially decreasing profile of γ_s, the entire length of the solid insulator is considered to be photoemissive at any time during the avalanche propagation.

RESULTS AND DISCUSSION

Simulation of electron avalanches was performed with an applied voltage of 30 kV and a modified electric field profile illustrated in Figure 1. The electron avalanche in Figure 3 has no photoemissive contribution from the solid insulator ($\gamma_s = 0$) and can be used for comparison with other simulated avalanches. The electron avalanches with a constant ($\gamma_s = 7\gamma$), a linearly decreasing (peak value of $\gamma_s = 7\gamma$) and an exponentially decreasing (peak value of $\gamma_s = 0.3\gamma$) profiles of γ_s are illustrated in Figures 4, 5, and 6, respectively. Comparison of Figures 3 and

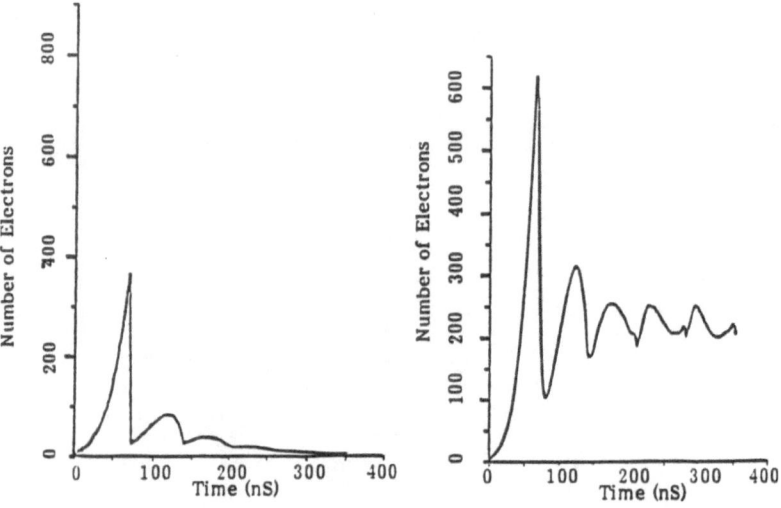

Fig. 3. An Electron Avalanche without Photoemissive Contribution from the Solid Insulator (Voltage = 30 kV; $\gamma_s = 0$) .

Fig. 4. An Electron Avalanche with Photoemissive Contribution from the Solid Insulator (Voltage = 30 kV; $\gamma_s = 7\gamma$ and constant).

4 indicates a significantly higher growth of secondaries with a constant γ_s profile. The obvious reason for this being the additional photoemissive contribution from the solid insulator. As against this, the linearly decreasing profile of γ_s (Figure 2(b)) has a minimal contribution to the electron avalanche (Figures 3 and 5). The most interesting electron avalanche was obtained with an exponentially decreasing profile of γ_s (Figure 6). An important feature of this electron avalanche is that the secondaries are not so distinct in nature. The primary and four secondaries are merged in such a way that it appears like a single avalanche with a relatively long transit time.

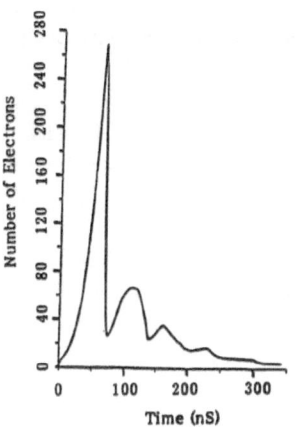

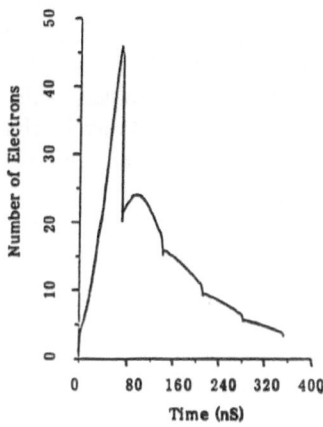

Fig. 5. An Electron Avalanche with Photoemissive Contribution from the Solid Insulator (Voltage = 30 kV; $\gamma_s = 7\gamma$ and linearly decreasing).

Fig. 6. An Electron Avalanche with Photoemissive Contribution from the Solid Insulator (Voltage = 30 kV; $\gamma_s = 0.3\gamma$ and exponentially decreasing).

Figure 7 illustrates an experimentally obtained electron avalanche near a Nylon surface in nitrogen gas at 101.3 kPa [9]. It should be noted that the number of initiating electrons in the simulation is '1' whereas the same number was $> 10^3$ in the experiment. The trace of the experimental avalanche (Figure 7) has been turned upside down to make the comparison with other figures easier. Although there is a difference in the applied voltage in Figures 6 and 7, the basic nature of the avalanches appears to be identical.

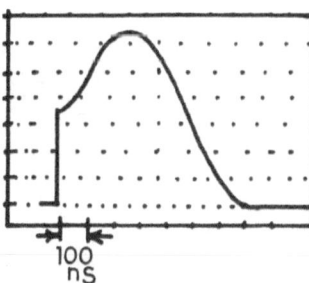

Fig. 7. An Electron Avalanche near a Nylon-Nitrogen Interface [9]
(Voltage = 17.3 kV; 53.3 μA/div.; 100 nS/div.).

Thus, a photoemissive contribution from a solid insulator to an electron avalanche in a gas seems to be possible. Assumption of an exponentially decreasing profile of photoemission resulted in an electron avalanche similar to the one that was experimentally obtained. This result is to be expected since the photon density at the tip of an electron avalanche should decrease exponentially on either side of the avalanche tip. Another interesting feature of an avalanche with an exponentially decreasing profile is that the value of γ_s is only 30% of the secondary ionization coefficient of nitrogen gas (γ). In the case of avalanches with constant and linearly decreasing γ_s profiles, the assumed values of γ_s are probably too high.

More experiments are needed to determine the spatial variation of photon density near a gas-solid interface. Precise knowledge of secondary emission coefficient (γ_s) and the mechanisms behind photoemission from a solid insulator will help expand the scope of this simulation.

REFERENCES

1. Sudarshan, T. S. and Dougal R. A., Mechanisms of Surface Flashover along Solid Dielectrics in Compressed Gases: A Review, *IEEE Trans. on Elect. Insul.*, Vol. EI-21, No. 5, Oct. 86, pp. 727-746.

2. Tanaka, M., Murooka, Y., and Hidaka, K., Nanosecond Surface Discharge Study using Computer Simulation Method, *CEIDP*, 1986, pp. 93-98.

3. Tanaka, M., Murooka, and Hidaka, K., Nanosecond Surface Discharge Development using the Computer Simulation Method, *J. App. Phys.*, 61(9), May 1987, pp. 4471-4478.

4. Yumoto, M. and Sakai, T., Photo Electron Emission from Dielectric Materials by Ultra-Violet-Ray Irradiation, *CEIDP*, 1980, pp. 75-81.

5. Knecht, A., Development of Surface Charges on Epoxy Resin Spacers Stressed with Direct Applied Voltages, *Gaseous Dielectrics II*, 1982, p. 356.

6. K. W. Lam, S. M. Mahajan, and t. S. Sudarshan, Simulation of Avalanches Near a Composite Dielectric, *Proceedings of 6th International Symposium on High Voltage Engineering*, New Orleans, 1989, Paper 46-05.

7. Mahajan, S. M., Electron and Ion Avalanche Growth Modifications Near a solid Insulator in Nitrogen Gas, Ph.D. Dissertation, University of South Carolina, 1987.

8. Verhaart, H.F.A., Avalanches in Insulating Gases, Ph.D. Dissertation, Eindhoven University, 1982.

9. Mahajan, S. M., Sudarshan, T. S., and Dougal, R. A., Avalanches Near a Dielectric Spacer in Nitrogen Gas, *Gaseous Dielectrics V*, L.G. Christophorou and D.W. Bouldin (Eds.), Pergamon, New York, (1987), p. 546

DISCUSSION

I. GALLIMBERTI: As the electron velocity perpendicular to the surface cannot be zero, direct interaction between electrons and the dielectric surface is possible (e.g., impact ionization, electron capture, electron delayed release, etc.). Can you comment about the relative importance of these direct processes with respect to the photon interaction that you have assumed in your model? Could these direct processes explain the similarity between your current records and those presented this morning by T. Teich for avalanches dominated by the ionization/attachment/ detachment processes?

S. M. MAHAJAN: Since the avalanches were simulated in dry nitrogen gas, the attachment/detachment processes are unimportant. Delayed electrons (due to conversion process) usually occur in moist gases and therefore, were not considered here. The cross section for electron impact ionization with a solid surface as compared to the one in gas would be very small. This leaves photoemission from a solid surface as the most likely mechanism which could be responsible for additional contribution to the avalanche in a gas. Detrapping of captured electrons by photons could explain photoemission. Finally, the proximity of photons in the avalanche to the solid surface makes it an important mechanism.

J. M. WETZER: When comparing the model with the real situation, it turns out that the real field is not parallel to the insulator. It consists of a surface charge field super imposed on the applied field. As a result the drift velocity is not parallel to the insulator. In fact the electrons collide with the insulator. In any case the velocity component along the insulator is smaller than the calculated drift velocity. Could you comment on how realistically the model describes the real situation, and whether, and how, you intend to improve the model?

S. M. MAHAJAN: The normal component of the electric field would affect the drift velocity and consequently affect the pulse width of the electron avalanche to a certain extent. However, the purpose behind these simulations was not to obtain an exact replica of the experimentally obtained avalanches. Instead, the basic mechanism of photoemission from a solid insulator can be observed by comparing the overall nature and comparable pulse widths of experimental and simulated avalanches. The model can be improved by (i) including an exact electric field profile at the interface (which requires surface charge measurements) and (ii) recording simultaneous growth of photons and electrons in an avalanche near a gas–solid interface.

IMPULSE SURFACE CHARGING AND FLASHOVER

Owen Farish and Ibrahim Al-Bawy

Centre for Electrical Power Engineering
University of Strathclyde
Glasgow G1 1XW, UK

INTRODUCTION

In compressed-gas-insulated equipment, the weakest point in the system is often at the interface between the gas (usually SF_6) and the solid spacers used to support the conductors[1]. The low dielectric strength is often attributed to the effects of surface charges, or to ionisation at high-field sites such as the gas-electrode-spacer "triple junction"[2-4]. Although there have been some studies of surface charge development under dc stress[5,6] relatively little is known about the way charge builds up on a surface and how the surface charge influences the breakdown process under impulse conditions. In the present work a study was made of impulse flashover of model spacers under conditions where:

- surface charge was allowed to accumulate as a result of repeated impulse stressing

- a fault was simulated by introducing an annular gas gap at the triple junction

- the spacer surface was precharged with a line charge or uniform charge distribution

- metal inserts were used to shield the triple junction and move the high-field site to the mid-gap region.

An important feature of the study was that, in all cases, the surface charge density was measured before and after each impulse-voltage application; the equipment was designed so that the complete surface of the spacer, including both triple-junction regions, could be scanned by a charge probe.

EXPERIMENTAL

The spacer assembly and the charging and scanning systems were contained in a glass vessel which could be evacuated to 10^{-2} torr and which had a maximum SF_6 working pressure of 2 bar. The PTFE spacers were 20mm diameter and 10mm long, and were held between 125mm diameter Bruce-profile electrodes. As shown in Figure 1 the upper electrode, and the outer part of the lower electrode, could be withdrawn to allow scanning of the entire spacer. This was achieved by bringing the charge probe to within 1mm of the surface and using a geared drive system to provide simultaneous rotation and translation of the spacer.

The capacitive charge probe consisted of a 0.7mm diameter rod mounted coaxially in an earthed shield of outer diameter 6mm. Charge measurement was accomplished

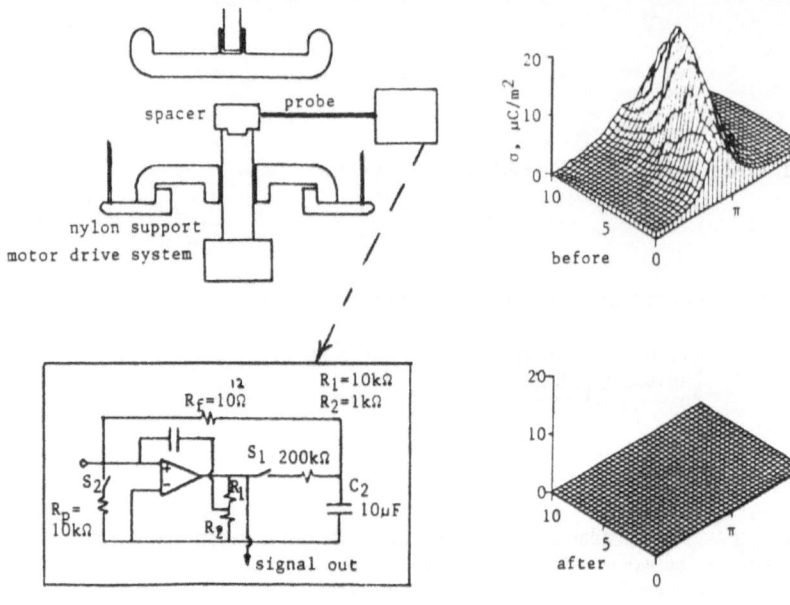

Fig.1· Electrode assembly.

Fig.2. Scan of spacer before and after charge removal by ac corona.

Fig.1· Electrode assembly.

Fig.2. Scan of spacer before and after charge removal by ac corona.

using a low-input-bias (~50fA) op-amp configured as a current integrator.

If required, a line charge or a uniform charge distribution could be deposited by using a corona needle with 5-15kV DC applied and moving the spacer in either a linear or spiral path. Charge deposited in this way, or introduced by impulse stressing, could also be removed using an ac corona source in the form of a wire ring arranged around the spacer. As shown in Fig· 2, the discharging technique was very effective and its use allowed a series of independent tests to be made on the same spacer.

The Marx generator which was used provided 0.15/20µs or 0.15/1500µs impulses with amplitudes up to 200kV. The minimum impulse breakdown level V_m was determined by applying at least 5 shots at each voltage level with voltage increments reducing to ~1% near the breakdown threshold. V_m was usually reproducible to within ~2%.

RESULTS AND DISCUSSION

Plain Spacer

With a plain spacer, the application of positive impulses of increasing magnitude resulted in deposition of negative charges with peak density near the cathode triple junction. There was a well-defined threshold at about 70% of the gas-only breakdown voltage at which the peak density increased suddenly to ~15µC/m². This behaviour was thought to be due to poor contact at the triple junction. A gas gap in this region will result in local field enhancement and field calculations showed that, with a PTFE spacer, the maximum enhancement is 2pu and the enhanced region extends over an axial distance of ~1.5 times the size of the contact defect. A rough estimate of the defect size was obtained by considering the conditions for streamer formation under these conditions and, for a pressure of 1 bar, the calculated gap was 80µm. The critical defect size at 4 bar, the pressure typical of GIS equipment, would therefore be ~20µm for similar onset conditions.

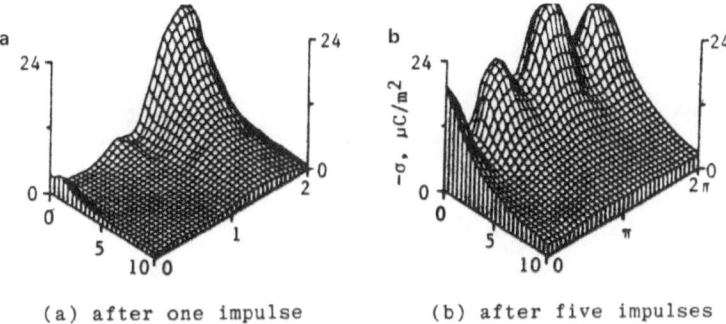

(a) after one impulse   (b) after five impulses

Fig. 3. Build-up of charge with repeated impulses.

Despite the early onset of discharge activity, the breakdown voltage with the plain spacer was typically within ~8% of the gas-only value for pressures up to 1.5 bar, indicating that the triple junction activity did not give rise to streamers able to cross the gap at fields less than ~90% of the limiting field strength for SF_6. Where charge was allowed to accumulate between impulses, each shot quenched the field enhancement at one section of the triple junction so that, after ~5 shots the discharge activity ceased (Fig. 3).

Spacer with a Gas Gap at the Triple Junction

The effect of a significant contact defect was investigated using a spacer with an annular groove at the cathode triple junction. The grooves were 2mm deep and 0.5-1.5mm wide. The calculated field distribution with these grooves is shown in Fig. 4.

For tests in which the charge was neutralised between impulse voltage applications, the reduction in strength was greatest with the largest gas gaps, increasing from ~3% for the 0.5mm groove to ~12% with a 1.5mm gap (Fig. 5). On the other hand, if charge was allowed to accumulate between shots, the greatest reduction (of ~30%) was recorded with the 0.5mm gap.

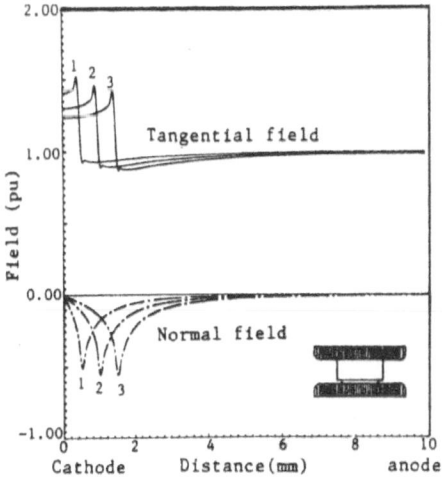

Fig. 4. Calculated field distribution
1:d=0.5mm; 2:d=1mm; 3:d=1.5mm.

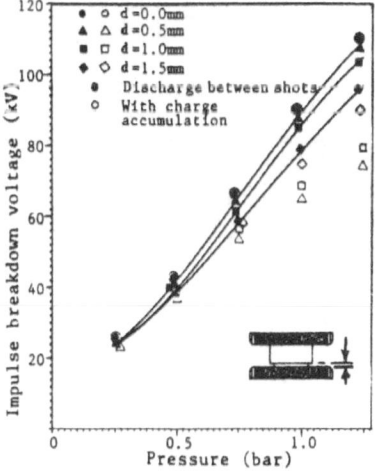

Fig. 5. Flashover characteristics of a spacer with a gas gap at the cathode.

307

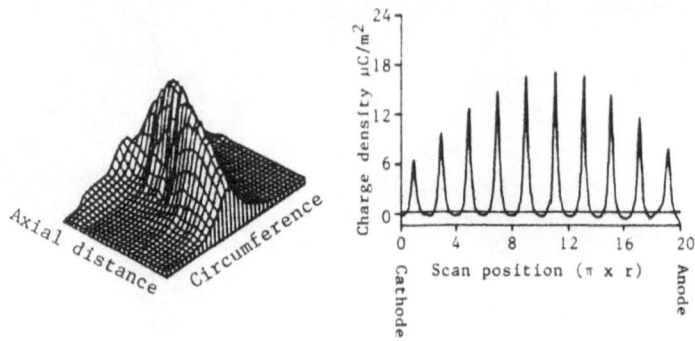

Fig. 6. Scan of linear charge distribution deposited
using a corona needle.

These results may be qualitatively explained on the basis of the computed tangential field distribution of Fig. 4. For the charge-free spacer the triple-junction streamers can propagate only when the field in the main part of the gap is close to the SF_6 limiting field of ~88.4kV/cm/bar. Any reduction in strength will be greatest for the most intense streamers. Since the ionisation activity depends on the area under the tangential field curve for $E > E_{lim}$ (with 1pu $\cong E_{lim}$ in this case) the reduction would be expected to be greatest for curve 3 in Fig. 4 (i.e. for the 1.5mm groove). When the charge was allowed to build up between shots, probe measurements showed that significant charge densities accumulated over the whole of the spacer. This would result in field enhancement which would allow streamer propagation, and hence breakdown, at a lower level (as observed). For these lower voltages E_{lim} would correspond to (say) 1.3pu and for this condition the largest area under the field curve, and hence the most intense streamer, occurs with the 0.5mm gap.

Precharged Spacers

Tests were carried out in which plain spacers were precharged with approximately linear charge distributions (like that in Fig. 6); with uniform monopolar charge of either polarity; and with bipolar charge distributions in which the deposited charge was approximately sinusoidal with the same polarity as the adjacent electrode (homocharge) or with the opposite polarity (heterocharge). For monopolar charge distributions there was a significance effect only for positive charge deposition, i.e. where the cathode field was enhanced. For this case, there was a threshold density of $\sim 25\mu C/m^2$ for deviation from the limiting field value, and the strength was reduced to ~60% of E_{lim} for a charge density of $70\mu C/m^2$ (Fig. 7). The worst case was with a heterocharge distribution, where peak densities of $\sim 10\mu C/m^2$ were sufficient to cause a reduction in impulse strength.

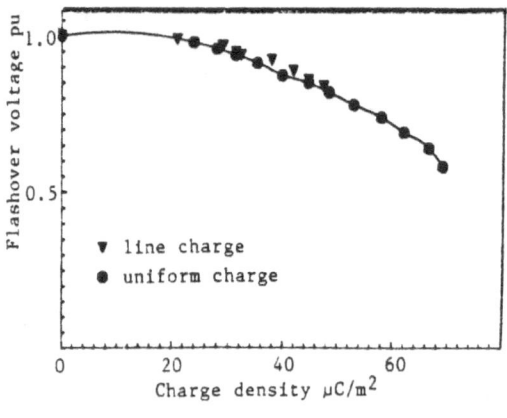

▼ line charge
● uniform charge

Fig. 7. Effect of line and uniform
charge distributions on
flashover of a Teflon
spacer in SF_6 at 1 bar.

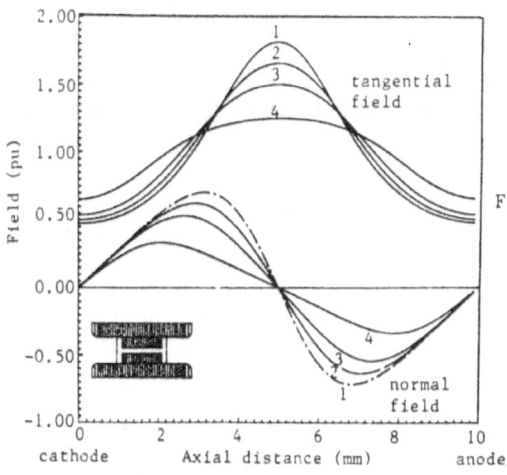

Fig.8. Field distribution for a spacer with metal inserts.
1:d=4mm; 2:d=3.5mm; 3:d=3mm; 4:d=2mm .

Spacers with Metal Inserts at the Triple Junctions

Tests were made on charge-free spacers with metal inserts at each end. The inserts were 2-4mm in depth and resulted in the field distribution shown in Fig. 8. One of the aims of this work was to use relatively large inserts to raise the midgap field in order to determine whether streamer inception and propagation could occur in that region.

The inserts were found to effectively shield the triple junctions, and no charging was observed at voltages below the breakdown level. Provided that the long-tailed (1500µs) wave was used, V_m was reduced to the calculated streamer inception level, indicating that conditions for inception and propagation were both satisfied at onset. With the shorter wave, V_m was ~20% above the calculated level, suggesting a lack of initiating electrons for midgap inception. This was confirmed by using a 5mCi caesium source to irradiate the gap.

One of the effects of the inserts is to cause a build-up of charge on the surface as a result of the normal component of field which they introduce. Thus, following a breakdown, positive charge was found near the cathode, and negative charge near the anode (Fig. 9). Clearly, if such charge accumulation occurred in practical systems, perhaps as a result of the effects of particles, the system might be weaker against a subsequent impulse stress than one without inserts.

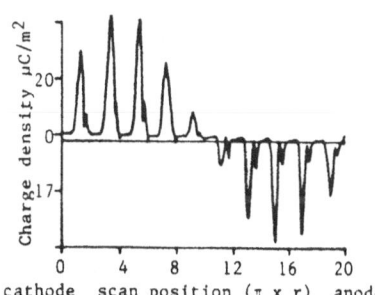

Fig.9. Charge deposited on a spacer with inserts following a breakdown.

CONCLUSIONS

For a plain spacer, the response to impulse stress is determined by conditions at the triple junction. For defects of a few tens of microns, discharge activity begins at about 70% of the limiting field strength but the breakdown level is only slightly affected.

With large (~mm) defects at the cathode triple junction the onset level is considerably reduced and charge can be deposited over most of the spacer surface. If charge is allowed to accumulate the impulse strength can be reduced by as much as 30%.

When controlled charging methods are used to create regions of high charge density the strength can be as low as 50% of the gas-only value, even for pressures of 1 bar, with the greatest reduction occurring for deposition of heterocharge.

Inserts can provide effective shielding of the triple junction. However, they introduce a normal field component which can attract surface charges. This may be detrimental in a system in which charges are produced as a result of microdischarge activity in the gas.

REFERENCES

1. J. R. Laghari, Spacer flashover in compressed gases, IEEE Trans, Vol. EI-20, No 1, 83-92 (1985).
2. Th. Stoop , J. Tom , H. F A. Verhaart and A. J. L. Verhage, The influence of surface charges on flashover of insulators in gas insulated systems, Fourth Int. Symp.on High Voltage Engineering, Paper No 32.10, Athens (1983).
3. K. Itaka and T. Hara, Influence of local field concentration on surface flashover characteristics of spacers in SF_6 gas, IEEE Int. Symp. on Elect. Insul., 56-60, Boston, USA (1980).
4. T. Nitta, Y. Shibuya, Y. Fujwara, Y. Arahata, H. Takahashi and H. Kuwahara, Factors controlling surface flashover in SF_6 insulated systems, IEEE Trans, Vol. PAS-97, No 3, 958-968 (1978).
5. K. Nakanishi, A. Yoshioka, Y. Arahata and Y. Shibuya, Surface charging on epoxy spacer at dc stress in compressed SF_6 gas, IEEE Trans., Vol. PAS-102, No 12, 3919-3926 (1983).
6. A. Knecht, Development of surface charge on epoxy resin spacers stressed with direct applied voltages', pp356-362 in "Gaseous Dielectrics III", L . G. Christophorou, ed., Pergamon Press, New York, (1982).

DISCUSSION

C. M. COOKE: You provided a very clear presentation of charge accumulations on spacers. Two questions about the method. How long does a "scan" take and have you observed any discharging of the surface due to the probe locally enhancing the field when it is near the charged region?

O. FARISH: With the present probe, a complete scan of the spacer took ~ 30 seconds. We observed no discharges due to the presence of the probe for the charge densities measured. The probe does cause an increase in the normal field (from a value of $\sigma/\epsilon_0(1 + \epsilon_r)$ to a value of σ/ϵ_0) but as long as the charge density is below $\sim 80 ~\mu C/m^2$ (for SF_6 at 1 bar) the field will not exceed the discharge threshold.

S. W. ROWE: When using the spacer with a slotted base (Fig. 4) the charge accumulation is uniform. Where are the streamers actualy situated (under the slot or at the edge of it)? Could this uniform charging be due to charge drift, from streamers "under" the slot, along the field lines, or do the streamers develop near the slot edge?

O. FARISH: We have not attempted to record the spatial distribution of the streamer activity around the slot. However, with the long–tailed wave there is time for a "series" of discharges so that even if the initial discharges are under the slot, the charge deposited would modify the field so that subsequent discharges are deposited to the edge and deposit charge on the surface. It is unlikely that the charge will drift around the circumference of the spacer. More probably, there are multiple discharges distributed around the cathode groove resulting in the observed quasi–uniform charge distribution.

K. NAKANISHI: The sinusoidal charging pattern shown in Fig. 9 of the text suggests that surface conduction exerts much influence to the charging pattern. Could you give us a comment on that?

O. FARISH: As there was no DC stress applied to the spacer prior to impulse voltage application, surface conduction should not be an important factor in the charging process. With inserts, even the gas ionization charge mechanism is largely suppressed because of the reduction of the triple–junction field. No charge is recorded under withstand conditions and sinusoidal pattern is recorded only after a breakdown, probably because the normal component of the field attracts heterocharge during the streamer stage of the breakdown.

PARTICLE-INITIATED PREBREAKDOWN CHARACTERISTICS

WITH AN INSULATOR STRESSED ON 60 H$_z$ VOLTAGE IN SF$_6$

H. T. Wang and R. G. van Heeswijk

Department of Electrical & Computer Engineering
University of Waterloo
Waterloo, Ontario, Canada
N2L 3G1

INTRODUCTION

The excellent dielectric properties of Sulfur hexafluoride (SF$_6$) have long been recognized. The use of compressed SF$_6$ as insulant as well as switching medium in electric power systems is now being extended to higher system voltages. It has been realized however that the presence of conducting contaminating particles alters the insulation characteristics, especially the performance of spacers in SF$_6$-insulated systems(Nitta et al., 1978; Cookson, 1981; Sudarshan et al., 1986). Considerable experimental investigations on paticle-initiated breakdown characteristics with and without the presence of an insulator under AC, DC, Impulse and their combinations have been carried out by many researchers(Wootton et al. 1979; Yamagiwa et al., 1987; van Heeswijk and Srivastava, 1985). However, there is relatively little information available on the particle-initiated pre-breakdown characteristics, especially with an insulator involved. It is therefore, essential to investigate these charactristics in order to gain some insight into the particle-initiated breakdown(or flashover) processes.

Particle-initiated breakdown is very complex and no theoretical comprehensive model or even a satisfactory explanation of the breakdown characteristics exists yet. Some mechanisms have been proposed to explain particle-initiated breakdown characteristics. They are: (1) the effects of increased ionization in the enhanced field at the particle tip(Cookson et al., 1972; Hara et al., 1976/1977); (2) the processes associated with rapid charge reversal at contact with an electrode in terms of "pulsed" mechanism(Cookson et al.; 1973; Cooke et al., 1977); (3) mechanisms based on the microdischarge between particle and electrode such as "trigatron effect"(Cookson et al., 1972,1973), electron emission from either the particle or the electrode depending upon the polarity(Chatterton et al.,1972); (4)mechanisms related to the dymanics of free particles such as "low density region or shock wave formation, or turbulent trace"(Cookson et al., 1972; Vilenichuk et al, 1980), "charge reversal" due to the accumulation of the charge with the same polarity of the electrode which the particle is approaching (Berger, 1974).

Most of the mechanisms proposed are based on the two gap breakdown process, neglecting the interaction between the microdischarge and the main gap breakdown. It is reported by Pfeiffer et al.(1980) that the discharge processes can propagate simultantaneously from both tips of a conducting particle. The particle-initiated breakdown process obviously needs to be investigated further. Moreover, the experimental results(Wootton et al., 1979; Wang and van Heeswijk, 1990; Berger, 1974)under ac and dc voltage, showed that for free moving particle breakdown the same important characteristic, the critical region(this term is thought to be more suitable than the critical distance) for which the breakdown(or flashover) shows the minimum values, can also be observed by using a fixed particle. So it may be concluded that particle-initiated breakdown mechanisms are not directly related with the dynamics of free particles. The movement of free particles can only increase the probability for a particle to be in the "critical region". So far there is no mechanism which can even give qualitative explanation(Cookson et al., 1981), especially for filamentary particles.

In the present work, a steel wire particle, 6.4 mm long and 0.35 mm diam. was fixed at a distance from the epoxy resin spacer or glued on its surface in order to study the particle-initiated prebreakdown characteristics and the influences of the spacer. The results are reported and discussed by considering the presence of surface or space charge.

EXPERIMENTAL

Two Al Brouce profile parallel-plane electrodes with diameter of 203 mm bridged by a cylindrical epoxy resin spacer, 40 mm height and 30 mm diameter were used. A specially designed subdivided bottom electrode, which forms part of a bridge circuit capable of effectively measuring only the prebreakdown current (both steady and pulsative components) initiated by a particle. The sensitivity of this measuring circuit is approximately $0.03\mu A$ (with $33.3K\Omega$ detecting resistor and the response time is less than $0.5\mu s$). A collimated photomultiplier(EMI9813B) can be used to examine the light emission from the particle. A more detailed description of the setup and procedure has been given elsewhere (Wang and van Heeswijk, 1990). The experiments were carried out in a steel test vessel filled with commercially pure SF_6 gas at pressure ranging from 50 kPa to 600kPa. The vessel is evacuated to a pressure of about 0.01 torr prior to pressurization with SF_6. The spacer and the particle were cleaned with ethanol and dried in dry air. During the experiment the spacer is continously checked for damage due to breakdown. The maximum current that occurred during about 1 minute was measured by using an OSC in storage mode. The polarity indicated in the figures is the instantenous polarity of the high voltage electrode.

RESULTS AND DISCUSSION

The particle-initiated prebreakdown current and light emission waveforms are strongly affected by the particle position, the presence of a spacer, gas pressure and applied voltage, etc.. In this experiment, five different locations of the particle were chosen, position A, B, C, D and E as illustrated in Figure 1.

Figures 2(a)-(e) show typical waveforms observed with a particle at position A or B. At the lower voltage and pressure of 50 kPa, the current waveforms for a particle at position A and B are similar and shown in Figure 2(a). As the applied voltage is increased, the rate of rise of the discharge current for position B is greater than that for position A[Figure 2(b)]. It can be seen that the discharge is in the form of glow with corona bursts superimposed. At the higher pressures of 100kPa and intermediate voltage, pulsative discharge occurs during the positive half cycle with strong intensity of light emission and small glow component with a large corona burst superimposed appears during the negative half cycle as shown in Figure 2(c). Similar current waveforms have been observed under these conditions for a particle at position B. Upon increasing the voltage, the current and the light waveform change to those shown in Figure 2(d). At the higher pressures of 300 kPa and above, the current shows corona bursts for a particle at position A as shown in Figure 2(e). At position B the current indicates a glow with burst larger than those for position A superimposed.

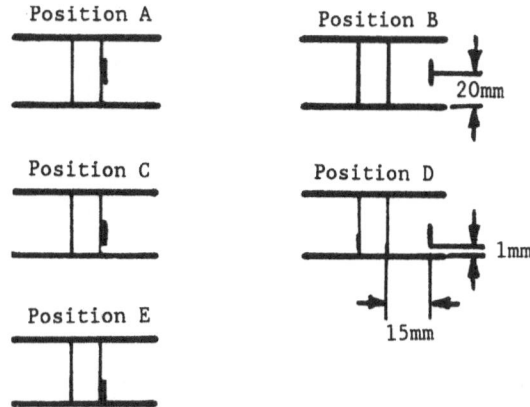

Fig.   1. Schematic of wire particle positions.

Figures 2(f)-(i) give the typical waveform when a particle is glued on a spacer at position C or is fixed at position D. At the lower pressure of 50kPa, for a particle at position C the first detectable current signal is due to small bursts. By increasing the voltage above the onset voltage, a pulse train is observed as shown in Figure 2(f). For a particle at position D, no such pulses were observed, but a rather steady glow discharge shown in Figure 2(h) occurred. At the higher pressure of 100kPa and above, with a particle at position C, the current is mainly due to corona bursts with a higher frequency during the positive half cycle. At position D, the discharge clearly shows corona bursts and glow during the positive half cycle and corona bursts at negative half cycle as shown in Figure 2(i).At higher pressure surface discharge is mainly pulsative in nature for both half cycles. Only at higher pressures and with particles fixed at position B and D is the corona during both half cycles similar to that observed for a highly nonuniform field gas gap. Figure 2(j) clearly shows with a particle touching the bottom electrode (position E), the polarity effect of the negative tip glow amd positive tip bursts during positive and negative half cycles, respectively.

Figure 3 shows the effect of the gas pressure on particle-initiated prebreakdown current. For all cases investigated, the current is suppressed by increasing the gas pressure. It should be noted that the polarity effects tends to be stronger with the increase of pressure.

Figure 4 shows the effect of a spacer on prebreakdown current at pressures of 50kPa. At the lower pressure the presence of the spacer suppresses the current. But at higher pressure, the case is reversed. It seems that an insulating surface has stronger effect on the diffuse-glow discharge than on the filamentary discharge processes.

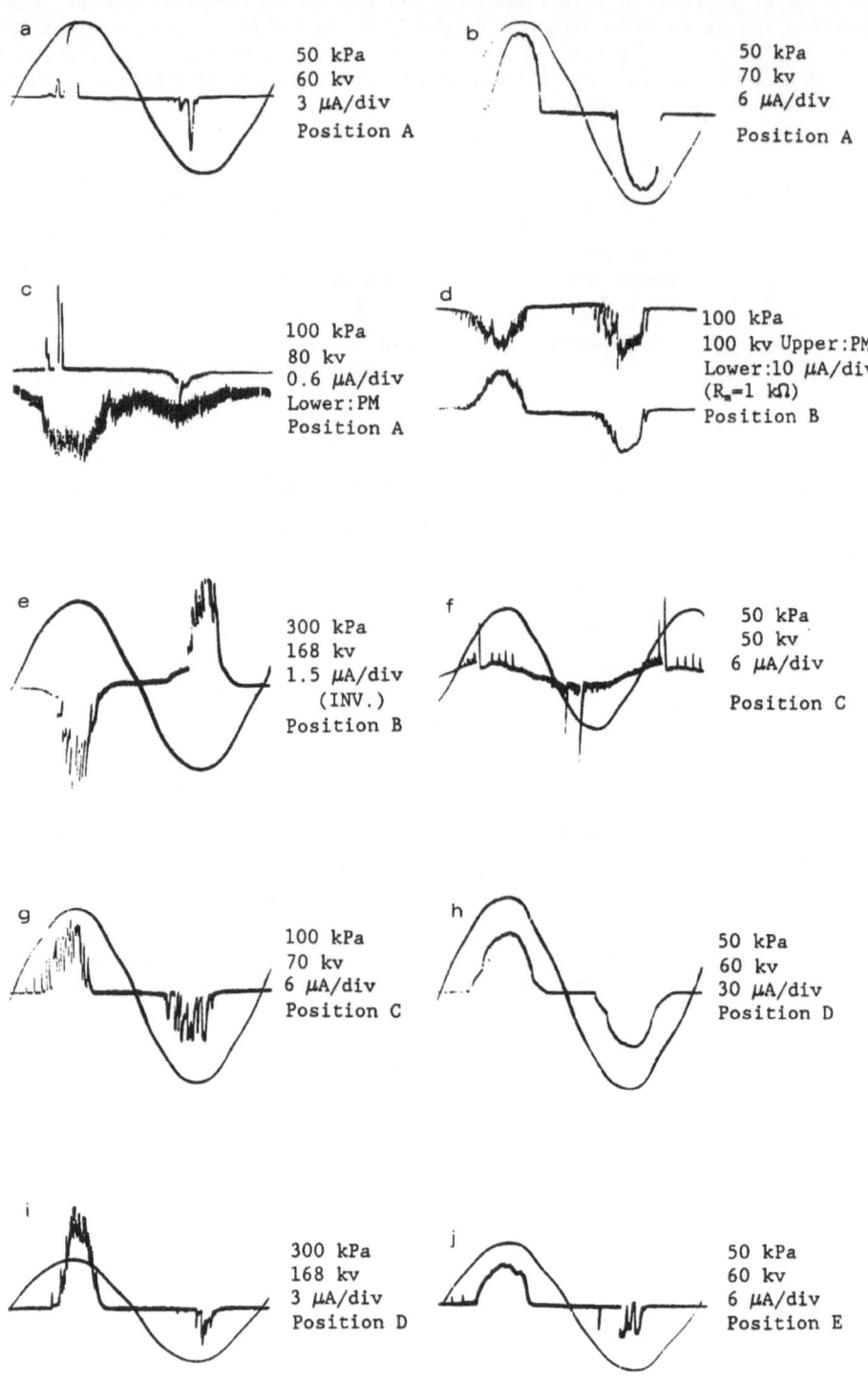

Fig. 2. Prebreakdown current and light oscillogrames in SF₆
(time base: 2ms/div, except e and f >2ms/div).

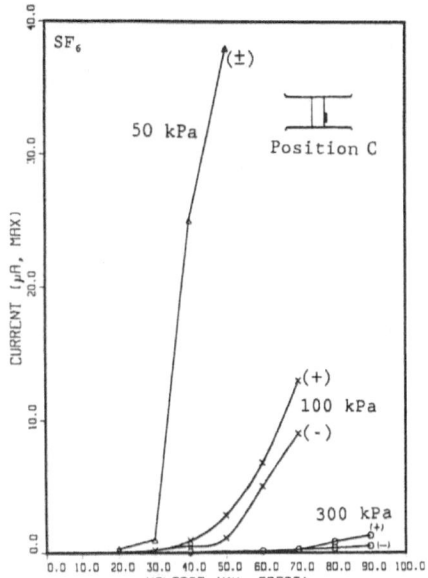

Fig. 3. The effect of the gas
pressure on prebreak-
down current.

Fig. 3. The effect of the gas
pressure on prebreak-
down current.

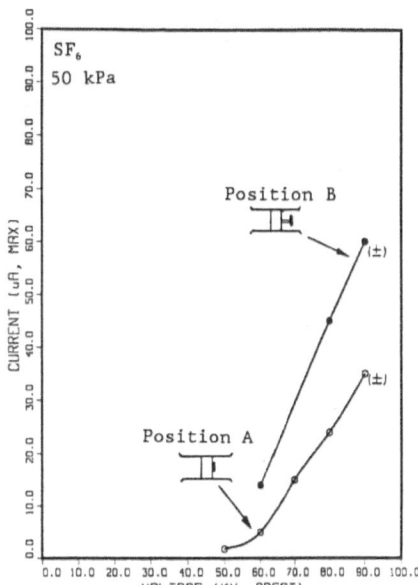

Fig. 4. The effect of a spacer
on prebreakdown current.

Fig. 4. The effect of a spacer
on prebreakdown current.

Figure 5 illustrates the effect of the particle position on a spacer. At the lower pressure of 50 kPa, there is no detectable polarity effect. It should be noted that a particle at position C initates a larger prebreakdown current than that at position E. On the other hand, at higher pressure of 300kPa, this is reversed. In addition, there is a strong polarity effect.

Figure 6 shows at the higher pressure of 300kPa, the effect of the particle position without the influence of a spacer (position B, D). It can be seen that the current with a particle at position B seems to coincide with the current at negative half cycle(negative tip faces the ground electrode) at position D. This probably indicates that by moving the particle closer to the ground electrode, the current in positive half cycle(positive tip facing the electrode) is enhanced. At the higher pressure the positive tip seems to be more active than the negative tip of the particle.

The foregoing implies that without the influence of a spacer(particle is located at position B and D) the particle-initiated prebreakdown current is similar to the prebreakdown current waveforms observed in highly nonuniform point-to-plane gaps(Hazel and Kuffel, 1976; Ibrahim and Farish, 1980) under dc voltage application. The presence of a spacer surface(position A) reduces the steady component of the discharge current at the lower pressure of 50 kPa (Figure 4), i.e., the surface charging process appears to play an important role. Moreover, the accumulation of homocharge on a spacer surface in front of the tips of a particle and therefore its effect on the discharge is somewhat random, depending on several factors[figure 2(c),(f) and (g)]. The surface charge in front of a particle tip can cause a back discharge, a

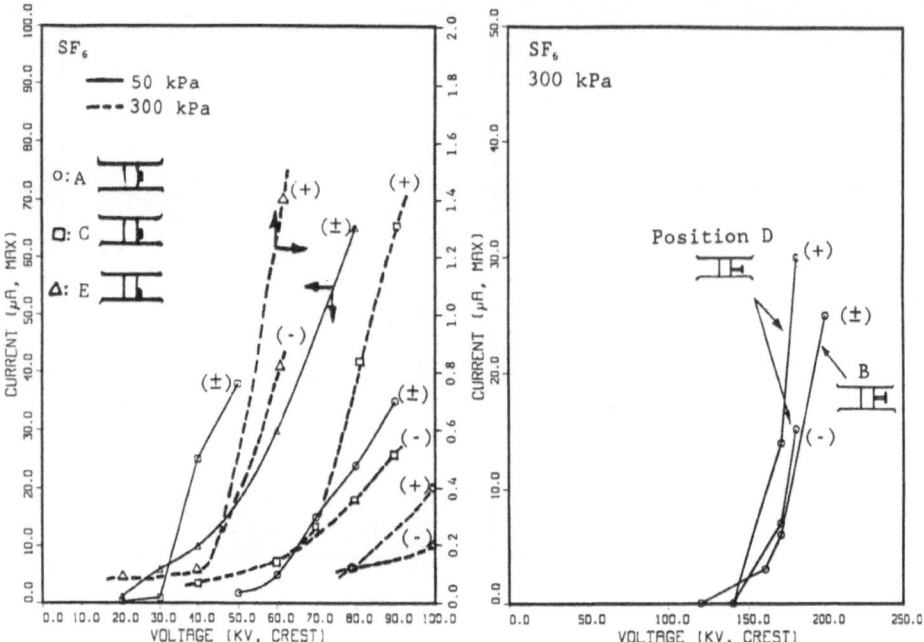

Fig. 5.The effect of the particle position on prebreakdown current.A particle on the spacer.

Fig. 6. The effect of the particle position on prebreakdown current. A particle 15 mm away from a spacer.

discharge at or near voltage zero[figure 2(f),(g) and (j)]. It was observed that at higher pressure of 300 kPa and above, the prebreakdown current is increased due to the presence of an insulating surface and the breakdown voltages are lower than that without the presence of a spacer.

It appears necessary to consider the effect of space charge in the microdischarge to interpret particle-initiated breakdown charateristics such as the "critical region". It is possible that the field enhancement at both tips of a particle due to the space charge in the microgap may be comparable, especially for a filamentary particle. The polarity effect of particle-initiated breakdown can be explained by considering the effect of space charge in the microgap. The microdischarge plays a crucial role in particle-initiated breakdown processes. The space charge in the microgap not only enhances the field at the particle tip facing the electrode, but also can enhance the field at the particle tip facing the main gap in pulsative manner.More quantitative work on clarifying the physical processes involved in the particle-initiated breakdown needs to be done in the future.

CONCLUSIONS

The results indicate that particle-initiated prebreakdown current charateristics are complex. At this moment, we can not explain quantitatively the why and how of the influence of the spacer on particle-initiated prebreakdown and breakdown processes. Nevertheless, some conclusions can be drawn:

a) Surface charge has a strong influence on particle-initiated prebreakdown current under ac voltage which depends on the gas pressure;

b) Particle-initiated prebreakdown current characteristics are particle position dependent even in uniform field;

c) Polarity effect becomes stronger with increasing gas pressure and decreasing the distance between the particle and electrode.

ACKNOWLEDGEMENTS

The authors wish to acknowledge the Natural Sciences and Engineering Research Council of Canada for financial support. The technical assistance of Mr. T. Weldon and Mr. R. Hinz is appreciated.

REFERENCES

Berger, S.K., 1974, Reduction of breakdown voltage in uniform and coaxial fields in atmospheric air through moving conducting spheres, 3rd International conference on Gas Discharges, 380-384.

Chatterton, P.A., Menon, M.M., and Srivastava, K.D., 1972, Discussion of "Effect of conducting particles on ac corona and breakdown in compressed SF_6", IEEE Trans. on PAS, 91:1337.

Cooke, C.M., 1975, Ionization, electrode surfaces and discharges in SF_6 at extra-high voltages, IEEE Trans. on PAS 94: 1518-1523.

Cooke, C.M., Wootton, R.E., and Cookson, A.H., 1977, Influence of particles on ac and dc electrical proformance of gas insulatd systems, IEEE Trans. on PAS 96: 768-777.

Cookson, A. H., 1981, Review of high-voltage gas breakdown and insulators in compressed gas, IEE Proc. pt.A, 128: 303-312.

Cookson, A.H., Farish, O., and Sommerman, G.M.L., 1972, Effect of conducting particles on ac corona and breakdown in compressed SF_6, IEEE Trans. on PAS, 91:1329-1338.

Cookson, A.H., Farish, O., 1973, Particle-initiated breakdown between coaxial electrodes in compressed SF_6, IEEE Trans. on PAS, 92:871-877.

Hara, M., and Akazaki, m., 1976/1977, A methode for prediction of gaseous discharge threshold voltage in the presence of a conducting particle, J. Electrostat., 120:223-239.

Hazel, R. and Kuffel, E., 1976, Static field anode corona characteristics in sulphur Hexafluoride, IEEE Trans. on PAS, 95: 178-186.

Ibrahim, O.E. and Farish, O., 1980, Negative-point breakdown and prebreakdown corona processes in SF_6 and SF_6/N_2 mixtures, 6th International Conference on Gas Discharges and their Applications, pt.1, 161-164.

Nitta,T., Shibuya,Y., Fujiwara,Y., Arahata,Y., Takahashi,H. and Kuwahara,H., 1978, Factors controlling surface flashover in SF_6 gas insulated systems, IEEE Trans. on PAS 97:959-968.

Pfeiffer, W. and Völker, P., 1980, Influence of conductive particles on the DC-voltage strength of spacers in compressed SF_6, in: "Gaseous dielectrics II, L.G. Christophorou, ed., Pegramon Press, New York, p.243.

Sudarshan, T. S. and Dougal, R. A., 1986, Machanism of surface flashover along solid dielectrics in Compressed gases, IEEE Trans. on EI, 21:727-746.

van Heeswijk, R.G. and Srivastava, K.D., 1985, Particle-initiated breakdown at a spacer surface in a gas insulated bus, Canadian Electrical Association, Montréal, Project report 154-T-343.

Vilenichuk, A.L., Pankratova, I.V., Titkova, V.G., Investigation of interaction of electromagnetic field with conducting particles in SF_6 field apparatus, 6th International Conference on Gas Discharges and their Applications, pt.1, 224-227.

Wang, H.T. and van Heeswijk, R.G., 1990, Experimental investigation of particle-initiated surface flashover in air and N_2 in uniform gap under 60 HZ voltage, <u>IEEE International Symposium on EI</u>, 244-247.

Wootton, R.E., Cookson, A.H., Emery, Farish, O., 1979, Investigation of high voltage particle-initiated breakdown in gas-insulated systems, EPRI report, EL-1007.

Yamagiwa, T., Ishikawa, T., Endo, F., Kamata, Y., 1987, Dielectric characteristics on spacer surface in SF_6 under lightning Impulse superimposed on AC voltage, <u>5th ISH</u>, paper 13:03.

DISCUSSION

M. GOLDMAN: Do dipoles on the surface have an effect on the particle–initiated breakdown?

H. T. WANG: I can not identify the existence of the dipoles on the insulating surface and their effect, based on our measurement. However, the dipoles can exist on the technically clean insulating surface, which may more or less modify the surface charging process and then influence particle–initiated breakdown.

CHAPTER 7: PULSED POWER SWITCHING AND LASER INITIATED BREAKDOWN

HIGH POWER GASEOUS OPENING AND CLOSING SWITCHES

M. Kristiansen[*]
Department of Electrical Engineering
Texas Tech University
Lubbock, TX 79409-4439

A. Guenther
Los Alamos National Laboratory
Los Alamos, NM 87545

G. Schaefer[**]
Polytechnic University
Farmingdale, NY 11735

INTRODUCTION

The total number of variants of gaseous opening and closing switches
is far too large to be dealt with adequately in a short review paper such
as this. Three recent books address much on this subject (Guenther et
al., 1987; Vitkovitsky, 1987; Schaefer, 1990). Switch types selected for
emphasis herein are: diffuse discharge opening switches, plasma (erosion)
opening switches, gas filled spark gaps, ignitrons, and back-lighted
thyratrons (pseudosparks) . The choice is not only an indication of the
authors' background and experience but reflects certain noteworthy recent
developments in the high power switching field. The detailed theory of
operation of each switch is not included, due to obvious space limitations. Instead we have provided the essential references in this regard
and have rather emphasized the salient recent developments and improvements for each of these switches.

Inductive energy storage, in spite of its obvious promise of much
higher energy storage density than capacitive systems, has never reached
the development and level of applications as has capacitive systems because of the lack of a true repetitive opening switch. There is no opening switch with the same wide range of parameters, ease of use, and
reliability as, for instance, spark gap or thyratron closing switches.
Two of the main differences in the operating mode of these switches explains to some extent why the opening switch is so much more difficult
than the closing switch to implement.

[*]Supported, in part, by the Texas Tech University Center for Energy Research.

[**]Our friend and collaborator Gerhard Schaefer of the Weber Research Institute, Polytechnic University of New York, passed away on September 17, 1989, at the age of 49. He was an active researcher in the gaseous dielectrics area particularly as it applied to laser and pulsed power technologies. Gerhard was known for his commitment and contributions to science as well as for his personality and sincerity. He will be both missed and remembered frequently. We dedicate this paper to him, based, in part, on his last major effort as editor of "Gas Closing Switches" in the series "Advances in Pulsed Power Technology" (Plenum Press, New York, NY).

First is the obvious problem of interrupting a discharge rather than initiating one. The second problem is a little more subtle. Repetitive closing switches typically conduct a relatively short time compared to the off time. A typical case may be 1-10 μs conduction time at pulse repetition rates of up to 1000 pps, which implies a typical duty factor of 0.1-1%. Opening switches, on the other hand, are on most of the time (i.e., during the inductor charging period) and only turn off for the short time of the output pulse. Hence they have typical duty cycles of ~ 99%. This means that they must have very low losses when closed and be able to dissipate large amounts of heat (i.e., they must be large and bulky) and yet be able to open very quickly. This conflicting requirement often leads to the concept of staged switches where typically 2-3 opening switches of successively less thermal capacity and faster opening speed are used in sequence.

In order to develop fast, repetitive opening switches using a gaseous conduction medium it is generally conceded that one desires a diffuse discharge system because of reduced current density. Much attention has also recently been given to plasma (erosion) opening switches. The brackets ( ) around one word indicates a disagreement as to the exact opening mechanism. We shall in the following refer to these switches generally as P(E)OS.

Diffuse Discharge Opening Switches

In a diffuse discharge switch, the switch medium is usually an externally sustained discharge. The most common sustainment method is electron beam pumping, but the pumping sources can also be microwaves, lasers, or other high intensity, high energy photon sources. Early reports on diffuse discharge opening switches were given by Hunter (1976) and by Koval'chuk and Mesyats (1976a, 1976b). A schematic of an experimental arrangement for investigating such switches is shown in Fig. 1. The gas between the electrodes is ionized by an external source (e.g., an e-beam) and becomes conductive. The reduced electric field, E/N, is kept in a range where self-ionization by the electrons in the discharge is negligible. When the external ionization source is turned off, the switch turns off by electron attachment and recombination processes.

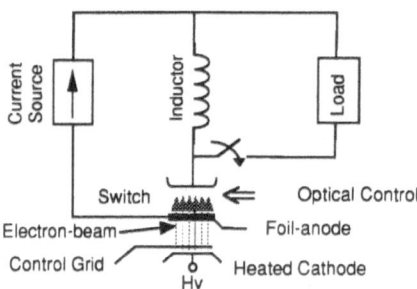

Fig. 1. Schematic of an externally controlled opening switch as part of an inductive energy storage system.

It was soon realized that the gases used in such switches should have certain special properties (Schoenbach et al., 1981, 1982, 1987; Christophorou et al., 1982):

322

1. For low values of the reduced field strength E/N, characteristic for the conduction phase, the electron drift velocity, w, should be large and the attachment rate coefficient, k_a, should be small in order to minimize losses.

2. With increasing E/N, characteristic of the opening phase of a switch in an inductive energy storage system, the attachment rate coefficient should increase and the electron drift velocity should decrease in order to support the switch opening process.

3. Additionally, the gas should have a high dielectric strength to hold off the expected high voltage across the switch when it opens.

In addition to the three switch gas criteria listed above there is also the concern for instabilities that might lead to an undesirable glow-to-arc transition, i.e., one wants the highest possible switch current density that does not lead to such a transition. Optical control methods were also considered but it is generally found that photons are too energy expensive at the present time to be used for the main sustaining source.

One possibility that was never investigated experimentally was to use electron beams as the main sustaining source and optically enhanced attachment together to decrease the switch losses (Schaefer et al., 1983). Various other "clever" photon-gas interactions were considered, such as vibrationally excited molecules with enhanced electronegativity (Bardsley and Wadehra, 1982), photo disassociation of molecules into fragments with enhanced electronegativity (Rossi et al., 1985), electronic excitation into vibrational excitation (Schaefer et al., 1988). A major contribution was the realization of the use of "gas engineering" to satisfy the three previously listed gas criteria (Schaefer, 1989). Hunter et al. (1984) investigated several gases with the general desired properties with regard to attachment rate and mobility vs E/N.

Plasma (Erosion) Opening Switch (P(E)OS)

The Plasma (Erosion) Opening Switch has received considerable interest in both the US and USSR over the past few years. The basic (erosion) opening mechanism has been described by Commisso et al. (1987). The basic idea is depicted in Fig. 2. Instead of using the classical Marx generator charging the capacitance of a PFL arrangement, the Marx generator charges up the inductance of the line via an opening switch. The switch is closed by injecting a plasma which then opens at some current level, transferring the inductive energy in the line to the load. It is suggested that the exact mechanism is either a plasma "erosion" mechanism or a plasma-dynamic (snowplow-like) process.

The key research issues appear to be the conduction time, the switch losses, and the opening time. These parameters basically control the voltage and power gain of the system. Both the US and the USSR laboratories (Comisso et al., 1986; Mesyats et al., 1987) have reported μs range conduction times. During a recent visit to the USA, G. Mesyats (1989) reported a voltage gain of 13 (500 kV to 6.5 MV) in a two stage switch. Bistritsky (1990) also reported a voltage gain of 3.5 in a combination switch including magnetic pulse compression.

The switch conduction losses or impedance determines how much of the current is diverted to the load (unless a peaking switch is also employed) during the charging cycle. It is obviously desirable to have a switch plasma with high electron drift velocity but it is also desirable to keep the plasma density low to improve the switch opening speed (Commisso et al., 1987).

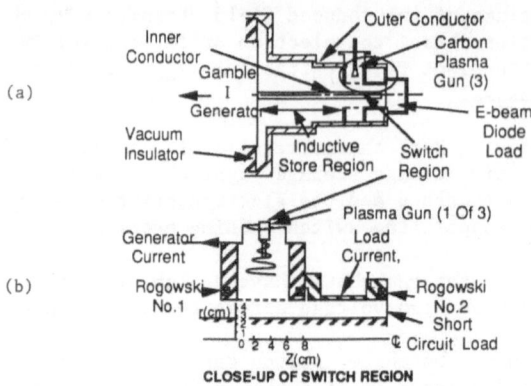

Fig. 2.   (a) Schematic of Gamble I P(E)OS experiment showing relative
          location of generator, storage inductance, P(E)OS region, and
          load.
          (b) Close-up of switch region showing current diagnostics,
          electrode polarity, and plasma gun orientation. Here the load
          is a short circuit.

GAS FILLED SPARK GAPS

     One of the main work horses of high power gaseous switches is the gas
filled spark gap. The number of possible configurations and parameter
ranges is very large, indeed. The parameters of main concern are usually:
self breakdown voltage, peak current, coulombs/shot, repetition rate,
lifetime, and trigger delay and jitter. The main aspects of these and
other features are described extensively in the book edited by Schaefer
(1990). Much of the recent research and development of spark gaps has
been devoted to improved triggering, increased repetition rates, increased
coulomb and peak current handling, and improved lifetime (reduced electrode erosion and insulator damage).

     Triggering of conventional trigatron type spark gaps has recently
been investigated by Peterkin and Williams (1989). They investigated the
basic breakdown mechanism in these switches and found that breakdown is
initiated by streamers from the trigger pin before the appearance of the
trigger arc. Laser triggering of large spark gap arrays to ensure good
synchronization of the switches has been discussed by Wilson and Donovan
(1989). They describe a KrF triggered, 5 MV spark gap with less than 3 ns
rms jitter in routine operation. Sub-ns jitter has been reported for more
optimally arranged systems (Williams and Guenther, 1990).

     The repetition rate of gas filled spark gaps primarily depends upon
the rate at which heat can be removed from the inter-electrode space.
Without extensive gas flushing the repetition rate is generally less than
100 pps, whereas high flush rates can extend the operation into the kpps
range, at least when operated in a burst mode (Buttram and Sampayan,
1990).   Recent work by Thayer et al. (1989) describes a "tuned" flow
system which utilizes the shock waves and unsteady flow after a discharge
event to reduce the gas flow by a factor of 2 and also reduces the power
losses in circulating the gas by more than a factor of 10 at higher repetition rates (> 1000 pps). The same authors also describe the use of a
two-phase flow system, where the high heat of vaporization of the liquid
droplets provides both gas and wall cooling. The large volume change due
to vaporization generates sufficient flow rates.

Several efforts to increase the peak current and coulomb handling capability of gas filled spark gaps have been reported (Harrison, 1990; Hammon, 1990). The rotating arc gap (RAG) of Harrison attempts to utilize the same principle as is used in vacuum interrupters, namely a slit electrode to provide a JxB effect which moves the arc around on the surface and does not allow it to anchor in one spot. In the Hammon switch the electrodes are water cooled and uses graphite electrode tips. The insulator is moved away from the immediate arc region to reduce the heat load on it. In both cases the goal is several hundred coulomb, near MA level operation.

A large data base with regard to insulator and electrode damage in intense discharge environmebts has been collected by Donaldson (1990) and Engel et al. (1989, 1990). All standard electrode materials, except graphite which does not have a molten phase, exhibit the erosion behavior shown in Fig. 3. This figure also summarizes the consequences of varying various spark gap parameters. Note that the general shape is independent of the erosion parameter (I_{peak}, $\int i \, dt$, $\int i^2 \, dt$, etc.). It is found that copper with various extruded refractory metal fibers (e.g. Nb) has the lowest volume erosion. In repetitive spark gaps and in surface discharge spark gaps, the insulator damage becomes a major issue. Two "goodness" parameters which seem to describe beneficial insulator characteristics are (Engel et al., 1990):

$$\textbf{Arc Melting Resistance,} \qquad AMR = (T_m + \frac{\Delta H_m}{c_p}) \sqrt{\rho c_p K_t},$$

where T_m = melting temperature ρ = density
ΔH_m = enthalpy of melting K_t = thermal conductivity
c_p = specific heat

and the voltage **Holdoff Degradation Resistance,** $HDR = \frac{-\Delta H_f}{M.W.}$,

where $-\Delta H_f$ = enthalpy of formation and M.W. = molecular weight. Some of the best insulator materials include reinforced alumina, others are industry proprietary.

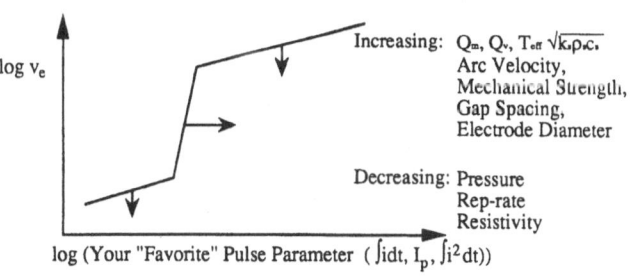

Fig. 3. Typical Erosion Curve.

IGNITRONS

Ignitrons utilize a mercury liquid pool cathode and a solid anode with a semiconductor (boron nitride) trigger electrode dipping into the mercury pool. Their main use has been in power rectification and welding control.

Since early 1987, a cooperative R&D program including Lawrence Livermore National Laboratory, Richardson Electronics/U.S.A., EEV/United Kingdom, and Texas Tech University has been carried out to redesign existing "Size E" ignitrons for high current, high coulomb switching. The development goals are peak current ratings of up to 1000 kA and simultaneous charge transfer rates of 250-500 C, having a lifetime of 500 shots or more. Studies of conventional "Size E" ignitrons (Kihara, 1987; Rei et al., 1989) revealed the following problems, in order of severity, that arise when increasing the peak current above 100 kA.

Ignitor problems such as wetting and breakage.
Erosion of anode and wall.
Premature self breakdown or prefires.
Balance and support of magnetic forces.
High forward voltage drop.

The approach to improve the tube performance was to replace the cylindrical shaped anode of the conventional tube with a cup anode and to reduce the anode-cathode (mercury) spacing drastically. This confines and stabilizes the plasma to the region underneath the anode, minimizes arc transfer to the wall, and reduces the forward voltage drop and therefore the dissipated energy. The ignitor is located in the center of the tube underneath the cup anode and is, thereby, protected from direct exposure to the main discharge.

Today the first commercial tube incorporating the new design, the NL-9000 made by Richardson Electronics, demonstrates simultaneous peak current and charge transfer ratings of 700 kA and 250 C, respectively (Richardson Electronics, 1989). Two years ago, the peak current rating of "Size E" tubes was 300 kA for comparable life expectancy. Figure 4 shows the peak current and coulomb capabilities of conventional "Size D" and "Size E" ignitrons as well as the data from the new NL-9000. Besides the increased peak current capability, the forward voltage drop of the NL-9000 has been lowered from approximately 400 V to 120 V at 400 kA for comparable current waveforms (critically damped sinusoid with 160 μs quarter period). It is important to note that the forward voltage drop increases with decreasing current risetime.

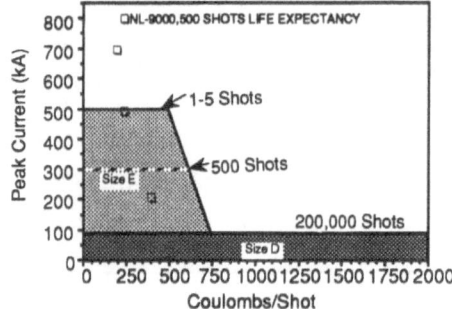

Fig. 4. Peak current and coulomb capabilities of conventional "Size D" and "Size-E" ignitrons compared to data from the new NL-9000. (NL-9000 data are taken from manufacturer's data sheet but originated from tests at LLNL).

BACK-LIGHTED THYRATRONS

Much work has been done in Europe on the so-called "pseudo-spark" switch (Christiansen 1990). An optically triggered version of this switch is called the back-lighted thyratron (BLT) and has been investigated and developed in this country by Kirkman and Gundersen (1990).

The basic switch structure is shown in Fig. 5. Laser or unfocussed UV radiation on the back surface of the cathode initiates a uniform, superdense glow discharge. The cathode is self-heated by ion bombardment during the closing phase, enabling the switch to operate in a glow mode, even at very high currents and short risetimes. Table I compares some of its characteristics with that of a conventional hydrogen thyratron. The BLT has been triggered using light from various UV sources, coherent and incoherent, and also via an optical fiber. Low delay and jitter (78 and 0.4 ns, respectively) and switch lifetimes in excess of 10^5 shots have been obtained in an unoptimized prototype. As little as 10 μJ of optical trigger energy was used at a wavelength of 222 nm.

The BLT is being developed commercially by Integrated Applied Physics, Arcadia, CA and is being posed as a competitor to conventional hydrogen thyratrons, one of its main advantages in this regard being the absence of an externally heated cathode.

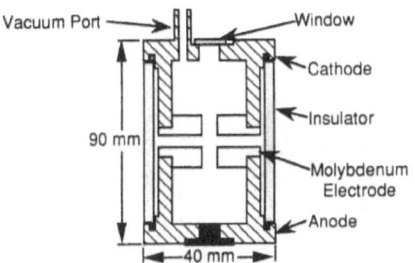

Fig. 5. BLT Switch structure. The switch is cylindrically symmetric with a central aperture, and the electrodes are fabricated from a refractory metal such as Mo or Ni. In this realization light enters through a window and is incident on the back surface of the cathode.

Table I

Comparison of High Power Hydrogen Thyratron and Pseudospark/BLT Devices that are Roughly the Same Size, Based on Data Obtained from Prototype Devices in Laboratory Experiments (Kirkman and Gundersen (1990))

Parameter	Hydrogen Thyratron	BLT/Pseudospark
Cathode	Externally Heated 50 A/cm^2	No External Heater 10,000 A/cm^2
Cathode Material	Oxide or Dispenser	Mo, Ni
Conduction Medium	Diffuse Glow	Diffuse Glow
Peak Current	10 kA	> 100 kA
dI/dt (A/sec)	2-10^{11}	> 10^{12}
Reverse Current	10%	$\approx$ 100%

327

REFERENCES

Bardsley, J.N. and Wadehra, J.M., 1982, private communication.

Bistritsky, V.M., Krasik, Ya.E., Ivanov, I.B., Mesyats, G.A., Ryzhakin, N.N., and Usov, Yu.P., 1990, Pulse Sharpening Using the Magnetic Core and Plasma Opening Switch, <u>1990 International Magnetic Pulse Compression Workshop</u>, Lake Tahoe, NV.

Buttram, M. and Sampayan, S. 1990, Repetitive Spark Gap Switches: G. Schaefer, ed., Gas Discharge Closing Switches, Vol. II, <u>in</u> "Advances in Pulsed Power Technology", A. Guenther and M. Kristiansen, eds., Plenum Press, New York and London.

Christiansen, J., 1990, The Pseudospark Switch, G. Schaefer, ed., Gas Discharge Closing Switches, Vol. II, <u>in</u> "Advances in Pulsed Power Technology", A. Guenther and M. Kristiansen, eds., Plenum Press, New York and London.

Christophorou, L.G., Hunter, S.R., Carter, J.G., and Mathis, R.A., 1982, Gases for Possible Use in Diffuse-Discharge Switches, <u>Appl. Phys. Lett.</u>, 41:147.

Commisso, R.J., Cooperstein, G., Meger, R.A., Neri, J.M., Ottinger, P.F., and Weber, B.V., 1987, The Plasma Erosion Opening Switch: A. Guenther, M. Kristiansen, and T. Martin, Vol. I, Opening Switches, <u>in</u> "Advances in Pulsed Power Technology", A. Guenther and M. Kristiansen, eds., Plenum Press, New York and London.

Donaldson, A.L., 1990, Electrode Erosion in High Current, High Energy Transient Arcs, Ph.D Thesis, Texas Tech University, Lubbock, TX.

Engel, T.G., Kristiansen, M., Smith, B.D., and Marx, J.N., 1989, High Performance Insulator Materials for High Current Switching Applications, <u>Proc. of 7th IEEE Pulsed Power Conference</u>, Monterey, CA, :336. IEEE Cat. No. 89CH2678-2.

Engel, T.G., Kristiansen, M., O'Hair, E., and Marx, J.N., 1990, Estimating the Erosion and Degradation Performance of Ceramic and Polymeric Insulator Materials in High Current Arc Environments, <u>5th Sypmosium on Electromagnetic Launch Technology</u>, Destin, FL.

Guenther, A., Kristiansen, M., and Martin, T., 1987, Opening Switches <u>in</u> "Advances in Pulsed Power Technology", A. Guenther and M. Kristiansen, eds., Plenum Press, New York and London.

Hammond, H.G., III, Frazier, G.b., Roth, I., Naff, J.T., and Sincerny, P, "High-Coulomb Spark Gap for Electric Guns", 5th Symposium on

Harrison, J., 1990, Private Commucication.

Hunter, R.O., 1976, Electron Beam Controlled Switching, <u>Proc. 1st IEEE International Pulsed Power Conference</u>, Lubbock, TX.

Hunter, S.R., Carter, J.G., Christophorou, J.L., and Lakdawala, V.K., 1984, Transport Properties and Dielectric Strengths of Gas Mixtures for Use in Diffuse Discharge Opening Switches, <u>Gaseous Dielectrics IV</u>, Pergamon Press, New York.

Kihara, R., 1987, Evaluation of Commercially Available Ignitrons as High-Current, High-Coulomb Transfer Switches, Proc. 6th IEEE Pulsed Power Conf. Arlington, VA, :581.

Kirkman, G. and Gundersen, M., 1990, The Back-Lighted Thyratron: G. Schafer, ed., Gas Discharge Closing Switches, Vol. II, <u>in</u> "Advances in Pulsed Power Technology", A. Guenther and M. Kristiansen, eds., Plenum Press, New York and London

Koval'chuk, B.M. and Mesyats, G.A., 1976a, Rapid Cutoff of a High Current in an Electron-Beam-Excited Discharge, <u>Sov. Tech. Phys. Lett.</u>, 2:252.

Koval'chuk, B.M. and Mesyats, G.A., 1976b, Current Breaker with Space Discharge Controlled by Electron Beam, <u>1st IEEE International Pulsed Power Conference</u>, Lubbock, TX.

Mesyats, G.A, Bugaev, S.P., Kim, A.A., Koval'chuk, B.M., and Kokshenov, V.A., 1987, Microsecond Plasma Opening Switches, <u>IEEE Transactions on Plasma Science</u>, PS-15:649.

Mesyats, G.A., 1989, Private Communication.

Peterkin, F.E. and Williams, P.F., 1989, Triggering of Trigatron Spark Gaps, <u>7th IEEE Pulsed Power Conference</u>, Monterey, CA, :559, IEEE Cat. No. 89CH2678-2.

Rei, D.J., Barnes, G.A., Gibble, R.J., Hinckley, J.E. Kreider, T.W. Waganaar, W.J., 1989, Design and Performance of the 10-kV, 5 MA Pulsed-Power System for the FRX-C Magnetic Compression Experiment, Los Alamos Report LA-11519-MS, UC-426.

Richardson Electronics Ltd., 1989, Tentative NL-9000 Data Sheet, National Electronics Div., La Fox, Illinois, 60147.

Rossi, M.J., Helm, H., and Lorents, D.C., 1985, Photoenhanced Electron Attachment in Vinylchloride and Trifluorethylene at 193 nm, <u>Appl. Phys. Lett.</u>, 47:576.

Schaefer, G. Schoenbach, K.H., Tran, P., Wang, J., and Guenther, A.H., 1983, Computer Calculations of the Time Dependent Behavior of Diffuse Discharge Switches, <u>Proc. 4th IEEE Pulsed Power Conference</u>, Albuquerque, NM, :714, IEEE Cat. No.83CH1908-3.

Schaefer, G., Giesselmann, M., Pashaie, B., and Kristiansen, M., 1988, CO_2-Laser Enhanced Electron Attachment in Externally Sustained Diffuse Gas Discharges Containing Vinyl Chloride, <u>J. Appl. Phys.</u>, 64:6123.

Schaefer, G, 1989, Basic Mechanisms and Applied Methods for Gas Discharge Control, Invited Paper, <u>Proc. International Conference on Phenomena in Ionized Gases</u>, University of Belgrade, Yugoslavia.

Schaefer, G., 1990, Gas Discharge Closing Switches <u>in</u> "Advances in Pulsed Power Technology", A. Guenther and M. Kristiansen, eds., Plenum Press, New York and London.

Schoenbach, K.H., Kristiansen, M., Kunhardt, E.E., Hatfield, L.L., and Guenther, A.H., 1981, Exploratory Concepts of Opening Switches, <u>Proc. ARO Workshop on Repetitive Opening Switches</u>, Tamarron, CO, :65, DTIC No. AD-A110770.

Schoenbach, K.H., Schaefer, G., Kristiansen, M., Hatfield, L.L., and Guenther, A.H., 1982, Concepts for Optical Control of Diffuse Opening Switches, <u>IEEE Trans. Plasma Sci.</u>, PS-10:246.

Schoenbach, K.H. and Schaefer, G., 1987, Diffuse Discharge Opening Switches: A. Guenther, M. Kristiansen, and T. Martin, Vol. I, Opening Switches <u>in</u> "Advances in Pulsed Power Technology", A. Guenther and M. Kristiansen, eds., Plenum Press, New York and London

Thayer, W.J., Lo, C.H., and Cousins, A.K., 1989, Evaluation of Innovative High Pulse Rate, Purged Spark Gap Concepts, <u>Proc. of 7th IEEE Pulsed Power Conference</u>, Monterey, CA, :548. IEEE Cat. No. 89CH2678-2.

Williams, P.F. and Guenther, A.H., 1990, Laser Triggering of Gas Filled Spark Gaps: A. Guenther, M. Kristiansen, and T. Martin, Vol. I, Opening Switches <u>in</u> "Advances in Pulsed Power Technology", A. Guenther and M. Kristiansen, eds., Plenum Press, New York and London.

Wilson, J.M. and Donovan, G.L., 1989, Improvements in Synchronization of the PBFA-II Accelerator with Laser-Triggered Gas Switches, <u>Proc. of 7th IEEE Pulsed Power Conference</u>, Monterey, CA, :197. IEEE Cat. No. 89CH2678-2.

Vitkovitsky, I., 1987, "High Power Switching", Van Nostrand Reinhold Co., New York.

DISCUSSION

F. SCHWIRZKE: What mechanism causes the steep increase in the erosion curve (Fig. 3)?

M. KRISTIANSEN: It is caused by total melting of the top surface of the electrode accompanied by ejection of molten metal (droplets).

R. T. WATERS: Is "ion starvation", which can cause sudden current chopping, a limiting factor in the current–carrying capacity of mercury–vapor ignitrons? A similar current chopping is reported in 10 kPa SF_6 discharges by Pfeiffer in the same session of this symposium. Does arc control by an external magnetic field have a useful role in limiting radial losses of charge carriers?

M. KRISTIANSEN: We have not seen any evidence of this in our work, where basically all data, except for some calibration shots, are taken at $I > 100$ kA.

R. J. GRIPSHOVER: At NSWC (Naval Surface Warfare Center) we have demonstrated 100 μs recovery of a high pressure (400 psi) hydrogen–filled undervolted spark gap operating at 50 kV and 170 kA for 5 μs pulses.

M. KRISTIANSEN: Yes, I referred to this work in my talk.

S. W. ROWE: The problem of erosion is a major one, not only for this type of switch but also for SF_6 circuit breakers. We use magnetic field induced arc rotation in order to constantly move the arc root and hence keep melting and droplet ejection to a reasonable level. Furthermore, contact materials used are similar, CuW being a frequently used one. Finally, any metal vapor generated is evacuated by careful design of gas flow permitting voltage withstand of tens of kV a few microseconds after current interruption.

M. KRISTIANSEN: I am not actively involved in circuit breaker design but it is my impression that the $\vec{J}x\vec{B}$ induced arc rotation is mostly used in vacuum interrupters. It is now also used or being investigated for high pressure spark gaps (Maxwell Laboratories, Inc.) and ignitrons (Texas Tech University). It should be pointed out, however, that the arc relation may not be as effective for some pulsed power applications where the conduction time may be in the submicrosecond range and the currents in the MA range. The μs recovery quoted is impressive if it pertains to gas filled, rather than vacuum, switches.

LASER TRIGGERING OF GAS FILLED SPARK GAPS

P.F. Williams

Department of Electrical Engineering
University of Nebraska-Lincoln
Lincoln, NE 68588-0511

INTRODUCTION

The laser-triggered spark gap was first reported by Pendleton and Guenther in 1965,[1] and has since been employed in a wide range of applications. The primary advantage laser-triggering offers over other triggering methods is precise synchronization of the gap closing time. High power (megavolt, megamp) gaps can be triggered with closure delays of ~10 ns, and R.M.S. jitter less than 1 ns. Further, under proper operating conditions, the closure delay is a weak function of the ratio of charging voltage to self-breakdown voltage, so that there is little drift in closure time with spark gap aging. Another advantage of laser-triggering is the essentially perfect electrical isolation of the trigger circuit from the switched high voltage circuit it provides.

Guenther and Bettis[2] have published several timely, comprehensive reviews of laser-triggered switching technology. In this paper I will concentrate on the physical mechanisms responsible for the laser triggering of undervolted spark gaps. In most laser-triggered gaps the laser beam enters the gap through a hole in the center of one electrode and traverses the gap along the gap axis. Two types of gaps are commonly used. In one type, first reported in 1965 by Guenther and Pendleton,[1] the laser beam is focused onto the surface of the opposite electrode where it produces a small plasma fireball. This fireball is responsible for triggering the gap. In the other type, first reported in 1980 by Rapoport, Goldhar, and Murray,[3] the laser is focussed to a point in the center of the gap, and may pass out of the gap through a hole in the opposite electrode. In this case the laser induces the optical breakdown of the gap fill gas, and the electrical breakdown of the gap is triggered by the occurrence of this optical breakdown. I will refer to these two types as *electrode-surface triggering*, and *volume triggering*, respectively. I will discuss both types, but will emphasize *electrode-surface triggering* because more is understood about the initial triggering mechanisms.

PHYSICAL TRIGGERING CONSIDERATIONS

A charged, undervolted spark gap in the hold-off (open) state is an example of a system in quasi-stable equilibrium. The system is stable to small perturbations such as thermal fluctuations in free electron density, but a "lower energy" state exists, and for a perturbation larger than some threshold value the gap will make the transition to this state, resulting in electrical breakdown, conduction, and switch closure. Triggering such a gap requires the introduction of a perturbation larger than this threshold, but useful triggering typically requires a perturbation substantially larger than this value to produce acceptable delay and jitter.

The physical basis for this threshold is of interest. Consider a slightly undervolted gap filled with a non-attaching gas such as N_2. The gap is stable to small perturbations such as the release of a few electrons near the cathode, even though the electron impact ionization avalanche gain is much larger than one. The reason for the stability is that the operative secondary emission processes at the cathode whereby these initial electrons reproduce themselves are very inefficient, and if the gap is undervolted the avalanche gain is not yet large enough to compensate. On the other hand, a self-sustaining glow discharge, once initiated, is possible at even lower applied voltages because of the creation of a space-charge-induced region of high field (the cathode fall) just in front of the cathode. This region greatly enhances the electron emission from the cathode, thereby reducing the avalanche gain required in the positive column for self-sustained operation. Apparently the size of the minimum perturbation needed to trigger breakdown is partly determined by the requirement that it be large enough to allow (either directly or through some transient intermediary mechanisms) the creation of such a region. After this point, a self-sustained glow-like discharge is assured. For typical spark gaps this discharge is usually unstable, and rapidly undergoes a glow-to-arc transition to close the switch.

For *electrode-surface triggered* gaps triggering occurs in several steps. The laser produces a small plasma fireball on one electrode, which initiates a streamer. This streamer propagates across the gap, bridging it with a free ionization density of $\sim 10^{15}$ cm^{-3}, well above the threshold for instituting a cathode fall. This transient glow-like discharge transforms rapidly into an arc. For *volume triggered* gaps, the laser directly produces a needle-shaped region of free ionization between the electrodes. This ionization produces very high fields in the regions between the needle and the electrodes, highly overvolting them. Breakdown of these regions occurs rapidly, and a transient glow-like discharge is probably instituted which soon transforms into an arc.

ELECTRODE-SURFACE TRIGGERING

Guenther and Bettis[4] first suggested that triggering in *electrode-surface triggered* gaps was due to the formation and propagation across the gap of a streamer, and they later presented evidence for the interaction of the propagating streamer with the laser. Their conclusions were supported by a large volume of experimental evidence showing the variation of closure delay and jitter as a function of experimental parameters such as charging voltage and laser power. More recently, Dougal and Williams[5][6] have published direct evidence for both this streamer mechanism, and the interaction between the propagating streamer and the laser.

As a result of this work, the following picture of the physical processes occurring in laser-triggered breakdown in this geometry has emerged. The focussed laser beam first produces a small plasma fireball on the struck electrode which serves as a seed to initiate a streamer. This streamer propagates across the gap, generally with laser assistance, and bridges the gap with a low conductivity channel of free ionization. This ionization density is well above the threshold for establishing a cathode fall region, and a filamentary, transient glow discharge is instituted. The conductivity of the channel then increases until a glow-to-arc transition occurs to form the spark channel and, finally, close the gap. I will briefly discuss each of these stages separately.

Plasma Fireball Formation

The focussed laser initiates a small plasma on the surface of the struck electrode, probably by desorption or evaporation and subsequent heating of material on the electrode surface. Once the plasma has formed it grows by direct heating of the gas at the plasma-ambient boundary, as demonstrated in Figure 1 which shows streak photographs of the formation and growth of this plasma.[5] The plasma surface moves with a speed of $\sim 2 \times 10^7$ cm/sec, much faster than any reasonable shock wave velocity, ruling out the possibility of growth through expansion of the super-heated gas inside the original fireball. Further, in the photos of Figs. 1b and 1c, secondary plasmas are seen to be formed above the initial plasma, presumably at hot spots in the focussed laser beam. These secondary plasmas are opaque to the incident laser, so that once formed, any plasma downstream will no longer be illuminated by the incident laser beam. The growth of the initial plasma ceased abruptly upon the appearance of the first secondary plasma, demonstrating clearly that the plasma growth is a direct result of interaction with the laser at the plasma boundary.

Streamer Formation and Propagation

After the formation of a volume of plasma on the struck electrode, the free electrons drift in the applied field, setting up a localized region of enhanced field, and launching a streamer. This streamer propagates across the gap (usually with assistance from the laser beam), bridging it with a resistive channel of free ionization. The importance of streamers to the triggering of breakdown in *electrode-surface triggered* gaps was first suggested by Guenther and Bettis.[4] Perhaps the most direct evidence linking streamers to the triggering mechanism in these gaps was obtained by Dougal and Williams,[5] who obtained high sensitivity streak photographs showing a very weak luminous front associated with the tip of a streamer propagating across the gap during the earliest stage of laser triggering.

Guenther and Bettis[4] were also the first to note the importance of the laser-streamer interaction in *electrode-surface triggered* gaps. They reported two operational regimes for their gap. For charging voltages above some critical value V_{CR}, (roughly 50% of the self-break voltage, V_{SB}) switching delay increased slowly with decreasing charging voltage, ranging from about 5 ns just below V_{SB} to about 8 ns at V_{CR};

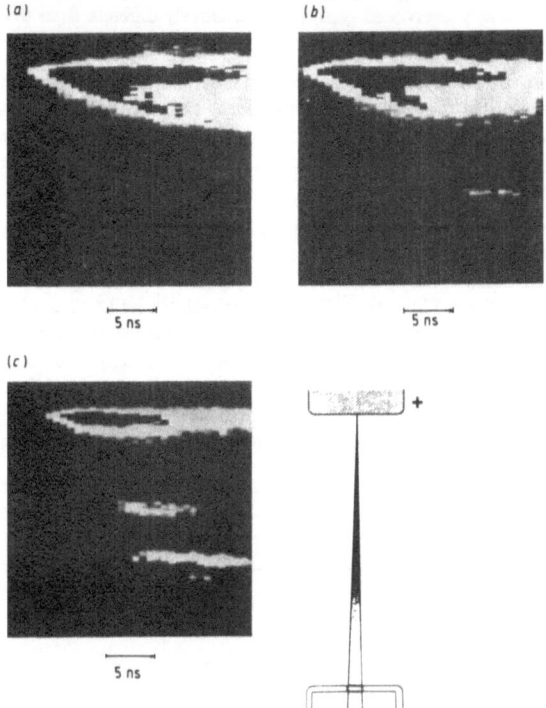

Fig. 1. Digitized streak photographs showing the formation of the plasma fireball produced by the triggering laser as it impinges on the struck electrode. The black islands centered in regions of high intensity are artifacts resulting from overloading the digitizer. Experimental conditions were: 800 Torr N_2; 100 mJ, 1.06 μm, 15 ns laser; brass electrodes, 5 mm gap spacing. Charging voltages were: (a) 54%; (b) 81%; (c) 98% of the static breakdown voltage, V_{SB}. The charging voltage in (a) was below the triggering threshold. [From Ref. 5]

whereas for voltages below this value delay increased rapidly with decreasing voltage, reaching ~40 ns at about 30% of V_{SB}. They suggested that a laser-streamer interaction was responsible for this behavior. As the charging voltage was decreased, the streamer propagation velocity slowed until at V_{CR} the laser pulse shut off before the streamer had finished crossing the gap. Unassisted propagation was then required to complete the bridging of the gap. It was assumed that the streamer propagates much slower without the laser interaction than with it, leading to the observed rapid increase in delay with decreasing charging voltage below V_{CR}.

This view was given strong support by open shutter photographs published later by the same authors showing the arc formed when the gap was 1) laser-triggered, and 2) when it was electrically triggered.[7] In the laser-triggered case, there was one straight channel (except near the electrode through which the laser entered the gap); whereas in the electrically-triggered case there were several crooked channels. The obvious interpretation of these photos is that in the laser-triggered case, the laser interacted with some stage of the nascent arc channel, guiding it. One electrode of the gap had a hole in it to allow injection of the laser beam into the gap region, and the arc was forced to deviate from the laser path in order to complete the circuit. In this region, channel growth was unguided, and the path therefore more erratic.

Since that time, Vaill et al.,[8] Bradley,[9] Koopman and Saum,[10] and Aleksandrov et al.[11] have shown, in experiments not directly related to laser-triggered gaps, that a laser beam can guide streamer propagation. Dougal and Williams[5][6] have published evidence showing directly the interaction between the laser and streamer in laser-triggered gaps. They described the results of a set of experiments conducted with a gap in which a small recess was drilled into the center of the struck electrode. When the laser was

333

focussed at the bottom of this electrode the plasma thus formed was initially in a nearly-field-free region, and did not initially form a streamer. After about 40 ns the surface of this plasma expanded to emerge from the recess and to experience the full applied field. At this point a streamer was probably formed, but the laser pulse had ended so that the propagation across the gap was without laser assistance. It was found that oscillograms of the gap current and streak photographs of the breakdown in this case were very similar to those found for untriggered, weakly overvolted gaps, and qualitatively different from the corresponding traces for conventionally-triggered breakdown.

Further direct evidence for the laser-streamer interaction is seen in the streak photographs of Fig. 1. Secondary plasmas form at isolated points in the focussed laser beam only when the applied voltage is sufficient to allow triggered breakdown. It is evident that the laser interacts with some product of the initial breakdown process which propagates toward the cathode at a speed of at least 5×10^7 cm/sec. A cathode-directed streamer is the only likely candidate.

It is clear that the properties of the streamer play an important role in determining the switching characteristics of a laser-triggered spark gap. Although the streamer mechanism was first suggested by Raether,[12] and by Loeb and Meek[13] in the late 1930's, the quantitative description of a streamer has remained elusive. Recently Dhali and Williams,[14] Wu and Kunhardt[15] and Dhali and Pal[16] have made significant progress in this direction by applying flux-corrected transport techniques to the numerical simulation of streamer formation and propagation. These calculations are fully two-dimensional (three-dimensional with cylindrical symmetry assumed), and take streamers to be solutions of the standard hydrodynamic particle conservation equations with the Townsend ionization process and other electron and ion creation and loss processes included. Figure 2 shows typical results for overvolted cathode-directed streamers.[14]

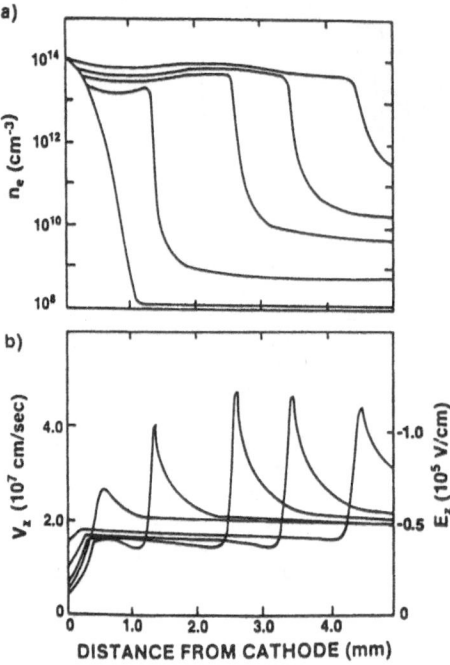

Fig. 2. Plots showing (a) on-axis electron density and (b) electric field and drift velocity for an anode-directed streamer in atmospheric pressure N_2. The curves correspond to times of t = 0.1, 1.0, 2.0, 2.5, and 3.0 ns after the start of the simulation. [From Ref. 14].

Although streamer behavior is complex, and not easily given to broad generalizations, several observations have emerged from this work. Making use of the effective field approximation developed by Raizer,[17] preliminary estimates of the effect of the interaction between the laser and the propagating streamer head were carried out by Dhali.[18] For conditions similar to those present in laser-triggered gaps, these calculations demonstrate that this interaction is an important effect in aiding the propagation of under-volted streamers, and clearly support the role of the laser-streamer interaction in laser triggering of spark gaps.

After the streamer has traversed the gap, there is a thin column of weakly ionized plasma connecting the electrodes. Assuming an ionization density of 10^{15} cm^{-3} and a cross sectional area of the channel of 10^{-4} cm^2, for atmospheric pressure N$_2$ the resistance is of the order of 100 kΩ per cm of channel length. For most applications the gap would be considered to be still open at this stage, and closure must await the creation of additional ionization to reduce the switch resistance. The creation of this additional ionization is responsible for a substantial portion of the closure delay and jitter of the switch, and also determines in part the current risetime of the switch.

The physical mechanisms operating during this phase are not as well understood as those in the earlier phases. Several mechanisms may be responsible for initially increasing the ionization density. For initial voltages within 10 or 20% of V_{SB} the Townsend coefficient is still appreciable, and some growth occurs due to this process. The exponential growth time constant for N$_2$ at 90% V_{SB} is about 6 ns, for example.[19] Cascade ionization processes may also be important.

Marode et al.[20] have proposed a different mechanism to explain ionization growth during this phase of over-volted point-plane breakdown. The mechanism they propose relies on heating of the translational motions of the neutral species of the gas, raising the kinetic temperature. This process decreases the gas density, and therefore increases E/N for a fixed E. Since the efficiencies of most excitation mechanisms increase rapidly with increasing values of E/N, this true heating effect leads to a runaway process and rapid increase of ionization. This mechanism requires the physical motion of gas molecules, and may not, therefore, be able to explain the very fast breakdown times observed in laser-triggered switching under some conditions.

The laser-plasma interaction is an important mechanism for laser-triggered gaps. As demonstrated in Fig. 1, the focussed laser interacts strongly with free electrons, rapidly creating brightly luminous plasmas. This observation is consistent with estimates of the strength of the interaction based on the effective field approximation. For example, for a typical Nd:YAG laser pulse used in laser triggering ($\lambda = 1.06$ μm, 50 mJ in a 10 ns pulse, focussed to a spot 0.01 cm in diameter) the effective field in 760 Torr N$_2$ is approximately 60 kV/cm.[17][21] This field corresponds to an exponential growth time of about 0.2 ns. The effective field falls quadratically with distance from the focus. For a 1 cm gap with a 20 cm focal length lens and a 1 cm beam diameter at the lens the effective field from the laser would fall to 2.4 kV/cm at the beam entrance electrode. This field adds in quadrature with the applied D.C. field. Even well away from the laser focus, the laser is expected to play a significant role in the evolution of the nascent spark channel.

The free electron density in the spark channel of *electrode-surface triggered* gaps has been measured by Crumley et al.,[22] Dhali et al.,[23] and by Najafzadeh et al.[24] Crumley et al.[22] and Dhali et al.[23] measured electron densities in a 6.5 mm gap filled with 400 Torr of H$_2$, using time and space resolved spectral measurements of the Stark broadening of the H$_\beta$ line. The gap was charged through a 50 Ω resistor to 5 kV, and it was triggered with 10 ns, 4 mJ pulses of 336.1 nm radiation from a N$_2$ laser. The electrical pulse width was 1.0 μs, and the peak electron density was about 10^{17} cm^{-3}, measured 0.1 μs after triggering. Najafzadeh et al.[24] measured electron densities in atmospheric air using an interferometric technique in an 0.5 mm gap charged to about 3 kV with a low-inductance capacitor system and triggered with a N$_2$ laser. A peak electron density of 6×10^{19} cm^{-3} was found, occurring 8 ns after triggering. The reasons for the differences between the densities measured by the two groups are not clear, but may be related to the different peak currents.

VOLUME TRIGGERING

A spark gap can be volume triggered by introducing the laser beam into the gap in either a longitudinal or a transverse geometry. I will discuss here only triggering with the longitudinal geometry because the transverse triggering mode is not very well developed. Taylor and Leopold[25] have described transverse triggering of a low-impedance rail gap using volume triggering.

Rapaport, Goldhar, and Murray[3] first reported the volume triggering of a spark gap in the axial geometry, and the technique was developed by a group at Sandia Laboratories.[26-29] The later stages of gap closure have been studied in extensive detail by a group at Mathematical Sciences Northwest.[30][31] For this technique, triggering requires the production of substantial free ionization along the axis of the spark gap through the optically-induced breakdown of the fill gas. Gap triggering results from the formation of this relatively high conductivity channel in the middle of the gap. Dielectric relaxation inside this channel causes rapid shielding of the interior of the channel from the applied field. Until the gap voltage collapses, this shielding of high conductivity regions unavoidably results in enhancement of the field in regions of lower conductivity, such as those away from the laser focus, or regions not illuminated directly by the laser, such as the short path between the laser beam and the edge of the hole in the entrance electrode. With sufficient field enhancement in these regions, electrical breakdown proceeds rapidly, bringing the ionization density up to a value similar to that in the originally higher conductivity regions. At this point, the gap is well past the early ionization growth regime, and breakdown proceeds, producing finally a glow-to-arc transition.

The physical mechanisms responsible for the initial optical breakdown are only partially understood. Woodworth et al.[28] found that the jitter in closure time was several ns when triggering with the 1.06 μm output of a Q-switched Nd:YAG laser, whereas the jitter was less than 1 ns when the 248 nm output of a KrF excimer laser was used. Similar low jitters were reported with the frequency-quadrupled output of the Nd:YAG laser at 266 nm. Dougal[32] found that the jitter in this geometry, using the 1.06 μm Nd:YAG laser output, when measured from the time of arrival of the laser pulse at the spark gap was several ns, in agreement with Woodworth et al.[28] When measured from the first appearance of emission from the optical breakdown from the region of the laser focus, however, the jitter was too small to be measured (< 1 ns). Recently Yoshida at al.[33] have published the results of a detailed study of the effect of U.V. laser preionization on CO_2-laser-induced optical breakdown in helium. They find that the U.V. irradiation plays an important role in providing the initial electrons needed for the inverse bremsstrahlung heating process driven by the 10.6 μm CO_2 radiation. Without U.V. pre-irradiation the delay and jitter in the optical breakdown increased substantially, but the characteristics of the final discharge were otherwise little affected.

These results are all consistent and clearly indicate that low jitter triggering in this geometry requires a two-step optical breakdown process. First, weak photoionization of the fill gas must occur, followed by a rapid ionization growth process. This latter process is probably inverse bremsstrahlung heating, a cascade heating process in which the free electrons are directly driven by the laser field and impart energy to the neutral atoms and molecules through frequent inelastic collisions. The details of the first process, however, remain unclear. Good results are obtained with KrF laser triggering at 248 nm in nominally pure SF_6 or SF_6/N_2 mixtures. Several laser photons are required to ionize either molecule, implying that in the absence of an accidental near resonance with a molecular energy level, multiphoton ionization should be a relatively weak process. Direct photoionization of some impurity in the fill gas seems more likely, but little experimental information is available on the point. Woodworth, Frost, and Green report, however, that the intentional introduction of small quantities of additives, chosen to have an ionization potential less than the energy of a laser photon to an otherwise pure N_2 fill produced triggering characteristics little changed from those of a gap with a fill of "pure" N_2, and inferior to those of a gap with a "pure" SF_6 fill.[26]

After the appearance of this initial free ionization produced by the laser, the field in the gap redistributes itself, with the field decreasing in the regions of higher conductivity, and increasing in the lower conductivity regions. Further ionization amplification occurs in the high field regions, either through direct electron impact ionization, or through a cascade process, until the ionization becomes more nearly axially uniform. At this point, if the gap voltage has not yet collapsed, further excitation occurs roughly uniformly along the axis, increasing the free ionization density, and resulting in the formation of a thermalized arc channel. The results of experiments reported by Woodworth, Adams, and Frost[27] are consistent with this model. They found that the delay to gap closure was a function of the axial length of the optical spark, with the delay decreasing as the optical spark was made longer. This model predicts this result, since the magnitude of the field in the low ionization regions, such as the short connecting paths between the region ionized by the laser and the edges of the holes in the electrodes needed for optical access, increases with increasing length of the high conductivity volume. Also consistent with the model are the results of experiments in which it was found that the delay decreased as the laser beam was moved off the axis of the entrance and exit holes and closer to the edges of the holes.

The group at Mathematical Sciences Northwest carried out an extensive study of the arc formation stage of switch closure in these gaps. This work is an excellent example of the value of a research program involving combined experimental and theoretical approaches.[30][31] The work also demonstrates the utility of numerical calculations in determining the important mechanisms causing switch behavior. In the experimental aspect of this program,[30] the neutral and the electron densities in a nascent laser-triggered arc channel were determined with excellent spatial and temporal resolution using an interferometric technique. The results clearly show that the arc channel starts as a narrow (~50 μm) channel of plasma and then grows primarily by hydrodynamic means to reach a diameter of about 1 mm in ~100 ns. Growth occurs due to expansion of the strongly heated core, forming after a short time a relatively low density core surrounded by a high density shell. The presence of this shell tends to confine the central plasma, limiting its expansion velocity to hydrodynamic values. Electron densities of the order of 10^{19} cm^{-3} are found in the central core, corresponding to essentially complete ionization of the gas.

Combined with this experimental effort was a numerical modeling program.[31] The model was one-dimensional, treating only the radial expansion of the arc column, and consisted of a fairly complete set of continuity equations for the free electrons and important molecular and atomic species. The model clearly demonstrated the formation of a shock front at the interface between the hot central arc core and the ambient gas, and it showed the role of this shock front in confining the core, both through the inertia of the shell and through the opacity of the shell to ionizing radiation from the core. The model further showed that there are three primary mechanisms for radial growth of the arc channel: 1) hydrodynamic expansion of the hot core; 2) electron avalanche growth initiated by photoionization and occurring primarily in the shell; and 3) electron avalanche growth initiated by thermal ionization by the hot neutral species in the shell and at the interface with the cool ambient gas.

Since the plasma in the hot central core is essentially fully ionized, little further reduction in switch resistance can be obtained though more heating. The model showed that after an initial period of rapid decrease in resistance, further decrease is primarily due to the radial expansion of the arc column. Although all three mechanisms for radial growth listed above are important, the expansion velocity is limited by the first. The authors therefore suggested that faster closure, with lower on-state resistance, will be obtained with a fill gas consisting of molecules with the lowest average atomic weight, all other factors being equal. They carried out experiments to test this prediction and found good agreement.

REFERENCES

1. W.K. Pendleton and A.H. Guenther, Investigation of a Laser Triggered Spark Gap, Rev. Sci. Inst., **36**, 62 (1965).

2. A.H. Guenther and J.R. Bettis, A Review of Laser-Triggered Switching, Proc. IEEE QE-3, 581 (1971); The Laser Triggering of High Voltage Switches, J. Phys. D **11**, 1577 (1978); Recent Advances in Optically Controlled Discharges, in *Digest of Technical Papers, 5th IEEE Pulsed Power Conference*, M.F. Rose and P.J. Turchi eds., (IEEE, New York, 1987), pp. 47-53.

3. W.R. Rapoport, J. Goldhar, and J.R. Murray, KrF Laser-Triggered SF_6 Spark Gap for Low Jitter Timing, IEEE Trans. Plasma Sci., **PS-8**, 167 (1980).

4. A.H. Guenther and J.R. Bettis, Laser Triggered Megavolt Switching, IEEE J. Quantum Elect. QE-3, 581 (1967).

5. R.A. Dougal and P.F. Williams, Fundamental Processes in Laser-Triggered Electrical Breakdown of Gases, J. Phys. D **17**, 903 (1984).

6. R.A. Dougal and P.F. Williams, Fundamental Processes in the Laser-Triggered Electrical Breakdown of Gases: Unconventional Geometries, J. Appl. Phys. **60**, 4240 (1986).

7. J.R. Bettis and A.H. Guenther, Subnanosecond-Jitter Laser-Triggered Switching at Moderate Repetition Rates, IEEE J. Quantum Elect. QE-6, 483 (1970).

8. J.R. Vaill, D.A. Tidman, T.D. Wilkerson, and D.W. Koopman, Propagation of High-Voltage Streamers Along Laser-Induced Ionization Trails, Appl. Phys. Lett. **17**, 20 (1970).

9. L.P. Bradley, Preionization Control of Streamer Propagation, J. Appl. Phys. **43**, 886 (1972).

10. D.W. Koopman and K.A. Saum, Formation and Guiding of High-Velocity Electrical Streamers by Laser-Induced Ionization, J. Appl. Phys. **44**, 5328 (1973).

11. G.N. Aleksandrov, V.L. Ivanov, G.P. Kadzov, V.A. Parfenov, L.N. Pakomov, V.Y. Petrun'kin, V.A. Podlevskii, and Y.G. Seleznev, Effect of a Laser-Induced Ionization Channel in a Long Discharge in Air, Sov. Phys. Tech. Phys. **22**, 1233 (1977). [Russian original in Zh. Tekh. Fiz. **47**, 2122 (1977)].

12. H. Raether, Die Entwicklung der Elektronenlawine in den Funkenkanal, Z. Phys. **112**, 464 (1939).

13. L.B. Loeb and J.M. Meek, The Mechanism of Spark Discharges in Air at Atmospheric Pressure, I, J. Appl. Phys. **11**, 438 (1940).

14. S.K. Dhali and P.F. Williams, Two-Dimensional Studies of Streamers in Gases, J. Appl. Phys. **62**, 4696 (1987).

15. C. Wu and E.E. Kunhardt, Formation and Propagation of Streamers in N_2 and N_2-SF_6 mixtures, Phys. Rev. A **37**, 4396 (1988).

16. S.K. Dhali and A.K. Pal, Numerical Simulation of Streamer in SF_6, J. Appl. Phys. **63**, 1355 (1988).

17. Y.P. Raizer, Laser-Induced Discharge Phenomena, Consultants Bureau, New York (1977).

18. S.K. Dhali, *Space-Charge Dominated Phenomena in Electrical Breakdown of Gases*, PhD. Dissertation (Texas Tech University, Lubbock, 1984).

19. J. Dutton, A Survey of Electron Swarm Data, J. Phys. Chem. Ref. Data **4**, 577 (1975).

20. E. Marode, F. Bastien, and M. Bakker, A Model of the Streamer-Induced Spark Formation Based on Neutral Dynamics, J. Appl. Phys. **50**, 140 (1979).

21. Y.P. Raizer, Optical Discharges, J. de Physique, **40**, C7-141 (1979).

22. R.J. Crumley, P.F. Williams, M.A. Gundersen, and A. Watson, Electron Densities in Laser-Triggered Discharges, in *Digest of Technical Papers, 2nd IEEE International Pulsed Power Conference,"* A.H. Guenther and M. Kristiansen eds., (IEEE, New York, 1979), pp. 119-121.

23. S.K. Dhali, P.F. Williams, R.J. Crumley, and M.A. Gundersen, Electron Densities in Laser-Triggered Hydrogen Sparks, IEEE Trans. Plasma Science **PS-8**, 164 (1980).

24. R. Najafzaden, E.E. Bergmann, and R.J. Emrich, Schlieren and Interferometric Study of a Laser Triggered Air Spark in the Nanosecond Regime, J. Appl. Phys. **62**, 2261 (1987).

25. R.S. Taylor and K.E. Leopold, U.V. Radiation-Triggered Rail Gap Switches, Rev. Sci. Inst. **55**, 52 (1984).

26. J.R. Woodworth, C.A. Frost, and T.A. Green, U.V. Laser Triggering of High-Voltage Gas Switches, J. Appl. Phys. **53**, 4734 (1982).

27. J.R. Woodworth, R.G. Adams, and C.A. Frost, U.V.-Laser Triggering of 2.8-Megavolt Gas Switches, IEEE Trans. Plasma Sci. **PS-10**, 257 (1982).

28. J.R. Woodworth, P.J. Hargis, L.C. Pitchford, and R.A. Hamil, Laser Triggering of a 500 kV Gas-Filled Switch: A Parametric Study, J. Appl. Phys. **56**, 1382 (1984).

29. C.A. Frost, J.R. Woodworth, J.N. Olsen, and T.A. Green, Plasma Channel Formation with Ultraviolet Lasers, Appl. Phys. Lett. **41**, 813 (1982).

30. W.D. Kimura, E.A. Crawford, and M.J. Kushner, Investigation of Laser Preionization Triggered High Power Switches using Interferometric Techniques, in *Conference Record of 1984 Sixteenth Power Modulator Symposium*, (IEEE, New York, 1984); W.D. Kimura, M.J. Kushner, E.A. Crawford, and S.R. Byron, Laser Interferometric Measurements of a Laser Preionization Triggered Spark Column, IEEE Trans. Plasma Sci. **PS-14**, 246 (1986); and W.D. Kimura, M.J. Kushner, and J.F. Seamans, Characteristics of a Laser Triggered Spark gap Using Air, Ar, CH_4, H_2, He, N_2, SF_6, and Xe, J. Appl. Phys. **63**, 1882 (1988).

31. M.J. Kushner, R.D. Milroy, and W.D. Kimura, A Laser-Triggered Spark Gap Model, J. Appl. Phys. **58**, 2988 (1985); M.J. Kushner, W.D. Kimura, and S.R. Byron, Arc Resistance of Laser-Triggered Spark Gaps, J. Appl. Phys. **58**, 1744 (1985); and M.J. Kushner, W.D. Kimura, D.H. Ford, and S.R. Byron, Dual Channel Formation in a Laser-Triggered Spark Gap, J. Appl. Phys. **58**, 4015 (1985).

32. R.A. Dougal, *Breakdown Processes in Laser Triggered Switching*, Ph.D. Dissertation, (Texas Tech University, Lubbock, 1983).

33. S. Yoshida, J. Sasaki, Y. Arai, and T. Uchiyama, Effect of U.V. Laser Preionization on CO_2-Laser-Induced Optical Breakdown, J. Appl. Phys. **58**, 4003 (1985).

338

DISCUSSION

E. E. KUNHARDT: What is happening between the laser and the streamer?

P. F. WILLIAMS: The interaction is probably inverse bremsstrahlung.

J. R. ROBINS: Could you discuss the different effects that depend on the laser wavelength?

P. F. WILLIAMS: For the triggering method in which the laser strikes a gap electrode, an infrared laser provides better triggering than does ultraviolet. This observation is consistent with the conclusion that the laser and streamer interact via an inverse bremsstrahlung mechanism. For the other principal triggering method in which the laser induces optical breakdown of the gas in the gap, an ultraviolet laser provides better triggering. I believe that the reason for this is due to an "initial electron" effect. An initial electron is required to initiate the optical breakdown, and ultraviolet light would be more efficient at inducing multiphoton ionization than would infrared light. This explanation is consistent with experimental results.

E. MARODE: What is the thickness of the streamer?

P. F. WILLIAMS: In a laser–triggered spark gap, I do not know. In a trigatron spark gap streamers are produced, but with no laser interaction, and these typically have a diameter of 1–2 mm. In a laser–triggered gap, I would expect the interaction with the focussed laser beam to cause the streamer shape to tend to conform with the dimensions of the beam in the gap. In most cases, this would imply a more narrow streamer channel than without the laser interaction.

A. H. GUENTHER: For a laser triggered gap, from interferograms, it is a fraction of a mm.

P. F. WILLIAMS: I believe these results are for the nascent arc channel, after some heating and ionization growth have occurred. It is possible that the current channel may have constricted somewhat as a result of these processes.

A NEW APPROACH FOR AN OPENING SWITCH FOR REPETITIVE OPERATIONS

W. Pfeiffer and D. Stolz

TH Darmstadt; Fachgebiet Elektrische Meßtechnik

6100 Darmstadt, Landgraf-Georg-Str. 4; FRG

ABSTRACT

The aim of the investigations described in this paper is the realization of an opening switch for repetitive operation based on a spark gap. SF_6 and SF_6-N_2-mixtures at low pressure in the range of 10 kPa were used. For these conditions an independent quenching of the current flow without external control could be obtained. Further investigations show that an admixture of nitrogen to SF_6 allow to influence the duration of the current flow. This effect seems to be very promising for the realization of an opening switch for repetitive operation. A repetition rate of 1 MHz and a breaking capacity of 100 MW seems to be possible at least for short-time operations.

INTRODUCTION

For many pulsed power applications inductive energy storage is very attractive, because of the high energy density as for instance described by Schoenbach et al[1]. The key problem of inductive energy pulse systems is the development of an fast and repetitive opening switch. The main demands to realize an opening switch for repetitive operation are summarized by Honig[3] :

1. The possibility to reclose on command for repetitive operation
2. Full recovery of the dielectric strength

Schaefer[4] has shown different approaches to realize an opening switch. These methods are based on external control[2,5]. The following described investigations demonstrate a special discharge type with a self quenching of the current, without any external influence.

EXPERIMENTAL SETUP

For the described investigations a high voltage, high power pulse generator was used. The pulse generator delivers square wave pulses with a maximum amplitude of 250 kV, a rise time of 3 ns and a pulse duration of 400 ns. This generator can also provide two or multiple square wave voltage pulses with a pulse duration of 400 ns each and an impulse pause of the same magnitude[6]. Furthermore optical investigations of the discharge development were performed by using a high speed photographic system. This system is based on a gated image intensifier tube, a low light level video camera and a digital frame store. With this system an exposure time of 1 ns, a luminous gain of 10^7 and a spatial resolution of 8 Lp/mm is realized[7].

EXPERIMENTAL RESULTS

In figure 1 the current course of the special discharge type is demonstrated in comparison with the typical current course of a high pressure discharge. This special behaviour of the current was detected

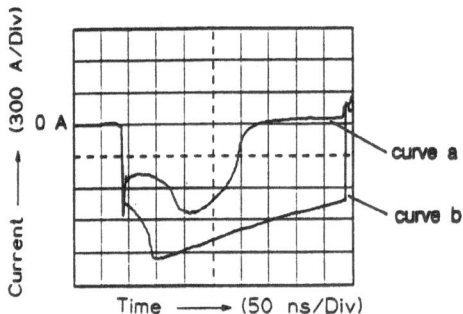

Fig. 1. Current course at the testgap for a breakdown voltage of 89.8 kV, curve a: Special Discharge type (p=10 kPa), curve b: Normal behaviour of the current (p=50 kPa).

both in the low pressure range of 10 kPa and for high over-voltages. For further analysis of the discharge phenomena optical investigations were performed. The exposure time T_{EXP} is measured from the first detectable current flow, which is in the range of 0.3 A. V_L marks the luminous gain of the optical diagnostic system as a relative value[8].

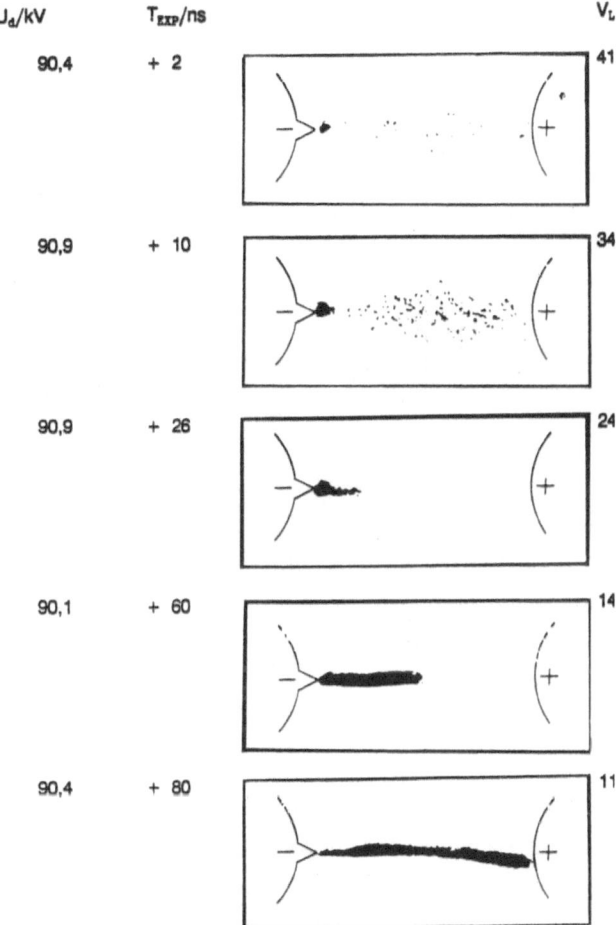

U_d/kV	T_{EXP}/ns		V_L
90,4	+ 2		41
90,9	+ 10		34
90,9	+ 26		24
90,1	+ 60		14
90,4	+ 80		11

Fig. 2.   Discharge development in pure SF_6 at a pressure of 10 kPa.

The optical investigation of the discharge development shows, starting with a diffuse luminous phenomenon, a very intense luminous spot in front of the cathode at an early moment of exposure (T_{EXP}=+2ns). This luminous spot at the cathode is the starting point of discharge channel towards the anode. For later moments of exposure a significant increase of the intensity of the discharge is detected. At the exposure time of $T_{EXP}=+60ns$ the discharge channel bridges nearly half of the gap distance. After additional 20 ns a continuous discharge channel is formed. Obviously the growth velocity of luminous phenomenon changes its speed. Simultanuously the discharge type changes from a diffuse discharge to a anode directed channel discharge. If the polarity of the voltage is

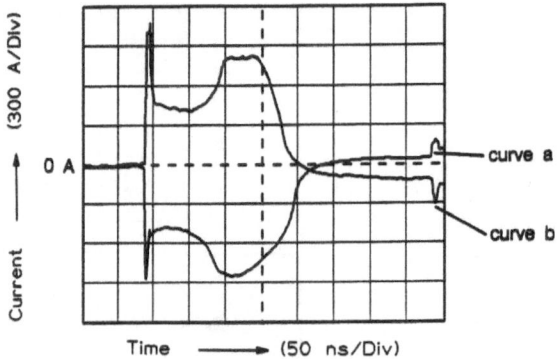

Fig. 3. Current Course of the special discharge type for negative (curve a) and positive (curve b) polarity of the overvoltage.

changed no significant influence on the behaviour of the current course is found, as demonstrated by figure 3.

Of particular interest with respect to the investigated discharge type is the admixture of nitrogen to SF_6. The percentage of the nitrogen admixture allows to influence the duration of the current flow as shown by figure 4. Thereby it is possible to vary the duration of the current flow in the range of 100 to 200 ns. If the discharge development of the SF_6-N_2 mixture is investigated no principle difference in the discharge development in comparison to pure SF_6 is detected.

CONCLUSIONS

The investigations describe a special discharge type which may be the base for an opening switch with high repetition rate. The measured current course showed a self quenching behaviour. The discharge type was found in the low pressure region of ≈10 kPa in connection with high overvoltage. The polarity of the voltage has no influence on this effect. A very

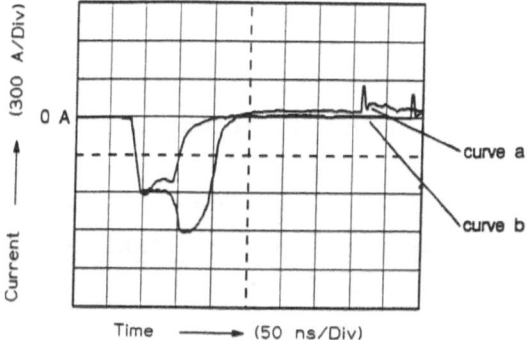

Fig. 4· Current course of the special discharge type for two different mixtures of SF_6 and N_2 (curve a: 25% SF_6-75% N_2; T_I=120 ns and curve b: 25% N_2-75% SF_6; T_I=160 ns) p=15 kPa and U_d=90.6 kV.

interesting phenomenon is detected, if a mixture of SF_6 and N_2 is investigated. With the admixture ratio of nitrogen the duration of the current flow can be varied. The discharge starts with a diffuse discharge and turns over in a directed channel discharge. During the development of the directed channel discharge the space charge in the discharge head causes an electric field of opposite polarity in front of the cathode, which has a screening effect. This field influences the electron dispensation process of the cathode, so that the discharge development is slowed down. For the self quenching behaviour a metastable intermediate stage of the SF_6 molecule is the main reason. During this intermediate stage electrons can be captured very easily , this results in a local charge carrier depletion.

ACKNOWLEDGEMENT

This research programm is sponsored by the German ministry of research and technology (under contract No. 13N5388/9). The authors are responsible for the contents.

REFERENCES

1. K. H. Schoenbach and M. Kristiansen, Diffuse Discharges And Opening Switches - A Review Of The Tamarron Workshops, in: 4th IEEE Pulsed Power Conference, Albuquerque/NM : 26 (1983).

2. G. Schaefer and K.H. Schoenbach, External Control Of Diffuse Discharge Switches, in: 5th IEEE Pulsed Power Conference; Arlington/VA :644 (1985).

3. E. M. Honig, Inductive Energy Storage, in: "Pulsed Power Short Course", Lublock/TX (1989).

4. G. Schaefer and K.H. Schoenbach, A Review Of Diffuse Discharge Opening Switches, IEEE Trans. On Plasma Science Vol. PS-14 56 : 574 (1986).

5. S. R. Hunter, J.G. Carter and L.G. Christophorou, Electron Transport Studies Of Gas Mixtures For Use In E-Beam Controlled Diffuse Discharge Switches, J. Appl. Phys. 58 (8): 3001 (1985).

6. M. Giesselman, W. Pfeiffer and J. Wolf, Short Time Optical And Electrical Diagnostics Os Pulsed N_2 and SF_6 Discharges, in: 6th IEEE Pulsed Power Conference :182, Arlington/VA (1987).

7. B. Lieberoth-Lenden, "Über ein Kurzzeitkammerasystem für den Subnanosekundenbereich mit einem getastetn Mikrokanal-platten-Bildverstärker und dessen Anwendung in komprimierten Isoliergasen", Fortschritt. -Bericht VDI Reihe 21 Nr. 12, VdI Verlag, Düsseldorf (1987).

8. W. Pfeiffer, D. Stolz and J. Wolf, Discharge Development And Formative Time Lags For Non-Uniform Fields in SF_6 and SF_6-N_2 Gas Mixtures, in: 9th Intern. Conf. On Gas discharges And Their Applications :315, Venezia/Italy (1988).

DISCUSSION

F. SCHWIRZKE: Did you take pictures of the plasma and the plasma channel at the time of quenching of the current?

W. PFEIFFER: This was not performed until now for the self–quenching discharge type. However, from other experiments during the recovery phase of pulsed discharges it is known that the after glow is rather long. For instance, during the 400 ns following a high current discharge channel the luminosity is not reduced greatly. Therefore, it is doubtful if useful information can be obtained from such experiments.

O. FARISH: Did you measure the voltage accross the gap? If so what was the waveshape?

W. PFEIFFER: We measured the voltage just in front of the gap related to the traveling wave and we measured the voltage behind the gap. There were two probes, one just before the gap, and one just after the gap.

S. W. ROWE: What is the origin of the switchover from diffuse to concentrated conduction mode? What is the contribution of the electrodes?

W. PFEIFFER: The reason for the transition from the diffuse discharge to a directed discharge with a narrow channel of high conductivity is the inherent instability of high pressure diffuse discharges. To some degree this can be influenced by a large volume preionization of the gap. However, the final transition seems to be only a matter of time.

E. E. KUNHARDT: The development to the state of high current is typical for electronegative gases in transmission line gaps. The voltage across the gap (not the cable charging or incident pulse voltage) after breakdown (i.e. first voltage transition) is determined by the critical field for the particular gas. After the gas heats, there is a second transition to the final conducting state where the voltage across the gap is a minimum. The quenching of the current, which is pressure dependent, may be similar to what happens in thyratrons when the limiting current is reached. This current is determined by the size of the grid hole and the heating of the gas leading to a lowering of the density in the channel and thus the available number of electrons that can carry the current.

I. GALLIMBERTI: If the voltage across the gap remains at about 200 kV, with a current flow of about 1000A, you should conclude that you have a transient discharge resistance of about 200 Ω, which could not correspond to a full conducting state of the channel. Can you comment on that?

W. PFEIFFER: The data in this paper are a bit different: an applied impulse voltage of $\sim$ 90 kV corresponding to a peak discharge current of $\sim$ 800A. The resulting impedance of 110 Ω corresponds rather precisely to twice the characteristic impedance of the coaxial lines being used. The impedance of the discharge channel therefore is on the order of 10 Ω or less.

A MACROSCOPIC GAS BREAKDOWN RELATIONSHIP*

Thomas H. Martin

Research Scientist of Pulsed Power
Sandia National Laboratories
Albuquerque, N.M.

INTRODUCTION

The design and construction of accelerators in the 100-terawatt class
have significantly increased the demands placed on gas and liquid
switching.[1] The 100-terawatt PBFA II (Particle Beam Fusion Accelerator)
uses 36 separate modules. Each module includes a Marx generator, a water-
insulated energy store, a laser-triggered 5-MV switch, a 5-MV charge water
pulse-forming line, and multiple self-breakdown megavolt water pulse
sharpening switches. The resultant electrical pulse formed by PBFA II has
subnanosecond syncronization in the main diode. During the analysis of
the PBFA switching data, a scaling law was found that has implications for
switch research and development.

Heating Phase Formula Description

The scaling law relates the time between the initiation of the
primary streamer and the final decrease of gas gap voltage (basically, the
formative time)[2] to the gas mean electric field. The relationship was
obtained from atmospheric breakdown data and includes gaps ranging from
sub-cm size to several-meter-long lightning simulator size.

The fast discharge (the primary streamer) closes a gap in a small
percentage of the total breakdown time. The heating phase, taking the
major portion of the time, heats up the gas to a temperature where the gas
discharge develops into an arc. To apply this concept, the transient
discharge current must be above the critical corona current.[2]

The scaling law is:

$$\rho\tau = 97800 \ (E/\rho)^{-3.44} \tag{1}$$

where ρ is the gas density in grams/cc, τ is the time to breakdown in
seconds, and E is the average electric field in kV/cm. The probable
values for the constants in the formula are 97800 ± 5000 and -3.44 ± 0.06.

---

*This work was supported by the U.S. Department of Energy under Contract
No. DE-AC04-76DP00789.

Gaseous Dielectrics VI, Edited by L.G. Christophorou and
I. Sauers, Plenum Press, New York, 1991

Comments Regarding the Formula

Three factors simplify gas breakdown analysis. First, as predicted,[2] the current is constant along the resistive filament that is formed. This constant current provides a reasonably uniform electric field. The heating times are independent of the original electrode structure and are determined by the mean electric field (V/d). Second, the formula indicates that time to breakdown scales inversely with the gas density if E/ρ is held constant. This fact has been noted before.[3] Third, the electrode spacing is unimportant since the initial filament heats at a rate determined by the average electrical field. The breakdown time is not determined by a transport phenomena between electrodes during heating.

The formula, initially derived from data by Felsenthal and Proud,[4] extends their data base by about three orders of magnitude in breakdown time. Mean field levels below dc breakdown, where most switching gaps operate, are included in the data.

Applications of the Formula Shown in this Paper

Data from electrically triggered gaps, laser-triggered gaps, and untriggered gaps are included in the approximations. The formula relationship appears valid for SF_6 and air or nitrogen over a wide range of parameters. Other gases such as oxygen, argon, helium, Freon 12, Freon 114, and Freon C318 obey this relationship over the data range available.[4] In addition, the predictive capability of this formula for the time to breakdown of mixtures of air and SF_6 will be shown.

DEVELOPING THE HEATING-PHASE RELATIONSHIP

Data Bases Used for this Approximation

The data used were obtained from several sources: Felsenthal and Proud (F&P),[4] Atomic Weapons Establishment (AWE),[5] Shuropat Trigatron (ST),[6] Wells and Martin Trigatron (WM),[7] Pendleton-Guenther Laser-Triggered Switch data (P&G LTS),[8] and T. Udo long-gap data (Udo).[9] An extended discussion of these data bases can be found in Reference 6. The data include trigatrons, laser-triggered gaps, highly field-enhanced gaps, self-breakdown gaps, constant field gaps, and megavolt lightning simulator gaps. Since the time to breakdown is independent of the electrode geometry, the average fields (V/d) were used from the data bases.

Graph and Empirical Relationship

The empirical approximation is shown in Figure 1. Several hundred data points from the above references define a power-law relationship. The data are presented in terms of the gas density to indicate applicability to other gases.

Observations from the Relationship

Sharp Points-Dull Points-Uniformly Graded Electrodes. The F&P data were taken using uniformly graded electrodes that were overvolted by a fast-rise voltage pulse (0.25 - 1.0 ns). The electrode surfaces were smooth and regular. The JCM (AWE) data were taken using a needle with a small radius point above a plane surface. The ST and WM data were derived from trigatrons using their trigger voltages and gap characteristics. The overlapping data were obtained using weakly triggered laser gaps with spherical electrodes by P&G. Finally, the data from Udo, using points or wires discharging to a ground plane, complete the breakdown selection.

Other Gas and Mixture Breakdown Delays. One intriguing aspect of this relationship is the possibility (not certainty) that the breakdown delay of different gases, not yet tested, could be predicted. F&P shows the widest collection of gases. However, their data do not continue below the dc breakdown of their test facility. The F&P data for nitrogen, helium, and SF_6 were compared with argon, air, and SF_6 data from AWE. The gas density was used instead of pressure, and the breakdown times are closely correlated for three orders of magnitude. The results are presented as Figure 2.

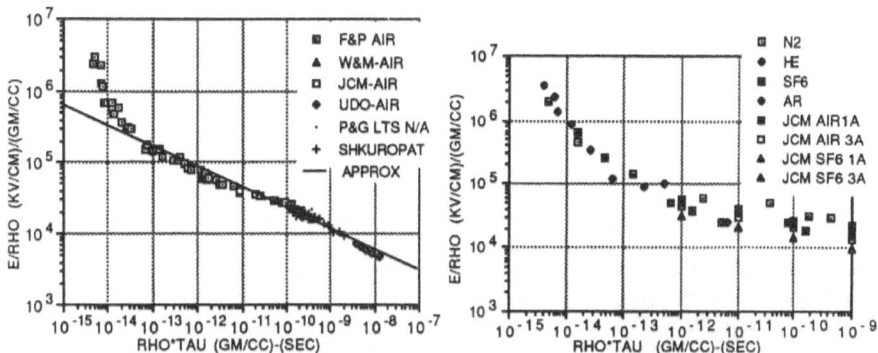

Fig. 1. Air heating data summary.

Fig. 2. Breakdown times for air, N_2, He, Ar, and SF_6.

Weak and Strong Triggering Modes. During the data analysis of triggered gaps, a concept of weakly and strongly triggered gaps was found useful. The data are consistent with the concept of a fast discharge closing a gap and starting a heating stage that completes the breakdown. The fast discharge can be initiated by a voltage pulse, a laser beam, or an injected electron beam. The energy requirements for the weakly triggered mode are small, and the breakdown time is simply determined by the field of V/d.

However, if the triggering mode modifies the field after the fast discharge is created, then the modified field must be used in the relationship. A classic case illustrating a different mode is a laser-created conducting filament that partially closes the gap while the heating phase is in process. The nominal gap field is changed by the laser-ionized region, and the gap breaks down in a shorter time due to a higher average field. This condition is the strongly triggered mode.

For instance, the strongly triggered laser gap that has a conducting laser filament of length x will have a breakdown time given by the modified E field in the relationship of[6]

$$E' = V/(d - x) . \tag{2}$$

Or, a trigatron gap with a trigger electrode gap of d_t, a trigger voltage of V_t, a main gap distance of d_g, and a main gap voltage of V_g will have two characteristic electrical fields that determine the breakdown delay.

The first is the trigger gap field which is

$$E_t = V_t/d_t . \qquad (3)$$

The second is the main gap after-trigger field that is

$$E_g' = (V_g + V_t)/d_g , \qquad (4)$$

where the trigger voltage can either aid or hinder the original gap field.[6,7] This concept explains the data taken by ST, where the application of a higher voltage trigger pulse does not shorten the breakdown time for the main gap. The trigger gap breaks down earlier, and the trigger voltage is removed from the main discharge. This, in effect, lowered the gap total electrical field during the heating phase.

In a similar manner, the relationship can be applied to a variety of either voltage-triggered gaps or to other energy-triggered gaps such as electron beams or lasers of weakly interacting wavelengths such as IR in SF_6 gaps.

Applications of the Relationship

SF_6 and Air Mixtures. Soviet experimental data[10] and the relationship's prediction are shown as Figure 3. An electron-beam trigger caused gap breakdown in a mixture of SF_6 and air. This particular gap is operating in the previously described weakly triggered mode. The trigger delay was measured. The upper breakdown curve was obtained from U.S. data for SF_6 and air mixtures. The curve approximates the data[10] by using a field enhancement factor of 1.3. The Soviet gaps were tested at a voltage of 1.45 MV, which was below dc breakdown. The lower data show the breakdown delay after the electron beam was injected. The lower curve was generated using the empirical formula along with the electric field shown by the upper curve. After subtracting out a constant of two nanoseconds from the data to allow for the rise time of the injected electron beam, the lower curve is well within the experimental error of the data.

This result is significant for three reasons. First, the empirical relationship predicted a result outside of the data boundaries that generated it, and, second, the results show that the electron-beam-triggered gap breakdown time can be approximated. Third, this result indicates that analyzing a switch run time in gas mixtures, using only the gas density and the mean electric field, is possible.

Trigger-Pin Experiment. Another prediction made by the approximation is the independence of the gap length on the breakdown delay. Only the

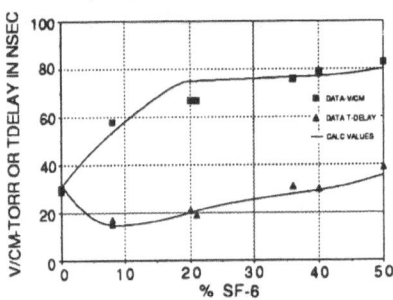

Fig. 3. e-beam triggered data
SF_6 and air mixtures·

density of the gas and the mean electric field are important. Gaps have the same time delay to breakdown regardless of their electrode spacing if their mean electric fields are the same.

To test the hypothesis of channels heating up with equal fields in equal times, an experiment was devised that had extreme sensitivity to this condition. Reference 7 describes a trigatron where the trigger pin was energized to cause breakdown. As indicated above, a trigatron is actually a switch containing two rather separate gaps, yet operating within the same gas envelope and initiated by the same fast discharge. If the trigger-pin voltage of the trigatron is monitored during breakdown, then two modes of gap closure are evident. One closure mode occurs when the trigger-pin discharge closes the trigger pin to ground and the voltage on the pin drops to zero. The second mode of closure occurs when the trigger pin closes to the high-voltage electrode. The main gap voltage then appears at the trigger pin. The difference between these two types of closure is easily detected. For two experiments, the mean fields in the gaps were determined using equations 3 and 4. The trigger voltage aided the gap breakdown, using a positive trigger on the positive electrode. The ratio of trigger gap distance to main gap distance was 1 to 4.7. A trigger voltage was chosen and held constant for each experiment. The gap voltage was adjusted until the trigger-pin monitor indicated a mode between the two extremes. This intermediate mode is the simultaneous breakdown between the trigger pin, ground, and the high-voltage electrode. The results are given in Table 1.

Table 1. Trigatron Mean Fields at Breakdown

Test	V_t kV	d_t cm	V_g kV	d_g cm	E_g kV/cm	E_t kV/cm
1	26	1.28	95	6	20.17	20.31
2	21.5	1.28	80	6	16.92	16.80

The mean main gap and trigger fields found in this experiment, by using the two equations, were within 1% of the same values. Thus, for this experiment with a 4.7 to 1 ratio of gap lengths, the breakdown times were equal for equal fields.

Three-Step Dimensional Derivation of Breakdown Time Delay. If the breakdown time is independent of the gap distance, then a different method of deriving a similar empirical relationship for any gas is possible. An illustration will be given for air.

First, the air Paschen curve is obtained. Next, the Paschen voltage is divided by the pd to obtain the reduced electric field vs. pd. Next, the relationship for the gas drift velocity (d/t) is divided into the pd at the same E/p, which gives the relationship of E/p vs. pt. This plot is shown as Figure 4 and compared to the F&P data.

The agreement of this derivation at pt values greater than 10^{-8} with the F&P uniform gap data is excellent for air. The differences between E/p for values of pt less than 10^{-8} are explained by the several-order-of-magnitude extension of the available data for the drift velocity.[11] This method of derivation, which uses the Paschen curve and the gas drift velocity, should be applicable to other gases.

This agreement was obtained through dimensional analysis and does not attempt to reconcile the results with the physics of the breakdown process.

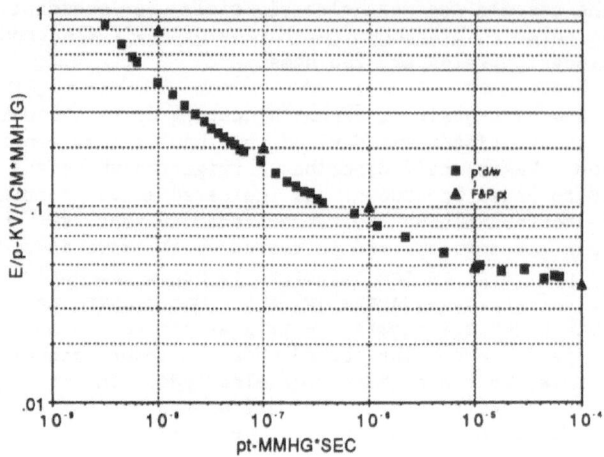

Fig. 4. Paschen derived air
breakdown delay.

SUMMARY

A method to predict several important gas parameters, using macroscopic
characteristics, has been shown.

REFERENCES

1. T. H. Martin, B. N. Turman, S. A. Goldstein, J. M. Wilson, D. L.
    Cook, D. H. McDaniel, E. L. Burgess, G. E. Rochau, E. L. Neau, and
    D. R. Humphreys, PBFA II, The Pulsed Power Characterization
    Phase, Proc. 6th IEEE Pulsed Power Conf. (Arlington, VA, 1987)
    IEEE Cat. No. 87CH2522-1, p. 225.
2. E. Marode, The Mechanism of Spark Breakdown in Air at Atmospheric
    Pressure Between a Positive Point and a Plane, I. Experimental:
    Nature of a Streamer Track, J Appl. Phys. 46:2005 (1975).
3. C. Cooke and A. Cookson, The Nature and Practice of Gases as
    Electrical Insulators, IEEE Tran. Elecr. Insul. EII-13:239 (1978).
4. P. Felsenthal and J. Proud, Nanosecond-Pulse Breakdown in Gases,
    Phys. Rev. 139A:1796 (1965).
5. J. C. Martin, "Dielectric Strength Notes," 15 (1967) first printed
    as SSWA/JCM/697/71, AWE internal publication.
6. T. H. Martin, An Empirical Formula for Gas Switch Breakdown Delay,
    Proc. 7th IEEE Pulsed Power Conf. (Monterey, CA, 1989) IEEE Cat.
    No. 89CH2678-2, p. 73.
7. J. Wells, Thesis for MS in Electrical Engineering (Dec. 1987), Univ.
    of New Mexico, Albuquerque, NM, T. Martin, Advisor.
8. A. H. Guenther, J Phys. D: Appl. Phys. 11:1577 (1978).
9. T. Udo, IEEE Trans. on Power Apparatus and Systems PAS-84:304 (1965).
10. B. M. Kovalchuk, V. V. Kremnev and Yu. F Potalitsyn, High Current
    Nanosecond Switches, Chapter 3, Figure 3-15.
11. J. Dutton, A Survey of Electron Swarm Data, J Phys. Chem. Ref. Data
    4:577 (1975).

DISCUSSION

E. E. KUNHARDT: Can you explain your results on the basis of a phase—like transition between a cold gas and a thermal plasma?

T. H. MARTIN: The fast or initial discharge probably initiates an electron—ion background. The current rapidly goes into a relatively small diameter channel which then heats electronically from energy supplied by the driving circuit. As the resistance of the channel falls, the current increases and the heating rate increases. During this time the voltage across the gap has not changed even though the channel resistance changes several orders of magnitude. The onset of final breakdown, where the gap resistance approximates the source impedance, likely occurs due to a phase change in the channel at about 3000° K, at this point the gas turns into a thermal plasma and becomes self sustaining as the resistance drops to a fraction of the circuit impedance and rapidly changes the gap voltage. Typical parameters for a cm size gap in atmospheric air are 10 A at 33 kV for 300 ns for an energy of about 100 mJ.

MITIGATION OF POTENTIAL HEALTH HAZARDS

OF TRANSMISSION LINE FIELDS

Imre Gyuk

Program Manager for
Electromagnetic Research
U. S. Department of Energy

THE PUBLIC CONCERN

Two years ago, the Wall Street Journal brought into public
prominence an epidemiological study investigating a possible
correlation between childhood leukemia and residential magnetic
field exposure. Since then, potentially harmful effects of
electric and magnetic fields have become a major public
concern. Articles varying from informative to alarmist to
downright scurrilous have appeared in almost every daily
newspaper. A series of reports published in the New Yorker have
been particularly notable in this respect. Public response
ranges from individual concern to organized citizens groups
opposing construction of new power lines. There have been law
suits concerned with issues such as the placement of power
lines near schools or the decrease of property values in the
proximity of transmission lines. Recently, the first suit of
'wrongful death' has been entered. The issue has become
politically sensitive and three congressional hearings have
been held on the subject in the last two years.

Actually, concern for possible health hazards of electric and
magnetic fields at power line frequencies antedates the public
furor by more than a decade. Following Russian reports of
potential hazards due to such fields, the Department of Energy
established a research program to investigate the problem. This
program has maintained an annual funding level of some 3
million dollars. The Electric Power Research Institute has
maintained a similar program, vastly accelerated in recent
years. Considerable knowledge about the interaction between
fields and biological systems has been developed. But
ironically, the same scientists on whose work the current
public concern is based, are now often reproached for not
having resolved the problem.

WHAT ARE ELF FIELDS?

We are dealing with both electric and magnetic fields. For very high frequency radiation, Maxwell's Laws show that the two are inextricably related to each other. At lower frequencies they become increasingly de-coupled and at extremely low frequencies (ELF), such as 60 Hz, they are virtually independent. We can talk about magnetic fields and electric fields separately and the word 'radiation' is inappropriate.

At high frequencies (e.g. x-rays), wavelengths are extremely short and the energy associated with each wave-packet is large. Such radiation will damage DNA and thus is quite obviously mutagenic. Lower frequencies, such as microwaves, have less energy - but there is still enough to 'cook' cellular protein. ELF fields cannot affect cells directly because the energy involved is too low. At 60 Hz the wavelength stretches from Washington to Los Angeles. The observed biological phenomena, therefore, require a process of collecting and amplifying incoming energy over both time and space - a process which is very little understood.

Electric fields are perceived by man, baboon and rat alike at about 15 kV/m. Magnetic fields are not perceived at all - although birds and some other animals may be sensitive to them.

Right underneath a 500kV transmission line maximum electric fields are about 5kV/m, magnetic fields are as high as 100 mGauss. At the edge of the right-of-way the fields will have dropped to 1-2 kV/m and perhaps 10 mG. Electric fields are easily shielded by earth, trees and houses. Magnetic fields are only very slightly attenuated.

Some epidemiology studies suggest that magnetic fields of 3 mG can be considered 'high' and potentially dangerous. But 30 mG are routine in New York City - in many office buildings fields as high as 1000 mG have been measured.

Inside residences, the observed fields are roughly 30% due to distribution (not transmission!) lines and 30% due to household appliances such as electric blankets, stoves, toasters,etc. The remaining contribution comes from the return neutral currents which often run through the plumbing.

HEALTH CONCERNS

Concern during the last decade has focussed on three areas of potential health effects: Neuro-physiology, Reproductive Problems and Abnormal Cell Growth. Although no unequivocal proof of a definite health effect has been given, these three areas surface consistently in epidemiological studies as well as work with animals and humans.

Stress due to long term exposure, changes in hormone balance, decreased memory capacity of animals exposed before birth and changes in heart beat frequency have been reported in laboratory experiments as examples of neuro-physiological effects.

Epidemiological work has revealed possible correlations of magnetic fields on childhood leukemia, brain tumor growth and (recently) male breast cancer.

Observed melatonin decrease of exposed rodents may be correlated with increased breast cancer rates. Animal experiments have shown increased rates of fetal malformations. Women using electric blankets have reported changes in menstrual cycle.

EPIDEMIOLOGY

Since human experimentation involving toxic agents presents certain difficulties, epidemiology can offer a useful alternative.

There have been some 40 epidemiology studies on potential health hazards of electric and magnetic fields. Roughly half of these are residential, half are occupational. Twenty studies are ongoing.

In general, risk factors for residential studies are in the vicinity of 2, while occupational studies yield higher risk factors (for example 8). However, in many cases, studies showing statistically significant correlation with exposure are matched by other studies which do not. Besides, the diseases involved are fortunately rare - the total number of cases is therefore orders of magnitude smaller than those involved in accepted correlations such as lung cancer and smoking.

Among the most often quoted studies, The Savitz Study investigated cases of childhood leukemia in Denver. He found that houses with fields of about 3 mG have a slightly higher correlation with the disease than those with only 1mG. While the study itself was carefully conducted, certain facts have to be kept in mind. The risk factor involved is only 2. The total number of cases is about 135. And there was no correlation to adult leukemia. While the study indicates potential effects, it is certainly not proof that leukemia is caused by power line fields.

A recent study of telephone office workers showed an increased risk of male breast cancer. A very rare disease, it usually occurs only once for a millon males. The John Hopkins study found 2 cases among four thousand workers. This may be very significant or it may be the statistics of small numbers.

Epidemiology studies are easier to understand by the general public than work on cellular biology. However, such studies also lend themselves to facile (and misleading) interpretations. They are, therefore, particularly prone to lead to alarmist reports in the press and to general confusion. Actually, a wide variety of causes lead to risk factors of approximately 2, without arousing any notable public concern. Secondary smoking and bottle feeding infants are among such causes.

During the last dozen years, extensive experimental work has been carried out. Research spans the entire spectrum from humans, primates and rodents to tissues cells and DNA. After the early screening studies, many of which were negative, effects have now been identified for a considerable number of systems.

It must be noted, that many of the observed effects are not very robust. Moreover, most studies have not been extensively replicated. In part, this may be due to the fact that appropriate exposure parameters have not yet been fully identified.

In human research, male volunteers are exposed for 6 hours to mixed electric and magnetic fields which they cannot perceive. Of some 50 blood, urine, physiological and psychological variables investigated, only three showed significant changes due to the fields: Changes in certain brain waves, a slowing of motor responses and, in particular, a slight slowing of heart rate (3 out of 70 beats per minute). These effects were consistent. However, they were present only at 9kV/m and 0.2 G and not at fields above and below these values. There appears to be a 'window' effect quite unlike the usual dose/effect relationship. Moreover, when fields were intermittent (on-off 4 times per minute) the effect becomes stronger. Both of these features had previously appeared in cellular work.

Melatonin is known to be an important hormone produced in the pineal gland and regulating thyroid gland, adrenal gland and reproductive organs. Reduction of melatonin production is strongly correlated with breast cancer in rats as well as human females. Exposure of rats to electric fields for 3 weeks has resulted in depression of daily melatonin production by some 50%. Continuing experiments have shown that this effect can be elicited with fields ranging from 3.5 kV/m to 120 kV/m. Currently, effects of mixed electric and magnetic fields are being explored. Preliminary results show that intermittent magnetic fields yield stronger effects.

Cellular work has shown varied responses to exposure such as irregular firing of neurons and reduced killing capacity of white blood cells (thus reducing effectiveness of the immune system). It has become clear, largely through DOE work, that the cellular basis of many bio-effects seems to be a disturbance of the flux of biologically important ions through the cell membrane. These ions, particularly calcium, serve as messengers telling cells how to respond to external stimuli. It was found, for example, that exposure causes changes in the intra cellular hormones ornithine-decarboxylase and parathyroid which are similar to changes caused by known cancer promoters.

Although extensive experimentation has revealed no direct effect of ELF fields on DNA, recent results reveal a more subtle effect. DNA transcription and translation into messenger RNA is apparently affected by exposure. Put simply, this means that DNA is not affected directly, but the way DNA works may be changed by fields.

All these effects are biological effects - they are not yet health effects. They do, however, present cause for future concern.

DOSE AND EFFECT

Most toxic agents found in the environment have a fairly simple dose/effect relationship. Basically this relationship is linear - at least for small doses. Twice the dose gives twice the effect. Often there is some saturation, when further increases in dose do not increase the effect.

The situation for bio-electromagnetic effects is completely different. Both human and cellular work indicate the existence of intensity 'windows'. An effect only seems to occur within a certain range of the parameter. Certain phenomena, in fact, appear to be restricted to narrow resonance-like bands. Non-linearity also seems to be characteristic of the frequency dependence. Furthermore, intermittency emerges as an important factor. The magnitude of the field may not be as important as fluctuations in exposure. Finally, other factors such as timing and even the local geo-magnetic field seem to play a role as well.

The non-linearity of the response to a field should perhaps come as no surprise when considering the necessity for a complex process of energy collection which is required at the cell surface.

While extremely interesting from a scientific point of view, this situation makes any evaluation of 'exposure' quite diffi-cult. An extensive program of measurement is not warranted at this time, since we do not yet know what to measure. Similar remarks apply to regulation and mitigation.

Finding the dose/effect relationship for biological effects must be the prime scientific objective. Once such a relation-ship is found, one will be in a much better position to look for possible health effects - either through epidemiology or through animal experiments.

BIOLOGY - HEALTH - RISK

A decade ago, a substantial number of scientists might have doubted whether ELF fields can interact with biological mecha-nisms. Today, the existence of 'biological' effects is accepted by a majority of scientists. However, such biological effects do not necessarily imply that there are 'health' effects. Experiments are made under carefully controlled laboratory conditions, which may have little relevance to realistic exposure environments. For example, small amounts of light, can essentially negate the effects of the melatonin experiment. Furthermore, body mechanisms are able to take care of most other biological perturbations. Proof of health effects will need extensive and costly animal experimentation. Until there is better understanding of the dose/effect relation such work will not be conclusive.

While biological effects can be considered as established, health effects of ELF fields must be considered as unproved.

Only if there are health effects, will the question of risk become relevant. Because of the apparent nature of the dose/-effect relationship, one might say that the special exposure conditions which result in effects might be comparatively rare. On the other hand, because electricity is virtually ubiquitous, one could say that even very small health effects will result in major risks.

REGULATION?

To date, seven states have passed regulations to limit fields in the vicinity of transmission lines. Naturally, all these regulations differ from each other. In general, the electric field inside the right of way is restricted to magnitudes less than between 7 and 11 kV/m. At the edge of the right of way the maximum field is set as 1 to 3 kV/m. One state, Florida, also restricts the magnetic field to less than 150 to 200 G. There are usually exceptions for short line segments or lower voltages.

U. S. regulation is in general considerably more stringent than international standards. For example IRPA and WHO guidelines followed by most of Europe require electric fields to be less than 5 to 30 kV/m while magnetic field limits are between 1 and 250 (!) Gauss.

Since there are no compelling scientific reasons for choosing field limits, regulation is based on legal or political arguments. For example, regulation may be designed so that fields of future transmission lines are no 'worse' than existing measured fields.

Because of the non-linear nature of the dose/effect relationship, the window effect in particular, there is always the danger that lower field limits will give the public a false sense of security. Regulation might actually put more people into the biologically active range of intensities.

MITIGATION?

The arguments against pursuing a vigorous program of mitigation are much the same as for regulation. To begin, we do not really know whether and to what degree there really are health effects. But more importantly, we really don't know what to mitigate against! If we were dealing with the usual toxicological situation, one could simply say that any reduction in field strength would be commendable. But because of window effects, reduction of field levels may be counter-productive. If it turns out that only certain frequencies are biologically effective then it may be much more important to decrease contributions of that specific component rather than lowering the 60 Hz amplitude in general.

A number of strategies for reducing field levels of transmission lines are available. None of them are attractive from an economical point of view.

One could simply circumvent the problem by using lines with lower voltage or current. But, of course, this would require more lines with associated higher costs and increased chance for litigation.

The right-of-way could be made wider or the poles could be made higher. Cost of land or equipment would obviously be greater.

Undergrounding would be aesthetically pleasing. In Europe this is practiced much more widely than in this country. But costs are considerably higher than regular lines, even though there would be some advantage due to lower maintenance costs. Most importantly, while electric fields are shielded out by the ground, magnetic fields are not. A person walking over or near the buried lines would be subjected to much higher fields than is now the case.

New configurations could be used. BPA, for example is already using the 'delta' configuration with a field decrease of some 20 to 50 % at the edge of the right-of-way.

Finally, new technologies could be introduced. Higher phase order systems or gas insulated transmission are among these. Because such systems would require extensive reconfiguration and construction, they will not be readily accepted by the utility industry unless there are other strong reasons for their introduction.

THE COSTS

Potential health effects of electric and magnetic fields are an extremely expensive proposition. In the past decade, there have been over a hundred legal challenges to construction of new transmission line projects throughout the U. S.. Health concerns figured prominently in most of these. In some cases, such as the Washington-Baltimore loop, it has been estimated that the delay has already cost the rate payer several billion dollars by having to buy more expensive power.

Regulation can be expensive too. Extrapolation from a recent Florida study indicates that incremental cost of establishing a 1.5 kV/m, 100 mG right-of-way would amount to some $ 5 000 million per year for the entire U. S.

Ultimately all such costs would have to be borne by the rate payer. By comparison, the cost of federal and private research programs on the biological effects of electric and magnetic fields are minute.

CONCLUSIONS

It is becoming apparent that there are biological effects of electric and magnetic fields with frequencies between 15 and 150 Hz. However, it is not clear whether these will lead to health effects.

Because of our limited knowledge of the exposure parameters involved and the non-linearity of the dose/effect relationship, there is at present no scientific basis for regulatory action. For the same reason, an extensive program of mitigation is not warranted at this time. There is, consequently, much room for research to improve our understanding of this complex area.

There is no real reason for immediate alarm. Health effects cannot be considered as proven. If they exist, epidemiology indicates that they are likely to be small. There is, however, reason for continued concern. Scientists, regulators and utility representatives should co-operate in resolving this important issue.

ELECTRIC AND MAGNETIC FIELDS: AN ENGINEER'S PERSPECTIVE

F. S. Young

Electric Power Research Institute
Palo Alto, California

ABSTRACT

Much has been written and discussed about electric and magnetic fields. Extensive programs are underway to discover whether these fields can affect our health.

But, engineers face some serious dilemmas in dealing with the electromagnetic field issue. A number of scenarios concerning what field characteristics might cause effects have been postulated. Until more evidence is available, the engineer does not really know what to fix, or even if anything needs to be fixed at all.

Presently, a program of research is underway to provide an electrical engineering characterization of field magnitudes, sources, time and space variations, and other important parameters. This work will be helpful to the medical research community in understanding the nature of fields. It will also help identify possible options for reducing field strengths, if that is deemed necessary.

INTRODUCTION

A discussion of the engineering aspects of this issue should be preceded by noting that although power systems have both electric and magnetic fields, one can generally associate the electric field with the voltage on the wire independently from the magnetic field which is associated with the current flowing in the wires. This paper deals principally with the magnetic field caused by current flow.

There are two important questions to this issue: (1) what are the medical and biological effects; and (2) what can engineers do about designing power systems that produce lower magnetic fields. To an engineer, at present, it is not completely clear what to fix.

There are a number of different hypotheses about what might cause effects from magnetic fields:

- Amplitude
- Time–integrated dose
- Thresholds
- Windows

With regard to the absolute magnitude of the magnetic field, some investigators report finding health effects at magnetic fields as low as 3 mG. Thresholds and windows are also factors. If the magnetic field strength is lowered, one can not be sure from present data that it is below the threshold or if it has been lowered to the middle of a window, causing more problems than if it had not been lowered at all. That is the general dilemma that faces the electrical engineer when looking at this problem.

Gaseous Dielectrics VI, Edited by L.G. Christophorou and
I. Sauers, Plenum Press, New York, 1991

The main engineering effort to date has been to develop instruments that will measure magnetic fields and then use them to develop a knowledge of the sources and characteristics of the fields. The results of this effort will show how strong the fields really are, what their variations are in time and space, and what electric facilities are responsible for producing the fields.

EPRI PILOT STUDY

A few years ago EPRI organized a pilot study with six member utilities to make measurements in the houses of some of their employees. EPRI examined nine houses at each of the six utilities for a total of 54 houses. The houses were not selected in a scientific way. Some were near transmission lines and others were not. There were three objectives in these studies: (1) to determine the sources of the magnetic field; (2) to test the instruments used to make the measurements; and (3) to collect initial data needed to develop computer software that could simulate magnetic fields in residences.

The pilot study showed that sources of the magnetic fields could be discerned inside a house. Generally, there were three important sources (leaving out appliances for the moment) of magnetic fields that were identified.

1. <u>Transmission lines</u>. A transmission line can produce a field inside the house. However, there are not many people that live near transmission lines.

2. <u>Distribution circuits</u>. The net current flow in the distribution circuits near a house can produce fields that can be measured inside the house. Net current flow is difficult to describe, because the distribution system is a complex configuration of wires, including high voltage primary feeders, secondaries, and neutrals. By taking the vector sum of all the currents in all the wires, the net current results in a field that can be measured in the house.

3. <u>Grounding system</u>. Because of the way the power system is grounded (for safety reasons) there is flow of current in the grounding system (often in water pipes). Several cases were observed where an appliance was turned on in one house, the current flowed via the water pipes through another house and then back through the neutral wire to the distribution transformer.

Figure 1 shows data from the 17 most interesting houses. The remainder had fields of 1 mG or less. The houses near transmission lines had the highest fields (14 mG). In some cases when a home owner modified the house wiring and ignored wiring codes, large loops were created resulting in high fields. With the exception of the homes near transmission lines the rest of the houses had fields generally less than 5 mG. The pilot study was successful in identifying the independent sources. For comparison purposes appliances such as television, portable heater, range, and heated floors produce fields greater than 5 mG.

The second objective of the pilot study was to test instruments to measure the magnetic fields. An example is an instrument called the STAR, a stationary three axis recording magnetic field meter. This instrument can be made part of another device called the VANA which holds the STAR, a digital compass, and a device that measures the distance, so that one can plot magnetic field profiles. Other instruments were developed to measure current flow adjacent to houses in circuits at the same time field measurements were made in the houses so that a correlation could be made between current flow and magnetic field.

The third objective of the pilot study was to collect initial data for the development of software. The transmission line is of symmetrical geometry which lends itself to simple analysis. ENVIRO is a software package developed previously by EPRI to permit calculation of the electric and magnetic fields and also permits calculation of other field dependent factors like surface gradients on conductors, audible noise, radio influence and other quantities related to the voltage and current flow. The distribution circuit, however, is much more complex, making the problem of developing software a more significantly difficult problem.

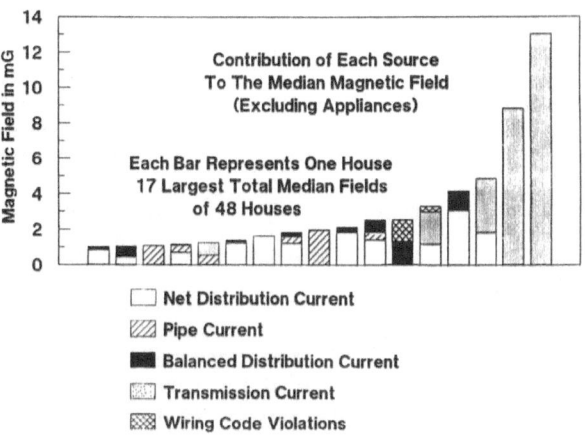

Fig. 1. Magnetic field strength survey of 17 houses in EPRI pilot study.

PRESENT STUDIES

Based on the results of the pilot study and on the development of instrumentation, EPRI is now launching a nationwide statistically–correct study of 1,000 utility customer homes that will cover 25 different utility systems. To date, the measurements have been concluded on the first three utilities. EPRI has also initiated a non–residential measurement program in offices, schools, factories, and utility facilities such as substations and power plants. One of the interesting results from that study relates to the way large buildings are grounded. The power system may be grounded to the structural steel, so that a person near a support steel column, may be exposed to a relatively high magnetic field occuring when current returns through that structural member.

In addition to the survey, EPRI has built a magnetic field research facility in Lenox, Massachusetts. It consists of seventeen simulated loads, one real house, a grounding system with water pipes, a 23–kV feeder and related components. The purpose of the house is to study various wiring configurations, determine influence of measures to reduce fields, train utility personnel in proper measuring techniques, and verify software.

FIELD REDUCTION OPTIONS

EPRI is now beginning additional studies to explore the options available to reduce strengths of magnetic fields emanating from existing and new facilities. But here is where the engineer's dilemma gets serious. Table 1 shows several options that could be applied if health effects are related to the field magnitude or to the duration of exposure. A different strategy would be used and different priorities would be applied.

Table 1. Priority of Approaches Based onMagnetic Field Magnitude
or on Magnetic Field Exposure,
(Based on Pilot Study).

Field Magnitude	Field Exposure
— certain appliances	— grounding system of residence
— transmission lines	— net current of distribution lines
— grounding system of residence	— tranmission lines
— net current of distribution lines	— certain appliances
— balanced current of distrib—ution lines	— balanced currents of distrib—ution lines

For example, if just field magnitude is of concern, then one might work on modifying appliances and transmission lines, but if the concern is exposure duration, the grounding system and the distribution circuits would be more important. If it is eventually found that magnetic fields really do cause cancer, then transmission lines are not likely to be the cause, but rather those devices that are around us every day.

The priority of this research work will be different for existing and new facilities. Table 2 lists research that must be conducted on existing systems and Table 3 outlines work that will be useful for future facility installations.

Table 2. Magnetic Field Reduction For Existing Sources.

— Reduce ground current
— residential/non—residential
— Modify transmission lines
— Reduce spacing upgrade
— Modify distribution substations in commerical buildings
— Modify appliances and industrial equipment

The reduction of ground current in residential neighborhoods and non—residential buildings must be done with caution since grounding is essential for safety. A premature change of the grounding system, could result in more electrocutions and fires.

Modification of transmission lines may take many forms. For example, in a double circuit transmission line normally the three lines are phased ABC (top to bottom) on the left side and ABC on the right side (Fig. 2). If phasing one of the sides is reversed so that the line is now CBA from top to bottom, there would be a 50% reduction in the magnetic field at the ground level. However, there would also be a reduction in the impedance of the circuit stressing the circuit breakers and also producing higher gradients on the conductors producing more radio and audio noise. Engineers must carefully consider these trade—off and associated costs.

Table 3. Magnetic Field Reduction For Future Equipment Installations.

— Optimize single/multi—circuit transmission line designs
— Reduce cost of underground transmission and dristribution
— Establish new grounding practices
— Redesign appliances and industrial equipment

EPRI does not plan to immediately take an active role in redesign of appliances and industrial equipment. The Institute is, however, available to assist in these efforts if needed. Presently several manufacturers are undertaking work in this area under their own initiative. For example IBM is working on improved video terminals, and two electric blanket manufacturers have modified wiring in their blankets to produce lower magnetic fields.

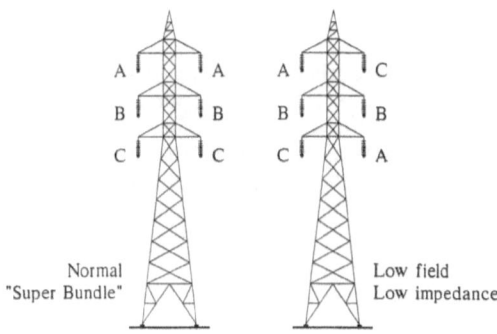

Fig. 2. Modification of transmission line phasing to reduce magnetic field strength.

Table 4. Exposure Reductions For Utility Workers.

— Identify high field areas
— Commercialize TOMCAT for hot line work
— Develop auto—surveillance techniques
— Modify maintenance practices
— Investigate shielding

Exposure of workers (Table 4) is also an important subject. TOMCAT is a remote manipulated robot that can work in hazardous locations where people would ordinarily be sent. Auto—surveillance refers to the explosion in sensor technology that is presently taking place. Development of platforms (robots) to carry these new sensors in to sensitive areas is also important.

Finally, there is the shielding problem. It is not clear how to effectively shield magnetic fields. Possibilities include foils, thick slabs of soft iron or possibly even some of the new technology, such as that used for the stealth bomber.

CONCLUSIONS

Magnetic field effects are not just a problem that utilities face, but a problem that all society faces. A large number of other groups, such as those listed below, must work towards the goal of minimizing the possible adverse health effects due to exposure to electric and magnetic fields:

- National Electric Code
- National Electric Safety Code
- American Waterworks Association
- American Gas Association
- Home Builders Association
- American Architectural Association
- Telephone Group
- National Electrical Manufacturers Association
- National Home Appliance Manufacturers Association
- Canadian Electric Association
- Labor Unions

The utilitites are taking a proactive role in this issue. Some of them are beginning to reduce the fields by using the ABC/CBA approach configuration, even before the scientific results on health effects are in. The benefits of this strategy are in terms of public perception and political and regulatory interests. The EPRI program, supplemented by the work of others, will help in this process.

DISCUSSION*

M. KRISTIANSEN: Do you have any comments on microwave interactions?

I. GYUK: The microwave situation is entirely different. Here, we are strictly talking about the region from 15–150 Hz, i.e. extremely low frequencies.

M. KRISTIANSEN: What about pulsed power?

I. GYUK: It is an interesting subject, but our particular program does not concern itself with that, although it may be of interest in the future. Systems such as magnetically levitated trains, superconducting storage units or the supercollider are also possible research problems.

O. FARISH: You mentioned risk factors of typically 2, but it seems that the case of childhood leukemia has attracted particular attention. Is that because the risk factor is greater? What is the reason?

I. GYUK: In a childhood leukemia situation in residential applications, we usually get risk factors of around 2. In certain other occupational situations we may get higher risk factors, up to seven, for example, for cable splicers, from a study at Johns Hopkins. Remember, these are classified purely according to their profession, where many of the workers use welding equipment, where there are toxic fumes, or where workers are in underground tunnels. So there are many factors to consider.

D. C. AGOURIDIS: <u>Comments on near fields:</u> (1) We cannot determine the value of the magnetic field from calculations and/or measurements of the electric field. (2) We cannot determine the value of the electric field from calculations and/or measurements of the magnetic field. (3) We must do each separately. <u>Comments on electric field coupling, magnetic field coupling and electromagnetic field coupling:</u> (1) We cannot determine the magnetically induced current/voltage from measurements and/or calculations of the electrically induced current/voltage. (2) We cannot determine the electrically induced current/voltage from measurements and/or calculations of the magnetically induced current/voltage. (3) We cannot determine the electromagnetically induced current/voltage from (1) and/or (2).

I. GYUK: In some of the experiments we used electric and magnetic fields simultaneously to try in some ways to simulate these simultaneous electric and magnetic fields. However, with respect to direct interaction of electric and magnetic fields, at high frequencies—light, for example — the electric and magnetic fields are totally interrelated, 50:50, electric and magnetic fields. As the frequency becomes lower and the wavelength becomes longer the electric and magnetic fields become more and more decoupled. When we go to low frequency 15–150 Hz, basically, the electric and magnetic fields have become totally decoupled and can be considered as independent, since the induced electric and induced magnetic fields have become so small that they really do not count.

E. E. KUNHARDT: In the experiment that you reported what is the effect of and variation in the geomagnetic field?

I. GYUK: The geomagnetic is roughly 500 mG DC and the variation is about ± 150 mG. It is also a vector quantity. Although we cannot adjust the geomagnetic field, at least we can record what the field is. It is possible that some of the diversity in results may be due to the local geomagnetic field. There may be a very narrow dependence, not on the absolute value, but on the component of the applied magnetic field that is parallel or perpendicular to the geomagnetic field.

F. S. YOUNG: We measure the three components of the magnetic field, so that we can get the magnetic field vector.

S. W. ROWE: As I understand it, the substance melatonin is produced on a day–night sequence. During the night, except for unusual situations, the people are immobile with respect to the magnetic and electric fields. Is the fact that they are immobile in the field a good thing or not?

I. GYUK: It depends on whether the most important factor is intensity or intermittency. If it is intermittency, then, the fact they are not moving inside the field is fine. If it is intensity and they are in the right "window", for example if there is a wire underneath the floor under a bed, then the field could have a greater effect than the field due to a transmission line outside the house.

F. S. YOUNG: I think that we are in the first inch of a mile in understanding what is going on.

*Of the two preceding papers.

RELATIVE POTENCY AS A MEANS OF EVALUATING ELF HEALTH RISKS

Clay E. Easterly and Larry R. Glass

Oak Ridge National Laboratory[1]
Post Office Box 2008
Oak Ridge, Tennessee 37831-6101

INTRODUCTION

In the 1970's, a variety of developments took place to heighten public and scientific interest in electromagnetic fields. Examples include the microwave beaming of the U.S. Embassy in Moscow, the U.S. Navy's large low-frequency antenna project (Project Sanguine), the intense magnetic fields used in proposed magnetic fusion reactors and the attempt by utilities to utilize very high voltage transmission lines to enhance conservation. During this time, biological studies of nonionizing electromagnetic fields were taking place, but no clear evidence of risks to public health was identified. Then came the surprising epidemiological finding of Wertheimer and Leeper (1979) suggesting that 60 Hz magnetic fields may be related to some childhood leukemias. Our particular interest at ORNL was how to interpret the available data with respect to human exposures to the nearly ubiquitous fields.

A review of the available data showed that consistent biological effects were difficult to identify. Neither *in vivo* or *in vitro* experiments were able to provide a consistent level or type of effect. Classical toxicological tests used in chemical risk assessment had not been performed with Extremely Low Frequency (ELF) fields but rather a wide range of mechanistic studies had been pursued. To evaluate the level of anticipated hazard or risk there was neither a mechanistic understanding nor a consistent phenomenological outcome. A risk evaluation normally requires one or the other of these two types of information. Two quite different approaches were pursued: *meta-analysis* and *relative potency*. The first of these is a method to combine data from similar experiments to enhance the relative statistical power of a collection

---

[1]Operated by Martin Marietta Energy Systems, Inc., under contract DE-AC05-84OR21400 with the U.S. Department of Energy.

of small sample size studies (Morris et al. 1989), and will not be discussed further. The second, relative potency, will be the focus of this paper.

RELATIVE POTENCY BACKGROUND

The roots of the present application of relative potency can be seen in pharmacology wherein drugs are compared in terms of relative doses required to produce the same therapeutic action. For example, if it takes two aspirin tablets to cure a headache and one tablet of a new drug, the potency of the new drug relative to aspirin would be 2. The relative potency framework is used to compare doses required to produce equal levels of effects. These relative comparisons in pharmacological uses are performed without making assumptions about biological mechanisms. The relative potency approach being pursued at ORNL is also based on observed biological response rather than hypothesized mechanism of action. The potency scale is constructed by expressing results as a function of an arbitrarily selected standard, usually a well studied toxic agent. A distinguishing feature is that it is "data intensive and model sparse;" as such it incorporates a broad spectrum of biological data without incorporating a host of assumptions to support a theoretical model. Thus, rather than focusing on identification of the single "best" animal experiment, as is done in "quantitative risk assessment" (U.S. EPA 1989), relative potency can utilize all available sources of data.

In order for this idea to be given a fair hearing in a scientific environment centered on the pursuit of mechanisms, it was important to demonstrate that a relative potency framework, as we envisioned it, could operate successfully without a detailed knowledge of mechanisms. The framework used has undergone continuous development in the Health Effects Group at the Oak Ridge National Laboratory for approximately a decade (Griffin et al. 1979; Walsh et al. 1982; Dudney et al. 1983; Jones et al. 1983, 1985, 1988; Watson et al. 1989; Glass et al. 1990). The results of the latter work, to be described next, demonstrate that this ranking can take place without explicit knowledge of mechanism of action; therefore, the use of relative potency in health risk assessments of electromagnetic fields is not bound by our lack of mechanistic knowledge. Application of this technique, however, may be limited at present because of the available data and how it is accumulated.

CONCEPTUAL PERSPECTIVE

The relative potency logic of addressing the issue of mechanisms is as follows. Chemical carcinogens, those chemicals recognized by the International Agency for Research on Cancer (IARC) as human or potential human carcinogens, represent a wide array of mechanism of action. That is, the means by which these chemicals produce cancer in animal systems is quite different from one chemical to another. If, by some technique, these carcinogens could be ranked relative to a reference compound of one mechanism of action, that ranking could be compared with a ranking using another reference compound which had a different mechanism of action. This process could be carried out for reference compounds having different mechanisms of action. If the rankings are dependent on mechanism, then the rankings based upon the different reference compounds would be different. If the rankings are independent of mechanism of action, then the technique could serve as a valuable model for future studies. Results of the exercise demonstrated that, in fact, the approach developed at ORNL, using a wide spectrum of long-term whole animal data is useful in providing a ranking of carcinogenic potency independent of the reference compound's mechanism of action (Glass et al. 1990).

The most appropriate next evaluation was one which allowed for the greatest diversity in biological responses in order to demonstrate the application of the relative potency approach to a diverse set of bioeffects endpoints such as are present in the ELF bioeffects database. The study was undertaken to determine if potency ranking as predicted with long-term whole animals could be replicated using short-term tests. The results of that study (Glass et al. 1990) demonstrate that a potency ranking on the basis of short-term tests provides essentially the same ranking of the carcinogens as the long-term whole animal studies (Glass et al. 1990).

RANKING STUDIES

The relative potency framework as utilized at ORNL involves a comparison of dose levels of different agents (chemicals) to give a common outcome in a common experimental protocol. These ratios of dose of reference agent to dose of test agent represent a distribution in sensitivity of the various test systems, which themselves represent a variety of species, strains, dosing methods, time course of dosing etc. Since nearly all of the biological test systems are "tuned" to be sensitive to some insult, the test systems should not be thought of as being individual surrogates for humans. Rather, they can be thought of as collectively representing the range of responses by living organisms. Not having information upon which to determine the underlying distribution of responses, the summary measure determined for use in this work is the median (Jones et al. 1985). The summary relative potency is then the median of the collection of relative potency ratios.

For evaluation of these concepts, we selected 52 carcinogens, primarily from the list of carcinogens published by the IARC. Long-term whole-animal data on tumorigenicity and short-term bioassay data were obtained from the Registry of Toxic Effects of Chemical Substances (RTECS). To avoid any potential bias, 6 chemicals from among the 52 were chosen to serve as reference compounds. These 6 reference compounds were chosen to represent different mechanisms of action: Direct acting---Propiolactone; Indirect acting---Dimethylnitrosamine; Epigenetic---Benzene, 2,3,7,8-Tetrachlorodibenzodioxin, and epichlorohydrin; and Metal-cadmium. Theorized mechanisms of action were taken from Klaassen et al. (1986).

In determining whether or not mechanism of action was important for ranking the long-term whole-animal tumorigenic studies, 6 rankings of the 52 chemicals were performed. These

375

6 lists were compared with one another, two-by-two, for consistency using the Spearman Rank Correlation test. All of the 15 pairs which were ranked had Spearman Rho values of 0.8 or greater and p-values were typically less than 0.0001 (Glass et al. 1990). These results provide a high degree of confidence that the relative potency framework, as used, can be expected to reliably rank toxicologic potency regardless of the mechanism of action of the agents being ranked.

Short-term tests involve bacteria, yeast, mammalian cells, and, sometimes, host animals. Despite this diversity, when relative potencies using data from short-term tests are evaluated for test chemicals with respect to a well studied reference chemical, a distribution of ratios is found similar to that for the whole animal tumorigenicity data. Summary relative potencies are again described by the median value. Using short-term data, the 6 reference chemicals were used to rank the 52 carcinogens, just as before. As might be expected, the range of responses (ratios) within the short-term assays was greater than within the whole-animal assays and therefore the median was less stable. As a consequence, when the 6 rankings of the short-term tests were compared with the 6 rankings of the tumorigenic tests using the Spearman Rank Correlation test, the correlations were lower than before. Correlation values were typically 0.4, and the p-values were higher, typically 0.01 to 0.05 (Glass et al. 1990). Considering the fact that the individual ratios for a given test and reference chemical often vary by factors of 1,000,000 or more, the consistency of Rho (which can vary from -1 to +1) which generally varied between 0.35 and 0.6 is remarkable. Only three p-values of the 15 two-by-two pairs were greater than 0.05. They were 0.16, 0.08 and 0.06. Again, given the consistency of results and the summary statistic (tested against a two-tailed distribution), the strength of the correlation is remarkable.

APPLICATION TO ELF ELECTROMAGNETIC FIELDS

On the basis of the similar ranking of the potency of the chemicals between whole animal data and the widely varying short-term data, a cautious step can be made toward analyzing effects derived by ELF electromagnetic fields, and for a variety of different biological endpoints.

Chronic exposure to 60 Hz electric fields have, since 1978, resulted in consistent reports of effects on synaptic transmission in rats (Jaffe 1978, 1979; Jaffe et al. 1980, 1981); this set of experiments will be used to demonstrate relative potency concepts. The actual endpoint measured is the postsynaptic compound action-potential height ratio at specific conditioning-test (C-T) intervals. Clearly, this type of endpoint is different from those typically considered in human health assessment. However, each measure of toxicity contributes to the summary relative toxicological potency. Electromagnetic field data to be used in the relative potency example are derived from the basic data of Jaffe 1978, 1979, and Jaffe et al. 1980, 1981. Similar measurements were made by Whitcomb and Santolucito (1976) who examined the effects of carbaryl on postsynaptic compound action potential height ratio at specific C-T intervals. Since reduction of enzyme (acetylcholinesterase) activity is a reliable measure of carbaryl toxicity (Fernandez et al. 1982), we will take carbaryl as the reference agent. Figure 1 depicts the percent depression brain acetylcholinesterase as a function of carbaryl dose (Mount et al. 1981). For the reference carbaryl data, Holtzman adult white male rats were used, while for the ELF field studies, Sprague-Dawley male rats approximately 3 months old were used. For our illustration, the experimental systems are assumed to be sufficiently similar with respect to the synaptic transmission to allow direct comparison.

The data are quite similar in that the C-T response was measured over the same inter-stimuli intervals, and, the spike height ratios relative to the conditioning stimuli are similar, to within about a factor of three. For both ELF and carbaryl data, when the interval between the two stimuli exceeds 800 ms, there is no difference from controls. Below 800 ms, both agents exhibit an increase in synaptic excitability. On this basis, the area under the two curves can be compared for an overall excitability response (Fig. 2). Since a dose-response relationship for 60 Hz electric fields has not yet been demonstrated, the isoeffect dose of carbaryl must be calculated. Thus, in order to obtain similar integrated responses, that is, for the dose of the reference agent carbaryl to produce an equal level of response as the 60 Hz exposure, a reduction of the carbaryl dose from that given by Whitcomb and Santolucito (1982) would be required. The original carbaryl dose was quoted as being 1/10 of an oral LD_{50}. The actual dose was not specified but an LD_{50} for the rat should be between 500 and 700 mg/kg. For the present example, the value of 600 mg/kg is chosen.

To reduce depression of acetylcholinesterase activity, and thereby lessen the excitability of the synapses would require a lower dose of carbaryl. The integrated excitability curves of Figure 2 indicates that a reduction of the carbaryl response by approximately 2.3 is required so that the carbaryl response could equal that of the ELF exposed system. From data in Figs. 1 and 2 the carbaryl isodose required to reduce by a factor of 2.3 the response from the 1/10 LD_{50} level (60 mg/kg, 24% acetylcholinesterase inhibition) is 4% of the carbaryl LD_{50} or 26 mg/kg. This dose provides one relative potency ratio.

In order to determine the summary relative potency of electric fields, a variety of endpoints will have to be analyzed similarly to the above analysis and the resultant equivalent doses of reference agents will themselves need to all be calibrated to a single agent. The median of this collection then would represent the potency of electric fields as represented by the different exposure regimes and test systems. This potency, expressed as equivalent dose of some well studied reference agent, can then be used to gain a perspective on the general toxicological potency of the electric fields.

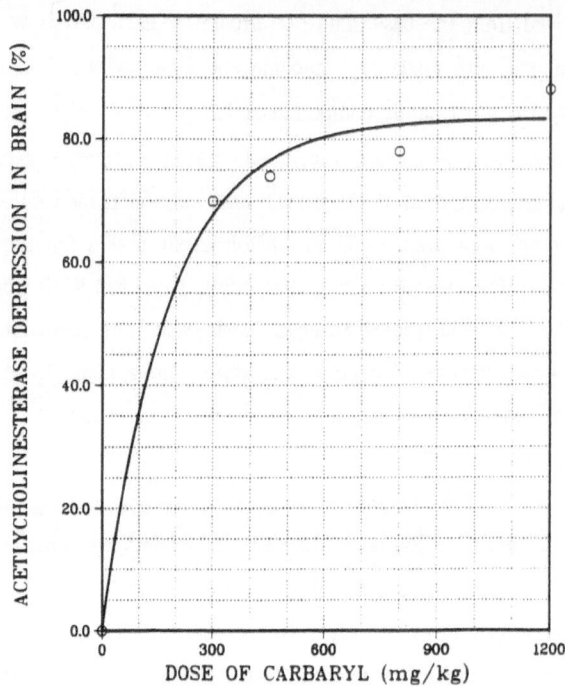

Fig. 1. Acetylcholinesterase depression in rat brain as a function of carbaryl dose.

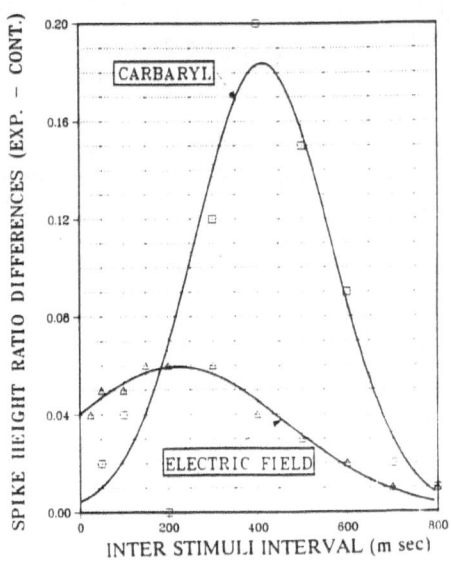

Fig. 2. Comparison of synaptic excitability in rats as a function of inter stimuli interval using carbaryl and electric fields .

DISCUSSION

There exists within society a need to understand, in a general way, the gross level of hazard posed by 60 Hz electromagnetic fields. Are they potent like dioxin or relatively impotent like saccharin? This question must be answered before more elegant questions are addressed such as: What is(are) the mechanisms of action? What is(are) the dose-response curves? Do fields act via a co-factor? At present, we do not know the important exposure variables, but this does not keep us from being concerned about the possible level of hazard for any arbitrary exposure. Because of this dilemma, we propose that, for the time being, all positive biological responses be lumped together as "exposed." As time goes on and we gain more information, this "exposed" category can be refined, first into classes like low, medium, and high, and later a more continuous measure.

On-going bioeffects experiments are mostly designed to search for mechanisms and results of these experiments do not easily lend themselves to conventional hazard assessment. Construction of a range of relative potencies clearly will take a considerable effort and will require that researchers perform positive controls during their experiments. If positive controls are used, resultant data can be incorporated into the relative potency framework. Our experience has shown that the median of relative potencies becomes stable with a small number of input points (usually less than a dozen).

Capabilities exist to derive gross relative toxological potencies of electromagnetic fields based on "exposed" and "not exposed" categories. Assistance from the experimental bioeffects community is required to provide positive control information in order to expand the number and diversity of experimental test systems making up a median relative toxological potency for 60 Hz electromagnetic fields.

REFERENCES

Anderson, E.L. (1983). Quantitative approaches in use to assess cancer risk. *Risk Anal.* 3(4):277-295.

Dudney, C.S., Walsh, P.J., Jones, T.D., Calle, E.E., and Griffin, G.D. (1983). On the use of relative toxicity for risk estimation. pp 499-513 in *Short-term Bioassays in the Analysis of Complex Environmental Mixtures III.* M.D. Waters, S.S. Sandhu, J. Lewtas, L. Claxton and S.N. Nesnow, eds., Plenum Press, New York.

Fernandez, Y., Falzon, M., Cambon-Gros, C., and Miljavila, S. (1982). Carbaryl triicompartmental toxicokinetics and anticholinesterase activity. *Toxicol. Lett.* 13: 253-258.

Glass, L.R., Jones, T.D., Easterly, C.E., and Walsh, P.J. (1990). Use of short-term test systems for the prediction of the hazard represented by potential chemical carcinogens. Oak Ridge National Laboratory Report, ORNL/TM-11413, Oak Ridge, TN.

Griffin, G.D., Jones, T.D., and Walsh, P.J. (1979). Chemical cytotoxicity: pp 723-732 in *Polynuclear Aromatic Hydrocarbons*, P.W. Jones and P. Leber, eds., Ann Arbor Science Publishers, Inc., Ann Arbor, Mich.

Jaffe, R.A. (1978). Chapter XIII. Neurophysiology, pp. 141-155 in *Biological Effects of High Strength Electric Fields on Small Laboratory Animals, Annual Report*, April 1977 to March 1978, U.S. Department of Energy, HCP/T 1830-03, NTIS, Springfield, Virginia.

Jaffe, R.A. (1979). Chapter XI. Neurophysiology. pp. 195-227 in *Biological Effects of High Strength Electric Fields on Small Laboratory Animals*, Interim Progress Report DOE/TIC-10084, eds. R.D. Phillips, L.B. Anderson, and W.T. Kaune, U.S. Department of Energy, NTIS, Springfield, Virginia.

Jaffe, R.A. Laszewski, B.L., and Carr, D.B. (1980). Chronic exposure to a 60-Hz electric field: effects on neuromuscular function in the rat, *Bioelectromagnetics* 2: 227-240.

Jaffe, R.A. Laszewski, B.L., and Carr, D.B. (1981). Chronic exposure to a 60-Hz electric field: effects on synaptic transmission and peripheral nerve Function in the rat, *Bioelectromagnetics* 1: 131-147.

Jones, T.D., Griffin, G.D., and Walsh, P.J. (1983). A unifying concept for carcinogenic risk assessments. *J. Theor. Biol.* 105: 35-61.

Jones, T.D., Walsh, P.J., and Zeighami, E.A. (1985). Permissible concentrations of chemicals in air and water derived from RTECS entries: a "RASH" chemical scoring system. *Tox. Ind. Health* 1(4):213-234.

Jones, T.D., Walsh, P.J., Watson, A.P., Owen, B.A., Barnthouse, L.W., and Sanders, D.A. (1988). Chemical scoring by a *Ra*pid *S*creening of *H*azard (RASH) method. *Risk Anal.* 8(1):99-118.

Klaassen, C. D. (1986). Principles of Toxicology, pp. 11-32 in: *Casarett and Doull's Toxicology. The Basic Science of Poisons*. C. D. Klaassen, M. O. Amdur and J. Doull Eds. Macmillan Publishing Co., New York.

Morris, M.D., Kimball, K.T., Aldrich, T.E., and Easterly, C.E. (1989). Statistical approach to combining the results of similar experiments, with application to the hematologic effects of extremely-low-frequency electric field exposure. *Bioelectromagnetics* 10: 23-34.

Mount, M.E., Dayton, A.D., and Oehme, F.E. (1981). Carbaryl residues in tissues and cholinesterase activities in brain and blood of rats receiving carbaryl, *Toxicol. Appl. Pharmacol.* 58: 282-296.

Walsh, P.J., Jones, T.D., Griffin, G.D., Dudney, C.S., Calle, E.E., and Easterly, C.E. (1982). Risk assessment approaches: General definitions, limitations, and research needs. *J.Environ.Sci. Health* A17:541-552.

Watson, A.P., Jones, T.D., and Griffin, G.D. (1989). Sulfur mustard as a carcinogen: Application of relative potency analysis to the chemical warfare agents H, HD, and HT. *Regul. Toxicol. Appl. Pharm.* 10: 1-25.

U.S. EPA (U.S. Environmental Protection Agency) (1989). *Workshop Report on EPA Guidelines for Carcinogenic Risk Assessment*, EPA/625/3-89-015, March 1989, Washington D.C. 20460.

Wertheimer N. and Leeper, E. (1979). Electrical wiring configurations and childhood cancer. *Am. J. Epidemiol.* 109: 273-284.

Whitcomb E.R. and Santolucito J.A. (1976). The action of pesticides on conduction in the rat superior cervical ganglion. *Bull Environ. Contam. Toxicol.* 15(5): 348-356.

INFLUENCE OF MEMORY ON THE STATISTICS OF PULSATING CORONA

R. J. Van Brunt and S. V. Kulkarni

National Institute of Standards and Technology
Gaithersburg, MD 20899

INTRODUCTION

It has been shown in the recent work of Van Brunt and Kulkarni (1990) that the well known pulsating negative corona (Trichel pulse) discharge in electronegative gases is a stochastic process in which memory effects play an important role. A complete understanding of this phenomenon cannot be achieved without information about these memory effects which are associated with the influence of negative-ion space charge and metastable species from previous discharge pulses on the initiation and growth of subsequent pulses. The purpose of this paper is to illustrate how information about memory can be obtained from measurements of various conditional discharge pulse-amplitude and pulse-time-separation distributions. The discharge phenomenon is represented here by a random point process corresponding to the set $\{q_1, q_n, \Delta t_{n-1}\}, n = 2, 3, 4, \cdots$ of pulse amplitudes, q_n, and time separations, Δt_{n-1}. Here Δt_{n-1} is the time separation between the nth and $(n - 1)$st discharge pulses.

MEASUREMENTS

The results reported here were obtained using a measurement system previously described (Van Brunt and Kulkarni, 1989) which allows a direct, "real-time" determination of the set of conditional and unconditional pulse-amplitude and pulse-time separation distributions: $p_0(q_n)$; $p_0(\Delta t_n)$; $p_1(q_n|q_{n-1})$; $p_1(q_n|\Delta t_{n-j})$, $j \geq 1$; $p_1(\Delta t_n|\xi)$, $\xi = q_n$ or Δt_{n-1}; $p_2(q_n|\Delta t_{n-1}, \xi)$, $\xi = q_{n-1}$, or Δt_{n-2}. These distributions are defined such that $p_0(q_n)dq_n$ is the probability that the nth discharge pulse, for arbitrary n, has an amplitude between q_n and $q_n + dq_n$ independent of previous pulse amplitudes or time separations; $p_1(q_n|\Delta t_{n-1})dq_n$ is the probability that the nth discharge pulse has an amplitude in the same range if this pulse is separated from the previous pulse by a fixed time separation Δt_{n-1}; and $p_2(q_n|\Delta t_{n-1}, q_{n-1})dq_n$ is the same with both Δt_{n-1} and q_{n-1} fixed.

The results reported here apply to a self-sustained discharge in a Ne/5%O$_2$ gas mixture at an absolute pressure of 100 kPa ($\sim$ 1 atm). Polished stainless-steel point and plane electrodes were used, where the point electrode served as the cathode for a dc gap voltage, V_a, of 6.5 kV. The point-to-plane gap spacing was 2.0 cm and the radius of curvature at the tip of the point electrode was 0.15 mm. The pulse amplitude is expressed in units of picocoulombs (see Van Brunt and Leep, 1981).

RESULTS AND DISCUSSION

Examples of the results obtained for the various conditional and unconditional distributions are shown in Figs. 1–5. Data for the different sets of distributions were obtained at different times so that the cathode surface conditions that apply, for example, to the results in Fig. 1

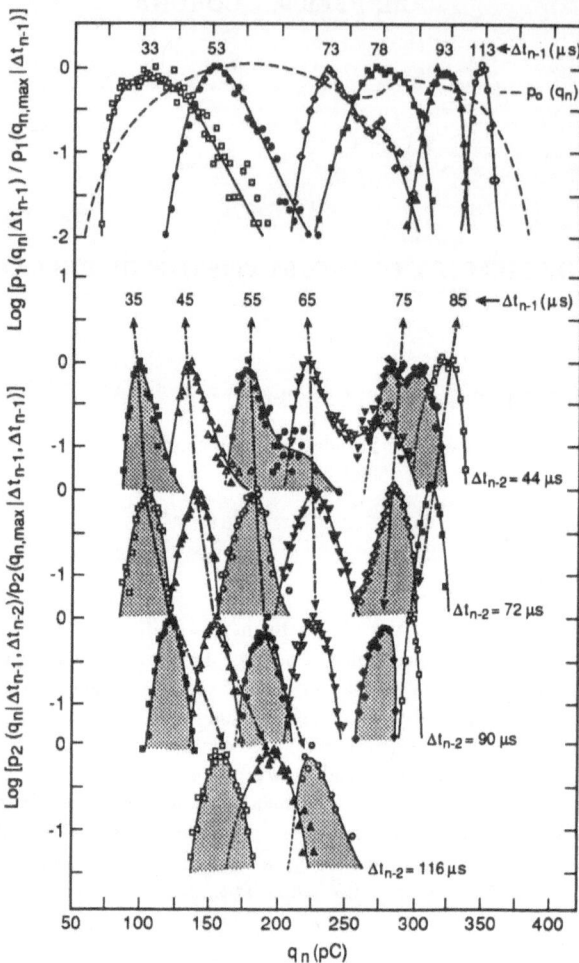

Fig. 1. Measured unconditional and conditional discharge pulse-amplitude distributions $p_0(q_n)$, $p_1(q_n|\Delta t_{n-1})$, and $p_2(q_n|\Delta t_{n-1}, q_{n-1})$ for the indicated fixed values for Δt_{n-1} and q_{n-1}. The distributions have been normalized to the maxima.

differ slightly from those that apply to Fig. 2. This accounts for the difference in $p_0(q_n)$ shown in these two figures. A detailed interpretation of the results presented here would go beyond the scope of this report. However, salient features of the data and certain important conclusions that can be derived therefrom should be noted. A reasonably complete discussion of the physical bases for the stochastic behavior of the Trichel-pulse phenomenon has been given by Van Brunt and Kulkarni (1990).

The fact that the second-order conditional distributions, p_2, differ from the corresponding first-order distributions, p_1, which in turn differ from the corresponding unconditional distributions, p_0, indicates unequivocally that the set of random variables $\{\Delta t_n, q_n, \Delta t_{n-1}, q_{n-1}, \Delta t_{n-2}, \cdots\}$ associated with adjacent pulses are not independent. For example, it is seen from Fig's 1 and 2 that q_n has a strong positive dependence on Δt_{n-1}. This dependence can be related to the influence of moving negative-ion space charge from previous pulses in suppressing the magnitude of the electric field at the cathode when the next pulse develops. It is also seen from Figure 1 that the amplitude, q_n, of a pulse can be either positively or negatively dependent on the amplitude, q_{n-1}, of the previous pulse. The sign of this dependence can be explained in terms of the competing effects of negative-ion space charge and metastable species in respec-

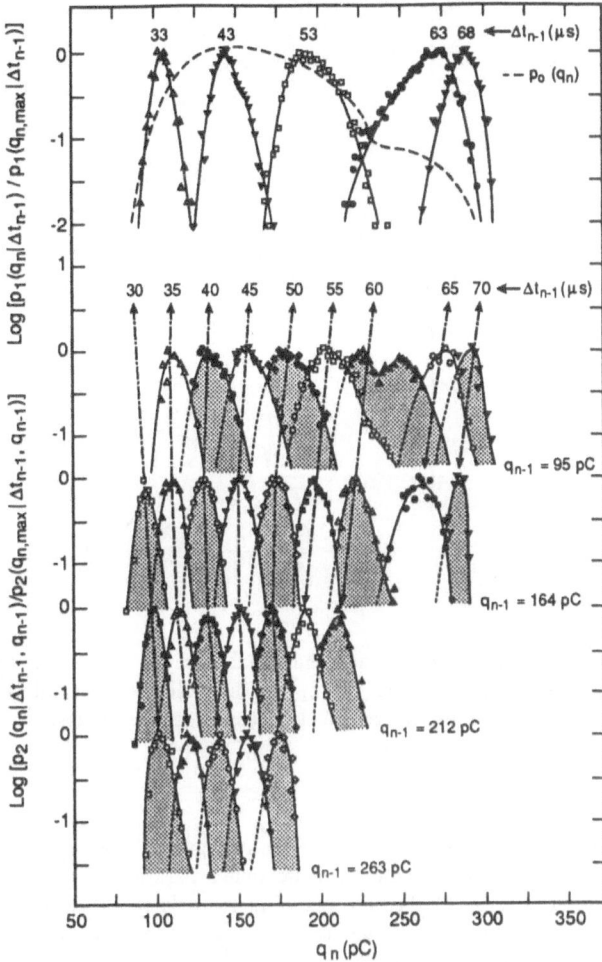

Fig. 2. Measured unconditional and conditional discharge pulse-amplitude distributions $p_0(q_n)$, $p_1(q_n|\Delta t_{n-1})$, and $p_2(q_n|\Delta t_{n-1}, \Delta t_{n-2})$ for the indicated fixed values for Δt_{n-1} and Δt_{n-2}. The distributions have been normalized to the maxima.

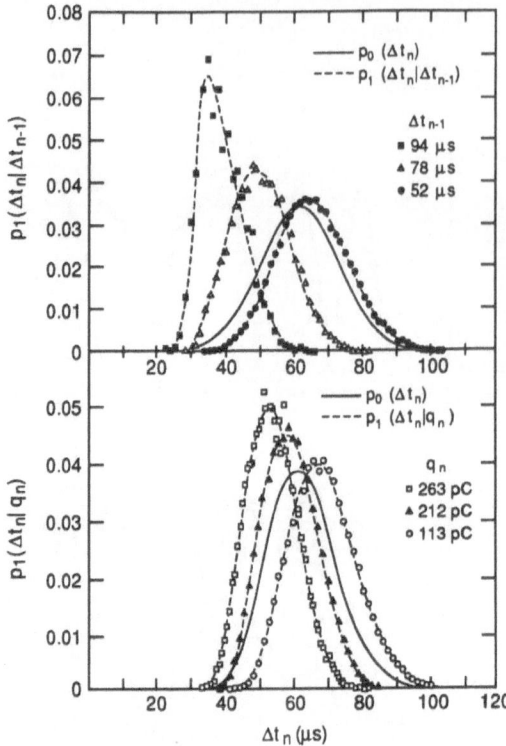

Fig . 3. Measured unconditional and conditional discharge pulse-time-separation distributions $p_0(\Delta t_n)$, $p_1(\Delta t_n | q_n)$, and $p_1(\Delta t_n | \Delta t_{n-1})$ for the indicated fixed values for q_n and Δt_{n-1}. The distributions have been normalized to the areas under the curves.

tively retarding or enhancing the growth of the next pulse. The negative dependence of Δt_n on q_n (and Δt_{n-1}) implied by the conditional time-separation distributions shown in Fig. 3 can be understood in terms of the influence of metastable species from the previous pulse in enhancing the probability for initiating the next pulse by ejecting electrons from the cathode surface during field-assisted quenching.

Because of the correlations among the amplitudes and time separations of successive pulses, the distributions shown in Figures 1–3 are all related. It can be shown, for example, from the law of probabilities that $p_0(q_n)$, $p_0(\Delta t_n)$, and $p_1(q_n | \Delta t_{n-1})$ are related by the integral expression

$$p_0(q_n) = \int_0^\infty p_0(\Delta t_{n-1}) p_1(q_n | \Delta t_{n-1}) d(\Delta t_{n-1}), \tag{1}$$

and the distributions $p_0(q_n)$, $p_0(\Delta t_n)$, $p_1(q_n | \Delta t_{n-1})$, $p_1(\Delta t_n | q_n)$, and $p_2(q_n | q_n, \Delta t_{n-1})$ are related by

$$p_1(q_n | \Delta t_{n-1}) = p_0(\Delta t_{n-1})^{-1} \int_0^\infty p_0(q_{n-1}) p_1(\Delta t_{n-1} | q_{n-1}) \times p_2(q_n | q_{n-1}, \Delta t_{n-1}) dq_{n-1}. \tag{2}$$

Equation (1) indicates that if q_n is dependent on Δt_{n-1}, then any externally-induced change in the time-interval distribution, $p_0(\Delta t_n)$, will necessarily be reflected as a change in the amplitude distribution, $p_0(q_n)$.

Since, as seen from the data in Figs. 2, 4, and 5, the profiles for $p_2(q_n | \Delta t_{n-1}, \Delta t_{n-2})$ and for $p_1(q_n | \Delta t_{n-j})$, $j = 2, 3, 4$ do not match the profile for $p_0(q_n)$, it can be stated that q_n

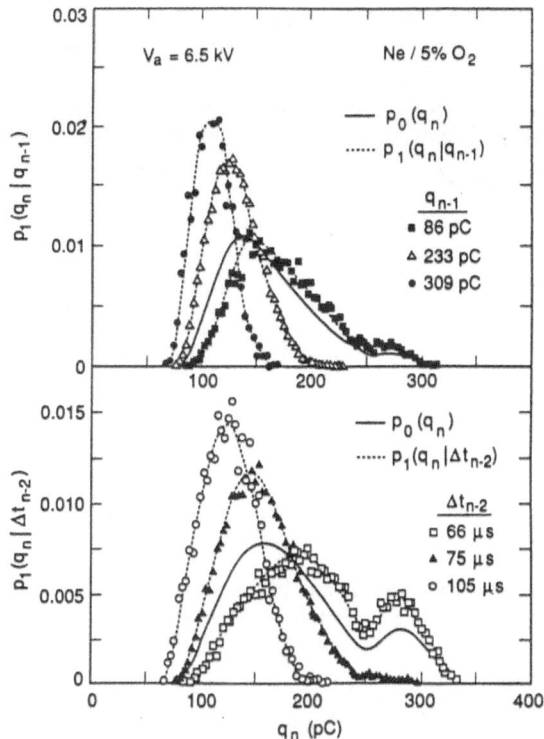

Fig. 4. Measured unconditional and conditional discharge pulse-amplitude distributions $p_0(q_n)$, $p_1(q_n|\Delta t_{n-2})$, and $p_1(q_n|q_{n-1})$ for the indicated fixed values for Δt_{n-2} and q_{n-1}. The distributions have been normalized to the areas under the curves.

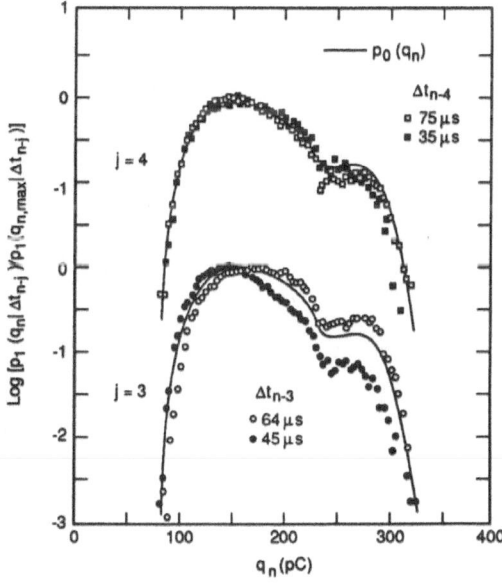

Fig. 5. Measured unconditional and conditional discharge pulse-amplitude distributions $p_0(q_n)$, $p_1(q_n|\Delta t_{n-3})$, and $p_1(q_n|\Delta t_{n-4})$ for the indicated fixed values for Δt_{n-3} and Δt_{n-4}. The distributions have been normalized to the maxima.

depends on $\Delta t_{n-j}, j > 1$, and therefore, the process is one for which memory extends back in time beyond the most recent event, i.e., the process is non-Markovian. This observation is consistent with results reported in the recent work of Steiner (1988). It can, in fact, be shown (Van Brunt and Kulkarni, 1990) that because of the relatively strong dependence of q_n on both Δt_{n-1} and q_{n-1}, it is possible for memory to propagate indefinitely back in time.

ACKNOWLEDGEMENTS

This work was supported in part by the U.S. Department of Energy, Division of Electric Energy Systems.

REFERENCES

1. Steiner, J. P., "Digital Measurement of Partial Discharge,," Ph.D. Thesis, Purdue University, W. Lafayette, IN (1988).

2. Van Brunt, R. J. and Leep, D., 1981, Characterization of Point-Plane Corona Pulses in SF_6, *J. Appl. Phys.* 52:6588.

3. Van Brunt, R. J. and Kulkarni, S. V., 1989, Method for Measuring the Stochastic Properties of Corona and Partial Discharge Pulses, *Rev. Sci. Instrum.* 60:3012.

4. Van Brunt, R. J. and Kulkarni, S. V., 1990, Stochastic Properties of Trichel-Pulse Corona: A Non-Markovian Random Point Process, *Phys. Rev. A.*, in press.

DISCUSSION

R. T. WATERS: What are the effects of gas type on your corona statistics?

R. J. VAN BRUNT: The kinds of stochastic behavior exhibited here for neon–oxygen gas mixtures are also observed for other mixtures like air, nitrogen–oxygen, sulfur hexafluoride–oxygen, and pure oxygen. This behavior is, in fact, ubiquitous and inherent to the Trichel–pulse phenomenon independent of gas mixture.

I. GALLIMBERTI: Did you do any covariance analysis on the stochastic process that could show how far the "memory" of one single event extends in time on the subsequent ones?

R. J. VAN BRUNT: We have carried out a covariance analysis in the sense that we used our data on conditional distributions to compute correlation coefficients to determine the degrees of correlation among the amplitudes and time separations of successive Trichel pulses. This is discussed in our recent paper (R. J. Van Brunt and S. V. Kulkarni, Phys. Rev. A, Oct. 15, 1990). This kind of analysis was also carried out in the work of J. P. Steiner (Ph.D. Thesis, Purdue University, 1988) which again showed that memory indeed propagates far back in time for this discharge phenomenon.

L. NIEMEYER: The interrelations between the probability distributions of different orders indicate that there is redundance. Can one devise, based on statistical theory, a systematic strategy of the sequence in which one has to measure the different distributions? As an example, if one wants to check for a memory effect, what are the necessary distributions to assess the existence/nonexistence of this effect?

R. J. VAN BRUNT: It is true that there is a certain redundancy built into our measurements in as much as we look at distributions that are related. We have done this to determine that the distributions are all self consistent and to determine which distributions might provide the most interesting information about the physical mechanisms of the process. In the design of practical partial–discharge measurement systems to assess stochastic behavior, one could use a subset of the distributions considered in this work. One could, for example, build a four channel system to simultaneously measure $p_0(q_n)$, $p_0(\Delta t_n)$, $p_1(q_n|\Delta t_{n-1})$ and $p_1(\Delta t_n|q_n)$ that might be adequate for some applications.

PULSED CORONA EXPERIMENTS

E.J.M. van Heesch, A.J.M. Pemen, J.W. van der Snoek and P.C.T. van der Laan

High-Voltage Group, Eindhoven University of Technology
P.O. Box 513, 5600 MB Eindhoven, The Netherlands

INTRODUCTION

With a recently modified setup for pulsed corona experiments we operate corona at reduced pressures (100-760 Torr) and obtain direct readings of pulsed corona currents. Pulsed currents of up to 93 A have been recorded. A flexible design of the high-voltage circuit allows easy adjustment of pulse parameters, DC-bias voltage and voltage polarities.

The reported experiments on pulsed corona show that completely new types of discharges can be produced, in particular when in addition to the voltage pulse a DC bias is used to remove the initial space charge. Features are: high intensity, excellent controllability, more homogeneous than DC corona and a strong interaction with the external circuit. These pulsed discharges should not be categorized in the same way as DC corona.

Direct pulsed current readings are obtained via a bridge circuit that subtracts the high displacement current caused by the pulsed voltage. A balun type transformer matches the balanced output of the bridge to the unbalanced input of a digital oscilloscope.

Previously the pulsed current was derived in a few cases from the voltage waveforms measured with and without DC bias applied (Ref.1). Indications from those earlier measurements of currents of about 50 A in corona at 400°C, 1 atm., are confirmed by present direct current measurements in air of 20°C and similar density, i.e. at a pressure of 330 Torr.

To allow problem-free operation of diagnostics and digital oscilloscopes during the intense interference pulse caused by the experiment, adequate EMC techniques, as developed by our group (Ref.3) are incorporated in the experimental setup.

EXPERIMENTAL SETUP

The corona vessel is a 1 m long, 0.29 m diameter steel cylinder with a 0.55 m long centered corona wire of 1 mm diameter. The wire is suspended from a 90 mm diameter tube supported by a GIS spacer used as high-voltage feedthrough.

The external circuit consists of a pulse source, a coupling capacitor and a DC bias source. Various types of circuits for the pulse source were built and tested (Ref.1). The circuit used presently, depicted in Fig.1, allows a wide range of pulse parameters: rise time 30 ns - 5 μs, decay 5 μs - 2 ms, peak value 5 kV - 50 kV, both polarities, repetition rate up to 20 Hz.

The switching element is a triggered sparkgap flushed with air. Capacitor C_1, charged via R_3, discharges via the gap into R_2. The corona wire is coupled to this RC discharge via capacitor C_2. A DC bias voltage (up to 60 kV) can be fed to the corona wire via resistor R_4.

Gaseous Dielectrics VI, Edited by L.G. Christophorou and
I. Sauers, Plenum Press, New York, 1991

Inside the corona vessel two separate cylinders (with capacitances C_a and C_b to high voltage) are installed for balanced current measurements (see next section). Each cylinder is connected to the center conductor of a coaxial cable with its braid attached to the vessel at the same location. Both cables are led to a passive bridge circuit which provides the actual discharge current.

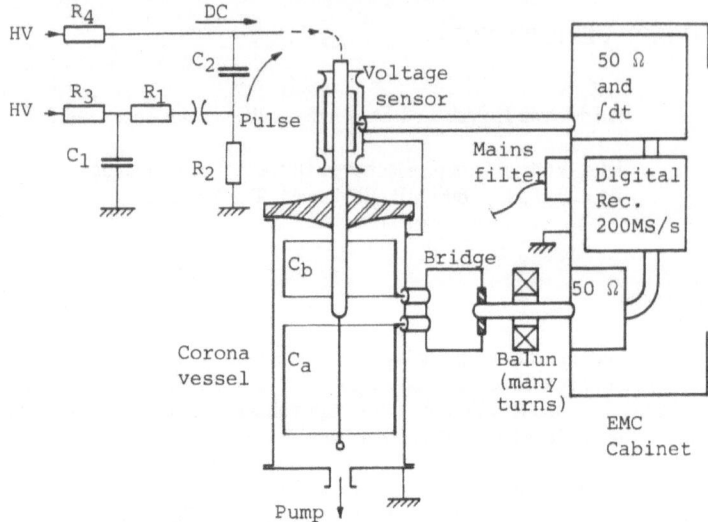

Fig. 1. Overview of the experimental setup; $C_1 = 1.2nF$, $C_2 = 8nF$, $R_2 = 20$-$90k\Omega$, $R_1 = 0$-$10k\Omega$.

A differentiating voltage sensor is mounted near the high-voltage feedthrough of the vessel. An integrator in a special measuring cabinet restores the original waveform. This Differentating/Integrating (D/I) technique is described in   Ref.2.

Electromagnetic Compatibility (EMC) of measuring equipment, pulsed high voltage and discharge are given special attention. Our EMC techniques are based on a philosophy that concentrates on the control of current loops. The transfer impedance between these loops and the sensitive apparatus is kept very low. The measuring cabinet (EMC cabinet) fits in this strategy.

CURRENT DETECTION

The External Current

The current I_e fed externally to the corona wire, crosses the space between wire and outer cylinder. Within the gap I_e has two components, the material current I_m and the displacement current I_d which is the time derivative of the total E-field flux, $\partial\Psi/\partial t$:

$$I_e = I_m + \epsilon_0 \partial\Psi/\partial t \tag{1}$$

The R.H. side of Eq. (1) can be taken at any surface enclosing the high-voltage electrode. The flux Ψ consists of a component Ψ_0, caused by the vacuum field and a component Ψ_p, caused by the space charge built up in the gap. The vacuum flux crossing the gap is directly related to the external voltage V_e and the vacuum capacitance C_0 of the gap:

$$\epsilon_0 \partial\Psi_0/\partial t = C_0 dV_e/dt \tag{2}$$

With this purely capacitive current I_{c0} of the vacuum field we rewrite Eq. (1):

$$I_e = I_{c0} + I_m + \epsilon_0 \partial\Psi_p/\partial t \tag{3}$$

As could be expected the external current equals a purely capacitive current plus a component resulting from the corona discharge, commonly referred to as the apparent current, (e.g. in Ref.4):

$$I_{app} = I_m + \epsilon_0 \partial \Psi_p / \partial t \qquad (4)$$

At a large dV/dt a considerable fraction of I_e in Eq.(3) flows in the gap as the capacitive current I_{c0}. In that case the E-field distribution is mainly determined by the capacitive currents and not by the charge carriers in the gap. The attainable maximum for the material current, before full breakdown, is expected to be of the order of the capacitive current.

Balanced Detection

To obtain a direct measure of the apparent current in pulsed experiments a detection is needed of the external current minus the purely capacitive current of the vacuum field. Basically I_e can be derived from the voltage across a small resistor in series with the ground lead of cylinder C_a in Fig.1. Similarly a measure for the capacitive current $C_0 dV_e/dt$ is found from the current in the ground lead of cylinder C_b. This cylinder has approximately the same capacitance as the corona cylinder but remains corona-free due to the much larger diameter of the high-voltage conductor.

Passive circuitry is used to subtract the two current signals. The problem of subtraction of two signals with equal polarities, a wide dynamic range and a large bandwidth was solved by means of a balun-type transformer. With respect to EMC it is especially attractive to use the balun since it is linear and operates also at high frequencies.

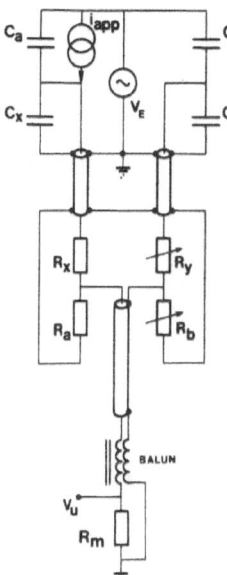

Fig.2.Circuit for corona current detection. The resistor R_m is the 50Ω cable termination at the output where V_u is measured.

A complication not mentioned yet is the presence of the large capacitances of both measuring cylinders to the grounded vessel (C_x and C_y in Fig.2). The bridge however can be tuned for deviations between the values of C_a and C_b as well as for deviations between C_x and C_y.
The circuit is balanced if we tune $R_a C_a = R_b C_b$ and $\tau_a = \tau_b$, where

$$\tau_a = (R_a + R_x)(C_a + C_x) \text{ and } \tau_b = (R_b + R_y)(C_b + C_y) \qquad (5)$$

In the frequency range for normal operation, we have $\omega M \gg R_b$ and $\omega \tau_a \ll 1$, and an output signal:

$$V_u = R_m R_a I_{app} / (R_m + R_a + R_b) \qquad (6)$$

MEASUREMENTS

Adjustment and Calibration of the Bridge Circuit

After an initial adjustment with 1 MHz and 5 MHz, sinewaves of 100 V at the high-voltage feedthrough, additional small corrections were made during pulsed operation at a low voltage level

(10 kV) and no DC bias, i.e. no corona. Fig. 3 demonstrates the quality of the adjustment. At about 40 kV pulse and 8 kV bias the corona process is delayed; therefore the residual interference and the current pulse are separated in time.

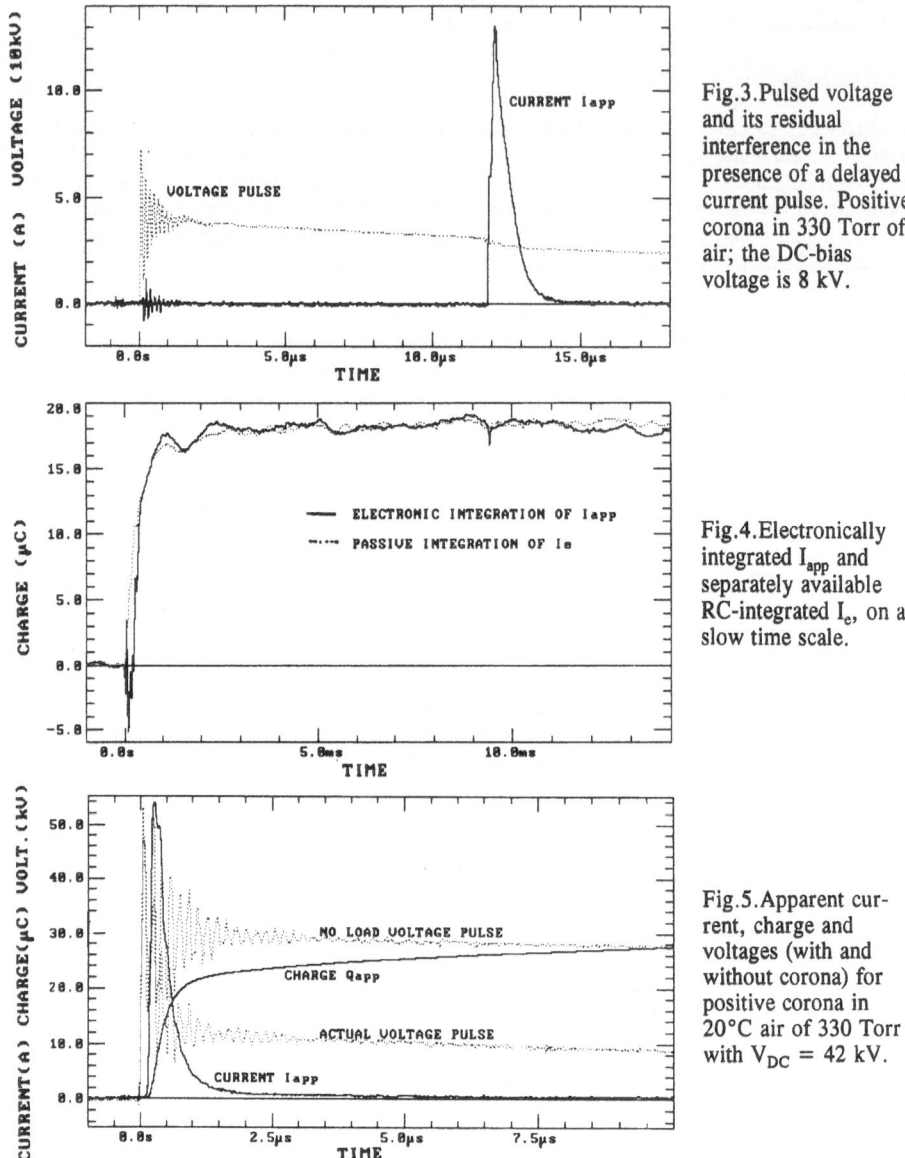

Fig.3.Pulsed voltage and its residual interference in the presence of a delayed current pulse. Positive corona in 330 Torr of air; the DC-bias voltage is 8 kV.

Fig.4.Electronically integrated I_{app} and separately available RC-integrated I_e, on a slow time scale.

Fig.5.Apparent current, charge and voltages (with and without corona) for positive corona in 20°C air of 330 Torr with $V_{DC} = 42$ kV.

A DC current source was connected between the two inputs of the bridge circuit to determine the calibration factor. Current source and bridge were made floating whereas the balance output was monitored with one side grounded. The resulting calibration was verified in several ways:
- Trichel pulse operation: The bridge output, I_{app} integrated per second, differed less than 7% from the separately available signal for the average (DC) corona current.
- Pulsed high voltage operation with high I_{app}: For a few cases the apparent charge was estimated from the dip in the voltage waveforms, according to the calculation described in Ref.1. The deviations from the integrated I_{app} were less than 30% for $t > 1\mu s$.
- Pulsed high voltage operation with moderate current values: Figure 4 shows that the I_{app} signal fed via a Walker integrator is almost identical to the separately available, low-pass RC-filtered signal for I_e.

Effects of Density, Polarity, Voltage and Risetime

An example of a high I_{app} is shown in Fig.5 together with the numerically integrated current and the simultaneous voltage pulse. With this high current, the apparent charge (2.75 10^{-5} C) represents a large fraction of the external charge in the circuit. The resulting drop in the applied voltage pulse is however not as complete as it is in Fig. 6, a case of real breakdown.

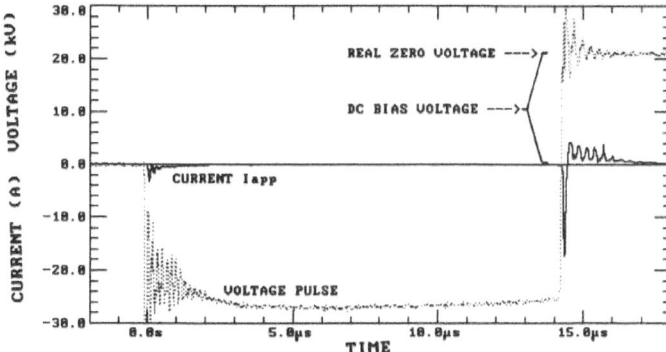

Fig.6.Corona current pulse followed by breakdown; negative corona in 20°C air of 200 Torr with V_{DC} = 20 kV.

Voltage, apparent current and charge were measured for a large number of conditions. Both for positive and negative corona, current and charge increase with pulse voltage and decrease with risetime and pressure. With negative corona the pulsed corona intensity does not increase with DC-bias voltages in excess of the inception level. The measuring results for negative corona are summarized in Fig. 7 where I_{app} is given as a function of the pressure; the parameter is the pulse amplitude and the pulse risetime and the error bars denote the (mostly small) variation with DC-bias voltage. For positive corona the peak I_{app} values continue to rise with DC-bias over the entire range of applied voltages (0.5 kV - breakdown). It turned out that the peak currents scale approximately with the sum of DC and pulse voltage, V_{tot}. The results for positive corona are given in Fig. 8-10 where I_{app} is plotted versus V_{tot}; the parameters are the pressure and the pulse risetime. A few exceptions on the V_{tot} dependence are: the extra line for 40 kV pulses in Fig. 8 and the scatter of points around the upper line in Fig. 9.

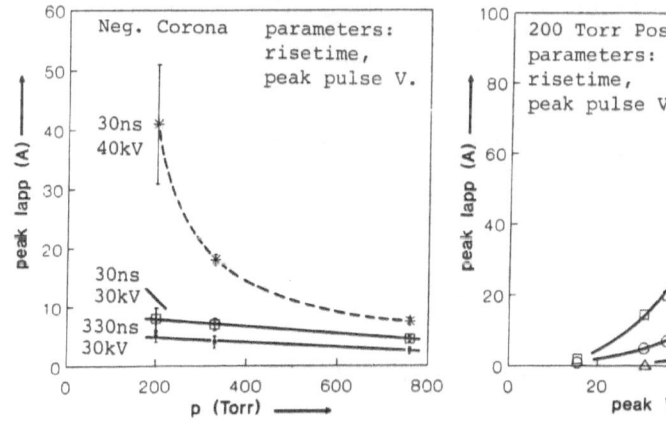

Fig.7. Variation of the peak current of negative corona with pressure; V_{DC} above inception, parameters are the risetime and the peak pulse voltage.

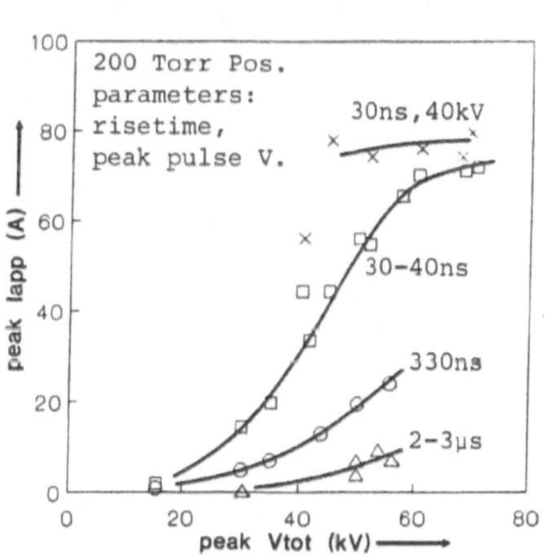

Fig.8. Variation of the peak current of positive corona with total voltage; various risetimes; p = 200 Torr; parameters are the risetime and the peak pulse voltage.

Existence Regions for Pulsed Corona

The experiments confirm an increased width of the existence region (voltage range between inception and breakdown) for pulsed corona at reduced pressures and at room temperature. In the present setup with pulses of less than 200 μs duration, the gain in width compared to DC voltage is 3.6 (pos.) and 1.7 (neg.) for corona in 200 Torr of air (the density at 840°C, 1 atm.). At 330 Torr (i.e. 400°C) these gains are 1.6 (pos.) and 1.2 (neg.).

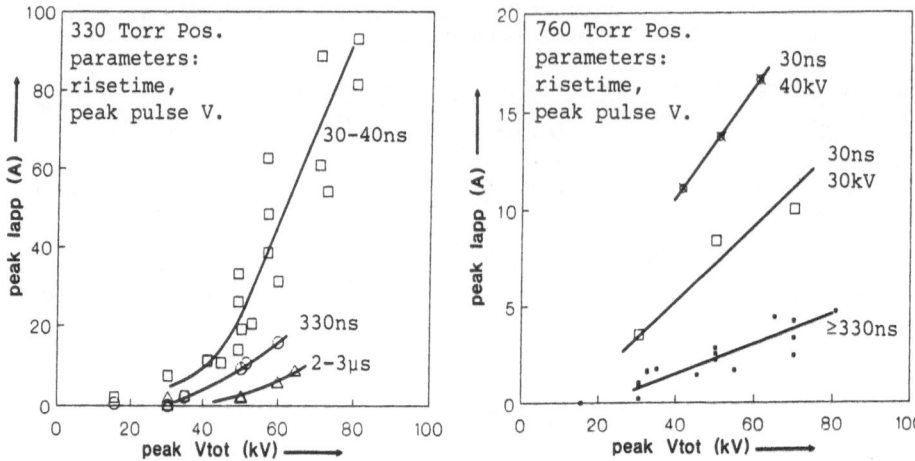

Fig.9. See Fig.8 caption; p=330 Torr.   Fig.10 See Fig.8 caption; p=760 Torr.

DISCUSSION

The measured values of Q_{app} are considerably higher than the original (vacuum) charge Q_0 present on the gap electrodes after the application of the steep pulses. The real injected charge is even higher, generally, since in Eq.(4) I_m and $\partial \Psi_p/\partial t$ usually have opposite signs at those locations where I_m flows. Figure 4 illustrates how the Q_{app} value is approaching the real Q after most of the charge has crossed the gap. Immediately after the current pulse, one would assume that no charge carriers have as yet completely crossed the gap (since we have no complete breakdown). Accordingly, at some surface around the corona wire, the term I_m (in Eqs. 1 to 4) remains zero. Therefore a large Ψ_p is needed to explain the large I_{app} and Q_{app}. A partial shorting of the gap by newly formed space charge could be looked at as an increase of C_0 with ΔC. A measured value $Q_{app}/C_0.V = 30$ would correspond to a space charge cylinder expanding from the wire (r = 0.5 mm) to a 12 cm radius; certainly a large fraction of the distance to the outer electrode (r = 14,5 cm).

Clearly, the high corona intensities are not easily explained. This interesting discharge behavior occurs both with negative and positive corona and is most pronounced for the shortest risetimes, in accordance with earlier remarks on E-field distribution and dV/dt.
Pulsed positive and negative corona are however differently affected by the DC bias voltage. Since DC bias above inception does not enhance pulsed negative corona, the highest negative corona intensities are found with the highest pulse voltages and a relatively low DC voltage. For positive corona, the sum of the two voltages is the important parameter within the range of applied voltage pulses.

REFERENCES

1. E.J.M. van Heesch, A.J.M. Pemen, and P.C.T. van der Laan, Pulsed corona existence up to 850°C. Sixth International Symposium on High Voltage Engineering, New Orleans, 1989, paper 14.07.
2. E.J.M. van Heesch, et al, Field tests and response of the D/I high voltage measuring system. Sixth International Symposium on High Voltage Engineering, New Orleans, 1989, paper 42.23.
3. M.A. van Houten, et al, General methods for protection of electronics against interference, tested in high-voltage substations, Proc. 8th International Symposium on Electromagnetic Compatibility Zurich, 1989, p. 429-434.
4. J.M. Wetzer and P.C.T. van der Laan, Prebreakdown currents, Basic interpretation and time-resolved measurements, IEEE Transactions on Electrical Insulation, Vol. 24, No. 2, p. 297-308, 1989.

DISCUSSION

R. T. WATERS: I appreciate the care with which these measurements of apparent current have been made. Have you performed any experiments to determine the spatial location of corona charge?

E. J. M. VAN HEESCH: Information regarding the spatial charge distribution would be very helpful in explaining the observed high corona intensities. To obtain such data, space– and time–resolved optical diagnostics need to be implemented in the setup for future experiments. With the various discharge models we can think of now, the one mentioned in the paper, for example, uses the ratio $Q_{app}/Q_o \approx 30$ to derive a cylindrical corona area extending from the wire to about 80% of the entire gap.

CURRENT STABILITY OF NEGATIVE CORONA DISCHARGES IN SF_6 AND

DELAYED SPARK BREAKDOWN

K. HADIDI and A. GOLDMAN

Laboratoire de Physique des Décharges (CNRS/ESE)

Plateau de Moulon,91192 Gif sur Yvette Cedex, France

ABSTRACT

Current stability of negative corona discharges in SF_6 is analysed in terms of gas and electrode surface evolution. It is found that the discharge behaviour is predominantly governed by the surface evolution of the high-field electrode and that there is a relation between its initial state and the spark breakdown current, implying the nature of the negative ions produced .

INTRODUCTION

It is known that the spark breakdown voltage of a corona gap is sensitive to the gas purity and to the surface characteristics of the high-field electrode. However, these factors are varying with time during corona exposure, and the resulting changes may in turn influence the discharge and its spark-over behaviour. The influence of such changes on some discharge characteristics at low current levels in SF_6 had already been studied by Van Brunt (1982) and by him and Leep (1982) with special emphasis on the effects of trace levels of water vapour, and by Fujimoto (1987) concerning charge trapping. On the other hand, Hughes and Hampton (1982) had shown that an improvement on the dielectric losses of compressed gas capacitors can be gained by submitting them to appropriate electrical stresses, while the use of repeated sparks, sometimes utilized as a conditioning process, leads on the contrary to increased losses even if good for other electrical characteristics.
A problem we tried to solve was to see how discharge-produced changes on the high-field electrode surface could affect the current stability, and if, under certain conditions, they could eventually lead to delayed spark breakdown at voltages initially too low for it. For this purpose we undertook exploratory test experiments with constant dc voltage applied to a point-to-plane gap, and saw that such a situation could be observed with the point operating as a cathode.
This paper reports results showing the influence of different preconditioning processes of the cathode point on this behaviour.

EXPERIMENTAL SET-UP AND PROCEDURE

The experiments were performed with a pair of copper electrodes in a 6 dm^3 stainless steel vessel. The high field electrode consisted in a rod ended by a hyperboloical cap of 1 mm curvature radius, used as a cathode, and the low-field electrode in a plane disk of 10 cm diameter, connected to earth through a 1.8 kohms resistance. The voltage drop across this resistor was used for determination of the mean total discharge current.

Almost all experiments were carried out with a 10 ± 0.5 mm gap length and with a negative constant voltage of 60 kV dc applied to the point electrode. The spark breakdown voltage measured at the beginning of a run under a rapidly increasing voltage was about 105 kV.

Each experiment was made with new gas (commercial SF_6 99.9% purity, filtered in order to have less than 0.1 ppmv O_2 and less than 0.4 ppmv H_2O) introduced to a pressure of 2 bars in the chamber, first evacuated to a pressure of about 10^{-9} bar with an ionic pump.

In all cases, standard polishing and cleaning procedures were used for the surface preparation of the point electrode, using abrasive paper and polishing paste for the polishing and ethanol in an ultrasonic cleaner during 30 min for the cleaning . In the foregoing text, experiments carried out only with these preliminary surface preparation procedures will be referred to as experiments of type A, while experiments made with complementary preconditioning procedures will be referred to as experiments of type B. The complementary preconditioning procedures used were mainly of three types :

* a baking at 300 °C during 16 hours in the pre-evacuated discharge chamber before filling it with SF_6 ;

* cathode sputtering by means of glow discharges also operated in the pre-evacuated discharge chamber with Ar/H_2 (70/30%) at a pressure of 10 mbar during 1 hour, as used by Avni et al.(1989) for the cleaning of iron surfaces before further surface treatments (After this cleaning procedure the experimental chamber had again to be evacuated to the vacuum level used as a standard for all experiments before SF_6 introduction);

* special polishing conditions consisting in not using polishing paste or in using the point electrode of a previous experiment with our standard cleaning procedure but without repolishing it.

All experiments were carried out during hours duration runs while continuously recording the mean total current with a time of response of 0.3 s.

Scanning microscopy and XPS spectrometry were used for morphological and chemical analyses of the electrodes before and after shortened runs of variable duration to get information on the kinetics of the surface evolution. Due to the small size and to the curvature of the point electrode, only qualitative informations could be got from XPS spectrometry.

RESULTS ON THE CURRENT STABILITY

Autostabilizing phase of the discharge

Strong effects of the preconditioning handling of the point electrode are observed as soon as the voltage is switched on.

1 . At corona onset, experiments of type A (without preconditioning the point electrode) give lower values for the corona onset current ($I_0 < 100$ μA) than experiments of type B (with a preconditioned point electrode) for which $I_0 > 100$ μA.

2.Immediately after corona is switched on the current generally decreases. In the case of experiments of type A, the period of current decrease and recovering lasts a time ($\Delta t)_{min}$ of about 1 to 3 hours depending on the I_0 value (Fig.1a) with a consistent current drop ($\Delta I)_{min} = I_0 - I_{min}$ very dependent on I_0 (Fig.1b).During this period, the discharge shows unstable both by its luminous emission, observed visually, and by its current fluctuations. These fluctuations are illustrated by the current record sample presented in Fig.2 and observable at the beginning of the evolution curve of Fig.3, typical for experiments of type A.

Preconditioning of the point electrode makes the stabilizing period shorter in time and less important in amplitude (Fig.1), while the discharge becomes rapidly quite stable in light emission and mean current.

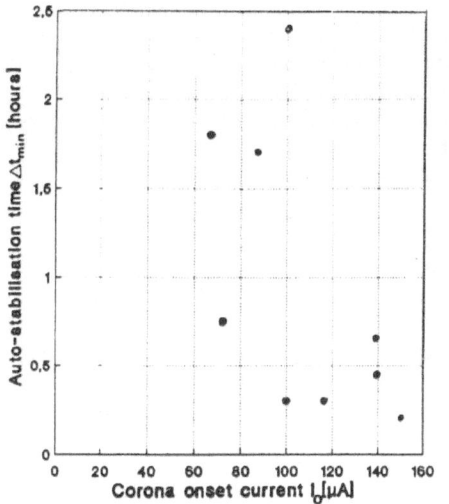

Fig.1a. Auto-stabilisation time
vs the corona onset current.

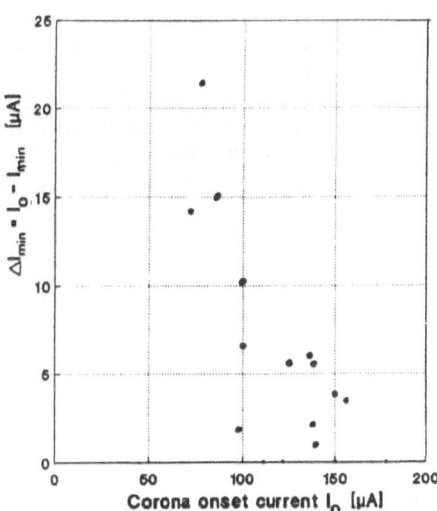

Fig.1b. Initial current drop vs
the corona onset current.

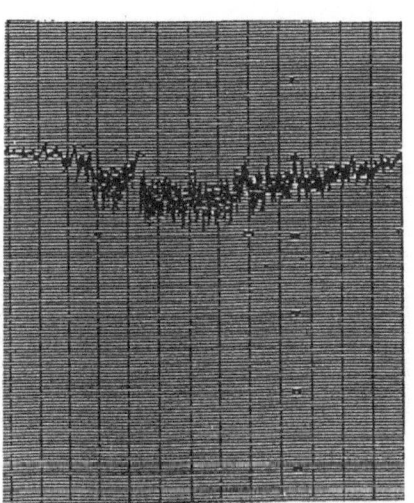

Fig.2. Current record of the auto-
stabilisation period of the discharge
(experiment of type A).

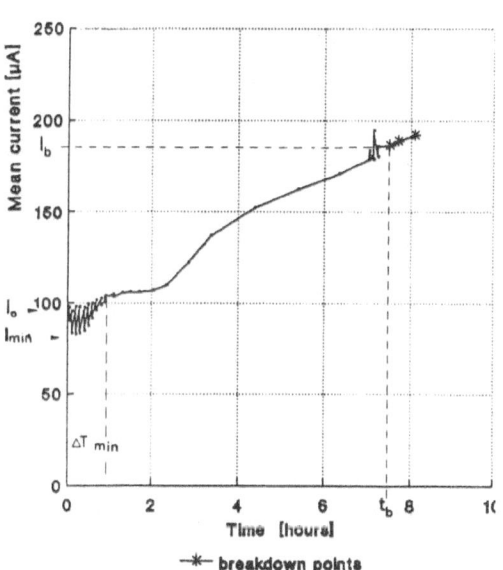

Fig.3. Mean current evolution
curve typical for an experiment
of type A (without preconditioning
the point).

Stabilized and increasing current phases

After the autostabilizing first period, often the current does not vary much
during a more or less long duration, depending on different conditions; then it increases.
In all experiments carried out under our working conditions, we never got stationary
currents during more than 4 hours, even if spark breakdown was not always achieved.

Figure 4a shows points corresponding to experiments with observed spark
breakdown. We can see that the time-to-breakdown t_b measured between corona onset
and spark breakdown is generally shorter ($t_b < 3$ h) for experiments of type B (with a

preconditioned point) than for experiments of type A (without preconditioning the point) for which they lie in a range of 7 to 21 hours. Spark breakdown occurs for a mean current I_b which shows a tendency to increase with the onset current I_0 (Fig.4b), mostly lying in a range of 175 to 190 μA to be discussed hereafter.

Figure 4 also shows points corresponding to experiments stopped without getting spark breakdown. Their position in abscissae obey to the rule : $I_0 < 100$μA for experiments of type A and $I_0 > 100$μA for experiments of type B. In ordinates, they are all located under the points corresponding to experiments with breakdown ; we may think that they concern experiments which might have led to spark breakdown if they could have been sufficiently prolonged.

Some test experiments have been carried out under lower stress conditions :
(i) An experiment at 42 kV began at $I_0 = 8$ μA; when stopped after 13 days, the mean current was at 50 μA, still slightly increasing.

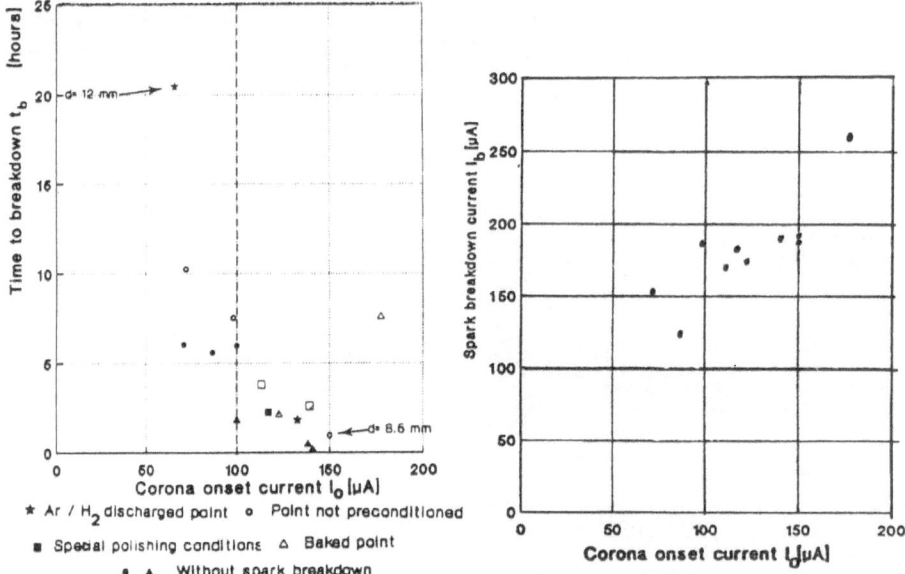

Fig.4a. Time to breakdown vs the corona onset current.

Fig.4b. Spark breakdown current vs the corona onset current.

(ii) Two experiments of type A were carried out always at 60 kV, but with a smaller and a greater gap length (d=8.5 mm and d= 12 mm instead of 10 mm). They gave values coherent with the above described behaviour, as it can be seen in Fig.4, i.e :
* For d=8.5 mm and V=60kV, i.e with an increased applied field (340 kV cm^{-1} at the point electrode instead of 325 kV cm^{-1}), was obtained a point with high I_0, high I_b and short time-to-breakdown t_b.

* For d=12 mm and V=60 kV, i.e with a decreased applied field (310 kV cm^{-1} at the point electrode instead of 325 kV cm^{-1}) was obtained a point with low I_0, and long t_b (or rather long time "without breakdown" since this experiment was stopped without getting breakdown).

402

DISCUSSION

First we can consider that our discharge acts primarily as a pseudo-glow, with a continuous current and current pulses observed oscillographically. We have seen that its life begins with or without a fluctuating phase of current, according to the preparation mode used for the high-field electrode. In other terms, the fluctuating phase of the experiments of type A (without preconditioning the point) should be regarded as an auto-conditioning phase. The fact that the current I_0 increases with the cleanliness of the point electrode while this first phase vanishes could be simply explained by a better area occupancy of the glow when a better level of cleanliness is achieved.

Now concerning the further evolution of the I(t) curves, in all experiments, with the point electrode either auto-conditioned by the discharge or preconditioned by other means, spark breakdown is observed to occur with a qualitatively similar evolution of current, as seen above, and a similar evolution of the point electrode surface characteristics, as it will be now seen. As illustrated by Fig.5 SEM micrography first a thin layer of very small grains ($\sim$ 1 µm diameter) settles on the electrode tip surface, then covered with islets ($\sim$ 10µm diameter),very dispersed during the stationary current phase and edge to edge when breakdown occurs. According to the XPS analyses, the surface grains as the islets should be formed by $(CF_2CHF)_n, (CF_2)_n$ and C_4F groups.

Fig.5. Point micrography during the current increase course.

We know from an energy shift observed on the XPS peaks that the so-formed layers have a bad electrical conductivity.How conduction proceeds through them becomes an actual question. It might be through pores and/or by microsparks across the layers.The informations on the nature and structure of the layers get from surface analyses and reported above, combined with the presence of pulses in the current signals observed oscillographically suggest the idea of microdischarges occurring in the interstices between the insulating islets and acting as a current amplifier like backdischarges in electrostatic precipitators (see Dancer et al.,1980; Cross, 1986).

For their part, changes produced by the discharge in the gas composition as well as those which may occur in the negative ion population could enter into play in the growth of current during the increasing current phase. At least on the first point, we can answer a priori negatively since no clear influence on the characteristics of the discharge should be observed neither by Casanova (1989) when he added SF_6 decomposition products to the gas in a controlled manner nor by ourselves when we made a gas renewal during experimental runs.

Let us now comment on our I_b spark breakdown current values. Knowledge on dc negative corona breakdown is still poor. However, it is known that negative SF_6 coronas remain localized at the cathode under most conditions (Farish et al., 1979), till spark breakdown which was seen to occur, at least at pressures lower than 1 bar, within some nanoseconds after the development of a streamer in the cathode region (Sigmond. et al., 1988). In these conditions, as in Sigmond et al. conditions (1982), our I_b breakdown current values should be identified with the maximum I_s saturation value of the unipolar ion current which may be conducted before bipolar phenomena like streamers must occur. Using the approximative expression given by Sigmond (1982) for this limiting I_s current, then we should write:

$$I_b = I_s = 2\mu_- \epsilon_0 V^2 / d$$

where V=10 kV, d=10 mm, ϵ_0=8.85 pF.m^{-1} for the gas permittivity and where μ_- is a mean mobility value for the negative ions governing the drift region space charge in the gap. By this way, we obtain μ_- values mostly lying in a 0.27 - 0.3 cm^{2}V^{-1}s^{-1} range with 0.19 and 0.4 cm^{2}V^{-1}s^{-1} as extremum values at both ends of this range. All these values enter in the range of values previously found by Waters et al. (1982) on the one hand and by Arson and Bortnik (1980) on the other hand, also under discharge conditions but on the basis of the I(V) characteristic shape under various field, gas pressure and gap length conditions. Since the physico-chemical surface state of the cathode working as the high-field electrode seems to be the only variable in our case, we should attribute the μ_- changes to its influence on the nature of the predominant negative ions into play, in the same way as an incidental influence of the cathode high-field electrode material was previously observed on ozone production (see Goldman et al., 1982). According for instance to Patterson data (1970), our mobilities should most generally correspond to SF_6^- ions (μ_- =0.27 cm^2 V^{-1} s^{-1}) and to negative SF_6 ionic fragments, perhaps SF_5^- for the 0.3 cm^{2}V^{-1}s^{-1} mobility and a not identified fragment for the 0.4 cm^{2}V^{-1}s^{-1}, with a possible incidence of detachment phenomena during the life of the negative ions across the discharge interval and also a possible incidence of an increasing electron production efficiency of the point cathode region which should finally make the corona current reach the unipolar saturation limit. Since our operating conditions were relatively clean ones, such high mobilities are not to be surprising and our lowest μ_- values should correspond to a larger contribution of heavy complex, eventually clustered, ions to be associated to the less clean cathode surface conditions.

At last, we may emphasize that at this stage, we have no actual knowledge about the breakdown mechanism itself and we do not know under what level of electrical stress delayed spark breakdown could be completely avoided in our operating conditions. Furthermore, we do not know how the gas purity and the material of the high-field electrode should act on the current evolution and particulary on the delayed spark breakdown phenomenon. This will need further investigations.

CONCLUSION

Correlations have been clearly established, linking to the initial surface state of the high-field electrode the onset current and the different evolutive parameters of the discharge behaviour till spark breakdown. They show that the negative ions into play and that the current evolution are together predominantly governed by physico-chemical surface phenomena on this high-field electrode.

ACKNOWLEDGEMENTS

The authors wish to thank Prof. R. S. Sigmond for his helpful suggestions and M. Goldman for his constant interest and the valuable discussions they had with him.

REFERENCES

Arson, A. G., and Bortnik, I. M., 1980, Mobility of ions in SF6, in: "Proc. 6 th Int. Conf. on Gas Discharges and their Applications", , 165-167.

Avni, R., Fuchs, R., and Polak, M., 1989, Cleaning of tool steels surface by RF plasmas of Ar and Ar + H2 gas mixtures, in: "Proc. 9th Int. Symp. on Plasma Chemistry," R. d'Agostino, ed., Pugnochiuso, Italy, 1110-1115.

Cross, J. A., 1986, Back ionisation in a negative point-to-plane corona discharge, J.Electrostatics, 18,327-344.

Casanova, J., 1989, Private communication.

Dancer, P., Goldman, M., and Le Fur, D., 1980, Non-destructive breakdowns in non-impregnated papers, J. Phys. D., 13, 449-454.

Farish, O., Ibrahim, O. E., and Kurimoto, A., 1979, Prebreakdown corona processes in SF6 and SF6/N2 mixtures, in: "Proc. 3th Int. Symp. on High Voltage Engineering", Paper 31.15.

Fujimoto, N., 1987, Conduction currents in gas-insulated switchgear for low level dc stress, in: "Gaseous Dielectrics V," L. G. Christophorou and D. W. Bouldin, ed., Pergamon Press, New York, 513-519.

Goldman, M., Lecuiller,M., and Palierne,M., 1982, Influence of the nature of electrode material on the production of corrosive species in a corona discharge, in: "Gaseous Dielectrics III", L. G. Christophorou, ed., Pergamon Press, New York, 327-331.

Hughes, R. C., and Hampton, B. F., 1982, Conditioning compressed gas capacitors, and variations in their loss angle, in: "Proc. 7th Int. Conf. on Gas Discharges and their Applications," London, 296-299.

Patterson, P. L., 1970, Mobilities of negative ions in SF6, J. Chem. Phys., 53, 696 -704.

Sigmond, R. S., 1982, Simple approximate treatment of unipolar space-charge-dominated corona: The Warburg law and the saturation current., J. Appl. Phys.,53, 891-898.

Sigmond, R. S., Hegerberg, R., and Baranov, V. V., 1982, Positive glow pulses and sparks in 1 cm point-to-plane gaps in SF6 up to one atmosphere, in:"Proc. 7th Int. Conf. on Gas Discharges and their Applications," London, 227-230.

Sigmond, R. S., Linhjell, D., Hegerberg,R., and Sigmond, T., 1988, DC Corona breakdown in SF6 and mixtures, in: "Nordic Symp. on Insulation", Trondheim,Norway.

Van Brunt, R. J., 1982, Effects of H2O on the behaviour of SF6 corona, in: "Proc. 7th Int. Conf. on Gas Discharges and their Applications," London, 140-142.

Van Brunt, R. J., and Leep, D. A., 1982, Corona-induced decomposition of SF6, in: "Gaseous Dielectrics III," L. G. Christophorou, ed., Pergamon Press, New York, 402-408.

Waters, R. T., Farish,O., and Ibrahim,O., 1982, Positive and negative mean ion mobilities in corona discharges in SF6 and mixtures, in: "Proc. 7th Int. Conf. on Gas Discharges and their Applications", London, 251-253.

DISCUSSION

L. G. CHRISTOPHOROU: Have you conducted these studies on other systems besides SF_6?

M. GOLDMAN: No, if we think about spark breakdown; yes, if we think about polymerization at high pressure. With fluorinated gases, it is possible to create an insulating layer in a corona discharge.

VERIFICATION OF DIRECT-CURRENT CORONA MODELS EMPLOYING

THE DEUTSCH APPROXIMATION

A. Bouziane, K. Hidaka,
M.C. Taplamacioglu and R.T. Waters

School of Electrical, Electronic & Systems Engineering
University of Wales College of Cardiff, U.K.

INTRODUCTION

When a voltage significantly in excess of the inception voltage is applied, the spatial distribution of the electric field and current density in direct-current corona is found to become independent of voltage. The normalised distributions $\overline{E} = E/E_{max}$ and $\overline{J} = J/J_{max}$ over a passive collector electrode thus share the invariant character of the original Laplacian electrostatic field. The field shape differs appreciably from the Laplacian however, and is naturally more difficult to compute. Nevertheless, it is significant that the distributions of $\overline{J}$ and $\overline{E}$ depend only upon the electrode geometry, and their computation is important from the practical aspects of power transmission and corona applications.

A number of authors have modelled corona using the Deutsch approximation (where the field direction is assumed to remain Laplacian), usually to simulate only the total corona current magnitude (see a review by Sigmond (1986)). One of the most successful of these approaches is that of Popkov (1949), and a later variation of this method by Sarma and Janischewskyj (1969). In the present paper this model has been implemented to calculate J and E distributions in paraxial (non-coaxial) cylinders, an electrode system specially designed for the present investigation, where the geometry can be varied from the coaxial case towards the wire-plane case. In addition to these calculations, new measurements of current density and electric field distributions are presented.

BASIS OF MODELS

General Equation and simplifying assumptions

The electric field E, current density J and the unipolar space-charge density and mobility ρ and μ are related by :

$$\nabla.E = \rho/\epsilon_o \qquad (1)$$

$$\nabla.J = 0 \qquad (2)$$

and

$$J = \mu\rho E \qquad (3)$$

which combine to give the differential equation which describes the distribution of potential in the ion drift region of the corona discharge :

$$\nabla . \; (\nabla \Phi \; \nabla^2 \Phi) = 0 \qquad\qquad (4)$$

These equations neglect variations of ion mobility and diffusion.

Although finite difference solutions of these equations can be attempted, their solution is aided by simplifying assumptions of the kind made by Popkov and Sarma and Janischewskyj :

1.   The shape of the electric field distribution over the surface of the corona electrode remains of the Laplace form, and has a magnitude equal to that existing at the corona inception voltage V_i (Kaptsov condition).

2.   The space charge influences only the magnitude of the electric field E, but not its direction, which throughout the discharge space remains that of the Laplacian field E_L (Deutsch approximation). If ξ is the field enhancement factor (a scalar function of position), this gives :

$$E = \xi \; E_L \qquad\qquad (5)$$

Of the above equations, only (1) and (2) are independent, and ξ and ρ can be chosen as the independent variables. The relationship between ξ and ρ is easy to find : from (1), (4) and (5)

$$\nabla . \; \left[(-\xi \; E_L) \; (-\rho/\epsilon_o) \right] = 0 = \nabla . \; \left[E_L \; \xi \, \rho \right] = E_L \; .\nabla \; (\xi \rho) \qquad\qquad (6)$$

The zero value of this scalar product shows that $\nabla \; (\xi \rho)$ is orthogonal to E_L, so that

$$\xi \, \rho = \xi_e \rho_e \qquad\qquad (7)$$

for any point on a Laplacian field line.   The boundary conditions at the electrode on the r.h.s. of (7) neglect any ionisation zone effects.   A value for ξ_e is easily obtained from the Kaptsov condition :

$$\xi_e = V_i / V \qquad\qquad (8)$$

Determination of ξ (eliminating ρ)

Consideration of equations (1) and (5) gives

$$\nabla . \; (\xi \; E_L) = E_L \; \nabla \; \xi = \rho/\epsilon_o \qquad\qquad (9)$$

For a coordinate s along a Laplacian field line where the original space-charge-free potential was $\emptyset$, then $\nabla \xi = d\xi/ds$ and $ds = - \, d\emptyset/E_L$, so that equation (9) can be integrated using equation (7) to obtain

$$\xi^2 = \xi_e^2 + (2 \; \xi_e \, \rho_e/\epsilon_o) \int_{\emptyset}^{V} d\emptyset/E_L^2 = \xi_e^2 + F \; (\rho_e, \, \emptyset, \, V) \qquad\qquad (10)$$

This integral involving the Laplacian field can be evaluated analytically for some geometries only, but can be computed numerically for any system.   Equation (10) can thus be used to calculate ξ if the charge density ρ_e at the appropriate position on the corona electrode can be determined.

Determination of ρ (eliminating ξ)

From equations (7) and (9)

$$E_L \nabla \xi = E_L \nabla \left[\frac{\xi_e \rho_e}{\rho} \right] = E_L \xi_e \rho_e \left[-\frac{\nabla \rho}{\rho^2} \right] = \frac{\rho}{\epsilon_0} \qquad (11)$$

Since $\nabla \rho = d\rho/ds = - E_L \, d\rho/d\emptyset$ we have

$$d\rho/\rho^3 = (1/\epsilon_0 \xi_e \rho_e) \, d\emptyset/E_L^2 \qquad (12)$$

$$\therefore \qquad 1/\rho^2 - 1/\rho_e^2 = (2/\epsilon_0 \, \xi_e \rho_e) \int_\emptyset^V d\emptyset/E_L^2 = F \, (\rho_e, \emptyset, V)/(\xi_e \rho_e)^2 \quad (13)$$

The same Laplacian integral can thus be used to calculate ρ if ρ_e is known.

Determination of ρ_e

We may now utilise the known applied voltage between the electrodes:

$$V = - \int_b^a \xi \, E_L \, ds = - \int_0^V \xi \, E_L \, (-d\emptyset/E_L) = \int_0^V \xi \, d\emptyset \qquad (14)$$

Popkov combined (10) and (14) so that

$$\int_0^V \left[\xi_e^2 + F \, (\rho_e, \emptyset, V) \right]^{\frac{1}{2}} d\emptyset = V \qquad (15)$$

In an iterative procedure to find ρ_e we could use the convergence condition

$$\left| \text{LHS (equation (15))} - V \right| < \epsilon \qquad (16)$$

In the Sarma/Janischewskyj model, equation (14) is again used in a less direct way : from equation (9)

$$\int_{\xi_e}^\xi d\xi = \xi - \xi_e = \int_a^s (\rho/\epsilon_0 E_L) \, ds = (1/\epsilon_0) \int_\emptyset^V (\rho/E_L^2) \, d\emptyset \qquad (17)$$

and this ξ can be used in equation (14) so that

$$V = \int_0^V \xi_e \, d\emptyset + (1/\epsilon_0) \int_0^V \int_\emptyset^V (\rho/E_L^2) \, d\emptyset \, d\emptyset \qquad (18)$$

$$\therefore \qquad F_i (\rho, V) = V - V_i \qquad (19)$$

The iteration on ρ_e will converge so that

$$\left| \text{LHS (equation 19)} - (V-V_i) \right| < \epsilon \qquad (20)$$

For initialisation of ρ_e for the convergences of equations (16) and (20) a mean value of ρ_m could be used as defined by Sarma and Janischewskyj :

$$\rho_m = (V-V_i)/ \int_0^V \int_\emptyset^V d\emptyset \, d\emptyset/E_L^2 \qquad (21)$$

A flow chart which shows the alternative computations of Popkov and Sarma and Janischewskyj is shown in Figure 1. Both procedures are found to yield equivalent results.

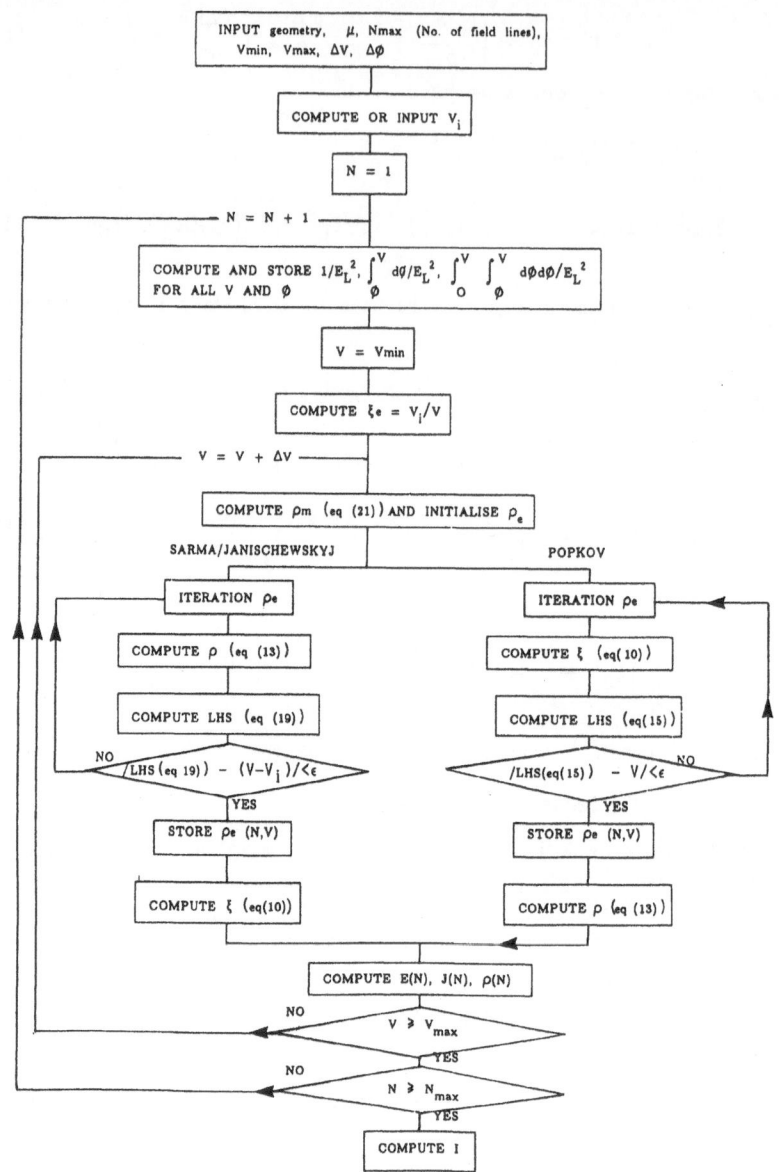

INPUT geometry, μ, Nmax (No. of field lines), Vmin, Vmax, ΔV, $\Delta\phi$

COMPUTE OR INPUT V_i

N = 1

N = N + 1

COMPUTE AND STORE $1/E_L^2$, $\int_\phi^V d\phi/E_L^2$, $\int_O^V \int_\phi^V d\phi d\phi/E_L^2$ FOR ALL V AND ϕ

V = Vmin

COMPUTE $\xi_e = V_i/V$

V = V + ΔV

COMPUTE ρ_m (eq (21)) AND INITIALISE ρ_e

SARMA/JANISCHEWSKYJ   POPKOV

ITERATION ρ_e ITERATION ρ_e

COMPUTE ρ (eq (13))   COMPUTE ξ (eq(10))

COMPUTE LHS (eq (19))   COMPUTE LHS (eq(15))

NO $/LHS (eq 19)) - (V-V_i)/<\epsilon$ $/LHS(eq(15)) - V/<\epsilon$ NO

YES   YES

STORE ρ_e (N,V)   STORE ρ_e (N,V)

COMPUTE ξ (eq(10))   COMPUTE ρ (eq (13))

COMPUTE E(N), J(N), ρ(N)

NO   V $\geqslant$ V$_{max}$

YES

NO   N $\geqslant$ N$_{max}$

YES

COMPUTE I

Fig. 1. Flow chart for computation by Deutsch models.

EXPERIMENTAL PROCEDURES

The paraxial electrode geometry and the design of the rectangular current-density/electric-field probe have been described in earlier papers (Bouziane et al· 1988, 1989). A wire of diameter 0.56 mm located within a cylinder of diameter 100 mm is arranged to be displaced from the axis of the cylinder. Results are presented here for a displacement Δ = 21 mm as typical for the range $0 \leqslant \Delta \leqslant 30$ mm. Experiments were performed in ambient air with positive voltage applied to the wire. The probe was of the biased-collector type, located flush with the inner surface of the cylinder. Both current density J(θ) and electric field E(θ) could be obtained as a function of angular position by a simple rotation of the outer cylinder. The reference angle θ = 0 was defined by the plane of the shortest wire-to-cylinder path. In addition to the J(θ) and E(θ) profiles, the total corona current per unit length was also measured.

410

COMPARISON OF RESULTS WITH DEUTSCH MODELS

Figure 2(a) shows the variation of total corona current per unit length with voltage, where the experimental points relate to Δ = 21 mm. The corona inception voltage for this system is 11.8 kV, and the current shows the usual square-law increase. The curves which accompany the experimental points are the output from the computer models; the computed current is proportional to the value assumed for the ionic mobility, and excellent agreement is found for a mobility of 1.8 cm^2/Vs. The same is true for the maximum current density at θ = 0 on the cylinder (Fig. 2(b)). The largest measured electric field is also at θ = 0, and Fig. 2(b) compares the experimental points with the computed curve. Both show a linear increase with applied voltage, although the measured field is 8 - 10% higher than the computed value, which unlike current density is not sensitive to the assumed mobility value. However, this computed value is, together with the computed current density, dependent upon the value assigned to the corona inception voltage which may be responsible for this relatively small discrepancy between the measured and computed field. If an increase of 5% in the value of inception voltage is inserted into the computation, then decreases of about 10% in J_{max} and 5% in E_{max} result.

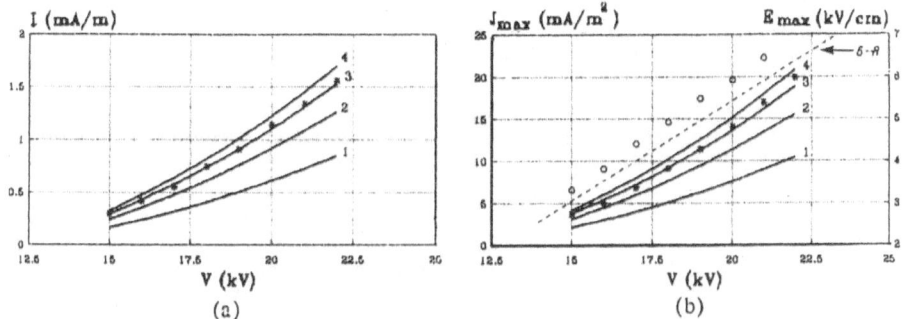

Fig. 2. Corona current and maximum current density and field at cylinder.
Measured values * (current, current density), $\circ$ (field).
Model curves 1-4 (current, current density) 5-8 (field) assuming ion mobilities 1, 1.5, 1.8, 2 cm^2/Vs.

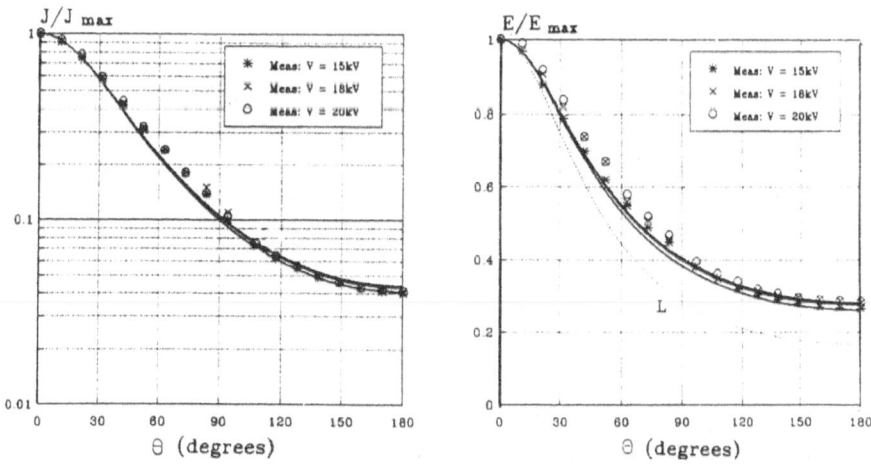

Fig. 3. Normalised profiles of current density and electric field - (m = 1.8 cm^2/Vs). L - Laplacian field profile.

Although this is satisfactory, perhaps more impressive than the prediction of maximum current and electric field is the performance of the Deutsch models in simulating the profiles of both current density and electric field. Fig. 3 shows the measured experimental points normalised with respect to maximum values together with the computed curves. Not only is there good shape agreement between measurements and model, but both manifest a voltage independence. The current density and electric field profiles are related empirically by

$$J/J_{max} = (E/E_{max})^{2.5}$$

The electric field profile is found to be broader than the original Laplacian field, which is also shown in the Figure. It is found that

$$E/E_{max} = (E/E_{Lmax})^{0.75}$$

CONCLUSIONS

Although it is clearly possible to define notional static charge distributions which would satisfy the Deutsch condition, such charge distributions do not comply with the current continuity condition incorporated in equation (4). The Poissonian field profile illustrated by the experimental points of Fig. 3 is thus quite different from the Laplacian profile implicit in the Deutsch approximation. For this reason, it is unexpected that a modelling procedure based upon that approximation is found to yield such excellent agreement with experiment.

The formulation of the model clarifies this apparent conflict. Although a flux line terminating at a given position on the collector electrode will have an origin at the active electrode which is significantly different in the Laplacian and Poissonian fields, the error in the computation of the integral $F(\rho_e, \emptyset, V)$ of equation (10) to obtain ξ is relatively small if the Laplacian line associated with the terminal point rather than the point of origin is chosen. Work is under way to quantify this error by comparative studies of the Deutsch model with finite-difference solutions for corona current in the paraxial system. Such errors are likely to become unacceptably large where interacting corona sources are involved : experiments in twin-wire configurations are also in progress.

ACKNOWLEDGMENT

Professor Hidaka wishes to thank University of Tokyo for study leave to enable him to undertake this research collaboration at Cardiff.

REFERENCES

Bouziane, A., Hartmann, G., Jones, J.E., Waters, R.T. and Zebboudj, Y. (1988). Improved instrumentation for field measurement in cylindrical coronas. 9th Int. Conf. Gas Discharges (Venice), 527-530.

Bouziane, A., Taplamacioglu, M.C., Hartmann, G. and Waters, R.T. (1989). Calibration of a linear probe for coaxial d.c. corona experiments. Proc. 6th Int. Symp. H.V. Eng. (New Orleans), Paper 42.01.

Popkov, V.I. (1949). On the theory of unipolar d.c. corona. Electrichestvo No.1, 33-43.

Sarma, M.P. and Janischewskyj (1969). Analysis of corona losses on D.C. transmission lines. Part I unipolar lines. IEEE Trans. PAS-88, 95-102.

Sigmond, R.S. (1986). The unipolar space charge flow problem. J. Electrostatics 18, 249-279.

FREQUENCY EFFECTS IN ALTERNATING CURRENT CORONA

D. A. Rickard and R. T. Waters
University of Wales College of Cardiff
P. O. Box 904
Cardiff CF1 3YH, UK

and

J. Dupuy
Laboratoire Génie Electrique
IURS, Université de Pau
Avenue de l'Université
F 64000, Pau, FRANCE

INTRODUCTION

Corona phenomena upon overhead high voltage lines have detrimental influence in terms of power loss and noise. Present design work relies heavily upon tests in experimental cages or on full scale lines. In this study a controlled laboratory experiment has been performed to measure a wide range of corona parameters under variable voltage and frequency of a.c. excitation. The results have then been used to validate an existing corona model.

TEST CAGE MEASUREMENTS

The test apparatus is shown schematically in Figure 1. An a.c. signal source and power amplifier are connected via the step up transformers T1 and T2 (overall ratio of 1:649) to the test cage. The magnitude of the high voltage is measured by both the electrostatic voltmeter E.V. and the capacitors C1 and C2, which also provide its phase. The test cage itself is of special construction as described in [1,2]. The central conductor (radius 92.5 μm or 156.5 μm) is coaxial with an electrode system consisting of two sets of interleaved grid wires (each of radius 1 mm) 5 mm inside a plain cylindrical sensor electrode of radius 100 mm. The measurement section has a length of 254 mm.

A bipolar bias voltage ($\pm$ Vb) is applied to the two grids, creating a strong d.c. electric field, which attracts the corona ions to the grid and prevents any reaching the sensor. Synchronous current measurement at the sensor and grid electrodes then allows the conduction and displacement components of the corona space charge current to be distinguished which is not possible with a conventional test cage. Modification of the experimental procedure from that used in [1 and 2] by utilising opto–isolators has allowed these simultaneous measurements. The displacement current corresponds to the rate of change of surface charge (field), and may be integrated to determine the absolute magnitude of the electric field. The U–Function meter measures the corona power loss from the voltage and total corona current.

Gaseous Dielectrics VI, Edited by L.G. Christophorou and
I. Sauers, Plenum Press, New York, 1991

Optical data from a photomultiplier have been used to measure the start and finish times of the corona in each half cycle.

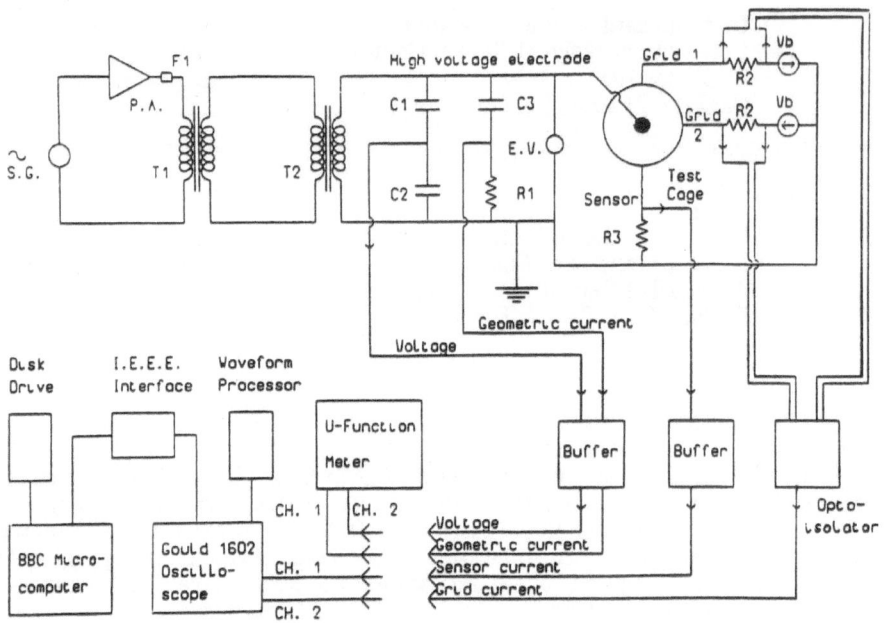

Fig. 1.   A. C. corona test cage system.

S.G. Signal generator; P.A. Power amplifier; F1 quick blowing fuse; T1 L.V. transformer (0.7/1 p.u.); T2 H.V. transformer (220/100 kV); E.V. electrostatic voltmeter; C1,C3 loss free H.V. capacitors; C2 standard capacitor box; R1 geometric current resistor (500 Ω typical); R2 grid resistor (3 kΩ typical); R3 sensor resistor (31.5 kΩ typical); V_b 1400 V d.c. power supply.

MODELLING THE A.C. CORONA PROCESS

The model adopted here, first proposed by Gary et al. [3], is based upon simplifying assumptions because the fundamental phenomena, determined by the Poisson and continuity equations, are intractable to exact solutions. Corona space charge is therefore assumed to be emitted as discrete infinitesimally thin rings at each time step in the cycle, if required to limit the electric field at the corona wire to the critical field given by Peek's Law (or similar). These rings then drift with constant mobility in the total electric field calculated at their spatial location unless neutralised at either electrode. Recombination although allowed for in [3] has been neglected in this implementation.

414

ACCURACY AND LIMITATIONS OF RESULTS

Experimental data have been collected for applied voltages of 25–100 Hz and 6–28 kV, the latter representing voltages up to approximately 5 times the inception voltage of the chosen wires. The data are reproducible for a given corona wire and have been collected under the same atmospheric conditions as far as is possible without a climatic chamber. Digital signal averaging has been used to reduce noise and the effect of any statistical variation between cycles.

The modelled results have been checked for their sensitivity to variations in the surface state coefficient of the corona wire, the ionic mobility values and the representation of the ionisation region.

The solutions for current, power loss and charge transport are particularly sensitive to the first two parameters whilst surprisingly changing the ionisation region, which has been modelled as in [3] and [4], has little effect. The corona critical times are not sensitive to any of the above variations.

Variation of the incremental time step for the solution also affects the results, although the solutions do converge as the time step is reduced corresponding to a closer approximation of the true continuous charge distributions by the rings. In practice longer computational times limit the smallest practicable time step to 1/400th of the cycle period.

The surface state coefficient has therefore been chosen to match the experimental and calculated inception voltages, whilst the positive and negative ion mobilities provide the other degree of freedom required to match the experimental data. These optimum values of 1.6 and 1.8 cm²/Vs are in the range of values quoted in the literature.

COMPARISON OF THEORETICAL AND CALCULATED WAVEFORMS

The overall qualitative agreement between these shown in Figure 2 is excellent, the major points to be noted being:

1. The total corona current equals the geometric current immediately prior to corona inception as the net current due to the space charge (but not its conduction and displacement components) has decayed to zero from the previous half cycle.

2. The displacement current at the outer electrode has a double peak. The first peak after the positive going voltage zero is caused by the cage filling with active corona ions after positive inception. When these start to reach the outer electrode the current rapidly decreases becoming increasingly negative. At positive extinction the space charge starts to decrease as it expands under its self field so the magnitude of the current decreases. The cycle then repeats after negative inception for the opposite polarity ions. Due to the effect of the space charge field the net electric field at the outer electrode lags the applied voltage field.

3. A clear partitioning of charge collection of opposite polarity ions occurs, separated by periods of zero charge collection. The charge arrival is detected a finite time after the inception time due to the transit time across the cage, and ceases when the overall electric field at the outer electrode reverses (although for the experimental results there appears to be a slight residual phase error between the points of field reversal and the starting times of the zero charge collection plateaux).

4. More negative than positive charge is collected, as expected from the higher mobility of these ions.

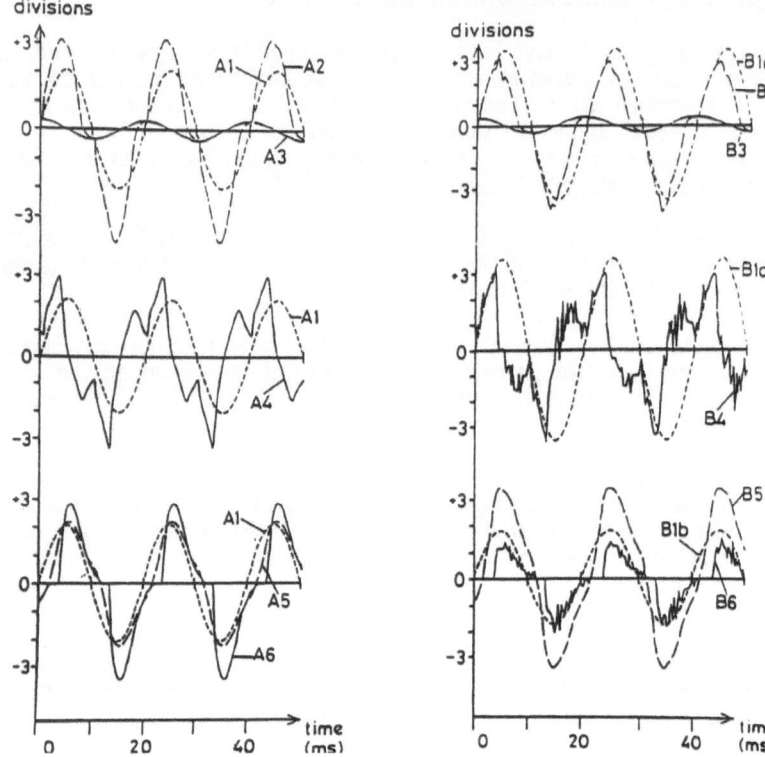

Fig. 2.   Experimental and calculated corona waveforms.   V=20 kV r.m.s., f=50
            Hz, wire radius=92.5 μm.

            Experimental data: A1 voltage 14 kV/div; A2 total current 56 μA/div;
            A3 geometric current 56 μA/div; A4 displacement current 56 μA/div; A5
            electric field 1.33 kV/cm/div; A6 charge transport rate 53 μA/div.

            Calculated data: B1a voltage 8 kV/div; B1b voltage 16 kV/div; B2 total
            current 51 μA/div; B3 geometric current 51 μA/div; B4 displacement
            current 51 μA/div; B5 electric field 0.76 kV/cm/div; B6 charge transport
            rate 102 μA/div.

VOLTAGE EFFECTS UPON CORONA

        The power loss and integrated total charge transport follow a square law fit
against voltage with frequency constant as is shown in Figure 3. The tendency of the
model to over–estimate quantities at lower voltages is likely to be due to the neglect
of recombination phenomena.

        The inception times become progressively earlier at constant frequency as the
voltage is increased and extinction times become later.   Arrival time of the first
corona ions at the outer electrode (the transition from the zero plateaux) occurs
progressively earlier due to the higher drift velocities and collection of the last ions
gets later because of the later extinction times at the higher voltages. These trends
are mirrored in experimental and calculated results.

At voltages slightly (less than approximately 50%) above the inception voltage no charge is collected at the outer electrode showing the spatial extent of the space charge to be voltage controlled.

The transit times of the first corona ions produced in each cycle, for positive and negative ions, may be estimated, by assuming that the first ions produced at corona inception are the first ions of that sign collected in each half cycle. Ion transit times match approximately, although surprisingly the experimentally determined time for positive ions is slightly less than that for the corresponding negative ions. (This is verified over the range of experimental data.) The converse is indicated by the model, where the above assumption is redundant, which is the expected result in view of the higher mobility of the negative ions.

A plausible explanation of the experimental result is that the first negative ions (or possibly electrons), recombining more rapidly than the equivalent positive ones, are completely neutralised before they reach the outer electrode. The first ions collected are therefore produced some time after negative inception reducing the transit time from that assumed. Alternatively the positive ions are initially injected further into the air gap, by a mechanism such as streamers. (There is little experimental evidence to support this supposition.)

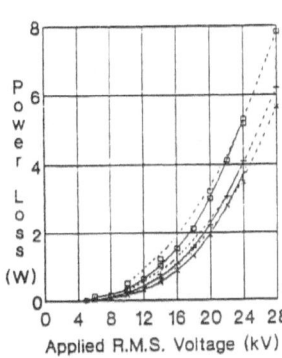
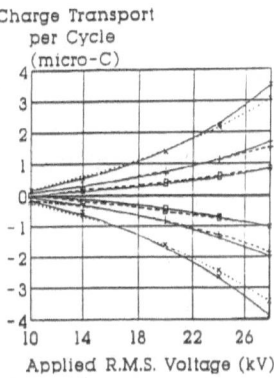

Fig. 3.   Power loss and positive and negative charge transport per cycle with voltage for 92.5 μm radius wire.

——x—— measured at 25 Hz.   — —x— — modelled at 25 Hz.
——+—— measured at 50 Hz.   — —+— — modelled at 50 Hz.
——□—— measured at 100 Hz.   — —□— — modelled at 100 Hz.

FREQUENCY EFFECTS UPON CORONA

A rise in power loss occurs with increasing frequency (Figure 4). This increase is nearly linear at lower voltages, but becomes quadratic at higher voltages where charge transport across the cage dominates.

Normalising the inception and extinction times with frequency indicates both times occur earlier in the cycle, due to the lower relative drift distances, as the frequency increases. The residual space charge from the previous cycle being closer to the active electrode at inception enhances the field there to a greater extent at higher frequencies. Similarly near to extinction the newly produced corona ions are closer to and of the same sign as the active electrode and therefore reduce the field there to a greater extent.

The voltage at which ions start to reach the outer electrode becomes lower as the frequency is reduced as the d.c. case of unipolar drift is approached.

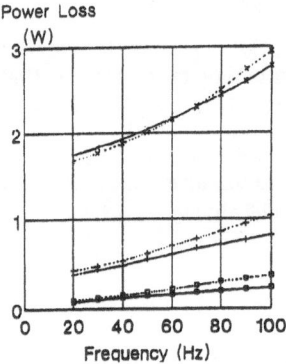

Fig. 4. Power loss with frequency for 156.5 μm radius wire.

—□— measured at 10 kV.	— —□— — modelled at 10 kV.
—+— measured at 14 kV.	— —+— — modelled at 14 kV.
—x— measured at 20 kV.	— —x— — modelled at 20 kV.

CONCLUSIONS AND FUTURE WORK

The results are in accord with the data over the complete range of test conditions, including the new charge transport measurements allowing greater confidence to be placed in both measurement and calculations. The charge transport measurements have allowed an estimation of the ionic mobilities in accordance with widely accepted values.

Tests for bundle conductors are currently in progress and an extension to mixed a.c. and d.c. voltages is planned. Modelling the former via an equivalent capacitance method or by considering the true two dimensional nature of the problem by a means such as charge simulation is then possible. The latter measurements will provide more data to test the existing model.

REFERENCES

[1] H. S. B. Elayyan, G. Hartmann, A. Robledo–Martinez, and R. T. Waters, Charge Injection and Loss in A.C. Corona, Proc. Roy. Soc. London, U.K. A410, 1477–1500 (1987).

[2] A. Bouziane, J. Dupuy, D. A. Rickard, and R. T. Waters, A Double Walled Test Cage to Measure A.C. Corona Charge Flow, Sixth International Symposium on High Voltage Engineering, New Orleans, U.S.A. Paper 40.021 (1989).

[3]   J. J. Cladé, C. H. Gary, and C. A. Lefèvre, Calculation of Corona Losses Beyond the Critical Gradient in Alternating Voltage, <u>IEEE Trans. P.A.S.</u>, Vol. 88, 1695–1703 (1969).

[4]   M. A. Al–Tai, H. S. B. Elayyan, D. M. German, A. Haddad, N. Harid, and R. T. Waters, The Simulation of Surge Corona on Transmission Lines, <u>IEEE Trans. P.W.D.</u>, Vol. 4, 1360–1368 (1989).

A MASS SPECTROMETER STUDY OF IONIZATION IN SF$_6$ CORONA: INFLUENCE

OF WATER AND NEUTRAL BY-PRODUCTS

I. Sauers and G. Harman
Oak Ridge National Laboratory
Oak Ridge, Tennessee 37831-6123

ABSTRACT

Mass identified ions sampled from positive- and negative-point-to-grounded plane corona discharges in SF$_6$ are reported. In general the mass spectra obtained from positive and negative corona depend significantly on trace contaminants, either preexisting in the gas or produced by the corona itself and accumulating with time. The major positive ions observed from positive corona were, SF$_5^+$, SF$_3^+$ and SF$_2^+$ fragments produced by electron impact of SF$_6$. Addition of water in the range 40-600 ppm in SF$_6$ resulted in the formation of clusters SF$_5^+$(H$_2$O)$_n$, SF$_3^+$(H$_2$O)$_n$, SF$_2^+$(H$_2$O)$_n$, SOF$_3^+$(H$_2$O)$_n$, and H$^+$(H$_2$O)$_n$. The major negative ions observed from negative corona were SF$_6^-$, SF$_5^-$ and F$^-$ without water addition. Addition of water resulted in the formation of the clusters SF$_6^-$(HF), F$^-$(HF)$_n$ and OH$^-$(H$_2$O)$_n$ where HF is an SF$_6$ discharge by-product. Other negative ions observed included SO$_2$F$^-$, SO$_2$F$_2^-$ and SOF$_5^-$ which are products of reactions of SF$_6^-$ with neutral by-products of SF$_6$ corona.

INTRODUCTION

The point-to-plane corona discharge in a dense gas has been modeled[1] by considering the total gas volume to be composed of three zones: (1) the glow region near the point electrode where electron impact processes lead to ionization, attachment, and dissociation; (2) the interelectrode (gap) region where ions and electrons drift and react with neutral gas particles; and (3) the main gas volume where slower neutral chemistry occurs. This paper concerns the ion-chemical processes of zone 2 and to a more limited extent to the electron-impact processes occurring in zone 1 in SF$_6$ gas. In general ions emerging from the glow region, where the density-normalized electric-field (E/N) is high, into zone 2, where E/N is low, interact with the gas, undergoing ion conversion through reactions which depend on the composition of impurities, the concentrations of which may be time dependent. In the measurement of ion drift velocities in SF$_6$ corona[2] the primary positive- and negative-charge carriers were assumed to be SF$_5^+$ and SF$_6^-$ respectively. We will show in this paper, however, that ions which exit the glow region of the discharge can be significantly different from those ions which arrive at the plane electrode. Data will be presented on the positively- and negatively-charged mass identified ions in SF$_6$ corona.

EXPERIMENT

Corona source and mass spectrometer

Corona discharges were produced in a point-to-plane geometry housed in a stainless-steel cell coupled directly to a differentially pumped quadrupole mass spectrometer having a mass

Gaseous Dielectrics VI, Edited by L.G. Christophorou and
I. Sauers, Plenum Press, New York, 1991

range 3-1000 amu. In the corona source the needle electrode was mounted to a linear motion vacuum feed-through so that the gap (nominally set to 2 mm) could be adjusted from outside the vacuum chamber. The high-voltage electrode could be biased to either positive or negative polarity with respect to the grounded plane electrode. A small aperture (30 μm diameter) in the plane electrode permitted the sampling of ions, produced either by the discharge or by subsequent reactions. Although it was possible to operate the corona discharge cell at SF_6 pressure in the range 1.3-40 kPa (10-300 Torr) with this aperture, the data reported here are for 6.5 kPa (50 Torr) SF_6 pressure. Over the above pressure range both the positive-ion mass- spectrum (PIMS) and the negative-ion mass spectrum (NIMS) became more complex with increasing pressure. By operating at the low end of the pressure range we were able to observe ions emerging from the glow region minimizing as much as possible the ion conversion due to reactions with trace contaminants. The PIMS and NIMS at higher pressures were consistent with those obtained at lower pressure but at longer gap or higher contaminant content. The SF_6 gas used was instrument grade with a purity of 99.99%. The results of two types of experiments are reported here: positive ions sampled from positive point-to-plane corona and negative ions sampled from negative point-to-plane corona. Typical voltage (on the needle electrode) and currents employed were 1450 V at 0.5 μA for positive corona and -875 V at 0.5 μA for negative corona. For the H_2O/SF_6 mixtures the voltage-current characteristic for positive corona depended on water content. In these cases the voltage was decreased to maintain constant current at 0.5 μA.

The needle electrode was made of either tungsten (W) or platinum-iridium alloy (Pt-Ir) with a tip radius less than 1 μm. The needle material was found to influence the NIMS at mass-to-charge ratio, m/z > 200. For example when the W needle was used, WF_6 was formed in the corona and this species subsequently entered the ion chemistry leading to formation of WF_6^- and WF_7^-. Data reported here cover mainly the m/z range 10-210 and are not affected by the electrode material for the discharge parameters described above. The Pt-Ir needle was used in all the H_2O/SF_6 experiments.

Interpretation of mass spectra

When the ions observed are simple fragments, i.e. formed by single bond breakage with no rearrangement as in the case of single-collision electron-impact, the identification of ions is straight-forward requiring only knowledge of the mass-to-charge ratio, m/z, of the ion mass peak. The PIMS in Fig. 1 illustrates a spectrum composed only of SF_6 fragments, SF_5^+, SF_3^+, and SF_2^+. However when the ions observed are formed as a result of reaction with trace constituents (some of which are unknown or are poorly characterized), the ion identities are subject to considerably more uncertainty. In this study where SF_6 is involved the natural isotopic abundances of sulfur (^{32}S-95.03%, and ^{34}S-4.18%) are used to aid in determining whether the ion contains S and the number of S atoms in the ion. The other two S isotopes (^{33}S-0.74% and ^{36}S-0.016%) are observed but generally not used for sulfur identification due to the small intensities. Thus for singly charged ions of mass, m, the number of S atoms, N, can be determined from the ratio of ion intensity at mass m+2 to the ion intensity at m by $i_{m+2}/i_m = [^{34}S]/[^{32}S]$ N = 0.044 N, assuming that other ions of mass m+2 are not present in significant amounts. For example the ions at m/z = 32 in Fig. 2 must be O_2^+ rather than S^+ because of the absence of an ion peak at m/z = 34. In some of the spectra, particularly for experiments involving addition of water vapor to SF_6, peaks are ascribed to ion clusters involving H_2O (18 amu separation between mass peaks) or HF (20 amu separation between mass peaks). Actually, the structure of the ions are not known. These ion species are written as a core ion (generally an SF_6 fragment) attached to one or more polar molecules as in the case of $F^-(HF)_n$ [n=0,1,...]. The $F^-(HF)_2$ ion could also be written as $HF_2^-(HF)$ or $H_2F_3^-$. Finally since only singly charged ions are reported here m/z is replaced by m (in amu) when ion mass is discussed. No correction has been made for mass discrimination due to either the sampling aperture or the quadrupole mass filter. Neglect of mass discrimination does not affect the conclusions made on ion identification and reaction pathways reported.

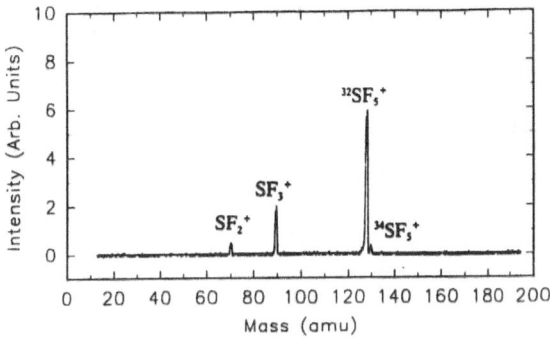

Fig. 1. Positive-ion mass-spectrum of SF_6 (P=6.5 kPa, T=28°C) for positive corona (V=1.45 kV, i_c=0.5 μA).

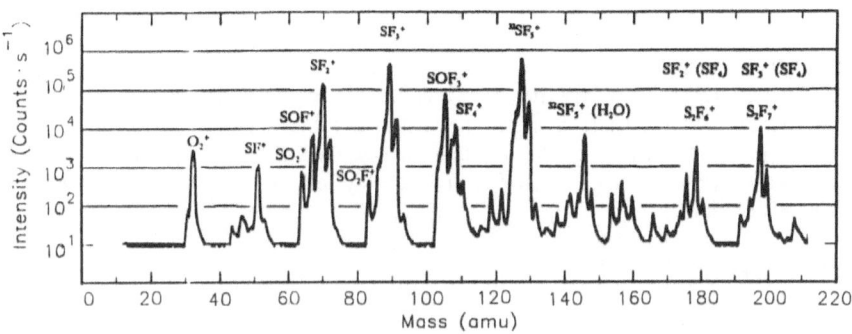

Fig. 2. Logarithmic plot of positive-ion mass-spectrum for positive corona (same conditions as Fig. 1) showing the influence of trace constituents of the gas (see text).

RESULTS AND DISCUSSION

A common feature in the mass spectra of ions formed in SF_6 corona is the influence of trace (ppm) constituents of the gas. Two types of impurities are considered, (1) impurities that are present in the gas prior to discharge initiation and (2) impurities that are created as a result of the discharge and whose concentration increases with time. Water which falls in the first group, is the primary contaminant considered while species such as SOF_4, SO_2, and SF_4, (neutral by-products of SF_6 corona) belong to the second group. Other SF_6 by-products such as HF may actually fall in both groups.

PIMS from positive corona

In positive corona (P=6.5 kPa, i_c=0.5 μA, d=2 mm, T=28 °C) under relatively dry conditions ([H_2O]<<50 ppm) the PIMS is relatively simple exhibiting only three SF_6 fragment ions SF_5^+ (127 amu), SF_3^+ (89 amu), and SF_2^+ (70 amu) as illustrated by the mass spectrum in Fig. 1. The fragment ion SF_4^+ is notably absent in the spectrum despite the low threshold energy (ϵ_{th} = 18.9 eV) which is below those of SF_3^+ (ϵ_{th} = 20.1 eV) and SF_2^+ (ϵ_{th} = 26.8 eV). In Fig. 2 we show a PIMS taken under similar discharge conditions where the ion signal is presented on logarithmic scale covering 5 decades of ion intensity. Above 10^5 counts s^{-1} (Fig.

Table 1. Comparison of calculated ion intensities relative to SF_5^+ with experiment for SF_6 corona.

	SF_5^+	SF_4^+	SF_3^+	SF_2^+	SF^+
Experiment[1]	1.0	0.011	0.33	0.083	1.5×10^{-3}
Calculated[2]	1.0	0.11	0.14	0.037	8.3×10^{-5}

[1]Relative to SF_5^+ intensity = 1.0

[2]Calculated from $\int_{\epsilon_{th}}^{\infty} f(\epsilon)\, \epsilon^{\frac{1}{2}}\, \sigma_i(\epsilon)\, d\epsilon / \int_{\epsilon_{th}}^{\infty} f(\epsilon)\, \epsilon^{\frac{1}{2}}\, \sigma_j(\epsilon)\, d\epsilon$ where

$i = SF_x^+$, x = 1-4, $j = SF_5^+$, $f(\epsilon)$ = electron - energy distribution - function[4] at E/N = 3.54 $\times 10^{-15}$ Vcm^2 and $\sigma_i(\epsilon)$ = partial dissociative ionization cross section[3]

(2) the fragments are the same as those shown in Fig. 1. Below 10^5 counts s^{-1} the fragments SF_4^+ (108 amu) and SF^+ (51 amu) are observed. In addition to these are the oxygen-containing fragments SO_2^+ (64 amu), SOF^+ (67 amu), SO_2F^+ (83 amu), SOF_3^+ (105 amu), and $SF_5^+(H_2O)$ (145 amu). Using cross section data for electron-impact ionization in SF_6[3] and an electron-energy distribution-function corresponding to the limiting density-normalized electric field, $(E/N)_{lim} = 3.54 \times 10^{-17}$ Vcm^2, applicable to the glow region[4] the relative ion intensities have been calculated and are presented in Table 1 and compared with experiment. The low intensity of SF_4^+ and the fact that no new ions of comparable intensity appear in the spectrum (Fig. 1) suggest that SF_4^+ is probably lost in a reaction such as the charge changing collision

$$SF_4^+ + SF_6 \rightarrow SF_5 + SF_5^+ \tag{1}$$

The intensity of SF^+ is considerably higher than expected possibly indicating that it is formed from neutral fragments (such as SF_4 or SF_2) in the discharge.

Two disulfur ions $S_2F_6^+$ (178 amu) and $S_2F_7^+$ (197 amu) were also observed (Fig. 2). One of these ions ($S_2F_7^+$) has been reported previously in spark discharges[5] and in radio-frequency plasma discharges[6]. These ions may be respective clusters of SF_2^+ and SF_3^+ with SF_4 where SF_4 is a major neutral by-product of SF_6 discharges[7,8].

Positive corona in H_2O/SF_6 mixtures

Water, when added to SF_6 in the concentration range 40-640 ppm, was found to alter dramatically the PIMS. Water vapor was introduced into the corona cell via a metering valve and its pressure was monitored by a capacitance manometer (0-10 Torr range) with a resolution of 10^{-3} Torr. The SF_6 gas was added after several minutes to allow the water-vapor pressure to reach equilibrium with the walls. A typical PIMS for 300 ppm H_2O/SF_6 is shown in Fig. 3. Ion clusters identified include $H^+(H_2O)_n$, [n=1-4]; $SF_3^+(H_2O)_n$, [n=0-2]; $SOF_3^+(H_2O)_n$, [n=0,1]; and $SF_5^+(H_2O)_n$, [n=0-2]. The dependence of some of the SF_6 ion fragment intensities (SF_5^+, SF_3^+, and SF_2^+) and the reaction product ion intensities (SOF_3^+ and $SF_5^+(H_2O)$) on water content is shown in Fig. 4. For $P_{H_2O} = 32$ mTorr (corresponding to 640 ppm) the SF_5^+ ion intensity has decreased to nearly 0. The major reaction product is SOF_3^+ which may be formed by

$$SF_5^+ + H_2O \rightarrow SOF_3^+ + 2HF. \tag{2}$$

A plot of the relative ion intensity of $SF_5^+(H_2O)$ to that of SF_5^+ in Fig. 5 indicate a linear dependence in the 0-640 ppm range of water concentration. Using the data in Fig. 5 as a calibration curve for water content in SF_6, the water concentration in the experiment reported in Fig. 2 is calculated to be 6 ppm, a reasonable value for this grade of SF_6.

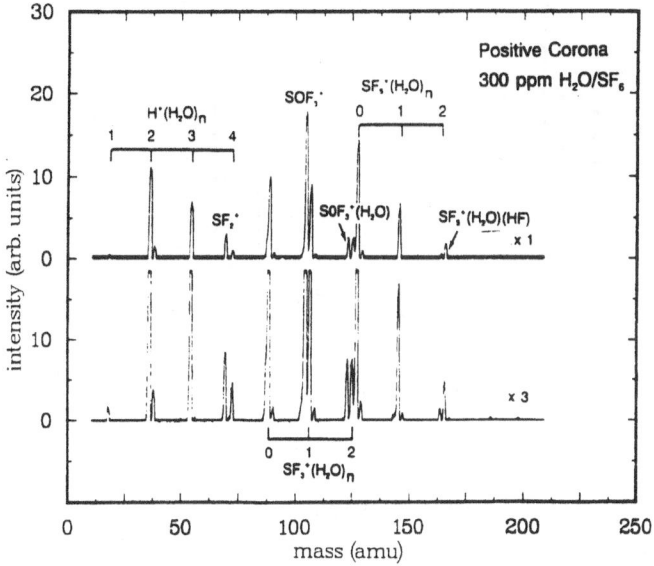

Fig. 3. Positive-ion mass-spectrum of positive corona in 300 ppm H_2O/SF_6 mixture at two sensitivities (denoted $\times$ 1 and $\times$ 3).

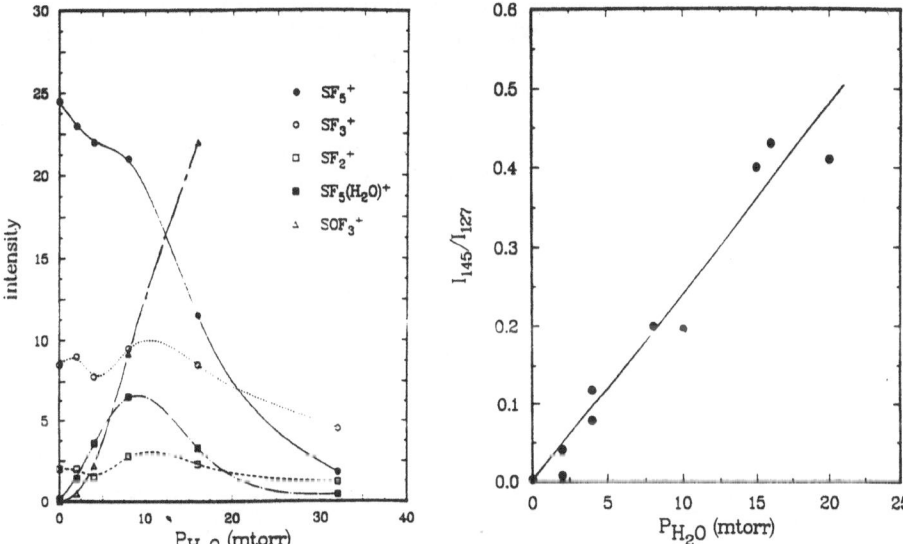

Fig. 4. Dependence of ion intensity on H_2O partial pressure for SF_5^+, SF_3^+, SF_2^+, $SF_5^+(H_2O)$ and SOF_3^+.

Fig. 5. Increase in the ion intensity ratio of $SF_5^+(H_2O)$ to SF_5^+, I_{145}/I_{127}, with water partial pressure P_{H_2O} yielding a water calibration curve for water content in SF_6 at the ppm level.

NIMS from negative corona in SF_6

For negative corona discharges in SF_6 under gas purity conditions similar to those used for the positive corona data, the NIMS (Fig. 6) contained more ion peaks involving impurities than did the positive ion spectra (Fig. 1). In addition to the parent negative ion SF_6^- (146 amu) the fragment ions SF_5^- (127 amu) and F^- (19 amu) are observed. Other peaks in the spectrum at masses 59, 143, and 166 amu are ascribed to $F^-(HF)_2$, SOF_5^-, and $SF_6^-(HF)$

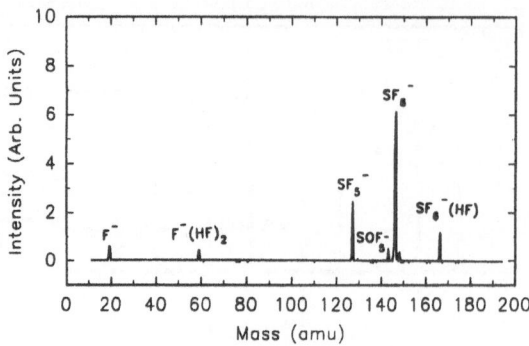

Fig. 6. Negative-ion mass-spectrum of SF_6 (P=6.5 kPa, T=28°C) for negative corona (V= - 875 v, i_c=0.5 μA).

respectively. The formation of these ions will be discussed further in the next two sections with regard to H_2O, SOF_4, and HF contamination in SF_6.

Negative corona in H_2O/SF_6 mixtures

The effect of water on the SF_6 NIMS is illustrated in Fig. 7 for 600 ppm H_2O/SF_6. The sensitivity is such that several peaks [SF_6^-, SF_5^-, and $F^-(HF)_2$] are offscale. Ion clusters observed include $F^-(HF)_n$, [n=0-4]; $OH^-(H_2O)_n$, [n=0-2]; and $SF_6^-(HF)_n$, [n=0,1]. In addition, SO_2F (83 amu), $SO_2F_2^-$ (102 amu), SOF_4^- (124 amu), SOF_5^- (143 amu), and $SOF_5^-(HF)$ (163 amu) are noted in the figure. The formation of these sulfur oxyfluoride ions are related to by-products and will be discussed in the following section.

The species HF may be present initially in SF_6 as a contaminant or formed as a product of the discharge. As an SF_6 by-product, HF may be formed in a fast process[9] in the glow region

$$F + H_2O \xrightarrow{\text{k}} HF + OH, \quad k=1.1 \text{ x } 10^{-11} \text{ cm}^3 \text{ s}^{-1} \tag{3}$$

or formed outside the glow in the main gas volume[10] by much slower reactions such as

$$SF_4 + H_2O \xrightarrow{\text{k}} SOF_2 + 2HF, \quad k=(0.9 - 2.6) \text{ x } 10^{-19} \text{ cm}^3 \text{ s}^{-1} \tag{4}$$

Therefore H_2O enters the negative ion chemistry either directly to form $OH^-(H_2O)_n$ or indirectly to form HF leading to negative ion clusters such as $F^-(HF)_n$.

Peaks shown but not labelled in Fig. 7 in the NIMS were also observed at 37 and 55 amu for which the identities are less certain. The ion at mass 37 may be written as either $F^-(H_2O)$ or $OH^-(HF)$. Similarly the ion at mass 55 may be $F^-(H_2O)_2$ or $OH^-(H_2O)(HF)$. The $F^-(H_2O)_n$ series seems more likely since it avoids the mixed cluster of H_2O and HF. Figure 8 shows the dependencies of most of the negative ion intensities on water content.

Various negative ions such as $(H_2O)_n^-$ and $O_2^-(H_2O)_n$ have been suggested[11,12] as the source ('seed' ion) for collisionally detached electrons in H_2O/SF_6 mixtures where H_2O was found to influence the discharge characteristics even at the ppm level. Neither of these two ion clusters were observed in the present experiments. The ion cluster $OH^-(H_2O)_n$ is observed, however, (see Fig. 7) and is a possible candidate for the seed ion.

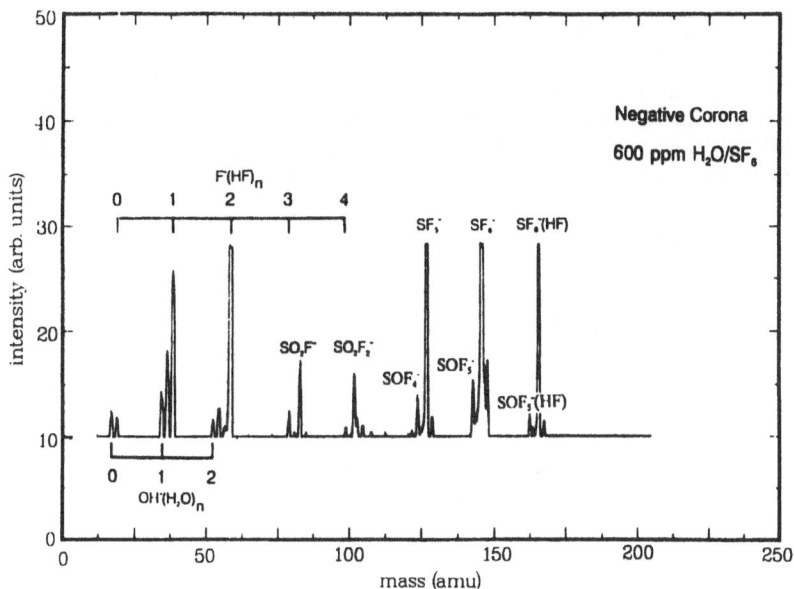

Fig. 7. Negative-ion mass-spectrum of negative corona in 600 ppm H_2O/SF_6. At this sensitivity some of the ion peaks [$F^-(HF)_2$, SF_5^-, and SF_6^-] are off-scale.

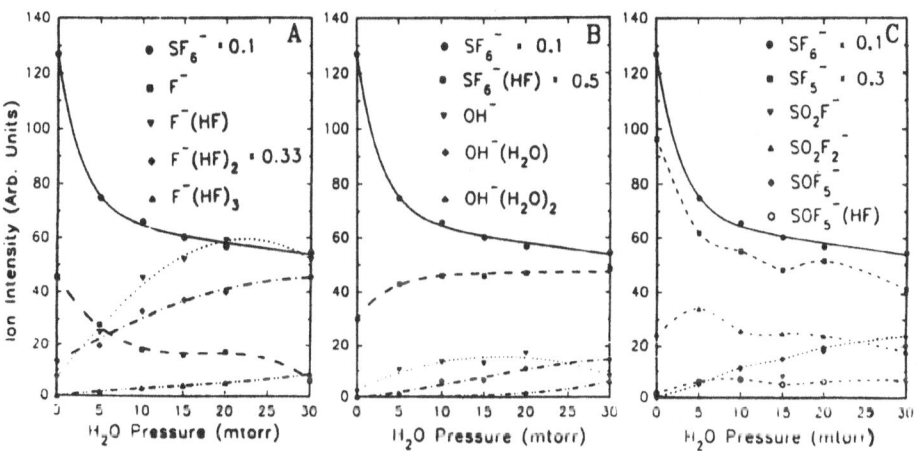

Fig. 8. Dependence of negative ion intensities on H_2O partial pressure for: (A) SF_6^-, $F^-(HF)_n$, [n=0-3]; (B) SF_6^-, $SF_6^-(HF)$ and $OH^-(H_2O)_n$, [n=0-2]; and (C) SF_6^-, SF_5^-, SO_2F^-, $SO_2F_2^-$, SOF_5^- and $SOF_5^-(HF)$.

Effect of SF_6 by-products on the NIMS

Many of the oxygen-containing ions observed in SF_6 and H_2O/SF_6 corona are formed via reactions of SF_6^- with neutral discharge by-products of SF_6. The ions SOF_5^- and SO_2F^- can be formed by

$$SF_6^- + SOF_4 \longrightarrow SOF_5^- + SF_5 \tag{5}$$

$$SF_6^- + SO_2 \longrightarrow SO_2F^- + SF_5 \tag{6}$$

Rate constants for reactions of the form

$$SF_6^- + M \longrightarrow SF_5 + MF^- \tag{7}$$

have been measured in electric fields[13] for M=SOF_4, SiF_4, SF_4, WF_6 and as a function of temperature[14] for M=SOF_4, SiF_4, SF_4, SO_2 and in all cases were found to be extremely large ($k > 10^{-10}$ cm^3 s^{-1}) approaching the collision rate. In Table 2 are listed several relevant species and their corresponding F$^-$ ion affinities. Ion conversion proceeds down the list where the last species is the most stable. Results of pulsed-electron-beam high-pressure mass spectrometer experiments[14] have shown that in reactions involving SOF_4 with SF_6^-, SF_6^- will efficiently convert to SOF_5^- as reaction time increases (by increasing gap or gas pressure). These authors also showed that when SiF_4 is present (formed by attack of HF on Si-containing insulators or vacuum greases), after long reaction time an equilibrium will be established between SOF_5^- and SiF_5^- due to the reversible reaction

$$SOF_5^- + SiF_4 \longleftrightarrow SiF_5^- + SOF_4. \tag{8}$$

In Fig. 9 we show the time dependence of the NIMS over the mass range 130-170 amu during a corona discharge which was run for 10 minutes. As the SOF_4 by-product increases with time, the relative intensities of SOF_5^- (143 amu) to SF_6^- (146 amu) increases. The same behavior occurs for their respective clusters to HF at masses 163 and 166 amu. As further

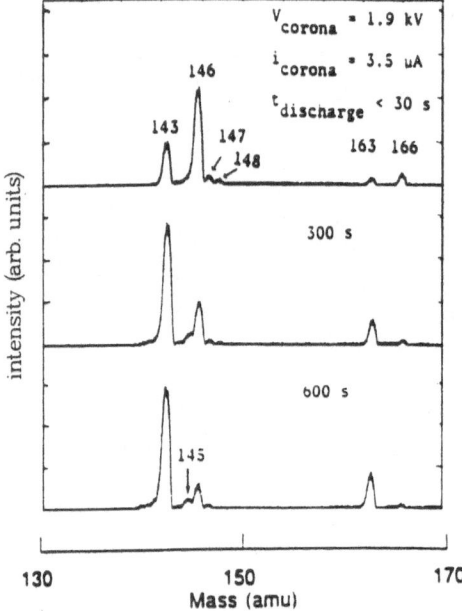

Fig. 9. Change in the negative-ion mass-spectrum with time, t, in the mass range 130-170 amu for negative corona (V=1.9 kV, i_c=3.5 μA) in SF_6 (P=6.6 kPa, T=28°C): (a) t<30 s, (b) t=300 s and (c) t=600 s. Ions and their masses shown are SOF_5^- [143], SF_5^- (H_2O) [145], SF_6^-[146], SF_5^-(HF) [147], $^{34}SF_6^-$ [148], SOF_5^-(HF) [163] and SF_6^-(HF) [166].

evidence of the importance of SOF_4 to the negative ion chemistry in SF_6 corona, we show in Fig. 10 what happens when a large amount (2%) of SOF_4 is added to SF_6, resulting in the conversion of negative charge carriers in the discharge to SOF_5^- and SOF_5^-(HF). From Table 2 it would also appear that HF reacts with SF_6^- according to

$$SF_6^- + HF \longrightarrow HF_2^- + SF_5 \tag{9}$$

No rate constants are currently available for this reaction however.

Table 2. Fluoride ion (F⁻) affinities for various neutral species relevant to SF₆ corona.

Species	F⁻ affinity[1] (eV)
SO_2F_2	1.55
SOF_2	1.62
SF_5	1.62-1.90
SF_4	1.9
SO_2	1.9
HF	2.01[2]
SOF_4	2.59
SiF_4	2.69

[1]Taken from Ref. 14 except for HF.
[2]Theoretical calculation from Ref. 15.

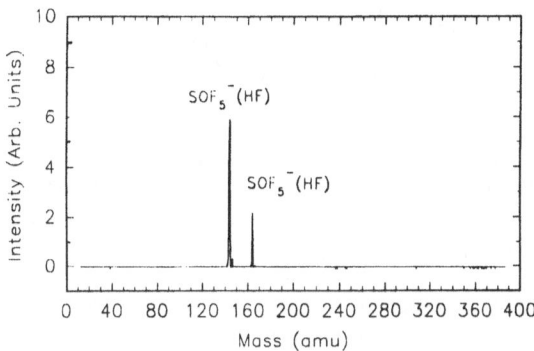

Fig. 10. Negative ion mass spectrum of negative corona in 2% SOF₄/SF₆ (P=6.6 kPa, T=28°C).

SUMMARY

Trace constituents in SF₆ gas such as H₂O, HF or various SF₆ by-products significantly alter both the PIMS and NIMS of SF₆ corona. At the lowest impurity levels attained in the present study the major positive and negative ions were SF₃⁺ and SF₆⁻ respectively. The effect of water is to produce cluster ions involving either H₂O or HF: $H^+(H_2O)_n$, $SF_x^+(H_2O)_n$, $SOF_3^+(H_2O)_n$, $F^-(HF)_n$, $SF_x^-(HF)_n$, $OH^-(H_2O)_n$, $F^-(HF)_n$, $SOF_5^-(HF)_n$. Neutral by-products of SF₆ corona were found to produce a time dependent effect on the NIMS. As SOF₄ increases with corona discharge time, the initial negative charge carriers composed primarily of SF₆⁻, SF₅⁻, and F⁻, convert almost completely to SOF₅⁻ and SOF₅⁻(HF).

ACKNOWLEDGEMENT

We gratefully acknowledge support by the Office of Energy Management, Electric Energy Program, U.S. Department of Energy, under contract DE-AC05-84OR21400 with Martin Marietta Energy Systems, Inc.

REFERENCES

1.   R. J. Van Brunt, J. T. Herron, and C. Fenimore, Corona-Induced decomposition of dielectric gases, in: "Gaseous Dielectrics V" (Proc. of the 5th Int. Symp. on Gaseous

Dielectrics) ed. L. G. Christophorou and D. W. Bouldin, Pergamon Press, New York, 163 (1987).

2. D. T. Blair, B. H. Crichton, F. J. Al-Kindi, and T. L. Sharma, Drift velocities of positive ions and negative ions in cylinder SF_6, J. Phys. D: Appl. Phys. 22:755 (1989).

3. T. Stanski and B. Adamcyzk, Measurements of dissociative ionization cross section of SF_6 by using double collector cycloidal mass spectrometer, Int. J. Mass. Spect. and Ion Phys. 46:31 (1983).

4. R. J. Van Brunt, Private Communication.

5. L. C. Frees, I. Sauers, H. W. Ellis, and L. G. Christophorou, Positive ions in spark breakdown of SF_6, J. Phys. D: Appl. Phys. 14:1629 (1981).

6. A. Picard, G. Turban, and B. Grolleau, Plasma diagnostics of a SF_6 radio frequency discharge used for the etching of silicon, J. Phys. D: Appl. Phys. 19:991 (1986).

7. I. Sauers, H. W. Ellis, and L. G. Christophorou, Neutral decomposition products in spark breakdown of SF_6, IEEE Trans. on Elect. Insul. EI-21(2):111 (1986).

8. R. J. Van Brunt, Production rates for oxyfluorides SOF_2, SO_2F_2 and SOF_4 in SF_6 corona discharges, J. Res. Nat. Bur. Stand. 90(3):229 (1985).

9. D. L. Baulch, R. A. Cox, R. F. Hampson, Jr., J. A. Kerr, J. Troe, and R. T. Watson, Evaluated kinetic and photochemical data for atmospheric chemistry: Supplement II, J. Phys. Chem. Ref. Data 13:1259 (1984).

10. I. Sauers, J. L. Adcock, L. G. Christophorou, and H. W. Ellis, Gas phase hydrolysis of sulfur tetrafluoride: A comparison of the gaseous and liquid phase rate constants, J. Chem. Phys. 83(5):2618 (1985).

11. R. J. Van Brunt, "Water vapor-enhanced electron-avalanche growth in SF_6 for nonuniform fields, J. Appl. Phys. 59(7):2314 (1986).

12. G. Berger, B. Senouci, B. Hutzler, and G. Riquel, The influence of impurities on the dielectric strength of SF_6 for positive polarity, in: "Gaseous Dielectrics V" (Proceedings of the 5th Int. Symp. on Gaseous Dielectrics) ed. L. G. Christophorou and D. W. Bouldin, Pergamon Press, New York (1987).

13. I. Sauers, Sensitive detection of by-products formed in electrically discharged sulfur hexafluoride, IEEE Trans. on Elect. Insul. EI-21(2):105 (1986).

14. L. W. Sieck and R. J. Van Brunt, Rate constants for F^- transfer from SF_6^- to fluorinated gases and SO_2. Temperature dependence and implications for electrical discharges in SF_6, J. Phys. Chem. 92(3):708 (1988).

15. M. J. Frisch, J. E. Del Bene, J. S. Binkley, H. F. Schaefle, III, Extensive theoretical studies of the hydrogen bonded complexes $(H_2O)_2$, $(H_2O)_2H^+$, $(HF)_2$, $(HF)_2$, $(HF)_2H^+$, F_2H^-, and $(NH_3)_2$, J. Chem. Phys. 84(4):2279 (1986).

DISCUSSION

J. CASTONGUAY: (1) Do you have an idea of the concentration of the SOF_4 by-product buildup that is responsible for the mass spectral charge shown in Fig. 9? (2) In the determination of the water content in your high grade SF_6 ($\sim$6 ppm), did you take into account the fact that some of the original moisture present in the gas was lost, adsorbed on the walls of the cell? Six ppm seem to me rather low. (3) How much do ion–molecule reactions contribute to final decomposition by–products compared to other processes?

I. SAUERS: (1) The SOF_4 concentration was not directly measured. However SOF_4 is known to be formed as the major by–product of SF_6 corona. Corona experiments that were made with 50 ppm SOF_4 in SF_6 resulted in SF_6^- to SOF_5^- ion intensity ratios similar to that shown in the t=300s spectrum in Fig. 9. This concentration is consistent with SOF_4 corona yields measured at higher pressures [Van Brunt, J. Res. Nat. Bur. Stand. 90 (3), 229–53 (1985)] and with a calculation based on the rate constant $k=2x10^{-10}$ cm^3 s^{-1} for reaction (5) in the paper. (2) It is possible that some of the water is lost to the walls. However, from the calibration curve (Fig. 5) determined in experiments where the water vapor was in equilibrium with the walls and the water vapor partial pressure was measured, the water content in SF_6 was much less than $\sim$ 40 ppm (in high purity SF_6). The main point is that the ratio of SF_5^+ (H_2O) to SF_5^+ ion intensities yields an instantaneous measure of the water content in SF_6 and that this measured content was consistent with the expected water content for this grade of SF_6. Another calibration curve (not shown in this paper) based on the ion intensity ratio SF_3^+ (H_2O) to SF_3^+, yields a water content of $\sim$ 7 ppm for the spectrum in Fig. 2. (3) In general ion–molecule reactions have not been found to contribute significantly to the final decomposition product formation rate for the major by–products SOF_4, SOF_2 and SO_2F_2. The F^- exchange reaction (reaction (5)) has been shown [Van Brunt et al., Plasma Chem. Plasma Proc. 8 (2), 225–246 (1988)] to decrease slightly the net charge rate–of–production of SOF_4.

CHAPTER 10: HIGH VOLTAGE DC INSULATION; UTILITY EXPERIENCE WITH GIS

HIGH VOLTAGE DC GAS INSULATION

- A WORLD-WIDE REVIEW

K. Nakanishi

Manufacturing Development Laboratory
Mitsubishi Electric Corporation
8-1-1, Tsukaguchi-Hommachi, Amagasaki, Japan

ABSTRACT

The paper reviews charge accumulation on supporting insulators and the behaviour of conducting particles which should be especially taken into consideration for dc gas insulation. For charge accumulation, the mechanisms and the dc breakdown properties of spacers at a polarity reversal are stated. The optimum designs of spacers for DC-GIS are presented. In the latter part of the paper, the behaviour of conducting particles at the situation which electric and magnetic fields coexist is studied. Ion wind which the partial discharges of the particles generate is discussed in relation with the structure of a particle trap

INTRODUCTION

Compressed SF_6 gas insulation has accomplished high reliability and compactness of the ac power apparatus since the apparatus insulated with excellent insulating gas is not exposed to the influence of contamination or any other external disturbances. Studies on application of gas insulation to high voltage dc power apparatus (Jaster et al., 1978; Matsumura et al., 1983) have been done to realize the benefits which the ac gas insulated equipment has enjoyed. The aim of the paper is to show the studies which have been performed to understand the discharge properties of SF_6 peculiar to dc electric fields and establish the insulation design for HVDC gas insulated equipment.

Surface charging has been a fundamental consideration in high voltage dc gas insulation since charges accumulated on solid insulators work to reduce the breakdown properties particularly at the polarity reversal of dc voltage. The breakdown properties along solid insulators with accumulated charges and the mechanisms on charge accumulation are discussed in the paper. I refer to the problems involved in capacitive probe measurement for charge distribution and briefly show a solution method for the problems. The shapes of the epoxy spacer which minimize the influence of charge accumulation are described.

Influence of metallic particles on HVDC insulation is another limiting factor in designing HVDC-GIS since compressed gas insulation is more susceptible to metallic particles at dc fields. In the paper, the behaviour of

Gaseous Dielectrics VI, Edited by L.G. Christophorou and
I. Sauers, Plenum Press, New York, 1991

Table 1. Conductor field at flashover (SF6 0.4 MPa).

Spacers / Voltage	Post - type 109/292mm	Tri - post 89/226mm	Conical-type 89/226mm
Ave. dc	>15.0kV/mm	13.4kV/mm	10.6kV/mm
Symmetric dc polarity reversal	>11.2kV/mm	10.8kV/mm	8.4kV/mm

metallic particles at dc electric fields together with magnetic fields is discussed. The influence of ion wind generated by partial discharges to the behaviour of the particles is also shown.

CHARGE ACCUMULATION ON/IN SOLID INSULATORS

Flashover Due To Charge Accumulation

When a corrugated post-type spacer was energized with dc voltages for long duration (10^3 minutes), the breakdown voltage decreased to a half value of the short-term flashover voltage as shown in Fig.1 (Mangelsdorf et al., 1980). The reduction of the flashover voltages was considered to be caused by the local field enhancement due to accumulated charges which increased with the passage of time. They reported that accumulation of homo-charges (charges with the same polarity as that of the nearest metal insert) happened in the bulk of the spacer.

The studies on surface flashover along spacers at dc fields performed in the US also indicated that the tri-post and conical designs exhibited a long term failure mode (Cooke et al., 1982). The flashover voltages at the polarity reversal of dc voltage were reported for post-type, tri-post and conical type spacers set in coaxial electrode gaps. The data summarized in Table 1 show that accumulation of charges worked to lower the dc flashover voltages to 75-80 % of the short-term breakdown values, depending on the insulator geometry.

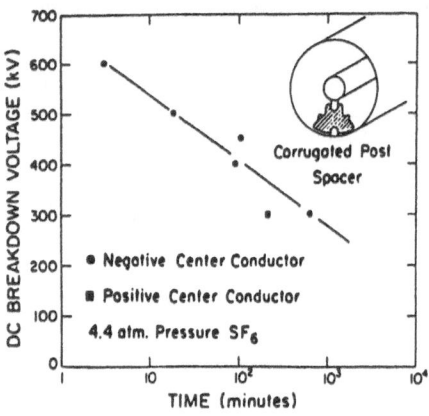

Fig.1. Time to breakdown of a corrugated post spacer due to charge accumulation at dc application.

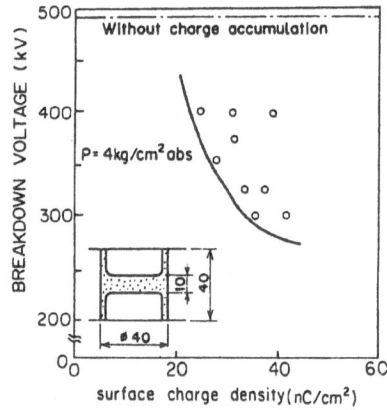

Fig.2. Relation between flashover voltage and residual charge density of model spacer at the polarity reversal.

434

Since accumulated charges distort the electric field along the spacer and reduce the withstand level particularly at the polarity reversal test, the influence of surface charge densities to the breakdown voltage was studied using a simple cylindrical insulator with metal inserts. In the tests, accumulation of hetero-charges (charges with the polarity opposite to that of the nearest metal insert) was measured before reversing the polarity of dc voltage. The reduction of the flashover voltages was noticed when the surface charge density accumulated at some spot on the surface was over 20 nC/cm^2 as shown in Fig.2 (Nakanishi et al, 1983). The cylindrical insulator with the charge density of about 40 nC/cm^2 resulted in showing the breakdown voltage of 300 kV although the insulator without charge accumulation could withstand 490 kV dc application.

Mechanisms On Charge Accumulation

In order to design the optimum shape of spacer which supresses accumulation of charge, the researchers have performed the studies to clarify the mechanisms on charge accumulation. They have proposed the following conduction paths via which charges were transported.

(1) bulk conduction
(2) surface conduction
(3) gas conduction

Mangelsdorf and Cooke (1980) proposed charge deposition via bulk conduction in the epoxy insulator. They accounted for homo-charge accumulation around the metal inserted in the epoxy, using field dependence of bulk conductivity or nonuniform conductivity due to inhomogeneity of the materials. The CIGRE paper (Cooke et al, 1982) also indicated that homo-charge accumulation was observed for toroidal-type spacer at the carefully cleaned condition such as highly buffing the adjacent electrodes. They considered that non-homogeneous insulator materials (tabular alumina filler and epoxy resin) lead to resistivity variation and charging. At particular, an abrupt change in resistivity in the structure between heavily sedimented regions which are resin-rich/filler-starved, and vice versa appeared to be influential to charging and the following breakdown flashover along the spacer. However, the tests performed in other areas of the world have shown the hetero-charge accumulation on the surface of spacers. Although the difference cannot be explained, it might be attributed to that between the insulator materials and the process of casting they used.

Charge accumulation through surface conduction was observed when the test was done with well buffened electorodes in a small and clean chamber. Hetero-charges were deposited locally on the surface of the epoxy sample as

Fig.3. Dust figure of residual charge distribution on untreated spacer.

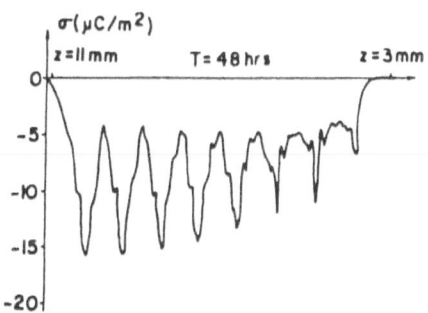

Fig.4. Spatio-temporal development of charges on a teflon spacer when stressed with 5 kV in SF_6.

435

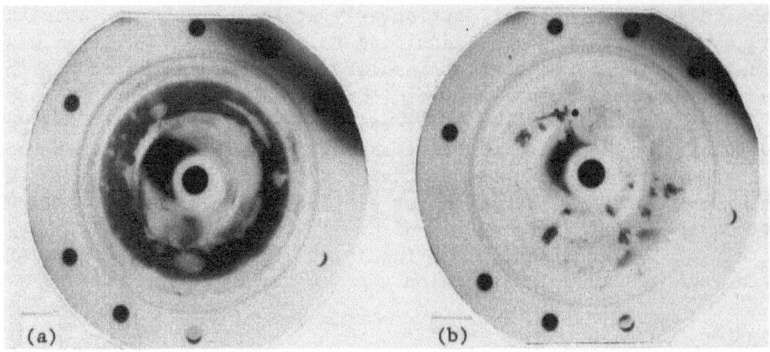

Fig.5. Dust figures of conical spacers with rough (a) and fine (b) finish
conductors set in a 60/140 mm dia. coaxial bus subjected to negative dc
150 kV for 5 hours.

shown in Fig.3. The phenomenon was accounted for by the dependence of surface
conductivity on the tangential electric field along the spacer surface. The
local charge accumulation suggested that the conductive layer such as the
greasy materials locally remained on the surface. When the thin surface layer
of the epoxy sample was removed by honing treatment, the hetero-charge was
deposited uniformly around the circumference of spacer surface and the magnitude of the charge density lowered to about 30 % of that of the untreated
epoxy (Nakanishi et al., 1982). The phenomenon can be explained by considering
that the surface with nonlinear resistivity was eliminated. When a teflon
spacer was stressed with dc voltages, hetero-charges periodically accumulated along the spacer as shown in Fig.4 (Bektas et al.,1988). Their data
showed that charge deposition was considerably influenced by the surface
conductivity of the material.

    In the mechanism of gas conduction, the charge carriers are considered
to be transported from the source to the surface of the spacer through gas
phase (Cooke et al., 1982; Knecht, 1982). They supposed that the normal electric field attracted the carriers until the field is eliminated by accumulation of the charges with the polarity opposite to that of the metal insert.
The charge carriers would be charged microparticles (Bargigia et al., 1987;
Fujimoto et al., 1987), ions produced by micro-discharges (Fujinami et al.,
1989) and negative gas ions which captured the electrons emitted from irregularities on the electrode. Therefore, this type of charge accumulation is
likely to happen in the industrially clean system in which the electrodes had
the practical surface finish. As shown in Fig.5(a), negative charges deposited
circumferentially on a concave part of spacer when the inner conductor with
the rough surface finish of 20-30 um was stressed with negative dc voltage
(Nakanishi et al., 1983). However, in the case of fine finish conductor (about
5 μm), only a small amount of charges was observed. The data suggested that
negative charging was caused by electrons field-emitted from the protrusions
on the electrodes. Although accumulation of positive charges on a convex side
was observed, the magnitude was low compared to that of negative charges.

Capacitive Probe Measurement

    A capacitive probe was used to measure the charges deposited on a spacer
by several researchers since charge distribution was quantitatively detected
by inserting the probe into a tank without breaking the gas. The principle
of capacitive probe technique is to detect the potential across a detective
capacitor which is in proportion to the charge density on a spacer. However,
Pedersen (1984) indicated that the probe technique was effective only in
limited situations and that it might have serious inaccuracies at charge

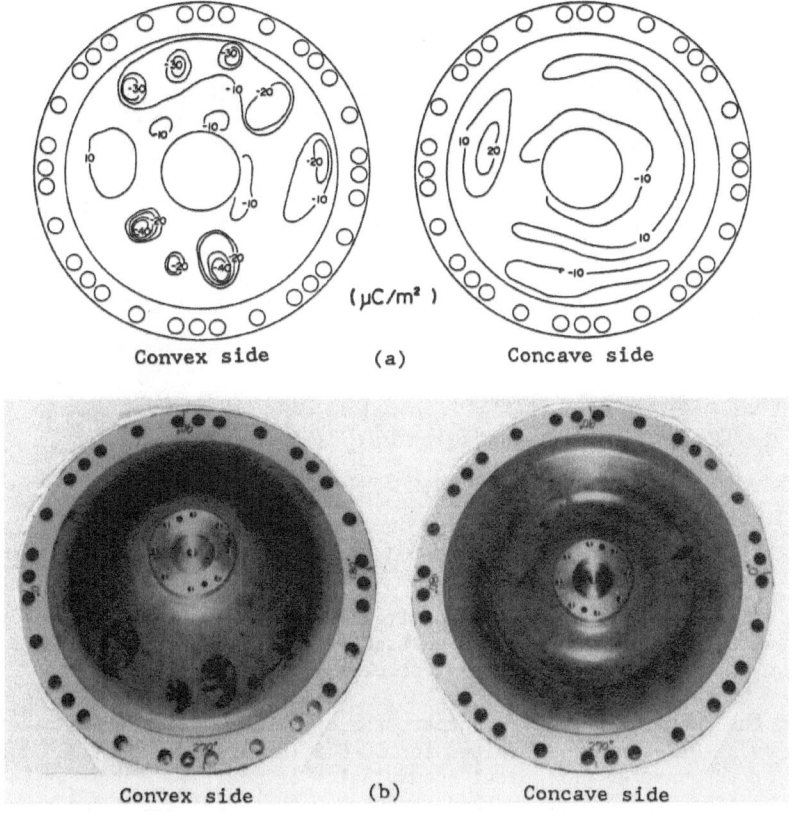

Convex side   (a)   Concave side

Convex side   (b)   Concave side

Fig.6. Surface charge distribution (a) analyzed by the computational method and the corresponding dust figures (b) on both sides of conical spacer.

measurement for a three-dimensional construction such as a spacer set in GIS. Conolly et al.(1984) stated the factors which generated the measurement errors, referring to the thichness of the dielectric and the calibration method.

Ootera and Nakanishi (1988) showed that the influences of the factors can be numerically computed, extending a three-dimensional field calculation method (surface charge method). They developed an analytical computation method for evaluating charge distribution on a conical spacer from the probe outputs. Figure 6 shows the charge distribution analyzed by the computational method and the corresponding dust figures on the convex and concave sides of the spacer developed for 500 kV DC-GIS. The method was proved to be effective when charge accumulation happened on the surface of insulators.

Optimum Design Of Spacers For DC-GIS

In the practically constructed system, the charges were supplied from the surface of the center conductor and moved along the electric field lines in the gas to the sheath. Therefore, the design of spacer which did not lie in the conduction paths of charged particles should be optimum for dc spacers. Since a disk-type spacer did not intersect the electric field lines, it was relatively free of charge accumulation and indicated small reduction of breakdown voltages at the polarity reversal test as shown in Fig.7. While, the spacer for AC-GIS which stands in the way of charged particles conduction was subjected to considerable amount of negative charges on the concave side of the spacer, leading to deteriorated flashover properties at the polarity

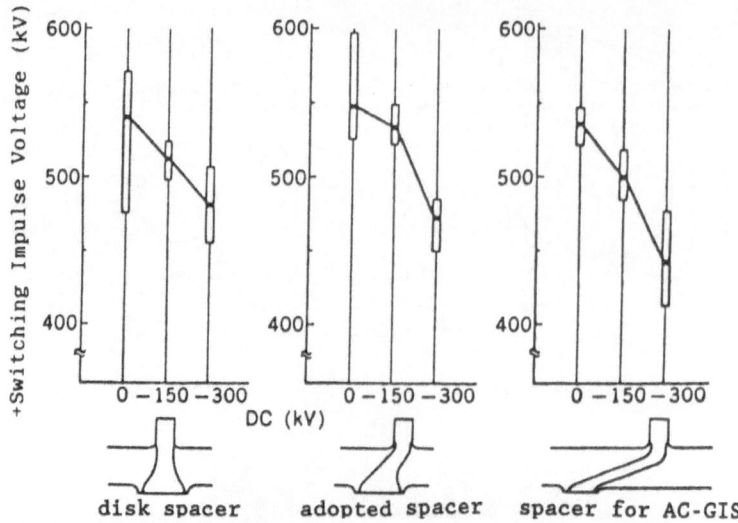

Fig.7. Flashover voltages of coaxial-type spacers at the polarity reversal
test with negative dc and the following switching impulse applications.

reversal. For instance, when the negative 300 kV was pre-stressed, reduction
of the switching impulse breakdown voltage reached to about a quarter of the
breakdown voltage without pre-stress process.

For the shape of post-type spacer, the spacer with ribs which have been
applied to AC-GIS should be avoided for DC-GIS (Hasegawa et al., 1983). Since
the spacer encountered periodical accumulation of positive and negative charges on the ribs, the charges worked to enhance the local field and reduce the
breakdown voltages to the considerably low level. A sphere-type spacer was
revealed to be one of best designs both by the experiments and by the theories based on the surface and the gas conduction mechanisms.

In selecting the design of spacer for DC-GIS, other important factors
such as the productivity of the spacer and the insulating properties at the
existence of conducting particles should be taken into account.

INFLUENCES OF CONDUCTING PARTICLES TO GAS INSULATION

The behaviours of the conducting particles peculiar to dc electric
fields, such as bouncing and hovering (Trump, 1962; Cooke, 1978) have been
found. Since their works, many researchers have studied the breakdown properties of SF6 gas caused by the particles left in the system. Since the
particles can give the more detrimental effects to compressed gas insulation
at dc fields than at ac fields, special care has been paid in designing the
high voltage dc power apparatus (Ouyang et al., 1982; Nitta et al., 1985). The
following two ways have been adopted to avoid the harmful effects of the particles to the system. One way is to place particle traps which scavenge the
conducting particles, utilizing their behaviour under electric stress. The
other is to give fair margin to the design stress, compared with the stress
applied to the ac apparatus.

The paticle trap has been placed in a long gas insulated bus which does
not include the moving equipments such as switchgears and disconnectors. It
is because cleaning and inspection works were difficult in the long bus without branch connections. In addition, having scavenged the particles before
service on site could improve the reliability of the system since the moving
parts which might generate the conducting particles in operation were not set

438

in the bus. However, in a complicated section in which switchgears, CT and PT were placed, the electric stress was limited to the level which was compatible with a certain carefully controlled system without traps.

The behaviour of conducting particles just at dc fields is not described here because they have been written in detail in many other publications. In the paper, I mention the behaviour of particles at dc fields together with magnetic fields generated by feeding dc current to a coaxial electrode system.

Magnetic fields hardly affected the motion of the particles which were non-magnetic materials such as copper and aluminium. However, for iron wire particles lying on the sheath, they turned to the direction of magnetic field lines when dc current was fed to the center conductor. Magnetic fields also worked to prevent the iron particles from lifting from the sheath when the magnetic material such as steel was used as the sheath. Figure 8 shows the effect of magnetic field on the lifting electric field for steel wire particles 0.2 mm in diameter and 5 mm long put on copper or steel sheath. As shown in Fig.8, when the material of sheath was copper, the lifting field decreased with the magnetic field and reached to about a half of that at the situation without magnetic field. However, the reduction was relatively small in the case of the steel sheath. The phenomenon can be accounted for by considering that the attractive magnetic force between the particle and the sheath which was generated by the magnetic field prevented the particles from levitating.

The principle of the particle trap which has been widely used in GIS is to move the particles by electric stress and confine them into very low stressed area. The metal sheet-type trap with slits was proved to be also effective in collecting the particles at dc fields as well as at ac fields. However, we often experienced that the particles confined in the trap escaped out of the side of the trap in a certain situation. The situation was strongly related with the behaviour of the particles which was peculiar to negative dc application.

As have already been reported, when the center conductor is energized with negative dc voltage, the particles sitting on the sheath are levitated

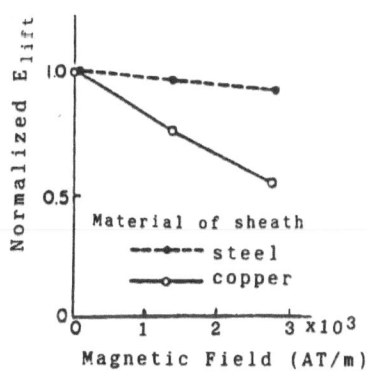

Fig.8. Effect of magnetic field on the behaviour of steel wire particle (0.2 mm dia., 5 mm long).

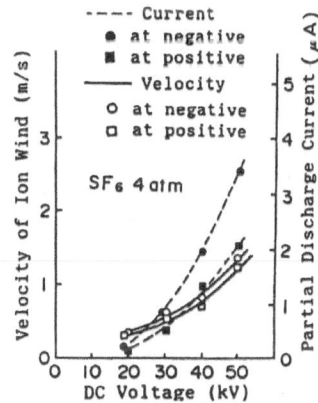

Fig.9. Dependencies of ion wind and partial discharge currents on applied voltage (0.2 mm dia.wire - 150 mm plane , gap ; 27.5 mm).

Table 2. Minimum Wind Velocities At Which Metallic Particles Begin To Move In A Pressurized Wind Tunnel

Shape of particle		Velocities at which a particle begins to move					
		Al sphere			Al wire (5mm long)		Al powder
Dia. of particle (mm)		⌀ 1	⌀ 2	⌀ 3	⌀ 0.2	⌀ 2.4	
SF₆ gas pressure	4 atm	1.1	1.1	1.3	0.6	—	0.8–1.0
	5 atm	0.9	1.1	1.3	0.5	0.6	0.8–1.0

at a certain voltage and reach to the conductor and are lingering there during voltage application. Then, dischaging and charging happen at the tip of the lingering particles. The ions produced by the discharges move along the electric lines to the direction of the sheath. The movement of the ions is accompanied with the wind of the gas which is referred to as ion wind. The velocity of ion wind was measured in the compressed gas, using a needle-plane model. The test results shown in Fig.9 shows that the velocity of ion wind produced by partial discharges reached to about 1.3 m/s at negative 50 kV application. In the practical system such as coaxial systems, the ion wind with the velocity of the same degree was confirmed when the particles were lingering at the inner conductor energized with the negative dc.

The minimum velocity of wind which the particles begin to move were measured, using a pressurized wind tunnel (Nakanishi et al., 1990). The results are indicated in Table 2. The data in Fig.9 and Table 2 show that the ion wind generated by partial discharges of the lingering particles around the center conductor has the force to blow the particles out of the trap. The particle trap with the side walls which can prevent both the ion wind from entering and the particles from escaping will be effective for DC-GIS.

CONCLUSIONS

Accumulation of charges which were transported in the bulk and on the surface of an insulator happened when the test was performed in a carefully cleaned system with well buffened electrodes. Surface charging due to gas conduction was observed in an indusrially cleaned condition. The large amount of charges accumulated via gas conduction covered the charging patterns which might be generated by other mechanisms. Therefore, we should design the physical configuration of the spacer and the electrodes for DC-GIS in a way that the surface of the spacer intersects the electric field lines in slant angle as possible.

When the sheath was made of the magnetic material such as steel, levitation of the iron particle was suppressed due to the attractive force between the two magnetic substances. The lingering particle at the inner conductor energized with negative dc generated the ion wind which could move the particles confined in the particle trap. The particle trap with the side walls was proposed.

REFERENCES

Bargigia, A., G.Mazza, A.Pigini, G.Rizzi (1987). Gaseous Dielectrics V, L.G. Christophrou and D.W.Bouldin, eds., Pergamon, New York, p. 621.
Bektas, S.I., O.Farish (1988). 9th International Symposium on Gas Discharges and their Applications, 295.

Connolly, P., O.Farish (1984). Gaseous Dielectrics IV, L.G.Christophorou and M.O.Pace, eds. Pergamon, New York, p. 405.

Cooke, C.M. (1978). Gaseous Dielectrics, L.G. Christophorou (Ed.), ORNL CONF-780301, p. 162.

Cooke, C.M. (1982a). IEEE Trans. Electrical Insulation, EI-17, 172.

Cooke, C.M., R.Nakata, M.Ouyang, S.J.Dale, T.F.Garity (1982). CIGRE paper No.15-14.

Fujimoto, N. (1987). Gaseous Dielectrics V, L.G.Christophorou and D.Bouldin, eds., Pergamon, New York, p. 513.

Fujinami, H., T.Takuma, M.Yashima, T.Kawamoto (1989). IEEE Trans. Power Delivery, Vol.4, 1765.

Hasegawa, K., Y.Shibuya, K.Nakanishi, Y.Arahata, A.Yoshioka (1983). Mitsubishi Denki Giho, Vol.57, 698.

Jaster, H., R.Nakata, N.Singh (1978). IEEE Trans. Power Apparatus and Systems, PAS-97, 828

Knecht, A. (1982). Gaseous Dielectrics III, L.G.Christophorou ed., Pergamon, New York, p. 356.

Mangelsdorf, C.W., C.M.Cooke (1980). IEEE International Symposium on Electrical Insulation, 146.

Matsumura, S., T.Tanabe, Y.Harumoto, S.Tominaga, H.Kuwahara, N.Nagai, T.Tada (1983). IEEE Trans. Power Apparatus and Systems, PAS-102, 2871.

Nakanishi, K., A.Yoshioka, Y.Shibuya, T.Nitta (1982). Gaseous Dielectrics III, L.G.Christophorou, ed., Pergamon, New York, p. 365.

Nakanishi, K., A.Yoshioka, Y.Arahata, Y.Shibuya (1983). IEEE Trans. Power Apparatus & Systems, PAS-102, 3919.

Nakanishi, k., T.Okamoto, K.Inami, Y.Arahata (1990). Trans. IEE of Japan, to be published in July, 1990.

Nitta, T., K.Nakanishi, Y.Arahata, M.Shimada (1985). 1985 International Conf. Properties and Applications of Dielectric Materials, Vol.1, 340.

Ootera, H., K.Nakanishi (1988). IEEE Trans. Power Delivery, Vol.3, 165.

Ouyang, M., M.A.Baker, T.F.Garrity (1982). IEEE Trans. Power Apparatus and Systems, PAS-101, 2194.

Pedersen, A.(1984). Gaseous Dielectrics IV, L.G.Christophorou and M.O.Pace, eds., Pergamon, New York, p. 414.

Trump, J.G. (1962). IEEE Trans. Nuclear Science, NS-14, 113.

DISCUSSION

A. H. COOKSON: For the DC support insulator, have you considered dissipating the charges with a coating or additive to the epoxy for the insulators of higher conductivity than the present epoxy materials?

K. NAKANISHI: A low conductivity coating or a low conductivity additive to the epoxy may decrease charge accumulation. However, the level of conductivity required for the significant reduction of charge density is quite high and it is almost impossible to control and maintain the conductivity and the uniformity of the material in industrial applications.

H. FUJINAMI: In charge accumulation on a GIS spacer, you mentioned three possible mechanisms, that is, by bulk, surface or gas processes. What do you think is the most important rule in an actual GIS spacer? Second, if the bulk charge exists, how do you measure the bulk charge?

K. NAKANISHI: (1) In industrial quality GIS, the presence of micro—dusts and field—emission from the conductor will be the most important sources of charges. In that condition, the gas conduction process will be a governing factor for charge accumulation on the spacer. (2) The measurement presented in my paper is good only for the charges on the surface of the spacer. If we evaluate the penetration depth of surface charges into the bulk, the depth will be of the order of one hundred microns. This means that the measurement is good for the charges accumulated on the surface of spacers.

P. F. WILLIAMS: The surface charge accumulation measured on a conical insulator was very inhomogeneous. Could you correlate these inhomogeneities with activity in the gap or some other cause?

K. NAKANISHI: The local charge accumulation on a conical spacer will be attributed to the localized sources of charge carriers on the electrode.

L. NIEMEYER: You have shown the case of a needle on the high voltage conductor causing the electric wind. Is this a stationary, stable configuration in a real system, i.e., do the particles cross the gap and stay on the high voltage conductor for a long time?

K. NAKANISHI: The particle continues to keep its position on the conductor during the voltage application. It is not repelled.

GAS-INSULATED SUBSTATION PERFORMANCE IN BRAZILIAN SYSTEM

H.J.A. Martins*   V.R. Fernandes** and R.S. Jacobsen**

*CEPEL - Rio de Janeiro-Brazil
**ELETROPAULO - São Paulo-Brazil

ABSTRACT

This work is based on a report developed in the Working Group 23-03 of CIGRÉ-Brazil [1], about gas-insulated substations performance in the Brazilian electric system from 1975 to 1989. The main points presented are about global characteristics of the installations and the occurrences during the whole operation period, with emphasis on this last topic. The American electric system is taken as a reference in order to compare performances between both systems [2].

INTRODUCTION

Although the SF_6 gas has been used as an electric insulating in GIS since 60's, in Brazil the first installations began the operation in July 1975 and December 1976, both provided by Delle Alsthom for CEMIG (145 kV, ST Centro) and ELETROPAULO (242 kV, ST Centro) respectively.

From then on, many other supplies were made, including the first 550 kV installation of the occidental world, which started the operation in 1978. At that time, as the technology was new for the country, some problems such as the necessity for qualified labor force for its assembly and maintenance and a strong dependence on the manufacturers with regard to components and spare parts, appeared immediately.

So far, the equipment has been totally imported, including the SF_6 gas, with no perspectives of immediate manufacture in the country.

In Brazil the GIS have been used mainly in big urban centers like São Paulo, Rio de Janeiro and Belo Horizonte, and in hydropower stations with an increase tendency for the 550 kV class (Itaipu, Tucurui, Foz do Areia).

Although there are some reports about specific problems in GIS, there is little published literature about countries experiences with the equipment. In Brazil a questionnaire covering many topics about the equipment was distributed to the utilities and high voltage special customers. All of them answered providing information to create a data bank on which this report is based.

Gaseous Dielectrics VI, Edited by L.G. Christophorou and
I. Sauers, Plenum Press, New York, 1991

GENERAL CHARACTERISTICS OF THE INSTALLATIONS

Nowadays eleven utilities and special high voltage customers use GIS in Brazil, with only one outdoor installation which has great problem of corrosion in the support structure (ST Terminal Sul, 145 kV, LIGHT - Rio). The equipments in their majority come from Europe, more specifically from France and Switzerland, with few furnishings coming from Japan too. Figure 1 shows the participation on the Brazilian market of many manufacturers in function of circuit-breaker bay. In this work, one bay is composed of three poles of a circuit breaker together with disconnectors, earthing switches, etc., through the text it will be used simply the word "bay".

In this figure one notes that Delle Alsthom has the majority of the supplies, mainly in the 145 kV level, while Brown Boveri stands out itself in the 550 kV level. One can also note that there are many different technologies involved in the system, so it is difficult for the utilities to handle or exchange experience among themselves, besides importing spare parts from different countries. The quantity of bays per voltage level is shown in Figure 1 where one can see the predominance of the 145 kV level (metropolis) and 550 kV level (hydropower station) is noted. Among the four 550 kV installations only Grajaú ST is located in a metropolis, although São Paulo reached the 362 kV level. It seems that due to the needs this level will be commonly used from now on.

OCCURRENCES

This term was adopted for all and every type of defect or failure, which had as consequence the disconnection of the equipment, or the necessity of maintenance without causing interruption in the electric energy supply.

The data bank considers for the equipment the defective component as well as the dates of manufacture, energyzing and occurrence. For the occurrence, its description, cause and solution adopted in order to clear it. The intention in analyzing these occurrences is to increase the exchange of information among the utilities, this will help in the solution of some common problems, like new specifications, maintenance of equipments of different technologies and improve the interchange of spare parts.

There is also the purpose of contributing to improve the quality of the equipment, identifying and examining possible causes of occurrences.

EQUIPMENTS VERSUS OCCURRENCES

The equipments were grouped as shown in Figure 2 where we can see the number of occurrences for each equipment, including the ones related to the erection and test phases.

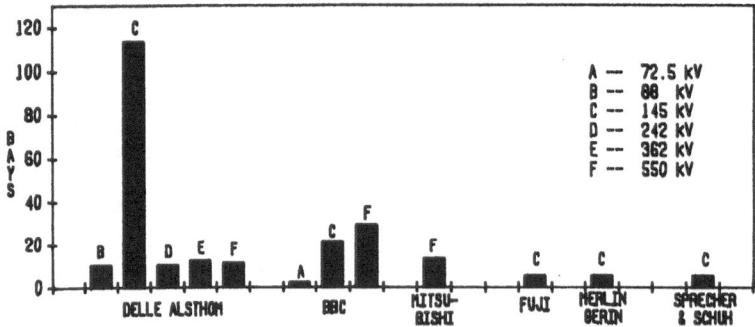

Fig. 1. Manufacturer participation on the Brazilian market.

444

The bus-ducts were the equipments in which the majority of the problems occurred. We understand a bus-duct as a composed of metal enclosure, conductor, solid insulators and accessories. Most part of the occurrences come from the 550 kV installations.

Following the bus-ducts, come the circuit-breakers, disconnectors and arresters with a great contribution coming from the 550 kV installations.

Discounting the occurrences relative to the erection and test phases during comissioning, there is a little decrease at values (except for the bushings), however, the shape of the graphic remains the same.

Although only two utilities have provided information about occurrences during assembly and erection tests, we believe that the number of them was big and in some cases unit ships that were tested in the factory were mounted with carbonizing paths of secondary flashovers across solid insulators as a result of the tests and inconvenient cleared of the compartment before sending to the utility. In some cases this situation was only discovered after some time of functioning and after a failure occurred.

Next, is shown for each equipment the index of occurrences per component. An auxiliary figure, in a circular shape, shows the ratio between occurrences of mechanical and electrical origin.

In the Bus bars (Fig. 3), the solid insulators stand out, being the majority of 550 kV installations. The main causes of failure are: bad quality in the manufacturing of the component, surface contamination due to particles and inapropriate transportation. To the gas were attributed the loss of its dielectrics characteristics in function of excessive humidity level, particle contamination and failure due to very fast transients. The main contribution of the 145 kV level was due to the conductors length, which was smaller than that stabilished in the original project, with some cases of disconnection of the conductor from the solid insulator.

In the Circuit-breakers (Fig. 4) the operating rod (550 kV) was the component that presented more failures of identical characteristics, revealing a project mistake.

Following the rods come the electrovalves (145 kV), with failure during the assembly in the factory.

The hydraulic operating mechanism, presents occurrences in all levels of voltage, being interesting to note that in the cylinders (72.5 kV class) the material used in the o'rings was deteriorated by the oil.

The double pressure SF_6 circuit-breakers attract the attention about the blast valve between the high and low pressure compartments, showing a premature consuming of the piece. Only two installations, of different utilities, use this type of equipment and both presented the same problem.

There are four occurrences involving the gas insulator, one of them with contaminated gas by particles.

There is a predominance of mechanical occurrences over those of electrical origin, more specifically dielectrics, what is coherent with the

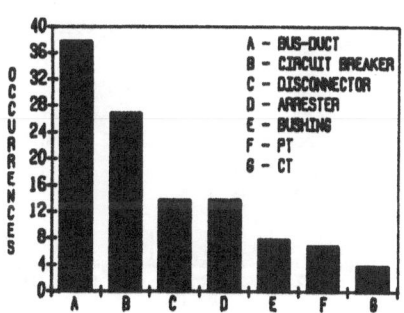

Fig. 2. Occurrences per equipment.

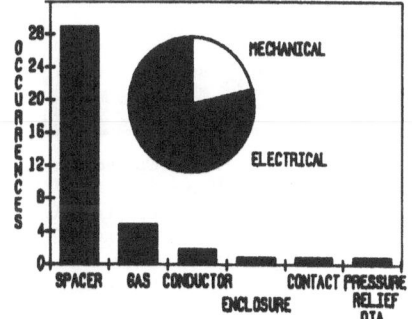

Fig. 3. Occurrences per bus-duct components.

characteristics of the equipment, and the majority of the mechanical occurrences come from the 145 kV installations.

In the Disconnectors (Fig. 5) the solid insulators were the ones to present more problems for many causes, such as: particle contamination on its surface, inconvenient test in an oil fluid cable connected to the installation, a screw forgotten in the inferior part of the compartment, opening operation of a charged bus-bar.

Two undue closing operations together with an earthing switch in a close position, provoking a phase-to-ground short-circuit, one of them with consequent explosion of the pole. Another explosion was caused by a defect in a remote command that failed during the opening operation and the moving contact did not complete its course.

There is not any register of occurrences due to high frequency surges when switchings were made up to the 242 kV level.

In the Bushings the causes of occurrences are: lack of interconnection between the base of the bushing and the GIS metal enclosure; different distributions of wholelocks in the flange of the base of the bushing and in the GIS flange of the metal enclosure; solid insulator contamined by particles and paper oil insulation deteriorated by high frequency surges due to switching of disconnectors (550 kV).

In the Potencial and Current Transformers the main occurrences are related to bad connection between the PT and the bus (145 kV), internal insulation deterioration due to very fast transients caused by disconnector switchings (550 kV), SF_6 gas leakage through the secondary bornes (145 kV), manufacturing defect in the winding (145 kV), superficial burning and cracking in a solid insulator of 550 kV (ring geometry).

All the Arrester occurrences come from Itaipu, where due to the lack of reliability of the SiC equipment the utility is changing them by ZnO arresters (66 indoor and 27 outdoor).

CAUSES

In order to understand the various causes that led the equipment to present failures, or any type of abnormality, all the occurrences were classified as shown in Figure 6. In this figure it is noted that the items manufacturing failure, assembling failure, failure in the projection and particle contamination distinguish themselves in relation to the others.

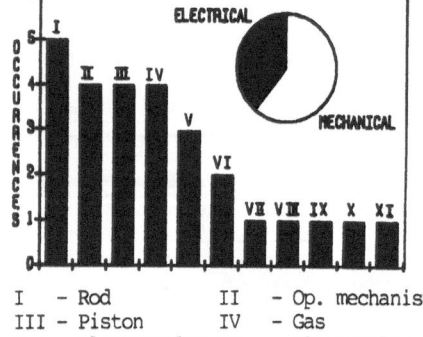

I   – Rod   II   – Op. mechanism
III – Piston   IV   – Gas
V   – Electrovalve VI   – Blast valve
VII – Spacer   VIII – Intake valve
IX   – Electrohydraulic command
X   – Hydraulic operating circuit
XI   – High pressure accumulator

Fig. 4. Occurrences per
        circuit-breaker components.

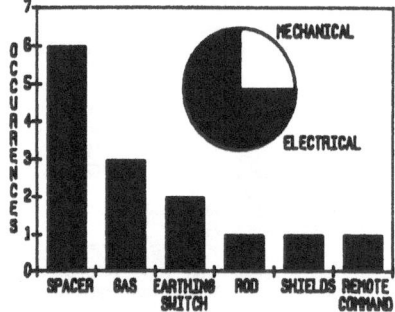

Fig. 5. Occurrences per
        disconnector components.

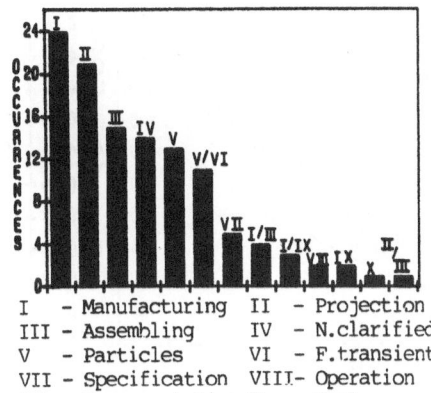 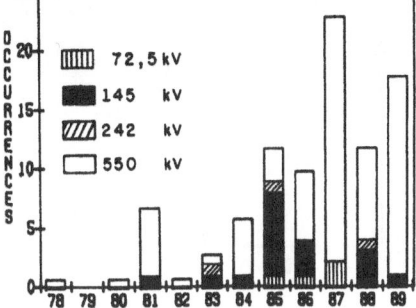

I   - Manufacturing   II - Projection
III - Assembling   IV - N.clarified
V   - Particles   VI - F.transient
VII - Specification   VIII- Operation
IX - Transportation   X   - Test

Fig. 6. Probable causes of occurrences.   Fig. 7. Occurrences per year.

Another worrying item refers to the particle contamination associated
to very fast transients (550 kV class).

It's important to note that due to some occurrences, there were
changes in the operation philosophy to installations like: restrictions to
disconnectors switching in 550 kV installations in order to prevent
interference in the control wire circuit, out-of-phase operations and
changes in the circuit-breakers of the original project (550 kV) involving
operating rods and shields.

EVOLUTION OF THE INSTALLATIONS VERSUS OCCURRENCES

At first, it is made an analysis of the occurrences concentrated per
year, in the period of 1975-1989. Figure 7 shows these values for each
year, also considering those during the assembly and tests. Analysing this
figure, it is noted that some years are in evidence. 1981 due to Grajaú GIS
(550 kV) and 1985 due to Guarulhos GIS (145 kV) along with Itaipu GIS (550
kV). To the subsequent years the occurrences are basically from Itaipu GIS.
It should be noted that proportionaly to the number of bays the major
number of occurrences come from the 550 kV installations.

To verify the situation of the Brazilian system, a comparison to the
USA system was made using the data from reference [2]. In order to permit
this comparison, the occurrences that happened during assembling and tests
as well as those occurrences that did not take the equipment immediatly out

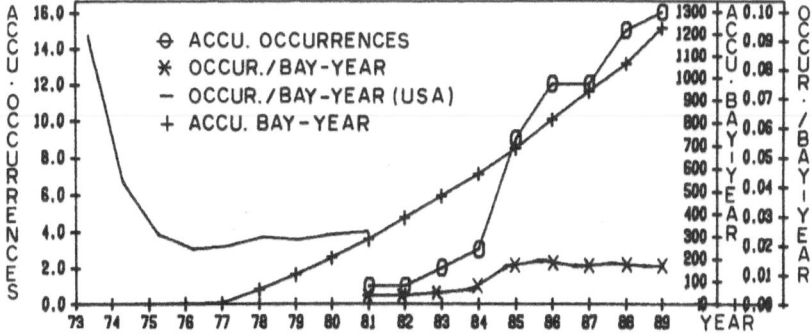

Fig. 8. Comparison between Brazilian and American systems (145 kV).

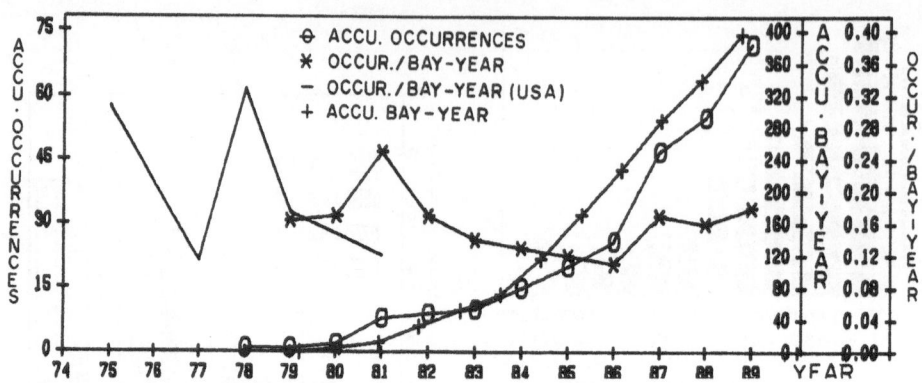

Fig. 9 · Comparison between Brazilian and American systems (550 kV).

of service were discounted.

Comparing the 550 and 145 kV installation indexes separately with the American system, as can be seen in Figures 8 and 9, the 145 kV installation index shows values under those of the American system, while the 550 kV installation index is in the same order as the American one.

CONCLUSIONS/SUGGESTIONS

- The Brazilian utilities should accompany the manufacturing stage and tests in foreign laboratories better. It is clear that some failures could be detected if a good accompaniment were made in some situations.
- The rated withstand power frequency voltage test, during 1 min., was the most used one, mainly on-site, in some cases together with a Partial discharge measurement. The Oscillating surge impulse test was restricted to the 550 kV installations on-site [1].
- The GIS installations of 145 kV class show a good performance when compared to those of the American system. The same conclusion is impossible to the 550 kV class installations, mainly considering that Itaipu has technology of a new GIS generation and the 550 kV occurrence index is increasing.
- A great number of occurrences of the 550 kV installations are due to very fast transients (switching circuit-breaker and/or disconnector) associated to contaminated particles.
- Although it was not mentioned as a occurrence, there are many SF_6 leakages in the 145 kV GIS, in some cases involving new installations.

ACKNOWLEDGEMENTS

We are grateful to the utilities, high voltage special customers and manufacturers who contributed providing information which permitted the consolidation of this work with special thanks to the colleague Rita de Cássia P.B. Martins (LIGHT - Rio) and Wladimir O. de Almeida (CEPEL).

REFERENCES

[1] Martins, H. J. A., Fernandes, V. R., and Jacobsen, R. S., 1988, "Gas-insulated substations performance in the Brazilian electric system", WG 23-03 Report, Cigré-Brasil (in Portuguese).

[2] Boggs, S.A., Chu, F.Y., and Mashikian, M.S., 1986, "Gas-Insulated Substation Reability: Present and Future Trends", EPRI EL-4422.

DISCUSSION

T. NITTA: The statistics you have shown in your presentation are quite different from what we have experienced in the past. Particularly the high failure rate at the spacer is contradictory to our experience. I would like to have your comment if the high number of spacer failures is common to all makes of GIS, or if it can be attributed to a particular make of GIS.

R. S. JACOBSEN: The spacers' failures are a generic problem, particulary in all installations of 550 kV.

F. Y. CHU: A comment on Dr. Nitta's question. According to the EPRI study, in which I happened to be one of the authors, spacers in GIS caused most of the failures. The EPRI study covered many makes of GIS by various manufacturers. Our results and the Brazilian results appear to agree that the spacer is a generic problem area, not just confined to one manufacturer. To Mr. Jacobsen, can you explain why the USA–Canadian experience as shown in Fig. 8 continued to improve and followed the classic bath–tub curve and that the Brazilian experience appeared to deteriorate as a function of time?

R. S. JACOBSEN: Theoretically you are correct, but the curve of the Brazilian occurrences, showed in Fig. 8, was based on the failures related by the utilities. Data from high voltage consumers would be necessary to analyse the causes.

GIS INSULATION AND DISCONNECTOR OPERATION

B. de Metz–Noblat and Y. Doin

Direction Technique
Merlin Gerin – Grenoble – France

INTRODUCTION

Four kinds of overvoltages are to be considered to design GIS insulation :

- lightning impulse
- switching impulse
- temporary overvoltages
- disconnector operation.

Up to now, the lightning impulse overvoltage was considered as the most severe and was used to determine the dimensioning of GIS. But, now, it is known that disconnector operation may lead to fast transient overvoltages which could be very severe for GIS insulation. The question is how to know accurately these fast transient levels and especially the wave-shapes of these overvoltages.

GENERAL

Computational studies, using EMTP, are currently done on complete GIS substations for insulation coordination purpose. These studies make the simulation of a lightning stroke on a high voltage line and consequently calculate the overvoltages distribution in the substation due to the lightning impulse.

The point now is to do similar calculations, using EMTP also, in order to simulate the fast transient overvoltages due to a disconnector operation in a substation. For these calculations, we need to know how to simulate the operation of the disconnector. This means that we must choose the values of different parameters such as arc resistance and duration of voltage collapse between contacts.

We have carried out tests on a disconnector in our high voltage laboratory and we have done at the same time calculations to simulate accurately these tests. From the confrontation of experimental results and calculations we deduce the values of the disconnector parameters. Then we can use these values for a computational study of a disconnector operation in a complete substation.

Gaseous Dielectrics VI, Edited by L.G. Christophorou and
I. Sauers, Plenum Press, New York, 1991

A test for disconnector operation has been defined in which a metal-clad disconnector feeded with a 1 per unit voltage closes on a trapped charge of − 1 p.u.

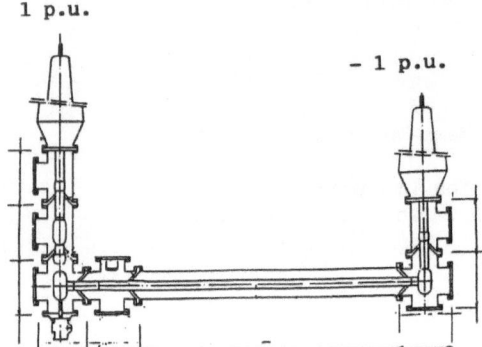

Fig. 1. Test of a disconnector closing.

Figure 1 shows the circuit used during this test. During the closing of the disconnector, a breakdown appears between contacts. At the time of the first spark, the potential of the floating contact which was previously of −1 p.u. increases up to the potential of the fixed contact with oscillations. Measurements have been done in the neighbourhood of the disconnector with field probes and Fig.2 shows the potential variation of the contact during the first spark.

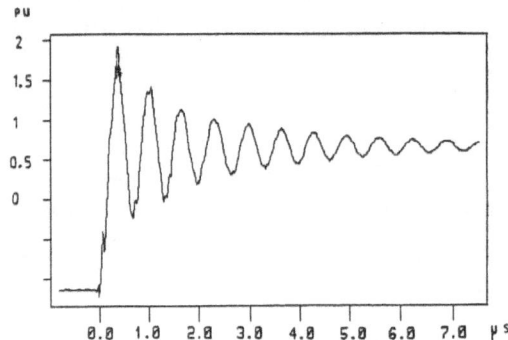

Fig. 2. Potential variation of the disconnector contact during the first spark.

Computerized calculations, with EMTP, of this test have been done with the following model of the disconnector (see Fig. 3).

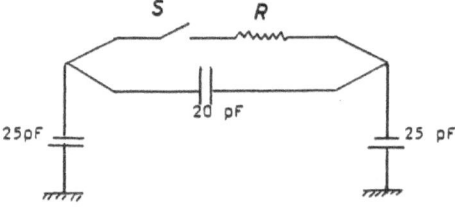

Fig. 3. Model of the disconnector.

On the figure 3, two 25 pF capacitances are intended to represent the spacers capacitances whereas the 20 pF capacitance is the intercontact disconnector capacitance. S is an ideal switch. Two adjustable parameters are :

      - R = arc resistance (2 Ω and 12 Ω)
      - t = duration of the voltage collapse between contacts
        from 0 to 15 ns

Results of these calculations are shown on Fig. 4.

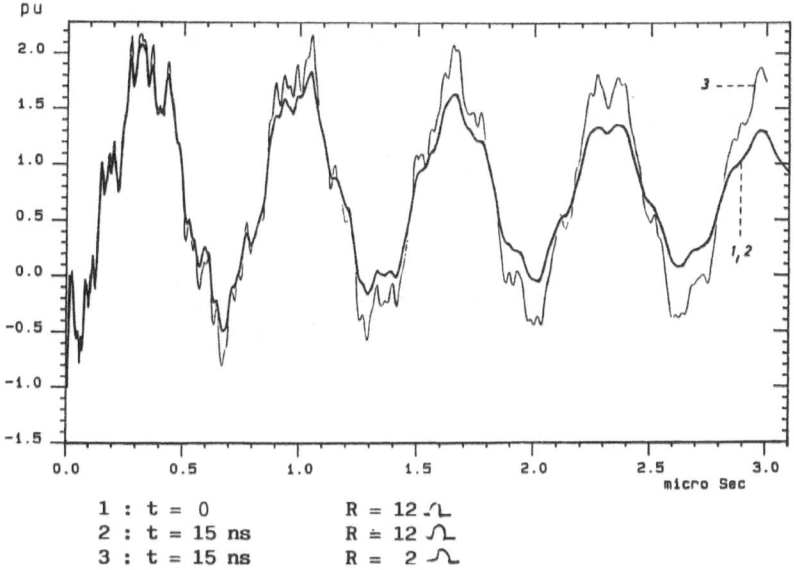

1 : t = 0   R = 12 Ω
2 : t = 15 ns   R ≐ 12 Ω
3 : t = 15 ns   R = 2 Ω

Fig. 4 .   Simulation of the disconnector test.

No influence of the voltage collapse duration could be noticed and the curves for t = 0 and t = 15 ns are approximately identical. From the comparison of the two curves (R = 2 Ω and R = 12 Ω), we can deduce a noticeable damping effect of the arc resistance R.

The best fitting between calculations and measurements is obtained with a 12 Ω arc resistance.
(Comparison of Fig. 5 = Simulation and Fig. 2 = measurement).

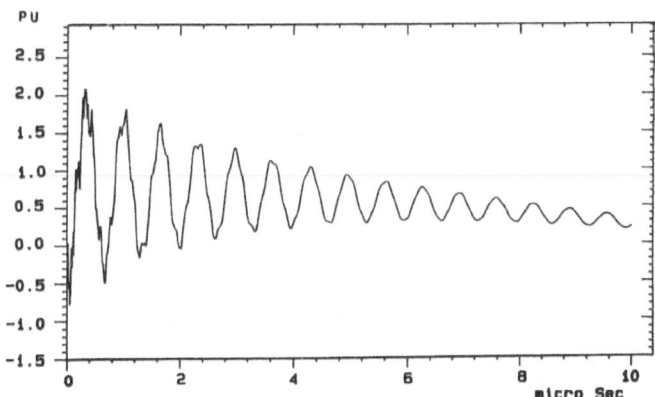

Fig.5 .   Simulation of the disconnector test (R = 12 Ω) .

HIGH VOLTAGE SUBSTATION

   The goal of this simulation is to ascertain the maximum value of
transient overvoltage in the substation due to the closing of a disconnector. We assume that this disconnector closes on a trapped charge of-1 p.u.
which is the worst case.

The main features of the modeling of the complete substation are:

   . GIS components (buses connections ...) considered as transmission
     lines characterized by their surge impedance and electrical length.

   . Spacers, disconnectors, circuit breakers, transformers represented
     with their equivalent circuit.

   . Disconnector spark modelled with a time to breakdown of 15 ns.

Transient voltages

   From a general overview of the French gas-insulated substations, a
typical substation design has been selected. On this configuration,

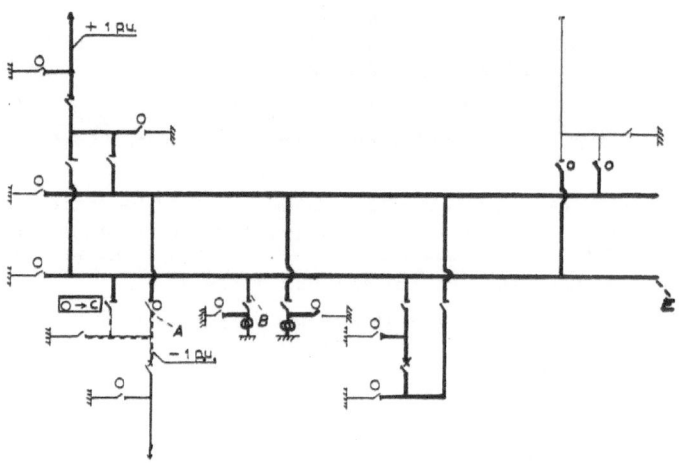

Fig. 6.   Configuration n° 1 : Line disconnector .

two different locations of the operating disconnector have been
selected as follows:

. Configuration n° 1 : HV line disconnector (Fig. 6)
. Configuration n° 2 : disconnector associated with coupling circuit
                       breaker (Fig. 7).

454

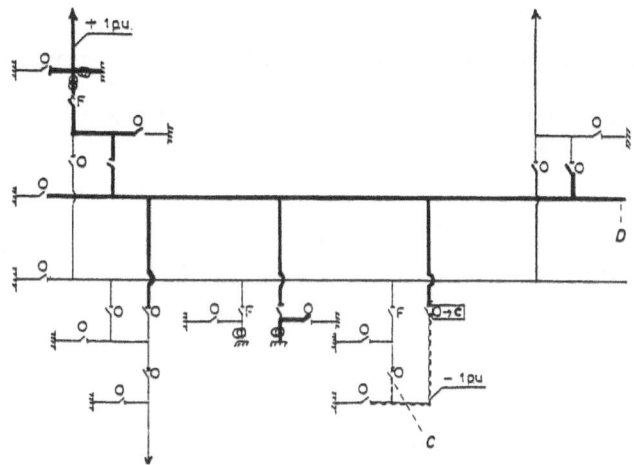

Fig. 7. Configuration n° 2 : Coupling disconnector.

These two disconnectors are closing on a short portion of busbar with a -1 p.u. trapped charge (dotted lines). The locations and the maximum values of the transient voltage are given in the following table :

	Config. 1		Config. 2	
Disconnector overvoltage p.u.	1,5		1,3	
Maximum overvoltage location (Fig.6,7)	A	B	C	D
Overvoltage value p.u.	1.8	1.85	1,65	2,05

An example of the calculated overvoltage at the point B inside the GIS is given on Fig. 8. The basic frequency of these fast transients is about 3 MHz with a superimposed frequency of 30 MHz.

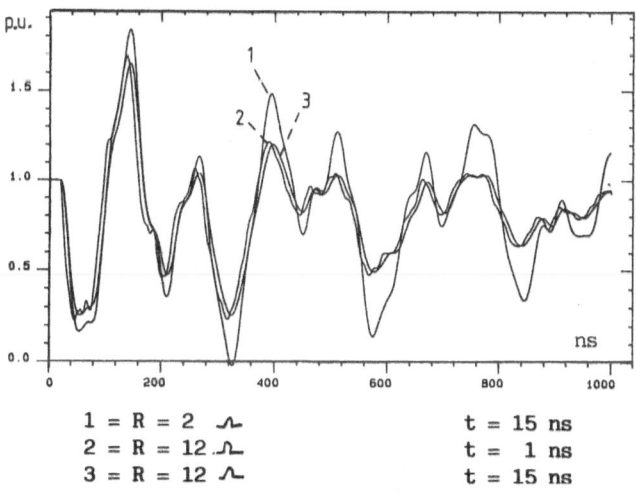

1 = R = 2 ⌁   t = 15 ns
2 = R = 12 ⌁   t = 1 ns
3 = R = 12 ⌁   t = 15 ns

Fig. 8 . Simulation of the closing of a disconnector. Config. n° 1.

455

Influence of busbar length

The influence of the busbar length has been studied on two variants :

Variant 1 : 120 m. extension of the busbar length at the right of
the coupling

Variant 2 : 120 m extension of the busbar length at the left of
the coupling .

The overvoltages are compared with the basis case in the following
table. We can see that the influence of the length is negligible
for the maximum value ($\leq$ 0.2 p.u.).

		Configuration 1			Configuration 2	
Location of overvoltage (Figs. 6 and 7)		A	B	E	C	D
Overvoltage Value p.u.	Basis case	1.8	1.85	1.7	1.65	2.05
	Variant 1	1.8	1.80	1.60	1.55	1.85
	Variant 2	1.65	1.35	1.7	1.65	2.00

CONCLUSION

From this work, we can point out that the overvoltages due to disconnector operation never exceed 2 p.u. with a basic frequency between
2 and 10 MHz.

Superimposed frequencies in the range of 30 MHz exist with smaller
amplitudes. The highest fast transient levels do not appear in the
disconnector itself but at the extremities of busbars.

Tests have been performed in our HV Laboratory with a circuit
simulating these fast transients (2 p.u. 2 MHz). Well designed
disconnectors withstand without failures these overvoltages and as
a conclusion, we can say that the lightning impulse remains the
most severe stress.

DISCUSSION

N. G. TRINH: I have two comments: (1) The withstand of GIS under fast transient is not very much higher than under lightning impulse when low breakdown probabilities are taken into account, as it is the case for disconnect switch overvoltages due to their higher frequencies of occurrences. (2) The ratio of 4 between lightning withstand and nominal voltage is decreasing at higher voltages.

Y. DOIN: The ratio between lightning impulse withstand level and nominal voltages is as follows:

V_{rated} (kV)	ratio: $LIWL/V_{rated}$
245	5
420	4
$\geq$ 550	3–4

The VFT crest voltages are not higher than 2 p. u., so we can deduce the margin between LIWL and VFT. This margin decreases with increasing rated voltages but remains above 1.

N. FUJIMOTO: The author concludes that VFT is not as critical as lightning impulse for GIS, based on his calculations of VFT magnitude and comparing it to the equipment BIL. Although for a GIS in "perfect" condition, the author's comments are probably correct, GIS is often full of a variety of defects some of which are very sensitive to VFT stress. If all such defects are not identified, some VFT–related problems may result. In these cases and as VFT represents "normal" service conditions, VFT may, in a practical sense, be more critical than lightning.

Y. DOIN: The amount of defects which are sensitive to VFT is not well known. Experiments made with long protrusions (see Riquel et al., "Insulation behavior of SF_6 gaps subjected to fast oscillatory overvoltages," Gas Discharges and Their Applications, 1988, Venice, p. 331) show that in their case VFT and LIWL are of the same order of magnitude. Particles may be sensitive, but the nature of the sensitivity must be investigated.

EXPERIMENTAL STUDIES ON THE PRESSURE RISE IN GIS BY INTERNAL ARCS

WITH VARIOUS MATERIAL PARAMETERS AND TEST ARRANGEMENTS

D. König

H. Schuhmann

Technical University of
Darmstadt, High Voltage Lab.,
D-6100 Darmstadt, Germany

Hoechst AG,
D-6230 Frankfurt/Main
Germany

INTRODUCTION

Considerable amount of work has been done in the past to study the effect of internal arcing in GIS on the pressure rise and the burn-through. This paper continues the earlier work[1] by members of CIGRE SC 23.03, where an attempt was made to relate the pressure rise to the design parameters of the GIS as well as stresses relevant to the power supply network. Although the agreement between the measured and calculated re-sults was fairly good, major deviations noticed in some cases could not be explained satisfactorily. This is assumed to be due to the poor quality of the measuring equipment then available in the high power laboratories and also due to unaccounted portions of energy during the investigations. Par-tial results from a research program intended to fill these gaps by applying modern digital measuring equipment, well-defined test arrange-ments and various contact materials[2,3] are reported in this paper.

ENERGY BALANCE

The pressure rise in a GIS-compartment with internal arc is mainly originated by the electrical energy fed into it[1,23] . Based on the assump-tion of energy balance and neglecting shock-wave effects, the pressure rise for a single-phase GIS-compartment can be calculated as:

$$\Delta p = (\kappa - 1) \cdot \frac{1}{V} \cdot k \cdot E_{arc} \qquad (1)$$

with Δp = pressure rise, κ = ratio of the specific heats
     V = volume of the k = factor characterizing that
        compartment   portion of E_{arc} relevant for
     E_{arc} = arc energy   the pressure rise.

The total electrical energy E_{arc} fed into the compartment is given by

$$E_{arc} = \int_{0}^{t_{arc}} u_{arc}(t) \cdot i_{arc}(t)\, dt \qquad (2)$$

with t_{arc} = arc duration; u_{arc} = arc voltage; i_{arc} = arc current.

Gaseous Dielectrics VI, Edited by L.G. Christophorou and
I. Sauers, Plenum Press, New York, 1991

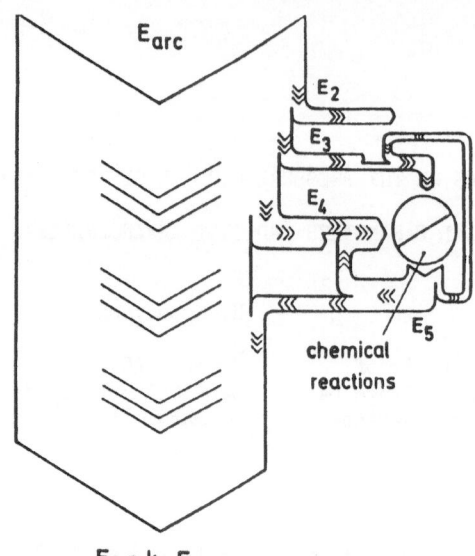

Fig.1. Energy flow diagram during internal arcing in GIS[1]
E_{arc} : total electrical energy input E_3: energy for vaporization of
E_1 : energy for heating the gas   the electrode material
E_2 : energy for heating the con- E_4: energy for SF_6-dissociation
          ductors and the enclosure E_5: energy for chemical reactions

The volume V is known. The ratio of the specific heats in case of SF_6 is $\kappa = 1.086$ and can be assumed to be nearly constant during arcing[1,2,4], while in case of air it is $\kappa = 1.4$. The arc energy E_{arc} can be measured. The only unknown quantity still needed for calculation of the pressure rise is the factor k. The investigations described in this paper concentrate on this factor.

The total energy balance in a system with internal arc is illustrated in Fig.1. Portions of energy heating up the enclosure and the main conductor are not available for heating the gas during the short time span of arcing. However, chemical reactions mainly between SF_6-decomposition products and vaporized material resulting from electrodes or insulating material have to be taken into account, which may contribute portions of energy relevant to the pressure rise. Up to now, no precise investigations of the role of the chemical energy have been published - except[2,3] - though some investigators had already presumed it to be essential under certain conditions[5]. The method applied to get improved information on the energy balance and portions relevant to the pressure rise is as follows:

The measured pressure rise Δp_m is compared with the calculated pressure rise Δp_c according to Eqn.(1) with the usual assumption[1] of k = 1.

$$\Delta p_c = (\kappa - 1) \cdot \frac{1}{V} \cdot E_{arc} \qquad (3)$$

The k-factor can then be defined as:

$$k = \frac{\Delta p_m}{\Delta p_c}. \qquad (4)$$

Thus, systematic variations of test parameters such as gap distance, electrode material and type of insulating gas will enable conclusions to be drawn with respect to energy balance and especially the role of chemical energy.

Table 1. Specific energy e_m (melting) and e_v (vaporization) (I) while changing the state of aggregation (II), during reaction with oxygen (III) and dissociated SF_6 (IV) and relevant differences of energy (V).

	Specific Energy kWs/g	Electrode Material		
		Al	Cu	Fe
I	e_m	0.9	0,6	1.3
	e_v	12,8	5,5	7.1
II	$e_m + e_v$	13.7	6.1	8,4
III	$e_{exo,air}$	31,0	2,4 (CuO) 1,3 (CuO_2)	6,7 (Fe_3O_4) 7,4 (Fe_2O_3)
IV	$e_{exo,SF6}$	25.6	2,02	5,69
V	$\Delta e = (IV - II)$	11.9	-4.08	-2.71

ENERGY FROM CHEMICAL REACTIONS

The main chemical reactions between metal vapor and relevant insulating gas components including the amount of exothermic energy are studied[2] based on published data[6]. Reactions between Al, Cu, Fe and O_2, in case of air, or S and F_2 in case of SF_6 as well as the dissociation energy of SF_6 have been considered. Table 1 contains the result of this investigation, quoted as specific values referred to the mass of burnt-out electrode material. The data about the specific energy for melting e_m and evaporation e_v were taken from published literature[7].

ARCING EXPERIMENTS

Experimental set-up

The arcing experiments have been performed in housings with different volumes (V = 5, 60, 102, 142 and 245 l), mainly parts of a 123 kV-GIS. The electrode arrangements were rod-rod, sphere-sphere, rod-plane and coaxial (Fig.2). The electrode materials were Al, Cu and Fe and the insulating gases SF_6 and air at an initial gas pressure in the range of 0.1-0.4 MPa.

The electrical energy was obtained by discharging a capacitor bank with maximum stored energy E_{max} = 2.4 MJ. The discharging took place via inductances, resulting in a damped oscillatory current with a frequency f = 50 Hz. The capacitor bank was charged to U_o = 25 kV, which resulted in a peak value of 20 kA at the first current peak. Arc current, arc voltage and pressure rise were measured with a transient recorder using analog optical data transmission systems to avoid EMC-problems. Arc power and arc energy were calculated with a PC. An example is given in Fig.3.

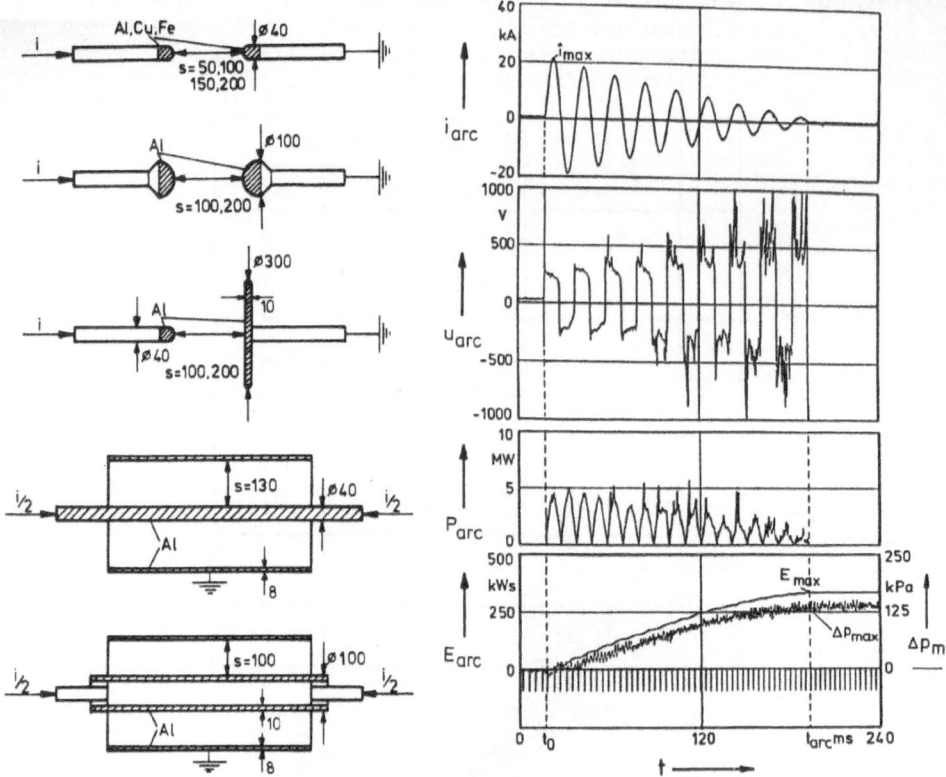

Fig. 2. Test arrangements.　　　Fig.3. Example for the evaluation
　　　　　　　　　　　　　　　　　　　　　　　　of an internal arc test.

Test results

Fig.4 shows the variation of the k-factor with gap distance for the rod-rod electrode configuration with the electrode materials Al, Cu and Fe and the insulating gas SF_6. The evaluation is based on 5 experiments under identical test conditions and shows the average, maximum and minimum values. The decay of k_{Al} with increasing gap distance is due to the peculiarity of the energy source[2]. The k-factors in case of aluminium are far greater than those of copper and steel. Factors k > 1 indicate that more energy is available for pressure rise than is fed into the system. For each arc test with electrodes of aluminium in SF_6 - including some in air - the amount of burnt-out electrode material Δm has been measured and evaluated in Table 2 with respect to the exothermic energy e_{exo} (average value), taking into account the data of Table 1. From that, a factor k_{exo} (average value) could be established, which indicates the amount of chemical energy contributing to the pressure rise. If a corrected factor $k_{Al}-k_{exo\ Al}$ is calculated and compared with the factors k_{Cu} and k_{Fe}, it is in the same order of about 0.6, including some scatter as shown in Fig.5.

While the described procedure makes sense in case of aluminium and SF_6, it does not apply in case of aluminium and air as indicated by Table 2. Even if an essential amount of chemical energy is available by reaction between oxygen and aluminium, it is not relevant for pressure rise. In any case, k-factors in the range of k ≃ 0.6 appear, as shown by Fig.6. These findings agree well with those published[7], where a k-factor k = 0.4 was measured in the low arc-energy range ≤ 100 kWs.

462

Table 2. Evaluation of k-factors for SF_6 and air with aluminium electrodes.

	SF_6				air	
s in mm	50	100	150	200	100	200
$\overline{\Delta m}$ in g	18.9	12.2	9.7	7.4	10.8	3.7
$e_{exo}-(e_m+e_v)$	11.9 kWs/g				17.3 kWs/g	
e_{exo} in kWs	225	145	115	88	187	64
e_{arc} in kWs	280	345	405	431	343	396
$\overline{\Delta p}_m$ in kPa	149	134	124	125	345	364
$\overline{\Delta p}_c$ in kPa	98.3	121	142	151	560	646
$\overline{k}$	1.52	1.1	0.87	0.83	0.62	0.56
$\overline{k}_{exo}$	0.8	0.42	0.31	0.21	0.55	0.16
$\overline{k} - \overline{k}_{exo}$	0.72	0.68	0.56	0.62	0.07	0.4

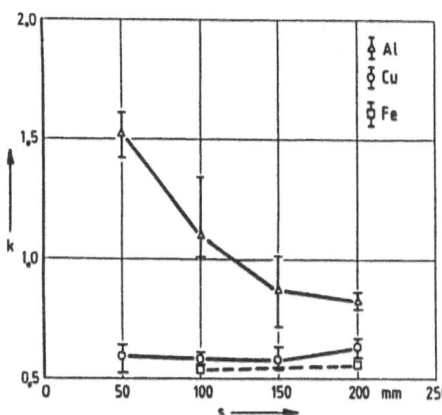

Fig.4. k-factor versus gap distance

Test arrangement: rod-rod, electrode material Al, Cu, Fe; Test parameters: SF_6, V=245 l, p_o=0.1 MPa, $\hat{i}_{max}$=20 kA.

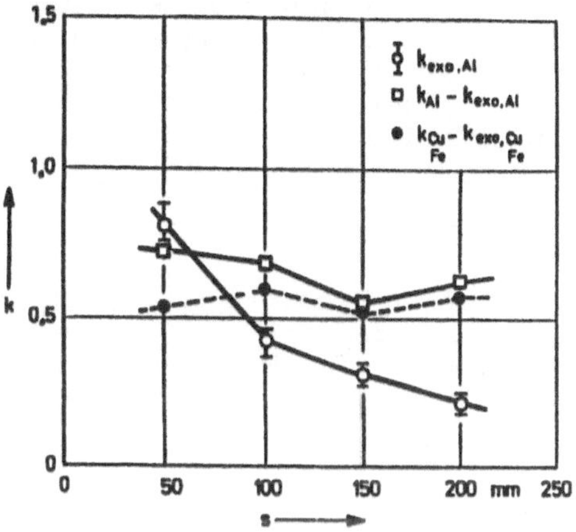

Fig.5. Factors $k_{exo,Al}$, $k_{Al}-k_{exo,Al}$ and $k_{Cu}-k_{exo,Cu}$ versus gap distance
Fe   Fe
Test arrangement and test parameters: see Fig. 4.

Interpretation and Conclusions

The evaluation of the test results as k-factors shows that in all the investigated material parameter combinations - except that of SF_6 and aluminium - a k-factor in the range of k ≈ 0.6 appears. This means that only 60% of the electrical energy put into the test vessel as E_{arc} according to Eqn.(2) is responsible for heating the insulating gas and increasing the gas pressure Δp (Fig.1). However, in case of SF and aluminium electrodes this factor can increase up to k ≈ 1.5 depending on the amount of burn-off. Since the burn-off depends on the arc movement, different contributions to the amount of chemical energy can be expected and, in fact, have been measured from arcs with and without stabilized arc roots. As shown by Table 2 the exothermic reactions between Al and dissociated SF_6 provide additional portions of energy inside the test vessel, which are relevant for the pressure rise.

463

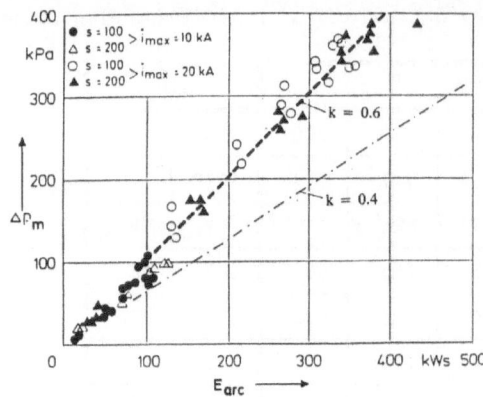

Fig. 6. Pressure rise versus arc energy at different gap distances and arc currents

Test arrangement: rod-rod, electrode material Al; Test parameters: air, $V = 245$ l, $p_o = 0.1$ MPa .

Even in the case of air and aluminium, exothermic reactions occur under arc conditions providing comparatively large portions of chemical energy, which, however, are not relevant to the pressure rise. Heat transfer mechanisms, which are different for air and SF_6, are supposed to be responsible for these findings. It can be assumed that the main portion of the exothermic chemical energy is converted to radiation energy. However, radiation energy is easily absorbed by heavy compound gases like SF_6, especially at certain ranges of wavelength[8,9], while simple gases like N_2 and O_2 are passed without appreciable energy transfer. Further investigations are needed to clarify the physical background more thoroughly.

ACKNOWLEDGMENT

This work was sponsored by the Deutsche Forschungsgemeinschaft (DFG). Bonn/Bad Godesberg, by Brown Boveri (BBC), now Asea Brown Boveri (ABB), Mannheim/Hanau and by the Max-Planck-Institut (MPI), München-Garching. The authors wish to express their thanks for the valuable support.

REFERENCES

1. D. König, Th. Facklam, Pressure Rise in Metal Enclosed SF_6-insulated High Voltage Switchgear of Single-phase Enclosure Type Due to Internal Arc, Electra 93, pp. 25-52 (1984).
2. H. Schuhmann, Investigations on the Pressure Rise in Compartments of Single-phase GIS by Internal High Power Arcs (in German), Thesis, Technical University of Darmstadt (1989).
3. D. König, H. Schuhmann, Estimation of Pressure Rise by High Power Arcs in GIS (in German), e&i 107, pp. 146-155 (1990).
4. Kali-Chemie, SF_6 (Dampftafel), Kali-Chemie, Hannover/Germany, (1979).
5. A. Kulsetas, A. Rein, P.A. Holt, Arcing in SF_6-insulated Equipment-Decomposition Products and Pressure Rise, A.I.M. Postes blindes isolés an SF_6, Liege, Contribution Nr. 22, (1979).
6. Gmehlin, Gmehlins Handbuch der anorganischen Chemie, System Nr. 9, 35, 59, 60, Teil B, Verlag Chemie, Weinheim/Bergstraße.
7. A. Dasbach, Investigations on the power balance of fault arcs with respect to the pressure stress of switchgear cubicles (in German), Thesis, RWTH Aachen (1987).
8. F. Hell, Grundlagen der Wärmeübertragung, 3. Auflage, VDI-Verlag, Düsseldorf (1982).
9. G. Herzberg, Molecular Spectra and Molecular Structure II. Infrared and Raman Spectra of Polyatomic Molecules, 4. edition, D. Van Nostrand Comp. Inc. (1949).

DISCUSSION

J. CASTONGUAY: Did you measure the amount of SF_6 decomposed during your test? Different electrode material vapors may decompose different quantities of SF_6, thus adding many more molecules or atoms to the gas, increasing the pressure.

D. KONIG: We did not measure the amount of SF_6 decomposed during the arc tests, since we concentrated on the precise measurement of the electrical quantitites and the pressure rise. In principle it must be possible to measure, what you are asking for, for the time after arcing, but not during arcing. However, I have some doubts, if this information would be helpful toward a better understanding of what happens during arcing, which is a fast transient process.

T. NITTA: The effect of arcing in GIS will be influenced not only by the total energy introduced by arcing but also by the duration of the arc. It is also influenced by whether the arc moves within the vessel or remains at a point. Did you include these factors in your discussion?

D. KONIG: Yes. As shown by eqn. (2), the total electrical energy E_{arc} introduced into the arcing vessel is dependent on the arc power, given by $u_{arc} \cdot i_{arc}$ and the arc duration, t_{arc}. The arc voltage u_{arc} as well as the k–factor according to eqn. (1) is influenced by the arc mode, i.e., whether the arc moves or is stabilized. In case of a stabilized arc the amount of burn–off will be high and (in case of aluminum electrodes and SF_6) pressure relevant chemical reactions will take place and increase the factor k. We studied different electrode arrangements with different tendencies for the arc to move or to remain stabilized. Our contribution takes these factors into account.

S. W. ROWE: In full scale switchgear the arc is not stable but can lengthen considerably, thereby modifying the energy input to the gas. (1) Can this effect be taken into account in these calculations? (2) Did you consider the vapor pressure due to melted metal to be capable of having a significant effect on pressure rise?

D. KONIG: (1) In case of an unstable arc in a full scale switchgear, where the arc lengthens considerably, the arc voltage, arc energy and the pressure increase proportionally. However, the "efficiency"–factor k, characterized by the percentage of pressure–relevant energy of the total arc energy (total electrical input energy), will be influenced mainly by phenomena happening in the range of the arc root. Our studies have concentrated on this particular aspect with the finding that stabilized arcs, which create high amounts of burn–off material, give high k–factors. It has not yet been investigated in general, if processes occurring in the arc column (where arc length will increase the arc energy) or in the arc roots (where burn–off and chemical reactions increase the factor k) are more effective with respect to the resulting pressure rise. (2) No, I do not think so. At comparable test conditions in the case of SF_6 and in the case of air, comparable amounts of material burn–off of aluminum electrodes were measured; however, the resulting k–factors were quite different. On the other hand, in case of Fe and Cu electrodes in SF_6 and in air and even in the case of Al electrodes in air k–factors in the same range were achieved. I assume that the main portion of the vaporized metal reacts rapidly with the insulating gas and its decomposition products.

ELECTROMAGNETIC INTERFERENCE WITH CONTROL EQUIPMENT BY

GIS SWITCHING SURGES

K. Nojima, H. Murase, S. Nishiwaki, N. Tanabe and S. Yanabu

Toshiba Corporation
Kawasaki, Japan

ABSTRACT

Electromagnetic interference (EMI) with control equipment near Gas Insulated Switchgear (GIS), which is caused during closing operation of Gas Circuit Breaker (GCB) in GIS, was investigated using a full–scale 66 kV GIS and a 1300 V thyristor unit. It is found that thyristor can be turned on by EMI from surges induced in earthing system during closing operation of GCB. It is also found that such mal–firing is caused by surges with frequency components over 10 MHz, which is peculiar to GIS system. Furthermore, it is experimentally proved that the mal–firing can be suppressed by eliminating bushing and connecting the tanks of high–voltage equipment directly to or via capacitor to power cable sheath.

INTRODUCTION

When Gas Insulated Switchgear (GIS) is operated, high–frequency surges are generated inside the GIS tank. It is known that these surges leak out into the earthing system through bushing or joint between GIS and power cable.[1,2,3] As for transient grounding potential rise which is one of the phenomena caused by high–frequency surges leaked out into earthing system, there have been reports including data measured on site.[3] But there are few studies conducted using actual control equipment about electromagnetic interference (EMI) due to surges leaked out into earthing system. In this study, a thyristor unit often installed near GIS, for example which are equipped in cycloconverter system near GIS, was used as a control equipment. Using actual thyristor unit and a full–scale 66 kV GIS, EMI which occurs during closing operation of Gas Circuit Breaker (GCB) in GIS was investigated. As a result, the frequency component of surges in earthing system which causes EMI was clarified and also the method to suppress this EMI was verified.

EXPERIMENT

To examine the characteristics of surges induced in earthing system during closing operation of GCB, and to examine the resultant EMI with the thyristor unit, the following experiment was conducted.

Gaseous Dielectrics VI, Edited by L.G. Christophorou and
I. Sauers, Plenum Press, New York, 1991

Experimental Apparatus and Method

The configuration of the experimental apparatus is shown in Fig. 1. This set—up simulates the circuit of substation system and consists of 66 kV GIS, transformer, and power cables. GIS consists of GCB and short gas—insulated bus. The power supply side of GIS is connected to charging power supply with a 100 m power cable.

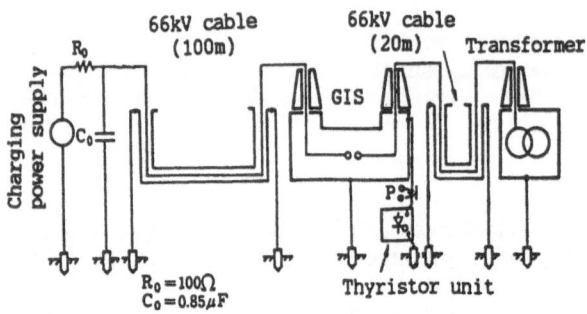

Fig. 1. Configuration of experimental apparatus.

The load side of GIS is connected to transformer with a 20 m power cable. The power cable is connected to GIS or transformer with a bushing. Power cable, GIS, and transformer are grounded with 100 mm² insulated wires about 5 m long. To simulate the capacitance in the circuit behind power supply side, capacitor (0.85 μF) is coupled in parallel with the charging power supply. The charging voltage is 54 kV (66 kV x $\sqrt{2}$ / $\sqrt{3}$). Using this experimental apparatus, the current flowing through the grounding wire of GIS tank during closing operation of GCB was measured at point P in Fig. 1, and the characteristics of surges induced in earthing system were examined.

In this experiment, EMI due to surges in earthing system was also examined. The circuit of the thyristor unit used in this experiment is shown in Fig. 2. The rating of the thyristor element is 4000 V, 2000 A. A diode—equipped circuit is put between the gate terminal of the thyristor element and the secondary side of the pulse transformer. The thyristor element and the pulse transformer were fixed on a metal frame but insulated from it. A 2 μF capacitor C_0 was connected between the anode and the cathode of the thyristor element to serve as a power supply. This capacitor was charged to 2000 V dc using a dc power supply. In actual field, the frame of thyristor unit is connected to earthing system. Therefore, surges in earthing system can propagate to the frame. To simulate this condition, the high—frequency current flowing through the grounding wire of GIS during closing operation of GCB was led to the frame of thyristor unit as shown in Fig. 1.

Through—type high—frequency current transformers were used to measure currents. Voltage between anode and cathode was measured using resistance—type potential divider. Measured signals of voltage and currents were converted to optical signals and transmitted to the observation room. This permitted highly reliable measurements, free of noise distortion.

RESULTS

Surges Induced in Earthing System. The current flowing through the grounding wire of GIS tank during closing operation of GCB was measured at point P in Fig. 1. The result is shown in Fig. 3(a)(b). Figure 3(b) is an enlargement of part A of Fig. 3(a). Evidently, the current waveform contains a relatively low frequency component, about 500 kHz, and very high frequency components, over 10 MHz. When the GCB was closed with power cable on the load side of GIS removed, the current waveform obtained did not show any low frequency component (around 500 kHz) any longer and only the frequency components over 10 MHz were observed.

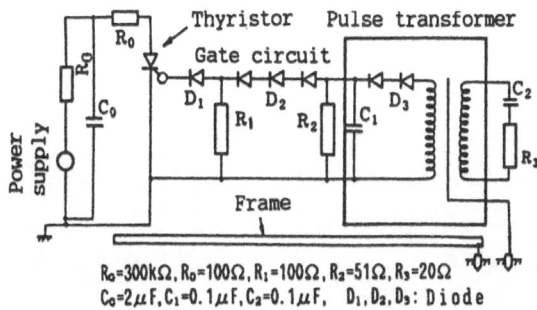

$R_0=300k\Omega, R_0=100\Omega, R_1=100\Omega, R_2=51\Omega, R_3=20\Omega$
$C_0=2\mu F, C_1=0.1\mu F, C_2=0.1\mu F,$ D_1, D_2, D_3: Diode

Fig. 2. Circuit diagram of thyristor unit.

This proves that the relatively low frequency component is related to power cable. The cause of high–frequency surges over 10 MHz is considered as follows. As shown by Yanabu et al.,[4] the rise–time of surges generated inside GIS during the operation of GIS is very short, about several nanoseconds. The waveform with short rise–time contains high–frequency components. The waveform of which rise–time is under 10 ns contains frequency components over 10 MHz. When such surges leak out to earthing system, surges containing frequency components over 10 MHz are induced in earthing system. In this respect, the frequency component over 10 MHz is the characteristic of surges in earthing system near GIS.

Electromagnetic Interference with Thyristor. When GIS grounding wire current with frequency components at about 500 kHz and over 10 MHz (shown in Fig. 3) was led to the frame of thyristor unit, the voltage between anode and cathode and the current induced in the gate circuit were measured. The results are shown in Fig. 4(a)(b). As mentioned previously, the thyristor and the pulse transformer were insulated from the frame. Nevertheless, the thyristor was turned on as shown in Fig. 4(a), indicating that the thyristor is mal–fired by EMI due to surges in earthing system. When the mal–firing occurred, current in frequency range over 10 MHz and duration about 1 μs was induced in gate circuit, as shown in Fig. 4(b). A current with 500 kHz component was rarely induced. Little is known about how thyristor responds to gate currents of such high frequencies and short duration.

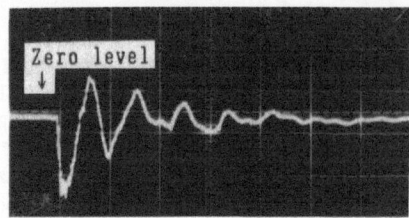 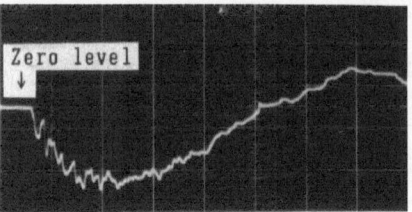

(a)Current through earthing wire of GIS

Vertical axis: 40 A/div
Horizontal axis: 2 μs/div

(b)Enlargement of A in (a)

Vertical axis: 40 A/div
Horizontal axis: 200 ns/div

Fig. 3. Current through earthing wire of GIS, measured during closing operation of GCB, with power cable on load side of GIS (measured at point P in Fig. 1).

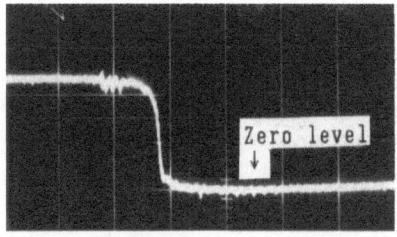 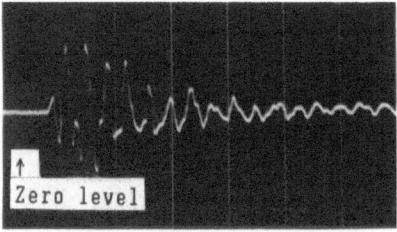

(a)Voltage between anode and cathode
Vertical axis: 870 V/div
Horizontal axis: 2 μs/div

(b)Induced gate current
Vertical axis: 4 A/div
Horizontal axis: 200 ns/div

Fig. 4. Voltage between anode and cathode(a), and induced gate current(b), measured during closing operation of GCB (with power cable on load side of GIS).

DISCUSSION

As the result of full–scale experiment, it is found that thyristor is mal–fired by EMI due to surges in earthing system. To keep the reliability of thyristor unit, it is necessary to suppress the EMI. To find the method for suppressing EMI, the frequency components contributing to mal–firing, which are contained in surges in earthing system, were considered. Using the result of this consideration, the method for suppressing EMI was discussed. Then the effect of the method was verified.

Frequency Components Causing Mal–Fire

To consider the frequency components causing mal–fire, high–frequency currents with peak value of 50–200 A and in frequency range of 500 kHz – 10 MHz were led to insulated lead wire laid on the frame of thyristor unit shown in Fig. 2. Then the current flowing through the insulated lead wire, the current induced in the gate circuit, and the voltage between anode and cathode were observed in the same manner as in the preceding experiment.

When the frequency of the current through the insulated lead wire was several 100 kHz, little current was induced in the gate circuit. When the frequency of current through the insulated lead wire was over 1 MHz, current was induced in the gate circuit. As the peak value of induced gate current rose to a certain level, turn–on occurred. The observed frequency dependency of the level of induced gate current is shown by the solid line in Fig. 5. The horizontal axis represents the frequency of the current through the insulated lead wire, and the vertical axis shows the ratio of the induced gate current to the current through the insulated lead wire. Also, minimum gate current levels necessary for turn–on are shown by the dotted line in Fig. 5. As frequency rises, the level of induced current and the current necessary for turn–on also increases; the induced current increases more rapidly. From these frequency dependencies, it is considered that if high–frequency surge over 10 MHz occurs in earthing system, the induced current rises over the current level necessary for turn–on and causes the thyristor to be mal–fired.

Suppression of Electromagnetic Interference

Method for Suppression. The previous considerations suggest that mal–fire could be prevented by suppressing high–frequency surges over 10 MHz in earthing system. Such high–frequency surges can be suppressed by preventing high–frequency surges generated inside GIS from leaking out. To do this, as shown by Buesh et al.,[1] the bushing must be eliminated and the GIS tank must be coupled directly to the power cable sheath. If this direct coupling is not practical, the connection can be made via capacitor. This capacitor, however, must have enough capacitance to be short–circuited practically to high–frequency surges over 10 MHz. Surges inside GIS also leak out to the earthing system through the joint between power cable and transformer. This leakage of high–frequency surges must also be prevented in the same manner.

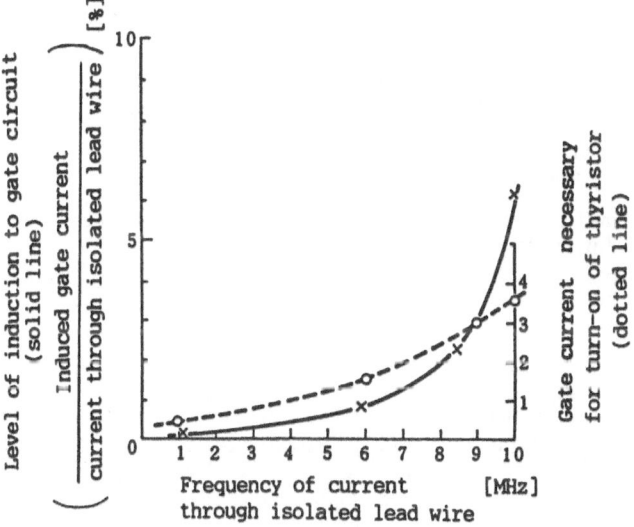

Fig. 5.   Frequency dependency of induction rate of gate current and gate current necessary for turn–on.

Effect of Suppressing Method. Remodeling the experimental apparatus shown in Fig. 1, the effect of the method for suppressing EMI, described in the previous section, was checked. A cable head was used in place of the GIS bushing unit. Also, the structure was adjusted so that the capacitors could be mounted within a short distance of about 20 cm on the circumference between GIS tank and power cable sheath. A steel tank simulating transformer tank was also used so that the direct connection of the transformer tank to cable sheath could be simulated. As with the structure of the experimental apparatus shown in Fig. 1, the grounding wire was connected to the earthing rod via frame of thyristor unit. To make checking for surge leakages from all high–voltage unit possible, not only the grounding wire of GIS, but also those of the power cable and the simulated transformer tank were connected to the frame.

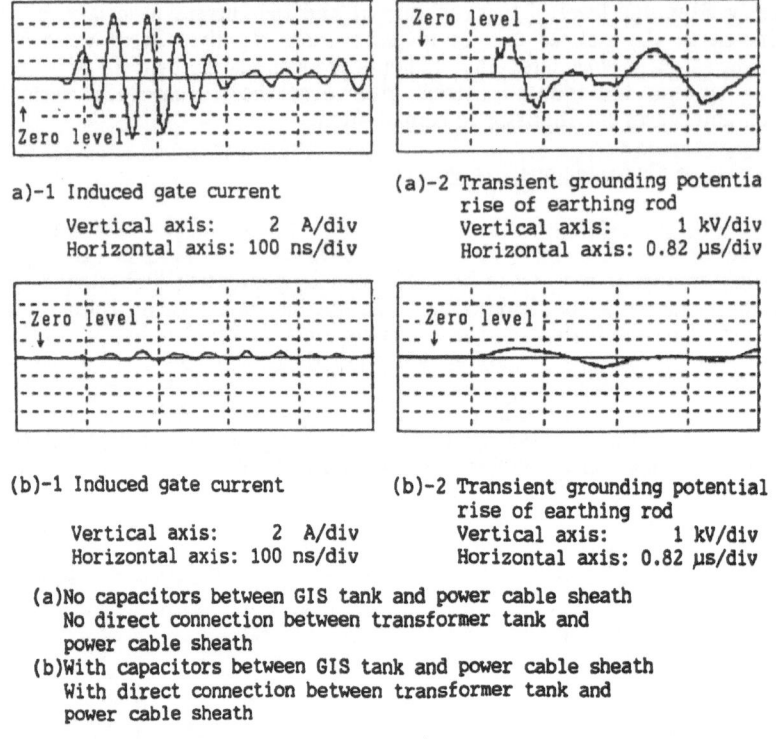

a)-1 Induced gate current

    Vertical axis: 2 A/div
    Horizontal axis: 100 ns/div

(a)-2 Transient grounding potentia rise of earthing rod

    Vertical axis: 1 kV/div
    Horizontal axis: 0.82 μs/div

(b)-1 Induced gate current

    Vertical axis: 2 A/div
    Horizontal axis: 100 ns/div

(b)-2 Transient grounding potential rise of earthing rod

    Vertical axis: 1 kV/div
    Horizontal axis: 0.82 μs/div

(a) No capacitors between GIS tank and power cable sheath
    No direct connection between transformer tank and
    power cable sheath
(b) With capacitors between GIS tank and power cable sheath
    With direct connection between transformer tank and
    power cable sheath

Fig. 6.   Induced gate current and transient grounding potential rise of earthing rod during closing operation of GCB.

With no capacitor attached between GIS tank and power cable sheath, and with simulated transformer tank not directly coupled to power cable sheath, GCB was closed. Then the induced gate current and the potential difference between the far–side reference grounding point and the earthing rod to which frame was connected were measured. The results are shown in Fig. 6(a)–1 and (a)–2. As shown in Fig. 6(a)–1, current with the frequency of about 10 MHz was induced in gate circuit. This current caused the thyristor to turn on. The initial part of the transient grounding potential rise waveform for the earthing rod, shown in Fig. 6(a)–2, shows a frequency component over 10 MHz, proving that surges of this frequency leaked out of the GIS. With 0.06 μF capacitors attached circumferentially to the cable heads as connectors between the GIS tank and the power cable, and with the simulated transformer tank directly coupled to power cable sheath, GCB was closed. The measured waveforms are shown in Fig. 6(b)–1 and (b)–2. As shown in Fig. 6(b)–1, the induced current in the gate circuit was very small. Nor was any turn–on caused. As shown in Fig. 6(b)–2, the transient grounding potential rise waveform for the earthing rod did not contain 10 MHz component, proving that surges in frequencies over 10 MHz no longer leaked out to earthing system from inside of GIS tank. All these prove that the suppressing method mentioned works as expected.

CONCLUSIONS

Using full–scale 66 kV experimental apparatus including GIS, transformer and power cable, the characteristics of surges induced in earthing system and their electromagnetic interference with thyristor unit were investigated. The obtained results are as follows.

1) Surges induced in the earthing system during closing operation of GCB in GIS have frequency components over 10 MHz, which are peculiar to GIS.

2) Surges in earthing system which contain frequency components over 10 MHz induce currents in the gate circuit of the thyristor unit, causing a mal–firing.

3) Surges over 10 MHz in earthing system can be suppressed by making the connection between power cable sheaths and the tanks of high–voltage equipment, such as GIS or transformer, to be short–circuited for high–frequency surges over 10 MHz. This connection can be a direct connection, or via circumferentially arranged capacitors. This method enables electromagnetic interferences with control equipment near GIS, such as thyristor unit, to be suppressed effectively.

REFERENCES

1. W. Buesh, H. Stephanides, and I. Heinemann, Attenuation of Fast Transients in GIS Earthing Systems, Cigre SC23, 23–10, 1988.

2. A. Welsch, Outcoupling and External Phenomena of Very Fast Transients (VFT) in GIS, Sixth International Symposium on High Voltage Engineering, 49.05, New Orleans, LA, USA, 1989.

3. J. Meppelink, K. Diederich, K. Feser, and W. Pfaff, Very Fast Transients in GIS, IEEE PWD, Vol. 4, No. 1, p. 223, 1988.

4. S. Yanabu, H. Murase, H. Aoyagi, H. Okubo, Y. Kawaguchi, Estimation of Fast Transient Overvoltage in Gas–Insulated Substation, IEEE, Summer Meeting SM628–0, 1988.

ABNORMAL PHENOMENA CAUSED BY CONTACT FAILURE IN 300KV GIS

Y. Mukaiyama*, F. Nonaka*, I. Takagi*
K. Izumi**, T. Sekiguchi**
A. Kobayashi*** and T. Sumikawa***

* Chubu Electric Power Co., Inc. — Nagoya, Japan
** Central Research Institute of Electric Power Industry — Tokyo, Japan
*** Toshiba Corporation — Kawasaki, Japan

INTRODUCTION

Recently, gas-insulated switchgear (GIS) has become the predominant switchgear of substations. Compared with air-insulated switchgear, GIS has two major advantages: smaller size and simpler maintenance work. On the other hand, because of the GIS's enclosed design, abnormalities are difficult to detect and identify.

To clarify the abnormality of contacts in GIS, contact failure simulation tests were carried out using an actual 300kV-2000A GIS, discovering faults, and observing deteriorative phenomena, from contact failure to arcing and ground fault by the authors [1].

The test method, conditions, and results are detailed in reference [1]. This paper deals with the deteriorative mechanisms of contact failures, the phenomena preceding contact failures, and how to detect these phenomena.

DETERIORATIVE MECHANISMS OF CONTACT FAILURE

Figure 1 shows the test circuit. The contact failure was simulated in disconnecting switch (DS) and the detachable bus (DB). The results of the tests are summarized in Table 1.

Contact Resistance Characteristics

The contact area at a contact failure can be divided into two parts: the contact part via an oxide coating and a small, limited part in contact with the metal. Contact resistance R in this condition is given by the following equation:

$$1/R = 1/Rm + 1/Rf \qquad (1)$$

where R: contact resistance; Rm: contact resistance of metal contact part; Rf: contact resistance of the contact part via oxide coating. Generally, $Rm < Rf$, and Rm nearly equals the concentrated resistance of the metal contact part. Thus, the resistance is given by this equation:

$$R = \rho/2a \qquad (2)$$

where ρ: resistivity of metal; a: equivalent radius of metal contact surface.

Gaseous Dielectrics VI, Edited by L.G. Christophorou and
I. Sauers, Plenum Press, New York, 1991

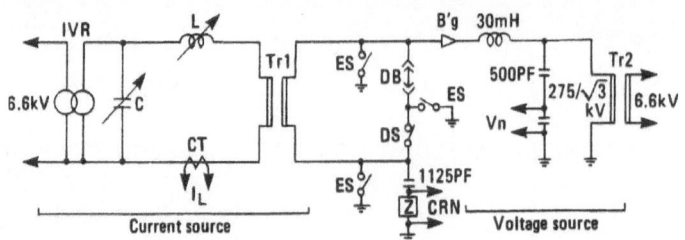

Fig. 1. Test circuit.

Table 1. Test results.

Test condition	Contact resistance before test	Phenomena in test	Gas analysis
Arc contact of DS	400-900 $\mu\Omega$	No abnormalities up to 750A. About 10 min after reaching 1000A, arcing (about 6.5s) between contacts, resulting in a ground fault.	SOF$_2$: 0.4-1.4 vol% SO$_2$: more than 25 PPM HF: more than 20 PPM
Main contact of DS without decomposed product	200-500 $\mu\Omega$	No abnormalities up to 1750A. 30 min after reaching 2000A, arcing (about 4.5s) and welded. Subsequently, 54-min test without abnormalities.	SOF$_2$: none SO$_2$: 1-4 PPM HF: more than 0.5 PPM
Main contact of DS with decomposed product	2 mΩ	The start of test was directly followed by arcing and subsequently by intermittent arcings until a continuous arcing and a ground fault occurred on the 6th day.	SOF$_2$: 1.4 vol% SO$_2$: more than 25 PPM HF: more than 20 PPM
Conduction to bolt in DB	2-9 mΩ	Test was started with one tightening-bolt. The start of supplying 300A was directly followed by intermittent arcing. Then, during repeated heat cycle up to 500A, single arcing was repeated, which turned continuous on the 13th day, resulting in a ground fault on the 14th day.	SOF$_2$: 0.33 vol% SO$_2$: more than 25 PPM HF: more than 20 PPM

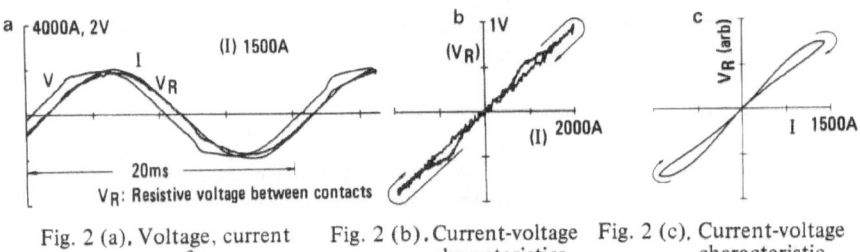

Fig. 2 (a). Voltage, current waveforms .

Fig. 2 (b). Current-voltage characteristics (measured) .

Fig. 2 (c). Current-voltage characteristic (calculated) .

In the test the contact resistance of the main and arc contacts of DS was nearly $500\mu\Omega$. Based on the assumption that there was only one contact point, equivalent radii of contact areas equivalent to $500\mu\Omega$ can be obtained as 0.02mm for the main contact and 0.05mm for the arc contact. These figures show that the metal contact area is very small.

Figure 2 (a) and (b) show the voltage waveform between contacts and the current-voltage characteristics measured during the test on the arc contact. The current-voltage characteristics show a rectangular form in the upward course of current. This rectangular form always appears in the upward course of current, irrespective of the polarity of current; it is not observed in the downward course of current.

It is considered that the contact resistance is given by the formula:

$$R \propto 1/\exp\left(-E/\kappa\, Tk\right) \tag{3}$$

where E: activation energy; κ: Boltzmann's constant; Tk: absolute temperature.

And, using heat capacity and heat resistance, contact temperature Tc is given by the formula:

$$Tc = Rt \cdot W \left[1 - \exp\left(-t/Ct \cdot Rt\right)\right] \tag{4}$$

where Rt: heat resistance; Ct: heat capacitance; W: heat flow, $= i^2 Rm$ (where $i = \sqrt{2}\, I \sin\omega t$; Rm: resistance of metal contact part; I: effective value of supplied current).

From equations (3) and (4), current-voltage characteristics can be calculated as shown in Fig. 2 (c). The results, similar to the actual test results, show that the rectangualr form appeared only in the upward course of current.

Arcing Mechanisms

It is assumed that whether current supply to a contact in failure will result in an arcing or not is determined by the balance between the energy injected into the contact and the energy radiated from it. The conditions resulting in an arcing are examined as follows.

DS arc contact. Considering heat transfer into the gas, the temperature rise in the DS arc contact is given by the equation:

$$\int \left[I^2 \cdot R(1 + \alpha \cdot Tf) - W \cdot S \cdot Tf\right] dt = C \cdot M \cdot Tf \tag{5}$$

where I: supplied current; R: resistance before temperature rise; α: temperature coefficient; Tf: temperature rise in arc contact; W: coefficient of heat transfer into gas; S: effective radiation area of arc contact; C: specific heat of arc contact; M: weight of arc contact.

Figure 3 shows the relation between supplied currents and temperature rises obtained from equation (5) when $R = 425\mu\Omega$. In actual tests ($R = 400\text{-}500\mu\Omega$), no arcing occurred with supplied currents up to 750A, but continuous arcing occurred in about 10 min with a current of 1000A. This result shows good agreement with the results of the calculation in Fig. 3.

The above result suggests that continuous arcing occurred due to melting down of the part of the arc contact, which was overheated because of extremely poor contact conditions.

DS main contact. Contact failure in the DS main contact caused an arcing for a short time, resulting in welding between contacts. This can be explained as follows: the contact parts were heated, and when their temperature reached the melting point, they melted, causing an arcing. Since the contact gap is small, the arcing did not grow continuous. Also, as the contact area was increased by melting, the contact resistance and the contact temperature decreased.

On the other hand, decomposed products were deposited on the contact, an intermittent arcing occurred, and this arcing became continuous. Here, the decomposed products pre-

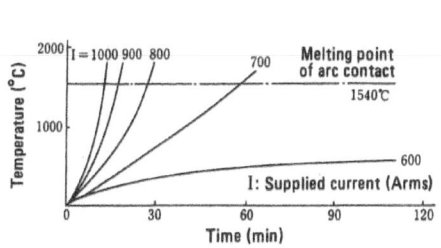

Fig. 3. Calculated temperature rise characteristics of arc contact in DS.

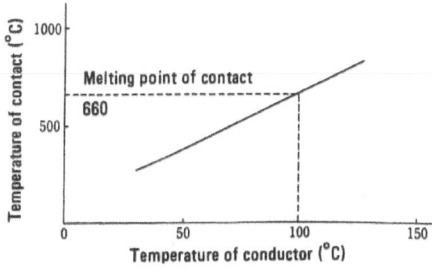

Fig. 4. Relation between conductor temperature and contact temperature

477

vented the increase of contact area and welding. So, the intermittent arcing repeated, growing continuous and resulting in ground fault.

Contact failure in DB. Using a heat conduction model, the relation between the conductor temperature measured by an infrared sensor and contact temperature can be calculated. The results are shown in Fig. 4. This result shows that if the conductor temperature exceeds 100°C, the contact temperature is expected to exceed the melting point. An oscillogram obtained in a long-term voltage-current test using a DB is shown in Fig. 5, where I_L: supplied current; Vn: applied voltage; T_1: tank surface temperature in DB; Ta: ambient temperature; Tc: conductor temperature; CNT: number of discharges above 10nC in 5 min.

In Fig. 5, except for arcing in the beginning of conduction, arcing occurs when the conductor temperature rises to around 120°-150°C ((A) and (B) in the figure). This suggests that arcings are caused by the melting of the conductor when heated.

In the test, arcing occurred during a constant current supply. At point (A), in Fig. 5, for example, the continued thermal balance might be destroyed due to a decrease in the ambient temperature at night, causing thermal contraction of the conductor, the contact resistance to jump up, and the conductor temperature to rise, resulting in an arcing. Other possible causes of the upset of thermal balance may be rapid current rises, mechanical vibration due to operation, and electromagnetic vibration in the GIS.

Relation Between Contact Resistance and Arcing

From the present test results, the relation between the initial contact resistance and the supplied current resulting in an arcing was obtained, shown in Fig. 6. Here, the relation

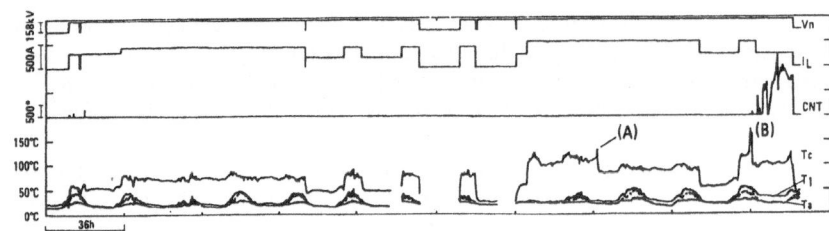

Fig. 5· Test results of long-term voltage-current test. *500 indicates 500 times of discharge in 5 min.

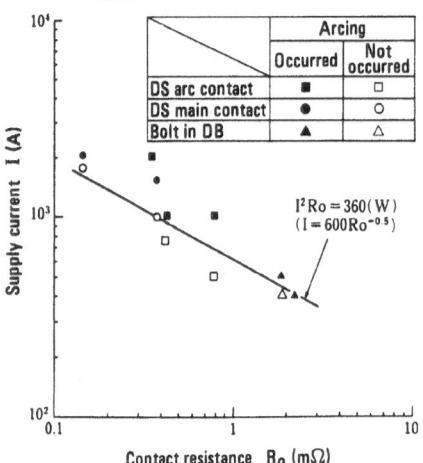

Fig. 6· Relation between contact resistance and current resulting in arcing·

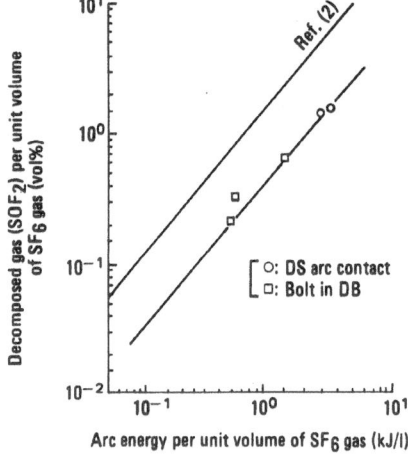

Fig. 7· Relation between arc energy and SOF_2 gas volume decomposed·

478

is very close to that given by the following equation with constant Joule-heat generation:

$$I \approx 600 \cdot R_o^{-0.5} \tag{6}$$

where I: current (A) resulting in arcing; R_o: contact resistance (mΩ) before test.

PHENOMENA AT ARCINGS

Decomposed Gas

When an arcing occurred, resulting in a ground fault, decomposed gases such as SOF_2, SO_2, and HF were detected. In very small or very short arcings, small quantities of SO_2 and HF were detected.

As the test results, the relation between arc energy and decomposed gas (SOF_2) is shown in Fig. 7 and expressed as

$$V(SOF_2) \approx 0.43 Earc^{1.1} \tag{7}$$

where $V(SOF_2)$: SOF_2 gas (vol%); Earc: arc energy per unit volume (kJ/l).

This relation showed a similar characteristic to that obtained with Al-Al electrodes [2].

PRECEDING PHENOMENA OF CONTACT FAILURES

Contact Failure Developments

From the test results, developments of a contact failure can be classified as follows:
(1) A contact failure causes the contact resistance to rise, causing the contact to be heated. If heat radiation exceeds heat generation, the contact remains stable.
(2) Heating increases, and when the temperature exceeds the melting point of the contact, an arcing occurs. The main contact of the DS turns stable after a short arcing, but the main contact of the DS, with decomposed products by arcing or contact failure in the DB, repeats arcings.
(3) As arcings repeat, the contact is heated and melted, resulting in a continuous arcing and a ground fault.
(4) Even after the contact is welded, the contact condition may be changed by ambient temperature drops, mechanical vibration, etc., causing the contact resistance to rise, resulting in an arcing.

Therefore, it is preferrable to detect abnormality at phase (1) or (2) above.

Detection of Preceding Phenomena

Detection of contact resistance rises. As described previously, there is a correlation between initial contact resistance and current resulting in arcing. Thus, the condition of the GIS can be checked by measuring the contact resistance.

Detection using tank temperature. Figure 8 shows the results of a long-term voltage-current test with the main contact of the DS with decomposed products. T_1 and T_4 represent the temperature of the tank top in the DB and the DS. Evidently, T_4 rose gradually, while defects grew in the contact failure part. Also, while there are obvious differences between T_1 and T_4 at night, without the effect of sunlight, there are slight differences between them in the daytime, with the effect of sunlight. Thus, if abnormalities in the contact failure parts grow, they can be detected by monitoring the tank top temperature at night.

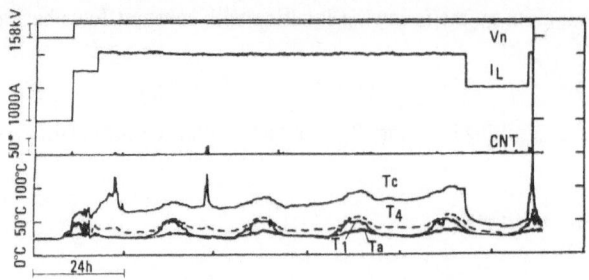

Fig. 8 . Test results in long-term voltage-current test.
*50 indicates 50 times of discharge in 5 min.

CONCLUSIONS

Using an actual 300kV-2000A GIS, contact-failure simulation tests were carried out, and contact failure developments and their preceding phenomena were studied. The conclusions are as follows.

Contact Resistance

There is a correlation between the contact resistance and the current resulting in an arcing; this correlation helps us decide the level of current to be fed without causing arcings.

Developments to Arcings

In the arc contact of the DS, the occurrence of arcing is determined by the heat capacitance of the arc contact. Once the arc contact melts down, the arcings become continuous, resulting in a ground fault. In the main contact of the DS, the contact turns stable because of welding caused by arcing. But in the main contact of the DS, with decomposed products, arcing becomes continuous, resulting in a ground fault. In the bolt of the DB, temperature rises cause the aluminum conductor to melt, resulting in an arcing or a ground fault.

Arcings can also be caused by thermal contraction of conductors, due to temperature drops at night, and by vibration in the GIS.

Decomposed Gases

If a continuous arcing occurs, resulting in a ground fault, gases such as SOF_2, SO_2, and HF are produced. These gases are associated with arc energy. In a very short arcing, small quantities of SO_2 and HF are produced.

Techniques for Detecting Contact Failure

(1) Currents resulting in arcings can be recognized by monitoring contact resistance.
(2) Contact failures which develop gradually can be detected by monitoring the temperature of the tank top at night.

REFERENCES

1. Y. Mukaiyama, I. Takagi, K. Izumi, T. Sekiguchi, A. Kobayashi, and T. Sumikawa, Investigation of Abnormal Phenomena of Contacts Using Disconnecting Switch and Detachable Bus in 300kV GIS, IEEE Trans. on Power Delivery, Vol. 5, No. 1, pp. 189-195 (1990).
2. F. Y. Chu, SF_6 Decomposition in Gas-Insulated Equipment, IEEE Trans. on Electrical Insulation, Vol. EI-21, No. 5 (October 1986).

INSULATION COORDINATION IN GAS INSULATED SWITCHGEAR

AGAINST LIGHTNING OVERVOLTAGES

T. Kawamura[*]   and   J. Ozawa[**]

    * University of Tokyo, Minato-ku, Tokyo, JAPAN 106
    ** Hitachi Research Laboratory, Hitachi Ltd., Hitachi-Shi
       Ibaraki-Ken, JAPAN 316

ABSTRACT

    This paper describes improved models for lightning overvoltage
analysis and example results by the models. Reduction of lightning impulse
withstand levels for 500 and 275kV gas insulated switchgear substations is
recomended. Newly developed high performance metal oxide surge arresters
are also presented.

INTRODUCTION

    An SF_6 gas insulated switchgear (GIS) substation offers such
advantages as compactness and economy compared with a conventional air
insulation substation. However, a dielectric breakdown fault in the GIS
sometimes causes a blackout with a long recovery time. Realizing
insulation rationalization, as well as improved insulation reliability, is
desired for the GIS. Then, technologies of lightning overvoltage analysis
and overvoltage suppression should be improved. For this, the insulation
level of the GIS will be reduced by considering the insulation
reliability.

LIGHTNING OVERVOLTAGE ANALYSIS

Lightning Stroke Parameters

    Peak amplitude and waveform of the lightning stroke current are
needed in advances to analyze lightning overvoltages in a GIS substation.
The peak amplitudes are described in terms of probabilities and some
cumulative frequency distribution curves of the peak amplitudes have been
obtained from observed data. However, the peak amplitudes proposed by the
Lightning Protection Design Study Committee of Japan have been used as the
standard values for each system voltage[1]. The most comprehensive waveshape

Gaseous Dielectrics VI, Edited by L.G. Christophorou and
I. Sauers, Plenum Press, New York, 1991

data for the lightning stroke current have come from the work of Berger and associates[2]. The front waves of the current show a concave shape like $1-\cos \omega t$ as shown by Anderson and Erikkson [3] from a computer simulation based on published data. The time to half value of the lightning stroke current is observed as about 70µs. The peak amplitude and the rate of current rise near the peak effect the critical flashover of an arcing horn of transmission lines[4].

Then in the present analysis, the ramp current wave with a front time of 1.0µs and a time to half value of 70µs was selected for the above reasons from the viewpoint of a realistic and simple approximation. The surge impedance of 400Ω was selected for the lightning stroke channel.

Multi-conductor Model of Transmission Line

Lightning overvoltages have usually been analyzed by using a single-conductor model, while an actual transmission line consists of a multi-phase circuit. For analysis of lightning overvoltage in a double circuit line, the line has been simulated by a multi-conductor model considering the mutual induction between conductors and the frequency dependency of the traveling wave on the line. Fig.1 shows an analysis circuit of a multi-conductor model by using the Semlyen setup in the EMTP (Electromagnetic Transients Program) developed by the Bonneville Power Administration. A substation is represented by equivalent resistances of average self surge impedance of the phase conductors. A back flashover to phase A is supposed. In the case of a single conductor model, lightning overvoltage on the phase conductor at the substation was 99.8% of that at tower No.1 for a lightning stroke to the tower. However, in the case of the multi-conductor model, the above value on phase A decreased to 61% as shown in Fig.2[5].

This decrease was investigated by the formula(1) based on the traveling wave theory in a two-conductor system.

$$V_1/e_1 = 2R_1 [(Z_2+R_2)e_1 - Z_{12}e_2] \cdot 100\% / [(Z_1+R_1)(Z_2+R_2) - Z_{12}^2] \quad (1)$$

where Z_1, Z_2, Z_{12}: self and mutual surge impedances of two conductors, i.e. a ground wire and a phase conductor; R_1, R_2: values of equivalent resistances representing a substation and an entrance tower; e_1, e_2, V_1: traveling wave voltages at the start points of two conductors and at the resistance R_1. The decrease ratio calculated with formula(1) and that analyzed for the multi-conductor model agreed within 5%. The decrease was

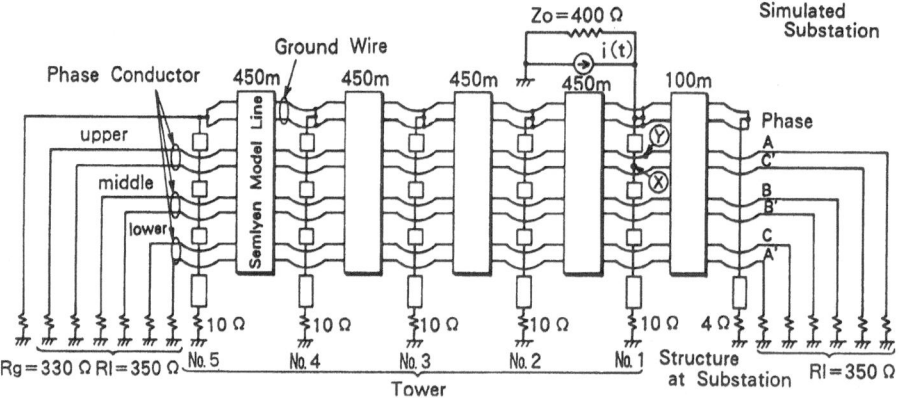

Fig. 1. Multi-conductor transmission line model.

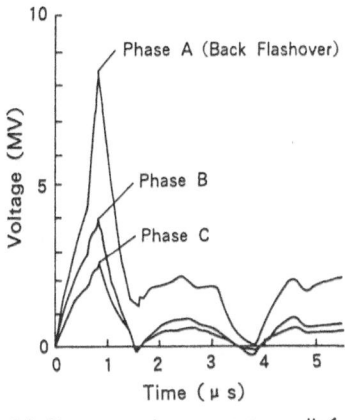

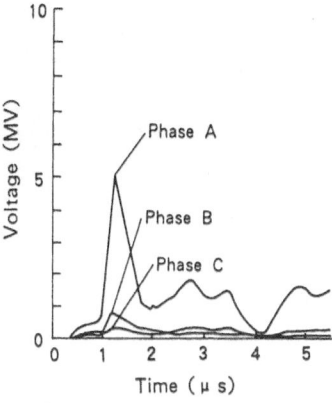

(a) Phase conductor at tower No. 1. (b) Phase conductor at substation.

Fig. 2. Lightning overvoltage waveform with back flashover for lightning stroke to tower No. 1 in multi−conductor system.

due to the effects of a traveling wave current flowing into the entrance tower and the electrical coupling between two conductors, i.e. a ground wire and a phase conductor.

As a result, lightning overvoltage analysis should be carried out by using a multi-conductor model, at least a two-conductor model, of a transmission line.

Transmission Tower Model

A transmission tower has usually been simulated by a single uniform distributed parameter line. A surge impedance of 100 to 115Ω, a traveling wave velocity of 210 to 240 m/μs and a surge attenuation coefficient of 0.8 to 0.9 have been used for lightning overvoltage analysis in Japanese systems. However, the surge response on the uniform tower model has not always been in agreement with that on an actual tower with phase conductors and ground wires. For the purpose of reducing this disagreement, a new multistory transmission tower model was developed as shown in Fig.3[6]. The new tower model consists of four sections divided at the upper, middle and lower phase crossarm positions. Each section consists of a lossless distributed line and lumped constants of a damping resistance shunted by an inductance. Damping resistances R_1-R_4 and inductances L_1-L_4 are calculated by formulas(2) and (3).

$$R_n = -2Z_T(\ln X)h_n/H \qquad (2)$$

$$L_n = R_n T \qquad (3)$$

where Z_T, X, h_n, H: the surge impedance, the attenuation coefficient, the section height and the total height of a tower; n: a constant from 1 to 4; T: the round trip traveling time of the tower.

The circuit parameters of the new tower model were determined from the measurement of voltages across the arcing horn on an actual 500kV double circuit transmission tower. Fig.4 shows the measured and calculated waveforms of the voltages across the upper phase arcing horn on the 500kV tower with ground wires. The surge impedances of the tower model were 220Ω for the upper three sections and 150Ω for the bottom. The calculated and measured results showed relatively good agreement.

483

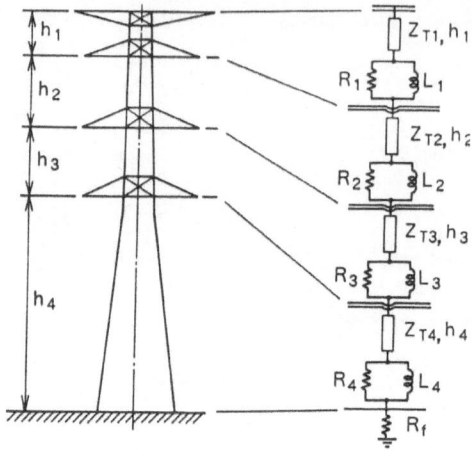

Fig. 3.   New multistory transmission tower model.

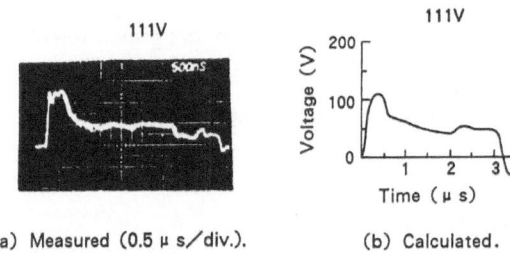

(a) Measured (0.5 μ s/div.).   (b) Calculated.

Fig. 4.   Measured and calculated waveforms of voltages across upper
phase arcing horn for the line with ground wires.

The new multistory transmission tower model is recommended for use in lightning overvoltage analysis on double circuit transmission lines.

Back Flashover Model of Arcing Horn

Back flashover of an arcing horn has usually been simulated by a switch with a fixed inductance. However, the back flashover model significantly effects the peak value and the rise rate of the wavefront of the lightning overvoltages coming into a substation. According to experiments on flashover phenomena for a long air gap, a large predischarge current prior to the flashover has been observed. A more reasonable back flashover model has been developed based on streamer and leader developement phenomena of the flashover in air. Fig.5 shows the new back flashover model with a time dependent inductance simulated in accordance with three stages such as a leader, a final jump and an arc discharge[7].

Leader velocity v is given by formula (4).

$$v = k_1 \cdot V_a^2/(D-2x) + k_2 \cdot (V_a \cdot i/(D-2x)) \cdot (x/D) \qquad (4)$$

where V_a: applied voltage; i: predischarge current; D: gap length of an

484

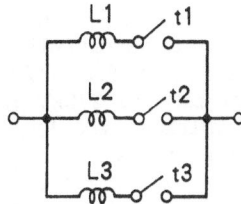

Fig. 5. New back flashover model (time dependent inductance model).

arcing horn; x: leader length; k_1, k_2: constants. Predischarge current is assumed to be proportional to the product of applied voltage V_a and the leader velocity v.

$$i = C_1 \cdot V_a \cdot v \qquad (5)$$

where C_1: a constant dependent on the leader length. Combining formulas (4) and (5) with a circuit equation, leader development characteristics and predischarge current are calculated. The equivalent inductances of the flashover model can be obtained from the voltage V_a and predischarge current i. The inductance of 1.0μH/m is used for an arc discharge phase.

HIGH PERFORMANCE SURGE ARRESTER

500 and 275kV high performance tank type MO (metal oxide) surge arresters have been developed to reduce the 500 and 275kV GIS insulation levels, based on economic considerations. Fig.6 compares the discharge

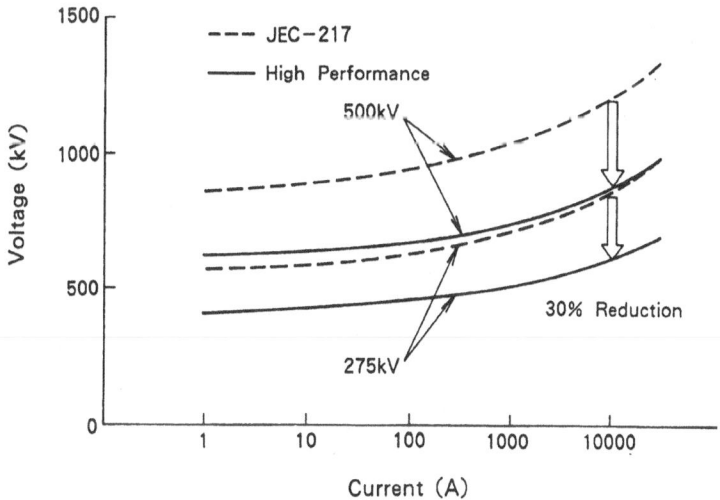

Fig. 6. Discharge voltage–current characteristics of 500kV high performance MO surge arrester.
JEC–217 : Standard of the Japanese Electrotechnical Committee.

voltage-current characteristics of the high performance arresters with those set by Japanese standard JEC-217-1984. The discharge voltage at 10kA was improved by lowering it from 1220kV to 870kV for the 500kV arrester and 851kV to 600kV for the 275kV arrester. Accompanying this reduction of the discharge voltage, the high performance arresters were required to have a higher energy absorption capability (about 250joules/cm^3) than the energy absorbed in AC temporary overvoltages. This high performance was realized by improvements in the composition of MO resistors and their manufacturing method.

LIGHTNING OVERVOLTAGE ANALYSIS

Condition of Analysis

Lightning overvoltages for the GIS substation have been analyzed by using the above lightning stroke condition, analytical methods and the characteristics of the high performance arresters[8],[9]. Lightning stroke currents of 150kA for the 500kV GIS substation and 100kA for the 275kV GIS substation were selected with a 1/70μs ramp wave. Substation configurations, such as a one line and one transformer configuration as shown in Fig.7(a), were set up for consideration of severity of overvoltages to a GIS and a transformer. Locations of high performance arresters were selected as the line entrance and transformer terminal from the viewpoint of a balance between overvoltage protection ability and economy.

Results of Analysis

Figure 7(b) shows typical results of analyzed lightning overvoltage in a 500kV substation with fifteen parameter combinations of gas bus length L_1 and cable length L_2. The maximum overvoltage in GIS was 1400kV at the section end of the GIS main bus and that at the transformer was

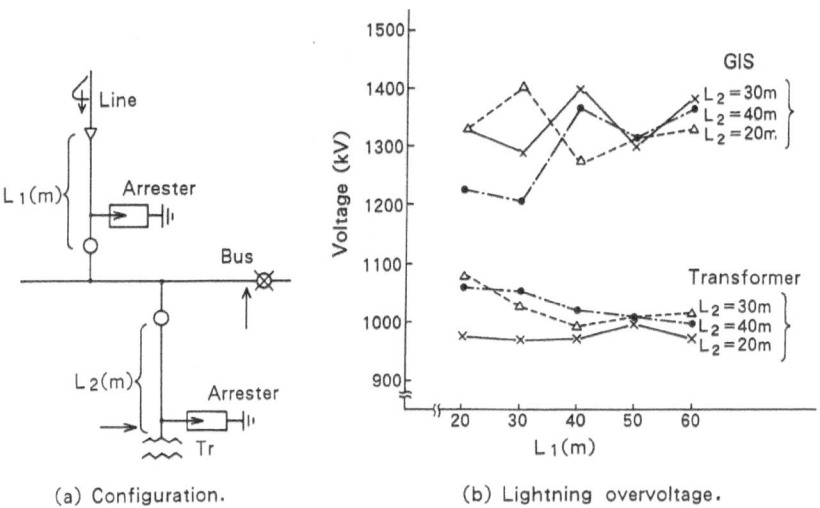

(a) Configuration.   (b) Lightning overvoltage.

Fig. 7. Analyzed lightning overvoltages in 500kV GIS substation with high performance MO surge arrester.

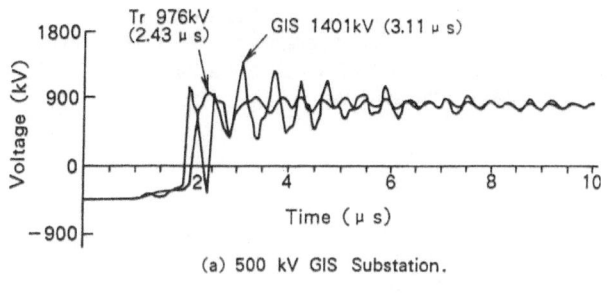

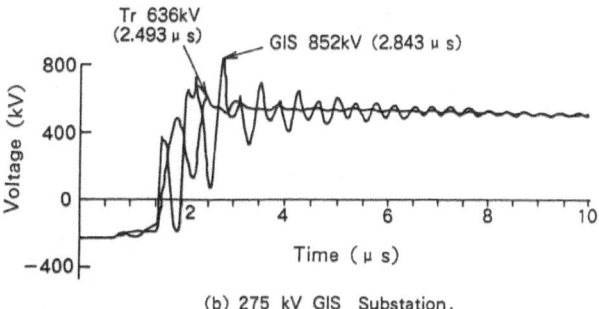

Fig. 8.   Analyzed lightning overvoltage waveforms.

1090kV. The analyzed waveforms have large oscillating voltages as shown in Fig.8, for the 500 and 275kV substations. However, the GIS insulation characteristics for such oscillating voltages were, for practical purposes, the same as those for the standard test voltage wave of 1.2/50µs. Therefore, LIWL (lightning impulse withstand level) values of 1425 and 950kV have been predicted for 500 and 275kV GISs, respectively, considering disconnector switching overvoltages and some safety margins[9].

GIS WITH REDUCED LIWL

Figure 9 shows an example layout of the newly installed 275kV GIS

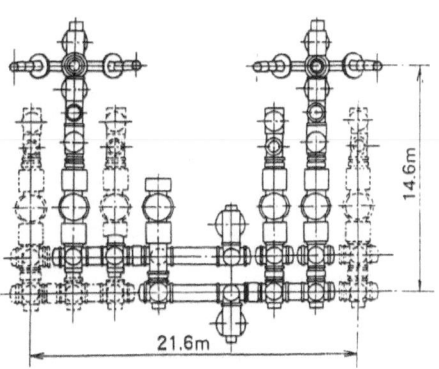

Fig. 9.   Example of 275kV GIS with reduced LIWL.

including the complete three phase common type equipment. Rationalized compactness and economy were achieved. The LIWL of 950kV, rated current of 2000/4000A, rated short circuit current of 50kA and rated gas pressure of 5bar are specified. A 500kV GIS with a reduced LIWL of 1425kV will be available in the near future.

CONCLUSIONS

Models of a transmission line, a tower and back flashover of an arcing horn were improved for lightning overvoltage analysis. The 500 and 275kV high performance MO (metal oxide) surge arresters with about a 30% reduction in discharge voltages were developed. As a result, 10-25% reduced LIWLs could be recomended for 500 and 275kV GIS substations.

REFERENCES

1. Lightning Protection Design Study Committee of Japan, Lightning Protection Design Guide-Book for Power Stations and Substations (in Japanese), CRIEPI Report No.175034 (1976)
2. K. Berger, R.B. Anderson, H. Kroninger, Parameters of Lightning Flashes, Electra No.41 (1975)
3. R.B. Anderson, A.J. Eriksson, Lightning Parameters for Engineering Application, Electra No.69 (1980)
4. J.G. Anderson,"Transmission Line Reference Book, 345kV and Above,"2nd Edition, Chapter 12, Palo Alto, California, EPRI (1981)
5. J. Ozawa, E. Ohsaki, M. Ishii, S. Kojima, H. Ishihara, T. Kouno, T. Kawamura, Lightning Surge Analysis in a Multi-Conductor System for Substation Insulation Design, IEEE Trans., PAS-104, No.8 (1985)
6. M. Ishii, T. Kawamura, T. Kouno, E. Ohsaki, K. Shiokawa, K. Murotani, T. Higuchi, Multistory Transmission Tower Model for Lightning Surge Analysis, IEEE/PES 89WM103-3PWRD (1989)
7. T. Shindo, T. Suzuki, A New Calculation Method of Breakdown Voltage-Time Characteristic of Long Air Gaps, IEEE Trans., Vol-104, No.6 (1985)
8. T. Kawamura, Y. Ichihara, Y. Takagi, M. Fujii, T. Suzuki, Pursuing Reduced Insulation Coordination for GIS Substation by Application of High Performance Metal Oxide Surge Arrester, CIGRE Paper 33-04 (1988)
9. Committee of Rationalization of Insulation Design, Rationalization of Insulation Design,(in Japanese) Report of The Society of Electrical Cooperative Research, Vol.44, No.3 (1988)

DISCUSSION

Y. DOIN: (1) What is the influence of the tower foot impedance on the lightning overvoltages calculated? (2) A value of 10 Ω is assumed for foot resistance. Is this value realistic? How do you choose this value and what is the probability of such a value for foot impedance?

J. OZAWA: The tower foot impedance may have complicated current and time dependent characteristics: linear or nonlinear, resistive, capacitive, or inductive types. I think inductive type impedance is preferable from the viewpoint of a conservative situation. But, up to now, the characteristics are not clear and depend on the tower locations. The tower foot resistance near the substation is usually less than 10 Ω. I think the value of 10 Ω is roughly realistic. We did not obtain the influence of the tower foot impedance.

S. W. ROWE: The design of the arrestor shown has a particularly interesting structure. Could you explain the reason why the two outer stacks have three ZnO resistors sandwiched between insulating blocks whereas the two inner stacks have alternate ZnO and insulating blocks? Is this to reduce inductance or to enable easy assembly?

J. OZAWA: The structure of ZnO resistor arrangement is similar to the structure of a non—inductive winding resistor to reduce the residual inductance of the surge arrester. Three ZnO resistors are stacked at both outer columns to have the same height as the connecting bars. This means the assembly reliability is also high.

IMPACT ON GIS DESIGN BY REDUCING THE RATED LIWL

Yasufumi MURAYAMA

TOSHIBA Corporation

Tokyo, Japan

ABSTRACT

This paper represents a contribution to the paper entitled "Insulation coordination in Gas Insulated Switchgear Against Lightning Overvoltage" of GIS Session. When the high performance surge arrester is applied , the maximum overvoltage appearing in GIS can be reduced and the rated Lightning Impulse Withstanding Voltage Level (LIWL) can be reduced. This paper introduces the influence on the design of 550kV GIS and the consideration to maintain the insulation reliability of GIS in the case of the reduction of LIWL.

INTRODUCTION

GIS has been widely employed for the substation switchgear due to the compactness and the higher reliability and so forth. In order to obtain the further compactness of the GIS, the reduction of the rated LIWL is considered effective. For that purpose, the high performance Metal Oxide surge arrester(MOA) has been developed and the residual voltage against lightning current can be reduced. The best allocation of the MOAs can be analyzed by using EMTP as shown in the paper(1).

As an example of the analysis, the rated LIWL of 550kV GIS is proposed to be reduced from 1,800kV to 1,425kV. The design of 550kV GIS can be also changed. Following comments are the example of the design change and the other parameters to be considered to maintain the insulation reliability.

550kV THREE-PHASE ENCAPSULATED BUSBAR

Figure 1 shows the configuration of 550kV GIS of Shin-Keiyo substation which has been put in service since 1989. The rated continuous current of main busbar is 8,000A and the three-phase

Fig.1. 550kV 8,000A GIS with three-phase encapsulated main busbar.

encapsulated design is employed for the busbar. The maximum
transportable component is the busbar disconnecting switch as shown
in Fig.2(a). The disconnecting switch is combined with main busbar
enclosure. The diameter of the main busbar is 1,800mm. When the rated
LIWL can be reduced to 1,425kV, the diameter can be reduced to 1,400mm.
The barrier spacer can be casted with the three-phase design as shown
in Fig.2(b). The reduction is contributive not only to compactness of
the disconnecting switch itself but also to the compactness of the total
layout of GIS.

(a) LIWL 1,800kV   (b) LIWL 1,425kV

Fig.2. 550kV 8,000A Busbar Disconnecting Switch.

OTHER DESIGN PARAMETERS

Although the LIWL is one of the important parameters to
determine the size of GIS, other parameters should be taken into account
in order to maintain the reliability. The most important parameter is
the insulation design for the practical metallic particles inside the
enclosure. When the metallic particle exists inside the enclosure, the
particle moves due to the electric field strength and the trapped
coulomb charge.

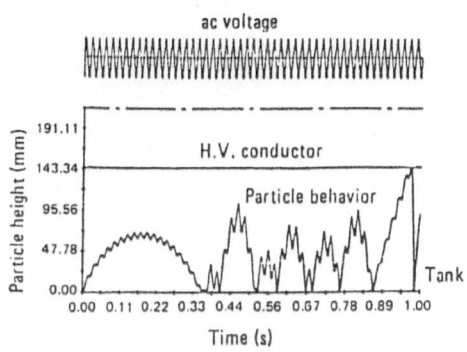

Fig.3. Computer-aided simulation
of the particle movement.

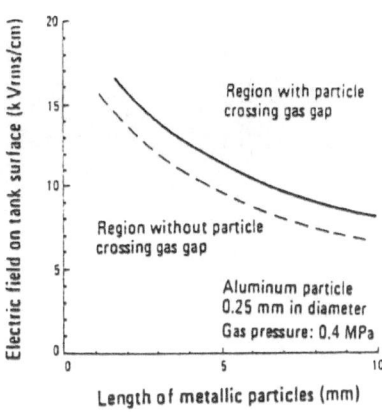

Fig.4. Relation between
particle length and crossing.

Figure 3 shows an example of the computer-aided simulation of the particle movement under power frequency voltage. The design is related to the design length of the particle. Figure 4 shows the relation between the length of the particle and the occurrence of the crossing. The "crossing" means the phenomena that the particle moves and reaches to the primary conductor.

When 5mm is considered as the design length of the particle, the electric field strength on the inner surface of the enclosure should be limited to less than 9 kV/mm in order to avoid the crossing phenomena. The size of enclosure is determined taking into account this electric field strength.

The temperature rise of the conductor and the enclosure should be also taken into account. The size and the material are considered.

DEVELOPMENT TEST

Three-phase busbar

As a typical result of the reduction of LIWL, the compact design of three-phase busbar has been developed. The insulation has been verified by phase-to-phase insulation test as shown in Fig.5. For one phase, AC nominal voltage is applied and for another one phase the lightning impulse voltage is applied with the reverse polarity. Figure 6 shows the temperature rise test by conducting three phase current.

High performance MOA

The impulse current residual voltage test is carried out as shown by using EMTP, the input data for the v-i characteristics of MOA shall be accurate. For that purpose, as the development test, a full pole column MOA is tested. As the result, the internal inductance is negligible, less than 5 micro-henry, and therefore the residual voltage against steep front wave is reliable.

Fig. 5. Dielectric test for
550kV three-phase busbar.

Fig. 6. Temperature rise test for
550kV three-phase busbar.

Fig. 7. Residual voltage test for full pole high performance MOA.

CONCLUSION

The rated LIWL is one of the most important parameters to
determine the size of GIS but another design parameters such as
temperature rise and design length of metallic particles should be also
taken into account to obtain the reliability of GIS.
If these conditions are satisfied, the reliability of GIS, even though
the rated voltage is 550kV to 800kV, can be ensured.

REFERENCE

(1) T. KAWAMURA and J. OZAWA : Insulation Coordination in Gas
Insulated Switchgear Against Lightning Overvoltages, These Proceedings,
p. 481.

DISCUSSION

K. FESER: There is an interesting development now in Japan to reduce the 1800 kV LIWL to 1425 kV LIWL on a 500 kV system. This information was mainly reported by a manufacturer. What are the opinions from the Japanese utilities on the reduction in LIWL?

M. HANAMURA: Under the joint cooperative development, the lightning overvoltage analysis appearing in GIS stations and the high—performance MOA have been accomplished. In order to ensure the insulation reliability, the practical margin between LIWL of the equipment and the maximum lightning overvoltage is maintained. Several tests have been carried out to confirm the good coincidence between calculation and field measurement. TEPCO believes the reduction of LIWL is effective to optimize GIS and the layout can become more compact with the reliability maintained.

TRANSIENT CURRENT AND VOLTAGE BEHAVIOUR DURING INTERRUPTION OF SMALL CAPACITIVE CURRENTS WITH GIS-DISCONNECTORS

Claus Neumann   Dieter König   Gerhard Imgrund

RWE Energie AG   Techn. Univers. of Darmstadt   VDMA,
Essen, Germany   High Voltage Laboratory   Frankfurt/M.,
                  Darmstadt, Germany   Germany

INTRODUCTION

In the past many investigations have dealt with switching of (small) capacitive currents with GIS-disconnectors. Most of the tests performed were done in laboratories with the aim to clarify the effects of very fast transient overvoltages (VFTO) (eg.,Refs. 1-5). With regard to these high-frequency phenomena the results gained by field tests agree well with those gained in the laboratory. In the laboratory comparable overvoltage factors and similar oscillations were obtained, when suitable test circuits have been applied.

However, in the range of low-frequency transients considerable deviations arise between field- and laboratory tests [6]. Particularly, field tests at load currents > 100 mA demonstrate that the arc in the gap does not extinguish after the decay of the high-frequency transients, but exists essentially longer. At the field tests arcing times up to 1 ms occurred after a reignition, while in the laboratory comparable load currents led only to arcing times of about 10 µs [7]. That means, in the current range mentioned the switching performance of a disconnector is not governed by the VFTO-phenomena, but by the reignition- and arc extinction behaviour. In the following this behaviour is investigated more thoroughly.

FREQUENCY REGIONS OF TRANSIENT VOLTAGES

The transient voltages when switching GIS-disconnector under network conditions may be classified into two larger regions and subdivided into two groups.

Fast and very fast transients (high-frequency oscillation, some MHz travelling waves, some 10 MHz).

Low-frequency transients (low-frequency oscillation, some kHz medium-frequency oscillation, some 10 MHz).

The travelling waves are generated by the voltage collapse after a restrike. They are superimposed on a high frequency oscillation in the MHz-range and damped after a few microseconds. The travelling waves are determined by the bus lengths on the source- and on the load side and their reflection and refraction conditions. The high-frequency oscillation is particularly affected by the external capacities on the source side.

Gaseous Dielectrics VI, Edited by L.G. Christophorou and
I. Sauers, Plenum Press, New York, 1991

The low-frequency transients follow after the decay of the high-frequency transients. They are of multiple frequency nature. The low-frequency oscillation of some kHz is superimposed by an oscillation in the range of some ten kHz. These multifrequency oscillations can be deduced to the characteristics of power transformers [8]. Such transformers have different more or less pronounced resonance frequencies which may be excited by the transient phenomena.

High voltage test transformers, however, generally possess only one distinct resonance frequency. Due to the high short-circuit inductance of such a transformer the resonance frequency is in the range of some 100 Hz.

TRANSIENT CURRENT INTERRUPTION DURING SWITCHING UNDER NETWORK CONDITIONS

The above mentioned higher frequencies in the low-frequency transients do affect the current interruption process after reignitions [7].

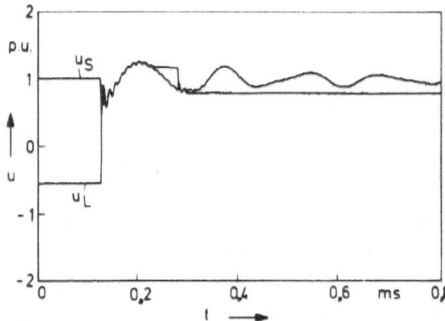

Fig. 1.
Low-frequency transients at a load current I_{LO} = 100 mA;

u_S source side voltage,
u_L load side voltage.

Fig. 1 shows the source side- and the load side voltage for a load current I_{LO} = 100 mA. In this case the basic oscillation of 7 kHz is superimposed by medium-frequency components. The arc extinguishes, if the medium-frequency and the low-frequency oscillation is decayed sufficiently. About 50 µs later a restrike occurs. This reignition takes place at an essentially reduced gap voltage corresponding to about 25 percent of the static dielectric strength. After the decay of the recurrent medium-frequency oscillation the current extinguishes at the next current zero.

Increasing the load current (I_{LO} = 360 mA) the arcing time becomes nearly 1 ms (Fig. 2). Extinction attempts can be detected at instant t_1 and t_2. The reignition voltage is less than 10 percent of the static dielectric strength. The arc does not extinguish until the low-frequency transients have dacayed sufficiently.

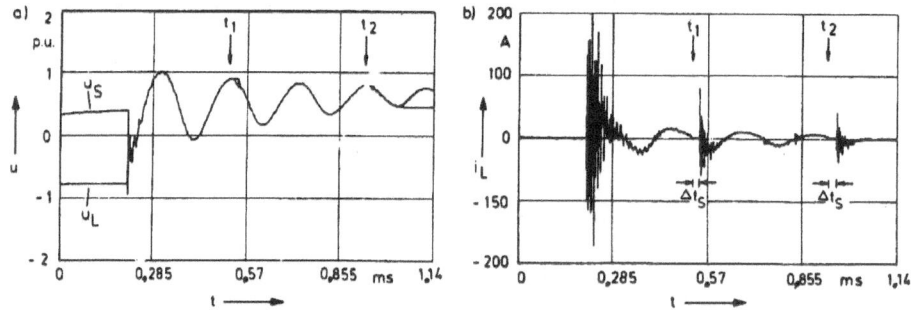

Fig. 2. Low-frequency transients at a load current I_{LO} = 360 mA.

a) source side voltage u_S and load side voltage u_L
b) transient current i_L .

498

The processes presented show that under network conditions the transient current is not interrupted after the decay of the high-frequency transients, but only if the low-frequency transients have decayed sufficiently early after the decay of the medium-frequency oscillation. This is caused by the unfavourable arc extinction- and recovery voltage conditions due to the higher frequency of the low-frequency transients. The arc cannot extinguish at current zero, until the rate of rise of the recovery voltage is small enough. If the rise of the recovery voltage is too steep after transient current interruption, reignitions will occur.

SIMULATION OF THE LOW-FREQUENCY TRANSIENTS IN THE LABORATORY BY A SYNTHETIC TEST CIRCUIT

The findings of the field tests presented were investigated more thoroughly in the laboratory.

Since the resonance frequency of HV-test transformers is in the range of some 100 Hz, the simulation of low-frequency transients in the kHz-range with long arcing times would be possible only by means of additional capacitances and inductances connected to the test transformer. Such a circuit, however, is unsuitable to produce higher capacitive load currents because the power of test transformers is usually limited.

Therefore a synthetic test circuit was developed which enabled simulations of low-frequency transients under network conditions [6, 8]. To this, the test transformer was replaced by a DC-voltage source. By variation of the cicuit elements the medium- and low-frequency transient current and the source side recovery voltage could be altered in amplitude and frequency.

ARC EXTINCTION AND REIGNITION PHENOMENA

The results of field tests demonstrated that the current interruption during the medium- and low-frequency transients was essentially influenced by the course of the transient current and of the recovery voltage after arc extinction.

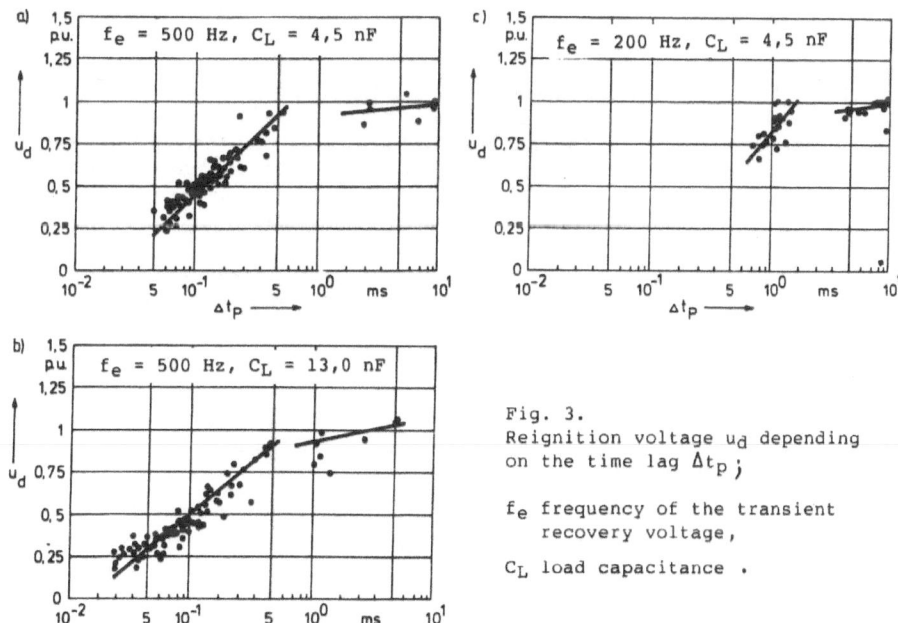

Fig. 3.
Reignition voltage u_d depending on the time lag Δt_p;

f_e frequency of the transient recovery voltage,

C_L load capacitance .

DYNAMIC BREAKDOWN VOLTAGE CHARACTERISTICS

The field test at load currents I_{LO} > 100 mA showed that in case of reignitions during the low-frequency transients the static dielectric strength can no longer be assumed. The prestress leads to a reduced breakdown voltage of the gap.

First fundamental findings could be gained by evaluation of the reignition sequence of a disconnector model with fixed contacts [8]. In these tests the rate of rise of the recovery voltage and the capacitive load current were varied. In Figs. 3 a – 3 c the reignition voltage U_d of the gap is recorded in dependance on the time lag Δt_p between two reignitions. In all diagramms one can find two regions. In the first region the breakdown voltage clearly depends on the prestress. This region can be defined as "dynamic" strength. In the second region the breakdown voltage is not generally influenced by the prestress. This may be characterized as "static" strength. Moreover one can recognize that the restrengthening of the gap depends on the transient frequency, i.e. the rate of rise of the recovery voltage, and on the current passing the arc channel in the gap.

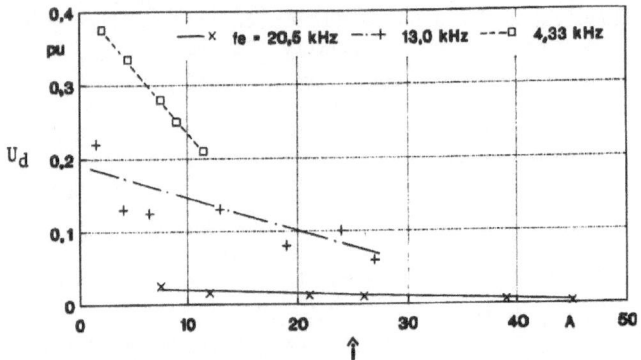

Fig. 4. Reignition voltage u_d depending on the crest value of the low-frequency transient current i and the frequency f_e of the transient recovery voltage.

The behaviour presented above could be investigated more thoroughly by switching tests in the synthetic test circuit. In Fig. 4 the reignitions during the low-frequency transients are evaluated. This Fig. shows the reignition voltage depending on the peak value of the low-frequency transient current and on different frequencies of the transient recovery voltage on the source side. One can recognize the strong dependance on the frequency, i.e. on the rate of rise of the recovery voltage, and the effects of the prestress by the low-frequency transient current. Therefore, particularly at higher amplitudes of the low-frequency transient current, i.e. at higher load currents, too, longer arcing times are to be expected.

ARC QUENCHING PHENOMENA

Apart from the reignition behaviour the arc quenching- and arc extinction process during the medium- and low-frequency transient current could be investigated by means of the synthetic test circuit. Three different arc quenching phenomena were observed (Figs. 5 a – 5 c).

Fig. 5 a shows a current interruption at natural current zero with a nearly constant current slope. In Fig. 5b a destinct current chopping can be detected. Fig. 5 c demonstrates that the current is set into oscillation before current zero subsequently leads to a current interruption. Both of the last mentioned arc quenching processes can be reduced to arc instabilities Which of the three arc quenching phenomena will occur, depends on the current slope of transient current. Additionally one can find an influence of the gas pressure.

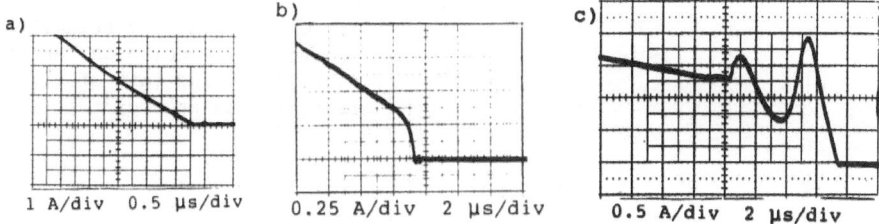

Fig. 5. Arc quenching phenomena;
  a) current interruption at natural current zero
  b) current chopping, c) current chopping with oscillations.

For a gas pressure of 3 bar one can take from Fig. 6 a stable arc behaviour up to a current slope $di/dt \gtrsim 0.4$ A/µs. At a current slope
< 0.4 A/µs current chopping occurs. A reduction of gas pressure to 2 bar
leads to current chopping at a current slope < 0.2 A/µs.
At a current slope < 0.1 A/µs the arc instabilities increase so much
that the current is set into oscillations. The current interruption takes
place in the current zero of the high-frequency current oscillation.

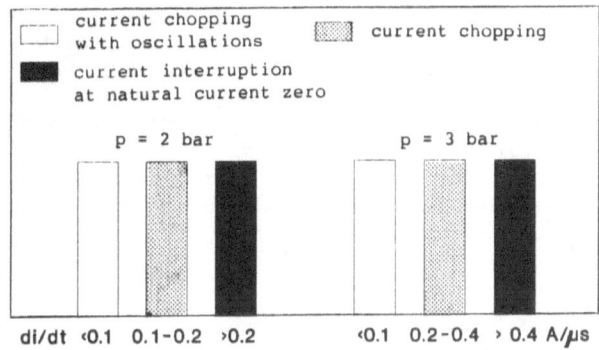

Fig. 6. Arc quenching phenomena depending on the current slope di/dt and
  the gas pressure p.

The arc quenching phenomena presented indicate that after a reignition
during the medium- and low-frequency transients the arc in the gap will not
burn as a straight channel, but produce unstable arc loops which may be
enlarged or short-circuited.

ARC EXTINCTION AND REIGNITION BEHAVIOUR

The arc extinction- and reignition behaviour can be characterized by the
current slope and the rate of rise of the recovery voltage. For this in
Fig. 7 the corresponding values of the first extinction attempt and of the
successful extinction are recorded. One can see that extinction attempts may
occur at a relatively high steepness of current and recovery voltage of
>1 A/µs and >1 kV/µs respectively, however they lead to reignitions.
A successful interruption of the transient current is only possible below a
certain steepness of current and recovery voltage. Based on the measuring
results a limiting curve is plotted in Fig. 7. A successful interruption
only takes place, if the transients have decayed so far that the steepness
of current and voltage is in the region beyond the limiting curve. Depending
on the current slope the arc extinction occurs at the normal current zero or
by current chopping. Current chopping always appears, if the current slope
at the last current zero has decreased so far that a stable arc cannot exist
any more.

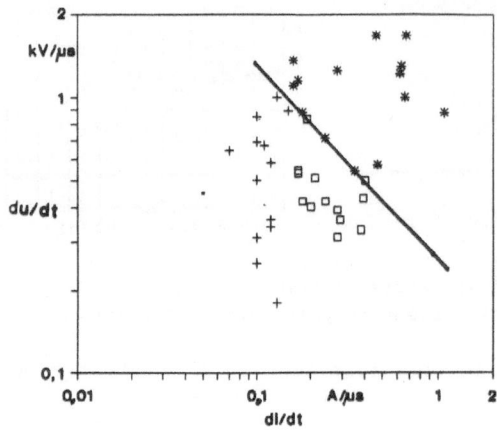

Fig. 7.
Arc extinction and reignition behaviour depending on current slope di/dt and rate of rise of recovery voltage du/dt.

*   arc extinction attempt

□   successful arc extinction

+   current chopping

.   current chopping with oscillations

CONCLUSIONS CONCERNING THE SWITCHING PERFORMANCE OF GIS-DISCONNECTORS

The investigations presented show that the switching performance of GIS-disconnectors under network conditions in the current range > 100 mA is influenced by dynamic processes during the medium- and low-frequency transients. The interruption of the transient current is governed by the current slope and the rate of rise of the recovery voltage. The phenomena determined at the arc extinction indicate arc instabilites. At arcing times of 1 ms and more arc migration and arc looping may occur.

The above mentioned switching case may be of special interest, particularly at higher voltage levels, where arc lengths of some cm appear. Therefore it is useful to clarify by further investigations, how far the phenomena described must be taken into account for the fundamental design of relevant equipment.

REFERENCES

1. Boggs, S.A.; Chu, F.Y.; Fujimoto, N.; Krenicky, A; Plessl, A.; Schlicht,D.: Disconnect switch induced transients and trapped charge in GIS, IEEE Trans. PAS-101, pp. 3593-3600 (1982).
2. Lalot, J.; Sabot, A.; Kieffer, J.; Rowe, S.W.: Preventing earth faulting during switching of disconnectors in GIS including voltage transformer, IEEE Trans. on Power Delivery, Vol. PWRD-1, No. 1, pp. 203-211 (1986).
3. Fujimoto, N.; Boggs, S.A.; Stone, G.C.: Mechanisms and analysis of short risetime GIS transients. CIGRE Symposium 05-87 Vienna, Paper 1010-03 (1987).
4. Boeck, W.; Witzmann, R.: Main influences on the fast transient development in gas insulated substations. 5. International Symposium on High Voltage Engineering, Braunschweig, 24-28 August, Paper 12.01 (1987).
5. Boeck, W.; Witzmann, R.: Critical GIS configurations with respect to VFT. 6. International Symposium on High Voltage Engineering, New Orleans, 28 August - 1 September, Paper 23.10 (1989).
6. König, D.; Imgrund, G.; Meppelink, J.; Schlicht, D.: Performance of GIS-disconnectors under laboratory and network conditions. Proceedings of the Int. Symp. on Gas-Insulated Substations, Toronto, Canada, Pergamon Press, pp. 75 - 85 (1985).
7. König, D.; Imgrund, G.; Neumann, C.; Maatz, K.; Schiweck, L.: Transients during switching of small capacitive currents with GIS-disconnectors in the 110-kV-network and their simulation in the HV-laboratory (in German). Elektrizitätswirtschaft, Vol. 85, pp. 131 - 138 (1985).
8. Imgrund, G.: Investigations of the switching performance and arc extinction behaviour of GIS-disconnectors in the 110-kV-network and in the HV-laboratory (in German). Thesis Technical University of Darmstadt, 1989.

FAST TRANSIENTS IN GAS INSULATED SUBSTATIONS - AN EXPERIMENTAL AND

THEORETICAL EVALUATION

P. F. Coventry and A. Wilson

The National Grid Research and Development Centre
Leatherhead, UK

ABSTRACT

The response of gas insulated substations to fast transients is being investigated in a programme at the National Grid Research and Development Centre, UK. Previously, workers in the area have relied on numerical models based on equivalent circuit or transmission line techniques. In the present work, a greater insight is obtained by using field theory to simulate the electrical stresses that occur under dynamic conditions. In particular, the field structure resulting from the interaction of a propagating pulse with an insulating barrier has been studied.

Instrumentation capable of measuring transient electric fields with nanosecond rise times have been developed and is used to support the theoretical models. These techniques have been evaluated both on a switchgear rig in an HV lab and also on site. High frequency couplers have been fitted at one of the National Grid's substations and measurements have been taken during switching operations.

Experimental and theoretical developments are described.

INTRODUCTION

The generation of fast fronted transients associated with disconnecter operation in gas insulated substations (GIS) is widely acknowledged (Luxa and others, 1988). The implications on substation reliability are being investigated in a programme of research in which experimental and theoretical techniques are applied.

In order to study fast transient phenomena, instrumentation permitting accurate measurement of waveforms is required. Its bandwidth should extend from below 50Hz to above 500MHz. Appropriate sensors were developed during this project.

The positions in a substation at which measurements may be performed are limited by practical constraints such as where sensors may be installed on hatch cover plates. To gain a fuller understanding of the conditions in a GIS, measurements are supported by mathematical models.

Gaseous Dielectrics VI, Edited by L.G. Christophorou and
I. Sauers, Plenum Press, New York, 1991

503

Transient propagation is widely studied using models based on equivalent circuit or transmission line analyses, and good agreement between calculated and measured waveforms may be obtained. In general, the overvoltage levels predicted appear to be within the capability of the GIS insulation and insufficient to explain failure. However, under the fast rates of change of field encountered, electrical stresses cannot be simply determined from an overvoltage level. In particular, where a transient encounters an insulating barrier, the speed of electromagnetic propagation changes on either side of the barrier surface. The resulting field structure is of interest.

In the present work, a computer code has been developed to simulate the electromagnetic fields of fast transients within a section of GIS, and the interaction with insulating barriers has been studied.

EXPERIMENTAL

It was considered that the requirement to record waveforms accurately over a wide frequency range would best be fulfilled using a specially designed capacitive coupler as the measurement sensor. The use of capacitive couplers has been described by other authors (Boggs and Fujimoto, 1984).

In the present work, couplers were evaluated by testing their response to a fast fronted square pulse in a laboratory GIS rig. The waveform was independently determined using a D-dot probe, having known response (Baum, 1986). This led to a succession of improvements in coupler design.

A high capacitance was required to reduce the low frequency cut-off. Initial experiments had shown that polythene terepthalate was a suitable dielectric. The minimum thickness that was found to be practical was 25μm. Coupler diameter was restricted in order to keep the lowest resonant frequency beyond the range of interest. This conflicted with the requirement for high capacitance, and an optimum diameter of 50mm was found. Machining the coupler electrodes reduced the effects of spaces between the dielectric and the electrode surfaces. Where a terminating resistor was used to permit the use of a buffer amplifier, the resistor was positioned to reduce the stray capacitance across it. The length of the connection to the buffer was kept as short as possible to raise the cut off frequency of the low pass filter formed by the capacitance of the connection and the terminating resistor. Couplers of this type gave satisfactory agreement with the D-dot probe in evaluations (Coventry and Wilson, 1990). They are used in laboratory measurements and have been installed on site.

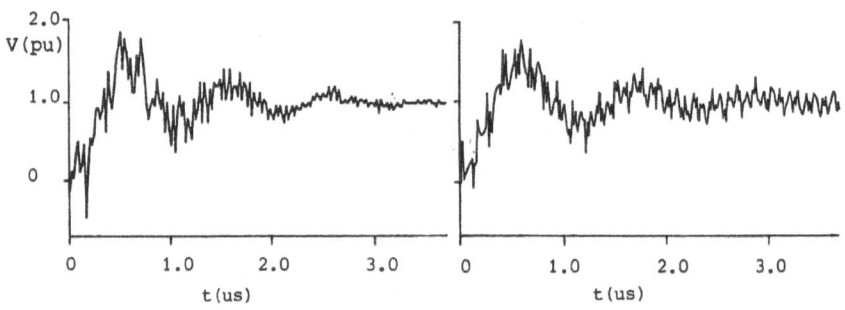

Fig. 1. Measured Waveform.   Fig. 2. Calculated Waveform.

One example of a waveform recorded during measurements on site is shown in Figure 1. The record was obtained on closing a circuit breaker during switching tests on a shunt reactor circuit in a 400 kV substation. The waveform exhibits an overvoltage of 60%, followed by a damped oscillation at about 1MHz which decays within about 3μs. Although there is evidence of further oscillation at higher frequencies, these were beyond the limits of the instrumentation connected during this test.

Part of the substation including the shunt reactor circuit was modelled by an equivalent circuit, using the EMTP code, in order to explain the measured waveform. As the wavelength corresponding to the oscillation of Figure 1 is about 300m, details such as elbows and barriers were neglected. The model produced the waveform shown in Figure 2, which reproduces the measured oscillatory behaviour. The results indicated that the oscillation arose from inductance associated with protection current transformers and the capacitance of the GIS busbar and reactor bushing. Damping was largely due to the source impedance.

THEORETICAL

Calculations described in the preceding paragraph demonstrate the point made in the introduction. Under defined conditions numerical models can lead to accurate explanations of measured waveforms. However for higher frequencies and where modes other than TEM are present they present limitations. Additionally fast transient breakdown is more likely to be produced by high stress rather than high voltages perse. A computer code has been developed to study stresses in a section of GIS under dynamic field conditions, thus supplementing the capability of circuit analysis programs.

A simple coaxial geometry has been considered, which consists of a central conductor, divided into two equal lengths, within an outer conductor. The geometry is illustrated in Figure 3. Symmetry about the axis is assumed, thus a solution in two dimensions is sought. Field equations are solved using a finite difference method.

For the initial field structure, a constant value of potential is assigned to the surface of each conductor, and the axial component of electric field at either end of the section is set to zero. The electric field structure is obtained by solving Laplace's equation.

The region separating the two lengths of central conductor is allowed to conduct, and a transient is generated. The subsequent electromagnetic field structure is obtained by solving Maxwell's equations.

Components of electric field tangential to conductor surfaces are zero for all time.

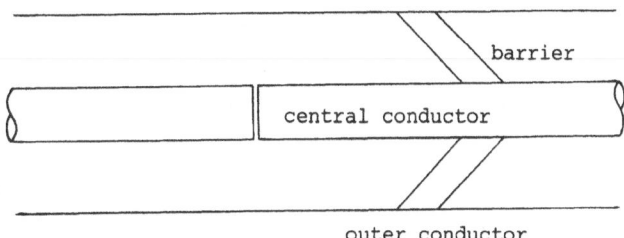

Fig. 3. Idealised GIS Geometry.

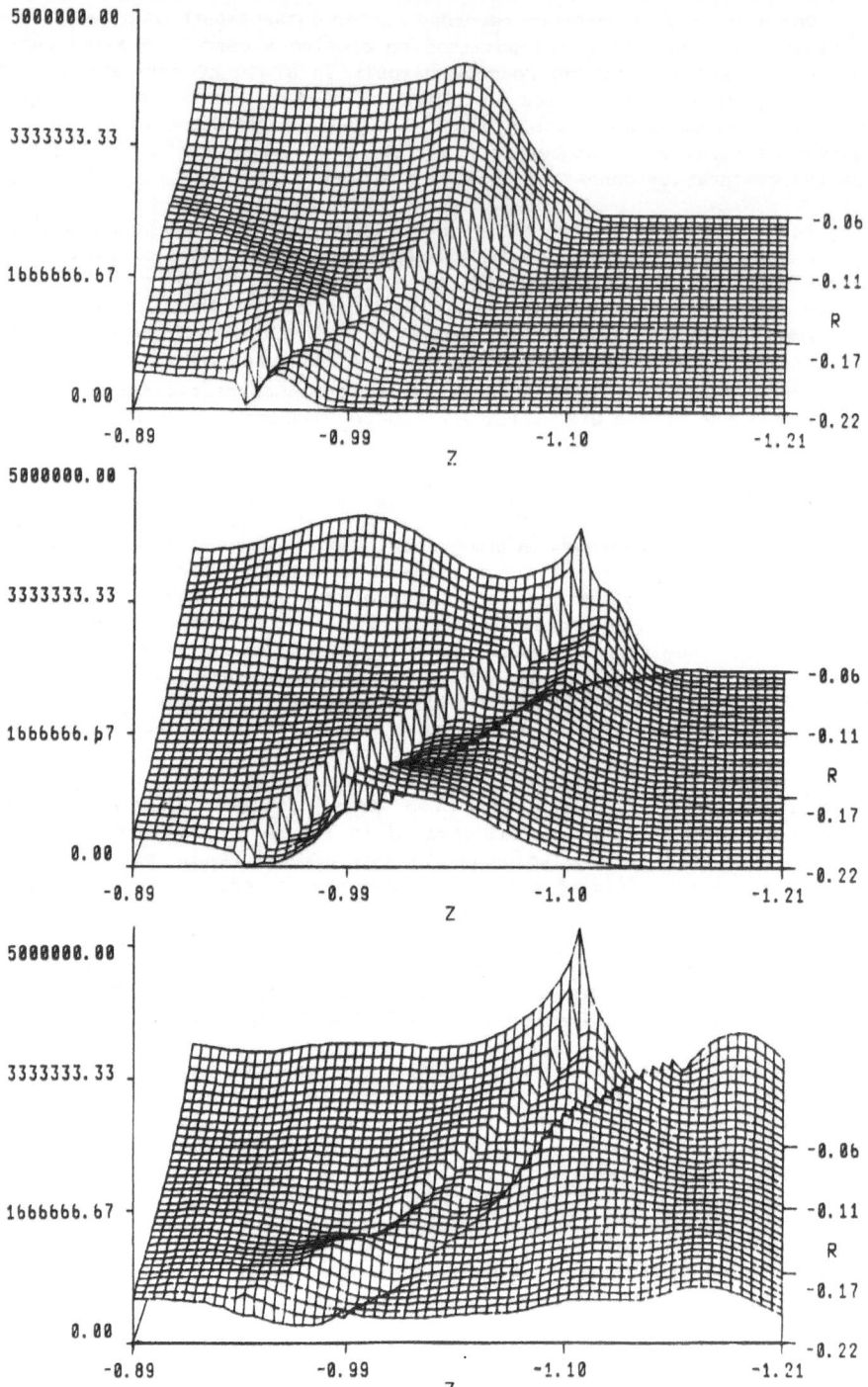

Fig. 4. Electric Field Intensity around Barrier.

Conical insulating barriers may be represented in the model. In the initial solution, the tangential component of electric field and the normal component of electric flux density are continuous across the barrier surface. A modification to take into account the presence of charge on the barrier surface is being implemented at present. In the solution of Maxwell's equations, the speed of electromagnetic propagation relevant to the barrier material is used for nodes inside the barrier.

The solution proceeds in a number of time steps. At each step, rates of change of magnetic and electric fields are calculated. Field values for the next time step are obtained by an iterative numerical integration.

Figure 4 shows a sequence of results at 0.4ns intervals for the interaction of a transient with a simple conical barrier. Results are plotted for a region around the barrier. The horizontal axes represent position and the vertical axis represents electric field intensity. The barrier surface was at 45° to the axis and its relative permittivity was 9. The influence of the barrier on the field structure is clear. The resulting behaviour depends on the steepness of the transient, but, in the cases studied, maximum field intensities were comparable with static conditions.

CONCLUSIONS

Experimental and theoretical techniques have been developed to support an investigation of fast transients in gas insulated substations.

A type of capacitive coupler has been developed for accurate measurement of waveforms covering a wide bandwidth. These couplers are used in the laboratory and have been installed on site.

A computer code has been developed to study stresses in a section of GIS under dynamic conditions. The code has been used to simulate the interaction of transients with insulating barriers. In the configuration tested, maximum field intensities were comparable with static conditions. However, a simplified model was used. Work is continuing to determine whether there are conditions under which the electric field is enhanced. The effects of charge can be modelled, and this is under investigation at present. Alternative, practical geometries will also be evaluated. The region separating the two lengths of central conductor was treated as a time dependent resistance, which gave control over the transient generated. Ideally, this would be replaced by an arc model.

ACKNOWLEDGEMENTS

The authors wish to acknowledge the contributions of their colleagues, D.S.Corben, R.Croxford, A.J.Harrin and R.J.Meats to this work. Advice on the use of D-dot probes was provided by Strathclyde University. The work was carried out at the National Grid Research and Development Centre and is published by permission of the National Grid Company.

REFERENCES

Baum, C.E.(1986). Electromagnetic Sensors and Measurement Techniques in J.E.Thompson and L.H.Luessen (Eds). "Fast Electrical and Optical Measurements", Vol. 1 NATO ASI series, Martinus Nijhoff Publishers, Dordrecht, pp. 73-144.

Boggs, S.A. and Fujimoto, N.(1984). Techniques and Instrumentation for Measurement of Transients in Gas-Insulated Switchgear. IEE Trans.EI, Vol. EI-19, 87-92.

Coventry, P. F. and Wilson, A. (1990). A Study of Fast Transients in Gas
  Insulated Substations. IEEE International Symposium on Electrical
  Insulation, Toronto.
Luxa, G., Kynast, E., Boeck, W., Hiesinger, H., Pigini, A, Bargigia, A,
  Schlicht, S., Wiegart, N. and Ullrich, L. (1988). Recent Research
  Activity on the dielectric performance of SF$_6$, with Special Reference
  to Very Fast Transients. CIGRE 1988, paper 15.06.

DISCUSSION

N. G. TRINH: Can you elaborate on the difference between the D—dot probe used and the capacitive coupler?

P. F. COVENTRY: The D—dot probe is a type of antenna developed for measurement of pulsed fields. It has a known response, which is determined by its geometry. However, since the signal obtained is proportional to the rate of change of electric field, the device is less sensitive to lower frequencies. The signal obtained by using the capacitive coupler is proportional to the electric field. In principle it is useful over a wide range of frequencies. However, good performance at high frequencies requires the effects of stray impedances and resonances to be minimized. To confirm this performance, comparison with waveforms obtained using the D—dot probe was made.

S. W. ROWE: The only way that the dynamic field for H.F. traveling waves could be significantly greater than the normal AC case (in this case) is by reflections off the spacer. In this case how is the spacer modeled?

P. F. COVENTRY: The barrier was modeled as a conical region in space having a relative permittivity of 9.

K. FESER: Very fast transients are a problem if we have sharp protrusions. Do you have practical insight that this is a problem?

P. F. COVENTRY: Thus far there is incomplete understanding of what is going on inside the GIS. I do not believe that the fields under dynamic conditions have been looked at before. Although sharp protrusions can lead to breakdown we cannot always be sure that this is the explanation.

CHARACTERISTICS OF ULTRA LOW LEVEL PARTIAL DISCHARGES IN GIS EPOXY INSULATORS *

N. Fujimoto, S. Rizzetto, J.M. Braun, G.L. Ford and G.C. Stone

Research Division
Ontario Hydro
Toronto, Ontario, Canada

ABSTRACT

The characteristics of partial discharges in full-scale, 138 kV epoxy insulators of the type used in gas-insulated substations were measured on a system with a sensitivity of 0.01 pC. The measurement system uses ultra-wideband concepts and also has the capability of x-ray irradiation to enhance detectability and defect location by the practical elimination of statistical time lag in small voids. The insulators tested included those made with typical production equipment but with known, controlled defects. By design, partial discharge levels were below typical acceptance limits in use today. Evidence suggests that many of these insulators would have failed prematurely despite the high probability of passing normal quality control tests. Over 100 insulators were studied with several defect types including a variety of voids and poor adhesion (delamination) of the epoxy to cast-in metal electrodes. Despite the controlled nature of the various defect types, a large variation in the characteristics of low level partial discharge activity was observed on specimens with nominal defects. In many cases, the observed patterns were unstable and changed as a function of applied voltage and time. Characteristics which may be related to the time-dependent properties of the void gas and the void wall surface, were observed.

INTRODUCTION

The detection of partial discharges (PD) is a commonly used technique for diagnosing and evaluating the condition of insulation systems. For solid epoxy spacer insulators found in gas-insulated switchgear (GIS), PD detection is often used as a quality control measure to detect those defective insulators which could ultimately lead to premature failure. The efficacy of these screening tests depends on the ability to detect PD with high sensitivity, especially for insulators of higher (300 kV and above) voltage classes. The detectability of defects by partial discharge measurement becomes increasingly more difficult for insulators of larger physical size [1].

The issue of PD detectability is being addressed within the context of investigations of the long-term reliability of GIS epoxy insulators in a research project co-funded by the Electric Power Research Institute. The project has undertaken to study the ageing characteristics of insulators with minor defects, which normally would not be detectable with PD measurement equipment in common use in production facilities, and to explore improvements to allow the detection of these defects [1]. In order to enhance the detectability of small void defects, a system which simultaneously applies high voltage and

* Supported in part by the Electric Power Research Institute under RP2669-1.

Gaseous Dielectrics VI, Edited by L.G. Christophorou and
I. Sauers, Plenum Press, New York, 1991

ionizing (x-ray) radiation during the electrical measurement of PD has been developed. The X-ray Induced Partial discharge Detection (XIPD) system uses x-rays to reduce the statistical time lag for discharges occurring within small voids, allowing PD to occur more consistently, at lower applied voltages and greatly enhancing the detectability of such small voids. By scanning the insulator with a narrow x-ray beam, the XIPD system also has the ability to locate defects (PD sources) within the insulator bulk.

The XIPD system uses ultra-wideband (UWB) techniques and has a detection sensitivity limit of about 0.01 pC in a near-factory environment. This paper provides a synopsis of some of the observations obtained with this system on over 100 individual insulators. The results suggest that a great deal of PD activity is often present at detection levels less than 0.5 pC and that these signals sometimes form specific patterns.

MEASUREMENT CONFIGURATION

Measurements were performed on an experimental post insulator (138 kV class) designed specifically for the purposes of the overall research project. The insulator consists of two mushroom-shaped electrodes spaced 50 mm apart and cast into an epoxy matrix. Clear, unfilled epoxies were used in addition to practical formulations with mineral fillers. In most cases, the insulators tested were manufactured with intentional defects for the purposes of investigation. These defects included several types of voids and minor delaminations at the epoxy/electrode interface. PD measurements were performed by detecting the current flow in the ground lead of the insulator. The PD signals were simultaneously displayed on a 1 GHz real-time oscilloscope and measured by a custom-fabricated, high frequency multi-channel analyzer (MCA).

PD Measurement Instrumentation

The XIPD system is based on UWB detection concepts throughout the measurement chain for optimal signal-to-noise ratio [2]. The custom-fabricated MCA is capable of responding to PD pulses with risetimes as short as 1 ns, durations as short as 1-2 ns and has a sensitivity of about 0.2 mV, which for the one type of PD pulse shape, corresponds to an apparent charge of about 0.01 pC. The MCA is configured to acquire accumulated counts of PD pulses resolved by magnitude, polarity and time of occurrence with respect to the phase of the applied high voltage. The system is configured for up to 256 phase windows and 16 logarithmically spaced magnitude channels covering 3 orders of magnitude for each polarity. The relatively low number of magnitude channels was a design compromise required for adequate response to the ns-wide pulses, necessary to achieve a high signal-to-noise ratio. The instrument can accept PD pulses with up to 50 kHz repetition rate, but since PD usually occurs at much lower rates, measurements were usually accumulated for a period of 5 seconds.

The display of PD counts in a phase-resolved manner has, in recent years, become popular as a means of evaluating the PD and external noise characteristics (for example, [3]. This form of data analysis was first implemented about fifteen years ago [4], but modern computer technology has facilitated its recent popularity. One of the principal attractions of the phase-resolved measurement is its ability to assist the operator (with expert knowledge) in segregating noise components of the signal from actual PD signals. In the present case, further noise discrimination is made possible by observing the pulse shape on the oscilloscope (as a result of filtering on the high voltage connection, external noise signals are usually highly oscillatory and not unipolar pulses, as described below). In addition, observing changes in signal behaviour when using a focussed x-ray beam also assists in identifying true PD signals. At a certain voltage range, PD can be modulated by the switching the x-ray beam on/off.

The data accumulated in the MCA are downloaded to a host computer which processes and displays the data in an appropriate form. Figure 1 shows a typical measurement of void discharge activity displayed several ways. This particular example shows a classic PD characteristic pattern with PD pulses occurring primarily in the first and third quadrants of the phase of the applied AC voltage. The pattern also clearly indicates a minimum PD level which is presumably linked to the minimum conditions for void dis-

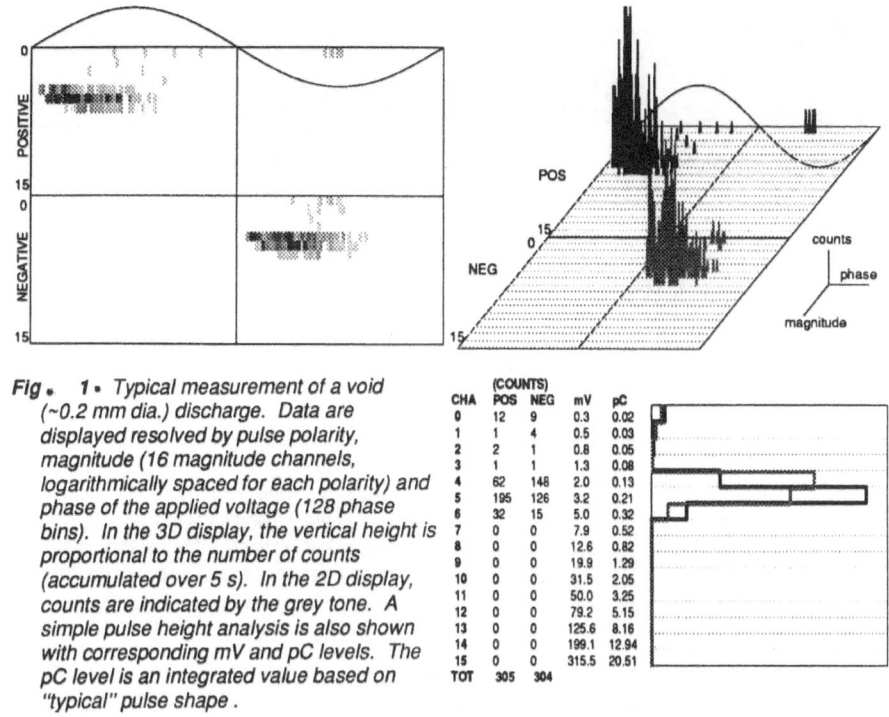

Fig. 1 · *Typical measurement of a void (~0.2 mm dia.) discharge. Data are displayed resolved by pulse polarity, magnitude (16 magnitude channels, logarithmically spaced for each polarity) and phase of the applied voltage (128 phase bins). In the 3D display, the vertical height is proportional to the number of counts (accumulated over 5 s). In the 2D display, counts are indicated by the grey tone. A simple pulse height analysis is also shown with corresponding mV and pC levels. The pC level is an integrated value based on "typical" pulse shape .*

CHA	(COUNTS) POS	NEG	mV	pC
0	12	9	0.3	0.02
1	1	4	0.5	0.03
2	2	1	0.8	0.05
3	1	1	1.3	0.08
4	62	148	2.0	0.13
5	195	126	3.2	0.21
6	32	15	5.0	0.32
7	0	0	7.9	0.52
8	0	0	12.6	0.82
9	0	0	19.9	1.29
10	0	0	31.5	2.05
11	0	0	50.0	3.25
12	0	0	79.2	5.15
13	0	0	125.6	8.16
14	0	0	199.1	12.94
15	0	0	315.5	20.51
TOT	305	304		

charge, which is in turn linked to size of the void and the void gas characteristics. The measurement in the figure was taken during x-ray application, which typically increases the total number of counts and compresses the variation in pulse magnitude and phase, leaving a more tightly distributed pattern clustered near the minimum PD level. These observations are, in general, consistent with the reduction of the statistical time lag for the void discharges. In many cases, measurement of PD at lower voltages was not possible without x-ray radiation.

PARTIAL DISCHARGE CHARACTERISTICS

The present measurement setup allows characterization of partial discharges in two ways. Since UWB techniques are used, direct measurement with an oscilloscope allows reasonable reproduction of the PD pulse waveshape. The accumulated counts recorded with the MCA allows visual interpretation of characteristic PD patterns when accumulated PD counts are displayed resolved by magnitude (pulse height) and phase (for example, figure 1).

PD Pulse Waveshapes

PD pulse waveshapes are generally known to have fast risetimes and short durations. PD pulses with sub-nanosecond risetimes and durations of 1-5 nanoseconds are typically recorded [1,2] and have been responsible for the development of the UWB measurement techniques. However, recent measurements of partial discharges from voids in small laboratory samples have indicated the existence of at least two distinct classes of PD pulse shapes. The first ("Type 1") is the classic fast, narrow pulse with sub-ns to ns risetimes and ns durations as discussed above. The second ("Type 2") PD pulse has a much broader shape (10-20 ns pulsewidth) and longer risetimes (several ns) and usually (but not always) a lower magnitude. Typical PD pulse shapes of these two categories are shown in figures 2 and 3. Similar pulse shapes were also measured on full-scale spacer insulators in the XIPD system, although stray inductances in the measurement circuit caused a slight degradation in risetime and an oscillatory response for the type 1 pulse.

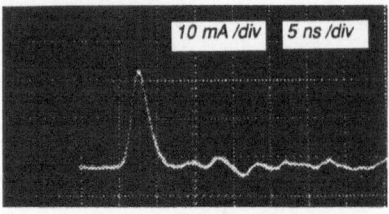

Fig. 2. *Typical "Type 1" pulse shape as measured on a laboratory sample with a single void in unfilled epoxy.*

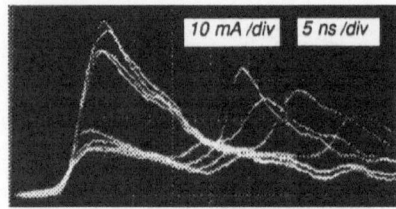

Fig. 3. *Multiple exposure oscillograph of a type 2 pulse showing progressive time indicating apparent separation of "component" discharges. Measurements were made with a void in filled epoxy in a small laboratory sample.*

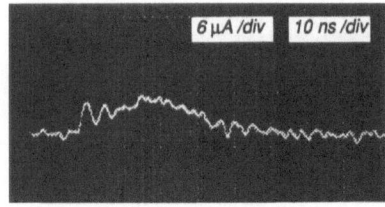

Fig. 4. *Measurement suggesting the apparent occurrence of a type 1 pulse immediately followed by a type 2 pulse. This example is from an XIPD sample with intentional poor adhesion between the HV electrode and epoxy.*

In many cases, overall PD pulse shapes suggest composite pulses possibly indicative of multiple discharge events within the void volume. Figure 3, for example, shows a type 2 PD pulse measured for a small laboratory sample with a single 4 mm dia. void in filled epoxy. As the applied voltage was increased, the PD pulse began to decompose into a pulse with two distinct elements resulting in the unusual shape indicated in the figure. In other cases, the measured waveform has appeared as a combination of the two types of pulses. Figure 4, for example, apparently shows a fast, type 1 pulse which is almost immediately followed by what could be a slower type 2 pulse. In this case, this type of pulse was observed both with and without x-rays and the relative magnitudes of these two components were erratic.

Other general observations include:

1)	Many of the tested insulators exhibited both types of PD pulse shapes. With the present measurement apparatus, the existence of a causal link between the two types of pulses cannot be determined. Although the case of figure 4 seems to indicate that one type of pulse "triggers" another, the measurement apparatus is not configured to determine if this idea is applicable to isolated PD pulses.

2)	As observed on the oscilloscope, the application of x-rays will often (but not always) change the relative distribution of the two types of PD pulses in favour of the type 2 pulse.

3)	Some insulators will exhibit only one type of PD pulse, either type 1 or type 2.

The two distinct pulse shapes indicate that at least two different discharge mechanisms are possible within voids. The physical mechanisms are probably related to the dynamics of charge movement and distributions on the void wall. In this manner, the discharge mechanism(s) is probably related to the material properties which affect the characteristics of the void wall, including epoxy formulation, filler type, void wall topography and void formation (void size and gas composition). Of the two broad classifications, one pulse type could be related to a gaseous discharge within the center of the void while the other type is a discharge involving the inside surface of the void wall, although this hypothesis has not been substantiated.

Phase-Resolved PD Measurements

The display of accumulated counts of PD pulses resolved by polarity, magnitude (pulse height) and phase of the applied AC voltage will typically reveal distinct patterns or "signatures". Unfortunately, present knowledge does not allow, in all cases, that the specific patterns are correlated with known physical processes or the defect type. Similar patterns are often (but not consistently) observed for a number of different defects, including several types of voids produced by different processes. One possibility is that the process of introducing controlled defects into the insulator has created other, inadvertent defects of a similar nature. Another possibility is that the observed patterns are related to fundamental discharge processes, which could be common, to some extent, for a wide range of defect types. The measured patterns are often unstable and change as a function of time and voltage making it difficult to correlate them with other observations. However, certain features within the patterns are frequently observed and are illustrated in figures 1, 5 and 6.

The pattern of figure 1 is the "classic" PD pattern of roughly equal populations of positive and negative counts which is often observed for many types of defects (including voids) over a wide range of voltages. The pattern of figure 5 is usually associated with discharges involving metal electrodes. In this case, these patterns were primarily observed for defects involving a delamination at the epoxy/high voltage electrode interface. The figure shows that the negative population is distinctive, extending over a large range of magnitudes and often dominates the positive counts. The bias to negative polarity counts and the observation that x-rays apparently have very little effect on this type of pattern suggest a strong role of the metal electrode, possibly through an electron emission mechanism.

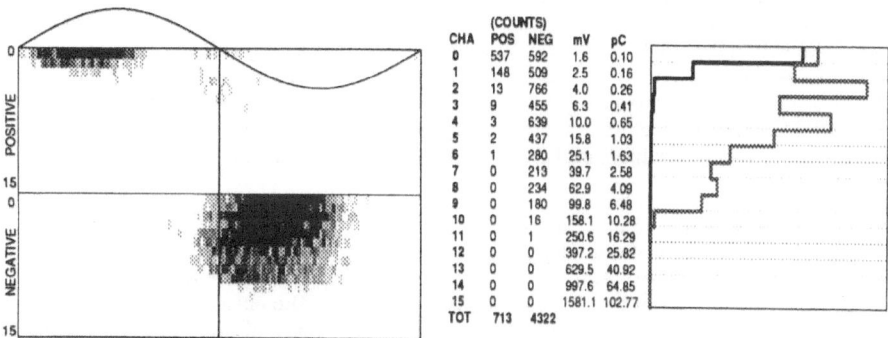

Fig. 5. *Example of commonly observed PD measurement pattern showing a strong polarity effect with the negative pulse height distribution extending over a wide range.*

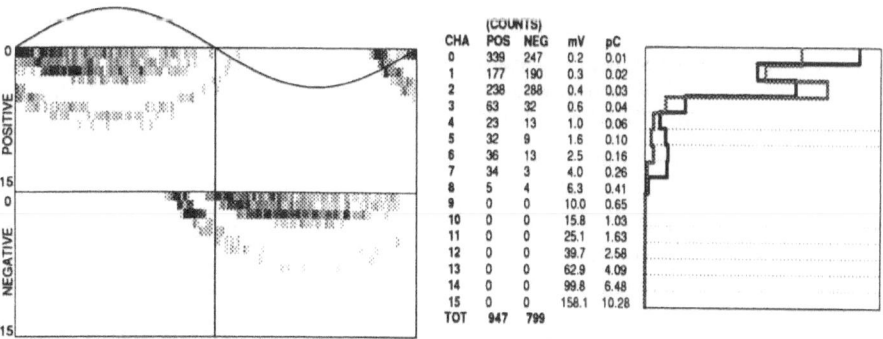

Fig. 6. *Example of commonly observed PD measurement pattern showing a well-defined curved pattern in addition to a second population of counts. These characteristics have been observed for a wide range of insulator and defect types.*

Figure 6 shows a pattern observed primarily for void discharges, although these have also been observed for other defect types (which possibly indicates the presence of a similar void in addition to the primary defect). The unique feature here is the curved pattern which can be seen easily in the figure, but is not evident on the pulse height analysis. This sample also shows two distinct populations of PD counts, the second of which appears similar to that of figure 1. However, the curved feature has on occasion been observed in isolation. In addition, the two components of the pattern will sometimes merge and blend into each other. These developments, in many cases, appear to be voltage, time and x-ray dependent. Possible physical processes which might give rise to such a pattern are not known. However, simultaneous observation of the associated PD pulse shapes, and the sensitivity to x-ray radiation indicate that the observed pattern is not an artifact of the measurement circuit.

Although three distinct patterns are shown in the figures, measurements are often made which display elements of all three patterns. Such observations make interpretation difficult, but may suggest, in some cases, the presence of multiple defects or possibly multiple discharge processes.

CONCLUSIONS

Measurements of ultra-low-level partial discharges have been made on full-scale epoxy insulators with a system using ultra-wideband measurement techniques and x-rays for partial discharge stimulation. The measurements indicate the existence of at least two general classifications of PD pulse shapes, which might be indicative of different physical mechanisms of discharges. In particular, the pulse-counting techniques for PD analysis and its display resolved by magnitude, pulse polarity and phase of the AC high voltage yield a great deal of information which, potentially, can be correlated with fundamental processes of the discharge. This information is contained in patterns displayed on conventional PD detector ellipses [5], but is much easier to interpret as a result of computer processing. Several common patterns of low magnitude PD have been observed, but there is insufficient evidence to conclusively correlate these to particular defect types. Many measurements exhibit elements of more than one of the patterns identified and in many cases, the patterns are unstable and change inexplicably with voltage, time and x-ray application.

The analysis of low magnitude partial discharges using UWB techniques and computer data processing is useful in displaying low level PD characteristics of defects which would normally pass quality control tests. However, substantially more research work is required before practical significance can be placed on any of the observed patterns. Further study is also necessary to assess the importance and hazard of these minor defects and the low level PD for the long term reliability of epoxy insulators. These issues are being addressed in the context of the overall research project.

REFERENCES

1. Fujimoto, N., G.C. Stone, H.G. Sedding and S. Rizzetto. Improved Partial Discharge Detection Methods for Epoxy Spacers in Gas-Insulated Switchgear. Sixth ISH, New Orleans, LA, USA (1989). Paper 15.04.

2. Boggs, S.A. and G.C. Stone. Fundamental limitations in the measurement of corona and partial discharge. IEEE Trans. on EI, Vol. EI-17, No.2, (April 1982).

3. Fruth, B., L. Niemeyer, M. Hässig, J. Fuhr and Th. Dunz. Phase Resolved Partial Discharge Measurements and Computer Aided Partial Discharge Analysis Performed on Different High Voltage Apparatus . Sixth ISH, New Orleans, LA, USA (1989). Paper 15.03 ·

4. Kelen, A. The Functional Testing of HV Generator Stator Insulation. 1976 Session of CIGRE, paper 15-03.

5. CIGRE Working Group 21-03. Recognition of Discharges. Electra, No. 11 (1969).

DISCUSSION

M. GOLDMAN: We made experiments in artificial voids of a few micron thickness and we found the same regions as you do. The physical mechanism between the two regions is the internal conductivity and/or wettability of the surface of the void.

N. FUJIMOTO: The voids we have studied range from <100 μm to a few mm in diameter. We intend further investigations into the void discharge mechanisms, and we agree that the void wall surface characteristics will play an important role in the discharge mechanism. Our preliminary results (mm–size voids) seem to indicate that the "fast" discharges correspond to a gas gap discharge in the center of the void and the "slower" discharges correspond to surface discharges on the void wall.

D. KONIG: Your interesting studies deal with the aging of epoxy resin post insulators made by different manufacturers, when stressed by high electrical field strengths. I feel, based on many years of experience, being responsible for the manufacturing of some hundred tons/year of epoxy resin components, the first step is to define the test specimen made by manufacturer A, B or C, without going more in–depth and without studies on the resulting internal structures. (1) Did you consider whether the manufacturing process as well as the kind of epoxy resin system, including the filler, and the measures of quality control during manufacturing have an essential influence on the internal structure of your test specimen? (2) How did you take this into acccount?

N. FUJIMOTO: (1) The intent of the overall project was to study insulators with minor defects, that is, defects which are realistic and credible but are just below the threshold of detection using standard, commercial partial discharge measurement methods. Such defects are believed to be responsible for many of the bulk insulation failures experienced in real equipment. While specific details vary from one manufacturer to another, each participating manufacturer has developed a procedure specific to his own manufacturing process and materials which generates the "minor" defect of interest. By this procedure, differences in quality control and, to some extent, materials are taken into account. (2) In general, the basic materials used in the manufacture of epoxy insulators do not vary a great deal from manufacturer to manufacturer and, as a result, we do not expect their electrical characteristics to vary significantly as well. We note that most manufacturers use very similar design stresses for their insulating components. The aging behaviour of epoxy insulators used at these stress levels depends more on the existence and nature of the minor defects described above, which is controlled by the quality control procedure. We would expect that the specific material properties might have some secondary influences but should not affect the validity of the results.

B. F. HAMPTON: (1) How was the discharge magnitude calibrated? (2) Does it refer to the apparent charge at the sample terminals? (3) If so, is it possible to estimate the actual charge in the void?

N. FUJIMOTO: The reported levels of partial discharges refer to the apparent charge, which is the signal available at the insulator electrodes. No specific calibration is performed. In its place, we measure the actual current waveform of the discharge ("apparent charge") and perform a simple integration to determine the charge level. For routine measurements, the charge is estimated by multiplying the peak voltage of the measured pulse with a constant factor, based on an assumed pulse shape. Unfortunately, this introduces some error if different pulse waveshapes occur. To estimate the actual charge in the void, I refer you to a formulation developed by Pedersen [G. C. Crichton, P. W. Karlson and A. Pedersen, IEEE Trans. on EI, 24, 335 (1989)] which relates void discharge to apparent charge, based on electrostatic field theory.

S. W. ROWE: For very small voids, the statistical time between creation of primary electrons can be hours, days or even much more. Furthermore, quenching could mean that the partial discharge dies out rapidly and only repeats once a day or week; also, treeing requires a minimum energy to enable atomic bond breaking before propagation can start. If the above is true, do you think that this type of micro—void can give a significant contribution to spacer breakdown?

N. FUJIMOTO: Unfortunately the void discharge to electrical tree transition is not very well understood. Probably there will be some threshold of void size (as a function of material charateristics and stress level) below which no significant aging effects are expected. We do not know what the threshold is at present, but this is the reason why we are conducting research in this area. We do know, however, that some "micro—voids" do cause premature failure and that some voids which may only discharge very sporadically at first may at some point in time suddenly "switch" into a mode of continuous discharge.

CHAPTER 12: RELIABILITY OF GIS/FAILURE MECHANISMS

FAILURE MECHANISMS OF GAS INSULATION AND AT GAS-INSULATOR INTERFACE:

THEIR INFLUENCE ON GAS-INSULATED EQUIPMENT

N. Giao Trinh

Institut de recherche d'Hydro-Québec (IREQ)
Varennes, Québec, CANADA J3X 1S1

ABSTRACT

This paper reviews the main mechanisms of failure in gas insulation and at the gas-insulator interface, and discusses parameters, such as voltage, field distribution, surface conditions, humidity, etc. which have a significant influence on these mechanisms. Emphasis will be on parameters which have a direct bearing on the design, operating conditions and maintenance of gas-insulated equipment.

INTRODUCTION

Since the introduction of SF_6-insulated equipment in the middle of the century, significant improvement has been achieved during the last two decades. SF_6-insulated equipment is now used either in gas-insulated substations (GIS) or in short links (GIL) between sections of high-voltage transmission systems up to the 800 kV AC and 500 kV DC. The overall experience with GIS has been good, as indicated by a CIGRE survey on GIS failure statistics[1] in 1982. Based on service experience with 3,100 breaker-bays and some 19,000 bay-years, a mean time between failure of more than 1,000 years was estimated for arcing faults in GIS over the voltage range between 120 and 420 kV AC. The failures involved breakdowns both of the gas gap and at the insulators. The causes of failure are numerous but often may be traced back to handling, installation and maintenance, quality control and the design of the failed components.

Research in the field of gas discharges has been most helpful in improving the performance of GIS components, first by explaining the failure mechanisms, second by identifying the main parameters affecting the breakdown processes, and third by developing adequate means for predicting and, hence, optimizing the dielectric performance of these components. This paper reviews the main failure mechanisms in SF_6 insulation and at the SF_6/insulator interface with an emphasis on parameters with a direct bearing on the design, operating conditions and maintenance of gas-insulated equipment.

PHYSICAL MECHANISMS OF BREAKDOWN IN THE GASEOUS DIELECTRIC

When SF_6 is the only insulating material separating two electrodes, the failure of the gap results from ionization of the SF_6 gas in an avalanche process which eventually builds up a conducting channel between the electrodes and causes breakdown. The physical mechanism leading to the gap breakdown is complex and depends on a number of parameters, including the physical properties of the gas, the gap configuration and the applied voltage stress.

Basic properties of SF_6

A considerable amount of published literature supports the fact that SF_6 not only is a superior insulating gas, with excellent thermal and arc quenching properties, it remains in the gaseous state over the practical range of temperatures, and is the gas least harmful to health and the environment. Recent efforts to find a better insulating gas[2] indicate that, when all aspects are taken into account, SF_6 remains the best candidate for gas-insulated power equipment.

From the dielectric point of view, SF_6 is a strongly electronegative gas. The single parameter characterizing its dielectric properties is the *effective ionization coefficient* $\alpha*$ expressed as[3]

$$\alpha* = \alpha - \eta = p \ C \ [(E/p) - (E/p)cr] \tag{1}$$

where α and η are respectively the ionization and attachment coefficients of the gas, p is the gas pressure (in atm.), C a constant, $C = 27 \ kV^{-1}$, and $(E/p)cr$ the limiting field normalized with respect to the gas pressure, $(E/p)cr = 89 \ kV/cm \ atm$.

Comparing the ionization potential of air and SF_6, the latter's excellent dielectric properties are mainly related to its strong *electron attachment*, which increases the limiting field to about three times that of atmospheric air. The large value of C in Eq. (1) implies a high rate of change in the effective ionization coefficient $\alpha*$ at field intensities exceeding the limiting field. The ionization activities, once initiated, are more intense, leading to rapid flashover of the discharge gap.

Breakdown in uniform fields

The basic process for the development of discharges in a gas, known as the *electron avalanche*, is field-induced ionization by electron collisions, whereby free electrons can accumulate energy from the applied field and ionize the gas molecules as they move across the gap. This causes an exponential multiplication of the number of free electrons according to

$$n(x) = \exp[\int (\alpha - \eta) \ dx] = \exp(\int \alpha* \ dx) \tag{2}$$

At low field intensities, especially when the conditions for generation of free electrons are not adequate, electron avalanches are *single events* which end naturally when they reach the low-field region or the anode.

Due to the action of secondary electrons, mainly photo-electrons produced in the gas or at the cathode, new avalanches are initiated and prolong the ionization activities in the gap. At field intensities exceeding the limiting field, these successive generations of electron avalanches cause a rapid transition into a *self-sustained streamer discharge*, which can be initiated from the cathode or in mid-gap and then propagates toward the electrodes to breakdown the gap.

The strong electronegative property of SF_6 rapidly transforms the free electrons produced in the avalanches into negative ions. Streamer propagation, as shown by Wiegart et al.[4], relies on intense ionization

518

activities in the strong space-charge field in front of the avalanche head, causing the charge centre to move forward. Such a propagation mechanism requires a field in the streamer trail comparable to the limiting field and differs considerably from the mechanism in air, where the streamer can propagate in near-zero fields.

Townsend criterion for breakdown in uniform fields: Assuming the secondary electrons are produced at the cathode, the following expression for the discharge current may be derived for a uniform-field gap[5]

$$\frac{i}{i0} = \frac{[\alpha/(\alpha-\eta)\ \exp[(\alpha-\eta)d]\ -\ \eta/(\alpha-\eta)}{1\ -\ \{\gamma\alpha/(\alpha-\eta)\}\{\exp[(\alpha-\eta)d]\ -\ 1\}} \tag{3}$$

where i0 is the current corresponding to the rate of free electrons produced at the cathode by an external source, and γ is a coefficient representing of the cathode's efficiency in producing secondary electrons.

A critical field intensity may be reached such that the denominator of Eq. (3) equals zero. The discharge current becomes undefined, implying the establishment of a self-sustained discharge, which in uniform fields leads rapidly to breakdown of the gap. The condition for breakdown thus defined is known as the *Townsend criterion*.

$$\exp[(\alpha-\eta)d] = \frac{1}{\gamma}\ (1 + \gamma - \frac{\eta}{\alpha}) \tag{4}$$

In a uniform field gap however, the breakdown field determined by the Townsend criterion is quite close to the limiting field, so that the breakdown conditions can be simplified to[5]

$$\alpha-\eta = 0 \tag{5}$$

which implies that breakdown will take place as soon as the field intensity in the gap allows the ionization (coefficient α) to exceed the attachment (coefficient η).

Breakdown in moderately divergent fields

In nonuniform fields, the breakdown usually follows a streamer discharge, initiated from the highly stressed electrode and propagating across the gap. Combined with the strong electronegative property of SF_6, a nonuniform field facilitates the accumulation of ion space charges in the low field regions, which modifies the local electric field and delays the breakdown process. Streamer discharges may then develop over a certain voltage range before breakdown occurs. The Townsend criterion modified for a nonuniform field, can be used to evaluate the *streamer onset voltage*.

$$\exp[\int_{0}^{z0} (\alpha-\eta)\ dr] = \frac{1}{\gamma}\ (1 + \gamma - \frac{\eta}{\alpha}) \tag{6}$$

Where z0 is the distance from the highly stressed electrode where the field intensity equals the limiting field.

In moderately divergent fields, breakdown usually occurs when the streamers successfully bridge the gap. The range of voltages for stable development of streamers is small, so the breakdown voltage is only slightly higher than the onset voltage of streamer discharges. Equation 6 is therefore a good estimate of the breakdown voltage.

<u>Streamer breakdown criterion</u>: In the streamer breakdown mechanism, breakdown of the gap is assumed to occur when the number of free electrons in an avalanche reaches a critical value Nc. The condition for the onset of breakdown was shown by Pedersen[6] to be

$$
Nc = \frac{N(z1)\ \exp[\int_0^{z0} (\alpha-\eta)\ dr]}{\exp[\int_0^{z1} (\alpha-\eta)\ dr]}
\tag{7}
$$

z1 is the distance from the cathode where the number N(z1) of free electrons in the avalanche is sufficient to assure an exponential growth according to Eq. (1). The above expression can be rewritten as

$$
\exp[\int_0^{z0} (\alpha-\eta)\ dr] = \frac{Nc}{kz}
\tag{8}
$$

where kz is the ratio of the actual number of free electrons in the avalanche N(z1) to the average avalanche size at z1.

The formally identical expressions of Eqs. (6) and (8) give rise to the familiar interpretation that the initial electron avalanche must be sufficiently large to ensure transition of the avalanche process into Townsend or streamer breakdown of the gap. Strictly speaking, the critical avalanche sizes defined by the right-hand sides in Eqs. (6) and (8) are not identical. However, experimental results indicate that the breakdown voltage in uniform and moderately divergent fields is relatively insensitive to the critical avalanche size, which is usually taken as $Nc = 10^8$. The breakdown criterion [*] may then be simplified to

$$
\int_0^{z0} (\alpha-\eta)\ dr = K = \ln(10^8) = 18.42
\tag{9}
$$

Relatively simple expressions can be derived for the breakdown field in practical SF_6-insulated gaps[3] of moderately divergent field

$$
R\ [\sqrt{Emax} - \sqrt{p}\ (E/p)cr]^2 = K/C
\tag{10}
$$

where Emax is the maximum filed at the electrode surface and R is the mean curvature at the surface of the highly stressed electrode surface.

For a coaxial conductor[8] of radii a and b, the following expressions can be derived for the critical avalanche length rc

$$
Ec\ rc\ \ln(rc/a) - Ec\ (rc - a) = K/C
\tag{11}
$$

and the breakdown voltage Uc

$$
Uc = Ec\ rc\ \ln(b/a)
\tag{12}
$$

where $Ec = p\ (E/p)cr$ is the limiting field.

(*) Difficulties in obtaining reliable data for the effective ionization coefficient α^*, combined with the unknown value of the critical avalanche size, have recently prompted the use of Paschen curves data in the application of the streamer criterion to breakdown in nonuniform fields in strongly electronegative gases[7]. The simplicity of Eq. (9) makes this form of the streamer criterion much more practical.

Figure 1 shows variations in the breakdown voltage of coaxial conductors in SF_6 (Ref. 9). It can be seen that over the range of pressures of interest, and for ideal electrode surface conditions, the calculated (Eq. 11) and measured breakdown voltages compare reasonably well.

Leader breakdown in nonuniform field

In a nonuniform field gap, the region of field intensities exceeding the limiting field is restricted to a small volume surrounding the highly stressed electrode. Both the nonuniform field distribution and the accumulation of ion space-charges in the low-field region of the gap tend to favor the development of streamers as a stable discharge mode at the highly stressed electrode[9]. The breakdown of the gap occurs as the result of the development of a *leader discharge*[10].

The phenomenon in SF_6 is very similar to that occurring in the breakdown of long air gaps, which has been extensively investigated[11] under switching-impulse voltages. The leader discharge was initiated from the highly stressed electrode and propagates across the gap in steps. At each step, a *streamer discharge* or *leader corona* is observed, initiated from the tip of the leader channel. The flow of the discharge current heats up the *streamer stem* and transforms it into a new section of the leader channel, thus extending it further into the gap.

As the voltage varies, leader development may be interrupted, and reignited several times before a *continuous leader* successfully develops in the gap. As the continuous leader approaches the ground plane, a point will be reached where the streamer discharge manages to bridge the remaining gap. The streamer-leader transition occurs rapidly through a process known as the *final jump*, and completes the breakdown of the gap.

Condition for leader breakdown in nonuniform field: The complex phenomenon at breakdown excludes practically all possibility of deriving a simple breakdown criterion for nonuniform field gaps. One approach consists in defining the most critical condition for the discharge gap and deriving empirical formulas relating the breakdown voltage (or field) to relevant parameters. The drawback of this approach resides in the large amount of experimental data needed for the derivation of the empirical formulas. Furthermore, empirical formulas are not reliable for extrapolating to conditions not covered by the experimental data, so that expensive tests on full-size models are required.

The preferred approach to predicting breakdown of nonuniform field gaps relies on the definition of a *breakdown model* which simulates the physical processes active at the instant of breakdown. For nonuniform field gaps, the most critical conditions are those prevailing during the development of the *continuous leader*. At the position of the *final jump*, breakdown will occur when the leader tip potential is adequate to cause *streamer breakdown* of the remaining gap. The breakdown voltage is then the sum of the continuous-leader inception voltage Ucl and the leader voltage drop ΔUl.

$$U = Ucl + \Delta Ul \tag{13}$$

For long air gaps, by calculating the field at the ground plane and equating it to the limiting field in air, Rizk[12] has derived the following expression for the U50% breakdown voltage as a function of the gap length.

$$U50\% = \frac{1830 + 59 \ d}{1 + \frac{3.89}{d}} + 92 \tag{14}$$

Figure 2, compares the calculated U50% with experimental data for long air gaps[12]. The same figure presents also the variations in the leader inception voltage as a function of the gap length. It may be seen that, whereas the leader inception voltage becomes saturated at long gaps, the breakdown voltage increases with the gap length.

In SF_6, Wiegart et al.[13] assumed that leader inception occurs when the charge produced by the streamer discharge reaches a critical value, which is a function of the gas pressure only

$$Qcr = 45 \ (p/p0)^{-2.2} \qquad (15)$$

where Qcr is the critical charge in nC and p0 is the reference gas pressure, p0=100 kPa.

A computer program was used to calculate the breakdown voltage of non-uniform field gaps subjected to step impulse voltages. The program starts with the calculation of the streamer inception voltage. The extension of the streamer discharge st, and the charge produced by the streamer Qs, were then evaluated and compared with the critical charge. The leader inception voltage was defined as the voltage at which the charge produced in the streamer equals the critical charge for the corresponding gas pressure. The leader length is then ls = d-st and the voltage drop along the leader channel can be estimated from experimental data for the average leader field, typically 2-3 kV/cm.

Typical results obtained by Wiegart et al. for different gap geometries are shown in Fig. 3, and compared with the experimental data. The agreement is quite reasonable, considering the simplification used by the authors in their model for leader breakdown.

FACTORS AFFECTING THE BREAKDOWN PROCESS

Although the breakdown of simple gap geometries may be determined with reasonable accuracy under ideal conditions, the same degree of accuracy has not been obtained in practical systems. This is due partly to their greater complexity, but also to the fact that several factors can affect the discharge development and breakdown processes. Since the breakdown path, even at an insulator/gas interface, is essentially through the gas, no fundamental difference is expected in the breakdown mechanism. However, the breakdown voltage varies considerably and can then be expressed as

$$Uc = k \ U0 \qquad (16)$$

where k is a gap factor, a function of both the gap geometry and the applied voltage, Uc and U0 the breakdown voltages of the gap under practical and ideal conditions.

The influence of different factors on the breakdown of gas insulation is discussed below.

Surface conditions

The effect of surface conditions on the breakdown voltage has been investigated by several authors. The findings point to the fact that the surface of practical electrodes has protrusions of various shapes and sizes, typically in the order of 25-100 μm. Because of the field enhancement, the local field at these protrusions may exceed the ionizing field in SF_6 and initiate the avalanche processes at voltages much lower than the calculated values. Pedersen[14] has shown that the effect of surface roughness can be accounted for calculating the breakdown field by using the streamer criterion with the local field enhancement.

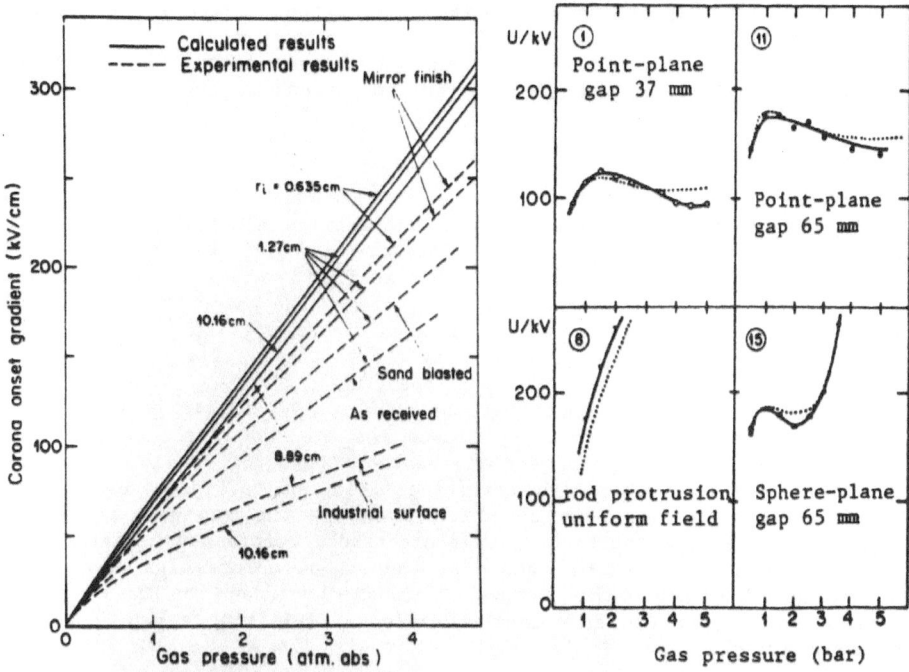

Fig. 1. Breakdown fields of coaxial
conductors[8].

Fig. 3. Leader-breakdown voltage
of nonuniform-field gaps[13].

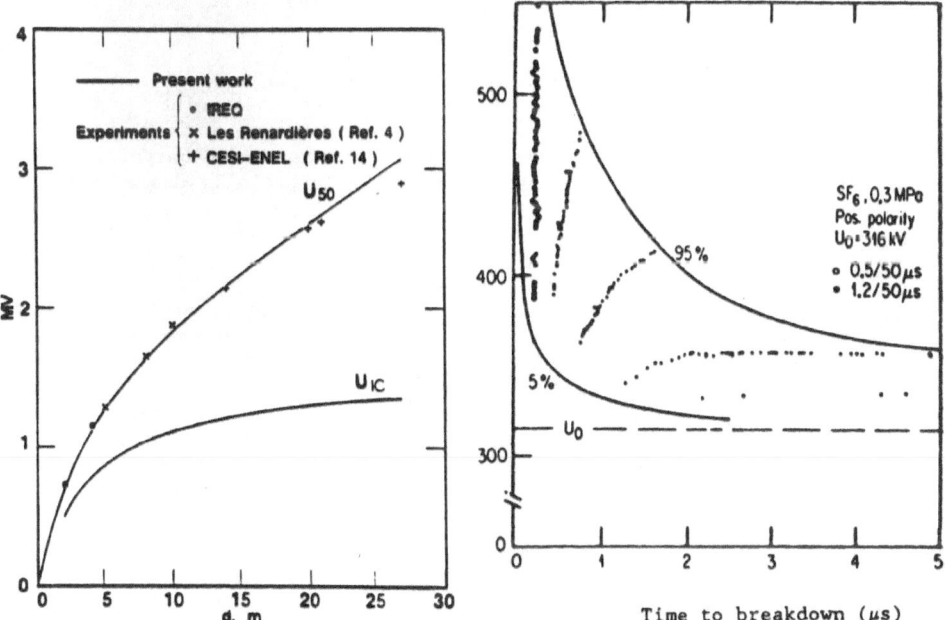

Fig. 2. Breakdown voltage of long
air gaps[12].

Fig. 4. Volt-time characteristics
of coaxial conductors[23].

The reduction in the breakdown field was found to be a function of the product, p h, of the gas pressure (in bars) times the height of the protrusions (in microns). A careful control of the surface finish at the electrode must be ensured to minimize the reduction in the withstand capability of the GIS. A typical value of the gap factor, k = 0.6, is representative for most practical surface conditions.

Particle contamination

The presence of solid particles of various sizes and shapes in GIS is now fully recognized. Under the effect of the applied electric field, these particles may be lifted and moved across the gas gap. The movement of a particle of simple geometry can be predicted from the general differential equation for the motion of a charged particle[15]. The height reached by the particle during its motion may be a significant portion of the inter-electrode gap, depending on its size and the voltage applied.

For small gaps, the mere fact that the conducting particle partly short-circuits the gap may effectively reduce the breakdown voltage. In long gaps, as the particle approaches the central conductor, field distortion at the particle may exceed the limiting field and initiates avalanches. Streamer discharge has been observed to bridge the gap separating the moving particle from the high-voltage electrode, before degenerating into full breakdown of the gap. Because of the random nature of the particle motion in the gap, the particle-induced breakdown voltage varies significantly under the same test conditions[15], yielding a gap factor as low as k = 0.25.

Stringent quality control must be ensured during manufacturing and installation to minimize the amount of particles left in the GIS. In addition, specially designed covers are usually installed around sliding connectors to collect particles detached from the contacts during thermal expansion and contraction cycles. Special test procedures have also been devised to condition the GIS prior to energization: particles are displaced towards natural or designed traps where their effect can be neutralized by the local near-zero fields.

For AC voltage applications, locations with inherently low field intensities serve as natural traps for free particles and most manufacturers do not provide specially designed traps in their GIS. The situation is more critical under DC voltages, where the unipolar field makes it more difficult to contain the particles. Specially designed traps are therefore required.

Insulator

Since insulators are inherent components of GIS, many studies have been conducted to evaluate their influence on the GIS withstand capability. Figure 1 compares the breakdown fields of a practical bus with the theoretical values of a coaxial conductor system of the same dimensions. It can be seen that the presence of insulators in a practical system reduces the breakdown field to about half the theoretical value. The effect of an insulator on breakdown can be related to field distortion, surface emission, particle contamination and surface charging.

Field distortion: Because of the high dielectric constant of epoxy resin, the presence of an insulator introduces a discontinuity in the insulating medium and disturbs the field distribution in the gas gap. The effect is especially pronounced at the triple junctions between the metal electrode, the insulator and the gas[16], where local field enhancement may cause early initiation of the discharge and facilitate the development of breakdown along the insulator interface.

Surface emission: Another effect of the insulator is the tendency of the discharge to develop along the interface, probably through secondary surface emission. Using a cylindrical insulator installed in a uniform-field gap, Pfeifer[17] has shown that the breakdown path deviated toward the insulator interface, even if the breakdown was initiated away from the insulator, and the cylindrical geometry of the insulator in this case did not modify the applied field.

Humidity: In practical GIS, it has been shown[18] that water is absorbed in aluminum and, particularly, epoxy insulators. When exposed to variations in ambient temperature, a continuous exchange of humidity between the gas, insulators and enclosures is observed, giving rise to seasonal humidity variations in the GIS. Several studies on the effect of humidity have shown that only liquid condensation of water at the insulator surface can affect its withstand capability[19], which has been observed to be as low as 25% of its value under dry conditions.

Particle contamination: The effect of a particle deposit on the insulator/gas interface[20] was found to be a function of its relative position with respect to the highly stressed electrode, and is most pronounced when the particle is located near the central conductor. Breakdown is then initiated from a discharge between the particle and the conductor and degenerates into full breakdown along the insulator surface.

Surface charge: The good insulating properties of epoxy facilitates the accumulation of electrostatic charges at the insulator. The most probable process is via ions produced in nearby partial discharges and transported to the insulator through the gas. The field enhancement caused by the gradual accumulation of surface charges eventually exceeds the limiting field and initiates the breakdown process. Surface charging is believed to play a predominant role during test for the volt-time characteristics of an insulator[21], where the charge deposit from one flashover may affect the breakdown process in the next voltage application.

The inherent presence of insulators in GIS has a predominant effect on GIS design and maintenance practices. Many precautions are usually taken by manufacturers to shield the triple junctions and reduce the risk of early initiation of breakdown from these locations. The tangential field distribution at the insulator/gas interface can be controlled by adjusting the insulator profile to force the breakdown path away from the insulator/gas interface, thus improve the GIS withstand capability[22].

Because particles collect most on horizontally-mounted cone- or disc-shape insulators, such a mounting position of the insulators is generally avoided in practical GIS.

The considerable variation in the humidity, combined with the wide range of temperatures to which GIS are subjected have set stringent limits on the acceptable humidity level in GIS, which must assure a dew point around $-10^{\circ}C$ to avoid formation of liquid water at the insulator interface[18]. The limits are usually more stringent in switchgear[19], due to the presence of arc by-products, which are more sensitive to humidity.

Time variations (short-time breakdown)

The condition for streamer breakdown expressed in Eq. (8) implies that

1- Some triggering electrons must be present in the high-field region to initiate the avalanche process.
2- Some finite time is required for the development of streamers, and its subsequent transition to breakdown of the gap.

For DC and low frequency AC voltages, these two conditions are usually satisfied, and the breakdown voltage is relatively independent of the time variation of the voltage. However, this is not necessarily true for impulse voltages, where the voltage time variations may be sufficiently rapid to influence the breakdown process. Breakdown of the gap then depends on the availability of free electrons in the high field region and on the probability that they can initiate the breakdown. The instantaneous breakdown voltage is usually higher than the static breakdown voltage U0, which implies a gap factor k>1.

The time to breakdown tb is decomposed into two components: tb = ts + tf. *The statistical time lag* ts is the time needed, once the applied voltage reaches the threshold value U0, for a free electron to appear in a favorable position in the gap to initiate an avalanche of critical size. *The formative time lag* tf is the time required for the avalanche-streamer transition to take place once a triggering electron was produced in the high-field region at time ts. A third time lag, corresponding to the streamer-arc transition, may also be distinguished, but it is much shorter than the other two, and is usually neglected in the evaluation of tb.

When an impulse voltage of a given shape and prospective peak is applied repeatedly to an insulation system, breakdown may occur at different points on the impulse wave, depending on the values of ts and tf. For a double exponential impulse voltage of prospective peak Up^{23}, the probability of having a time to breakdown less than or equal to td is

$$P(tb \leq td) = P(Up, td) = \int^{td} f(ts) \ Pc[(td-ts)|ts] \ dts \qquad (17)$$

Where f(ts) is the density function of the marginal distribution of the statistical time lag and Pc[(td-ts)|ts] is the conditional probability of the formative time lag.

When the breakdown points are plotted in a volt-time plane (Fig. 4), they are distributed over a finite area, and represents the volt-time characteristics of the insulation system[23]. It can be seen that most breakdown points are located within two boundary curves of constant joint time-lag breakdown probabilities of 5 and 95%. Similar behaviors were also observed with insulators[21], which in addition, also reduce the critical breakdown voltage and flatten the volt-time curves towards shorter times to breakdown.

The relatively flat volt-time characteristics of SF_6 insulation combined with the generation of very fast transient voltages in GIS tend to impose lower and more precise protection levels than in conventional substations. SF_6-insulated metal-oxyde arresters are therefore being used more and more for the surge protection of GIS.

CONCLUSIONS AND FUTURE WORK

Substantial knowledge has been obtained in recent years on the failure mechanisms in gas insualtion and at the insulator/gas interface. Their influence on the design, operating and maintenance practices of GIS is significant and has contributed to the good performance experienced with GIS so far. However, additional work is needed to provide a more in-depth understanding of the discharge mechanism, especially in nonuniform fields and under very fast transient voltages, where the breakdown model is still to be improved.

Electrostatic charging of insulators is another aspect of GIS insulation which is not fully understood and adequately controlled. Many in-service failures occurring at nominal operating voltages might be explained by the long term accumulation of electrostatic charges on the insulators. However, the charge accumulation processes are still to be clarified and their effect on the GIS withstand capability evaluated.

DC application of GIS has been modest, although its feasibility was demonstrated up to the 500 kV level. Surface charging of insulators and particle contamination are two identified problems in DC GIS. Highly resistive epoxy used in the bulk or as a surface coating, would prevent elctrostatic charging of the insulators and development of more effective particle traps would counter the problem of particle contamination.

ACKNOWLEDGEMENT

The author would like to thank Mrs. L. Régnier, Hydro-Québec, for the linguistic review of the text.

REFERENCES

1 . CIGRE Working Group 23-03, 1982, The State of International Development and Experience with SF_6 Gas-Insulated High Voltage Switchgear.

2 . Christophorou, L.G., Sauers, I., James, D.R., Rodrigo, H., Pace, M.O., Carter, J.G. and Hunter, S.R., 1984, Recent Advances in Gaseous Dielectrics at Oak-Ridge National Laboratory , IEEE Trans., Electrical Insulation, EI-19, pp. 550-566.

3 . Nitta, T. and Shibuya, Y., 1971, Electrical Breakdown of long gaps in Sulfur Hexafluoride , IEEE Trans., Power Apparatus & Systems, PAS-90, pp. 1065-1071.

4 . Wiegart, N. , Niemeyer, L., Pinnekamp, F., Boeck, W., Kindersberger, J., Morrow, R., Zaengl, W., Zwicky, M., Gallimberti, I. and Boggs, S.A., 1988, Inhomogeneous Field Breakdown in SF_6, IEEE Trans., Power Delivery, PWRD-3, pp.923-946.

5 . Geballe, R. and Reeves, M.L., 1953, A Condition on Uniform Field Breakdown in Electron-Attaching Gases, Phys. Rev., 92, pp. 867-868.

6 . Pedersen, A., 1989, On the Electrical Breakdown of Gaseous Dielectrics, 1989 Whitehead Memorial Lecture, IEEE Trans., Elect. Insul., EI-24, pp. 721-739.

7 . Pedersen, A., McAllister, I.W., Crichton, G.C. and Vibholm, S., 1984, Formulation of the Streamer Breakdown Criterion and its Application to Strongly Electronegative Gases and Gas Mixtures, Archiv. fur Elektrotechnik. 67, pp. 395-402.

8 . Trinh, N.G. and Vincent, C., 1978, Bundled-Conductors for EHV Transmission Systems with Compressed-SF_6 Insulation, IEEE Trans., Power Apparatus & Systems, PAS-97, pp. 2918-2206.

9 . Van Brunt, R.J. and Misakian, M., 1982, Mechanisms for Inception of DC and 60-Hz AC Corona in SF_6 , IEEE Trans., Elect. Insul., EI-17, pp. 106-120.

10. Niemeyer, L., Ullrich, L. and Wiegart, N., 1989, The Mechanism of Leader Breakdown in Electronegative Gases, IEEE Trans., Elect. Insul., EI-24, pp. 309-324.

11. Les Renardières Group, 1977, Research on Long Air Gap Discharges at les Renardières, Electra. No. 35, pp. 31-151.

12. Rizk, F.A.M.,1989, A Model for Switching Impulse Leader Inception and Breakdown of Long Air Gaps, IEEE Trans., Power Delivery, PWRD-4, pp. 596-606.

13. Wiegart, N., 1985, Corona Stabilization to Testing of Gas-insulated Switch-gear, Final report to CEA contract No. 153 T 310, 1985.

14. Pedersen, A., 1975, The Effect of Surface Roughness on Breakdown in SF_6, IEEE Trans., Power Apparatus & Systems, PAS-94, pp. 1749-1754.

15. Cooke, C.M., Wootton, R.E. and Cookson, A.H., 1977, Influence of Particles on AC and DC Electrical Performance of Gas-Insulated Systems at Extra-High-Voltage, IEEE Trans., Power Apparatus & Systems, PAS-96, pp. 768-777.

16. Takuma, T. and Kawamoto, T., 1984, Field Intensification near various Points of Contact with a Zero Contact Angle between a Solid Dielectric and an Electrode, IEEE Trans., Power Apparatus & Systems, PAS-104, pp. 2486-2494.

17. Giesselmann, M. and Pfeiffer, W., 1984, Influence of Solid Dielectrics upon Break-down Voltage and Predischarge Development in Compressed Gases, Gaseous Dielectrics IV, L.G. Christophorou and M.O. Pace (Eds.), Pergamon, New York, pp. 431-436.

18. Chu F.Y. and Braun, J.M., 1989, Assessment of Moisture in Gas-insulated Substations, Final report to CEA contract No. 217 T 424.

19. Nitta, T., Shibuya, Y., Arahata, Y., Takahashi, H. and Kuwahara, H., 1978, Factors controlling surface flashover in SF_6 gas insulated systems, IEEE Trans., Power Apparatus & Systems, PAS-97, pp. 959-968.

20. Rizk, F.A.M., Eteiba, M. Trinh, N.G. and Vincent, C., 1980, Influence of a Conducting Particle attached to an Epoxy Spacer on the Breakdown Voltage of Compressed Gas insulation", Gaseous Dielectrics II, L. G. Christophorou, (Ed.), Pergamon, New York, pp. 250-255.

21. Trinh, G.N., Mitchel, G. and Vincent, C., 1988, Influence of an Insulating Spacer on the V-t Characteristics of a Coaxial Gas-Insulated Cable - Part I: Study on a Reduced-scale Coaxial Conductor, IEEE Trans., Power Delivery, PWRD-3, pp. 16-24.

22. Trinh, N.G., Rizk, F.A.M. and Vincent, C., 1980, Electrostatic Field Optimisation of the Profile of Epoxy Spacers for Compressed SF_6 Insulated Cables, IEEE Trans., Power Apparatus & Systems, PAS-99, pp. 2164-2174.

23. Rizk, F.A.M. and Eteiba, M.B., 1982, Impulse Breakdown Voltage Time Curves of SF_6 and SF_6-N2 Coaxial Cylinder Gaps, IEEE Trans. Power Apparatus & Systems, Vol. PAS-101, pp. 4460-4471.

DISCUSSION

S. W. ROWE: Do you think that in a reasonably short time the model can be improved sufficiently enough to allow the precision that is required for improving our own designs?

N. G. TRINH: I believe that breakdown models will be more and more capable of predicting the dielectric performance of complex insulation systems under a variety of operating conditions. However from the standpoint of utility using GIS, it is of prime importance to us to be sure that the dielectric performance of the GIS conforms to the needs, and this can be obtained only through adequate tests on the equipment, both at the factory and on–site. This is particularly so due to the fact that unlike other power equipment, GIS has to be installed on–site from a large number of components under conditions often less favorable than those in the factory.

Y. DOIN: For designing GIS we do not know if we are in divergent fields or in homogeneous fields. For modeling we do need to know when to use streamer or leader processes. Could you comment on this?

N. G. TRINH: The exact discharge model to be used in evaluating the breakdown voltage in practical GIS is very much dependent on the gap configuration. In uniform and slightly divergent field gaps under clean conditions, the streamer breakdown model seems to be adequate. In non–uniform fields like point–to–plane gaps, the leader breakdown is most likely to predominate. For the intermediate situations, probably both breakdown models may be tried and selected for the most conservative results.

PREVENTION OF BREAKDOWN DUE TO OVERVOLTAGES

ACROSS INTERRUPTION OF GIS ENCLOSURE

J.M. Wetzer, M.A. van Houten and P.C.T. van der Laan

High-Voltage Group
Eindhoven University of Technology
P.O. Box 513, 5600 MB Eindhoven, The Netherlands

ABSTRACT

Voltage transients generated in GIS by the operation of circuit breakers or disconnect switches induce high voltages across interruptions of the enclosure. In this paper we discuss such interruptions and their necessity. We present fast measurements of voltage and current at the GIS/cable-interface of a 150 kV substation, where the enclosure is interrupted to permit current measurement. Circuit breaker operation causes a single voltage transient across the interruption with an amplitude up to 40 kV, a duration of 200-300 ns, and a risetime of about 5 ns. Disconnect switch operation causes multiple transients, each having a smaller amplitude and a longer risetime. The voltage across enclosure interruptions is adequately suppressed, and breakdown is completely eliminated, when the interruption is bridged with resistors. The resistors do not affect the current measurement.

INTRODUCTION

During switching events, a GIS-installation is a concentrated source of high frequency electromagnetic power. Part of the hf-power couples out at interruptions of the enclosure. This causes interference and may induce considerable voltage differences in secondary circuits or grounding structures. Sometimes, incorrectly, the term "transient ground potential rise" is introduced. Interruptions of the GIS-enclosure are found at the interfaces with overhead transmission lines and HV-cables, and sometimes between different GIS-sections. In this paper we discuss such interruptions and their necessity. In particular we investigate a GIS/cable-interface where the enclosure is interrupted to permit current measurement. The voltage across the interruption is measured with a differentiating/integrating measuring system with a time-resolution of 1 ns. We compare different measures to reduce the overvoltage, and to eliminate sparking.

INTERRUPTIONS OF GIS-ENCLOSURES

Interface with overhead transmission line. At the transition from a GIS-installation to an overhead transmission line, an interruption of the GIS-enclosure cannot be avoided. The large hf-currents which flow in the external circuit or in the GIS-building during switching events, can be reduced by means of proper grounding structures and by introducing useful hf-losses with resistive connections[1].

Interface with HV-cable. At GIS/cable-interfaces, the cable sheath is often insulated from the GIS-enclosure to allow cathodic protection, or to permit current measurement. During switching, high overvoltages occur across this insulation. Protective devices proposed in the literature to protect the insulating flange against sparking are capacitors, metal-oxide varistors and spark gaps[2]. When an interruption is made to permit current measurement, we can separate the 50 Hz and the hf return currents, and thereby reduce drastically the overvoltage, without affecting the (low frequency) current measurement (see next section).

Insulated enclosures. Sometimes, it is (naively) assumed that no 50 Hz current will flow in the metal enclosure, except under fault conditions, when different sections of a (single phase) enclosure are insulated with respect to each other. However, even for 50 Hz, induced eddy currents flow at the inside surface and, in opposite direction, at the outside surface[1]. We compare the dissipated power for continuous and insulated enclosures (see Fig.1). Each section of the insulated enclosure is connected to the grounding mesh of the substation at one single point. The three continuous enclosures are interconnected at the ends by wide metal plates. The wide metal end connections shown in Fig.1 (right) keep the magnetic field low outside the enclosure; the currents then tend to flow in the inside skin only[1]. The dissipated power per square meter of enclosure is shown in Fig.2 as a function of d/δ (enclosure thickness over skin depth).

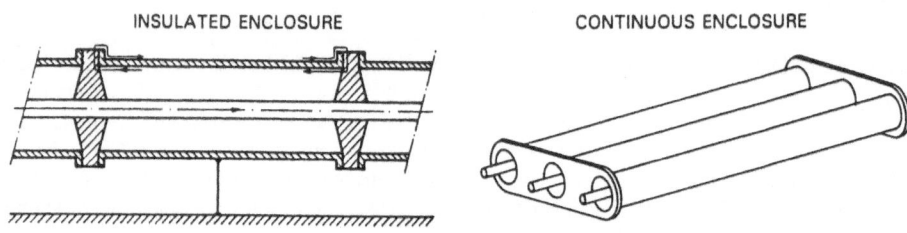

Fig. 1. Insulated and continuous GIS-enclosures.

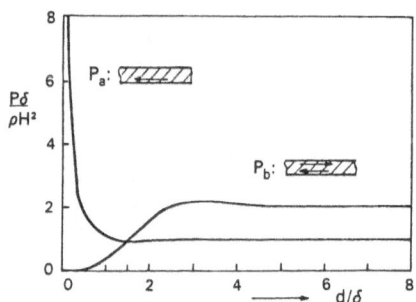

Fig. 2. The dissipated power per square meter of enclosure for continuous (P_a) and insulated (P_b) enclosures versus d/δ.

For iron or steel, a continuous enclosure is preferable because d/δ ≈ 6-8 at 50 Hz. For aluminum enclosures d/δ ≈ 0.65-1.2 at 50 Hz. Considering only dissipation in the enclosure, insulated enclosures would be desirable. The magnetic fields which escape at interruptions, however, cause additional dissipation in nearby structural steel as a result of eddy currents. Furthermore, a continuous enclosure has superior EMC properties and is therefore advisable also for aluminum enclosures[1].

MEASURING SYSTEM FOR GIS/CABLE INTERFACE

In a 150 kV GIS-substation of the PNEM (Power Company Province Noord-Brabant) current transformers are installed around each incoming or outgoing (single phase) HV-cable. The connection between GIS-enclosure and cable shield is interrupted to permit current measurements. The current return is provided by a lead making a detour around the transformer. A schematic view is given in Fig.3. Voltage transients generated by switching events cause high voltages, and subsequent breakdown, across the interruption. To avoid damage to critical insulation cylindrical spark gaps are installed.

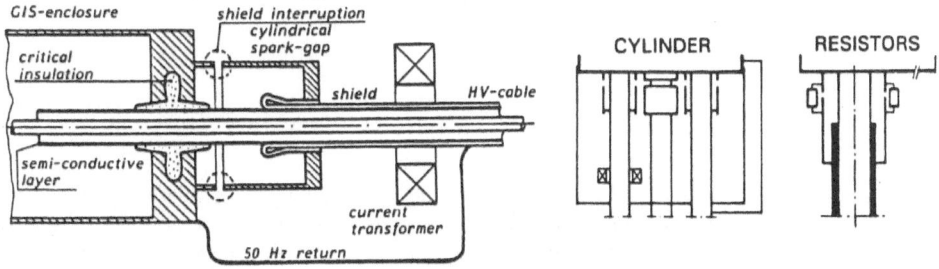

Fig. 3. Schematic view of GIS/cable-interface.
- left: original interface, only one cable shown
- middle: modification with metal cylinder around current transformers
- right: interruption bridged with resistors

The voltage across the gap is measured with a differentiating/integrating system. The differentiator consists of a capacitive sensor and a terminated 50 Ω coaxial cable. For maximum bandwidth, a passive integrator is used. Figure 4 shows the capacitive sensor, its location with respect to the spark gap, and the housing which contains the terminating resistor and the integrator. The measuring system has a clean stepresponse with a risetime of 1 ns. Special attention was paid to the hf-design of the different components[1]. Occasionally, current measurements have been performed with a 30 MHz Rogowski coil as the differentiating sensor.

The voltage waveforms have been recorded with a single-channel Tektronix digitizer (600 MHz bandwidth). At a lower bandwidth, simultaneous recordings of the voltage waveforms of three phases, as well as occasional current measurements, have been performed with two digital Nicolet oscilloscopes (sampling frequency 200 MHz). The recorders were placed within EMC cabinets. Copper tubing surrounded the signal cables. The grounding and shielding measures taken, made it possible to place the electronic equipment only a few meters away from the sparking interface. A more detailed description of the measuring system and EMC measures is presented elsewhere[1].

We have investigated two modifications to reduce the overvoltage and to eliminate all sparking (Fig.3). First the connection between cable shield and GIS-enclosure is given the form of a compact metal cylinder around the three current transformers. This reduces the discontinuity in the wave-impedance and provides a smaller inductance. Secondly, the interruption is bridged with a ring of resistors with a total resistance of 30 Ω. This provides a compact, low impedance hf-current return, without affecting the low frequency current measurement.

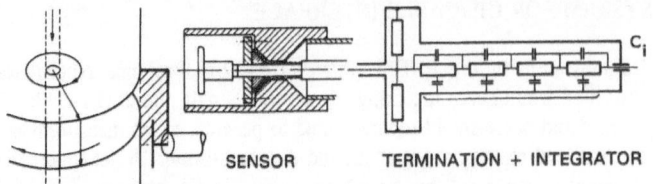

SENSOR   TERMINATION + INTEGRATOR

Fig. 4. D/I-measuring system.
      left: location of sensor with respect to spark gap,
      middle: capacitive sensor, the disk diameter is 30 mm,
      right: housing containing terminating resistor and passive integrator. The integration
                capacitor C_i is of the feedthrough type, the other capacitors indicate stray
                capacitance.

RESULTS AND DISCUSSION

After a number of initial tests, two characteristic switching events were selected for a systematic series of experiments. In the first event (CB) an open ended HV-cable is energized by a circuit-breaker. In the second event (DS) a small, floating, part of the GIS-installation is energized by a disconnect switch. Typical waveforms of the voltage across the spark gap at the GIS/cable interface, measured on a μs-timescale, are shown in Fig.5.

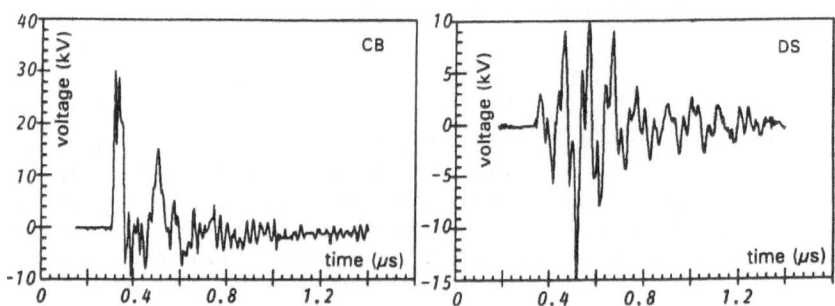

Fig. 5. Typical measured waveforms of the voltage across the gap during operation of
       circuit-breaker (CB) or disconnect-switch (DS).
       Recorder: Nicolet 4094C, sampling frequency 200 MHz.

Figure 6 shows voltage waveforms, measured during CB-operation, at three different spark gap settings, and recorded on a ns-timescale. For d = 2 mm the spark gap breaks down in the first front of the waveform, for d = 5 mm at the top of the voltage waveform. At breakdown, the voltage collapses within 1 ns. For d = 10 mm, breakdown occurs after the first 100 ns. Waveforms recorded during DS-operation show a longer risetime and a smaller amplitude.

Typical current waveforms measured during both switching events are shown in Fig.7. During CB-operation the spark gap breaks down once, and the transient current dies out within one 50 Hz period. DS-operation is characterized by multiple breakdowns, corresponding to a transient behavior during about 0.5 seconds.

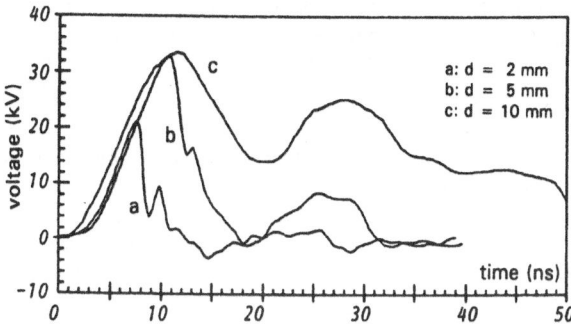

Fig. 6. Measured waveform of the voltage across the gap during operation of circuit-breaker for three different spark gap distances. The waveform measured at d = 10 mm corresponds to the waveform of Fig.5 (CB). Recorder: Tektronix 7912AD, bandwidth 600 MHz.

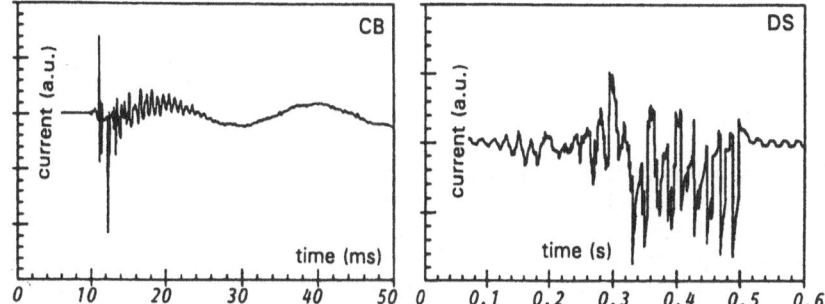

Fig. 7. Typical measured current waveforms during operation of circuit-breaker (CB) or disconnect-switch (DS).

For each voltage waveform we have determined the peak voltage and the 1/e-time of (the envelope of) the waveform. Measurements have been carried out for the original interface and for both modified interfaces (Fig. 3). The results for CB-operation are summarized in Table 1. The cylinder around the current transformers provides a reduction of both amplitude and time duration. The resistors across the interruption provide a less pronounced reduction of the overvoltage duration, but a significant reduction of the amplitude. Decisive however is the breakdown behavior. Both for the original interface and for the interface equiped with cylinder, the spark gap breaks down upon each switching event (CB and DS). With resistors installed, sparking is completely eliminated.

It can be easily verified that the resistors do not affect the 50 Hz current measurements for which the enclosure interruption was made in the first place. Even if the resistance is chosen a factor of 100 lower (300 mΩ) the error is smaller than 0.1 %. A smaller resistance will further reduce the overvoltage amplitude. The ring of resistors should not introduce additional inductance. It was therefore made out of 24 parallel branches of two series connected resistors each. The dissipated energy in the resistors is small: 10 mJ per transient at CB-operation and 2 mJ per transient at DS-operation. If we account for the repetitive behavior at DS-operation, the dissipated energy remains low: 100 mJ per (DS) switching event. The resistors should however be able to withstand the pulsed voltages applied. We therefore used 2W carbon composite resistors surrounded by shrink sleeve.

After a considerable number of switching events the resistors did not show any damage, or even temperature rise. Rings of resistors are now permanently installed across the interrupted enclosure.

TABLE 1. Summary of the Results for Circuit Breaker Induced Transient Voltages Across GIS/Cable-Interface.

DESIGN	maximum peak voltage (kV)	averaged peak voltage (kV)	1/e-time (ns)	sparking
Spark gap	39.7	29.0 (40%)	280 (20%)	yes
with cylinders	29.6	24.6 (20%)	65 (20%)	yes
with resistors	15.9	13.3 (25%)	200 (20%)	no

CONCLUSIONS

1. Piecewise insulated GIS-enclosures have no significant advantages over continuous enclosures. Continuous enclosures have much better EMC-qualities.
2. During circuit breaker operation, the voltage across interruptions of GIS- enclosures shows amplitudes up to 40 kV, an 1/e-time in the order of 200- 300 ns, and a risetime of about 5 ns. Waveforms measured under disconnect switch operation have smaller amplitudes and longer risetimes.
3. Circuit breaker operation causes a single voltage transient across enclosure interruptions, whereas disconnect switch operation causes multiple transients.
4. The voltage across enclosure interruptions is adequately suppressed, and breakdown is completely eliminated, when the interruption is bridged with resistors. In case the interruption serves to permit current measurement, the resistors do not affect the current measurement.

ACKNOWLEDGEMENT

The cooperation and support of the PNEM (Power Company of the Province Noord-Brabant) is gratefully acknowledged. The authors thank F.M. van Gompel, P.F.M. Gulickx, R.G. Noy and A.J.W.A. Oerlemans, as well as the PNEM employees, for their assistance during the experiments.

REFERENCES

1. M.A. van Houten, Electromagnetic Compatibility in High-Voltage Engineering, Ph.D. thesis, Eindhoven University of Technology (1990).
2. N. Fujimoto, S.J. Croall and S.M. Foty, Techniques for the protection of gas-insulated substation to cable interfaces, IEEE Trans.on Power Delivery, Vol.3, No.4, pp.1650-1655 (1988).

DISCUSSION

N. FUJIMOTO: As the authors are aware, we have adopted the use of ZnO materials for spark suppression at GIS/cable interfaces as a result of our specific requirements (Ref. 2 of paper).   Could the authors comment on the relative effectiveness of ZnO versus resistive materials for spark suppression? In addition, the authors could have chosen the ZnO solution for their problem as well. What were the reasons why resistors were selected over ZnO arresters for this specific application?

J. M. WETZER: The situation referred to in Ref. 2 of this paper concerns a GIS/cable interface, insulated for cathodic protection. In that case a nonconductive component (ZnO arrestor, capacitor) is required. The situation presented in this paper allows conductive components.   Resistors were chosen because: (1) they introduce damping by dissipation, which reduces the effect of reflected waveforms; (2) ZnO arrestors introduce higher frequency currents because of their strongly non linear characteristics resulting in a higher level of interference; and (3) with resistors the voltage can be reduced to lower values.

A. H. COOKSON: We made analyses of single point and cross–bonded, compressed–gas–insulated transmission lines which used an insulated enclosure joint. This would be for potential applications of long lengths of cable. Typically, this could reduce the total losses by 15%. For solidly grounded GITL systems the enclosure losses are typically 30% of the total losses, but with the insulated enclosure there are eddy current losses in the enclosure and conductors. The analysis indicated that for lengths of GITL the economics of lower losses might make this system attractive although so far it has never been used.

J. M. WETZER: I appreciate your comment. Indeed, insulated aluminum enclosures may have lower losses (as we have shown) if eddy current losses are limited. In our paper we include EMC–properties, and from that point of view continuous enclosures are superior. It would be interesting to include in the economic evaluation also the investments made to prevent damage and interference to secondary equipment, as well as the actual damage, related to poor EMC–properties of GIS equipment. According to our experience and understanding, proper EMC–measures may result in considerable savings.

S. W. ROWE: The use of cylindrical carbo–ceramic resistors are probably best for this application as the inductance can be reduced to very low values. Have you studied the effects on the overvoltages of varying the resistance value?

J. M. WETZER: We used rings of resistors as a provisional, quick–fit solution for experimental purposes. For a permanent solution a more compact (low inductance) solution is preferable. Your suggestion will serve this purpose. Due to the limited availability of the substation we did not study the effect of the resistance value on the overvoltage. However, much lower resistance values are allowed, and will reduce the overvoltage drastically.   The limit is about 100–500 mΩ because at much lower values the resistance will affect the power frequency current measurement.

THE APPLICATION OF INFRARED ABSORPTION SPECTROSCOPY

IN GAS-INSULATED EQUIPMENT DIAGNOSTICS

C.S. Vieira[1], J.R. Robins, H.D. Morrison and F.Y. Chu

Ontario Hydro Research Division

Toronto, Canada

ABSTRACT

Gas analysis is a powerful tool in diagnosing the internal conditions of SF_6 insulated equipment. Results of the analysis can be used to alert maintenance personnel for the formation of excessive toxic and corrosive byproducts in the equipment. Although many analysis techniques have been developed, they lack the required sensitivity and the capability of in-situ measurements to detect the formation of specific discharge byproducts in arcs, sparks and corona. This paper describes a gas diagnostics technique to detect discharge products in low concentrations by high resolution infrared absorption spectroscopy. S_2F_{10} and SOF_2 in sparked SF_6 have been observed by the infrared technique in the region of 825 cm^{-1} and 808 cm^{-1} respectively, thus confirming the observations by gas chromatography-mass spectrometry. The applicability of infrared techniques for GIS diagnostics in the field is discussed.

INTRODUCTION

The major incentive for using gas analysis as a diagnostic tool for gas-insulated switchgear (GIS) is that the technique itself is non-invasive and requires no retrofitting of complex equipment to the GIS. Among the many analytical techniques developed for GIS gas analysis, Chu (1987), infrared techniques offer the potential of in-situ measurement of specific molecular species at low concentrations. However, before infrared techniques can be widely applied for field measurements, the following requirements must be met. First, as SF_6 and its discharge products have rich infrared absorption spectra from 5 μm to 25 μm, the possible interference from the strong SF_6 background absorption and various molecular absorption lines makes interpretation of results difficult. It is essential to have prior knowledge of the byproduct's reference absorption spectra in regions with little interference for identification purpose. Second, simple and low cost instrumentation must be available for field use. The present generation of infrared equipment may be suitable for general purpose laboratory application, but field application will require further development of a simple device.

This paper describes the application of infrared techniques in the detection of S_2F_{10} and SOF_2 in sparked SF_6. The researchers Sauers (1989) and Olthoff et al.(1990) have used GC and

---

[1] Visiting Scientist, Permanent Address: CEPEL, Electrical Energy Research Centre, Caixa postal 2754, Rio de Janerio, Brazil.

Gaseous Dielectrics VI, Edited by L.G. Christophorou and
I. Sauers, Plenum Press, New York, 1991

539

GC-MS techniques to detect S_2F_{10} down to sub-ppm levels. However, the GC-MS technique depends on a surface catalysis reaction to convert the S_2F_{10} to SOF_2 on the separation membrane for detection by the mass spectrometer. Infrared techniques offer the opportunity of direct and in-situ detection of the specific species without secondary reactions if the reference spectrum is known and if there is no interference from background gases.

EXPERIMENTAL

The sparked SF_6 was prepared in a 1.35 litre stainless steel discharge cell fitted with 0.5 cm diameter spherical stainless steel electrodes separated by a 1 cm gap. Industrial grade SF_6 was used at a pressure of 765 torr. A 100 kV power supply was connected to the discharge cell. Breakdown of the SF_6 insulated gap occurred at 45 kV. The discharge rate was maintained at about 10 kHz with a supply current of 100 µA. The estimated energy expended in a typical run of 10^7 discharges is about 200 J.

Although the discharge cell was fitted with infrared transparent windows for direct absorption spectroscopy, the initial analysis was carried out with a 10 cm path length glass cell with NaCl windows. After sparking was completed, the gas was transferred to the sample cell for analysis. To minimize line broadening effects, the pressure of the sample cell was set at 100 torr. The time between sparking and analysis ranged from 5 minutes to several hours. The analyses were carried out with a grating infrared spectrometer (Pye Unicam) or a Fourier transform infrared spectrometer (Mattson), having resolutions of 2.5 cm^{-1} and 0.25 cm^{-1} respectively. Wavelength accuracy of the grating spectrometer is ± 1.5 cm^{-1} and the system was calibrated with known ammonia absorption lines.

Initial scans covered the wavelength regions of 400 cm^{-1} to 2000 cm^{-1}. As discussed in the next section, the region of interest is in the 800-850 cm^{-1} region. Spectra of background gases consisting of SF_6 and impurities such as H_2O were obtained as references. Reference spectra of samples consisting of S_2F_{10} in Ar, S_2F_{10} in SF_6, and SOF_2 and SO_2F_2 in SF_6 were obtained with the glass cell prior to the experiment. The S_2F_{10} samples were supplied by the Oak Ridge National Laboratory.

IDENTIFICATION OF S_2F_{10}

Low resolution infrared spectra of S_2F_{10} in the entire infrared region have been obtained by Wilmshurst and Bernstein (1957), Dodd et al.(1957), and Jones and Ekberg (1980). Argon matrix infrared spectra of S_2F_{10} have been observed by Smardzewski et al.(1976). Many S_2F_{10} absorption bands coincide with the absorption bands of SF_6, and SOF_2 and SO_2F_2, the dominant discharge products. This is not surprising because most of the absorption is from the S-F stretch in these molecules. Becher and Massonne (1970) observed a peak in the 825 cm^{-1} region when he sparked SF_6 repeatedly and he assigned that peak as S_2F_{10} absorption. Chu (1987) obtained a strong infrared absorption peak at 560 cm^{-1}, however, that peak coincides with absorption lines of SOF_2.

Table 1. Literature Observations of the S_2F_{10} Peak in the 825 cm^{-1} Region

Reference	Material	Wavelength (cm^{-1})
Edelson (1952)	pure S_2F_{10}	827
Wilmshurst and Berstein (1957)	pure S_2F_{10}	826
Dodd et al (1957)	pure S_2F_{10}	826
Jones and Ekberg (1980)	pure S_2F_{10}	824.5
Sauers (1990)	pure S_2F_{10}	823
Becher and Massonne (1970)	sparked SF_6	825
this work	sparked SF_6	825

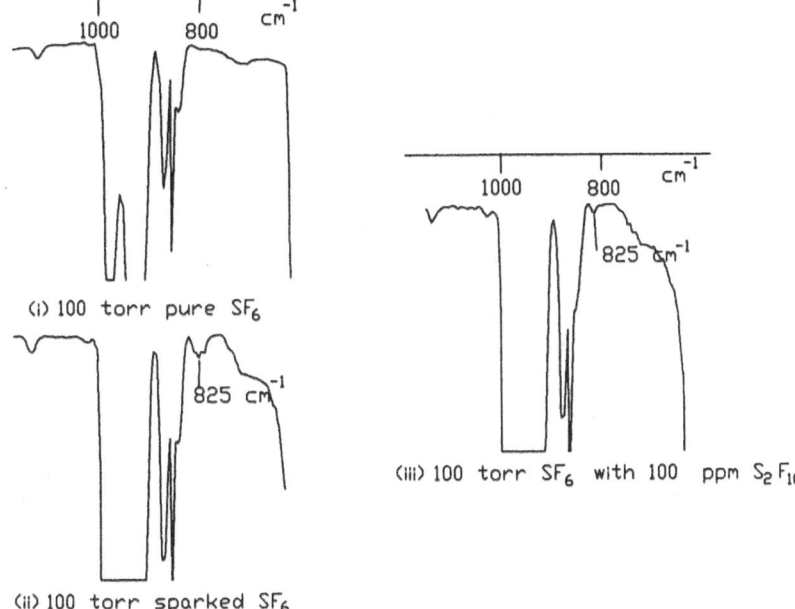

(i) 100 torr pure SF$_6$

(ii) 100 torr sparked SF$_6$

(iii) 100 torr SF$_6$ with 100 ppm S$_2$F$_{10}$

825 cm^{-1}

Fig. 1. Spectrum of SF$_6$, a sample of sparked SF$_6$, and SF$_6$ containing ~100 ppm S$_2$F$_{10}$.

In our initial search of an interference-free region, we have decided that the 820 cm^{-1} region is the most promising one to investigate. Samples subjected to 10^6 to 10^7 discharge pulses were analyzed with the grating spectrometer and then with an FT-IR to improve the resolution. Figure 1 shows the spectra of pure SF$_6$, a reference sample of a synthetic mixture of ~100 ppm (an estimate based on the degradation rate of a 1000 ppm mixture in a stainless steel cylinder over time, Sauers (1989)) of S$_2$F$_{10}$ in SF$_6$, and sparked SF$_6$ with an energy input of ~ 0.14kJ/l. The distinct absorption band at 820-825 cm^{-1} is clearly seen in the S$_2$F$_{10}$ reference sample and sparked SF$_6$ sample. Confirmation of the observed peak as S$_2$F$_{10}$ absorption peak is obtained by comparing the sparked SF$_6$ spectra with previous infrared spectra of pure S$_2$F$_{10}$ samples, Table 1.

The observed band at 825 cm^{-1} has been assigned as the v_6 band of S$_2$F$_{10}$ by Jones and Ekberg (1980). This band and the v_9 band at 937.8 cm^{-1} are the two strongest absorptions of S$_2$F$_{10}$. Unfortunately, the 937.8 cm^{-1} absorption of S$_2$F$_{10}$ coincides with a strong SF$_6$ band which makes direct observation of this band impossible.

Figure 2 shows the high resolution FT-IR spectra of the region between 805 cm^{-1} and 840 cm^{-1}. The strong absorption line at 808.5 is due to SOF$_2$ absorption. This is confirmed by comparison with the spectrum of pure SOF$_2$. The simultaneous observation of this line and the 825 cm^{-1} band permits investigation of the formation and breakdown kinetics of the various SF$_6$ discharge products. Initial analyses indicate that the SOF$_2$ peak starts to grow as a function of time after sparking while the S$_2$F$_{10}$ peak starts to decay. Further experiments will clarify this observation.

Figure 3 shows the strength of the S$_2$F$_{10}$ and SOF$_2$ absorption peaks as a function of discharge energy. Without the knowledge of the absorption strength of the v_6 band, it is impossible to convert the peak intensity to absolute number of absorbing molecules. Quantitative information will have to wait for the availability of a calibrated sample. At present there is no estimate of the sensitivity of the present technique.

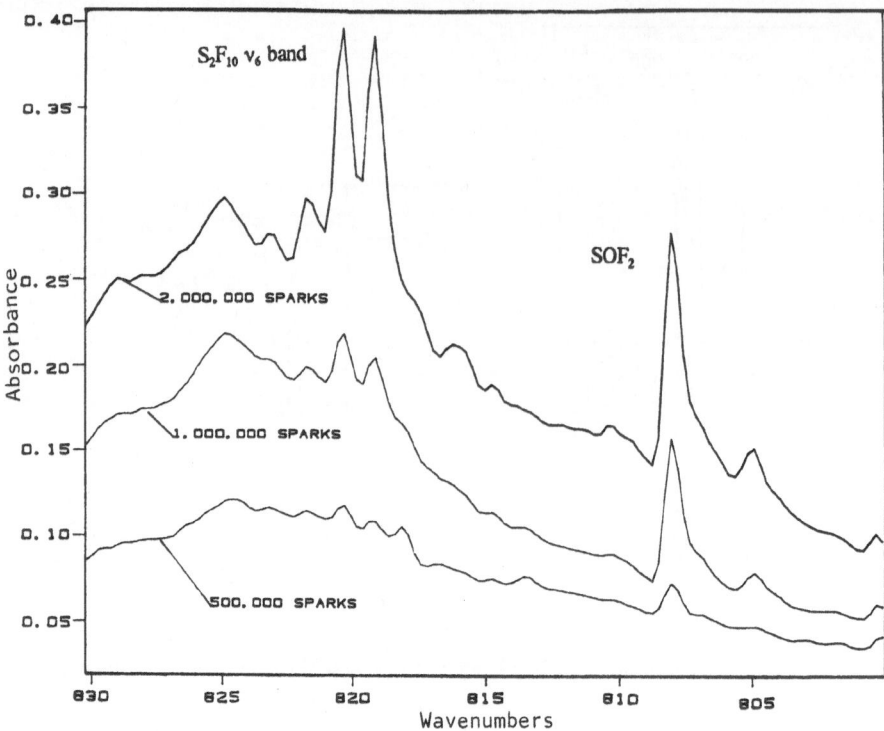

Fig. 2. High resolution FT-IR spectrum of sparked SF_6 in the region 805 to 830 cm^{-1}.

DISCUSSION

The positive identification of S_2F_{10} formation in sparked SF_6 subjected to discharge energy of about 0.1 kJ/l by infrared technique complements results obtained by GC or GC-MS methods. Due to the instability of S_2F_{10} and the complex mass spectrum causing difficulties in the interpretation of results, the infrared technique offers an independent confirmation.

For practical applications in GIS diagnostics, we plan to apply the infrared technique to detect the possible formation of S_2F_{10} in corona and arc SF_6 samples. S_2F_{10} formation in corona has been detected by high sensitivity GC-MS techniques, Sauers (1989); however, the formation of S_2F_{10} in arc samples remains to be confirmed. Although S_2F_{10} has been detected, in ppb levels, in arced samples by GC with cryogenic trapping, Janssen (1987), it is pointed out that in arc discharge, the temperature should be too high for S_2F_{10} to exist. At above 200°C, S_2F_{10} decomposes rapidly to SF_5 and subsequently forms other stable products by reaction with moisture and air. In corona or spark where non-equilibrium conditions dominate, the formation of S_2F_{10} is more likely than that in an arc. The availability of an infrared technique for in-situ measurement of S_2F_{10} formation in the neighbourhood of an arc discharge will help to clarify this controversial issue. Since many gas-insulated switchgear have experienced arc discharges either as a result of circuit breaker operation or of high current faults, the formation and stability of S_2F_{10} in an arc is an important practical issue. We plan to build an arc discharge cell equipped with the appropriate infrared optics for direct S_2F_{10} measurements. The proximity of the SOF_2 line at 808 cm^{-1} will serve as a marker since SOF_2 is the dominant byproduct formed in an arc discharge.

To address the issue of health and safety, the infrared technique must be able to measure quantitatively the concentration of S_2F_{10} down to 10 ppb level which is the ceiling limit of this toxic gas. This can be achieved by increasing the absorption path length or improving the signal-to-noise ratio of the source or detection electronics. With the availability of tunable diode lasers in the infrared region, detection sensitivity can be greatly improved. A lead salt diode laser operating at 823 cm^{-1} has been custom-made to probe the high resolution

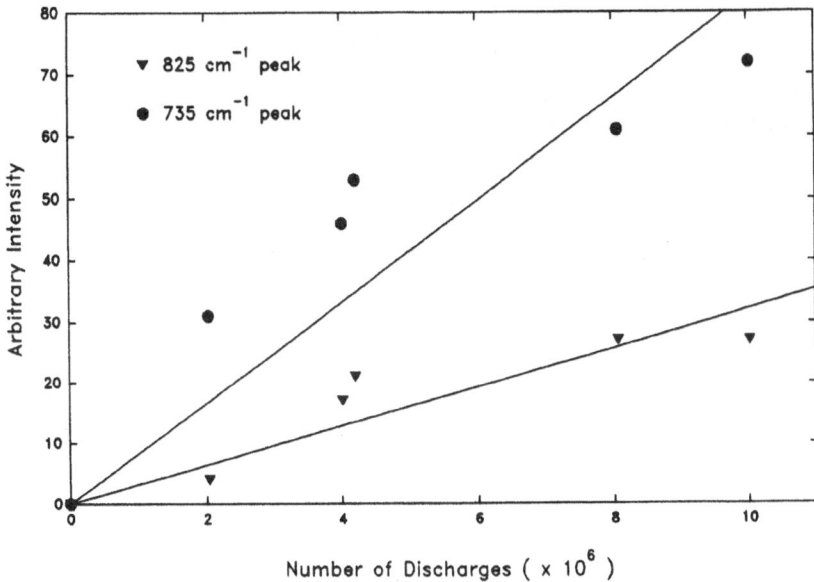

Fig. 3. Intensities of the S_2F_{10} peak (825 cm^{-1}) and the SOF_2 peak (735 cm^{-1}) as a function of number of discharges.

spectroscopy of sparked SF_6. Preliminary investigations show that a 0.005 cm^{-1} resolution can be achieved in that region. Since the tuning range for single mode operation is only 0.6 cm^{-1}, it will be time consuming to establish the complete high resolution spectra of S_2F_{10} as this heavy molecule will have complex vibrational-rotational absorption lines. However, once the wavelength of a strong absorption line is established, the laser can be made to operate at that wavelength for subsequent measurements with greatly improved sensitivity. Separate laser heads can be made to operate at regions of interest such as the absorption wavelengths of SOF_2 and SO_2F_2.

CONCLUSIONS

We have applied infrared absorption spectroscopy techniques to detect the formation of S_2F_{10} in sparked SF_6 samples. The observation of the S_2F_{10} band at 825 cm^{-1} with little interference from the background gas SF_6 and other byproducts offer a direct way to carry out GIS diagnostics in-situ. The observation also confirms previous results of S_2F_{10} formation in sparked samples by gas chromatography-mass spectrometry techniques. The simultaneously detection of S_2F_{10} and SOF_2 permits future investigation of the formation and stability of S_2F_{10} in corona and power arc discharges.

ACKNOWLEDGMENT

This work is partially supported by a consortium including the Electric Power Research Institute, Canadian Electrical Association, Empire State Electric Energy Research Corporation, Bonneville Power Administration, and Tennessee Valley Authority. The sample of pure S_2F_{10} was supplied by Dr. I. Sauers of the Oak Ridge National Lab. The FT-IR work was carried out by Dr. Otto Hermann of Ontario Hydro.

REFERENCES

Becher, W. and J. Massonne, 1970, Contribution to the study of decomposition of SF_6 in electric arc and sparks, *ETZ-A*, 91:605.

Chu, F.Y., 1987, Gas analysis as a diagnostic tool for gas insulated equipment , *in*: "Gaseous Dielectrics V," L.G. Christophorou & D. Bouldin, ed., Pergamon Press, New York.

Dodd, R.E., Woodward, L.A., and Roberts, H.L., 1957, Molecular vibrations of group 6 decafluorides, *Trans. Farad. Soc.*, 53:1545.

Edelson, D., 1952, The infrared spectrum of disulfur decafluoride , *J. Am. Chem. Soc.*, 74(1):262.

Janssen, F.J.J.G., 1987, Decomposition of SF_6 by arc discharge and the determination of the reaction product of S_2F_{10}, *in*: "Gaseous Dielectrics V," L.G. Christophorou & D. Bouldin, ed., Pergamon Press , New York.

Jones, L.H., and Ekberg, S.A., 1980, Vibrational spectrum and potential constants for S_2F_{10}, *Spectrochim. Acta*, 36A:761.

Olthoff, J.K., Van Brunt, R.J., and Herron, J.T., 1990, Catalytic Decomposition of S_2F_{10} and its implication on sampling and detection from SF_6, *in*: "Proc. 1990 Int'l Symp. on Electrical Insulation," Boston.

Sauers, I., Votaw, P.C., and Griffin, G.D., 1988, Production of S_2F_{10} in sparked SF_6, *J. Phys. D: Appl. Physics*, 21:1236.

Sauers, I. et. al., 1989, Production and stability of S_2F_{10} in corona discharge, *in*: "Proc. 6th Int'l. Symp. on High Voltage Engineering," New Orleans.

Sauers, I., 1990, private communication.

Smardzewski, R.R., Noftle, R.E., and Fox, W.B., 1976, Argon matrix raman and infrared spectra for SF_5Cl, SF_5Br and S_2F_{10}, *J. Mol. Spec.*, 62:449.

Wilmshurst, J.K., and Bernstein, H.J., 1957, The infrared and raman spectra of disulphur decafluoride (S_2F_{10}), *Can. J. Chem.*, 35:191.

DISCUSSION

J. CASTONGUAY: Do you have any idea of the amount or rate of formation of S_2F_{10}?

J. R. ROBINS: No, we do not. We have not tried to quantify how much we are producing. The amount of S_2F_{10} is probably above 100 ppm.

J. CASTONGUAY: You did not show the IR (infrared) spectrum of SF_4 in your table. Have you looked at SF_4?

J. R. ROBINS: We have not looked at the IR spectrum of SF_4 yet.

S. SUZER: Have you taken into account exchange reactions on the NaCl windows?

J. R. ROBINS: We have run regular backgrounds of the cell. We have seen a general decrease in transmission but no identifiable peaks.

DISULFUR DECAFLUORIDE (S_2F_{10}): A REVIEW OF THE BIOLOGICAL PROPERTIES AND OUR EXPERIMENTAL STUDIES OF THIS BREAKDOWN PRODUCT OF SF_6

G. D. Griffin, M. S. Ryan, K. Kurka, M. G. Nolan, I. Sauers and D. R. James

Health and Safety Research Division
Oak Ridge National Laboratory
Oak Ridge, TN 37831-6101

ABSTRACT

The toxicity of S_2F_{10} (disulfur decafluoride) to whole animals and to cells is discussed. The strong toxicity of S_2F_{10}, comparable to the toxicity of phosgene in some species, suggested the possible use of S_2F_{10} as a warfare agent. Exposures to as little as 0.1 ppm for 18 h have produced lung irritation in rats. Cell culture studies in our laboratory have shown S_2F_{10} to be by far the most toxic product we have so far identified in laboratory samples of electrically-decomposed SF_6. The significance of these toxicological studies for electrical utility personnel depends on (1) the presence and amount of S_2F_{10} found in actual applications and (2) the effectiveness of clean-up methods applied to faulted gas equipment.

INTRODUCTION: RELEVANCE TO SF_6-INSULATED EQUIPMENT

Although disulfur decafluoride (S_2F_{10}) has many interesting chemical and biological properties, it is, at best, a laboratory curiosity to scientists and engineers concerned with practical applications of electrical energy systems, unless it can be shown to be an actual byproduct of SF_6 decomposition (and present in "significant" amounts) in real-world electrical utility operations. If, in fact, such demonstrations are forthcoming, then issues regarding its degree of hazard to occupational workers, methods for its efficient removal, sensitive techniques to analyze for its presence, etc., need to be raised and addressed. Currently, we are unaware of definitive work which has identified and quantified S_2F_{10} in actual utility samples. Indeed, investigations for the presence of S_2F_{10} in such samples must rely on sophisticated analytical techniques, since the analytical requirement is to detect and quantify low (to very low) concentrations of a particular molecular species in an abundant milieu of a very similar molecular species (SF_6). Nevertheless, studies from a few laboratories have produced evidence that S_2F_{10} may be produced and accumulate as a result of electrical discharge in SF_6, under certain conditions and experimental designs. Thus, Pettinga (1985) demonstrated the presence of S_2F_{10} at very low concentrations (50-100 ppb) following a power arc burnthrough experiment. Work in the laboratory of I. Sauers has demonstrated that: (1) electric spark decomposition of SF_6 in a spark chamber can produce S_2F_{10} ($4\text{-}37 \times 10^{-11}$ mol/J) (Sauers et al., 1988a; 1988b; 1990); and (2) corona decomposition of SF_6 can produce highly significant yields of S_2F_{10} (2-15 μmoles/Coulomb) (Sauers et al. 1989; 1990; Olthoff et al. 1990a).

PHYSICAL/CHEMICAL PROPERTIES OF S_2F_{10}.

It is not the purpose of this discussion to provide an extensive overview of the physico-chemical properties of S_2F_{10}. Nevertheless, a few properties relevant to subsequent discussion will be mentioned. S_2F_{10} has a melting point of -53° C, and boils at 30.1° C (Renshaw and Gates, 1946; Eibeck and Mears, 1980). Its vapor pressure at 25° C is 675 mm Hg (Renshaw and Gates, 1946). It is quite insoluble in water, but soluble in a variety of organic solvents (e.g., acetone) (Cotton and Wilkinson, 1980; Renshaw and Gates, 1946). The compound can be decomposed by heating; below 200° C this decomposition is very slow, but proceeds rapidly at higher temperatures (Benson and Bott, 1969). S_2F_{10} does not react with common laboratory strong alkalis or acids. Activated charcoal catalytically decomposes S_2F_{10} (the stated products of this decomposition are SF_6 and SF_4) (Renshaw and Gates, 1946). In the S_2F_{10} molecule, each S is octahedral, and surrounded by 5 fluorines. While the individual S-F bonds are shorter by about 0.2Å than expected (thus stronger than anticipated for an S-F single bond), the S-S bond is unusually long (2.21Å as compared to 2.08Å expected for S-S) (Cotton and Wilkinson, 1980). The weakest bond in the S_2F_{10} molecule is thus apparently the S-S bond, and its breaking accounts for the decomposition products produced by heating or charcoal catalysis. S_2F_{10} is stated to be an oxidizing agent (Renshaw and Gates, 1946). Presumably, the chemistry underlying this assertion must involve the S-S bond.

ANIMAL TOXICOLOGY

One of the most interesting (and potentially important) properties of S_2F_{10} is its strong toxicological action. Apparently this particular property was recognized rather early after its discovery (in 1934 by Denbigh and Whytlaw-Gray) as it was considered as a candidate chemical warfare agent for use in World War II (see Renshaw and Gates, 1946). A further feature making S_2F_{10} attractive for military application was its insidious nature, as it provided little warning of exposure. (It did not produce lacrimation or skin irritation at toxicologically significant concentrations.) Pure samples are stated to be odorless and non-irritating to the respiratory tract, at least following brief exposures to concentrations up to 0.2 mg/L of air (Renshaw and Gates, 1946). On the other hand, some individuals have described an odor similar to SO_2 upon exposure to commercial preparations of S_2F_{10} (Renshaw and Gates, 1946). Whether these samples contained small contaminating concentrations of other sulfur-containing compounds was not determined.

The study of S_2F_{10} as a candidate warfare agent resulted in animal toxicology evaluations, and these data provide the most extensive extant source for the whole-animal toxicity of this compound. In Table 1 these toxicity data are considerably condensed from the data presented in the National Defense Research Committee (NDRC) report (Renshaw and Gates, 1946). We present only the LC_{50}'s for 1 and 10 min. exposures, although 30 min exposure times were also tested. It was found that the LCt_{50} (i.e., air concentration of S_2F_{10} x time of exposure, which produced lethality in 50% of the animals so exposed) did not significantly vary over the range of 10-30 min. This suggests significant detoxification over these short times of exposure was not occurring. Although the data in Table 1 do not provide evidence of this feature, the NDRC report indicates that there was a narrow range of S_2F_{10} concentration between that causing no death and that producing 100% mortality. The species differences in sensitivity to S_2F_{10} toxicity do not appear to be very marked, except in the case of the rhesus monkey. Animals exposed to lethal doses gave no indication during the course of exposure of impaired respiration or respiratory irritation. The majority of the animals died in the time period between 3 and 20 h after exposure. The initial symptoms were respiratory distress, which progressed to convulsions and death. The pathology seen in animals exposed to S_2F_{10} was consistent with classifying it as a pulmonary irritant. Death resulted from anoxia (lack of oxygen) due to a vigorous pulmonary edema (lungs filled with fluid) and hyperemia (blood in the lungs). Interestingly, no effects on other tissues attributable directly to S_2F_{10} were seen.

Table 1. Animal Toxicity of S_2F_{10}.

A. Renshaw and Gates, 1946

Species	Exposure Time (min)	LC_{50} (ppm)[1]
Mouse	1	133-209
	10	9.5-19
Rat	1	218
	10	19-28
Guinea Pig	10	38-57
Rabbit	10	38-57
Monkey	10	86

B. Greenberg and Lester, 1950 (Rats)

S_2F_{10} Concentration	Exposure Time (h)	Toxic Effect
10 ppm	1	hemorrhages in lung
1 ppm	1	severe lung congestion
0.1 ppm	1	no effect
1 ppm	16	lethal lung hemorrhages
0.5 ppm	18	severe lung lesions
0.1 ppm	18	lung irritation
0.01 ppm	18	no effect

C. O'Neill et al., 1980

Species	LCt_{50} (ppm x min)[2]
Mouse	120
Rat	127
Guinea Pig	412

[1]LC_{50} = concentration of a substance in air which causes death of 50% of animals exposed for the indicated period.

[2]LCt_{50} = concentration of a substance in air times duration of exposure, the product causing 50% mortality.

The purpose of these military toxicology studies was to evaluate S_2F_{10} as a chemical warfare agent. On the basis of the results observed, S_2F_{10} was considered to be possibly as toxic as phosgene. For some animal species, S_2F_{10} demonstrated a stronger acute toxicity than phosgene. In the case of the monkey, however, S_2F_{10} was only ~1/10 as toxic as phosgene. If the monkey is considered to be the best animal model for humans, then S_2F_{10} may be significantly less toxic than phosgene. Such an evaluation is very tentative, considering the limited data and the lack of understanding of the basic mechanism of S_2F_{10} toxicity.

Greenberg and Lester (1950) carried out S_2F_{10} toxicity studies in the rat which apparently form much of the basis for the setting of the Threshold Limit Value (TLV) proposed by the American Conference of Governmental Industrial Hygienists (ACGIH) for S_2F_{10}. The purity of the S_2F_{10} used in this study was not stated; the supplier was Allied Chemical Co. Exposures to 5% S_2F_{10} concentrations resulted in animal death within a few minutes. Exposure to 1780 ppm produced death at 1 h of exposure. Groups of animals (very small groups) were then exposed to various low concentrations of S_2F_{10}, and animals were autopsied immediately following exposure termination or 24 h later (see Table 1). Exposure to 10 ppm or 1 ppm for 1 h produced lung lesions in rats as detected immediately after exposure, although after 24 h recovery, no lung pathology was seen. Longer exposure times (18 h) were then tested. Of six rats exposed to 1 ppm all died within 16 h, and autopsy revealed extensive lung damage (edema and lung hemorrhages). Animals exposed to 0.5 ppm or 0.1 ppm survived the 18 h exposure, but autopsies immediately following the exposure termination showed significant lung damage from 0.5 ppm S_2F_{10} and a generalized lung irritation (i.e., generalized pinkness in the lungs) from 0.1 ppm. These authors confirm the NDRC observations that S_2F_{10} did not produce eye or nose irritation in any of their animal groups, and therefore they also emphasize the insidious nature of the S_2F_{10} hazard. They concluded that human exposure to S_2F_{10} should not exceed 0.01 ppm (10 ppb).

The accuracy with which these authors were able to prepare these very low concentrations of S_2F_{10} could not be determined, nor is any indication given of attempts to analyze the S_2F_{10} concentrations in the exposure chamber air. In the supporting documentation for the ACGIH establishment of the 10 ppb ceiling TLV for S_2F_{10} (ACGIH, 1986), the Greenberg and Lester study is cited along with only one other study, a report of work by Saunders et al. (1953). In the Saunders study, the S_2F_{10} was dissolved in a lecithin-saline emulsion and administered intravenously. We do not find this study to be particularly relevant to possible exposures to degraded SF_6 gas. Finally, we are aware of only one other study relevant to animal toxicity of S_2F_{10}, which appeared in abstract form in 1980 (see Table 1, work of O'Neill et al.). There is insufficient information provided to evaluate the purity of the S_2F_{10} gas used, the accuracy of the dilutions prepared, etc. The authors again remark on the insidious nature of S_2F_{10}, and note that over a time period of exposure up to 18 h, the toxicity appears to be proportional to the product of S_2F_{10} concentration x time of exposure.

OUR EXPERIENCE WITH BIOLOGICAL PROPERTIES OF S_2F_{10}: IN VITRO CELL CULTURE STUDIES

Over the past 4 years, we have carried out extensive investigations of the toxic action of S_2F_{10} using a cell culture system. S_2F_{10} mixtures in SF_6 have been made and analyzed by gas chromatography, and concentrations of S_2F_{10} from 5-5000 ppm have been tested against 3 separate cell lines. These data have been reported in detail elsewhere (Griffin et al., 1989a; 1989b; 1990) and will be only briefly summarized here. We find S_2F_{10} to be literally orders of magnitude more toxic than other SF_6 breakdown products (e.g., SOF_2, SO_2F_2, SOF_4, SiF_4, HF, etc.) in our cell culture systems (Fig. 1A). We find that the slope of the S_2F_{10} cytotoxicity vs. concentration curve is very steep (much more so than any other breakdown product we have tested), so small changes in S_2F_{10} concentration produce large changes in cytotoxic effect (Fig. 1B). This further means that the difference in S_2F_{10} concentration between no effect and 100% cell killing is small. Note that this same phenomenon was also observed in whole-animal studies (Renshaw and Gates, 1946). We have observed that the S_2F_{10} content of spark-decomposed SF_6 samples accounts for essentially all the cytoxic effect produced by these samples, and that destruction of the S_2F_{10} content by heating essentially eliminates the cytotoxicity of these sparked SF_6 samples. The different cell lines we use do show a somewhat different sensitivity to S_2F_{10}, although the sensitivity does not vary dramatically. We also find that the length of time of exposure is critical in determining S_2F_{10} cytotoxicity, and we have confirmed the previously cited results from animal studies, i.e., that the toxicity of S_2F_{10} can be expressed as a product of concentration and time. (Longer exposures to lower concentrations are equivalent to shorter exposures to higher concentrations.) To illustrate this, consider the following cytotoxicity data for one cell line (CHO = Chinese Hamster Ovary). Based on concentration vs. response curves for 1 h of exposure, the concentration of S_2F_{10} that kills 99% of the cells is 600 ppm. The LCt_{99} is therefore 600 ppm-h. Exposure of the cells for 24 h to 25 ppm produces $\geq$ 99% cell killing. Thus, the LCt_{99} for these particular conditions is also 600 ppm-h. It appears that the cytotoxic response is directly related to time of exposure, at least over the range of 1-24 h. The minimum concentration of S_2F_{10} which has shown cytotoxicity in our system is 25 ppm (24 h of exposure). This is a considerably higher concentration than was found to be minimally effective in producing toxicity in the whole animal (e.g., 0.1-1 ppm in the work of Greenberg and Lester). There are, of course, many differences between whole animal and in vitro systems. Nevertheless, we think that we can substantially increase the sensitivity of our cell culture system to the cytotoxic activity of S_2F_{10}. It is virtually certain that a degree of protection from the lethal effects of S_2F_{10} is afforded to our cells by: (1) being bathed (albeit periodically) in aqueous tissue culture medium; and (2) being bathed in fluid containing animal serum. Remember that: (1) S_2F_{10} is very insoluble in aqueous solution; the gas may therefore not be able to penetrate the thin liquid film overlying the cells. Simply slowing the rotation of the tubes in the roller drum in our exposure system may provide more effective exposure; and (2) animal serum has been shown to have significant protective effects for cells exposed to a variety of agents (e.g., Rasmussen, 1984). It would be a simple matter for us to carry out our exposures in serum-free conditions.

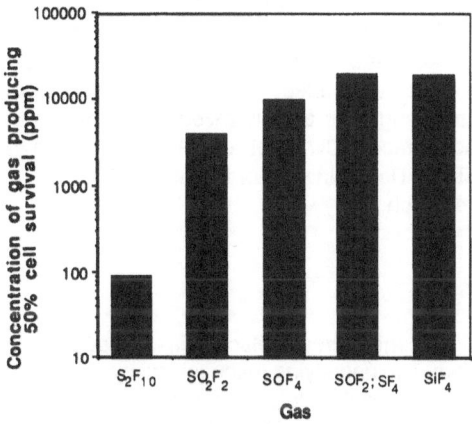

Fig. 1A. Comparative cytotoxicity of various SF_6 decomposition products.

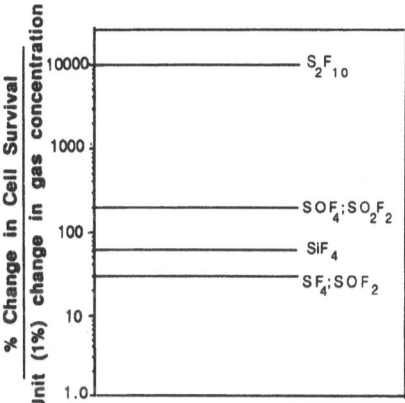

Fig. 1B. Incremental change in cell survival as a function of incremental change in gas concentration for various SF_6 decomposition products

EVALUATION OF DECOMPOSITION PRODUCT REMOVAL METHODS

The use of absorbers/scrubbers to remove corrosive and toxic decomposition products of SF_6 has been considered for at least 40 years. It was realized early on that toxic species such as SOF_2 should be removed to prevent possible exposure to maintenance workers. Also, such products as HF needed to be eliminated due not only to its toxicity but also to its highly corrosive effect on insulator materials and metals. Absorbers are now installed directly in some equipment and are used in gas reclaimer carts and separate filter/purifiers. A review of the literature concerning absorber studies appears in the EPRI Report EL-1646 (Baker et al., 1980).

A number of studies have indicated that the best materials for removal of decomposition products such as SOF_2, SO_2F_2, and SO_2 are activated alumina (porous Al_2O_3), soda lime (NaOH + CaO), and molecular sieve material (zeolite, containing Al_2O_3, SiO_2, and an alkali metal such as Na). These materials also act to absorb moisture. Activated charcoal has also been studied as an absorber (Baker et al., 1980). Commercial gas carts have internal purification cartridges to clean up SF_6 when it is being removed while equipment is serviced. For cases in which SF_6 is badly decomposed, separate purifiers have been developed to provide additional cleanup capacity. (Commercial devices are manufactured by companies such as LIMCO Manufacturing Corp., Glen Cove, NY and Cryoquip Corp. Murrieta, CA). As an example, one such system (LIMCO Series 2000 Rechargeable SF_6 Gas Filter-Purifier) contains four chemical beds separated by diffuser plates which remove different species, including one each for HF, moisture, oil vapors, and sulfur fluoride compounds. In addition, particle filters remove solid products down to 2 microns in size. The LIMCO filter and an experimental one assembled by Allied Chemical Co. were tested for their ability to filter out decomposition products from a 13.1 kA arc between Al electrodes with a duration of four cycles (Baker et al., 1980). The Allied unit contained soda lime, activated alumina, and a molecular sieve. The amounts of SOF_2 and SO_2F_2 present after flowing the gas through the scrubbers were each less than 100 ppm which was the detection limit of the analysis. In their product literature, LIMCO claims that their system can achieve a final concentration for those two species of 10 ppm by volume. It should be noted that in the EPRI study S_2F_{10} was not detected in arced samples or reference samples. This could be due to the use of Poropak Q column which has recently been shown by Olthoff et al. (1990a) to be unable to separate S_2F_{10} from SF_6. Hence, it is not known how effective these type of absorber materials are for S_2F_{10}. It is unlikely aqueous alkali (soda lime) alone would be very effective in absorbing/reacting with S_2F_{10}, due to its lack of solubility in aqueous solution. The studies relating to its use as a warfare agent indicate activated charcoal was a suitable agent for decomposing S_2F_{10}. If this decomposition is related to a surface catalysis effect [work from the laboratories of I. Sauers and R. Van Brunt suggests a generalized phenomenon of decomposition of S_2F_{10} on, at least, metallic surfaces (Olthoff et al., 1990b)], then other materials such as alumina may be as effective. Determination of the efficacy of various scrubber materials for S_2F_{10} awaits further research.

CONCLUDING COMMENTS

Whether S_2F_{10} does, in fact, exist to a significant degree in electrically-stressed SF_6 in utility settings is a question still to be answered. The underlying mechanisms by which this compound produces its toxic effects and the reasons it should be much more toxic than compounds of similar structure (i.e., SF_4) remain to be elucidated. Indeed, we have obtained preliminary evidence that $S_2F_{10}O$ shows no cytotoxicity in our cell culture system, even when tested at concentrations up to 5,000 ppm. Thus, the particular molecular structure of S_2F_{10} must be determinative in regard to its toxicity. Also, a better understanding of the chemistry of S_2F_{10} might serve to suggest appropriate amelioration techniques. Finally, understanding the molecular basis of the physiological/biochemical effects of S_2F_{10} may have applicability extending beyond the relatively narrow and focused interest in this particular toxic agent. The knowledge gained from such studies may open doors to other heretofore obscure or misunderstood areas of toxicology.

ACKNOWLEDGEMENTS

We acknowledge the patience and assistance of Mrs. Julia Cooper in preparing this manuscript. This research is sponsored by the Office of Energy Management, U.S. Department of Energy under contract DE-AC05-84OR21400 with Martin Marietta Energy Systems, Inc.

REFERENCES

American Conference of Governmental Industrial Hygienists (ACGIH)(1986). Documentation of the Threshold Limit Values and Biological Exposure Indices, 5th Ed. American Conference of Industrial Hygienists, Inc., Cincinnati, OH. p. 545.

Baker, A., R. Dethlefsen, J. Dodds, N. Oswalt, and P. Vouros (1980). Study of Arc By-Products in Gas-Insulated Equipment. EPRI EL-1646.

Benson, S. W. and J. Bott (1969). The kinetics and thermochemistry of S_2F_{10} pyrolysis. Int. J. Chem. Kinetics 1, 451-458.

Cotton, F. A. and G. Wilkinson (1980). The group VI elements, S, Se, Te, Po. Advanced Inorganic Chemistry, 4th Ed. John Wiley & Sons, New York, p. 522.

Eibeck, R. E. and W. Mears (1980). Fluorine compounds inorganic: sulfur-sulfur fluorides. Kirk-Othmer Encyclopedia of Chemical Technology Vol. 10, 3rd Ed. John Wiley & Sons, New York, pp. 799-811.

Greenberg, L. A. and D. Lester (1950). The toxicity of sulfur pentafluoride. Arch. Indust. Hygiene and Occupat. Med. 2, 350-353.

Griffin, G. D., I. Sauers, K. Kurka and C. E. Easterly (1989a). Spark decomposition of SF_6: chemical and biological studies. IEEE Transactions on Power Delivery 4(3), 1541-1551.

Griffin, G. D., K. Kurka, M. G. Nolan, M. D. Morris, I. Sauers and P. C. Votaw (1989b). Cytotoxic activity of disulfur decafluoride (S_2F_{10}), a decomposition product of electrically-stressed SF_6 In Vitro 25(8), 673-675.

Griffin, G. D., M. G. Nolan, C. E. Easterly, I. Sauers and P. C. Votaw (1990). Concerning biological effects of spark-decomposed SF_6. IEE Proceedings A (in press).

Olthoff, J. K., R. J. Van Brunt, J. T. Herron, and I. Sauers (1990a). Sensitive detection of trace S_2F_{10} in SF_6 Anal. Chem. (submitted).

Olthoff, J. K., R. J. Van Brunt, J. T. Herron, I. Sauers, and G. Harman (1990b). Catalytic decomposition of S_2F_{10} and its implications on sampling and detection from SF_6-insulated equipment. Conference Record, 1990, IEEE Symposium on Electrical Insulation, Toronto, Canada, June 3-6, pp. 248-252.

O'Neill, J.J., R.Leech, P. Beeton, M. Longhnane, and R. Raffa (1980). Sulfur decafluoride: a toxic contaminant of sulfur hexafluoride. Toxicol. Lett., Special Issue 1 p. 100.

Pettinga, J. A. J. (1985). Full-scale high-current model tests on busbar constructions for GIS. Proc. of CIGRE Symp. High Current in Power System (Brussels) pp. 506-585.

Rasmussen, R. E. (1984) In vitro systems for exposure of lung cells to NO_2 and O_3. J. Toxicol. Environ. Health 13, 397-411.

Renshaw, B. and M. Gates (1946). Disulfur decafluoride. Chemical Warfare Agents and Related Chemical Problems Parts I-II, Office of Scientific Research and Development, National Defense Research Committee, Washington, D.C. NTIS no. PB-158508, pp. 24-29.

Sauers, I., P.C. Votaw and G. D. Griffin (1988a) Production of S_2F_{10} in sparked SF_6. J. Phys. D. Applied Physics. 21, 1236-1238.

Sauers, I., P. C. Votaw, and G. D. Griffin (1988b) Production of S_2F_{10} by SF_6 spark discharges. Ninth International Conf. on Gas Discharges and their Applications. Tipografia B.G.M. snc. Padova, pp. 592-594.

Sauers, I. M.C. Siddagangappa, G. Harman, R. J. Van Brunt, and J.T. Herron (1989). Production and stability of S_2F_{10} in SF_6 corona discharges. Proc. of Sixth Internat. Symp. on High Voltage Engineering (New Orleans) 1, paper 23.08.

Sauers, I., G. Harman, J. Olthoff and R. J. Van Brunt (1990). S_2F_{10} formation by electrical discharges in SF_6: comparison of spark and corona. In L. G. Christophorou and I. Sauers (Eds) Gaseous Dielectrics VI Pergamon Press, New York, p. 553.

Saunders, J. P., M. M. Shoskes, M. R. DeCarlo, and E. C. Brown (1953). Some physiological effects of disulfur decafluoride after intravenous injection in dogs. Arch. Indus. Hygiene and Occupat. Med. 8, 436-445.

DISCUSSION

S. W. ROWE: Firstly, thank you for this very clear and unbiased presentation. (1) Probably the most important factor in practical installations would be a short term exposure due to an incident, or opening of equipment on short term laboratory work. What concentration level would be appropriate under these conditions? In your experiments you clearly employ pure S_2F_{10} samples. (2) What precautions do you employ for handling? (3) You state that S_2F_{10} is odorless, do you know the man who did the "sniffing"?

G. D. GRIFFIN: The first point is a significant one; i.e., a short—term exposure enables one to tolerate a higher level of S_2F_{10} without adverse biological consequences. There is no really satisfactory answer to this at present, given our lack of knowledge of even gross toxicology of S_2F_{10}, much less detailed understanding of toxic mechanisms. The observations of Renshaw and Gates (1946) that the product of concentration times time is constant over short times of exposure (10–30 min.) suggests that higher concentrations might be tolerated if the exposure time is reduced proportionately. It is still unclear, however, that this would really gain one all that much if S_2F_{10} is toxic/irritant for humans at the levels Greenburg and Lester (1950) found in rats. (2) Our S_2F_{10} is exclusively carried out in a chemical fume hood. When we remove the S_2F_{10} gas following cell exposure, it is collected by vacuum, and the collection flask has a charcoal filter between the flask and the pump. (3) The work of Renshaw and Gates (1946) states that pure samples of S_2F_{10} are odorless and nonirritating to the respiratory tract when breathed briefly at concentrations of 0.2 mg/L. This certainly implies a human(s) sniffed the gas. The identity of these "sniffers" is not revealed.

J. CASTONGUAY: Can you comment on this observation? In 1950 when toxicology tests were made, there was no sensitive analytical technique for analyzing S_2F_{10} at the 0.1 or 0.01 ppm level. Since recent experiments show that S_2F_{10} may decompose rapidly, the toxic effect may occur at concentrations (much) lower than thought.

G. D. GRIFFIN: This is an important and excellent observation. The paper of Greenberg and Lester does not provide information by which we can evaluate the accuracy of their S_2F_{10} dilutions, or even the purity of the initial S_2F_{10} sample from which these dilutions were made. As Dr. Castonguay correctly points out, there is the additional uncertainty associated with decay of S_2F_{10}. In the absence of accurate analysis of S_2F_{10} concentrations in the experimental animal exposure chambers, one can probably only consider the reported concentrations to be "ball park" figures, if that good.

L. G. CHRISTOPHOROU: Since $S_2F_{10}O$ is nontoxic is it possible to transform S_2F_{10} to $S_2F_{10}O$?

G. D. GRIFFIN: This is an excellent and intriguing question. Certainly some means of chemical reaction by which S_2F_{10} might be destroyed or modified to a less toxic compound could offer advantages. Catalytic decomposition to SF_5 free radicals, followed by reaction with free radical scavengers is certainly one possibility. As Dr. Christophorou points out, conversion to $S_2F_{10}O$ might be another. Unforturnately, we know too little regarding the chemistry of S_2F_{10} to know whether this is at all feasible. Potentially a peroxide might be capable of introducing the oxygen between the S–S bond of S_2F_{10}, but whether this reaction, if it went at all, would proceed at a reasonable rate is unknown to me.

S_2F_{10} FORMATION BY ELECTRICAL DISCHARGES IN SF_6: COMPARISON OF

SPARK AND CORONA

I. Sauers and G. Harman
Oak Ridge National Laboratory
Oak Ridge, Tennessee

J. K. Olthoff and R. J. Van Brunt
National Institute of Standards and Technology
Gaithersburg, Maryland

ABSTRACT

Among the SF_6 by-products of electrical discharges that have been investigated S_2F_{10} is probably the least understood (physical, chemical, and biological properties) and the most toxic. Its production in electrical discharges has been controversial because the presence of this chemical has been reported by only a few groups. We report on the yields of S_2F_{10} in two types of discharges: spark and corona. The S_2F_{10} yields for corona and spark were 2-4 μmol/C and 0.04-0.37 nmol/J respectively for experiments where the water content was low. For both types of discharges we have found that S_2F_{10} formation is dependent on the presence of moisture. For corona discharges model calculations based on known sulfur-fluorine chemistry are shown to yield reasonable agreement with experimental data. We show S_2F_{10} is formed in electrical discharges that occur in compressed-gas insulated equipment and address questions concerning effects of moisture and surface conditions.

INTRODUCTION

In addition to the desirable electrical properties of sulfur hexafluoride (SF_6), SF_6 is nontoxic and chemically inert (below $\sim 500\,^\circ C$). However, decomposition of SF_6 will occur when SF_6 is exposed to breakdown conditions including corona, spark, and arc. The major decomposition products (SOF_2, SF_4, SOF_4, SO_2F_2) have been examined in previous studies [1-3] and in some cases the production mechanisms are understandable in terms of interactions with impurities such as water or oxygen.

Probably the least understood SF_6 by-product is S_2F_{10} which can be formed by the reaction of SF_5 radicals

$$SF_5 + SF_5 \longrightarrow S_2F_{10} \tag{1}$$

where SF_5 is formed by electron-impact induced fragmentation of SF_6. Prior to 1985 this compound had only been observed by a few investigators [4,5]. Failure to detect S_2F_{10} in electrical discharges might be due to several factors including the use of inappropriate detection techniques such as mass spectrometry alone, or gas chromatography in which the column is not suitable for either separating or transmitting S_2F_{10}, and/or there is a lack of detection sensitivity. In addition, problems associated with low thermal and chemical stability, failure to use proper reference standards, and inappropriate sampling techniques can lead to significant errors in detection and identification of S_2F_{10} in SF_6. It is known that S_2F_{10} undergoes rapid unimolecular decomposition above about 200° C. At lower temperatures, S_2F_{10} has been found to undergo catalytic decomposition on surfaces[6]. As a result of its relatively low thermal stability, formation of S_2F_{10} in high-pressure SF_6 discharges has, been largely dismissed in the past. However, recent work has indicated that S_2F_{10} is formed in all major types of electrical discharges that can occur in practice, i.e., corona,[7] spark,[8] and arc.[9].

The importance of establishing the existence of S_2F_{10} in SF_6 discharges stems from its high degree of toxicity as indicated by the low ceiling TLV (threshold limit value) of 10 ppb (parts per billion) as set by the American Conference of Governmental and Industrial Hygienists. Occupational Safety and Health Administration (OSHA) has adopted this limit but has stayed enforcement until suitable detection techniques down to the TLV have been developed. Thus even at S_2F_{10} concentrations significantly lower than that of other SF_6 by-products, S_2F_{10} may be important in controlling biological and health effects. This was evidenced by the earlier cytotoxicity studies, [10] in which the major by-products of spark discharges excluding S_2F_{10} could not account for the total cytotoxicity in "in vitro" cell culture assays.

In this paper we present the results of a collaborative quantitative study of the corona and spark induced yields of S_2F_{10} formation in SF_6 and examine some of the factors that influence formation and destruction of S_2F_{10}.

EXPERIMENT

Corona discharge experiment

Sulfur hexafluoride was subjected to corona discharges in a negative point-to-grounded plane geometry. Corona experiments were performed both at National Institute of Standards and Technology (NIST) and Oak Ridge National Laboratory (ORNL) using three different discharge cells: 3.7 liter at NIST and 1.1 and 0.2 liter at ORNL. Discharges were run at constant current, i_c, and data were obtained on the S_2F_{10} production rate measured in moles per total charge collected $q = i_c \cdot t$, where t is the total discharge time. A detailed description of the NIST discharge cell and method of corona-discharge data analysis is given in Van Brunt [1]. Except where noted the electrodes were stainless steel.

Spark discharge experiment

Spark discharges were produced by repetitively discharging a 0.1 μF capacitor into a small 70-ml stainless-steel cell with concentric cylindrical electrode geometry. The spark energy was determined by $E_s = 1/2 \; nCV^2$ where n is the number of sparks, C is the capacitance of the charging capacitor and V is the voltage prior to breakdown. Since E_s is an upper limit to the injected energy, the energy-normalized yield is a lower limit. In the past, diagnostics have indicated that most of the energy is injected into the cell. However recent data of Derdouri et al [11] suggest that not all of the stored energy went into the spark in their spark experiment. Because of the strong dependence of S_2F_{10} production on conditions such as moisture and surface contamination, possible errors in spark energy cited above are not significant.

Method of gas analysis

All gas analyses were made using samples taken from the discharge cell by syringe and injected into a gas chromatograph (GC). Analyses were performed on two different GC systems: GC-MS (mass spectrometer) at NIST and GC-TCD (thermal conductivity detector)

at ORNL. The GC parameters were similar in both systems as previously specified [6,7,12].
Calibration was made with reference samples containing S_2F_{10} in either SF_6, N_2, or Ar.
The S_2F_{10} was synthesized at Clemson University and provided to us in liquid form with
greater than 99% purity determined by infrared absorption spectroscopy and GC-MS. The
reference samples were prepared at ORNL in a gas mixing vacuum manifold. Maintaining
reliable reference samples was not trivial since S_2F_{10} was found to deteriorate on surfaces
in the presence of moisture. A study of S_2F_{10} decomposition on surfaces has been reported
recently [6]. Due to the slow decay of S_2F_{10} in reference mixtures, new reference samples
were prepared periodically for use in quantitative analysis of S_2F_{10} in SF_6. Because of errors
due to unstable reference mixtures, preliminary corona yields reported by Sauers et al [7] were
overestimated in the NIST 60 μA data (presented in Fig. 4 of ref. 7). By taking proper
account of S_2F_{10} decay in the reference sample, the apparent current dependence originally
observed was eliminated as discussed in the following section.

The GC-MS technique employed for S_2F_{10} detection has been described in detail [12] and
only the most salient points of the method will be given here. As a general rule, a gas
sample of SF_6 containing a mixture of trace components is passed through a GC where the
components are separated both spatially and temporally. Each component passing into the
mass spectrometer is ionized by a 70 eV electron beam and one or more positive ions
characteristic of that species may be used to monitor the relative concentration of that
species. Because of the similarity of the S_2F_{10} and SF_6 electron-impact positive-ion mass
spectra [13], the monitoring of ions characteristic of S_2F_{10} is severely hindered by background
contributions from SF_6 ions, thereby reducing S_2F_{10} sensitivity to about 50 ppm. However,
it has been found that when the S_2F_{10} containing effluent from the GC flows through a
heated metal tube (T> 150° C) incomplete decomposition of the S_2F_{10} occurs to form SOF_2
possibly by surface processes such as:

$$S_2F_{10} \longrightarrow SF_5 + SF_5 \qquad\qquad (2)$$

$$SF_5 + H_2O \longrightarrow SOF_2 + 2HF + F \ . \qquad\qquad (3)$$

Thus the S_2F_{10} content can be monitored by observing the ion signals from SOF_2 (e.g.
SOF_2^+, m/z=86 or SOF^+, m/z=67). Because the ion background signals are significantly
lower at these m/z ratios than at m/z corresponding to SF_6 ion fragments, the S_2F_{10}
detection sensitivity is improved by several orders of magnitude to below 10 ppb. This
method of analysis was employed for the corona experiments made using the NIST 3.7-l cell.
The GC-TCD used for the analyses of spark and corona in the ORNL experiments was
applied in the conventional manner with sensitivity of ~25 ppm.

RESULTS OF CORONA EXPERIMENTS

"Clean" Cell, no water added

In Figs. 1 and 2 are shown the data obtained from the 3.7-l corona discharge cell (p=2 atm,
T=23°C, stainless-steel electrodes, 1 cm gap spacing). The data points in Fig. 1 are a
composite of three experiments, corresponding to one experiment at 60 μA, and two at 40
μA. The 60 μA data supercedes the data reported earlier [7]. There is no indication of a
dependence on i_c of the charge rate-of-production, r_q, of S_2F_{10}. This set of data gives a
production rate of $r_q=2.4$ μmol/C. However there are factors including surface
contamination and humidity that can either increase or decrease the net S_2F_{10} charge rate-
of-production from this value. Interestingly the least squares fit of the data in Fig. 1 does
not pass through the origin, indicating a non-linear production rate at short times. This is
shown more clearly by the data bounded by the dashed lines and replotted in Fig. 2. These
data were obtained from a 40 μA experiment at short elapsed times and analyzed by GC-
MS, after modification for increased sensitivity. This data set indicates a steeper initial rise
corresponding to a somewhat higher initial production rate (~4 μmol/C).

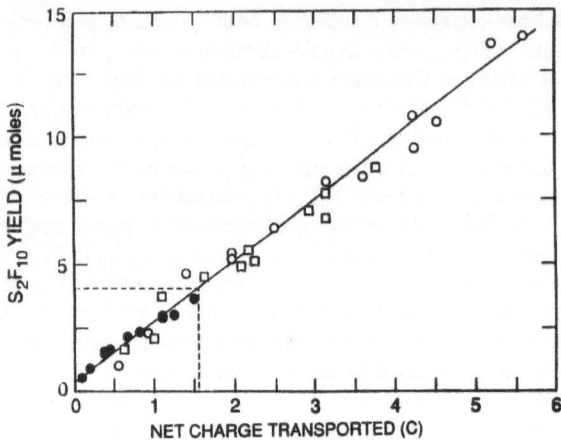

Fig . 1 . Production of S_2F_{10} in corona discharge of SF_6 at 200 kPa (3.7-*l* cell) as a function of the net charge transported: $i_c=40$ μA, open squares and closed circles; $i_c=60$ μA, open circles. Data in the region delineated by the dashed lines are replotted in Fig. (2).

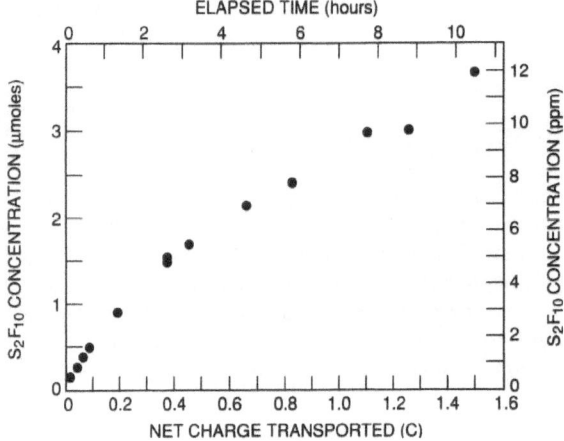

Fig . 2 . Production of S_2F_{10} in corona discharge of SF_6 at 200 kPa (3.7-*l* cell) at short times, corresponding to S_2F_{10} concentrations in the low- and sub-ppm range. The actual elapsed time during which the discharge was active is shown on the top scale.

Effects of contaminated surface

When the electrodes were not cleaned between experimental runs, that is, the SF_6 gas plus gaseous by-products were pumped out and the cell refilled with SF_6, the S_2F_{10} yield increased as shown in Fig. 3. Thus $r_q >10$ $\mu mol/C$ were obtained. The initial run (Run 1) yielded a charge rate of production within a factor of two of that shown in Fig. 1, but made

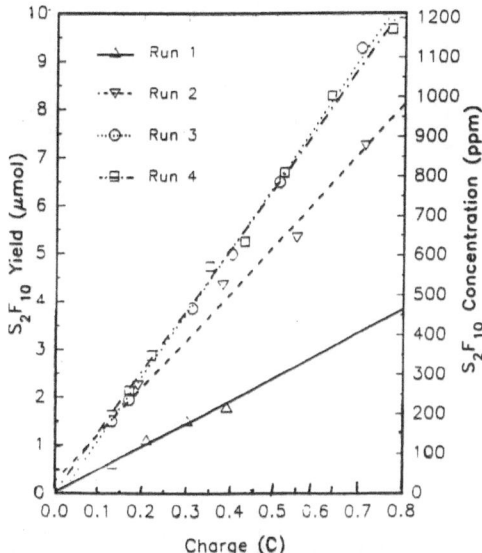

Fig. 3. Production of S_2F_{10} in corona discharge of SF_6 at 100 kPa (0.2-l cell) as a function of total charge for consecutive runs (Run 1-4) showing effects due to electrode contamination.

with a 0.2-l cell and tungsten (W) "point" electrode for i_c=24 μA. It is not known at this time what causes this increase in S_2F_{10} production, but it could be an effect associated with the buildup of solid contaminants on the electrode surface. It should be noted that the differences in S_2F_{10} concentrations shown in Figs. 1 and 3 are due to the differences in cell volumes used for the two experiments.

Influence of moisture: SF_6/H_2O mixtures

It was previously shown that adsorbed moisture will increase the consumption of S_2F_{10} on surfaces [7]. Gas-phase water can also influence the net yield of S_2F_{10} in SF_6 corona. Figure 4 shows a comparison of data taken with the 1.1-l corona chamber (P=2 atm, T=23°C, stainless-steel electrodes, 1 mm gap spacing) under relatively dry conditions with data obtained after introduction of water vapor prior to the run, for i_c=20 μA. The water concentration was determined from the partial pressure of water after introducing water into the corona cell by syringe and allowing sufficient time (~1 h) for the water vapor to reach equilibrium with the walls of the chamber. Yields obtained after accumulation of q=2 C total charge were substantially lower for water concentrations of 600 and 1200 ppm as shown in the figure. The three data sets represented by the open symbols were for experiments where the cell was "dry" meaning only that the cell was pumped down to 0.1 Pa prior to the run. The non-linear behavior for q>1 C could be due to electrode contamination as indicated by the data in Fig. 3, while the initial charge rate-of-production at q=1 C is similar to that exhibited in Fig. 1 and curve "Run 1" in Fig. 3. Because of the lower net S_2F_{10} yield (r_q<1 μmol/C) and the low sensitivity for the GC-TCD (25 ppm) used for these experiments, no data are shown below q=1C for the SF_6/H_2O experiments.

The above result appears contrary to previous data which showed a correlation between increased charge rate of production of S_2F_{10} with increased water as measured by GC-TCD using a Porapak T (Supelco) column [7]. The results of reference 7 were obtained in a 0.2-l corona discharge cell and W "point" electrode. It was pointed out in that work that the GC peak areas for H_2O content were not always consistent with the amounts of water added. However, as will be shown in the next section, decomposition of S_2F_{10} on the walls can account for lower yields. The results shown here and in reference 7 suggest that it may be necessary to consider the combined counteracting effects associated with electrode

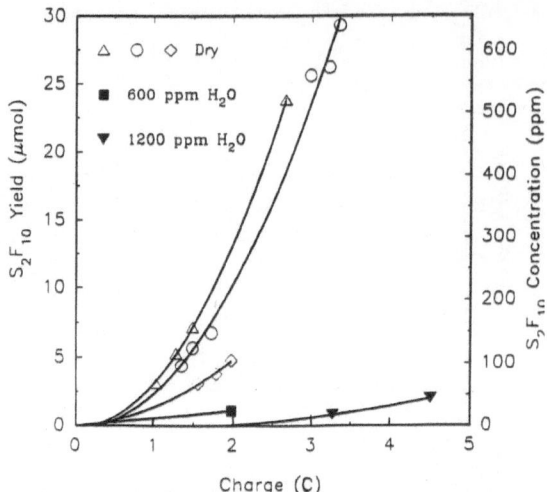

Fig. 4. Production of S_2F_{10} in corona discharge of SF_6 at 200 kPa (1.1-l cell) and $i_c = 10\ \mu A$ as a function of total charge, showing the effect of water addition: open symbols correspond to the "dry" case (no water addition); closed square, 600 ppm water; closed triangle, 1200 ppm water. Differences in the three "dry" curves are probably due to small but different amounts of adsorbed water.

contamination which increases S_2F_{10} production and adsorbed water which decreases S_2F_{10} production.

RESULTS OF SPARK EXPERIMENTS

The energy rates-of-production for spark discharges (see Table 1) exhibited considerably more scatter than the rates obtained for corona, falling in the range 0.04-0.37 nmol/J. The experimental conditions covered the range 9.7-30 kJ total energy. In a 10 kJ experiment, when 600 ppm water was added to SF_6, the S_2F_{10} yield dropped from 160 ppm to below detection limits (in this case 25 ppm using GC-TCD). As in the case of corona discharges, the S_2F_{10} yields probably depend on both surface conditions and on adsorbed trace moisture.

DISCUSSION

Model calculation of SF_6 corona

Previous model calculations using a chemical kinetics code[14] have achieved reasonable success in accounting for the charge rates-of-production of SOF_2, SOF_4, and SO_2F_2. In assessing the S_2F_{10} reaction scheme, reactions (4)-(8) were considered to be most important to S_2F_{10} formation:

$$SF_6 \xrightarrow{\text{corona}} SF_5 + F \tag{4}$$

$$SF_5 + SF_5 \longrightarrow S_2F_{10} \tag{5}$$

$$SF_5 + SF_5 \longrightarrow SF_4 + SF_6 \tag{6}$$

$$F + H_2O \longrightarrow HF + OH \tag{7}$$

$$SF_5 + F \longrightarrow SF_6 \tag{8}$$

Table 1. Comparisons of sulfur-oxyfluoride and S_2F_{10} by-product production rates from SF_6 discharges

Discharge Type

Species	Corona[a]		Spark[e]	Arc[g]
	$(\mu mo\ell/C)$	$(nmo\ell/J)$	$(nmo\ell/J)$	$(nmo\ell/J)$
SOF_4	50	0.90	0.2	0.2-100
SOF_2	32	0.54	1-3	100-600
SO_2F_2	14	0.25	0.02	
S_2F_{10}	$2-4^{[b]},(3.5\pm1.4)^{[c]},(1-12)^{[d]}$	0.05 [b]	$0.04-0.37$ [f]	$5.5-11 \times 10^{-5}$ [h]

(a) NIST data: P=200 kPa, i_c=40 μA (see Ref 1)
(b) Present data (see Fig 1), P=200 kPa, 3.7-l cell
(c) Present data (sec Fig 4), P=200 kPa, i_c=20 μA, 1.1-l cell
(d) Model calculation: $[H_2O]$=50-400 ppm (see Fig. 5)
(e) Data from Ref. 3
(f) Present data; range probably due to adsorbed water and varying electrode contamination
(g) Various sources and conditions (taken from Ref. 2)
(h) Yield determined from Refs. 9,15 (see text).

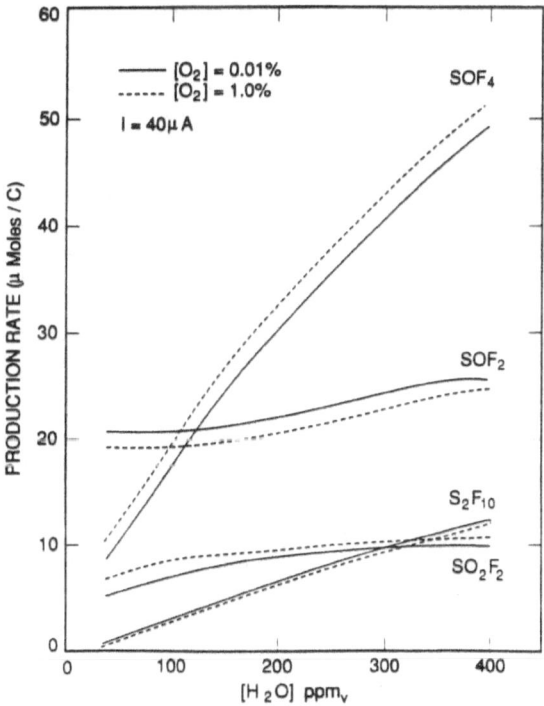

Fig. 5. Calculated charge rates-of-production of the oxyfluorides and S_2F_{10} as a function of H_2O content in 200 kPa SF_6 for a 40 μA discharge and the different indicated O_2 concentrations.

In this scheme, the presence of H_2O enhances the production of S_2F_{10} by reaction with F (reaction (7)) which suppresses reformation of SF_6 (reaction (8)) allowing SF_5 radicals to combine (reaction (5)) to form S_2F_{10}. According to this argument an increase in the gas-phase water concentration $[H_2O]$, would lead to an increase in the charge rate-of-production of S_2F_{10}. In Fig. 5 we show the results of model calculations for S_2F_{10}, SOF_2, SOF_4, and SO_2F_2 production rates as a function of gas-phase water concentration. In this model, only gas-phase reactions are considered, i.e., the model does not include loss of S_2F_{10} due to reactions occurring on surfaces, the rate of which depends on water as well.

The rate of S_2F_{10} production can be expressed by

$$\frac{d[S_2F_{10}]}{dt} = r_t - k_d[S_2F_{10}] \tag{9}$$

where $[S_2F_{10}]$ is the S_2F_{10} concentration, $r_t = r_q i_c$ is the initial rate of S_2F_{10} formation as determined for example by reactions (4)-(8), and k_d is a rate constant for S_2F_{10} decay which can depend on time and in some unknown way on the conditions of the surface on which S_2F_{10} decomposition occurs. Assuming a constant k_d, $[S_2F_{10}]$ will approach an equilibrium value, $[S_2F_{10}]_{eq} = r_t/k_d$. If we use $r_q = 3$ μmol/C (typical of the data reported here) and $i_c = 20$ μA then $k_d > 2.8 \times 10^{-5}$ would account for the low net yield shown in Fig 4. In previous work [6,7] S_2F_{10} was found to decay exponentially with rate constants falling in the range 0.001-2×10^{-5} s^{-1} depending on the size of the chamber, amount of adsorbed water, and on the temperature. Since the previously measured [6,7] decay rates are slower than this, a possible explanation for the faster decay during discharges is that the S_2F_{10} decay rate via surface reactions is higher at lower S_2F_{10} concentration ($<<$ 25 ppm).

Comparison of Corona and Spark yields

In Table 1 we summarize the results of S_2F_{10} production measurements from SF_6 in corona and spark discharges. In either case S_2F_{10} production rates are lower than for the other sulfur oxyfluorides listed. However the corona-induced S_2F_{10} charge rate-of-production is only an order of magnitude below that of the sulfur oxyfluorides for "clean" conditions where the water and oxygen contents are low. In the case of spark discharges, the yield of S_2F_{10} is lower relative to the most abundant by-product SOF_2. Included in the table for comparison are the production rates for the same sulfur oxyfluorides produced by arc discharges. The arc rate for S_2F_{10} production was determined from the S_2F_{10} concentration reported by Pettinga [9] in a power arc burn-through experiment and arc energy was taken from the related work of Janssen [15]. In either the spark or corona cases S_2F_{10} formation is significant enough to merit consideration in any evaluation of the potential health effects of decomposed SF_6.

CONCLUSIONS

Results of measuring S_2F_{10} production from both corona and spark discharges in SF_6 are reported here. Although the S_2F_{10} yield falls below the yields for SOF_2, SOF_4, and SO_2F_2, it is significant in light of the relatively high toxicity of S_2F_{10}. The results presented here demonstrate that during a continous low-level corona discharge, it is possible for the concentration of S_2F_{10} in an enclosed chamber containing SF_6 to build up to levels far in excess of the TLV within a matter of minutes. The charge rate-of-production of S_2F_{10}, $r_q = 2.4$ μmol/C, found for clean electrodes and relatively dry conditions can increase substantially under contaminated conditions. Under wet conditions, the net yield depends significantly not only on the formation rate of S_2F_{10} in the discharge, but also on the rate of decay as a result of contact with chamber walls. The larger the cell the smaller is the effect of degradation of S_2F_{10} after it is formed. An understanding of the influence of water on the S_2F_{10} formation rate requires additional experimental investigation. Results from model calculations of SF_6 by-product formation which show reasonable agreement with experimental data on sulfur oxyfluorides, also indicate an increase in S_2F_{10} yield with increasing gas-phase water concentration. It is not clear from the present data whether a model which neglects surface reactions involving S_2F_{10} is sufficient to adequately account for the low S_2F_{10} yield reported here when water is added to the discharge cell.

ACKNOWLEDGEMENTS

The authors would like to thank Dr. D. D. DesMarteau for his assistance in providing us with S_2F_{10} and to Dr. M. C. Siddangagappa for assistance in the early part of this work. This work was supported by the Office of Energy Management, Electric Energy Program, U. S. Department of Energy, under contract DE-AC05-84OR21400 with Martin Marietta Energy Systems, Inc. and under interagency agreement with NIST.

REFERENCES

1. R. J. Van Brunt, Production rates for oxyfluorides SOF_2, SO_2F_2, and SOF_4 in SF_6 corona discharges, J. Res. NBS 90(3):229 (1985).
2. I. Sauers, H. W. Ellis, and L. G. Christophorou, Neutral decomposition products in spark breakdown of SF_6, IEEE Trans. on Electrical Insulation, EI-21(2):111 (1986).
3. F. Y. Chu, SF_6 decomposition in gas-insulated equipment, IEEE Trans. on Elect. Insul. EI-21(5):693 (1986).
4. W. Becher and J. Massonne, Contribution to the study of the decomposition of SF_6 in electric arcs and sparks, ETZ-A 91(11):605 (1970).
5. B. Bartakova, J. Krump and V. Vosahlik, Effect of electric partial discharge in SF_6, Electroteck. Obzor. (Prague) 67:230 (1978).
6. J. K. Olthoff, R. J. Van Brunt, J. T. Herron, I. Sauers, and G. Harman, Catalytic decomposition of S_2F_{10} and its implications on sampling and detection from SF_6-insulated equipment, Conference Record of the 1990 IEEE International Symposium on Electrical Insulation, pp 248-252 (1990).
7. I. Sauers, M. C. Siddagangappa, G. Harman, R. J. Van Brunt and J. T. Herron, Production and stability of S_2F_{10} in SF_6 corona discharges, Proceedings of the Sixth International Symposium on High Voltage Engineering, paper 23.08 (1989).
8. I. Sauers, P. C. Votaw, and G. D. Griffin, Production of S_2F_{10} in sparked SF_6, J. Phys. D: Appl. Phys. 21:1236 (1988).
9. J. A. J. Pettinga, Full scale high current model tests on busbar constructions for GIS, Proc. of the CIGRE Symposium on High Current in Power Systems, pp. 506-511 (1985).
10. G. D. Griffin, I. Sauers, K. Kurka, C. E. Easterly, Spark decomposition of SF_6: chemical and biological studies, IEEE Trans on Power Delivery, 4(3):1541 (1989).
11. A. Derdouri, J. Casanovas, R. Grob, and J. Mathieu, Spark Decomposition of SF_6/H_2O Mixtures, IEEE Trans. on Elect. Insul., 24(6):1147 (1989).
12. J. K. Olthoff, R. J. Van Brunt, J. T. Herron, and I. Sauers, Sensitive detection of trace S_2F_{10} in SF_6, Analytical Chemistry (submitted, 1990).
13. J. K. Olthoff, R. J. Van Brunt, and I. Sauers, Electron-Energy Dependence of the S_2F_{10} Mass Spectrum, J. Phys D: Appl. Phys. 22:1399 (1989).
14. R. J. Van Brunt, J. T. Herron, and C. Fenimore, Corona-induced decomposition of gaseous dielectrics, in "Gaseous Dielectrics V," (Proc. of the 5th International Symposium on Gaseous Dielectrics), ed. L. G. Christophorou and D. W. Bouldin, pp. 163-173 (1987).
15. F. J. J. G. Janssen, Decomposition of SF_6 by Arc Discharge and the Determination of the Reaction Product S_2F_{10}, in "Gaseous Dielectrics V," (Proc. of the 5th International Symposium on Gaseous Dielectrics), ed. L. G. Christophorou and D. W. Bouldin, pp. 153-162 (1987).

DISCUSSION

J. CASTONGUAY: (1) How do the rates of formation of S_2F_{10} in various test setups (cell volume, electrode gap, and partial discharge current) compare with each other? (2) Is the rate of formation of S_2F_{10} influenced by the nature of the electrode? If so, is the same trend observed for the other more usual by–products SOF_2 and SO_2F_2?

I. SAUERS: (1) Although we observed that the charge rate–of–production of S_2F_{10} in SF_6 corona depended on water vapor and solid contamination, under clean, dry conditions we observed no systematic variations in the S_2F_{10} formation rate for all cell volumes and discharge currents used. Also, we could not discern any dependance of S_2F_{10} formation on electrode material where we used stainless steel, tungsten and Al. (2) No systematic investigation has been made of the dependence on electrode material for the other SF_6 by–products such as SOF_2 or SO_2F_2.

IMPROVEMENT OF THE RELIABILITY OF GIS BY DIELECTRIC DIAGNOSTICS

T.Nitta   M.Sakai   S.Sakuma

Mitsubishi Electric Co. Mitsubishi Electric Co. Mitsubishi Electric Co.
Headquarters R&D   Itami Works   Itami Works
Tokyo,Japan   Switchgear Dep.   Development Dep.
                        Hyogo, Japan   Hyogo, Japan

ABSTRACT

The analysis of the faults and the failures of gas insulated equipment for
A.C. power systems suggested a clear interrelation between the failure rate, the
level of quality control and the stress on the surface of enclosures of the
equipment. It also suggested that some means of diagnostics significantly
improve the reliability of the equipment. The diagnostic means are highly effective to reduce the risk of faults in service.

INTRODUCTION

The collected data of serious faults of gas insulated equipment in five
years period from 1981 to 1985 in Japan are summarized in Table 1 (Denki kyodo
Kenkyu Report Vol. 44, 1988). The number of involved units with the rated voltage from 33kV up to 550kV are 10,795 total at the end of 1985, which consist of
3,157 GIS, 329 Hybrid GIS and 7,309 GCB.

Table 1. Serious Fault Data for Gas Insulated Equipment.

Type	Cause	No. of faults
Dielectric faults	Low gas pressure Lightning surge Particle contamination	1 (4%) 8 (32%) 3 (12%)
	Total	12 (48%)
Thermal faults	Poor contact	4 (16%)
Mechanical faults		9 (36%)

Here, a serious fault means a fault in which some parts of electricity
supply are lost due to the incident. Out of 25 serious faults, 12 are dielectric
faults, 4 thermal faults and 9 mechanical faults, as shown in the Table 1. About
half of all the serious faults are dielectric faults. Three faults are caused by
particle contamination which occurred in the conditions where the electrical

Gaseous Dielectrics VI, Edited by L.G. Christophorou and
I. Sauers, Plenum Press, New York, 1991

stresses are well below the critical stress of the insulating media. One fault is caused by low gas pressure due to human error. Eight faults are caused by lightning surges which are assumed to occur with conducting particles under imp/AC superposed conditions, even if lightning arresters are mounted.

To make the problem clearer, a study on the failures in the factory test on 2,139 units of gas insulated equipment at different rated voltages is supplemented as follows. In the factory tests, A.C. breakdown voltages are measured raising applied voltages up to the A.C. withstand voltage for the tested equipment. Total of 20 dielectric breakdowns are recorded. Those gas insulated equipment are classified into 3 groups depending on lightning impulse withstand level (LIWL) or the ratio of lightning impulse withstand voltage over A.C. operating voltage of the equipment. The lightning impulse withstand voltage and the short-duration power frequency withstand voltage of an equipment are standardized depending on the operating voltage of the system on which the equipment is to be applied. Those insulation levels are represented in Fig.1, as the ratios to the maximum operating voltage for the three groups of gas insulated equipment. The ratios for the rated impulse voltage are 6.1, 4.5 and 4.0 for group A, B and C, respectively. The ratios for the rated short-duration power frequency voltage are 3.4, 2.8 and 2.4 for group A, B and C. Both the ratios decrease in the order of A, B and C.

Out of the 20 failures, 10 failures are for 124 units classified in group C, 9 failures are for 161 units classified in group B and 1 failure is for 1,854 units in group A. The rate of the failure is higher in the order of C, B and A.

The breakdown voltages measured in the test are much lower than design volage. The cause of the failures is assumed to be attributable to a defect of insulation such as particle contamination and the electrical stresses on the surface of enclosures. The posibility to detect such defect by diagnostic techniques under A.C. operating conditions is discussed in this paper.

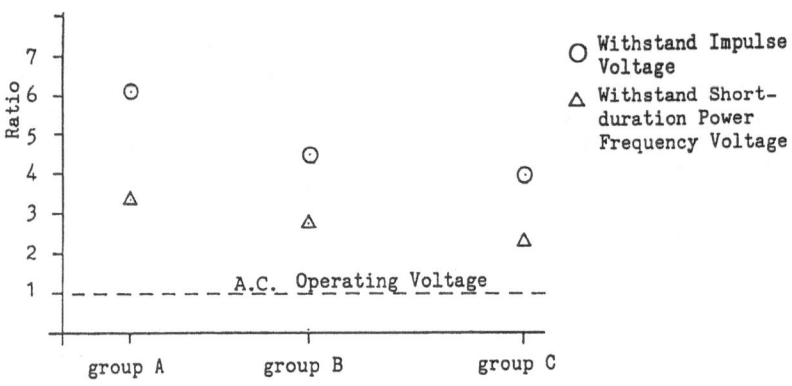

Fig.1. Ratio of Withstand Voltage to the A.C. Operating Voltage.

ANALYSIS OF THE FAILURES

The maximum electrical stress of the GIS on the inner conductor at the rated impulse voltage has been designed almost equal in all of the three groups. This means that the electrical stress on the surface of the enclosure under A.C. operating voltage in the field is higher in the order of C, B and A. The failure rate of each group is higher in the same order. It is interesting to recognize that the rate of the breakdown failures is related to the electrical stress on the surface of enclosure E, not the maximum stress on the conductor.

The histogram of breakdown data and the probability density of failure f(E) calculated by making use of the above data in the A.C. power frequency tests are shown in Fig.2. In the statistical analysis, a normal distribution on the value of E for breakdown probability of the three groups of GIS is assumed. Based on the calculation results in Fig.2, the breakdown probability is 1.6%, 5.7% and 6.1% for group A, B and C, respectively. It is found that the higher the electrical stress E on the surface of enclosures, the higher the breakdown probability.

As the electrical stresses for group A, B and C are differet under the A.C. operating conditions which are denoted as A, B and C in Fig.2, the probability density curve may be converted into three curves for group A, B and C as a function of the ratio of the electric stress E to that of the A.C. operating voltage E0 as shown in Fig.3. The curves suggest that the breakdown probability for group C is higher at a given ratio of E/E0. In other words, group C is much more vulnerable to temporary overvoltages in operation.

The lower the ratio of the rated impulse withstand voltage to the A.C. operating voltage, the higher the electrical stress on the surface of the enclosure under A.C. operating voltage, i.e, the curve is shifted to the left. And, if a reduced lightning impulse withstand level is applied in designing a GIS and the maximum stress st the lightning impulse withstand voltage is kept constant without any countermeasures, A.C. breakdown probability under A.C. operating conditions will increase.

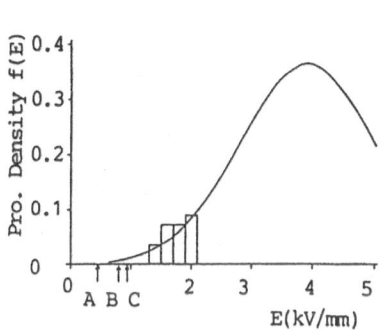

Fig.2.Distribution Curve versus
Electrical Stress E .

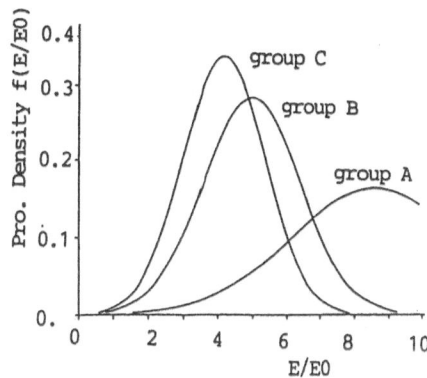

Fig.3.Distribution Curve versus
E/E0 .

INFLUENCE OF BREAKDOWN VOLTAGES BY PARTICLE CONTAMINATION

The breakdowns due to lightning surges in Table I also suggenst the existence of some kinds of insulation defect on the surface of enclosures, since the breakdown voltages are much lower than their basic impulse levels. If we consider the very short duration of the surge voltages, the defect must be active under A.C. operating conditions. It is interesting to discuss whether it can be detected by diagnostic techniques.

Fig.4 is an example to show the effect of the conducting particles on breakdown voltages with free aluminum particles sitting on the enclosure, measured under the superposed lightning impulse voltages on the pre-applied A.C. voltage. In the measurement, the A.C. pre-stresses at the surface of the enclosure are kept at the value between 0.85kV/mm and 1.0kV/mm. The breakdown voltage decreases sharply when the length of the conducting particle exceeds a critical length of 13mm in the example (Hattori et al.,1984).

There is another report which shows that the superposed impulse/AC breakdown voltage is reduced sharply when the A.C. pre-stress is high enough for free conducting particles to cross over the insulating gap (Yanabu et al., 1988). The relation between the length of particles and the crossing electrical stress E on the surface of the enclosure is shown in Fig. 5. It is interesting to note that the particles, whose length is more than 13mm and diameter is 0.25mm, crosses over the insulating gap when the A.C. pre-stress exceeds 0.85kV/mm corresponding to the knee point in Fig. 4. The values of E for the three groups of GIS in Fig. 1 are also given in Fig. 5.

The critical length of the free conducting particle for the three groups of GIS is getting smaller from group A to group C. The conducting particles larger than this critical length must be eliminated completely in the quality control of the production of GIS. This means that elimination of particles in quality control is severer in the order of C, B and A.

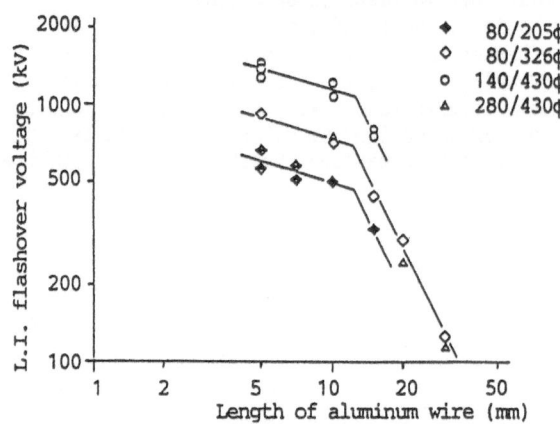

Fig. 4. Superposed Impulse/AC Flashover Voltages in SF6 Gas.

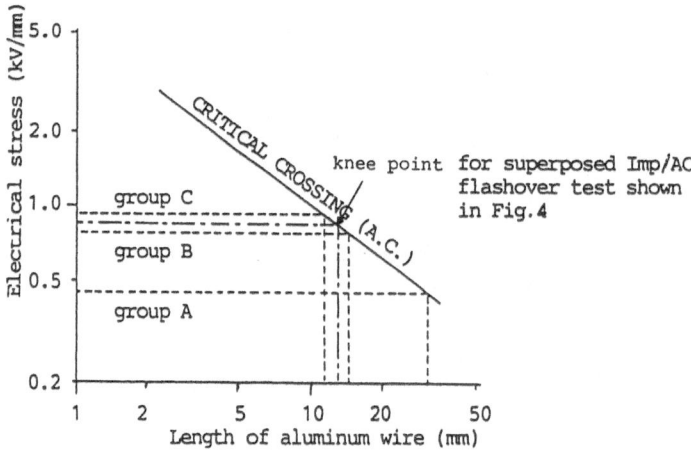

Fig. 5. Electrical Stress on the Surface of the Enclosure for Crossing.

DIAGNOSTIC DETECTORS FOR GAS INSULATED EQUIPMENT

Diagnostic detectors have been developed to detect some anomalies which may lead to the dielectric breakdown of gas insulated equipment (Denki Kyodo Kenkyu Report Vo.42). Even though the GIS is quality-controled not to have any of those harmful defect, still they may be introduced in the operating conditions of the GIS. Those diagnostic detectors are designed to detect those defects introduced afterward. Partial discharges at the defect generate lights, electro-magnetic waves, high frequency currents, mechanical vibrations, chemical compounds under A.C. operating conditions and these quantities have been studied as possible items to be detected in the diagnostic techniques.

The purpose of diagnostic techniques is the improvement of reliablity of equipement, and many people consider that the type of sensors to be mounted inside the enclosures are not suitable since they may interfere the quality of the GIS. Also the devices with external antennae to detect electro-magnetic wave may not be practical, because it is much influenced by the electro-magnetic waves from outside. Electrical potential detectors and mechanical vibration detectors have been recommended by many researchers of the technique.

One practical detector measures the electrical potential difference gene-rated by partial discharges between adjacent enclosures which are insulated each other (Kawada et al.,1988). In actual substations, high levels of noise are in the frequency range of 10MHz or lower, although several peaks are also observed in the frequency range between 10MHz and 20MHz. The noise level is quite low in the frequency range of 20MHz or higher. On the other hand, the pulses generated by partial discharges are very steep of the order of nanoseconds. Therefore, high sensitivity detection of the pulses can be obtained by selection of the tuning frequency at an appropriate range higher than 20MHz. This type of detect-or has the sensitivity of minimum 50pC of partial discharges under the actual field conditions. Fig.6 shows the relation between the length of conducting particles and the apparent electric charge of partial discharges to be measured, when wire type particles are located inside of the cylindrical configuration of gas insulated equipment under A.C. voltages (Hattori et al.,1984). From the test results, the discharge magnitude of the length of 5mm is around 50pC.

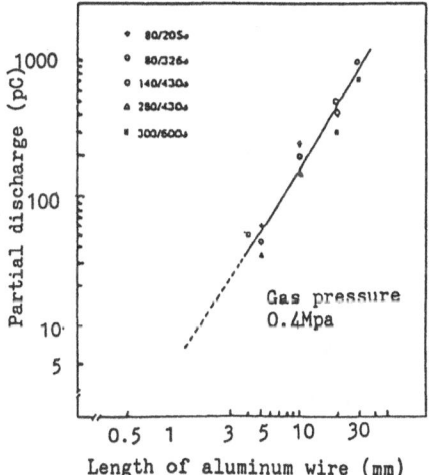

Fig. 6. Partial Discharge Magnitude versus Length of Aluminum Wire.

For mechanical vibration detectors, the partial discharges generate high frequency vibrations at the wall of the enclosure and the vibration waves propa-gate in gas insulated equipment. The mechanical vibration is detected by piezo-electric accelerometer. The frequency of the vibration produced by partial dis-charges is in the range of 5 to 40kHz and the vibration attenuates as it propa-gates in the enclosures. Acceleration produced on the enclosure in the frequency range of 10 to 30kHz is sampled at a time interval of about 10μs for a period of one cycle synchronously with source voltage. The sampling is repeated 64 times in one data collection, and the data are averaged in the microprocessor. By this averaging process, the acceleration in synchronism with the voltage source is selectively detected. The sensitivity of the accelerometer is basically determined by the continuous background noise, which is in the order of a few of $10\mu G$ in normal GIS. Fig.7 shows the relation between the magnitude of discharges and the vibration generated on the enclosure (Sakuma et al.,1989). Mechanical vibration detectors are able to detect more than 50pC partial discharges.

Both type of detectors are effective to detect the harmful conducting particles with margin.

The attenuation for electrical pulses to propagate in enclosures is very small but that for mechanical vibrations is significant. Because of this reason, electrical potential detectors are suitable to apply on-line supervision systems since a large area can be supervised by one detector. On the other hand, mechanical vibration detectors are suitable to apply off-line application to locate the defect when something wrong is detected by electrical means.

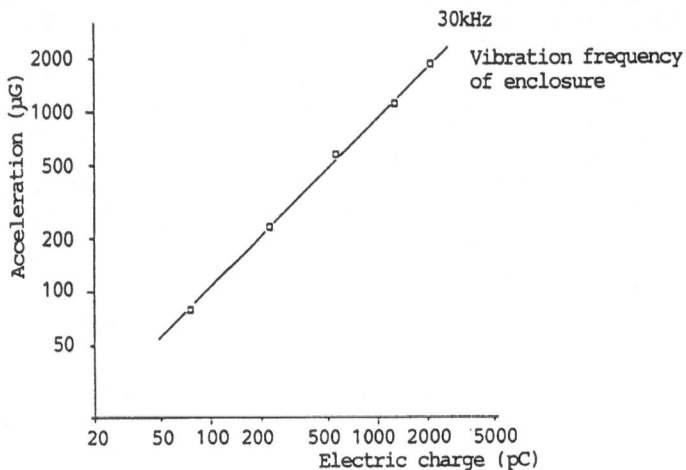

Fig.7. Electrical Charge versus Acceleration .

The result of these analysis shows that harmful conducting particles which may lead to GIS faults can be detected by a diagnostic means under A.C.operaing conditions. A sticking sharp edges may also have simular influence to the free conducting particles, in case all the sticking sharp edges will generate partial discharges higher than those accompanied with free conducting particles. And we may conclude that almost all the defects which may lead to dielectric faults of GIS can be detected by those diagnostic means.

CONCLUSION

When a better performance of lightning arresters or some other voltage limiting means permit to apply a reduced lightning impulse withstand level to a GIS, the reliability of the GIS may be endangered if no countermeasures are applied. In that case, electrical stresses on the surface of the enclosures under A.C.operating condition will increase to enhance the effects of conducting particles on the surface of enclosures. To cope with the problem, the following measures should be taken to maintain the reliabilty of gas insulated equipment;

(1) Better quality control to limit the size of conducting particles,
(2) Adoption of diagnostic techniques.

Practically and economically these are realistic solution to realize a a compact gas insulated equipment.

REFERENCES

1. Denki Kyodo Kenkyu Report, Vol. 42, No. 3 (1988).

2. Denki Kyodo Kenkyu Report, Vol. 44, No. 2 (1988).

3. T. Hattori, M. Honda, H. Aoyagi, N. Kobayashi, and K. Terasaka, A Study on Effects of Conducting Particles in SF_6 Gas and Test Methods for GIS, IEEE, 84 WM 155-8 (1984).

4. H. Kawada, K. Ando, T. Meda, M. Tamura, Y. Murakami, and T. Nitta, Application of Diagnostic Techniques to Gas Insulated Switchgears, Japanese Experience and The Future, CIGRE 23-05 (1988).

5. S. Sakauma, S. Hirogane, and S. Tada, Diagnostics For Gas Insulated Switchgears, EIM-89-5/HV-89-5 (1989).

6. S. Yanabu, N. Kobayashi, H. Ookubo, H. Aoyagi, and S. Matsumoto, A Theoretical Study of Metallic Particle Motion and Particle-Initiated Breakdown in GIS, T.IEE Japan, Vol. 108-B, No. 4 (1988).

DISCUSSION

A. E. D. HEYLEN: In Figure 1 was the impulse voltage of the lightning kind or of the switching kind?

M. SAKAI: Lightning.

DETECTION OF PARTIAL DISCHARGES DURING A LIGHTNING SURGE

B.F. Hampton*, O. Farish* and S. Larigaldie**

*Centre for Electrical Power Engineering   **ONERA
University of Strathclyde   Meudon
Glasgow G1 1XW, UK   France

INTRODUCTION

The wings and integral fuel tanks in modern aircraft may be fabricated from sheets of resin-bonded carbon fibre, producing a light and strong structure. The sheets are joined by coating the mating surfaces with a flexible sealant, and fixing them together with titanium screws.

Both the carbon fibre sheet and the sealant are resistive, and there is some uncertainty over whether discharges of sufficient intensity to ignite the fuel could occur in the wing tanks if they were struck by lightning. Means of detecting any discharge activity in sample joints carrying a current surge were therefore sought.

At first, optical methods using photomultipliers and high speed cameras were used. Subsequently the diagnostic techniques developed previously to detect partial discharge activity in GIS under AC voltages were adapted for this similar role under a lightning surge, and the main purpose of the paper is to describe these.

LIGHT EMISSION FROM A JOINT

In experiments carried out at Aerospatiale, Paris, a test joint made from sections of carbon fibre panel was screwed together and subjected to a through current of 95 kA peak and 4/10 μs waveshape.

When a poor contact had been created deliberately by loosening the nuts on the screws, discharges occurred and were recorded by an Imacon high-speed image-intensifier camera. A record of a discharge developing around one of the nuts was taken at a rate of 2×10^6 f/s, and is shown in Fig. 1.

This technique enabled the performance of a joint which was discharging heavily at high currents to be examined, and much useful information was gained. However at low currents the discharges were confined to the region around the shank of the screw, and were not visible. Attention was therefore turned to the UHF technique.

Gaseous Dielectrics VI, Edited by L.G. Christophorou and
I. Sauers, Plenum Press, New York, 1991

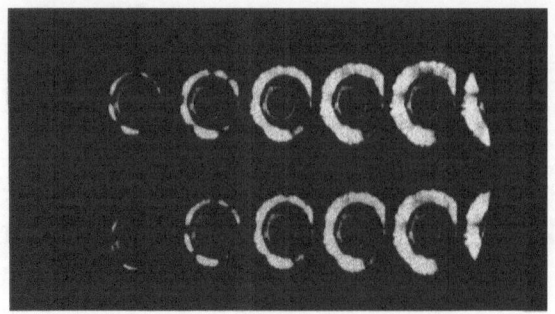

Fig.1. Discharges around a loose nut.

PARTIAL DISCHARGE DETECTION AT UHF

Principle

Success had been achieved in using the UHF technique to detect partial discharges in GIS (see, for example, Hampton 1990). The principle of the method is that the partial discharge pulse has a nanosecond rise-time, and the very high frequencies it contains excite the bus chambers into multiple resonances. Those in the TE and TM modes are particularly useful, and in a GIS reach their peak values in the range of 800 - 1200 MHz. Those measurements had been made using alternating voltages, but there seemed no reason why the technique should not be used equally well with a lightning surge if interference from the generator spark gap could be rejected. There seemed every possibility that this could be achieved, because the surge would have a rise-time of only a microsecond, and not contain the ultra high frequencies being recorded.

The test arrangement is shown in Fig. 2. It was housed in two aluminium chambers, both of size 0.8 x 0.6 x 0.4 m. One of them was simply for screening the surge generator. This consisted of a 20μF capacitor, which could be charged at up to 10 kV and discharged via a manually closed spark gap into the joint.

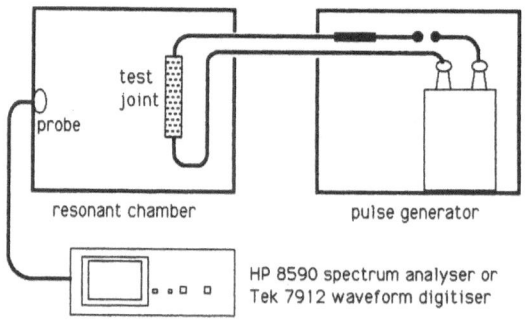

Fig.2. Test arrangement.

The test joint was suspended in the other chamber, which was expected to be set into resonance by any discharge activity. The resonances would be picked up by a 50 ohm D-dot probe mounted in the side wall of the chamber. This was used without an integrator, because it was more useful to achieve the maximum sensitivity that record the true waveshape of the signal.

Resonances in the Chamber

In a preliminary experiment the chamber was excited into resonance by a small repetitive discharge in a sphere gap energised by a separate, DC supply. The resonances were recorded on a 1.8 GHz Hewlett-Packard spectrum analyser, type 8590, and are shown in Fig.3.

The lowest calculated resonance in a higher-order mode for this chamber is the TE_{011} at 308 MHz, which is close to the peak recorded at 310 MHz. Other calculations enabled a further 27 resonances to be identified, including the following:

Table 1. Resonant Frequencies in the Chamber.

Mode	Calculated Resonant Frequency	Observed Resonant Frequency
TE_{011}	308	310
TE_{111} & TM_{111}	396	400
TE_{312} & TM_{312}	869	870
TE_{352} & TM_{352}	1482	1480

Resonances occurring below the higher-order mode cut-off at 300 MHz are in the TEM mode, but it has been found that these are generally of lower amplitude and less useful for diagnostic purposes.

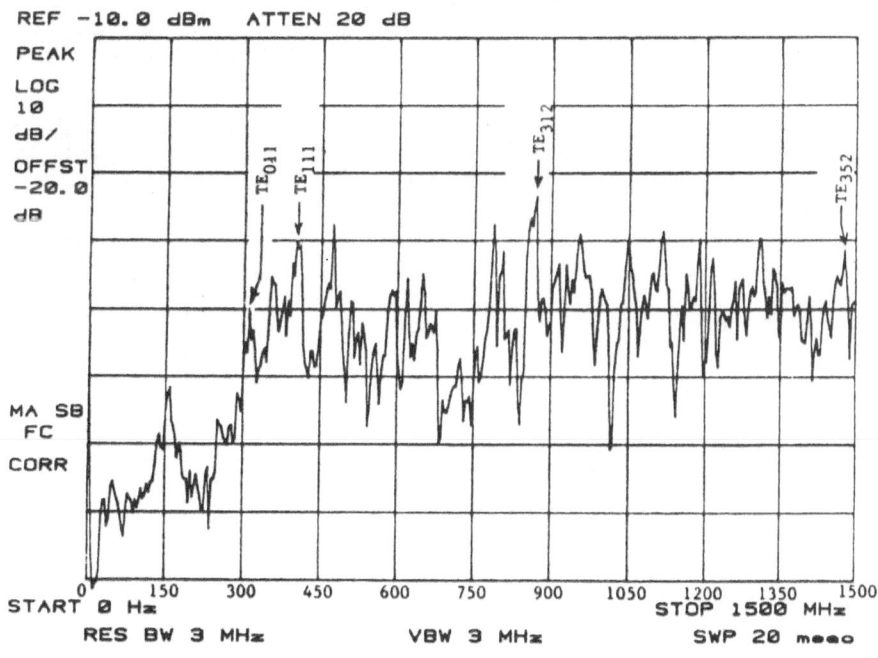

Fig.3. Resonances in the chamber.

PD Detection Under Surge Current

The spectrum analyser has a wide bandwidth and is an ideal instrument for recording repetitive signals. However it is unsuitable for single shot work, when measurements must be made with an oscilloscope. The instrument used was a Tektronix 7912 waveform digitiser, which was limited by its 500 MHz bandwidth and so could capture the signal only in the 300 - 500 MHz window. Nevertheless, it enabled the principle of the measurement to be established.

It will be recalled that the 7912 digitiser uses a scan converter to achieve a high writing speed, but the digitiser has only 512 data channels. These are insufficient to record a 500 MHz waveform on the 5 μs/cm timebase appropriate for the lightning surge, and the result is that only the envelope of the high frequency signal can be seen. Another characteristic of the instrument is that it detects the limits of blooming of the trace, removing these later by an 'average-to-centre' process. In the present case this could not be used without removing the signal, and so even a zero signal appears as two parallel lines. Nevertheless, the presence of a UHF signal is easily recognised once these limitations are appreciated.

JOINT TESTS

Background Interference

To check that the probe did not pick up any interference from the generator spark gap, the carbon fibre joint was first replaced by a copper conductor. The surge current of 5 kA peak passed through it is shown in Fig. 4, and the probe output in Fig.5. No significant interference was detected.

Joint Sample

The first sample was made by taking two sections of carbon fibre in their 'as-received' condition, and screwing them together tightly. The lowest current which could conveniently be passed through the joint was 0.5 kA, and even at this level the probe signal of Fig. 6 was recorded. Clearly the joint was discharging, although there was no visible sign of this. Some minor improvements were then made to the joint, and on re-testing at the same current the discharge level was seen to have fallen slightly (Fig.7).

In subsequent tests the joint resistance was lowered quite significantly, and currents up to the maximum of 36 kA obtainable with the present arrangement could be passed without discharges occurring in the joint or any significant interference being recorded.

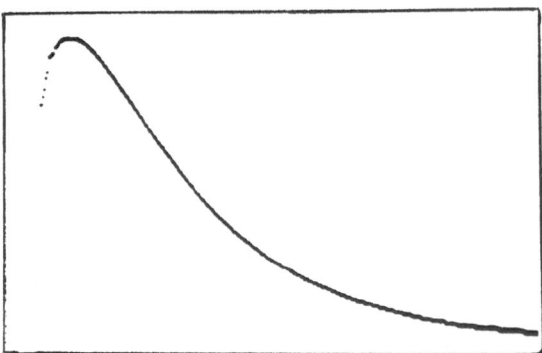

Fig.4 · Current pulse (5kA peak, 1/15μs waveshape).

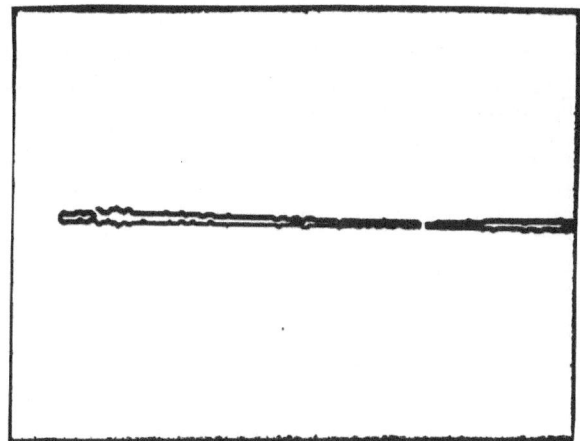

Fig.5. Background level.

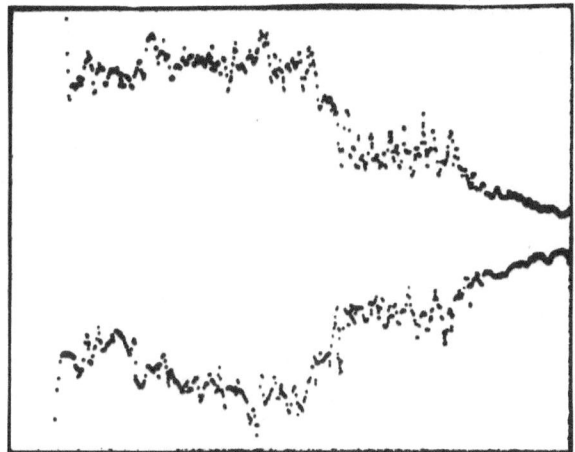

Fig.6. Discharges in joint A.

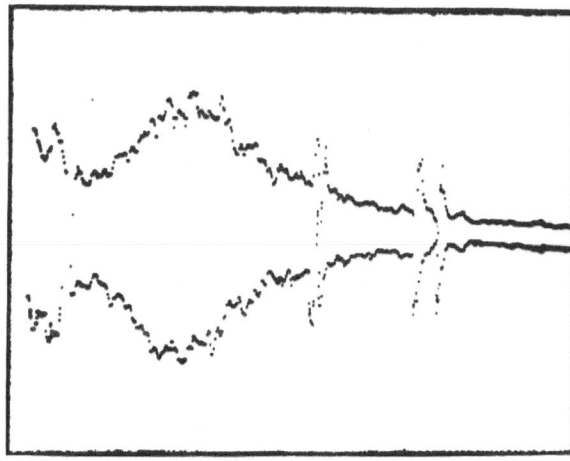

Fig.7. Discharges in joint B.

575

CONCLUSIONS

The UHF technique enables partial discharges, probably as low as 10pC, to be recorded during a lightning surge of some tens of kiloamperes. This suggests the possibility of making similar measurements in a GIS.

The 1 min AC overpotential test is commonly used when commissioning GIS, and is an effective means of detecting free particles and disconnected electrodes. This is especially so when it is used in conjunction with UHF diagnostics.

Other conditions such as a damaged or misplaced shield might be revealed only by a higher-voltage impulse or oscillatory switching surge. A disadvantage of these is that if a breakdown occurred during the test, the larger amount of stored energy could cause unknown damage. However, breakdowns could probably be prevented if the partial discharge level were monitored on each successive shot, and the test stopped when any activity was detected.

REFERENCE

B.F. Hampton, T. Irwin and D. Lightle, Diagnostic Measurements at Ultra High Frequency in GIS, CIGRE, Paris (1990).

DISCUSSION

N. FUJIMOTO: Previous arguments have stated that needle protrusions will result in corona under power frequency stress and, therefore, are detectable by the UHF technique. What, then, are the advantages of applying the UHF technique during lightning impulse application? Do you anticipate an improvement in sensitivity? If so, can you estimate the relative sensitivities to defects for the UHF technique applied during power frequency at impulse voltages?

B. F. HAMPTON: Any protrusion which discharges at the peak power frequency voltage can be detected by the UHF method. Smaller protrusions which may start to discharge only at a much higher voltage need to be detected during a lightning surge. So far we have not made any measurements in GIS under surge voltages and connot estimate the relative sensitivities.

L. NIEMEYER: When you do PD measurements with your coupler under AC do you have background noise problems due to, for example, air corona impulse noise coupled into the system through the bushings? How does such noise affect or limit the detection sensitivity level of the system and which means of discrimination are available against externally introduced noise?

B. F. HAMPTON: The interference of air corona coupled through the bushings is at relatively low frequencies, say up to 100 MHz. The PD signal itself often peaks in the range of 800–1000 MHz, well above the interference.

Y. DOIN: In order to apply PD measurements in on–site tests, we need a calibration of the sensor. The maximum level will be relative to the sensitivity of the coupler used for measuring. How can we manage with this?

B. F. HAMPTON: We calibrate our couplers using a discharge cell containing a 2 mm diameter ball moving in a metal dish. This gives a reproducible signal at a realistically low level.

S. W. ROWE: What are the possibilities of improving on–site testing of GIS with respect to defect detection by using either reduced pressure or another gas to promote discharge activity?

B. F. HAMPTON: These could be very good approaches, and we have done too little work in these areas to be sure at present.

M. GOLDMAN: Can you see some difference in your spectrum between different types of corona, point–plane in air, SF_6, or with dielectric materials?

B. F. HAMPTON: The spectrum is the same in all cases because the resonant peaks depend on the physical dimensions of the GIS chamber. The differences between the various sources appear in their point–on–wave occurrence, and this enables them to be identified.

PHASE RESOLVED PARTIAL DISCHARGE MEASUREMENTS

IN PARTICLE CONTAMINATED SF$_6$ INSULATION

Lutz Niemeyer, Berndt Fruth and Hartmut Kugel

ABB Research Center
C–5405 Baden, Switzerland

ABSTRACT

A study is presented on partial discharges from conducting particles in compressed SF$_6$ insulation under AC stress. A reference experiment was carried out with various diagnostic techniques including a phase resolved partial discharge analyzer. An interpretation of the measurements in terms of discharge mechanisms is given, based on which approximate scaling laws are derived which relate the PD data to the experimental parameters and to the particle characteristics.

INTRODUCTION

Conducting particles are known to reduce the breakdown level of gas insulated systems and to cause partial discharges (PD) below the breakdown level. The detection of such PD has become an important tool for quality control during manufacturing and for insulation monitoring in service.[1] In order to be of optimal use, PD measurements require adequate signal acquisition and evaluation techniques and a physically based interpretation concept. As a contribution to these two issues we present in this paper:

(1) a reference experiment in which a conducting particle on an electrode is exposed to a uniform background field;

(2) a diagnostic system which directly measures the discharge pulses from the particle and analyzes them with respect to charge, position on the AC phase, and frequency of occurrence;

(3) a tentative concept of the physical processes underlying the discharges;

(4) scaling laws relating the measured PD characteristics to the experimental parameters and to the particle characteristics.

EXPERIMENTAL

The most critical type of contamination in compressed gas insulation systems are conducting particles which may be electrostatically lifted and come to sit perpendicular on an electrode. This worst case configuration was chosen for the reference experiment shown in Fig. 1. A protrusion representing the particle is mounted on one electrode of a uniform field gap to which an AC voltage U_0 is

Gaseous Dielectrics VI, Edited by L.G. Christophorou and
I. Sauers, Plenum Press, New York, 1991

applied. The protrusion is connected to the electrode via a terminated 50 Ω cable that serves as a fast current measuring shunt. As shown in Fig. 1a the surface of the protrusion is coated with insulating material except for its tip. The coating is metallized on its outside with a shield which is connected to the electrode. By this arrangement the flow of displacement current to the protrusion is largely suppressed so that the current entering the cable shunt is very closely equal to the true current injected into the discharge from the tip. For very fast pulses, a signal correction is required because of the parasitic capacitance from the protrusion to ground. The shunt signal can be fed to fast oscilloscopes (TEK A7704 or DSA 602) or to a phase resolved partial discharge analyzer (PRPDA)[1] which evaluates the pulse charges in a multi–channel analyzer and stores them together with the AC phase at which they have occurred. The PD pulses thus acquired over an extended time period are processed by a computer and are displayed in a three dimensional diagram with the coordinates frequency of occurrence, charge, and phase. Such diagrams allow to study the phase correlation of the pulses together with their statistical characteristics.

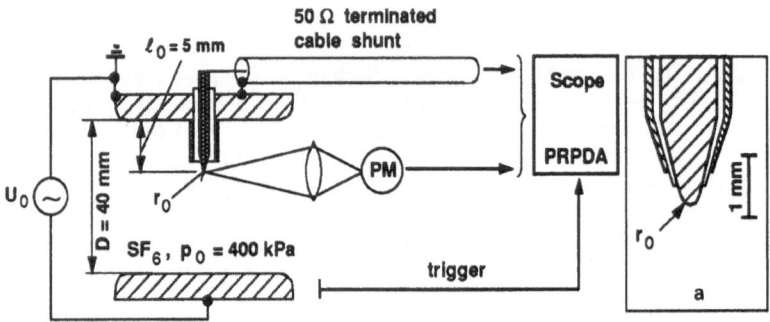

Fig. 1. Reference experiment. a: details of the protrusion tip.

DISCHARGE MECHANISMS

Qualitative Discussion

Partial discharges from point electrodes in compressed SF_6 have been thoroughly studied by Van Brunt and co–workers[2,3] with the following results:

(1) The discharges consist of sequences of pulses, the shape, charge, and repetition rate of which strongly depend on the experimental parameters. For negative polarity, a quasicontinuous background discharge may be observed in addition to the pulses.

(2) The pulse characteristics are strongly polarity dependent.

(3) For negative polarity the discharge inception voltage can be correctly predicted by the streamer criterion.

(4) For positive polarity the generation of initiatory electrons is critical for discharge inception which leads to higher inception voltages than those predicted by the streamer criterion.

These general features are also observed in the AC experiment described here, as can be seen from the oscillograms in Fig. 2. Figure 2a shows the discharge current during a full AC period. During the positive halfwave a randomly spaced sequence of pulses occurs each of which consists of an initial short peak in the ns range (Fig. 2d) followed by a tail of merging smaller pulses (Fig. 2b). During the negative halfwave a quasicontinuous current signal is observed (Fig. 2a), which, at higher time resolution (Fig. 2c), shows strong fluctuations with superimposed short pulses.

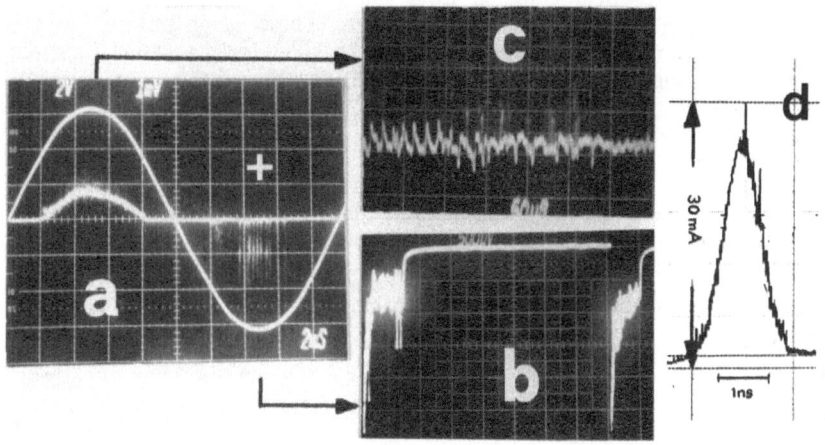

Fig. 2.   Partial discharge oscillograms obtained for p_0 = 400 kPa, ℓ_0 = 5 mm, r_0 = 100 μm and E_0 = 40 kV/cm.   Scales: 20 μA/div., 2 ms/div. a: discharge current over full AC cycle. b: sequence of positive discharge pulses. c: time resolved negative discharge.   Scales for b and c: 10 μA/div., 50 μs/div. d: initial portion of positive streamer burst.

These observations are tentatively interpreted as schematically illustrated by Fig. 3.   Within an overcritical field zone E_{cr} close to the tip of the protrusion, ionization processes can occur if initiatory electrons are provided.

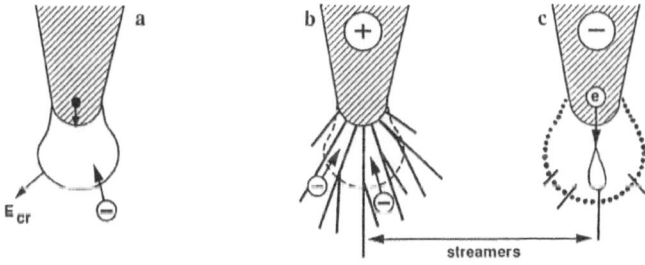

Fig. 3.   Schematic representation of discharge processes.   a: first electron generation, b: streamer corona at positive polarity, c: quasicontinuous discharge and streamer burst at negative polarity.

At positive polarity initiatory electrons are produced by field detachment from negative ions. Hence, a discharge is triggered whenever a negative ion appears in the critical volume E_{cr} (Fig. 3a) and undergoes a field detachment process.   The discharges therefore come in randomly spaced pulses. If the streamer criterion is exceeded the discharges develop into streamer coronae (Fig. 3b) which inject a charge pulse of typically ns duration[4] (Fig. 2d). The ionic space charge left by the corona reduces the electric field and provides detachable negative ions. These drift into the high field region and detach additional primary electrons thus starting secondary discharges which give rise to the pulse tails observed in Fig. 2b.

At **negative polarity** field emission from the particle tip provides abundant electrons thus allowing a fast sequence of discharges. These merge and produce the quasicontinuous discharge current observed in Fig. 2d. The short occasional current spikes are in the ns range and are interpreted as negative streamers.

Scaling Laws

Particles occurring as contamination in gas insulation systems are usually small in the sense that the electric background field, in which they are embedded, does not vary appreciably over their length. It is therefore practical to formulate the discharge characteristics in terms of the (undisturbed) background field E_0 at the particle location[5] and of the particle dimensions, namely, length ℓ_0 and tip curvature radius r_0.

As the background field E_0 is short–circuited by the particle the voltage drop $V_0 = E_0 \cdot \ell_0$ is redistributed ahead of the particle tip where it provides the field enhancement driving the discharge processes. With this concept the following scaling laws can be derived:

The background field E_0^{str} at which streamer inception takes place[6,5] is

$$E_0^{str} = (E/p)_{cr} \cdot p_0 \cdot f(p_0 \cdot \ell_0, \ell_0 / r_0) \tag{1}$$

where $(E/p)_{cr} = 89$ V/mPa is the pressure reduced critical field and p_0 the gas pressure. The function f decreases monotonically with $p_0 \cdot \ell_0$ and ℓ_0 / r_0.

The charge Q_c injected into the streamer corona has been shown[4] to scale approximately as

$$Q_c = [g \cdot 2\pi\epsilon_0 / (E/P)_{cr}] \cdot (\ell_0 \cdot E_0)^2 / p_0 \tag{2}$$

where $g \sim 0.1$ is a geometry factor and ϵ_0 the absolute permitivity of vacuum.

An upper limit for the average current of the negative quasicontinuous discharge can be estimated by assuming space charge limited ion flow. Then according to Sigmond[7] one obtains for point–to–plane gaps of distance D

$$I \approx 2\mu_i \epsilon_0 U^2 / D \tag{3}$$

where μ_i is the mobility of the negative ions[2]

$$\mu_i = C_\mu / p_0 \qquad C_\mu \approx 6 \text{ m}^2\text{Pa/Vs}$$

U the applied voltage, and D the gap distance. We can modify this relation for the uniform field gap with a protrusion by introducing the background field $E_0 \sim U/D$ and by accounting for the deviation from the point–to–plane geometry by a correction factor $\phi(\ell_0/D)$ so that

$$I \approx 2 C_\mu \epsilon_0 \phi(\ell_0/D) E_0^2 D / p_0 \tag{4}$$

We thus obtain a current proportional to the square of the background field and to the gap distance D and inversely proportional to the gas pressure p_0. A rough estimate of the geometry factor is obtained as $\phi \sim 0.6 \, \ell_0/D$ by a comparative evaluation of the electric flux from the electrode tip with the help of numerical field calculations.

It should be kept in mind that eqs. (2) and (4) are only order–of–magnitude estimates and have been derived primarily as a guidance for understanding the major parameter dependencies.

EXPERIMENTAL RESULTS AND DISCUSSION

In addition to the oscillograms shown in Fig. 2, Fig. 4 shows a three–dimensional PRPD plot obtained for the same conditions as the oscillograms in Fig. 2. The pulse response of the PRPD was set such that only the fast streamer burst pulses were acquired (> 1 MHz band width). It is seen that on both half waves the discharge activity is approximately centered to the field maxima. The slight asymmetries have not yet been studied in detail and are believed to be caused by space charge effects due to drifting ions. The negative streamer bursts are more frequent than the positive ones but have smaller charges in agreement with the oscilloscopic measurements and with the interpretation given earlier in the paper.

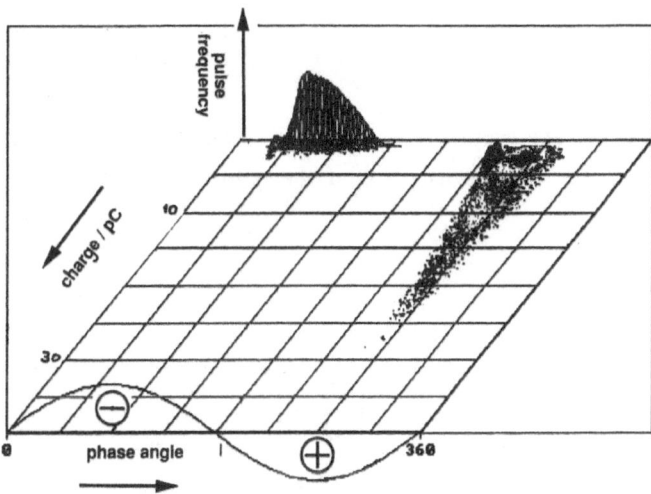

Fig. 4.   Three–dimensional PRPDA diagram.   Same parameters as in Fig. 2. Background field $E_0 = 40$ kV/cm.

The dependence of the discharge activity on the applied background field E_0 is represented in Fig. 5. For negative polarity, the average quasicontinuous current I is plotted (Fig. 5a) whereas for positive polarity the maximum pulse charge $\hat{Q}_c$ is given in Fig. 5b. Both diagrams also contain the inception field according to eq. (1) (arrows) and the scaling relations eqs. (2) and (4) as broken curves. It is seen that the measured inception fields are higher than the predicted ones, the discrepancy being larger for positive polarity in agreement with point (4) under "Qualitative Discussion." Both the negative quasicontinuous current I and the positive pulse charges are found to be of the theoretically estimated order of magnitude and show the predicted scaling with E_0^2.

CONCLUSIONS

The study of partial discharges from a protrusion in a uniform field has led to the following results:

(1) The discharges show a strong polarity asymmetry. At positive polarity, streamer bursts are dominant whereas at negative polarity a quasicontinuous discharge is the strongest phenomenon.

(2) Approximate scaling laws derived from a physical discharge concept relate the discharge intensity to the experimental parameters and are confirmed by the experimental data.

(3) A more precise discharge modeling will be required for an exact quantitative description of the experimental data.

(4) The PRPDA technique has been demonstrated to be a useful tool for a comprehensive analysis of the pulsed component of partial discharges.

The following conclusions may be drawn with respect to the interpretation of PD measurements in compressed gas insulated systems:

(a) The strongest and therefore easiest–to–detect discharges are positive streamer bursts. For sharply pointed particles of $\lesssim 10$ mm length and under typical GIS pressures around 400 kPa they may incept at several 10 kV/cm. At typical AC design fields in the $50 - 100$ kV/cm range pulse charges up to several 100 pC can be expected.

(b) The pulse charges are proportional to $(\ell_0 E_0)^2$ and are therefore a very sensitive measure of the particle length ℓ_0 if the position of the particle, i.e., E_0, is known.

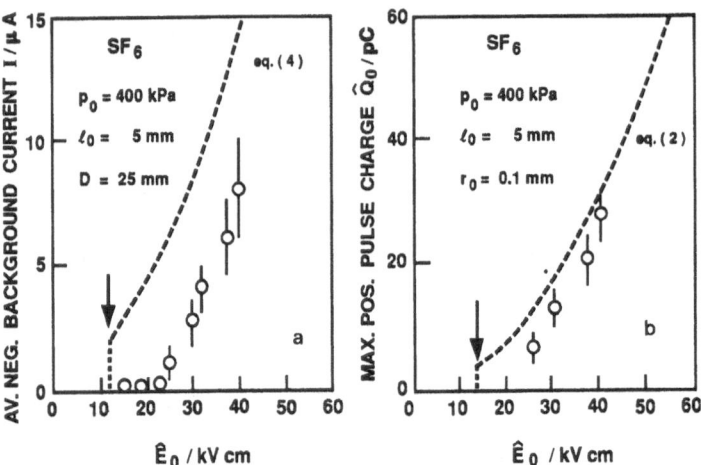

Fig. 5.   Characteristics of positive and negative partial discharges in dependence of the applied background field E_0. a: negative average continuous current I. b: maximal charge Q_c of positive streamer bursts. Broken curves are estimates according to eqs. (2) and (4).[8]

(c) The polarity asymmetry in conjunction with phase sensitive detection and the different background fields at both electrodes may be used to localize the position of a particle.

REFERENCES

1. J. Fuhr, M. Haessig, B. Fruth, and T. Kaiser, PD Fingerprints of Some High Voltage Apparatus, IEEE Symp. on El. Insul. June 3–6, Toronto (1990).

2. R. J. Van Brunt and M. Misakian, Mechanisms for Inception of DC and 60 Hz AC Corona in SF_6, IEEE–EI, Vol. EI–17, No. 2 (1982) p. 106–120.

3. R. J. Van Brunt and D. Leep, Characterization of Point–Plane Corona Pulses in SF_6, J. Appl. Phys. 52,11 (1981) p. 6580–6600.

4. L. Niemeyer, L. Ullrich, and N. Wiegart, The Mechanism of Leader Breakdown in Electronegative Gases, IEEE–Trans. El. Ins. Vol. EI–24 (1989) p. 309–324.

5. L. Niemeyer, Leader Breakdown in Compressed SF_6: Recent Concepts and Understanding, Paper of this conference.

6. C. Cooke, Ionization, Electrode Surfaces and Discharges in SF_6 at EHV, IEEE–PAS 94 (1975).

7. R. S. Sigmond, Simple Approximate Treatment of Unipolar Space–Charge–Dominated Coronas J. Appl. Phys. 53 (1982) p. 891–898.

DISCUSSION

R. J. VAN BRUNT: This is just a comment that when considering a repetitive pulsating discharge such as discussed here, one can expect memory effects to play a significant role. In particular, the ion space charge moving away from the point electrode from previous pulses will affect the field in which subsequent pulses develop. Therefore, correlations between not only successive pulses, but also successive bursts can be expected. A method for assessing such memory effects is available [see R. J. Van Brunt and S. V. Kulkarni, Rev. Sci. Instrum. 60, 3012 (1989)].

O. FARISH: You have interpreted the "burst" of current recorded under AC positive–point conditions on the basis of individual avalanches occurring around the point. We made some studies of DC corona in SF_6 [Farish and Ibrahim, ISH, Milan, (1979)] which showed that, at any instant, there is only a single filamentary discharge at the point, whose initial formation is accompanied by the large pulse at the beginning of the "burst", and that the subsequent "chain" of pulses is associated with restriking of this channel at a frequency of ~ 1 MHz. This channel dies out after 10–100 μs, and is followed by a new large pulse. For a slow rising AC waveform, this type of discharge will be more likely to occur than be "brush" type of streamer discharge which occurs for impulse, and which you have used in your modeling. Do you have any photographic or other evidence of the spatial appearance of the corona, and would the correlation be better if a single streamer channel was considered?

L. NIEMEYER: We have rough indications that the "abundance" of discharge channels occurring in the positive corona burst is strongly dependent on the experimental parameters as in the time separation between individual pulses. Although we have not yet sufficient experimental evidence, we expect both the frequency of discharge pulses and the complexity of each discharge pattern to increase, for example, with voltage. Conversely, when reducing the voltage, we would expect a transition into the regime described by Farish and co–workers. Further work is planned to quantify this conjecture.

R. T. WATERS: Recent work on AC corona modeling in air has been reported by Rickard et al. at this Symposium (see page 413). This is a development of an earlier method in the 1970's. The new work has shown that the magnitude and phase of the corona current is well simulated. An application to AC corona in SF_6 systems would seem to be now indicated.

L. NIEMEYER: Although air corona and SF_6 corona have many qualitative features in common their quantitative difference is substantial. The major difference is caused by the streamer propagation mechanism which implies critical field drop along the streamer in SF_6 and a field drop much lower than the critical one in air. This affects, among others, the corona structure and the major scaling laws relating the corona characteristics to the gap parameters.

AN ELECTRO OPTIC METHOD FOR TRANSIENT

FIELD MEASUREMENT IN GIS

Ian Chalmers, Walter Johnstone, Graham Thursby and Paul Coventry*

University of Strathclyde   *Nat.Grid R &D
Glasgow UK   Leatherhead UK

INTRODUCTION

It has been demonstrated[1,2] that compressed gas insulated systems (GIS), because of their compact coaxial geometry, can have transient overvoltages created within them of amplitudes up to 3 times the normal system voltage and rise times as short as a few nanoseconds. The occurence of such overvoltages is known to be associated with a combination of disconnector switch operation and the existence of residual charge on isolated sections and it has been suggested that the resultant travelling waves can stimulate resonance at microwave frequencies depending upon the geometry of the system. The overvoltages themselves are highly undesirable in that they can cause damage to parts of the system employing solid dielectric insulation such as voltage transformers and gas barriers.

The accepted method of monitoring these transients with the required high bandwidth has been by using electrical probes[3] either of the conical or planar D-dot type which are essentially C-R dividers or by straightforward capacitance couplers. The output from D-dot probes has to be integrated to reproduce the original E field and obviously the fidelity of the system is only as good as the integration involved. Also it must be borne in mind that any electrical method can be seriously upset by "noise" and by transient rises in the potential of the GIS enclosure; therefore it would clearly be an advantage if the electric field could be measured directly using a procedure which was immune from pick-up.

Optical methods of electric-field measurement exploiting the Pockels effect have been widely reported[4,5] and are known to offer the advantages of direct measurement and high bandwidth (>1GHz). This paper reports the design, assembly and characteristics of a 1GHz-bandwidth electric-field measurement system which comprises a lithium niobate Pockels cell addressed by 20m-long optical fibre leads. The system is intended ultimately for specific application in GIS-installation monitoring and the optical-fibre address arrangement is intended to provide flexibility in terms of siting the signal recovery equipment remote from the GIS system. By siting the instrumentation in a well screened enclosure, for example, it is possible to achieve complete isolation of the electronic measurement stages from the electrically "noisy" measurand environment.

Gaseous Dielectrics VI, Edited by L.G. Christophorou and
I. Sauers, Plenum Press, New York, 1991

SYSTEM DESCRIPTION

The general layout of the optical measurement system is shown in Figure 1. Its heart is the Pockels cell transducer consisting of a lithium niobate crystal (25mm x 9mm x 9mm) and a quarter waveplate sandwiched between two polarisers as shown in Figure 2. Lithium niobate in its crystalline form is electro optic and birefringent which means that it can be arranged to modify the polarisation of transmitted light in response to an applied electric field[6,7]. In general,

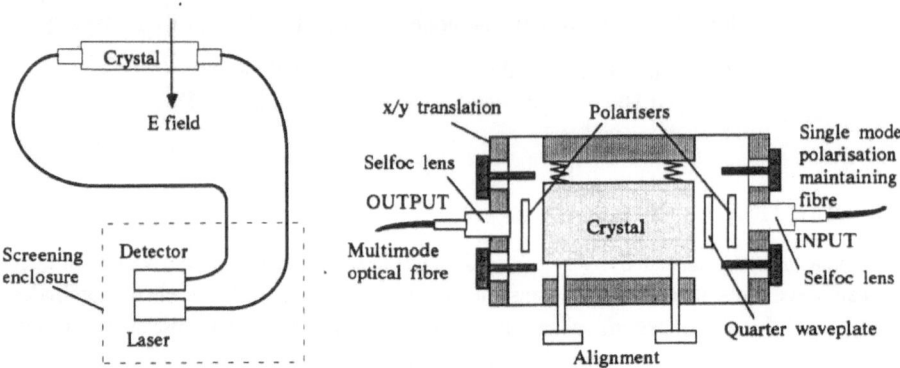

Fig.1. General layout.   Fig.2. Crystal mounting arrangement.

by sandwiching the crystal between crossed polarisers, the transmitted intensity varies sinusoidally with applied field as in the generalised response function shown in Figure 3. For the specific system reported here the incident light is polarised parallel to the x axis of the crystal and propagates along the z axis. The electric field is applied normal to the propagation direction along the x axis of the crystal. In this configuration the crystal presents no static birefringence to the incident light and hence for zero applied E field the polarisation of the incident light is not modified as it traverses the crystal. With no quarter wave plate, the system would then be automatically biassed at a peak or a trough on the raised sinusoidal response curve depending upon whether the polarisers were aligned parallel or orthogonal. By inserting the quarter wave plate between the input polariser and the crystal, the response curve is then biassed at the mid point of its linear section (point A in Figure 3). For the configuration described here, the bias point is stable and temperature independent. For most other configurations, however, the bias point determined by the static birefringence of the crystal is arbitrary and highly temperature sensitive.

The light source is a 0.82μm semiconductor laser. Its output is linearly polarised and is aligned using an elasto optic modulation technique[8] such that the input light polarisation is parallel to one of the major axes of the polarisation-maintaining optical fibre. This provides a stable linear polarisation at the transducer which can then be readily aligned to the input polariser. Collimation of the output from the optical fibre and collection of the transmitted light through the transducer are achieved using graded index (Selfoc) lenses. At the remote end of the system the transmitted light from the transducer is detected by a silicon avalanche photodiode with a bandwidth of 2GHz. Since the required information is in the form of optical intensity only, it is sufficient to use multimode fibre between the transducer and the detector.

SYSTEM ANALYSIS AND DESIGN

The most important characteristics of the system are the dynamic range, the sensitivity (ie the minimum detectable electric field) and the bandwidth. In the design process these specifications were critically analysed.

Dynamic range and sensitivity

For the present Pockels cell configuration with a peak transmitted light intensity, I_o, the response function (transmitted intensity I_T as a function of the phase shift ø induced by the applied field, between the orthogonal polarisations in the crystal) is given by[6]

$$I_T = I_o \sin^2 ø/2 \tag{1}$$

The phase shift ø is itself given by

$$ø = 2\pi n_o^3 r_{22} E_x L / \lambda_o \tag{2}$$

where n_o, r_{22}, E_x, L and λ_o represent respectively the ordinary index of lithium niobate (=2.286), the applicable electro optic coefficient of lithium niobate (= 3.35 x10^{-12} m/V at high frequency and 6.2 x 10^{-12}m/V for DC to audio frequency[7]), the applied electric field (V/m), the crystal length (m) and the free-space wavelength (m).

Using equations 1 and 2, the applied electric field E_π required to induce a π phase shift between the two orthogonal polarisation states in the crystal and hence to modulate the transmitted intensity from a peak to a trough in the response curve may be calculated for a given crystal length L.

For λ_o = 820nm and for high frequency fields (>10^7Hz),

$$E_\pi = 10245/L \text{ (V/m)} \tag{3}$$

The system is required to operate on the linear part of the response curve and thus the operating range (with a maximum permisible deviation of 5% from linearity) is $0.34E_\pi$ centred on the mid point (A in Figure 3). Hence from equation 3 , the maximum measurable E field can be derived as

$$E_{max} = 1741/L \text{ (V/m)} \tag{4}$$

Analytical studies[2] have demonstrated that overvoltages of up to three times the normal operating voltage can be generated in a typical GIS installation. For the NEI Reyrolle YG1 system which is designed for an operating voltage of 400kVand in which the inner and outer diameters are 127mm and 636mm respectively, the maximum transient field at the inside surface of the enclosure may be calculated as 2 MV/m. Any crystal with a dielectric constant of ε placed at this position would then be subjected to a field of 2/ε MV/m. Since the dielectric constant of lithium niobate is 32, the maximum field in the crystal may be calculated as approximately 62kV/m. Hence from equation 4 the optimum crystal length for high frequency fields is 28 mm. The crystal employed in the present work was, in fact, 25mm long.

To optimise the sensitivity (ie the minimum detectable field) and the dynamic range, the optical power launched into the system from the laser section must be adjusted such that the detector just saturates when the maximum field (62kV/m in this case) is applied. The linear operating range of the response curve is from $0.25I_p$ to $0.75I_p$ where I_p is the peak optical intensity transmitted by the system. Hence to ensure maximum sensitivity, the laser drive current must be adjusted such that the saturation intensity of the detector (I_s) is 0.75 I_p. If the system is biassed at the mid point of the linear portion of the response curve (ie at 0.5I_p) then, for the optimum crystal length, application of the maximum applied field induces a linear change in intensity from 0.5I_p to 0.75I_p (ie from 0.66I_s to I_s) Hence the optimum transducer responsivity, R (the gradient of the linear part of the response curve) is given by

$$R = 0.34 \ I_s/E_{max} \ \ W/Vm^{-1} \tag{5}$$

If it is assumed that the minimum detectable change in the incident optical power is the noise equivalent power (NEP) of the detector, then the minimum detectable electric field change (E_{min}) which generates a signal-to-noise ratio of 1 is given by

$$E_{min} = (NEP)/R \tag{6}$$

The optical detector used in the present work saturates at an incident power of 11.6 μW and the NEP is 0.05μW. Hence using equations 5 and 6, the responsivity and minimum detectable GIS field are respectively 1.9μW/MVm^{-1} and 26kV/m (about 0.8kV/m in the crystal) for a maximum GIS field of 2MV/m.

Bandwidth

The bandwidth of the optical system depends upon the temporal response of the transducer and also upon the bandwidth of the detector itself (2GHz). The linear electro optic effect is very fast with response times in the sub picosecond regime. Hence the bandwidth of the transducer is determined by the photon transit time (τ) of the crystal. Obviously this is determined by the length of the crystal and the speed of light and for the optimum length of 28mm, τ may be calculated as 183ps indicating a bandwidth of approximately 1GHz. The bandwidth is thus limited by the transducer although clearly the bandwidth could be increased by using a shorter crystal at the expense of sensitivity and dynamic range.

SYSTEM CHARACTERISATION

The intensity response of the optical system was determined by applying DC fields to electrodes directly in contact with the crystal. Figure 4 shows the measured response curves for both positive and negative applied fields. It is interesting to note the asymmetry with respect to the polarity of the applied field; this reflects the importance of aligning the z-axis of the crystal precisely with the incident beam. From the response curve of Figure 4, E_n is 230kV/m (from equation 2, the calculated theoretical value is 214kV/m).

To assess the system response to a transient free-space field, the transducer was placed between two parallel plane electrodes 50mm apart to which was applied the output voltage from a coaxial cable generator. The voltage appearing across the plates consisted of a long series of individual pulses each 30kV in amplitude, with a rise time of about 20ns and a duration of about 1μs. The pulses were separated one from the other by about 5ns although, after a few pulses, due to high frequency attenuation in the pulse forming cable, the waveform essentially resembled a DC step voltage. This then subjected the system to a fairly onerous test in terms of frequency. Figure 5 shows oscillograms of the applied field (upper trace) as measured by a coaxial capacitance divider and the response of the optical system (lower trace) for three time-base settings. Figure 5a clearly shows that the rising portion of the field, although delayed due to the 20m length of the optical fibre, is faithfully reproduced. In future work the risetime will

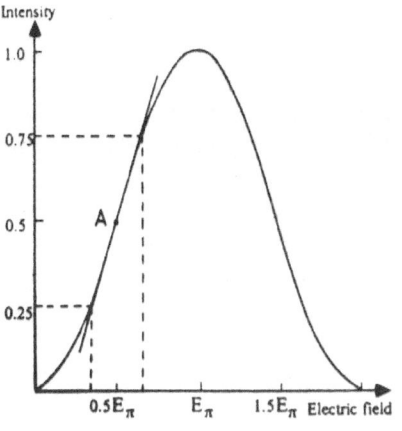

Fig.3. Generalised response curve.

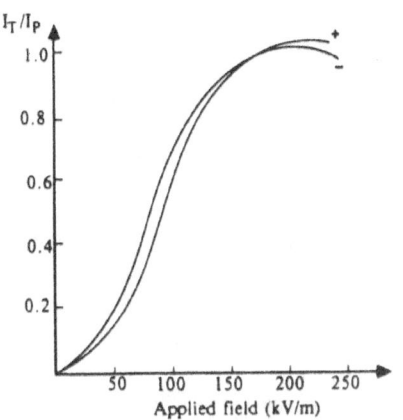

Fig.4. Measured Response.

Fig.4. Measured Response.

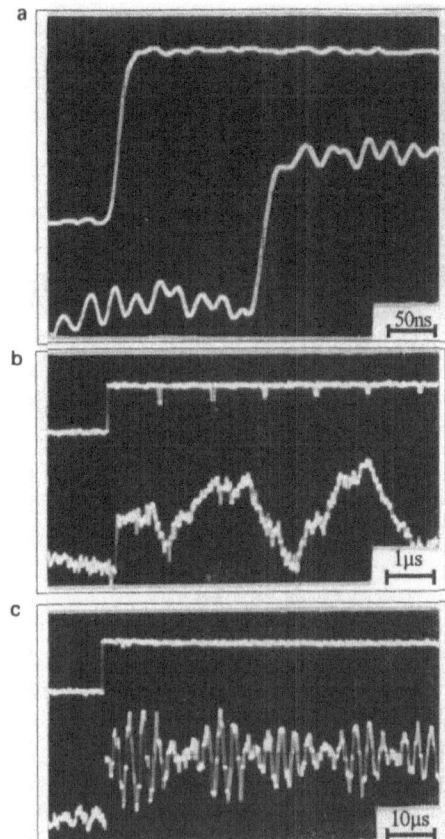

Fig.5. Oscillograms of applied voltage
and signal from optical detector.

be reduced by utilising pressurised SF$_6$ in the switch. The magnitude of the optical signal suggests an applied field in the crystal of 16.5kV/m whereas the value crudely estimated from the applied gap voltage of 30kV is 22.5kV/m.

The obvious noise on the signal merits some discussion. Initially it was believed that this noise was due to insufficient screening of the detector. Accordingly the detector was placed in a Belling Lee screened-room enclosure and the persistence of the noise then demonstrated that it was due to instability in the laser output probably arising from optical feedback into the laser cavity as a result of reflection from the optical fibre input. It should be possible to eliminate this by better design of the launch optics. By operating the system with the laser de energised, it was shown that the electrical noise was greater than 10dB down on the laser noise level.

From Figures 5b and c, it is clear that the broadband applied transient field has induced two dominant resonances in the optical system at frequencies of 3.3×10^5Hz and 8.3×10^4Hz approximately. These are attributed to the transverse and longitudenal acoustic modes of the crystal which (perhaps unfortunately) is piezo electric as well as electro optic. In any practical device however, these resonances could readily be filtered out if necessary.

CONCLUSIONS

A 1GHz bandwidth transient free space electric field measurement system based on a lithium niobate Pockels cell has been realised and assessed. The use of optical fibres allows isolation of the sensitive detection equipment from the hostile environment of the measured field. An analysis has been carried out for the system response and has been used to design the Pockels cell for optimum performance in terms of dynamic range, sensitivity and bandwidth for a GIS application. Absolute accurate calibration of the system has not yet been attempted and may be difficult to achieve although the response appears to be in reasonable agreement with that derived from theoretical considerations both for DC fields and for transient free-space fields. In terms of using the device in a GIS application, accurate calibration and, in fact, temperature drift, should it be present, might not be important since there will always be present a well defined reference field due to the operating voltage on the system.

It is intended that more work will be done using different profiles of applied voltage pulse to explore the bandwidth limitations of the system to compare with the predicted 1GHz. This will be achieved by sharpening the switching gap performance by employing presurised SF_6. Additionally the problem of the noise on the laser signal will be addressed.

ACKNOWLEDGEMENTS

The authors are indebted to the United Kingdom National Grid Co plc who sponsored the work.

REFERENCES

1. L. C. Campbell and A. Aked, Disconnector operations in gas insulated substations, CIGRE Paper 33-09, (1988).
2. B. F. Hampton and T. Irwin, Monitoring of GIS at ultra high frequency, Proc. 6th Int. Symposium on High Voltage Engineering, New Orleans, Paper 23.02, (1989).
3. S. J. MacGregor, O. Farish and I. D. Chalmers, The switching properties of SF_6 gas mixtures, Proc. 7th IEEE Conf. on Pulsed Power, Monterey, CA, P1.08, 510, (1989).
4. H. M. Hertz, Capacitively coupled KD*P Pockels cell for high voltage pulse measurement, J.Phys. E. Sci. Instrum., 18, 524, (1988).
5. E. A. Ballik and D. W. Liu,Measurement of high voltage pulses employing a quartz Pockels cell, IEEE J. Quantum Electronics- Quantum Electronics Letters, QE 19, 7, 1166, (1983).
6. J. Wilson and J. F. B. Hawkes, "Optoelectronics: An Introduction", Prentice Hall, International (1983).
7. A Yariv, "Optical Electronics" Holt Saunders, New York, (1985).
8. S. L. A. Carrara, B. Y. Lim and H. J. Shaw, Elasto optic alignment of birefringent axes in polarisation -holding optical fibre, Opt.Lett., Vol. 11, 470, (1986).

DISCUSSION

S. W. ROWE: What would the performance of this type of equipment be with respect to mechanical vibration?

W. JOHNSTONE: Two sources of noise would arise from vibration. (1) Mechanical misalignment of the optics leading to intensity modulation. This is reduced by ensuring fixed, stable assemblies. (2) Modulation via the piezo electric, electro optic effect. It is difficult to say at this point without details of the vibrational levels.

ANALYSIS OF SF$_6$ DISCHARGE BY OPTICAL SPECTROSCOPY

*Veysel Zengin, Şefik Süzer, Ali Gökmen and **Ahmet Rumeli
*Chemistry and **Electrical Engineering Departments
Middle East Technical University, 06531 Ankara, Turkey
                        and
M.Sezai Dinçer
Gazi University, Electrical Engineering Department
Maltepe, 06570 Ankara, Turkey

ABSTRACT

Optical emission spectra of SF$_6$ and CF$_3$H and mixtures of them containing nitrogen have been examined in the 250-600 nm range under positive and negative high voltages. In SF$_6$ discharge nitrogen impurity lines dominate the spectrum although features due to F and S atomic and ionic emission and a new molecular emission band around 420-500 nm are observable. It is found that the polarity of the electrode has a profound effect on the characteristics of the emission and favors nitrogen in negative polarities. The same effect, to a lesser degree, is observable in CF$_3$H discharges as well.

INTRODUCTION

Analyses of the emitted electromagnetic radiation from gaseous discharges have long been employed to investigate and/or quantify the mechanism of the various physical and chemical processes occuring in discharges[1-10]. Of the many gases investigated, nitrogen (or mixtures containing it) is the most thoroughly studied one. It is also encountered in various discharges when gases contain the unwanted nitrogen or air impurities. This is especially the case in SF$_6$ discharges since its excitation cross section is very small in comparison with that of nitrogen. Hence, the emitted light in the UV-Visible region is completely dominated by nitrogen even in samples of SF$_6$ containing less than 1% nitrogen[9].

In previous studies, however, no observation was reported on the polarity dependence of the spectral features of the main species or impurities. In this work, we report the spectral observations of the emitted light from SF$_6$, CF$_3$H and their mixtures with nitrogen at both high pressure (corona) and low pressure (glow) discharges under positive and negative high voltages.

EXPERIMENTAL

The experimental set-up is shown in Figure 1. The light output from the high voltage d.c. discharge in a pseudo point-plane geometry is admitted into the entrance slit of a 25cm optical monochromator (CVI Digikrom 240) via a small window on the electrode. Photomultiplier output is recorded via a 12 bit ADC to an IBM PC for data manipulation. SF$_6$ and CF$_3$H, both with a quoted purity of 99.5 %, are used as supplied.

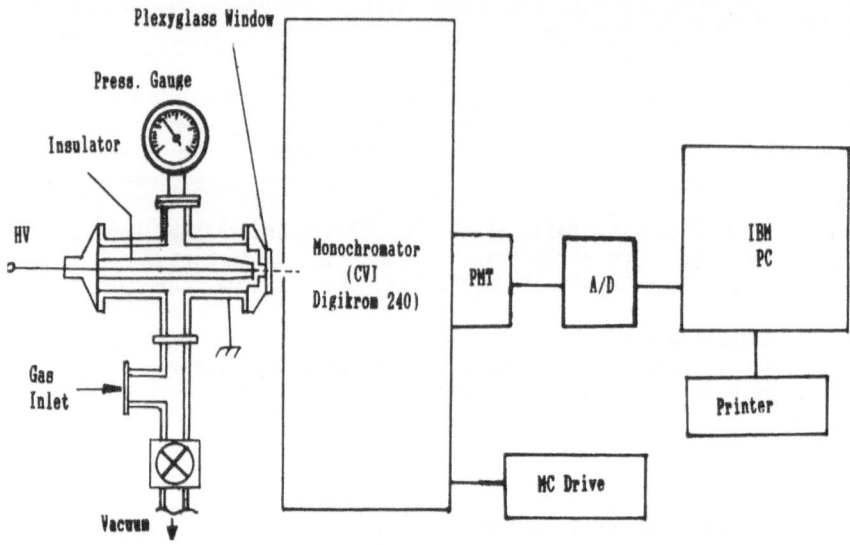

Fig. 1. Schematic diagram of the experimental set-up.

RESULTS AND DISCUSSION

As was reported earlier[9,10], emission spectrum of SF_6 is dominated by the impurity features mainly due to nitrogen both at high and low pressures and under both positive and negative high voltages. Features due to atomic and ionic F and S can also be seen. In corona discharges at pressures larger than 300 mbar, a new emission band around 420-500 nm spectral region is observed under positive high voltage as shown in Figure 2 (b). Figure 2 (c) displays the expanded region between 420-520 nm. The emission is clearly due to a molecular species with an average vibrational spacing of 290 cm^{-1}. This band disappears completely when the polarity of the electrode is changed to negative as shown in Figure 2 (a) and now only the nitrogen features are observable.

In order to get a better understanding of the processes occurring in positive and negative corona discharges, CF_3H is also subjected to the same test. Now, the emission is due mainly to CF_3 and CF_2 as was reported earlier [11-13]. Figure 3 (b) shows the emission spectrum of CF_3H under positive corona discharge. Changing the polarity of the electrode to negative causes again the features to weaken as shown in Figure 3 (a). Figure 4 shows the spectrum of a mixture of CF_3H with approximately 20 % nitrogen under (a) positive and (b) negative corona discharges. Similar to the case of SF_6, nitrogen features dominate the spectrum especially at negative polarity.

From the spectra we can deduce a qualitative order for the emission cross-section of these three molecules. Accordingly, N_2 has the largest one, CF_3H comes second and SF_6 has the smallest excitation/emission cross section. This order can not be explained by comparing the first ionization or excitation energies of these three molecules which are very close to each other[14-19]. The strong dependence on the polarity of the electrode, together with the diverse behaviour of these three gases suggest that second order processes, like collosional excitation, are main factors affecting the emission of light in the UV-Visible region.

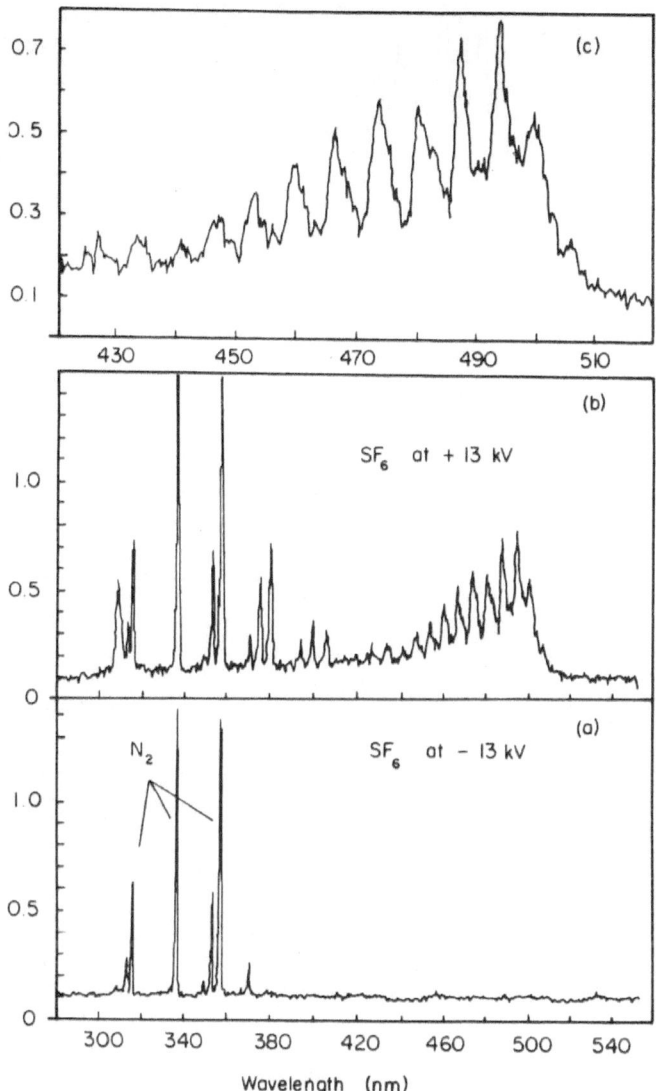

Fig.2. Emission spectrum of 300 mbar SF_6 (a) at −13 kV and (b) at +13 kV corona discharge. The expanded spectrum of the positive corona is shown in (c).

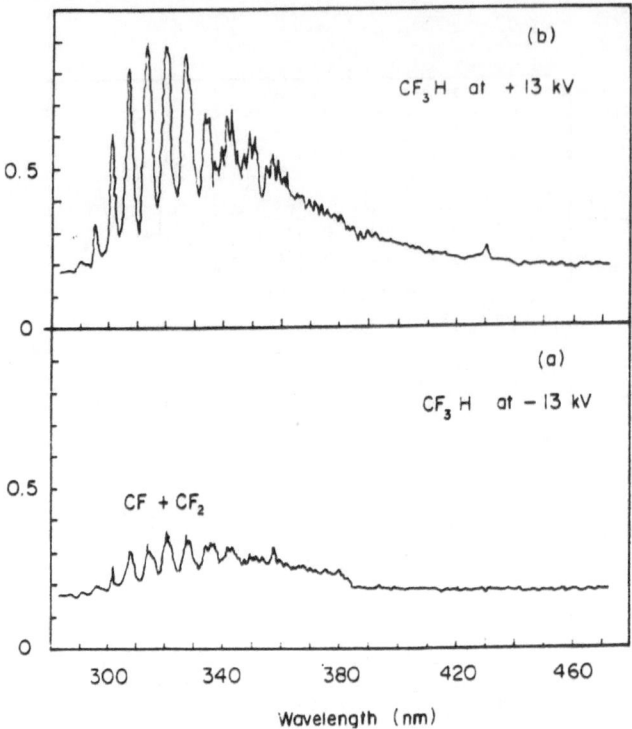

Fig. 3. Emission spectrum of 600 mbar CF_3H (a) at -13 kV and (b) at
(b) at + 13 kV.

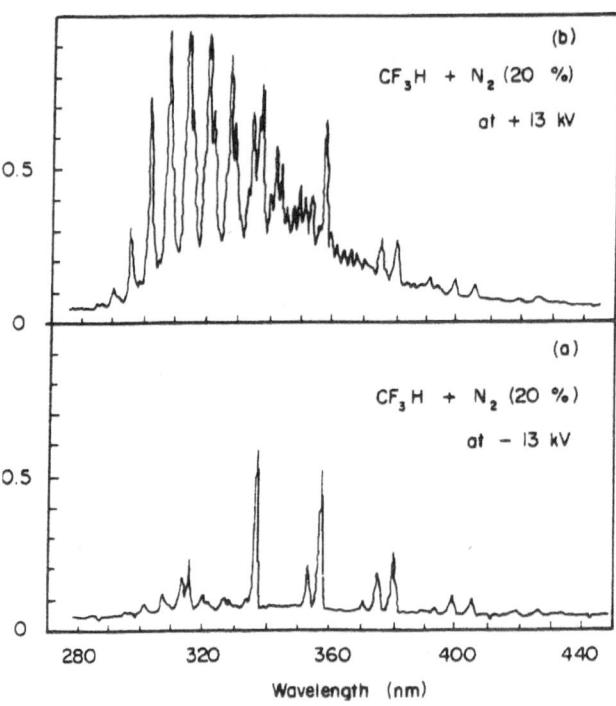

Fig.4. Emission spectrum of 600 mbar mixture 80% CF_3H + 20 % N_2
(a) at -13 kV and (b) at +13 kV.

ACKNOWLEDGEMENT

This research is supported by the Research Fund Project AFP-89-01-03-05 of the Middle East Technical University, Ankara, Turkey.

REFERENCES

1. J. Janca, V. Kapicka and A. Talsky, Application of N_2 Band Spectra in Diagnostics of Non-Isothermic Plasma, Proc. 1st Int. Conf. Gas Discharges, IEE London 1970, 138-142.
2. A. D'Alessio, F. Beretta and M. Diana,Disequilibrium Effects in Nitrogen Arc Plasma, Proc. 2nd Int. Conf. Gas Discharges, IEE, London 1972, 352-4.
3. I. Sander, Thermalization of Spark Channels in Nitrogen, Proc. 2nd Int. Conf. Gas Discharges, IEE, London 1972, 401-2.
4. N. Ikuta and K. Kondo, A Spectroscopic Study of Positive and Negative Coronas in N_2,O_2 Mixt., Proc. 4th Int. Conf. Gas Discharges, IEE, Swansea 1976, 227-30.
5. J.M. Baronnet, J. Rakowitz, P. Fauchais and J.F. Coudert, Selective Excitation of Nitrogen by Argon Atomic Metastable, Proc. 4th Int. Conf. Gas Discharges, IEE, Swansea 1976, 354-7.
6. W. Pfeiffer and A. Leitl, Spectroscopic Investigation of Breakdown Development in SF_6, Proc. 6th Int. Conf. Gas Discharges, IEE, Edinburgh 1980, 16-19.
7. H. Dienemann, SF_6 High Current Discharge, Proc. 6th Int. Conf. Gas Discharges, IEE, Edinburg 1980, 71-3.
8. G. Buchet, R. Haug and J. Maftoul, Time-Resolved Spectroscopy of a Free Burning Arc, Proc. 6th Int. Conf. Gas Discharges, IEE, Edinburg 1980, 116-7.
9. T.H. Teich and R. Braunlich, UV Radiation From Electron Avalanches in SF_6 with Small Admixtures of Nitrogen, in Gaseous Dielectrics V, L.G. Christophorou and M.O. Pace eds., Pergamon Press, New York 1984.
10. T.H. Teich and R. Braunlich, UV and Other Radiation from Discharges in Artificial Air and Its Constituents, Proc. 8th Int. Conf. Gas Discharges, IEE, Oxford 1985, 441-4.
11. M.M. Millard and E. Kay, Difluorocarbene Emission Spectra from Fluorocarbon Plasmas and Its Relationship to Fluorocarbon Polymer Formation, J. Electrochem. Soc. 129:160 (1982).
12. H.A. van Sprang, H.H. Brongersma, F.J. De Heer, Electron Impact Induced Light Emission from CF_4, CF_3H, CF_3Cl, $CF_2Cl_2$and $CFCl_3$, Chem. Phys. Lett. 35:51 (1978).
13. I. Tokue, T. Honda and Y. Ito, Electron-Impact-Induced Light Emission from $CXCl_3$ (X=H,F,Cl.Br), Chem. Phys. 140:157 (1990).
14. R. Bleekrode and W. van Benthem, Spectroscopic Investigation of High-Current Hollow-Cathode Discharges in Flowing Nitrogen at Low Pressures, J. Appl. Phys. 40:5274 (1969).
15. D.J. Burns, F.R. Simpson and J.W. McConkey, Absolute Cross Section for Electron Excitation of the Second Positive Band of Nitrogen, J. Phys. B 2:52 (1969).
16. A.W. Johnson and R.G. Fowler, Measured Lifetimes of Rotational and Vibrational Levels of Electronic States of N_2^*, J. Chem. Phys. 53:65 (1970).
17. I. Gallimberti, J.K. Hepworth and R.C. Klewe, Spectroscopic Investigation of Impulse Crona Discharge, J. Phys. D 7:880 (1974).
18. D.J. Burns, D.E. Golden and D.W. Galliardt, Electron Excitation of the $E^3\Sigma_g^+$ State of N_2 and Subsequent Collisional Deactivation and Energy Transfer to the $C^3\pi_u$ State, J. Chem. Phys. 65:2616 (1976).
19. S. Stokes and A.B.F. Duncan, Electronic Transitions in Methyl Fluoride and in Fluoroform, J. Am. Chem. Soc. 80:6177 (1958).

CHAPTER 14: REPORTS OF DISCUSSION GROUPS

GROUP DISCUSSION ON FUTURE DIRECTIONS FOR STUDIES ON

GASEOUS DIELECTRICS

PANELISTS: R. J. VAN BRUNT (CHAIRMAN),
O. FARISH, M. GOLDMAN, D. KÖNIG, I. SAUERS,
J. E. THOMPSON, J. DE URQUIJO AND J. M. WETZER

R. J. VAN BRUNT: First, I will introduce the panel members: Prof. Dieter König, Technical University, Darmstadt; Dr. Isidor Sauers, Oak Ridge National Laboratory; Dr. J. de Urquijo, Instituto de Fisica, Mexico; Dr. Max Goldman, CNRS, France; Dr. Owen Farish, University of Strathclyde, U.K.; Dr. J. M. Wetzer, Eindhoven University of Technology, The Netherlands; and, finally, Prof. Jim Thompson, University of New Mexico.

Each panelist will give a short presentation on areas for future research and problems to be considered, and after that the session will be opened for discussion. We will begin with Prof. König.

D. KONIG: By way of introduction I am an engineer, with more than ten years experience in industry (development, testing) and more than ten years experience in education and research in cooperation with industry.

There are four points that I would like to make.

(1) <u>There is a need to continue research in gaseous insulation and dielectrics.</u>

This recommendation should be made, since it appears that a shift towards solid state research has taken place in our physical science departments. Most of the fundamental studies on gaseous dielectrics are still mainly performed in high–voltage engineering institutions. Relevant studies on gaseous dielectrics have nearly completely died out at industrial research centers and in testing laboratories.

If research work is not continued at a certain level, existing knowledge will disappear in a short time. If it fails to interest young people to this field, it will have no future. The use of gaseous insulation, mainly atmospheric air and SF_6, is a fundamental part of our electrical power distribution system and those who have to obtain missing basic knowledge on gaseous dielectrics (people from the utilities and industry) suffer more and more.

(2) <u>Fields of interest from the standpoint of a high voltage engineer with respect to different power apparatus (improvement of conventional apparatus).</u>

We need improved and deeper fundamental knowledge on the function of gaseous systems used for insulation and switching. I feel that there is a gap in our knowledge with respect to special switching tasks as performed by GIS–disconnectors (and also air–disconnectors) or fuses, i.e. switching equipment, which have to operate at small load currents.

(3) <u>Basic investigations with repect to possible new applications of new gaseous dielectrics.</u>

We should encourage the chemical industry to develop new dielectrics which are not dangerous to humans, to the environment, and to materials. Since fluorocarbons will disappear in industrial production in the near future, substitutes are urgently needed, some of them for interesting power engineering applications, such as for combined insulating and boiling–cooling systems. These promising systems are already partly used in thyristor–guided drives miniaturized by forced cooling, and for rapid–city transportation systems. Prototypes of these new chemicals are recently available, but their dielectric performance is still unknown.

We should remember the tremendous push towards useful technical applications, which was given by the gift of SF_6, offered by the chemical industry. New factories for GIS have been built which provide work for thousands of people and nearly ten thousand articles on SF_6 problems have been published by researchers. We should state that gaseous dielectrics are part of the discipline of materials science, and progress can be expected from new gaseous materials as well as from other promising solid state materials.

(4) <u>Study of gaseous dielectrics in special environments (e.g., space).</u>

Special environments are characterized, for instance, by low pressure gas or "vacuum" (low to ultrahigh) and/or the presence of metallic and non–metallic (which may be organic or inorganic) layers of different wall–thickness on the electrodes. Outgassing effects, which produce gaseous dielectrics, should be studied. There is a need to increase the voltage range for power supplied in satellites. However, the insulating problems are still widely unsolved. This might be a subject of cooperation with people dealing with vacuum problems. A "dirty" vacuum may be understood as an environment containing small amounts of gaseous dielectrics.

I. SAUERS: There are two areas that I would like to present to this discussion group. These areas will only be discussed briefly since I mainly would like to hear what other people have to say on these topics. Both of these areas are subjects in which I am currently conducting research, so I am particularly interested in finding out how other researchers in this field feel about them as subjects for future work.

The first topic is the need for improved detection techniques for hazardous gases that may be found in GIS. I do not mean to imply that there is a significant hazard now, but rather how we develop the detection techniques that will be needed in the future particularly in light of the fact that they may be required by regulatory agencies. The second item is the need for a better understanding of ion chemistry especially under discharge conditions. I will make a few comments on each of these two topics.

<u>Improved detection techniques for hazardous gases in GIS</u>

There is a future trend of increasing regulatory requirements to ensure occupational safety and environmental compliance, placing increasing demands for improved detection techniques of hazardous materials. For SF_6 under discharge conditions, many of the gases generated as a result of SF_6 decomposition range from toxic (SOF_2) to highly toxic (S_2F_{10}). Development of techniques that can be applied to SF_6 by–products will not only be desirable from the point–of–view of diagnostics but mandatory if maximum permissible exposure levels are enforced. For example, recently basic research in this area has been successful in regard to S_2F_{10} detection through the collaborative efforts of Oak Ridge National Laboratory (ORNL) and the National Institute for Standards and Technology (NIST). Through that work the importance of defining sound gas sampling procedures has, also, been demonstrated.

Specific areas of interest include:
- improved gas analysis instrumentation
- improved gas sampling procedures
- improved understanding of gas–surface interactions
- improved data base.

Gas sampling has recently been found to be a non–trivial problem due to instabilities resulting from interaction of the gas with the surface. For example, one of the by–products of SF_6 discharges, S_2F_{10}, not only decays in a stainless steel chamber, but its decay rate depends on the size of the chamber. Another by–product of SF_6, SOF_2, has also been found to react on the walls of the storage container.

Gas–surface interactions are not only important for the development of detection techniques but also for understanding what is going on in electrical discharges. Dr. Goldman will elaborate futher on the subject of gas–surface interactions. With regard to an improved data base, there are many gaps in our knowledge of cross sections and rate constants that can be applied toward improved gas kinetic models.

The second topic that I want to present deals with the ion chemistry of gaseous dielectrics, particularly SF_6.

Need for better understanding of ion chemistry under discharge conditions.

Is there a need to understand ion interaction/conversion processes? The answer is yes.
Some reasons:
- collisional detachment
- electrode modification due to ion–surface interactions.

Are there other reasons?

We have shown in our paper at this conference (I. Sauers and G. Harman, A Mass Spectrometer Study of Ionization in SF_6 Corona: Influence of Water and Neutral By–products, p. 421), that the ion chemistry can be quite complex and highly dependent on impurities and other species which increase in density with time due to the discharge itself. This is why I am emphasizing that measurements should be conducted under actual discharge conditions.

What about positive ions?

Are there advantages to knowing more about the nature of the positive ions in electrical discharges, especially in the case of SF_6? Dr. de Urquijo will have further comments on this subject.

J. DE URQUIJO: I would like to talk about the need for more accurate, mass–identified ion transport data for electronegative gases, and, in particular, for SF_6 which is the most widely used gas in the electrical industry today. The great importance of having accurate electron swarm parameters is obvious, but it is also true that this gas, mostly used under conditions where negative ions, formed by both electron impact and ion–molecule reactions, plays an important role in discharge development. Therefore, a thorough understanding of the complex discharge scheme of SF_6 is highly desirable. This implies the identification of the various ions involved in the discharge developement, as well as the determination of their relevant transport properties as described by the coefficients of mobility and diffusion, the ion conversion rates, their recombination coefficients, and possibly, their velocity distribution over as wide E/N and N ranges as possible. Ion transport research on this gas has a long history, covering about four decades. Apparatus and analytical techniques have been systematically improved. However, it appears that in a number of experiments on the determination of swarm and transport parameters, many processes occur simultaneously, thereby complicating the interpretation and derivation of the relevant swarm parameters (e.g., ionization, electron attachment and detachment, electron and ion drift, ion conversion reactions). Recent reviews of the research on ion transport in SF_6 indicate that most of the work related to this matter has suffered from experimental difficulties regarding the identification of the drifting ions. In fact, most experiments carried out after 1970 have relied upon the only mass–analyzed data of Patterson, both inside and outside the E/N range that was covered (5–140 Td). It has not been until very recently that drift tube–mass spectrometer techniques, developed in the mid 1960's for the study of mostly atomic systems, have now been used for the study of ion transport in this gas. Thus, mobility and longitudinal diffusion data are now available for SF_5^-, SF_6^-, SF_5^+ and SF_3^+ in SF_6 over an E/N range extending up to about 500 Td.

Drift tube–mass spectrometers (DTMS) capable of working at high gas pressures have reached a state of maturity; recent DTMS also allow measurements up to E/N ~2000 Td, and novel techniques for measuring ion velocity distributions could be adopted for further research on this gas.

M. GOLDMAN: Among the subjects which, in my opinion, merit special interest for future studies are:

(1) <u>SF_6 stability in the pressence of metal surfaces</u>

Until now, much attention has been paid to SF_6 decomposition products from discharges, especially arc discharges, and to the effects of SF_6 gas impurities and of these by–products on the dielectric behavior of SF_6 insulated equipment, but much less interest has been paid to the possible importance and role of SF_6 by–products which may arise from gas–surface interactions. In high–pressure SF_6 (atmospheres at ambient temperature) such interactions may occur even from a simple gas–metal contact.

First indications of SF_6 reactivity on metal surfaces can be simply obtained from thermodynamical data listed in the table below, which shows that reactions between fluorine and zirconium, aluminum, or copper (to a lower extent) are possible.

Standard heat of formation of SF_6 and metallic fluorides

Compound	ΔH° (k J mole^{-1})	Compound	ΔH° (k J mole^{-1})
SF_6	- 1095	NiF_2	- 665
AlF_3	- 1300	CrF_2	- 757
$AlF_3, 1/2 H_2O$	- 1492	CrF_3	- 1108
CuF_2	- 530	ZnF_2	- 785
$CuF_2, 2H_2O$	- 1147	ZrF_2	- 961
FeF_2	- 743	ZrF_3	- 1463
FeF_3	- 1016	ZrF_4	- 1860

This does not mean that such reactions will occur or will occur rapidly; their occurrence will depend on their activation energy and reaction rates.

From experimental data already available from photoelectron spectroscopy (XPS) analyses made on metal surfaces after exposure to SF_6 atmospheres, we have found that:

- fluorine is selectively fixed by dissociative absorption, generally in good agreement with the thermodynamical expectations;
- fluorine fixation generally increases with gas purity;
- pre–absorbed oxygen limits fluorine fixation which may be interpreted as a consequence of a competitive behavior between oxygen and fluorine, also explaining the preceding point;

604

- fluorine fixation is also influenced by the polishing and cleaning procedures used for surface preparation; reactions between superficial traces of organic groups from the polishing and cleaning agents on the one hand and fluorine liberated underneath by dissociative absorption of SF_6 on the other hand will easily take place, if not hindered by the above mentioned oxidation effects on the material laid bare between its surface cleaning and its exposure to SF_6.

In connection with the fixation on the surface by fluorine originating from the absorption of SF_6 molecules, the remaining SF_6 fragments must be liberated in the gas. So the surface plays a role in producing SF_6 by–products. According to some results already available from mass spectrometric analyses made by us, it seems that the species involved in the gas composition evolution resulting from gas–surface interactions are mainly HF, SO_2, H_2O, CS_2 and F_2.

How all of these gas and surface changes subsequently affect the dielectric behavior of SF_6–insulated equipment is still unknown, but to illustrate that such effects may be of importance, we can simply refer to the paper presented at this conference by K. Hadidi and A. Goldman (Current Stability of Negative Corona Discharges in SF_6 and Delayed Spark Breakdown, p. 399) which shows how preconditioning or autoconditioning of the high–field electrode, at least in the case of negative corona discharges, can play a major role in the discharge stability up to breakdown, even under low stress discharge conditions.

Thorough investigations should be carried out on the different aspects of the problem, in particular on the mechanisms and kinetics of gas–surface interactions without and with electrical stress and on subsequent effects on the insulating properties of practical systems, taking into account the possibility of improving them by an optimized choice of materials and surface preparation for the metallic elements of the systems.

(2) Applications of gaseous dielectrics for purposes other than insulation

This point might be discussed in "Group Discussion on Industrial Outlook for Gas Dielectric Needs and Uses", p. 615. While some new applications may only need a transfer to practical uses of basic knowledge already acquired, others may need stimulation or assistance by basic research. It is towards these areas that I want to draw your attention; for instance, to the following fields dealing with detection and diagnostic techniques:

- induced changes in the electrical behavior of the discharge for gas traces or solid particles detection;
- plasma chemical reactions produced by the discharge for different purposes such as detection of gas traces or neutralization of toxic species;
- discharge–induced changes in surface properties (e.g., for surface diagnostic purposes).

O. FARISH: I would like to discuss the following three items:

(1) SF_6 Insulation
- Conditions for direct breakdown from fixed protrusions in quasi–uniform fields
- Partial discharge characteristics

(2) New dielectric gases/mixtures
- high intrinsic strength
- non–uniform field performance
- voltage–time characteristics
- trigger delay–jitter
- switching characteristics

(3) Gas–Solid Interface
  - surface and bulk charging mechanisms
  - charge migration–charge diffusion
  - surface tracking on contaminated insulators
  - discharge surface interaction
    - dielectrics
    - electrodes .

In the case of SF_6 insulation the main application is GIS. Over the next few years we will see increasing interest in diagnostics. The question is what basic studies are required to support these new developments. In the case of protrusions, for floating particles there are techniques such as the Ultra High Frequency (UHF) technique that Dr. Hampton has discussed, but the situation with fixed protrusions is not so clear. One thing to look at are different geometries to investigate different configurations of protrusions on a quasi–uniform field surface.

We heard at this conference about discharges in voids. We need useful information on partial discharges with respect to the characteristics of various flaws such as particles, points, and voids. Basic studies are needed to help identify these in GIS, either by UHF or other techniques, to determine the nature and size of the flaws in the system.

In the area of new gases/mixtures, we need to pay attention to the performance of new gases/mixtures under non–uniform field conditions. We also need to look at the V–t characteristics. In addition we might consider the combined effects of high electric stress and very fast transients on gas breakdown.

The last topic deals with the question of the gas–solid interface. We need to know more about the surface charging mechanisms such as when spacers become charged. This knowledge can aid in the development of surface treatments and the formulation of new materials.

Perhaps the most important topic is the study of surface tracking mechanisms on insulator surfaces that are contaminated by particles or charge. This could lead to the development of new materials that are resistant to damage caused by flashover.

Finally, work is needed on discharge surface interactions for dielectrics and electrodes, particularly for high current discharges. Applications include plasma processing and pyrolysis for destruction of toxic waste.

J. M. WETZER: In order to discuss future directions for studies on gaseous dielectrics we have to address the question of what basic research is necessary, desirable and possible. A key question is whether the breakdown threshold can be made higher. A major problem is the complicated breakdown process itself.

General Comments

  - The parameters involved are non–linear, for instance α/p is a steep function of E/p, particularly for gases where α is initially low (e.g., in SF_6).
  - The variety of collisional processes in technical gases is large; we have to consider many types of neutrals and ions in addition to the electrons. To obtain a correct list of the important processes with their cross sections requires much careful laboratory work.
  - In inhomogeneous fields and at higher voltages field distortion by space charge is of decisive importance in the breakdown process. Breakdown, then, depends on interrelated processes (mathematically on a closed set of equations), namely, initial ionization, space–charge buildup, field enhancement (in 3–D), more ionization and drift of electrons, often to a region in space where the field is lower. Whether breakdown occurs depends on the precarious balance of these processes.

- In technical geometries the field geometry is always inhomogeneous; not only in the overall sense, but also on a small scale as a result of local irregularities of the electrode surfaces.

Necessary Studies

<u>Breakdown studies</u> in real situations remain necessary to provide interesting cases for more detailed study, such as described below, or confirmation of predicted values of the breakdown voltage.

<u>Cross–section measurements</u> with beams, electron or ion beams, can provide necessary data on effective cross sections. Information can be obtained on specific processes.

<u>Avalanche studies</u> have the advantage that the avalanche is a pre–breakdown phenomenon quite close to the real breakdown process. These studies can be done in different forms:

(1) "Slow avalanche methods" with a time resolution of at best 1 μs can be used to obtain information on slower fundamental processes between mainly ions and neutrals. The problem is how to distinguish between the various ion species.

(2) "Fast avalanche methods" with a time resolution down to about 1 ns can be used to obtain information on the fast processes. These processes are almost invariably caused by the electrons. Since the electrons are necessary for the most important ionization process (α), it is highly interesting to follow the growth of the electron population.

<u>Further discussion on the fast avalanche method</u>

- Fast initiation is necessary, for instance, by short laser pulse illuminating the cathode.
- Fast measurements are necessary. This is not only a matter of the electronics, it is also very much dependent on the input circuit. It is amazing how important the coupling capacitor is to partial discharge studies, and, in contrast, how much the input capacitor and the coupling capacitor are neglected in avalanche studies.
- Digital recording of the avalanche current is necessary for efficient data handling.
- Computer models are necessary to simulate avalanche current waveforms and to derive avalanche parameters from experiments.
- Ramo–Shockley effects should be accounted for in homogeneous and inhomogeneous fields; Ramo–Shockley effects are very important in inhomogeneous fields. As argued above the breakdown process is more complicated in this case, but in addition to that the interpretation of the measured pre–breakdown or avalanche current is much less straightforward in inhomogeneous fields.
- Collision processes on a very short time scale do not have to be separately treated if they do not affect the electron energy distribution in the avalanche head.

Future Directions

Future research could benefit from a close coupling of beam studies and avalanche studies. Beam studies provide the necessary information on cross sections, whereas avalanche studies translate this information to electron growth. Furthermore, avalanche studies should be extended to incorporate inhomogeneous fields and space charge buildup in order to obtain a better insight into the first stages of breakdown. This approach can yield additional valuable information for the study of corona, in particular pulsed corona, and its applications.

J. E. THOMPSON: I will briefly present some ideas about future directions, applications, and needs associated with the insulating, conducting, and plasma characteristics of gas dielectrics. I have broken this topic into three application areas: (1) pulsed power associated with the delivery of high power at short delivery times; (2) possible industrial/commercial applications of gaseous dielectrics and gaseous conductors; and (3) space applications. There is an explicit need for basic models and parameter diagnostics to improve understanding.

In the category of pulsed power needs, in the context of switches, there is a need for switches that will transfer large amounts of charge and have long life. The technical aspects in accomplishing these goals are to have electrodes that have long lifetimes and this would imply that there is a need to study interactions of the electrodes with the gas conductor.

Pulsed power needs/applications (basic understanding/measurements)

Switches
- High Q, long life (electrodes, diffuse discharge, surface interactions);
- Fast recovery for repetitive operations (recombination, electrode life, cooling);
- Opening switches (plasma models, measurements, optical effects, new ideas);
- Light controlled (on/off capability);
- Fast switches.

There is a need for opening switches. What we would like to do is to store energy inductively rather than capacitively. Good ideas associated with opening switches are needed. Also there is a need to find ways to use light for controlling switches. It is important to have switches that can operate in the nanosecond regime and, in some applications, sub–nanosecond time scales.

Other topics associated with pulsed power include insulation designed for high fields but for a short time. Some insulation systems that work well for short times may not work for long times. A vacuum (or low pressure gas) is a good example of a good insulator over a short time, but not over a long time.

Other needs
- Insulation designed for very high E but short time ($V_{bd}(t)$);
- Short–pulse insulator–surface flashover ($V_{bd}(t)$, role of surface, electrode emission, gas);
- High frequency (1–16 MHz) bulk gas and surface insulation for high power rf–microwave devices—concepts, models—measurements;
- Atmospheric ionization/conduction (due to a need for transmitting power from one place to another in the atmosphere);
- Plasma ion sources (model basic plasma–gas measurements);
- Electromagnetic launchers (measurements, diffuse moving arc, lifetime);
- Use of "supercomputer" codes.

There is a need for insulating systems that will work from 1–16 MHz. An example is a high power rf–microwave system.

Commercial/Industrial Applications

The following are a few examples of commercial/industrial applications and concerns.

- Switches and insulators for high T, high radiation environment
- Utility applications
- Lightning and other transient protection
- Effluent emmision scrubbing
- Plasma waste decompositon and/or management
- Plasma–based chemistry/separation technology
- Predictive diagnostics (passive and active)
- Safe dielectrics

- Plasma Processing
  - semiconductor etching/cleaning
  - metal–surface hardening/corrosion resistant
  - surface polymerization/films
  - surface coatings
  - metal refinishing/melting
  - welding/machining .

There is a need for switches that can operate in a high temperature, high radiation flux environment such as in fission reactors. It is also possible to utilize high power discharges to, for example, decompose polychlorinated biphenyls (PCBs).

Space

Space represents another opportunity area for dielectric research. I wish to draw attention to the following:

- Higher voltages will be needed for operating power and for pulsed power
- Gas systems for ΔT, radiative, ionizing, outgassing environment
- Extreme reliability (basic models, earth/space environments)
- Low pressure or vacuum based switches .

Finally, I have, as did Professor König earlier, a few remarks to make with regard to education:

- Interest in engineering/science is decreasing
- Interest in high voltage dielectrics is decreasing even faster
- Substantial number of retirements are about to occur in the field
- Will probably need more technologists in this area in the future
- There is basically a shortage of trained people.

R. J. VAN BRUNT: I have been asked to be a spokesman for the theorists. So, I will make a few comments and suggestions about some of the challenging problems from a fundamental point of view. In a number of papers presented at this meeting (by L. Niemeyer, E. Kunhardt, and I. Gallimberti, for example), there has been suggested a need for more theoretical work: modeling and investigations into the fundamental limitations to our knowledge about the behavior of electrical discharge phenomena. The electrical discharge itself is a non–equilibrium phenomenon and can be thought of as a hydrodynamic instability or chaotic behavior. There is not enough support for these investigations. One fundamental question that can be asked, since the electrical discharge is a statistical process, is: How much can we know about the discharge process?

Other areas for fundamental investigation are:
- Fractal descriptions and chaos
- Instabilities leading to glow–to–arc and streamer–to–leader transitions
- Non–stationary behavior in glow discharges
- Factors that control stochastic behavior of discharge processes (partial discharges)
- Breakdown probabilities (what are the controlling factors?)
- Leader propagation models
- Models and simulations (we cannot even do simple avalanche simulations without reasonable models)
- Discharge development in non–uniform fields
- Space–charge effects
- Discharge–induced chemistry (kinetics models).

The message that I want to give is that there is a need for good theoretical support for many of the activities represented here.

I will now open the session for discussion.

M. KRISTIANSEN: What can you tell me about the availability of computer–based data bases for cross sections, rate coefficients, and other basic data? How can I obtain this information?

R. J. VAN BRUNT: There are people at NIST who are working on keeping the atomic data center at Boulder active. If people are interested in cross–section data, they can be obtained. There is also a data base that is being developed for plasma processing. A lot of work has been done recently on SF_6, for example, by Dr. John Herron who runs the data center at NIST. Other data bases are also available from Purdue and EPRI. Data bases tend to be focused on a particular problem. Professor Jack Dutton has been involved in a data base of cross sections.

J. DUTTON: We had exactly the same discussion on data bases at the last meeting and indeed at the one before that. The situation is still that there is a perceived need among the gaseous dielectrics community for critically evaluated data on cross sections, reaction rates, and swarm coefficients. The provision of such data, however, requires a great deal of scientific effort and funding. There is a willingness within the community to provide the effort but we have, for the most part, failed to persuade the funding agencies that this is a priority area. In this context, I was pleased to learn recently that the data analysis center at JILA is, up to now, maintaining its data base on swarm coefficients and has recently produced an up–dated swarm bibliography.

I. GALLIMBERTI: In Europe over the last ten years there has been a European collaboration called the European Group on Gas Discharges. Within this group we started a normalization of our data base. Unfortunately the last application for funding was not supported, but we think that we have a chance of continuing this project in the future. It is clear that without support it is very difficult to go on with the normalization, the transfer of information from laboratory to laboratory. For the moment there are many scattered data bases in each laboratory but at this stage we do not have a normalized data base.

M. KRISTIANSEN: We have developed two data base systems on bibliographies (one on electrode erosion and one on vacuum breakdown) each containing 1000–2000 references which can be cross searched by place, or author, and is available on dBase IV.

J. DUTTON: Is this available outside of the United States?

M. KRISTIANSEN: The one on erosion, which contains our own annotations, is available, but the other one on vacuum breakdown which contains abstracts, is not, due to a copyright problem at present. Copyrights are becoming more of a problem now in scientific communications.

R. J. VAN BRUNT: What other kinds of data do researchers in gaseous dielectrics need?

T. TEICH: We need all kinds of reaction coefficients, including coefficients for excitation to particular excited states and for dissociation.

R. J. VAN BRUNT: One of the big challenges in the atomic and molecular physics community is the measurement of dissociation coefficients. Any modeling of decomposition processes must take into account knowledge of how molecules break apart.

E. MARODE: Another important quantity is the amount of kinetic energy released in dissociation processes, which contributes to heating of the gas in electrical discharges.

I. GALLIMBERTI: I agree that we need cross—section data in order to model electrical discharges.   In some cases transport data have to be obtained from cross—section data.   It is important that the cross—section data are consistent with the transport data. Coefficients are important for all processes, including dissociation and excitation, and for processes involving metastable states.   We also need reaction rate coefficients for all the decomposition products.

M. FORYS: To improve the GIS performance there is a need for seeking electron acceptors (or their mixtures) having high cross section for electron capture over a higher range of energy than for SF_6 processes, and carrier gases with higher efficiency for slowing down electrons than nitrogen (e.g., CO_2, hydrocarbons).   It seems also useful to look for some three—body processes where increase in the carrier gas pressure can increase the electron capture rate constant without increasing the concentration of more expensive and hazardous electron acceptors.   New data for electron capture by Van der Waals complexes possibly could be taken into account.

R. J. VAN BRUNT: You are right.   There are a number of different mixtures such as those discussed by Dr. Wetzer which take into account synergism and three—body interactions. The main problem is, often, acceptability for some applications.

O. FARISH: The CO_2 mixtures have been looked at.

S. W. ROWE: (1) For most industrial situations we use the same gas to interrupt very high currents, so in those circumstances the thermal properties are also very important. Looking at dielectric properties alone may not be acceptable for industry. (2) For some experimental conditions S_2F_{10} has been found in low concentrations. However, the excellent work done on toxicology and detection leaves one main question open. Is S_2F_{10} really generated in significant quantities in real engineering situations? In my opinion this is priority number one for future work. Clearly an answer to this question should be sought by the independent experts in the next two or three years so as to supply users, manufacturers and other specialists with the answer.

I. SAUERS: Thus far S_2F_{10} has only been observed in laboratory experiments.   It remains to be seen whether S_2F_{10} will be found in actual equipment.   There is work going on now at ORNL, NIST and Ontario Hydro to develop techniques for detecting S_2F_{10} in the presence of SF_6.   Progress has been made toward significantly improving detection sensitivity. We certainly do not have all the answers yet, but progress is being made along these lines. One of the points that I brought up earlier is that you are going to have to satisfy regulatory requirements anyway, so it will be necessary to have these techniques available.   From experiments done at ORNL and at NIST and from information found in the literature it appears that S_2F_{10} is produced in all types of electrical discharges in SF_6 and there are indications that S_2F_{10} can be found in small amounts in cylinder SF_6 itself, presumably formed during the manufacture of SF_6.   The approach that we are taking is to improve the detection sensitivity, establish appropriate sampling proceedures and then to survey equipment in a systematic way. A final point that I would like to make is that occasionally you hear the comment that SF_6 gas from a system has been sampled and no S_2F_{10} was found. It is important to note that the method used for gas analysis should be shown to be appropriate for S_2F_{10} detection.   A method that might be appropriate for other SF_6 by—products may not be appropriate for S_2F_{10}.

I. GALLIMBERTI: I would like to make a comment on the S_2F_{10} issue.   We have done some experiments in collaboration with an Italian manufacturer that produces SF_6, in order to answer the question of whether S_2F_{10} is found in equipment.   It is difficult to answer this question because we have observed enormous fluctuations.   It is not only a matter of adding good detectors, but it is also a matter of properly defining the conditions. The problem of S_2F_{10} formation is very complex. It depends very much on thermodynamic and hydrodynamic processes, arc discharge formation and so on. As you change the current in switchgear, you also change the result of the analysis. For this reason we are not getting reproducible results. So, in addition to having good detectors and good sensitivity, we need to standardize the way in which we are making the measurements in order to understand what is going on.

J. CASTONGUAY: In reference to PCBs environmental problems, regulations, and concerns, fears followed the development of sensitive analytical techniques since with such techniques PCBs were found "everywhere". If this is to happen for S_2F_{10} and SF_6, it may be necessary to change SF_6 in equipment in 10–15 years to other dielectric gas(es). Research on alternative gases offerring higher "security" should be pursued. What should be done?

R. J. VAN BRUNT: I would like to comment on that since I have been directly involved in S_2F_{10} work. For a number of reasons one could question the toxicity level of S_2F_{10}, until more investigations are done. We seem to be dealing with very low concentrations of S_2F_{10} and we can now detect these low concentrations of S_2F_{10} in SF_6.

J. R. ROBINS: Is SF_6 a greenhouse gas? Is SF_6 a strong IR absorbing material? Does it build up in the upper atmosphere?*

S. W. ROWE: I cannot respond as an expert but I have seen several papers on SF_6, used for example as a flue gas tracer for atmospheric studies, which indicate that SF_6 does not contribute to global heating and that it is not a greenhouse gas. I have also read a recent CIGRE report that says the same thing.

R. J. VAN BRUNT: As I understand it, the reason SF_6 is not considered a greenhouse gas is because there is not very much of it in the atmosphere at present, but it is building up at a rapid rate. Because of its increased use, and since it is not destroyed very rapidly, this may be a concern in the future. I do not believe there is a definitive answer now.

D. KONIG: The work that Dr. Rowe mentioned was done by CIGRE and in the report is information on handling of SF_6 by–products. In the introduction there is an explanation of SF_6 structure, physical properties, by–products (such as S_2F_{10}) and a discussion of the greenhouse effect. I should mention that this is not original work but a collection of available information. I think that the answer given by Dr. Rowe is correct for the time being. If there is a contribution of SF_6 to the greenhouse effect, it is very small and not one of the dominant effects. Since you often hear rumors about SF_6 as a greenhouse gas, it is important to transfer this knowledge to the public to clear up misconceptions. The CIGRE report will be published as a guide on SF_6 handling and it is up to the user to decide what is important and what is not.

R. T. WATERS: One of the perennial difficulties in the scientific study of gas discharges, since the discharges are field driven, is in making good measurements of the field. I would make a plea that a future direction should be in the measurement of field distortion both in the pre–breakdown phase and during the breakdown phase.

---

* See discussion on the subject of SF_6 in the atmosphere on pages 624-626.

F. SCHWIRZKE: Breakdown is initiated by the formation of micron–size plasma spots (cathode spots) on surfaces. The locally enhanced electric field and the current density determine the breakdown process. The electric field changes rapidly due to changing pressure gradients, the dynamics of the plasma, and sheath effects. Measurements of the electric field distribution in the cathode spot seem to be difficult. A model of the cathode spot formation must also agree with the observed erosion pattern.

E. MARODE: Emphasis should be given to the following topics:
- Discharge studies
  - · Ways to produce high pressure diffuse discharge
  - · Studies on lightning protection and active lightning rods
- Dust: High and low pressure dust behavior in discharges
- Coupling between discharge phenomena and neutral hydrodynamic phenomena such as expansion phenomena and flows and turbulent gas mixing
- Gas chemistry: hydrocarbon gas compounds chemistry, with special reference to surface deposition and combustion ignition.

R. TOBAZEON: There is a need for the scientific community to have available an up–to–date book covering the present status of knowledge on gases (such as Meek and Craggs, 1978). Could a multipartner project be initiated in order to produce this book in a reasonable time?

R. J. VAN BRUNT: This comment goes along with the one made earlier on the need for a good data base.

GROUP DISCUSSION ON INDUSTRIAL OUTLOOK FOR GAS DIELECTRIC

NEEDS AND USES

PANELISTS: A. H. COOKSON (CHAIRMAN), P. C. BOLIN,
F. Y. CHU, A. DIESSNER, T. ISHII AND J. J. PACHOT

A. H. COOKSON: When this conference began several years ago, what we wanted to do was to get a working relationship between the people concerned with the research side and the people concerned with the applications side. This has always been a driving force for this conference and one of the ways we have done this is with the panel discussions. We have a strong panel consisting of three panelists from the utilities and two panelists from the manufacturers and all of the panelists have a research background. Each of the panelists will outline their views and then the session will be opened for discussion. The panelists include Frank Chu, section head on science in the research division at Ontario Hydro, with experience with GIS systems and arc burn–through; Arthur Diessner, manager of the high voltage laboratory at Siemens (Berlin), concerned with the design and testing of gas–insulated substations and also with research experience at MIT on SF_6 breakdown; Jim Pachot, chief R&D engineer at Bonneville Power, concerned with installation of SF_6 equipment including bus and circuit breaker, SF_6 research, and problems dealing with fish and wildlife; Mr. T. Ishii, manager of substation facilities and hydroelectric generation and transmission with Tokyo Electric Power, and co–author of two papers at this symposium on gas breakdown; and finally Phil Bolin, manager of gas products with Mitsubishi Electric in Pittsburgh, with experience in designing gas–insulated transmission lines and substations. We will begin with Phil Bolin.

P. C. BOLIN: Use of a gas dielectric superior to air has made possible GIS with several fundamental advantages: small, reliable, maintenance–free, safe, and economic. Utility and industrial companies needed such features in densely populated areas, leading to a conversion from air–insulated substations to gas–insulated substations in many countries and volume production by many manufacturers. Examples include: China Light and Power, and Mitsubishi Electric Production.

However, in North America, GIS has been much less used, and the perception is that experience is discouraging:

- Market is 10% of the size it should be, due to:
  · Perceived high cost of GIS vs AIS
  · Many very real problems with both early and recent GIS
- North American Failure Rate
- Market Status
- Perceptions Regarding GIS.

The outlook for GIS use in North America is poor until these perceptions are overcome. Some of the answers are technical in nature and those involving the gas dielectric are being answered in more and more detail as reported at meetings such as this. Others are experience related, dealing with failure modes, corrective measures, reliability data, and diagnostic techniques. These are also reported at this symposium. Progress in these areas are needed and will help. Mitsubishi Electric is undertaking a program to apply these locally:

- GIS engineer visits installed GIS once every year or so to:
  - Monitor conditions using non–intrusive diagnostic techniques (acoustic, gas analysis, etc.)
  - interview customer's personnel to obtain documentation experience
  - resolve any complaints/problems
  - issue written report of conditions.

In this way local data supporting the advantages of GIS are collected.

T. ISHII: Since the first application of SF_6 gas–insulated equipment (66 kV GIS) in 1969 in Japan, more than ten thousand GCB and GIS units have been put into service and their voltage class has been extended to 500 kV. This gas insulation technology is applicable to other power apparatus such as power transformers, neutral grounding registers and GITL. As for a gas–insulated transformer, 154 kV class has been in practical use since 1989, and 275 kV class is planned to be put into service in 1991. As the main equipment, GIS is described below.

The following is the Japanese utilities' applications, considerations, or experience, and their recent efforts on improving compactness and reliability and future expectations.

Main Reasons to Choose GIS

– High reliability

The reliability of GCB and GIS is at a satisfactory level in Japan. Recent research of the Electric Cooperative Research Institute reports that the five–year average failure rate of GCB and GIS from 1981 to 1985 is 0.56×10^{-2} (number/units–year) for major failure and 1.26×10^{-2} (number/units–year) for minor failure. These values are 1/2 to 1/10 of that of other type breakers such as air–blast breakers and oil–immersed breakers.

– Compactness

Since it is difficult to obtain sufficient space and to make an outdoor substation compatible in a metropolitan area, many underground substations including 275 kV class have been constructed. For underground substations, compact equipment is able to reduce the construction cost, that is, it can minimize the civil cost by limiting the excavation volume. Also, for 500 kV substations located in mountain areas, the space is limited due to the influence on the natural environment caused by large developments.

Comparing the required area for 500 kV switchyard, GIS needs only 8% of the area of a conventional air–insulated substation. As for the volume, this reduction ratio is about 3%.

– High compatibility for salt pollution problem

– High aseismic ability

– Less maintenance

Recent R&D Efforts

- <u>Reducing the GIS rated insulation level with the benefit of the high performance surge arrester</u>

GIS dimensions are determined by such design factors as the dielectric strength for LIWL, the dielectric strength for disconnector switching surge with free metallic particles, and the temperature–rise limits of enclosures and conductors.

To reduce GIS dimensions, these design factors should be reviewed while maintaining the present high reliability. One of the effective methods is in the reduction of LIWL by application of newly developed high performance surge arresters.

The newly developed metal oxide surge arrester has superior voltage–versus–current characteristics. For example, the impulse withstand voltage is 870 kV crest at 10 kA which is approximately 80% of the present value. This new arrester has been put into service since 1988 with a satisfactory operating experience.

In the study of lightning surge analysis using EMTP for 550 kV GIS substation, we have adopted lower lightning impulse withstand level (LIWL), 1425 kV, than the present one, 1800 kV. The reliability and margin between the LIWL and the overvoltage level are maintained the same. Due to the reduction of LIWL from 1800 kV to 1425 kV, a three–phase common enclosure type bus duct diameter can be reduced by 78%.

- <u>Diagnostic System for GIS</u>

The reliability of GIS has been proven through a satisfactory operating experience. Much higher reliability of GIS is still expected because society has become increasingly dependent on stable power supply, especially due to the recent prevalence of devices that are sensitive to voltage variations such as computers, requiring fewer electrical failures in the bulk transmission grid. Predictive maintenance is one of the methods to reduce electrical failures.

In order to achieve this, an integrated diagnostic system has been applied. The system continuously monitors the GIS conditions and can detect the symptoms of a failure. Before the major fault, the system will be switched and the reliability can be maintained.

In this system, partial discharges are monitored as signs of insulation failures by measuring a high frequency current to the graded shield in insulating spacers on the grounded side, as a sensor electrode.

- <u>Study for UHV (1,000 kV) GIS</u>

The Tokyo Electric Power Company has been planning to upgrade the transmission voltage to 1,000 kV which will take place within the first few years of the twenty–first century. To achieve this, we have been studying the design of UHV substations since 1970, and have obtained the following results for switchyards.

UHV substations will be constructed in mountain areas. In order to avoid the impact on the natural environment of mountain areas, GIS might be adopted because the installation space is limited to 70,000 m² (50,000 m² for a 500 kV substation). Basic equipment arrangement is considered by taking into account the experience of 500 kV substations.

Expectations for R&D on Gaseous Dielectrics

— <u>Further compactness toward optimization of UHV GIS</u>

— <u>Control a disconnector switching surge to reduce the effect on the secondary circuit</u>

— <u>Field data sampling to evaluate the expected lifetime</u>

A. DIESSNER: A few of the uses of gaseous dielectrics are:
  — GIS for AC, including incorporated and connected equipment
  — Circuit breakers
  — Transformers, including measuring transformers
  — GIS for DC (not in large numbers)
  — Electrostatic generators.

The following comments apply only to GIS and circuit breakers for AC.

The stresses during operation are:
  — Operating voltage: AC, including temporary overvoltages
  — Switching surges
  — Lightning surges
  — Very fast transients (mainly by disconnector switching)
  — Trapped charges: DC (mainly by disconnector switching).

VFT and trapped charges are caused by switching operations, mainly by disconnectors.

There are deficiencies or concerns with the present systems. The behavior of gases under good conditions (clean gas, good electrodes, clean insulator surfaces) is well known for technical purposes. The main question presently is: Which diagnostic methods are available for testing if we really do have "good" conditions?

  — HV testing: AC, impulse, and DC; e.g., AC and DC voltages are sensitive to free conducting particles; lightning impulse voltages are sensitive to needles.

  — PD detection: Which frequency range is best suited for detection and how can we identify a defect?

  — Acoustic detection: e.g., dancing particles can be identified.

  — Chemical detection: e.g., SF_6 dissociation products can be detected.

Does aging of the insulating materials used in GIS occur?
  — Gas: no
  — Insulator surface: some
  — Solid: yes

In conclusion, research and development is needed in the fields of diagnostics, fault detection, and aging of insulator surfaces.

F. Y. CHU: There are three aspects of the utility perspectives on GIS applications that I would like to consider: technical, cost, and environment.

Technical

- Reliability is the key factor

- Despite research effort, improved design, and cleanliness, utilities still experienced occasional failures

- Utilities will continue to demand problem-free, easily maintained GIS in critical applications.

<u>R&D Needs</u>

- Aging mechanisms

- Improved diagnostic techniques
  - Noninvasive
  - Monitoring
  - Early warning.

- Testing philosophies
  - Weed out and detect defects before GIS is put into service
  - Relate basic discharge processes and mechanisms to test techniques
  - Pros and cons of various test techniques.

Cost

- GI substation is competitive to AIS in many applications

- "The EMF factor may provide an incentive for increased use of GITL"
  - To what extent?
  - Huge cost hurdle to overcome; for example, underground transmission cost is ten times the overhead transmission line cost
  - GITL has to compete with other underground systems.

<u>R&D Needs</u>

- Cost competitive GITL

Environment

- Increased public concern on strange and exotic chemicals
- Do SF_6 or other dielectric gases cause ozone depletion?
- Does the release of SF_6 contribute to the greenhouse effect?
- Can utilities discharge SF_6 into the atmosphere?

<u>R&D Needs</u>

- Hard data and evidence is needed to answer the above questions
- Continued research on SF_6 decomposition leading to the establishment of safe proceedures.

There is a bright outlook for GIS in the next 25 years. The cost, environment, and some technical points are factors that need to be addressed.

J. J. PACHOT: During the past twenty-five years the electric utility industry has progressed in a somewhat erratic fashion from a typically air-insulated extra-high-voltage transmission system to one now using increasing amounts of enclosed gas-insulated equipment. In fact, several utilities are now committed to the use of gas-insulated equipment for virtually all new substations. This progression from open-air insulation to enclosed-gas insulation has been prompted by a reduction in available space and by the increasing reliability and reduced cost of new equipment. Future increases in its use may be dictated by either the real or perceived adverse effects from electric and magnetic fields.

During this period of development we have seen numerous technologies that, in the beginning, appeared to have tremendous potential for revolutionizing the utility industry. However, for one reason or another, these ideas never achieved maturity as operating equipment on an electric utility system.

Notable among these failures are the vacuum insulated, cryoresistive cable; the evaporative–cooled, gas–insulated cable; the 345 kV flexible gas cable, and very recently, a smaller 138 kV flexible gas cable.

The vacuum–insulated, cryoresistive cable design, which circulated liquid nitrogen in the conductor for cooling and used a vacuum as both the thermal and the electrical insulation, still has many appealing features. Unfortunately, the system which was assembled for full scale testing developed nitrogen leaks in the terminations and the project ran out of funding before the problems were solved. In this case additional funds and the involvement of persons knowledgeable in the ducting of liquid nitrogen might have altered the destiny of this system.

The evaporative–cooled, gas–insulated cable used liquid SF_6 inside the conductor for cooling. As the SF_6 evaporated during the cooling process, it was vented into the insulating area. This project was terminated when it was determined that no utility was interested in a cable that could deliver 8000 amps at 500 kV. Such a system, when used at its economic operating point, could result in some system instability if a failure occurred.

The 345 kV flexible gas spacer cable is a study in how not to develop a cable. From this project we have a fantastic flexible cable machine that can fabricate both the conductor and enclosure in diameters approaching 400 millimeters. However, too few resources were devoted toward the development of a suitable spacer. Finally, when a prospective customer was found, the manufacturer, for unknown reasons, priced the cable so high that it could not compete with other technologies. Due to the lack of a market and the high cost of storing the machine in a ready–to–operate state, it was eventually dismantled and moved to an unused storage area in Texas. There is presently no one in the United States that can set this machine up to produce a cable.

The recent failure of a 138 kV flexible gas cable project was due primarily to the incompatible personalities involved and the lack of a cohesive program to develop the product. More information regarding this project should be available after its termination.

There are several lessons to be learned from these failures:
   – Make sure that there is a market for the product before you start developing
   – Address the system in its entirety from the beginning
   – Be sure that the appropriate persons and experience are included on the project team
   – Make sure that support for the project, including funding, is adequate
   – Work together as a team and meet frequently to avoid interface problems
   – Do not market the equipment until it is fully tested.

Keep in mind that the perceived reliability of the system is most important in gaining acceptance by the electric utility industry.

This brings us to the identification of problems on presently operating gas–insulated equipment. These problems can be divided into four categories: mechanical, electrical, thermal, and contamination. Several of these problems may fit into more than one of these categories. For this panel I will limit my discussion to our experience with gas–insulated equipment on the Bonneville Power Administration system.

620

Over the past fifteen years we have had many problems with the mechanical linkages in the actuating mechanism of gas–insulated circuit breakers. These failures frequently leave one pole of the breaker in either the open or closed position while the other two poles operate satisfactorily. Some of these failures have also resulted in a flashover in the breaker.

Tracking and eventual electrical failure of insulators in GIS is another significant concern. This problem is especially of importance in our older systems. However, these problems may be due to aging effects which have not shown up in the newer installations. Contamination in gas–insulated equipment is still of concern both in newly installed systems and in older operating devices.

This last winter we had a significant reduction in gas density in a 500 kV gas insulated circuit breaker, located out–of–doors in the state of Montana. This reduction in density was caused by liquefication of the SF_6 during a long period of cold weather. Although there was no flashover, there was considerable concern over the ability of the breaker to withstand a switching surge.

Transient generation during switching operations is still a major problem in GIS. While it does not appear to be harmful to humans, it has caused damage to, and the inadvertant operation of other equipment. This is due to the coupling of the transients into adjacent control circuits. Generations of toxic by–products due to faults or partial discharge continues to represent a problem in the handling of both the contaminated gas and the equipment itself during maintenance or repair.

Finally, one of the biggest problems incurred by my utility is our inability to rapidly and accurately locate faults in GIS, especially the larger systems having significant bus runs.

While this list is not complete, it does represent the majority of our concerns. It also serves to identify those areas where we need to assign additional resources to mitigate problems in future gas–insulated equipment. The following is a list of areas where additional research and development is recommended.

Spacers have come a long way in the past fifteen years, both in the development of better shapes and better materials. However, we must continue to look at new materials and shapes to improve their performance in future systems. We must also work to reduce flaws such as voids and contaminants. Better factory testing is needed to "weed" out inferior spacers.

While a few problems have been reported for joints, the numbers appear to be small and limited to one or two manufacturers. However, we still should look for ways to obtain uniformly excellent joints that minimize wear and generate no contamination that can reach the insulating medium.

Contamination control must always be an important consideration in the design, fabrication, and installation of GIS. We must continue to address improved designs both to eliminate particle generation and to trap any particles that enter the insulating medium.

Testing and conditioning of GIS is especially important to minimize the high incidence of start–up failures. We are presently taxing the capabilities of existing power supplies for AC conditioning. Longer lengths of gas–insulated bus will require new methods of conditioning or larger power supplies.

Fault location has been very difficult in the past. We need to look at new fault location techniques, especially those that operate in real time and record the fault location during the first occurrence. The present need to incur additional flashovers to locate a fault tends to cause additional damage and contamination.

We need to improve our repair techniques. This means that the design of the GIS equipment should address the possibility of future repair and should accommodate the repair in a timely manner.

We are presently expecting to fund a project which will investigate the detection of toxic by–products in SF_6, especially S_2F_{10}. This project is also expected to identify proper gas handling techniques for contaminated gas. We need to get this project underway as soon as possible.

Finally, we need to look at the heating and circulation of gaseous dielectrics to avoid liquefication during extremely cold weather.

Now we come to directions for future research. This is not research intended to correct existing problems. It is research to obtain alternative gas–insulated equipment to be used on future utility installations. This list includes:

- Compact gas–insulated capacitors
- DC gas–insulated systems
- Gas–insulated equipment using high temperature superconductors
- High–speed grounding switches using gases that become conductive when illuminated by a laser
- Circuit breakers that use gases that become insulating when illuminated by a laser
- Development of standardized and modularized GIS components.

A. H. COOKSON: You have heard a very broad cross section of the status of the industry with respect to gas–insulated substations and gas–insulated equipment and with respect to reliability problems which we have in the U.S. There is a very strong appeal from the panelists to find non–intrusive methods for getting a diagnostic technique either for finding where the problems are going to occur or where they have occurred. There were also appeals for more work on the insulator and whether there can be better and more reliable designs. We also heard appeals to look at the very low probability of breakdown as it applies to the gas dielectric system. It is very easy to make 50 % probability measurements, but the very low probability is more difficult. Jim Pachot has given us a list of potentially new areas where gas dielectrics could be used. We are at a key point where the industry has reached a maturity phase and there is a need for new gas dielectric systems to improve the reliability of the system. With that, I would like to open the discussion and invite comments, questions, and views.

N. G. TRINH: (1) Will 800 kV GIS be available from most of the manufacturers? (2) Can probability tests, which require larger number or longer duration of voltage applications, be considered for GIS? Also can accelerated aging tests be considered to provide some information on the expected life of GIS at the time of purchase? (3) At 800 kV oil–paper cables are no longer economic. Future technology relies on the use of PPLP insulated cable. Some comparative studies between PPLP and SF_6 cables may be useful.

P. C. BOLIN: (1) Mitsubishi Electric has developed and tested 800 kV GIS so it is available from a design viewpoint, but it is not in routine production and therefore the price will be high relative to the lower voltage levels. (2) Low probability failures are being investigated on a component basis, but accelerated aging tests on complete GIS assemblies are not often undertaken. On the other hand, we should not forget that there is now more than 20 years of service experience with GIS. If enough manufacturers and users take the time and effort to carefully and thoroughly monitor the condition of this in–service GIS, we should be able to develop very good data on aging. We are now starting to see experience reports which are based on large numbers of unit–years so that parameters such as failure rates and types can be derived with a reasonable level of statistical reliability. These results are being used to guide efforts to improve reliability and indicate diagnostic techniques. As these techniques are applied to in–service equipment we can expect to obtain good information on aging of the equipment. This is probably a more effective way of studying aging than starting an accelerated aging test, but it does require cooperation among many organizations. Some of this is underway under the banner of reliability. Perhaps we could ensure appropriate attention to aging by making it a specific sub–topic and suggesting tests and examinations which bear directly on possible aging processes. (3) 800 kV SF$_6$–insulated cables of the simple rigid design are commercially available with well known characteristics. I expect the difficulty of a comparison with 800 kV PPLP cables will be that little is known about the production cost (selling price) and characteristics of an 800 kV PPLP cable.

A. DIESSNER: (1) 800 kV GIS has been developed a few years ago. However, the world–wide demand for it is presently very limited. (2) Low probability tests using large numbers of voltage applications can be useful for type tests of components. Real GIS equipment should not be endangered by excessive high voltage testing, but additional diagnostic methods, e.g., partial discharge measurements, should be used. In accelerated aging tests we must apply some models using "educated guesses". Such tests are, again, useful only for components. Real designs will always need some safety factors which will give confidence that the expected life will be some 50 years. (3) From the technical point–of–view, 800 kV SF$_6$ cables can be built. However, even more important are environmental and economic aspects. Here again, the world–wide demand is very limited.

F. Y. CHU: I think it may not be practical to apply accelerated aging test to the GIS as a system. Some aging tests can be applied to individual components such as spacers, for example, but it will take enormous efforts to conduct aging tests on the system.

J. J. PACHOT: (1) Several manufacturers have already fabricated and tested 750 kV AC equipment, so it is probably already available. (2) Higher and longer duration tests should be considered for installed GIS. This would help reduce our high rate of "start–up" failures. While accelerated aging and low probability tests on components at the factory might help, they do not correct the majority of the problems which occur between the time the equipment leaves the factory and is finally assembled for operation. (3) Several utilities have already performed economic and technical studies comparing GIS to paper–oil and PPP cables, including Bonneville Power, Philadelphia Electric, and Public Service Electric and Gas of New Jersey to name a few. However, most of these studies were at 500 kV.

T. NITTA: (1) I would like to hear the comments of the panelists on the two areas of techniques which are considered to be effective in improving the reliability of gas–insulated equipment. These are testing, particularly on–site, and diagnostic or monitoring techniques. Can we improve the reliability by just applying rigorous tests on–site? The test itself may deteriorate the quality of the equipment. High voltage testing on–site is a considerable amount of on–site work inside the equipment. (2) Monitoring technology is only valid for very high quality equipment. If you have alarms every day, you cannot live with that condition. The quality of GIS has to be high enough by itself to limit the alarms at most to once in a few years.

P. C. BOLIN: High voltage site tests (AC or impulse) are typically easy to perform in North America because most GIS here have SF$_6$–air bushings. Therefore, the equipment does not need to be opened. There is, of course, concern with starting problems which might otherwise never have appeared by applying a high voltage stress. This is balanced against the benefit of finding problems through the test which would, if not detected, lead to in–service failure. In the long run, as our understanding of GIS failure modes and diagnostic techniques improves, it should be possible to assure, at a gentle high voltage test, diagnostics good enough to show potential internal problems. (2) I agree completely that the objective must be high quality equipment which would, with even the most sensitive diagnostics, indicate a possible problem (alarm) not more than once every two years. I believe that most in–service GIS meets or exceeds this required quality level. The diagnostic challenge is to determine which installations do not achieve this level. Perhaps the near term focus of diagnostic efforts should be those installations which have had in—service problems. This would provide more information on the efficiency of the diagnostic systems and may improve overall reliability. In this sense I disagree that diagnostics should only be applied to high quality equipment. Diagnostics should be used to force an upgrading of GIS of marginal reliability.

J. J. PACHOT: (1) Reliable GIS is obtained by using a good design, extremely good factory testing and quality control while the components are still at the manufacturer's facility. Unfortunately, between the time the materials leave the factory and are assembled for operation, there are numerous possibilities to incur damage, moisture, improper installation, and contamination. Pre—energization testing is necessary to determine the condition of the completed system and possibly to move contamination into traps or other low—field regions. Unfortunately, the resonant AC test sets are difficult to move and we are presently reaching the limits of their operation. We as a utility would prefer longer high voltage AC testing, if it were possible, to minimize our high incidence of "start—up" failures. (2) Real—time monitoring of the system, which includes fault location, is the ultimate technique in continuously determining the condition of the GIS. We are presently looking at this for other equipment such as transformers, CTs, PTs, circuit breakers, and other gas—insulated equipment. This will allow us to correct a problem before significant damage occurs.

F. Y. CHU: I think Dr. Nitta's question refers to the reliability of the diagnostic sensor. I agree that the sensor has to be very reliable so that we do not get false alarms. But if the alarms are real, we should do something or plan to do something to correct the situation to prevent an in—service fault.

A. DIESSNER: (1) Reliability of GIS depends on many details of the design as well as on the production procedures. Quality control is important at various steps of assembly. GIS shipping units are tested in the factory. However, most GIS require major assembly work on—site, and this step is checked by site testing. Our experience has shown that this method can eliminate some defects which would have caused breakdown during operation. (2) Monitoring of GIS requires high quality GIS as well as of the monitoring system itself. Alarms (possibly erroneous ones) occurring every day certainly cannot be accepted.

B. F. HAMPTON: Reliability is centainly the most important requirement for GIS, especially if it serves a nuclear generating station with very high outage costs. Good reliability comes from sound design, manufacturing and installation techniques. Testing does not give reliability, but is necessary to detect any flaws which have crept in. But sound testing does not mean using higher voltages, which may generate undesirable transients, should breakdown occur. Rather it means using lower voltages together with diagnostic techniques to detect and identify any fault condition. Faults due, for example to sparking from floating stress shields can develop after the GIS has been in service for some time. It is therefore most desirable to maintain the diagnostic monitoring while the GIS is in normal operation. In the U.K. we have had a great deal of success using the UHF technique pioneered there, and are presently developing a version of this which will enable continuous monitoring of an unattended GIS and will automate reporting of any fault which might be developing.

J. OZAWA: Hitachi Ltd. recently used the oscillating impulse generator for confirmation of the insulation withstand capability of 500 kV GIS for on—site testing in Japan.

L. G. CHRISTOPHOROU: How much SF_6 is released into the environment per year, and how much of this comes from GIS?

L. NIEMEYER: Frank Chu has previously raised the issue of SF_6 and its effect on the atmosphere, i.e. stratospheric ozone depletion and global warming (greenhouse effect). The following order—of—magnitude data allow for a comparative assessment of the share that SF_6 contributes in comparison to the other man–made hologenated gases (CFC) released into the atmosphere.

The present total annual production rates are approximately:

Compound	Annual Production Rate (t/year)	Reference
CFC	1,000,000	F. Sherwood Rowland, "Chlorofluorocarbons, stratospheric ozone, and the antarctic ozone hole," <u>Environ. Conser</u>. 15, 101 (1988)
SF_6	5,000 – 8,000	SF_6 producer information: Kali–Chemie: < 10,000 t/year (1989). Allied Signal: 6,800 t/year. Both companies say that about 80% of the production goes into electrical equipment.

The average atmospheric trace concentrations of these gases into the atmosphere extrapolated to 1990 [V. Ramanathan et al., "Climate–chemical interactions and effects of changing atmospheric trace gases," <u>Rev. of Geophysics</u> 25, 12441–82 (1987)] are:

Compound	Concentration in the atmosphere (ppb vol.)
CFC	∼ 1.7
SF_6	∼ 0.0015

These data correspond approximately to the release rates. The SF_6 concentration is roughly consistent with the integral production since 1970.

The mechanism of stratospheric ozone depletion is characterized by a catalytic decomposition cycle [F. Sherwood Rowland: "Chlorofluorocarbons, stratospheric ozone, and the antarctic ozone hole," <u>Environmental Conservation</u> 15, 101 (1988)] of the form:

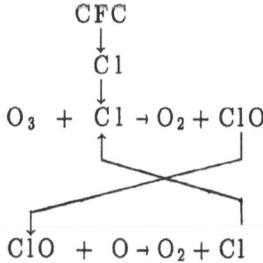

where chlorine (Cl) acts as a catalyst that is recycled. A similar cycle with fluorine (F), though possible in principle, does not occur because F is rapidly scavenged by the reaction $F + H \rightarrow HF$ [R. S. Stolarski and R. D. Rundel: "Fluorine photochemistry in the stratosphere," <u>Geophys. Res. Letters</u> 2, 443 (1975)] . Hence, SF_6, which does not contain chlorine, cannot contribute to stratospheric ozone depletion.

The mechanism of global warming (greenhouse effect) is based on the infrared absorption of the CFC molecules which contributes about 20 − 30% to man−made global warming. SF_6, having infrared absorption properties similar to the CFC, contributes roughly according to its relative concentration, i.e. about 10^{-3} of the CFC.

Contrary to CFCs, which are almost completely released into the atmosphere, SF_6 is increasingly being recycled by its major users, the electrical utilities. This recycling is to be encouraged and will reduce the SF_6 contribution further in the future by a factor of about 10^{-2} to 10^{-3} which is due to mainly leakage and operation errors.

Y. MURAYAMA: In order to maintain the reliability of GIS, good cooperation between utilities and manufacturers is essential. To improve the reliability, the cause of the failures should be analyzed and eliminated. Toshiba supplied 800 kV GIS which has been operated with satisfactory experience. The reliability can be achieved even for high voltage GIS, once the quality assurance procedure has been established. The understanding and cooperation of utilities are advisable to obtain high reliability.

R. A. HARTHUN: At the beginning of this panel discussion, it was mentioned that aging of most insulating materials typically occurs in GIS, and is related to bushing formulations, spacer formulations, and so forth. Do any of the panelists recognize new research into alternative organic/inorganic materials for use in GIS? One possible area concerns the use of the numerous composite−type materials now on the market.

P. C. BOLIN: The only materials being intensively studied for solid insulation in GIS are highly filled epoxies. In the past many materials have been tried for the solid support insulator of GIS, but none has made it into commercial service. There is some use of fiber reinforced epoxy composites for SF_6−air bushing cylinders, interrupter chambers and operating rods, but these are not under as high an electrical stress as the support insulators. To change the highly stressed solid dielectric material of the solid support insulators from what has been used for more than 20 years would mean starting over in terms of aging questions. So far, filled epoxy GIS insulators have not shown a measurable aging effect in service as long as the continuous operating AC stress level is kept below about 5 kV/mm rms, so there is little incentive to look for a better material.

A. DIESSNER: For high reliability, requiring failure probabilities for a single GIS insulator on the order of 10^{-4} after 50 years of operation, we are reluctant to change materials which have shown good performance in GIS for more than 20 years. Furthermore, long duration tests have clearly shown that production technology is at least as important as the insulating material itself. On the other hand, if new properties of the insulators are required for advanced designs, new materials must be applied, but only after thorough testing of various electrical, mechanical, thermal, technological and economical aspects of real−size insulators.

J. J. PACHOT: We need to continuously assess the value of new compounds to determine their suitability for use in GIS. Therefore, manufacturers must maintain a rigorous R&D program to continuously improve their systems. In addition to high dielectric strength a suitable spacer material should have a coefficient of expansion that matches the other materials used such as aluminum or steel. It must have a low dielectric constant, preferably 1, to match the gas and must be easy to work with and mold without shrinking and forming voids. Some present R&D on polyethylenes indicates that only long term aging at slightly elevated voltages is suitable to determine the actual suitability of a material. This means that we need to test numerous samples similar to the way paint is tested.

F. Y. CHU: The present formulation of GIS spacers consists of epoxy and filler materials. I do not know whether that belongs to the composite type. The material itself has proven to be very reliable. Preliminary results from aging studies show that the root cause of aging is from defects such as voids, debonding between metal and insulator interface. If these defects can be eliminated during the manufacturing process, GIS spacers can be very reliable.

D. C. Agouridis
Oak Ridge National Laboratory
Building 3500, MS-6010
P. O. Box 2008
Oak Ridge, TN 37831-6010

J. J. Bazley
Babcock & Wilcox
P. O. Box 11165
Lynchburg, VA 24506

A. P. Bitouni
Oak Ridge National Laboratory
Building 4500S, MS-6122
P. O. Box 2008
Oak Ridge, TN 37831-6122

J. P. Boeuf
CPAT - University P. Sabatier
118 Route de Narbonne
31062 Toulouse, Cedex
FRANCE

P. C. Bolin
GIS Product Manager
WM Power Products
512 Keystone Dr
Warrendale, PA 15086

A. F. Borghesani
Università Degli Studi di Padova
Dipartimento di Fisica
"Galileo Galilei"
35131 Padova-Via F.
Marzolo, 8
Padova
ITALY

D. W. Bouldin
Department of Electrical Engineering
The University of Tennessee
Knoxville, TN 37996

D. Bradley
Bonnville Power Administration
ELEL, Division of Laboratories
P. O. Box 491
Vancouver, WA 98666

J. G. Carter
Oak Ridge National Laboratory
Building 4500S, MS-6122
P. O. Box 2008
Oak Ridge, TN 37831-6122

J. Castonguay
Science des Materiaux
Institute de Recherche
 de l'Hydro-Quebec
C.P. 1000
1800 Montee Ste-Julie
Varennes, Quebec
CANADA J3X 1S1

I. D. Chalmers
Department of Electronic
 and Electrical Engineering
Royal College Building
204 George Street
University of Strathclyde
Glasgow G1 1XW
SCOTLAND, UK

R. L. Champion
Dept. of Physics
College of William and Mary
Williamsburgh, VA 23185

L. G. Christophorou
Oak Ridge National Laboratory
Building 4500S, MS-6122
P. O. Box 2008
Oak Ridge, TN 37831-6122

F. Y. Chu
Ontario Hydro Research Division
800 Kipling Avenue
KR-128
Toronto, Ontario
CANADA M8Z 5S4

C. M. Cooke
High Voltage Research Laboratory
Dept. of Electrical Engineering
  and Computer Science
155 Massachusetts Avenue
Massachusetts Institute of Technology
Cambridge, MA 02139

A. H. Cookson
Westinghouse Sc. & Tech. Center
1310 Beulah Road
Pittsburgh, PA 15235

R. W. Corell
National Science Foundation
1800 G Street, N.W.
Washington, DC 20550

P. F. Coventry
National Grid
Research & Development Centre
Kelvin Avenue
Leatherhead
Surrey KT22 7ST
UNITED KINGDOM

S. J. Dale
Oak Ridge National Laboratory
Building 3147, MS-6070
P. O. Box 2008
Oak Ridge, TN 37831-6070

P. G. Datskos
Oak Ridge National Laboratory
Building 4500S, MS-6122
P. O. Box 2008
Oak Ridge, TN 37831-6122

A. J. Davies
University College of Swansea
Singleton Park
DPT Physics
Swansea SA28PP
WALES, UK

A. Denat
Laboratoire d'Electrostatique et
  de Matériaux Diélectriques, C.N.R.S.
25 Avenue des Martyrs
BP 166X 38042 Grenoble Cedex
FRANCE

A. Diessner
Siemens AG-Schaltwerk Hochspannung/TVH
Postfach 140/D-1000
Berlin 13
GERMANY

Y. Doin
Volta
Merlin Gerin
38050 Grenoble
Cedex
FRANCE

J. G. Driggans
Tennessee Valley Authority
3N 54A Missionary Ridge Pl.
Chattanooga, TN 37402-2801

Th. Dunz
ASEA Brown Boveri Ltd.
Corporate Research
CH-5405 Baden-Dättwil
SWITZERLAND

J. Dupuy
Laboratoire Genie Electrique
IURS
Université de Pau
Avenue de l'Université
F 64000, PAU
FRANCE

630

J. Dutton
Department of Physics
University College of Swansea
Singleton Park
Swansea SA2 8PP
WALES, UK

C. E. Easterly
Oak Ridge National Laboratory
Building 4500S, MS-6101
P. O. Box 2008
Oak Ridge, TN 37831-6101

F. E. Evans
4768 Woodside Avenue
Hamburg, NY 14075

H. Faidas
Oak Ridge National Laboratory
Building 4500S, MS-6122
P. O. Box 2008
Oak Ridge, TN 37831-6122

O. Farish
Department of Electronics and
  Electrical Engineering
204 George Street
University of Strathclyde
Glasgow G1 1XW
SCOTLAND, UK

N. Femia
Instituto di Ingeneria Elettronica
Università di Salerno
84081 Baronissi
ITALY

K. Feser
Universität Stuttgart
Breitscheidstraße 2
D-7000 Stuttgart 1
GERMANY

M. Forys
Department of Chemistry
Agricultural & Teachers University
Ul. 3, Maja 54
08110 Siedlce
POLAND

M. F. Fréchette
IREQ
1800 Montee Ste-Julie
Varennes, Quebec
CANADA J3X 1S1

N. Fujimoto
Electrical Research Department
Ontario Hydro Research
800 Kipling Avenue, KR-151
Toronto, Ontario
CANADA M8Z 5S4

H. Fujinami
High Voltage Section
Central Research Institute of
  Electric Power Industry
2-11-1 Iwato-kita, Komae-shi
Tokyo 201
JAPAN

I. Gallimberti
Dept. of Electrical Engineering
Padova University
6/A via Gradenigo
35100 Padova
ITALY

A. Garscadden
Wright-Patterson Air Force Base
Advanced Plasma Research Group
Power Division
Aero Propulsion and Power Laboratory
Dayton, OH 45433-6523

V. H. Gehman, Jr.
Pulsed Power Technology Branch
Code F45
Naval Surface Warfare Center
Dahlgren, VA 22448-5000

G. A. Gerdin
Dept. of Elecectrical and
  Computer Engineering
Old Dominion University
Norfolk, VA 23529

M. Goldman
Laboratoire de Physique
  des Decharges
CNRS/ESE
Plateau du Moulon
91190 Gif-sur-Yvette-Cedex
FRANCE

G. D. Griffin
Oak Ridge National Laboratory
Building 4500S, MS-6101
P. O. Box 2008
Oak Ridge, TN 37831-6101

R. J. Gripshover
Naval Surface Warfare Center
Code F45
Dahlgren, VA 22448

A. H. Guenther
Los Alamos National Laboratory
MS-A110
Los Alamos, NM 87545

I. Gyuk
Dept. of Energy, CE-143
Office of Energy Management
Forestal Bldg., MS 5E-052
1000 Independence Ave SW
Washington, DC 20585

R. N. Hamm
Oak Ridge National Laboratory
Building 4500S, MS-6123
P. O. Box 2008
Oak Ridge, TN 37831-6123

B. F. Hampton
CEPE, Dept. of Electronics
  & Electrical Engineering
University of Strathclyde
204 George Street
Glasgow G1 1XW
UNITED KINGDOM

M. Hanai
High Voltage Engineering Group
Toshiba Corporation
2-1, Ukishima-cho, Kawasaki-ku
Kawasaki 210
JAPAN

M. Hanamura
Substation Facilities Division
Tokyo Electric Power Company
1-1-3, Uchisaiwaicho, Chiyodaku
Tokyo, 100
JAPAN

G. Harman
Oak Ridge National Laboratory
Building 4500S, MS-6123
P. O. Box 2008
Oak Ridge, TN 37831-6123

R. A. Harthun
Cooper Power Systems
11131 Adams Road
Franksville, WI 53072

T. Hasegawa
Hokuriku Electric Power Company
15-1, Ushijima-cho, Toyama
Toyama, 930
JAPAN

A.E.D. Heylen
Department of Electronics and
  Electrical Engineering
The University of Leeds
Leeds LS2 9JT
UNITED KINGDOM

H. Hiesinger
Technical University of Munich
High Voltage Institute
Arcisstrasse 21
D-8000 Muenchen 2
GERMANY

J. H. Hughes, III
Babcock & Wilcox
P. O. Box 785
Lynchburg, VA 24505

S. R. Hunter
GTE
Sylvania Lighting Design Center
100 Endicott St.
Danvers, MA 01923

T. Ishii
Hydro Generation &
  Transmission Department
Tokyo Electric Power Company
1-1-3, Uchisaiwaicho, Chiyodaku
Tokyo, 100
JAPAN

R. S. Jacobsen
Electropaulo-Electricidade
  De Sao Paulo S.A.
Av. Brigadeiro Wis Antonio, 1827
150 Andar - CEP 01317
Sao Paulo - SP
BRASIL

D. R. James
Oak Ridge National Laboratory
Building 4500S, MS-6123
P. O. Box 2008
Oak Ridge, TN 37831-6123

W. Johnstone
Department of Electronic
  and Electrical Engineering
Royal College Building
204 George Street
University of Strathclyde
Glasgow G1 1XW
SCOTLAND, UK

C. M. Jones
Oak Ridge National Laboratory
Bldg. 6000, MS-6368
Post Office Box 2008
Oak Ridge, TN 37831-6368

S. V. Kaye
Oak Ridge National Laboratory
Building 4500S, MS-6124
P. O. Box 2008
Oak Ridge, TN 37831-6124

L. E. Kline
Research and Development Center
Westinghouse Research Laboratories
1310 Beulah Road
Pittsburgh, PA 15235

D. König
Technical University Darmstadt
High Voltage Laboratory
FB 17, Landgraf-Georg-Str. 4
D-6100 Darmstadt
GERMANY

M. Kristiansen
Department of Electrical Engineering
Texas Tech University
Pulsed Power Laboratory
Lubbock, TX 79409-4439

E. E. Kunhardt
Weber Research Institute
Polytechnic University of New York
Route 110
Farmingdale, NY 11735

S. M. Mahajan
Tennessee Technological University
5004 Electrical Engineering
Cookeville, TN 38505

E. Marode
Laboratoire de Physique des Decharges
CNRS/ESE
Plateau du Moulon
91190 Gif-sur-Yvette
FRANCE

T. H. Martin
Sandia National Laboratories
Pulsed Power Systems
Department 1250
Albuquerque, NM 87185

I. W. McAllister
Physics Laboratory II
Building 309B
The Technical University of Denmark
DK 2800 Lyngby
DENMARK

D. L. McCorkle
Department of Physics
The University of Tennessee
Knoxville, TN 37996

J. McCoskey
Reynolds Industries
Electronic Product Divsion
3070 Skyway Drive, Suite 301
Santa Maria, CA 93455-1116

H. Mochizuki
Chubu Electric Power Co.
2-3-24 Yokota Atuta-Ku
Nagoya, 456
JAPAN

Y. Murayama
Toshiba Corporation
1-6, Uchisaiwai-cho 1-Chome
Chiyoda-Ku, Tokyo 100
JAPAN

K. Nakanishi
Manufacturing Development Laboratory
Mitsubishi Electric Corporation
1-1 Tsukaguchi-Hommachi 8-Chome
Amagasaki, Hyogo 661
JAPAN

T. Namera
Chubu Electric Power Company
1, Toshin-cho, Higashi-ku,
Nagoya, 461
JAPAN

J. B. Neilson
Powertech Laboratories, Inc.
12388 88th Avenue
Surrey, British Columbia
CANADA

A. G. Netto
Electropaulo-Electricidade
 De Sao Paulo S.A.
Av. Hove De Julio, 4939 - 100
Sao Paul 01417
BRASIL

C. Neuman
RWE Energie AG, Essen
Dept. E-G
Kruppstr. 5
D-4300 Essen 1
GERMANY

L. Niemeyer
ABB Research Center
CH5405 Baden
SWITZERLAND

T. Nitta
Mitsubishi Electric Corporation
Laser and Plasma Physics Dept.
Central Research Laboratory
8-1-1, Tsukaguchi-Honmachi
Amagasaki, Hyogo 661
JAPAN

J. K. Olthoff
National Institute of Standards
 and Technology
Building 220, Room B344
Gaithersburg, MD 20899

J. Ozawa
Hitachi Research Laboratory
Hitachi, Ltd.
4026 Kuji-cho, Hitachi-shi
Ibaraki-ken 319-12
JAPAN

M. O. Pace
Department of Electrical Engineering
The University of Tennessee
Knoxville, TN 37996

J. J. Pachot
Chief R&D Engineer
Bonneville Power Administration
905 NE 11th Ave, Box 3621
Portland, OR 97232

J. C. Paul
No. TEC/EE/DA/78
Department of Electrical Engineering
Tripura Engineering College
Tripura 799055
INDIA

A. Pedersen
Physics Laboratory 2
DTH - Building 309B
The Technical University
DK-2800 Lyngby
DENMARK

W. Pfeiffer
Institut für Hochspannungs-und
  Meßtechnik
Technische Hochschule Darmstadt
6100 Darmstadt, Schloßgraben 1
GERMANY

L. A. Pinnaduwage
Oak Ridge National Laboratory
Building 4500S, MS-6122
P. O. Box 2008
Oak Ridge, TN 37831-6122

L. C. Pitchford
CPAT - University P. Sabatier
118 Route de Narbonne
31062 Toulouse, Cedex
FRANCE

Y. Qiu
Electrical Engineering Department
Xi'an Jiaotong University
Xi'an 710049
PEOPLE'S REPUBLIC OF CHINA

D. A. Rickard
University of Wales College of Cardiff
P. O. Box 904
Cardiff
WALES, UK

L. L. Riedinger
The Science Alliance
The University of Tennessee
South College
Knoxville, TN 37996-1528

J. R. Robins
Ontario Hydro
800 Kipling Avenue, KR143
Toronto, Ontario M8Z 5S4
CANADA

X. Rong
Hochspannungslabor der
Universität Stuttgart
Nielsenstr. 18
7302 Ostfildern 2
GERMANY

S. W. Rowe
Merlin Gerin Company
Research Dept., USINE A2
38050 Grenoble Cedex
FRANCE

M. S. Ryan
Oak Ridge National Laboratory
Building 4500S, MS-6101
P. O. Box 2008
Oak Ridge, TN 37831-6101

M. Sakai
Mitsubishi Electric Company
Itami Works, Amagasaki-City
Hyogo-Pref. 661
JAPAN

I. Sauers
Oak Ridge National Laboratory
Building 4500S, MS-6123
P. O. Box 2008
Oak Ridge, TN 37831-6123

J. P. Sawyer
Oak Ridge National Laboratory
Building 4500S, MS-6122
P. O. Box 2008
Oak Ridge, TN 37831-6122

E. Schade
ABB Corporate Research
CRBP
CH-5405 Baden
SWITZERLAND

F. Schwirzke
Physics Dept., Code 61SW
Naval Postgraduate School
Monterey, CA 93943-5000

T. Sumikawa
High Voltage Switchgear Dept.
Hamakawasaki-works
Toshiba Corporation
JAPAN

S. Suzer
Chemistry Dept.
Middle East Technical University
06531 Ankara
TURKEY

M. Suzuki
Shikoku Electric Power Co.
2-5, Marunouchi, Takamatsu
Kagawa
JAPAN

T. H. Teich
High Voltage Engineering Group 1
Swiss Federal Institute of Technology
Physikstr. 3, CH-8092 Zurich
SWITZERLAND

J. E. Thompson
Department of Electrical
 and Computer Engineering
College of Engineering
University of New Mexico
107 Farris Engineering Center
Albuquerque, NM 87131

R. Tobazéon
Laboratoire d'Electrostatique
et de Matériaux Diélectriques
CNRS - 25 Avenue des Martyrs
166X - 38042 Grenoble - Cedex
FRANCE

N. G. Trinh
Cables & Insulation
Institute de recherche
 d' Hydro-Quebec
1800 Montee Ste-Julie
Varennes, Quebec
CANADA J0L 2P0

J. de Urquijo
Instituto de Física, UNAM
P. O. Box 139-B
62191 Cuernanaca, Mor.
MEXICO

R. J. Van Brunt
National Institute of
 Standards and Technology
Building 220, Room B344
Gaithersburg, MD 20899

E. J. M. Van Heesch
High Voltage Group
University of Technology
Eindhoven
P. O. Box 513
5600 MB Eindhoven
THE NETHERLANDS

H. T. Wang
Dept. of Electrical Engineering
University of Waterloo
200 University Avenue, West
Waterloo, Ontario
CANADA N2L 3G1

R. T. Waters
Department of Physics
 and Electrical Engineering
University of Wales Institute
 of Science and Technology
Cardiff CF1 3NU
UNITED KINGDOM

J. M. Wetzer
High Voltage Group
Dept. of Electrical Engineering
Eindhoven University of Technology
P. O. Box 513, 5600 MB Eindhoven
THE NETHERLANDS

W. R. White
Bonneville Power Administration
PO Box 491
Vancouver, WA 98666

T. B. Worzyk
ABB HV Switchgear AB
Box 701, S-771 01 Ludvika
SWEDEN

P. F. Williams
University of Nebraska-Lincoln
Dept. of Electrical Engineering
209N Walter Scott Engr. Center
Lincoln, NE 68588-0511

F. S. Young
Electric Power Research Institute
3412 Hillview Ave.
Palo Alto, CA 94303

Participants (left to right)

Third Row: V. H. Gehman Jr, P. F. Williams, R. Gripshover, R. A. Harthun, S. R. Hunter, E. J. M. Van Heesch

Second Row: H. Faidas, T. B. Worzyk, W. Pfeiffer, O. Farish, J. G. Carter, T. H. Martin, M. Kristiansen

First Row: L. A. Pinnadawage, K. Nakanishi, J. B. Neilson, J. C. Paul, P. G. Datskos, J. M. Wetzer, A. H. Guenther

639

Participants (left to right)

Fourth Row: D. R. James, L. G. Christophorou, J. P. Sawyer

Third Row: G. Harman, R. L. Champion, E. Schade, J. Dupuy, I. D. Chalmers, R. S. Jacobsen, A. G. Netto

Second Row: J. A. Cripps, B. W. McConnell, I. Sauers, L. Niemeyer, S. Suzer, S. W. Rowe, J. Dutton, I. Gallimberti, P. F. Coventry

First Row: N. F. Cardwell, J. E. Carrington, E. E. Kunhardt, E. Marode, M. S. Ryan, D. A. Rickard, A. E. D. Heylen, S. M. Mahajan

640

Participants (left to right)

Fourth Row: T. Hasegawa, T. Namera, I. W. McAllister, T. Sumikawa, X. Rong, R. T. Waters

Third Row: P. C. Bolin, M. O. Pace, Th. Dunz, H. Hiesinger, W. R. White, H. Fujinami

Second Row: R. Tobazeon, A. F. Borghesani, N. Fujimoto, N. Femia, B. F. Hampton, A. J. Davies

First Row: Y. Qiu, A. H. Cookson, A. Pedersen, J. J. Pachot, H. Mochizuki, M. Hanai, D. Denat

641

Participants (left to right)

Third Row: J. R. Robins, M. Forys, M. F. Freshette, J. Olthoff, J. Castonguay, T. Teich, R. J. Van Brunt

Second Row: A. Diessner, L. E. Kline, M. Goldman, A. Garscadden, N. G. Trinh, J. McCoskey

First Row: C. Neumann, D. König, F. Schwirzke, J. de Urquijo, Y. Doin, D. L. McCorkle, A. Bitouni

642

Participants (left to right)

Third Row: K. Feser, F. Y. Chu, I. Gyuk

Second Row: M. Suzuki, J. Ozawa, Y. Murayama, C. E. Easterly, G. D. Griffin, W. Johnstone

First Row: H. T. Wang, K. Nakanishi, T. Ishii, T. Nitta, D. C. Agouridis, M. Hanamura, M. Sakai

AUTHOR INDEX
(Italics Indicate Paper Authorship)

SUBJECT INDEX

Arc(s), 209, 459, 461, 477, 500
Avalanche, 74, 297, 607

By-products, 19, 421, 545, 553

Cathode spots, 209
Charge
  accumulation, 286, 434
  detection method, 285
  distribution, 215
  measurement, 84, 288
Corona
  alternating current, 413
  discharges, 171, 399
  frequency effects, 413
  in gases, 171, 383
  in liquids, 171
  in mixtures, 426
  in SF_6, 247, 399, 421, 553
  mass spectrometric study, 421
  models, 407
  pulsed, 383, 391
  statistics, 383
  surface effects, 399

Diagnostics, 509, 539, 563, 566, 615-624
Dielectric strength, 91, 193, 233
Disconnectors, 231, 446, 451, 497
Dissociation
  collision induced, 5
  cross sections, 123
  processes, 121

Electrical breakdown
  delayed, 353, 399
  effects of humidity on, 81
  electron emission processes, 152
  gas/liquid comparison, 159
  impulse, 164, 239, 247, 261, 267
  in gases, 49, 74, 91, 151, 187, 193,
    231, 247, 279, 351, 517

  in liquids, 159, 183, 187
  in space, 61, 609
  in vacuum, 151, 187
  laser-initiated, 331
  leader, 49
  macroscopic relation, 349
  mechanisms, 580
  non-uniform field, 61, 239, 247
  particle-initiated, 54, 313
  statistics, 82
  time delay, 353
  under electric and magnetic fields, 187
Electrical discharges
  development, 81, 129
  diffuse, 322
  effect of electrode, 285
  fractal description, 137
  hollow-cathode, 145, 201
  in air, 81
  in gas mixtures, 101, 231
  in SF_6, 239, 247, 553
  pseudo spark, 109, 145
  space charge effects, 101
  Townsend, 102, 114
Electromagnetic interference, 467
Electron
  attachment, 9, 19, 27, 35, 43, 73, 89, 221
  capture, 43
  density profiles, 206, 218
  detachment, 1, 73, 215
  drift velocity, 179
  emission, 151, 211, 263
  excitation, 95
  impact dissociation, 121
  impact ionization, 89, 95, 221
  in liquids, 165, 179
  initiating, 50, 262
  localization, 27
  mobility, 29, 263
  scattering, 19

649